14th International Conference on Hydroinformatics (HIC 2022)

IOP Conference Series: Earth and Environmental Science
Volume 1136

Bucharest, Romania
4 - 8 July 2022

ISBN: 978-1-7138-6991-7
ISSN: 1755-1307

Printed from e-media with permission by:

Curran Associates, Inc.
57 Morehouse Lane
Red Hook, NY 12571

Some format issues inherent in the e-media version may also appear in this print version.

Institute of Physics
Dirac House, Temple Back
Bristol BS1 6BE UK

Phone: 44 1 17 929 7481
Fax: 44 1 17 920 0979

techtracking@iop.org

Additional copies of this publication are available from:

Curran Associates, Inc.
57 Morehouse Lane
Red Hook, NY 12571 USA
Phone: 845-758-0400
Fax: 845-758-2633
Email: curran@proceedings.com
Web: www.proceedings.com

TABLE OF CONTENTS

Preface of the Proceedings of 14th International Conference on Hydroinformatics – HIC 2022

Andrei-Mugur Georgescu[1] **and Sanda-Carmen Georgescu**[2] **(editors)**

[1] Department of Hydraulics, Sanitary Engineering and Environmental Protection; Technical University of Civil Engineering Bucharest, 020396 Bucharest, Romania
[2] Department of Hydraulics, Hydraulic Machinery and Environmental Engineering; University "Politehnica" of Bucharest, 060042 Bucharest, Romania

andrei.georgescu@utcb.ro, carmen.georgescu@upb.ro

Abstract. The present volume contains the Proceedings of the 14th International Conference on Hydroinformatics – HIC 2022, held from the 4th to the 8th of July 2022 in Bucharest, Romania. The conference was jointly organised by the Technical University of Civil Engineering Bucharest and the University "Politehnica" of Bucharest.

1. Introduction

Hydroinformatics focuses on the application of information and communications technologies, to address problems in the water industry, water resources management, and other fields connected to the aquatic environment. New trends, like the digital transformation of urban water systems, smart water networks and smart cities demand multidisciplinarity and interoperability, to ensure a sustainable management of water resources. As the Hydroinformatics scientific community continuously grows, its interests encompass a wide range of topics, from the numerical modelling to the development of new sensors and tools, the interpretation of increasing amounts of data, the analysis and management of complex water systems, the integration of models, data and controls for practical applications.

The International Conference on Hydroinformatics (HIC) is organized since 1994, every 2 years, by the IAHR/IWA Joint Committee on Hydroinformatics [https://www.iahr.org/index/committe/15].

After a 4-years gap due to the COVID-19 pandemic, the **14th International Conference on Hydroinformatics – HIC 2022** was jointly organised in Bucharest, Romania, on 4-8 July 2022, by the Technical University of Civil Engineering Bucharest and the University "Politehnica" of Bucharest. The conference was hosted at the University "Politehnica" of Bucharest.

2. HIC 2022 relevant details and statistics

Adding to the COVID-19 pandemic, the war in Ukraine imposed new travel restrictions to the existing ones, so the HIC 2022 conference experienced, for the first time, a new hybrid format that was most challenging for the hosts, with a total number of 208 participants, out of which 70% were present onsite and 30% online.

Nevertheless, judging on opinions expressed during or after the conference, by the members of the Hydroinformatics community that participated onsite (146 attendees onsite, out of which 69 from abroad), HIC 2022 offered a frame for direct exchanges after a 4-years gap, namely, a good opportunity for knowledge integration among scientists, practitioners and technology experts, to tackle the complex challenges of Hydroinformatics in a changing world.

The conference-logo is presented in figure 1, together with an image of the conference webpage [https://hic2022.utcb.ro/].

Figure 1. HIC 2022: conference-logo (upper frame); webpage home window (lower frame) [https://hic2022.utcb.ro/].

HIC 2022 was structured around the following 6 main topics: Technologies for Water Management and Monitoring; Big-Data, Knowledge and Water Data Management; Digital Transformation of Urban Water Systems; Digital Approach in Water Engineering; Hydraulic and Hydrological Modelling; Water – Energy Nexus.

HIC 2022 received 165 papers from 39 countries; 110 papers were accepted for presentation. The conference contributions were presented in parallel sessions from the 5th to the 7th of July – they were divided into 22 sessions, based on the main topics. A plenary session was held on the 4th of July, after the Opening Ceremony. We acknowledge here the outstanding lectures given by the HIC 2022 Plenary Speakers:

- Professor Kalanithy Vairavamoorthy (IWA Executive Director), lecture title *How Hydro-informatics Can Help the Water Sector Navigate the New Normal*;
- Professor Orazio Giustolisi (Technical University of Bari, Italy), lecture title *Digital Transformation in Water Distribution Networks: from Digital Twin to Digital Water Service Concept*;
- Professor Anton Anton (Technical University of Civil Engineering Bucharest, Romania), lecture title *Digitization Path of Hydropower and Water Supply in Romania*.

All HIC 2022 submissions were peer-reviewed through a double-blind review process (using the EasyChair Smart CFP environment [https://easychair.org/cfp/HIC2022]), by at least two international experts, and decisions were taken based on research quality; 34 reviewers (members of program committees) were involved in this process. We are very pleased to report that the quality of the submissions turned out to be very high. A total of 62 papers, out of 110 accepted papers, were selected to be published as full papers in the present volume, which is around 56% paper acceptance rate. Most reviewers provided detailed and constructive comments, which were valuable for the authors to continue improving their papers, even if their submissions were not selected for the conference proceedings.

Given the high-quality works done by authors and program committee members, we are confident that the HIC 2022 Proceedings capture the current state of the art research in Hydroinformatics and will have significant impact to the research on longer term.

3. HIC 2022 Program Committees

We acknowledge here the invaluable assistance of all program committees' members.

3.1. Conference Leading Team

Chairs:

- Andrei-Mugur Georgescu, Technical University of Civil Engineering Bucharest, Romania
- Lăcrămioara Diana Robescu, University "Politehnica" of Bucharest, Romania

Members:

- Gabriele Freni, University of Enna "Kore", Italy
- Philippe Gourbesville, Polytech Nice Sophia, Université Côte d'Azur, France
- Sanda-Carmen Georgescu, University "Politehnica" of Bucharest, Romania.

3.2. Conference Steering Committee

Members:

- Orazio Giustolisi, Technical University of Bari, Italy
- Dragan Savić, KWR Water Research Institute, The Netherlands
- Duan Chen, Changjiang River Scientific Research Institute, China
- Maria C. Cunha, University of Coimbra, Portugal
- Ibrahim Demir, Iowa University, U.S.
- Raziyeh Farmani, University of Exeter, U.K.
- Gabriele Freni, Kore University of Enna, Italy
- María Pilar García-Navarro, University of Zaragoza, Spain
- Philippe Gourbesville, Polytech Nice Sophia, Université Côte d'Azur, France
- George P. Karatzas, Technical University of Crete, Greece
- Gwo-Fong Lin, National Taiwan University, Taiwan
- Ole Mark, Krüger A/S, Denmark
- Miguel Rico-Ramirez, University of Bristol, U.K.
- Dimitri Solomatine, UNESCO-IHE Institute for Water Education, The Netherlands.

3.3. International Scientific Committee

Members:

- Jean-Luc Achard, Grenoble National Polytechnic Institute, France
- Anton Anton, Technical University of Civil Engineering Bucharest, Romania
- Luigi Berardi, University G. d'Annunzio of Chieti-Pescara, Italy
- Sandor Ianos Bernad, Romanian Academy – Timisoara Branch, Romania
- Bruno Brunone, University of Perugia, Italy
- Epsică Chiru, ARA – Romanian Water Association, Romania
- Xin Dong, Tsinghua University, China
- Ioana Făgărășan, University „Politehnica" of Bucharest, Romania

IOP Conf. Series: Earth and Environmental Science 1136 (2023) 011001 doi:10.1088/1755-1315/1136/1/011001

- Zoran Kapelan, Delft University, The Netherlands
- Branko Kerkez, University of Michigan, U.S.
- Daniele Biagio Laucelli, Technical University of Bari, Italy
- Florentina Moatar, National Research Institute of Science and Technology for Environment and Agriculture (INRAE), France
- Frank Molkenthin, Brandeburg University of Technology, Germany
- Sebastian Muntean, Romanian Academy – Timisoara Branch, Romania
- Ilinca Năstase, Technical University of Civil Engineering Bucharest, Romania
- Konstantinos Papatheodorou, International Hellenic University, Serres Campus, Greece
- Olivier Piller, National Research Institute of Science and Technology for Environment and Agriculture (INRAE), France
- Dušan Prodanović, University of Belgrade, Serbia
- Liang Qiuhua, Loughborough University, U.K.
- Gabriel Racovițeanu, Technical University of Civil Engineering Bucharest, Romania
- Maurizio Righetti, Free University of Bozen-Bolzano, Italy
- Nicolas Riviere, INSA Lyon, France
- Dan Niculae Robescu, University „Politehnica" of Bucharest, Romania
- Sandra Soares-Frazão, Louvain School of Engineering, Belgium
- Romeo Florin Susan-Resiga, „Politehnica" University of Timisoara, Romania
- Nicolaos Theodossiou, Aristotel University of Thessaloniki, Greece
- Ken Thomas, Waterford Institute of Technology – WIT, Ireland
- Lydia Vamvakeridou-Lyroudia, KWR Water Research Institute, The Netherlands
- Emmanouil Varouchakis, Technical University of Crete, Greece.

3.4. Local Organising Committee

Honorary Committee:

- Mihnea Cosmin Costoiu, Rector of University „Politehnica" of Bucharest – UPB
- Radu Văcăreanu, Rector of Technical University of Civil Engineering Bucharest – UTCB
- Jorj-Mădălin Mihailovici, Country Manager VEOLIA Romania
- Aurel Presură, General Director S.C. RAJA Constanța S.A.
- Ilie Vlaicu, President of ARA – Romanian Water Association

Members:

- Alexandru Aldea, UTCB
- Mircea Bițoiu, Association of Energy and Electrical Engineers – AEEE
- Florin Bode, Technical University of Cluj-Napoca
- Diana-Maria Bucur, UPB
- Florian Burnar, Partnership Association for European Funds and Projects, APPFE
- Costin Ioan Coșoiu, UTCB
- Cristiana-Verona Croitoru, UTCB
- George Darie, UPB
- Alexandru Dimache, UTCB
- Mihaela Amalia Diminescu, UPB
- Georgiana Dunca, UPB
- Dragoș Găitănaru, UTCB
- Matei-Răzvan Georgescu, UTCB
- Cristina Sorana Ionescu, UPB
- Nicolae Marcoie, "G. Asachi" Technical University of Iasi – TUIASI
- Horia Necula, UPB
- Otilia Oproescu, UTCB
- Vlad Florin Pîrăianu, UPB
- Bogdan Popa, UPB
- Carmen Anca Safta, UPB
- Mihnea Sandu, UTCB
- Adrian Stuparu, "Politehnica" University of Timisoara
- Nicoleta Octavia Tănase, UPB
- Daniela Teodorescu, UTCB
- Eliza Isabela Tică, UPB
- Daniel Toma, TUIASI
- Bianca Ungureanu, AEEE

- Daniela Elena Gogoașe, UPB
- Constantin Radu Gogu, UTCB
- Iulian Iancu, UTCB
- Grigore Ursu, UTCB
- Alexandru Cezar Vlăduț, UTCB
- Elena Vulpașu, UTCB

Acknowledgments
On behalf of the HIC 2022 Leading Team, we express our sincere gratitude to all authors and presenters. We are also grateful to all members of the HIC 2022 Steering Committee, International Scientific Committee and Local Organising Committee, for their valuable support.

Last but not least, we are grateful to all student-volunteers, from the University „Politehnica" of Bucharest (members of Association of Energy and Electrical Engineers, and of Students' Association of Energy Engineering Faculty), as well as from the Technical University of Civil Engineering Bucharest, for their enthusiastic and efficient support to the HIC 2022 Registration and Information Desks.

The joint effort of everyone contributed to the great success of HIC 2022 – one of the first scientific conferences organised face-to-face in 2022 in Bucharest, with a great number of participants that travelled to Romania from abroad.

IOP Conf. Series: Earth and Environmental Science 1136 (2023) 011002 doi:10.1088/1755-1315/1136/1/011002

Peer Review Statement

All papers published in this volume have been reviewed through processes administered by the Editors. Reviews were conducted by expert referees to the professional and scientific standards expected of a proceedings journal published by IOP Publishing Publishing.

- **Type of peer review:** Single Anonymous
- **Conference submission management system:** Easychair
- **Number of submissions received:** 165
- **Number of submissions sent for review:** 110
- **Number of submissions accepted:** 62
- **Acceptance Rate (Submissions Accepted / Submissions Received × 100):** 37.6
- **Average number of reviews per paper:** 2.13
- **Total number of reviewers involved:** 34
- **Contact person for queries:**
  Name: Andrei-Mugur Georgescu
  Email: andrei_georgescu2003@yahoo.com
  Affiliation: Technical University of Civil Engineering Bucharest

IOP Conf. Series: Earth and Environmental Science 1136 (2023) 012001 doi:10.1088/1755-1315/1136/1/012001

A complete digital water experience to support real leakage management planning

Daniele Laucelli[1], Serena Spagnuolo[2], Antonio Rinaldi[3], Gianluca Perrone[4], Luigi Berardi[5] and Orazio Giustolisi[1]

[1] Department of Civil, Environmental, Land, Building Engineering and Chemistry, Technical University of Bari, Bari, 70126, Italy
[2] Acquedotto Pugliese S.p.A., via S. Cognetti 36, Bari, 70121, Italy
[3] IA.ING S.r.l., viale M. Chiatante 60, Lecce, 73100, Italy
[4] STUDIO AC3 Ingegneria S.r.l., Via Pertini 2, San Ferdinando di Puglia, 76017, Italy
[5] Department of Engineering and Geology, University "G. D'Annunzio" of Chieti Pescara, Pescara, 65127, Italy

danielebiagio.laucelli@poliba.it

Abstract. In the next years digital transition of Water Distribution Networks (WDNs) will represent a long-standing challenge for triggering innovations in the WDNs management tasks. The ongoing impulse for innovation, boosted also by the economic resources for post-pandemic recovery, is opening novel opportunities for technicians and researchers in the context of Digital Water. This work reports the key steps where novel concepts and tools of Digital Water have been used to support consultant companies in designing WDN leakage management actions in a real WDN managed by Acquedotto Pugliese, in Southern Italy. The Digital Water Services (DWSs) were used to support key phases of the workflow, including validation of topological data and integration of georeferenced connection to private properties; calibration of advanced hydraulic model exploiting mass-balance in pressure-driven analysis; the application of strategies for DMA design and pressure control; and the planning of pipe rehabilitation. WDNetXL platform was used to develop WDN Digital Twin and exploit the DWSs. Applying the Digital Water experience to multiple WDNs managed by the same provider also enabled planning the best courses of actions accounting for investments vs. expected leakage reduction at global scale.

1. Introduction

The concept of digital transition in WDN management has still not a global consensus and talking about "digital water" is intuitively referred to the integration of the latest Information and Communication technology (ICT) and Internet of Things (IoT) capabilities to manage data on water systems to facilitate visualization and analysis of complex information. In WDN analysis and management, the digital transition is expected to integrate advancements in data acquisition, storage and representation with the latest methodologies developed from technical-scientific research based complex network theory, artificial intelligence and multi-objective optimization. The main purpose of such transition is providing technicians and water managers with effective and transferable products that implements efficient processes to face various technical issues. The paradigm for digital transformation based on *digital water services* [1] has been introduces to integrate the concept of

Digital Twin, tailored for WDN, with well-established algorithms and methodologies, to support solving WDN life-cycle management issues.

Indeed, the interest for WDN management has significantly increased in the last two decades mainly due to the increasing concerns about asset deterioration and water losses. In such a context many initiatives have been adopted worldwide supported by massive investments on WDN asset management, which further increased due to post-pandemic recovery plans. For instance, the European "Next generation EU" Programme is funding actions in Italy [2] to support both sustainable water usage and digital transition. On one hand, all these investments represent unprecedented opportunities for water companies to improve the quality of WDN service. On the other hand, the digital transition in WDN sector can provide processes and products to support technicians in taking effective and efficient decisions on asset management planning based on structured, verifiable and replicable workflows. While doing so, the digital transition should also assist the re-qualification of information and data available in current form.

Acquedotto Pugliese S.p.A. (AQP) is one of the biggest water supply companies in Europe managing WDNs in about 300 cities in Southern Italy. More than one decade ago AQP started a process for asset management aimed at reducing water losses while improving service performances of urban WDNs. The latest initiative on this side was completed in 2021 and was aimed at planning asset management actions to reducing water losses and improve service performances. The main activities aimed at District Metering Area (DMA) design, pressure control and planning pipeline replacement. An essential pre-requisite for AQP was that all actions should be measurable, verifiable, repeatable and scalable in order to be applied on various size WDNs following a structured approach.

IA.ING s.r.l. and STUDIO AC3 Ingegneria S.r.l. as technical consultants for AQP on this call, tuned to IDEA-RT s.r.l., an innovative start-up company, spin-off of the Technical University of Bari, as a unique international provider of innovation in the field of water network systems management. This contribution presents the key steps where novel processes and products of digital water have been used to support consultant companies along the workflow for planning actions including validation of topological data and integration of georeferenced connection to private properties; calibration of advanced hydraulic model exploiting mass-balance in pressure-driven analysis; the application of strategies for DMA design and pressure control; and the planning of pipe rehabilitation. Each phase is supported by specific Digital Water Services (DWSs) developed as plug-ins in QGIS environment and based on data structure and modelling capabilities available in the WDNetXL platform [3]. The network serving the city of Modugno is used to show the complete digital water experience.

2. Materials and methods

To achieve effective and efficient planning of asset management aimed at the reduction of real volumetric losses, a strategy for DMA planning, pressure control, hydraulic monitoring and rehabilitation was developed. Such a methodology is firmly based on advanced hydraulic modelling of each WDN as well as on the innovative paradigm for model calibration based on water balances [4]. In fact, the innovative hydraulic model overcomes the limitations of classical modelling (e.g., EPANET-based software packages) and allows accounting for volumetric leakages as a function of deterioration and average pressure for each single pipe (e.g. [5]. Besides other innovations, such model goes beyond classical pressure-driven analysis (e.g., based on Wagner's model of water demands [6]) allowing different types of private connections (i.e., direct, fed by private tanks, or in multistorey buildings) at single user level (i.e., based on georeferenced meters). Figure 1 summarizes the workflow starting from the available topological and hydraulic data.

The topological data were enhanced by integrating the information coming from the Geographical Information System (GIS) data used in AQP with the hydraulic models (i.e., in .inp file format) from previous design activities. Such data is the basis for the structured procedure for DMA design [7] including two steps: (i) the *topological segmentation* into virtual districts to identify the candidate positions for meters and closed gates; (ii) optimal *hydraulic DMA design* to allocate meters and closed gates that might reconfigure water paths to reduce pressures and leakages, along with optimal setting of pressure reduction valves.

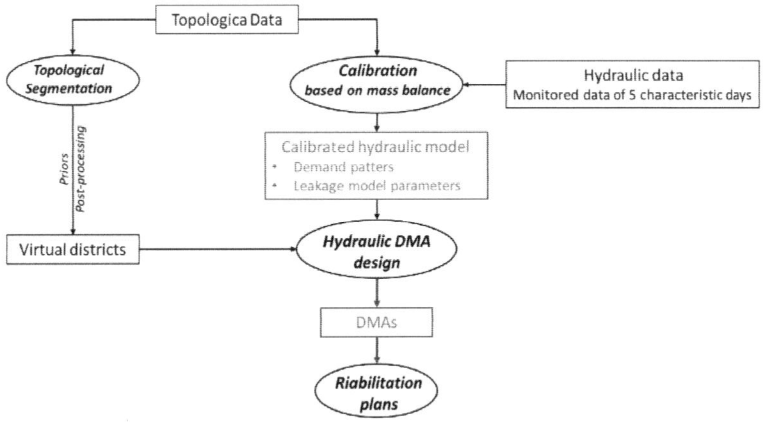

Figure 1. Asset management planning workflow supported by digital water experience.

The same topological data together with the hydraulic data relating to the consumption of single consumers and flow/pressure measurements monitored by AQP, were used to perform the innovative calibration based on mass balance approach [8]. To get a robust model for next planning activities, the calibration was based on measurements from five real daily cycles in 2019 representative of characteristic hydraulic status of the system, i.e., working/holiday, winter/summer conditions New Year's Day. The calibration results provided the parameters of the leakage model at single pipe level and the demand patterns consistent with each characteristic daily cycles used.

3. Results and discussion

3.1. Acquisition and re-qualification of topological data and consumption data

The main data used to build the hydraulic model of each WDN was provided by AQP as:

- geodatabase extracted from the company GIS including length, diameter and material of pipes; valves along the pipes;
- georeferenced meters at private connections;
- recently replaced pipelines;
- data monitored at flow and pressure meters in 2019;
- consumption database for each user for the year 2019, consistently with other monitored data.

Since the topology of the network represents the domain in which its hydraulic functioning takes place, the following analyses were carried on using GIS data and available models:

- Identification and correction of possible major errors in the GIS data;
- Topological simplification of small pipes representing isolation valves which are implemented in the WDNetXL model as features of each pipe object. The analysis of GIS data showed that almost all pipes less than 1m long coincide with valves in the geodatabase (i.e., vaults);
- Identification of double nodes in GIS data, i.e., with a distance between them of less than 0.1m, which cause many disconnections within the network. The network connectivity was automatically corrected and synchronized using one of the functionalities in the WDNetXL import utilities that joins two existing nodes into a single node which restores the hydraulic connectivity. Figure 2 highlights in magenta the position of the duplicated nodes for the network of Modugno;
- Identification of possible existing districts, as a starting point for next DMA design. For the Modugno network, the closed gate valves lead to the identification of pseudo-districts, each

fed by a reservoir (i.e., sub-networks). Since no flow meters are reported, closed or partialized valves do not identify actual DMA but only allows pressure control on a heuristic basis, not supported by hydraulic modelling. Figure 3 shows the automatic identification of existing districts using the WDNetXL tool;

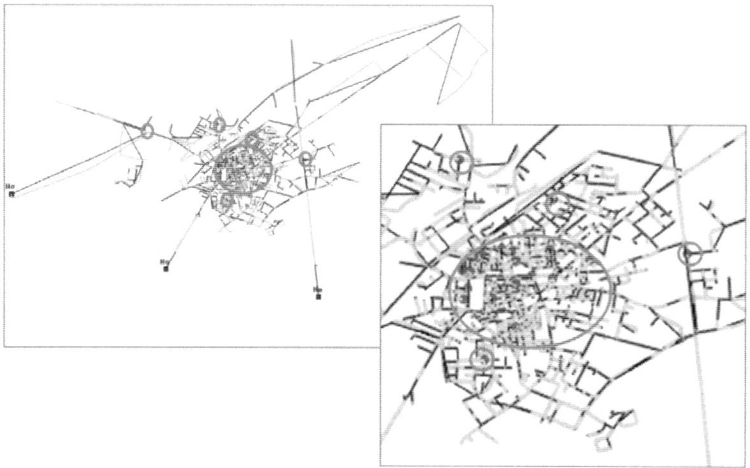

Figure 2. Identification of double nodes in GIS data.

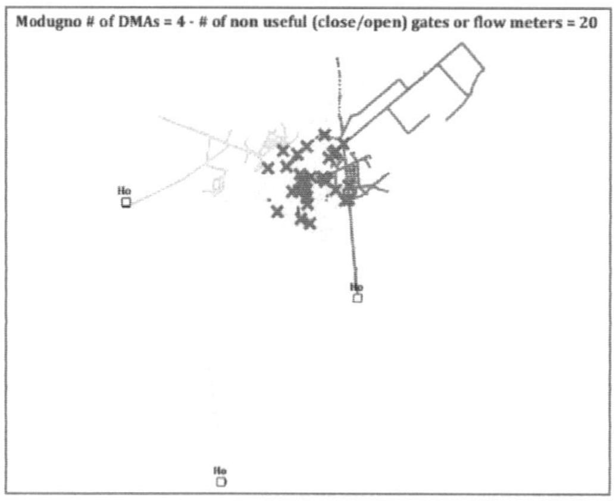

Figure 3. Closed gates and pseudo-districts in Modugno network.

- Import of water consumption data of georeferenced water meters at consumers' connections. Such data were imported into a separate table from the hydraulic model in the WDNetXL system. Such data are recalled only during the hydraulic simulation phase and returns the detailed hydraulic status at each individual user connection. Such a representation avoids the aggregation of different types of users in model nodes and allows storing individual information on the type of contract (e.g., commercial, industrial, household) and the service status (e.g., "active" or "inactive") to perform detailed hydraulic modelling. In addition, it is ready to accommodate detailed data that are expected to be available in the near future (e.g., the height of the buildings, the number of floors, the presence of private storage volumes), consistently with the digital transition trend. It is worth noting that using georeferenced works also helps in identifying pipelines not yet reported in the GIS or in previous models. Figure 4 shows the single users imported in the GIS interface of WDNetXL system, where digital water services are used.

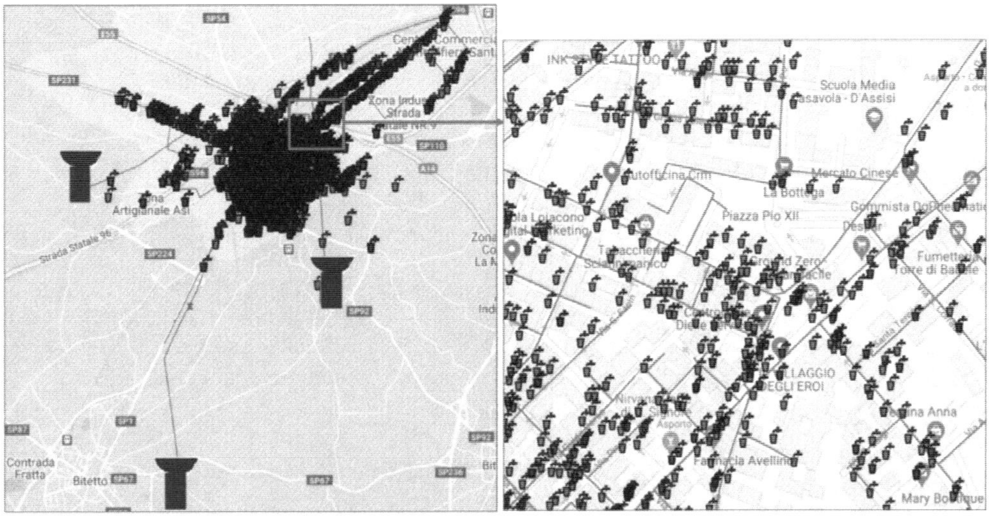

Figure 4. single users in Modugno network.

3.2. *Model calibration based on mass balance approach*

The calibration of the advanced hydraulic model, which includes the volumetric leakages at the single pipe level as a function of pressure and asset deterioration, was performed using an innovative approach, based on the mass balance concept which allows to identify (i) the parameters of the leakage model and (ii) a demand pattern for each monitored district [4][8]. The strategy also allows the simultaneous calibration of the hydraulic resistances of the most relevant pipes.

The procedure uses flow and pressure measures collected over the year 2019 by AQP, with reference to five days (24-hours operating cycles) representative of different operating states of the system: weekdays and winter holidays, weekdays and summer holiday and New Year's Day (i.e., atypical functioning). Therefore, the adopted methodology allows more robust modelling than using a limited number of filed measurements on specific days, usually not synchronous with the water consumption data. It is worth to remark that data of the year 2020 represent an anomalous behaviour of the system due to the Covid-19 pandemic, therefore not useful to support robust design.

Although the Modugno network is fed by three tanks, the inflow measurements were available from the reservoirs [A] and [B] in figure 5. In the absence of inlet measures for a portion of the network, the calibration strategy, based on the possibility of separating the model-based leakage components from stochastic consumers' water consumption in the mass balance, allows getting consistent results. In fact, the calibration assumed a single demand pattern for the entire Modugno network and the same values of leakage parameters for similar pipelines. In more details, it was assumed that the propensity to leak of each pipe is based on the diameter and number of private connections [8]. The pressure measurements at the PRV (red circle in figure 5) were used as a boundary condition for calibration.

The search for calibration solutions, used a DWS named *DigitalWaterMass_Calibration*, which explores the search space using a population-based algorithm. The results file are also available in Excel format and provides: the values of leakage model parameters for each individual pipe; the average user consumption data corrected during the calibration for each of the five days to account for the stochastic day-by-day variation of water consumptions; the unique demand pattern for the five days; the comparison between the measured network flow rate and that of the calibrated model; the values of the linear leakage indicator (i.e. M1a according to the Italian regulation [9] in [m^3/km/day]) calculated for each of the five days. Figures 5 and 6 show the user interface of the *DigitalWaterMass_Calibration* and the patterns of water demand and M1a leakage index of the unique DMA which can be calibrated based on available measurements.

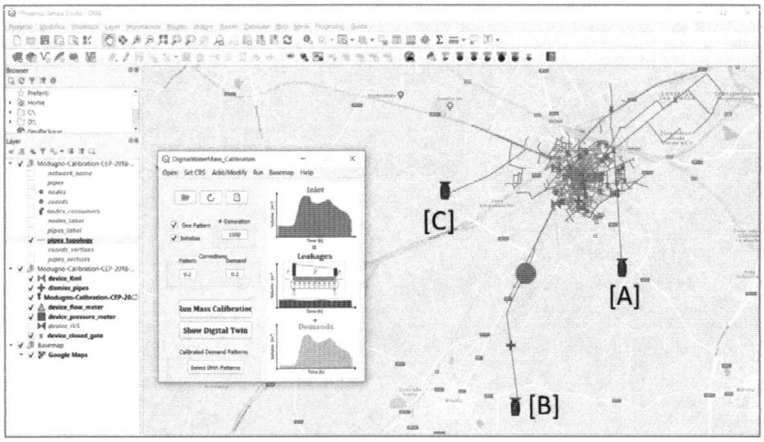

Figure 5. *DigitalWaterMass_Calibration* interface for Modugno network.

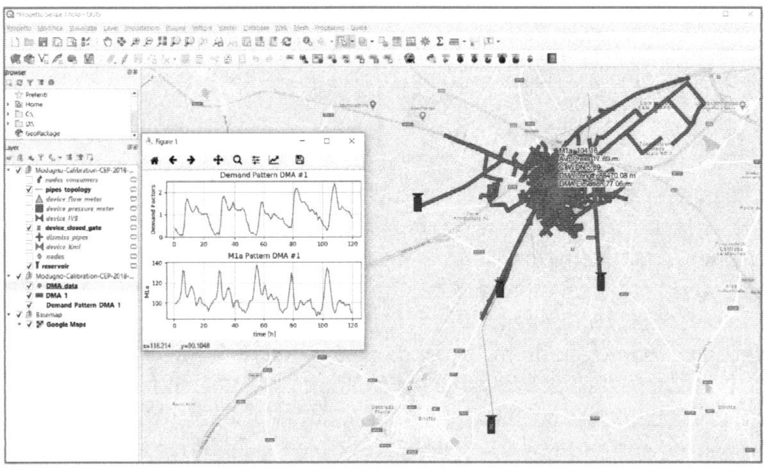

Figure 6. *DigitalWaterMass_Calibration* interface: patterns of consumers' demand and linear leakage index M1a [m^3/km/day].

3.3. Topological segmentation

As mentioned above, the DMA design for leakage management is a two-stage process including topological segmentation and the hydraulic DMA design. The topological segmentation is formulated as a bi-objective optimization that maximizes the infrastructure modularity index tailored for WDNs (e.g., [10]) and minimizes the number of conceptual cuts that separate segments/modules from each other, i.e., candidate positions for meters or closed gates in the next hydraulic DMA design.

It should be noted that the topological segmentation is independent of the hydraulic calibration of the model and, therefore, is performed in parallel as reported in figure 1.

The topological segmentation process allows assuming some *prior* position of vaults that can be useful for the subsequent design of the DMAs. Such position may come from existing vaults/meters or from technical judgement about improving WDN hydraulic functioning.

The DWSs used for the topological segmentation phase are *DigitalWaterVirtualDMA_Design* and *DigitalWaterSegment_Viewer*. The locations of *prior* conceptual cuts are integrated into the Digital Twin of the network as a basis for the segmentation service.

The segmentation of the Modugno network was performed using the segment *length* as weight for infrastructure modularity index, i.e., the topological segmentation has created districts like each other in terms of length. The DWS also allows selecting the *elevation* as weight, meaning that districts are uniform in terms of geodetic elevation within them. Considering the size of the Modugno network, a minimum length of 5 km for each district was set as a further constraint.

The DigitalWaterVirtualDMA_Design service provided two Pareto fronts of optimal segmentation solutions. Each of the solutions is nested in the one with the lowest number of virtual districts, i.e. it is possible to move from one solution to another with more districts on the Pareto front by adding conceptual cuts. This property of the segmentation process increases the flexibility while selecting the basis for next hydraulic DMA design. Figure 7 shows interface of *DigitalWaterVirtualDMA_Design* service along with the list of solutions (i.e., groups of layes).

The *DigitalWaterSegment_Viewer* service allows to view all the optimal segmentation solutions provided in the Pareto front and to move some conceptual cuts (i.e., vaults) to cope with possible restrictions/constraints. This action generates a new group of layers with the updated segmentation solution that can be used in the subsequent hydraulic DMA design phase. Figure 8 shows interface of *DigitalWaterSegment_Viewer* and the segmentation selected as basis for next hydraulic DMA design in Modugno network comprising 63 conceptual cuts (i.e., vaults for meters or sectioning valves).

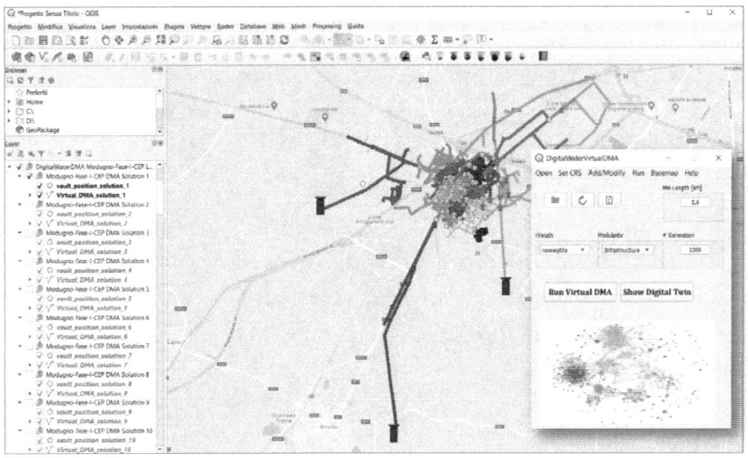

Figure 7. *DigitalWaterVirtualDMA_Design* interface.

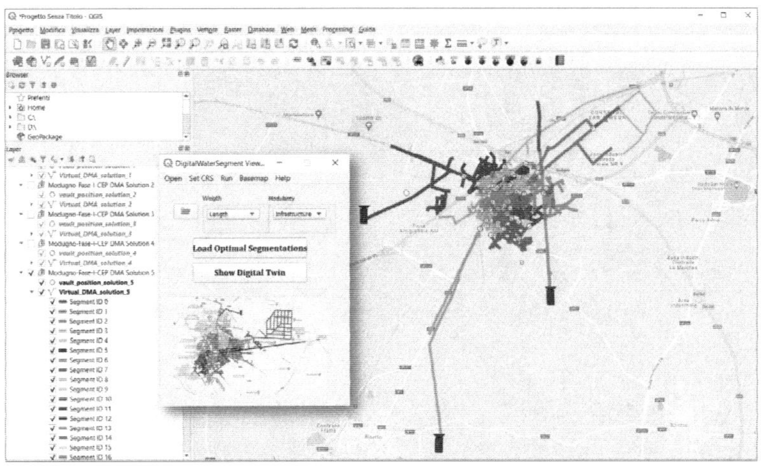

Figure 8. *DigitalWaterSegment_Viewer* interface and solution used for hydraulic DMA design.

3.4. Hydraulic DMA design

The hydraulic DMA design phase is based on solving a multi-objective optimization to identify the following design variables:

- closed sectioning valves: aimed at reducing volumetric leakages by re-configuring water paths in the WDN, in compliance with the minimum pressure for correct service at users;
- flow meters at the edge of the designed DMAs, in the minimum number possible;
- the optimal set-point value(s) of the pressure reduction valves if any is assumed.

All hydraulic analyses use the demand patterns and the leakage model parameters got from the hydraulic model calibration as discussed above.

In the case of Modugno, 13 closed gates were assumed *a priori* to allow effective pressure control using two pressure reduction valves (PRV), one of which is controlled by a remote sentinel node in the network. These design hypotheses allow a robust and rational pressure control in the network, while guaranteeing the minimum pressure of 6m water column for correct service at all users, as required by AQP.

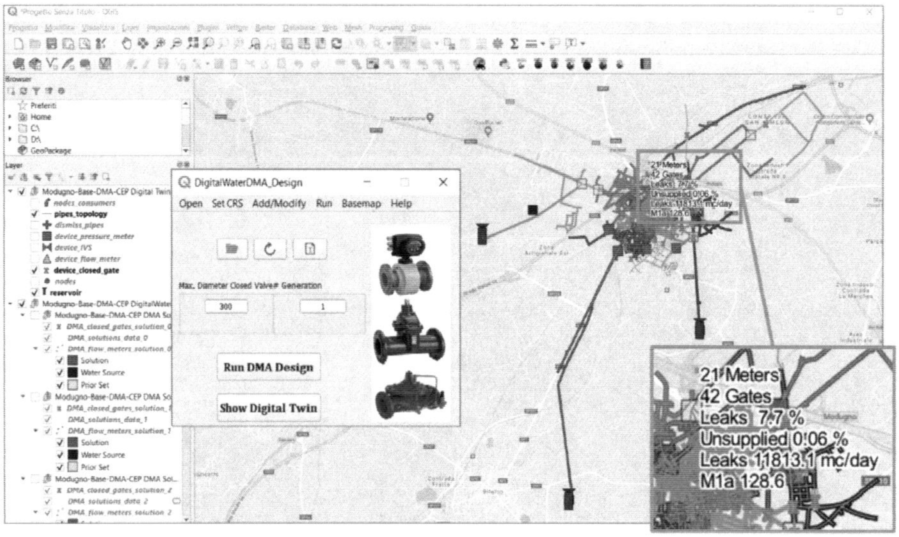

Figure 9. *DigitalWaterDMA_Design* interface with an example of solution.

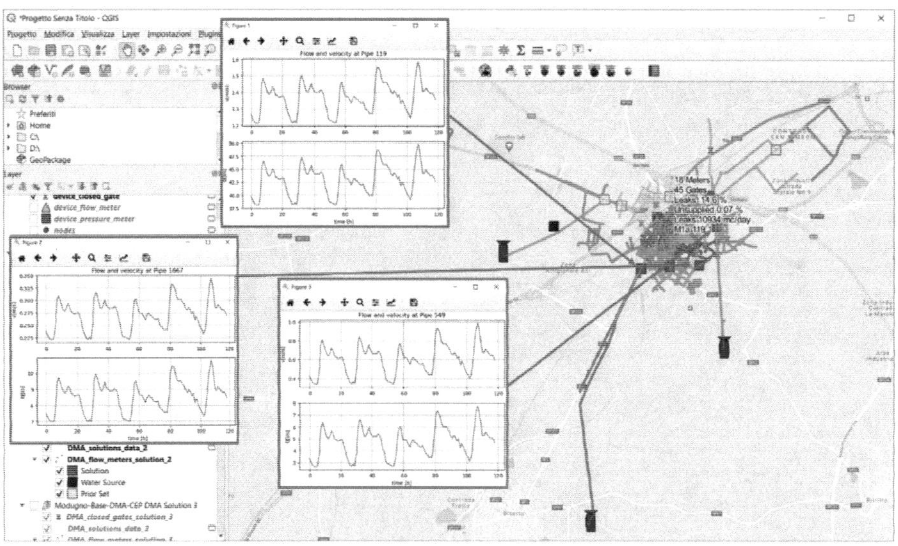

Figure 10. *DigitalWaterSDMA_Analyzer* interface with metrological information on tree flow meters.

The DWS to support the design and comparison of alternative DMA configurations are *DigitalWaterDMA_Design* and *DigitalWaterDMA_Analyzer*. As reported in figure 9, the optimization provides a Pareto front of optimal DMA configurations, in terms of: the number of closed sectioning valves and flow meters at the edge of the districts, any unsupplied demand due to pressure deficit, the percentage and volume of volumetric losses, and value of the linear leakage indicator M1a.

The solutions are directly arranged on different groups of layers in the QGIS environment to support the selection of the most effective DMA configuration in agreement with AQP. The same DWS allows selecting some "perspective" flow meters to be installed later, for example to activate nested DMA allowing step tests for leakage detection.

The *DigitalWaterDMA_Analyzer* also allows visualizing solutions and supports the metrological evaluations of the flow meters to plan. Figure 10 shows three examples of flow and velocity pattern visualization, based on advanced modelling, to support the selection of water meters to be activated. This is crucial to avoid meters where inversions of flow or very small water velocities, may occur.

3.5. *Planning pipeline rehabilitation*

Replacement of deteriorated pipes must be done by optimizing the investment cost while reducing volumetric leakages, as a global management indicator. While doing so the extended hydraulic consequences of pipe replacements have to be evaluated to avoid the contrasting effect of increased water losses, caused by the significant greater hydraulic conductance of the new pipes which increases the operating pressures in the portions of the system downstream. Therefore, the choice of the pipes to be replaced should follows the best cost/benefit ratio, i.e., investments vs. reduction of volumetric leakages, and depends on the new hydraulic regime obtained coming from DMA design and pressure control.

The DWS called *DigitalWaterRehabilitation* supports the evaluation of pipe replacement plans. The position of the devices, flow meters and closed valves, as well as the set-point of the PRVs, corresponding to the chosen DMA configuration, are included in the Digital Twin of the network.

The *DigitalWaterRehabilitation* service highlights pipes on different colours according to the linear leakage indicator M1a (see figure 11). This enables user to manually select pipes based on the value of M1a. Alternatively, it is possible to load external optimal rehabilitation plans, such as those generated by the WDNetXL Design module. Once pipes are selected the "Run rehabilitation" button returns the expected changes in WDN hydraulic performances based on advanced hydraulic modelling.

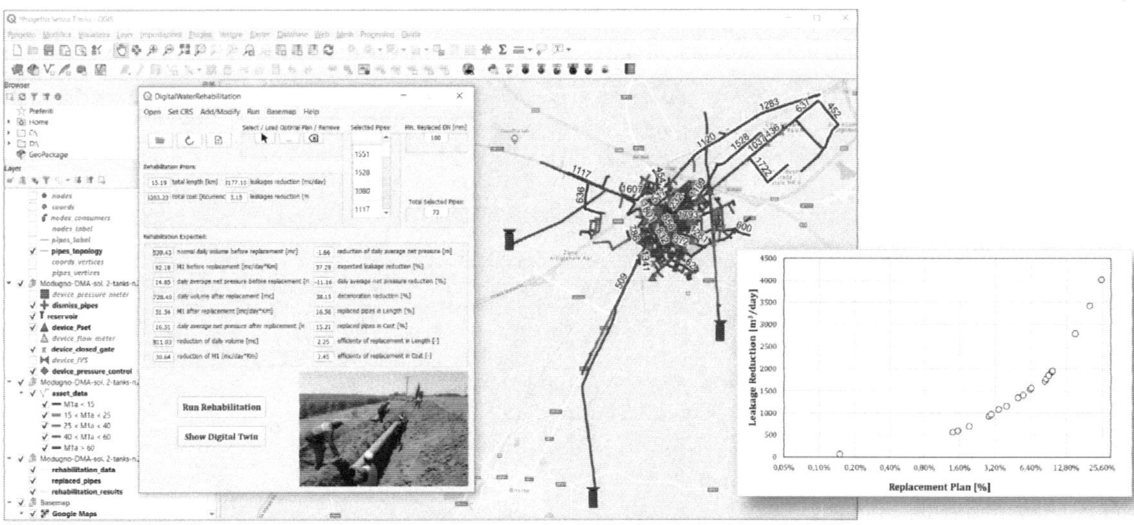

Figure 11. *DigitalWaterRehabilitation* interface and expected results for replacement cost of about 15% of total network cost.

For Modugno network, which is highly deteriorated, various replacement plans was obtained based on different investment levels up to 25% of the replacement cost of the entire system. Such optimal plans include pipes with the maximum efficiency in terms of ratio between the cost of intervention and the expected reduction in leakages. Figure 11 shows the results obtained by applying the rehabilitation plan corresponding to a cost of replacing the pipes equal to about 15% total network replacement (pipes are highlighted in yellow. This way it is possible to evaluate the reduction of leakages achievable for various investment level, thus supporting investment also in conjunction with multiple WDN managed by the same water provider.

4. Conclusions

This contribution demonstrates the potentiality of digital transition to WDN asset management using digital water services developed to support specific tasks within the workflow to plan leakage management actions. Such services cannot substitute the technical skills of engineers but allows integration of advanced topological and hydraulic analyses, artificial intelligence and efficient optimization tools to provide robust and consistent alternatives for design.

Acknowledgments
This paper was partly financed by European Union – PON Ricerca e Innovazione 2014-2020 E FSC, call D.D. 13 July 2017 n. 1735, Project "Energidrica – Efficienza energetica nella gestione delle reti idriche"(code ARS01_00625).

References

[1] Ciliberti FG, Berardi L, Laucelli DB and Giustolisi O 2021 Digital Transformation paradigm for asset management in water distribution networks *2021 10th Int. Conf. Energy and Environment (CIEM)* Bucharest pp 1–5 https://doi.org/10.1109/CIEM52821.2021.9614864
[2] *Call REACT-UE* https://ponir.mit.gov.it/documentazione/gestione-e-attuazione/manifestazioni-di-interesse-react
[3] *WDNetXL-WDNetGIS* 2021 http://www.idea-rt.com/digital-water-latest-innovation
[4] Berardi L, Simone A, Laucelli D and Giustolisi O 2017 Feasibility of mass balance approach to Water Distribution Network model calibration *XVIII Int. Conf. Water Distribution Systems, WDSA2016, Procedia Engineer.* **186** 551–558 https://doi.org/10.1016/j.proeng.2017.03.269
[5] Germanopoulos G 1985 A technical note on the inclusion of pressure-dependent demand and leakage terms in water supply network models *Civil Eng. Syst.* **2** 171–179 https://doi.org/10.1080/02630258508970401
[6] Wagner JM, Shamir U and Marks DH 1989 Water distribution reliability: simulation methods *J. Water Resour. Plan. Manage.* **114** 276–294 https://doi.org/10.1061/(ASCE)0733-9496(1988)114:3(276)
[7] Laucelli D, Simone A, Berardi L and Giustolisi O 2017 Optimal design of district metering areas for the reduction of leakages *J. Water Resour. Plan. Manage.* **143**(6) 04017017-1-12 https://doi.org/10.1061/(ASCE)WR.1943-5452.0000768
[8] Berardi L and Giustolisi O 2022 Calibration of design models for leakage management of water distribution networks *Water Resour. Manag.* **35** 2537–51 https://doi.org/10.1007/s11269-021-02847-x
[9] ARERA 2017 *Regolazione della qualità tecnica del servizio idrico integrato ovvero di ciascuno dei singoli servizi che lo compongono* Delibera 917/2017/R/IDR (in Italian)
[10] Giustolisi O and Ridolfi L 2014 A novel infrastructure modularity index for the segmentation of water distribution networks *Water Resour. Res.* **50**(10) 7648–61 https://doi.org/10.1002/2014WR016067

Digital Water Services using Digital Twin paradigm

Francesco G. Ciliberti[1], Luigi Berardi[1], Daniele B. Laucelli[2] and Orazio Giustolisi[2]

[1] Department of Engineering and Geology, University "G. D'Annunzio" of Chieti Pescara, Pescara, 65127, Italy
[2] Department of Civil, Environmental, Land, Building Engineering and Chemistry, Technical University of Bari, Bari, 70126, Italy

francesco.ciliberti@unich.it

Abstract. In the last years, the digital transition concept has spread all over public and private life as a process designed for improving problem solving in industry by the combination of models, information, and connectivity technologies. In the Water Distribution Networks (WDNs) management sector, the innovations in the areas of ICT/IoT, virtual representation of infrastructure elements in GIS/BIM platforms and the advancements in hydraulic modelling offer the opportunity to open a new era for water engineering. Nonetheless, a unique consensus about the digital transformation meaning in WDNs management is still missing. In this paper a recent paradigm for the digital transformation for WDNs is exposed and further extended, starting from the concept of Digital Twin for WDNs management. It encompasses the virtual representation of features and devices of the network integrated with advanced hydraulic modelling and Artificial Intelligence for supporting WDNs management tasks. The WDN Digital Twin is used within tools, named Digital Water Services (DWSs), working on GIS platforms. DWSs are conceived to support distinct phases of WDNs management by improving engineering awareness on technical decisions. Several DWSs, previously adopted for real WDNs management and planning, are here presented, as support tools for technicians and water utilities for achieving short-term and long-term management tasks.

1. Introduction

The advent of the 4th industrial revolution, Industry 4.0, gave significant advancement in technology and innovations that conducted companies and firms to adopt digital technologies integrating Information and Communication Technologies (ICT), Internet of Things (IoT) and Internet of Services (IoS). The ability to connect and control their systems through a cyber-physical environment has enabled to evaluate and support their actions across the value chains of their products [1]. Although several authors have different visions on that phenomenon, they all agree to identify the "digital transformation" as a process aimed at triggering enhancements a system through the combination of information, computing, and connectivity technologies [2].

However, in many areas of industry, there is not a clear consensus about the meaning of the "digital transformation" of systems and services. In the Water Distribution Networks (WDNs) sector, the first steps for the application of the digital transformation concept have been taken only recently. [3] identified the drivers of the "digital transformation" in the advancements in ICT and IoT, such as fast communication networks (e.g., Narrowband-IoT, LPWAN protocols, 5G networks), sensors, big data gathering and analysis, and geographic and building information systems (e.g., BIM, GIS, WEBGIS services), which have been introduced in water systems over the last 20 years. Therefore, few researchers and companies borrowed concepts and approaches of "digital transformation" from other industrial contexts, developing Cyber-Physical Systems (CPS) [4] and Digital Twins (DTs) [5, 6] as digital environments to support water utilities achieving sustainable, effective, and efficient management of the systems [7]. WDNs are complex infrastructure systems, which interact with anthropic activities. As such, differently from other industrial processes, they have many constraints coming from the surrounding environment, socio-economic activities and/or from the interactions with other infrastructures. In addition, the spatial extent of WDNs represents a challenge by itself since it may span from urban to territorial scale, differently from industrial applications.

In recent years, the changes of regulatory contexts provided further motivations to exploit the potentialities of digital transition. For example, in Italy [8] the Regulatory Authority for Energy, Networks and Environment (ARERA) introduced novel penalty/reward mechanisms based on service performances indicators, mainly related to leakages, reliability and water quality. Therefore, water utilities are seeking for effective processes and products to support effective and efficient solutions. In addition, the proposed approaches should be replicable and scalable, for those water utilities managing multiple networks; and flexible to cope with currently available funds. Such innovations are definitively becoming part of an industrial plan for water utilities to plan actions in the short (operational), medium (tactical) and long (strategic) time horizon.

Implementing a mere "digital representation" of such systems into "black boxes" tools that provide "optimal" solutions would turn into limited transfer innovations to engineers and decision makers. The idea of developing a unique digital system able to answer every possible technical need, minimizing the interaction with end users is simply not applicable in the WDN sector since current WDNs management practices in many water utilities worldwide strongly depends on the human factor in terms of knowledge of a specific network by technicians who daily work on that system.

On this premises, any innovation to support WDN management and planning is credible and acceptable by water utilities only if it preserves the key role of technicians/engineers that have to use their expertise and deep knowledge of the system in any phase of complex workflows. Therefore, the digital transition in WDN sector is expected to provide effective and efficient tools [9], using advanced computing capabilities and innovative procedures, to make easier each specific task of technicians, leaving them the chance to evaluate each technical alterative.

In the last century, technical-scientific research provided several advancements in various areas of analysis and management of WDNs, including physically based modelling (e.g. [10, 11]) optimization (e.g. [12]), data mining (e.g. [13]), complex network analysis (e.g. [14]), also tailoring concepts and tools from other research areas. Such advancements provided algorithms, methodologies and procedures that stand behind the ongoing "digital transformation" process and can foster the interactions between technical and scientific communities. All such aspects make the "digital transformation" in WDNs an unprecedented opportunity to develop innovation, also delivered as processes/products, to proceed side by side with technical world to accomplish classical and novel technical tasks.

In this work, a recently proposed paradigm for the digital transformation for asset management in WDNs [15] has been further exploited by developing a wide set of Digital Water Services (DWSs) suited to support various tasks along common workflows in WDNs management. Such services are implemented as plugins in QGIS environment using the same Digital Twin of the WDN. The DWSs are completely interoperable with the WDNetXL platform, working on Microsoft Excel, and share the same data structure and core functions. Nonetheless, DWSs are conceived to expose just the key information needed to take effective decisions for each single task, thus maintaining the key role of technicians.

2. Digital Twin

The Digital Twin (DT) technology firstly conceptualized by Grieves [5] and applied in the industry to prototype efficiency [16] and by NASA [6], is a virtual representation of a product, such as an object or a system that comprises its characteristics by means of models, information, and data across multiple life cycle phases.

DT technologies allows representing a "virtual replica" of a real system, by integrating collection and analysis of data from the real system, to get realistic analyses of that system under different scenarios and support informed decision making. In industry, DT technologies are applied for increasing efficiency, accuracy of the whole production system and consequently for gaining economic benefits in the productions [17, 18]. From such perspectives, the DT is considered as the candidate technology for sustaining the digital transformation in various industries [16].

The DT concept, as conceived by [5], consists of three parts: the physical layer, representing the reality; the digital layer, which contains all information gathered from the physical layer, and the connections between the two layers. The connections are represented by data, automatically collected by sensors or stored into database, that flow from the physical layer to the digital layer, and by results and decisions, based on analyses using the digital layer, that modify the physical environment.

In the WDNs context, as reported by [3][15], the physical layer is represented by the network, its real layout comprising pipes, sources, customer demands, devices (e.g., pumps, isolation valve system, flow, pressure control valves, etc), the hydraulic status (including the control of devices) and the technical constraints which are known by the water company.

All elements are translated into the digital layer, as a virtual representation of each geometric feature of the network integrated with the hydraulic model which should provide a representation of the physical behaviour of the system. In that sense, GIS stands as the main developing environment for the DT of WDNs: it enables to create relations between geographic features of the network, data and hydraulic modelling, as well as it allows visualization and immediate verification of information by the WDN experts.

The digital transformation paradigm of WDNs, reported by [15], couples the DT representation of the whole WDN with methodologies and algorithms, either data-driven or physically-based, which support various specific tasks of technical relevance for WDN analysis, management and planning. Such framework is exploited into a set of Digital Water Services (DWSs). After importing the DT of the network, each DWS allows performing specific analyses that integrate in different processes advanced hydraulic modelling with Artificial Intelligence (AI) and/or Complex Network Theory (CNT) metrics. The results are visualized as layers in the QGIS map canvas, stored as ESRI® shapefiles and, whenever needed to support further decisions, in Excel spreadsheets.

The advanced hydraulic model in all DWS uses WDNetXL platform core functions [19], which overcomes some limitations of classical modelling package, like those based on EPANET [20]. This is actually a pre-requisite to get reliable results and avoid misleading decisions, thus fostering the digital transition. Some key advancements of such model are summarized as follows:

- The model allows defining demand at single consumer level in a database which is separate from the hydraulic model graph; this representation enables to keep the actual database of consumers constantly updated both in terms of active/non-active connections and defining a pressure-driven model for each type of connection (i.e., direct from the network, free orifices, connection though private local tanks, multi-storey buildings). Unlike other models, any change of single or multiple consumers does not require updating demand data lumped at model nodes and hydraulic results are exposed at single connection level;
- Background and unreported leakages, which amounts to the major volumetric effects at annual scale, are modelled at single pipe level as a function of pipe pressure and deterioration;
- The simulation of variable level tanks does not suffer from the known instabilities due to uncoupling energy and mass balance equation, thus providing a more realistic representation of their actual behaviour;
- All kind of hydraulic devices (i.e., isolation valves, minor head-losses, pressure control valves and pumps) are modelled consistently with any type of control based on either time scheduling or hydraulic status variables.

3. Digital Water Services

The DWSs here reported have been designed for supporting and validating each phase of the WDNs management workflow, which spans from operational to tactical and strategic planning horizon.

Figure 1 summarizes the areas of WDN management where DWSs have been developed so far. They start from supporting the analysis of WDN topological domain and hydraulic model calibration (up-left box) as functional to build and validate the WDN DT. The DWSs supporting the analysis and design of Isolation Valve System (IVS), i.e., allowing the isolation of pipe segments for planned or unexpected pipe interruptions, are in the up-right box and tackles WDN mechanical reliability/vulnerability analysis. The analysis of abnormal scenarios due to pipe burst and subsequent interruptions, as of direct relevance to support operational activities, is supported by DWSs in the bottom left-box. Finally, various DWS have been developed to support asset management in terms of leakage reduction planning by pressure control and District Metering Areas (DMAs) design (tactical horizon), and pipe replacement (strategic horizon).

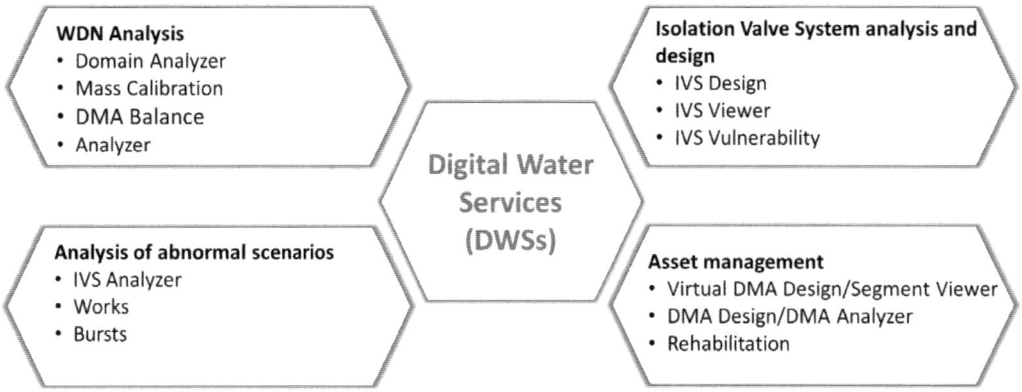

Figure 1. Block diagram of DWSs classification.

3.1. WDN Analysis

The DWSs reported in this group (figure 2) allows calibrating the hydraulic model of the network and evaluating the WDN behaviour, either applying the hydraulic modelling or analysing its topological features.

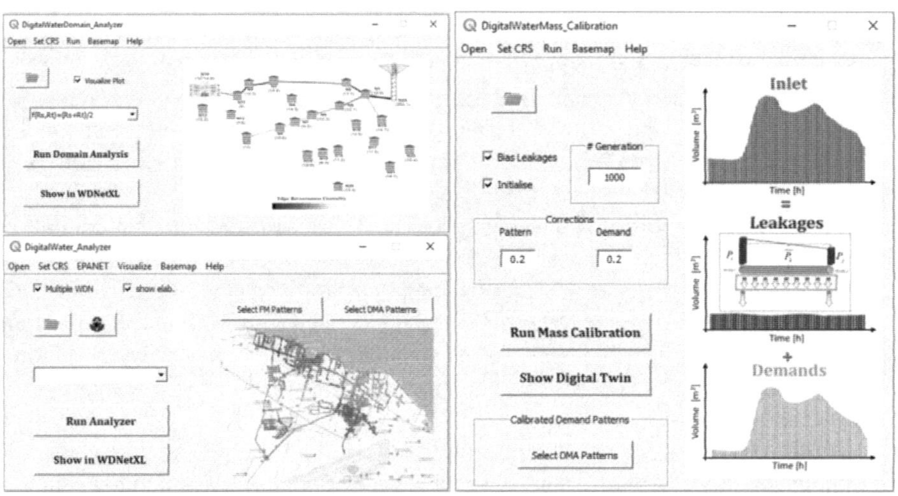

Figure 2. Graphical User Interfaces (GUIs) of *DigitalWaterDomain_Analyzer*, *DigitalWaterMass_Calibration* and *DigitalWater_Analyzer*.

IOP Conf. Series: Earth and Environmental Science 1136 (2023) 012002 doi:10.1088/1755-1315/1136/1/012002

The *DigitalWaterMass_Calibration* service allows users to calibrate the hydraulic model embedded in the digital layer of WDN assuming the mass-balance strategy [21] based on separating deterministic i.e., volumetric leakages) from stochastic (i.e., water requests of consumers) components of water outflow. It uses time series of water inflow, pressure and tank-levels collected over multiple days with different hydraulic functioning, even at single DMA level. The service returns the representation of the hydraulic model of the network and visualizes the calibrated demand pattern(s) (figure 3). Further analysis involving the estimation of the calibrated DMAs demand patterns can be carried out with the *DigitalWaterDMA_Balance*.

The *DigitalWaterDomain_Analyzer* service identifies the most relevant pipes/nodes of the WDN from the topological point of view based on WDN-tailored metrics taken from CNT [22, 23]. Such analysis complements the hydraulic simulation and enables further understanding WDN functioning. The *DigitalWater_Analyzer* service performs the hydraulic analysis of the WDN and returns the key data on demand patterns and pressure-dependent leakages at each DMA level; it also provides information on expected flow/velocity patterns at assumed flow meters, thus enabling decision based on metrological evaluations (e.g., removing flow meters where flow inversion is expected).

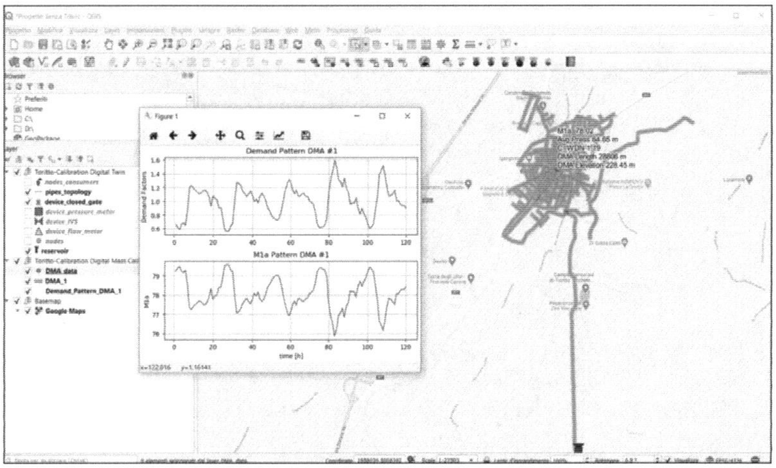

Figure 3. Visualization of the results of the *DigitalWaterMass_Calibration* service.

3.2. Isolation Valve System analysis and design

The *DigitalWaterIVS_Design* (figure 4) allows performing the optimal Isolation Valve System (IVS) design of the network by minimizing the risk of pipes disconnections and unintended isolations [24], based on multi-objective evolutionary optimization strategies.

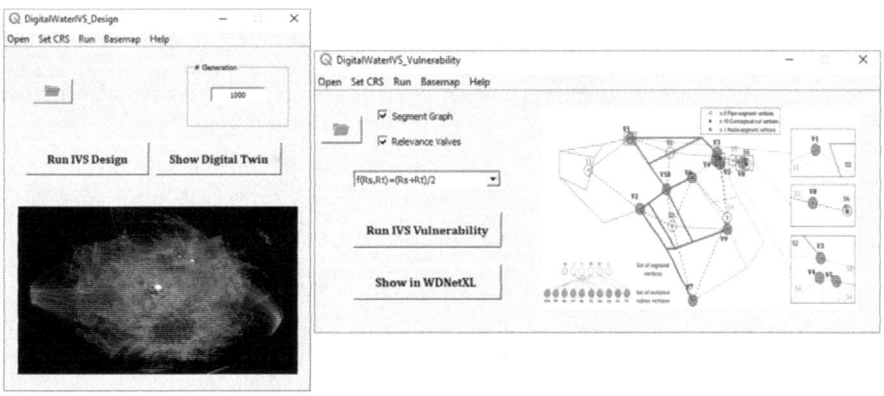

Figure 4. GUIs of *DigitalWaterIVS_Design* and *DigitalWaterIVS_Vulnerability*.

The *DigitalWaterIVS_Design* returns a set of IVS solutions, which are visualized on map canvas and stored into the digital layer of WDN. The *DigitalWaterIVS_Viewer* allows importing and visualizing solutions from previous optimal IVS design runs.

At last, the *DigitalWaterIVS_Vulnerability* (figure 5) applies a IVS reliability strategy proposed by [25] for mapping the spatial distribution of valves relevance in term of WDN reliability, based on WDN-tailored CNT metrics, and it allows supporting the identification of the most impacting failure scenarios and enabling the WDN management planning for IVS inspections and retraining.

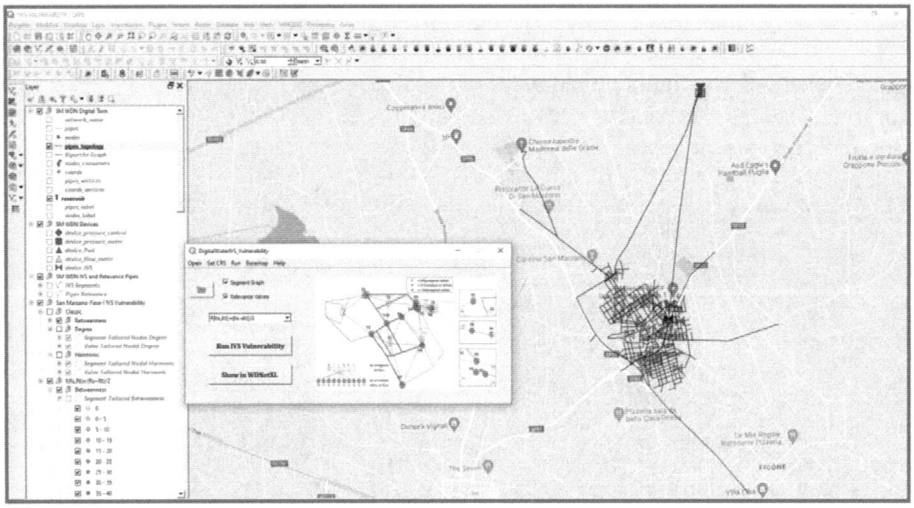

Figure 5. Visualization of the results of the *DigitalWaterIVS_Vulnerability* service.

3.3. Analyses of abnormal scenarios

The *DigitalWater_Works* and *DigitalWater_Bursts* services (figure 6) allow simulating the hydraulic behaviour of the WDNs when a failure pipe occurs and while isolation valves are closed for maintenance works. *DigitalWater_Works* results report the set of pipes and the customers involved in the failure events, the unsupplied water along the network as attribute of the water meters of the digital layer.

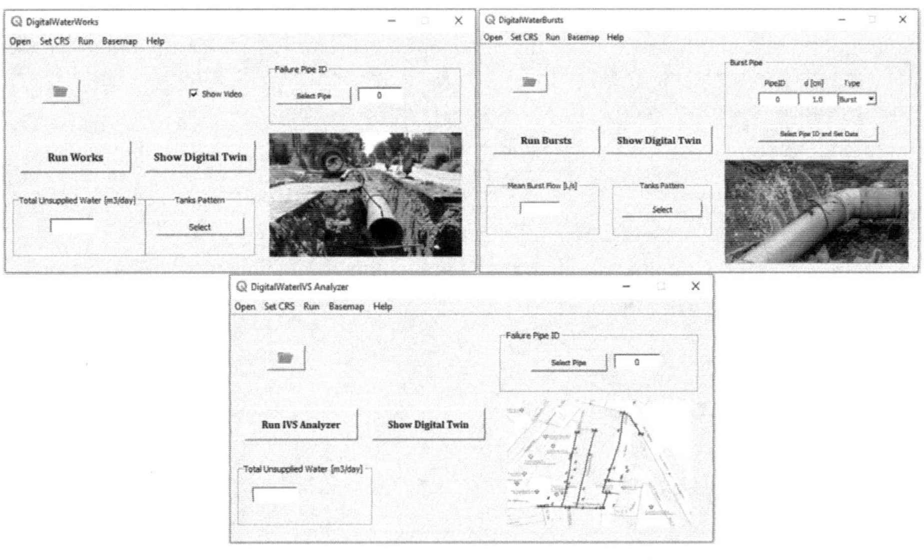

Figure 6. GUIs of *DigitalWaterWorks, DigitalWaterBursts* and *DigitalWaterIVS_Analyzer.*

The *DigitalWater_Bursts* enables the hydraulic analysis by defining geometric parameters of the burst and the type of pipe which is manually selected on the map canvas. It returns the unsupplied unsupplied demands to each customer. The *DigitalWaterIVS_Analyzer* (figure 7) allows detecting the portion of the IVS shutdown to allow work on a failed pipe and the valves to be closed.

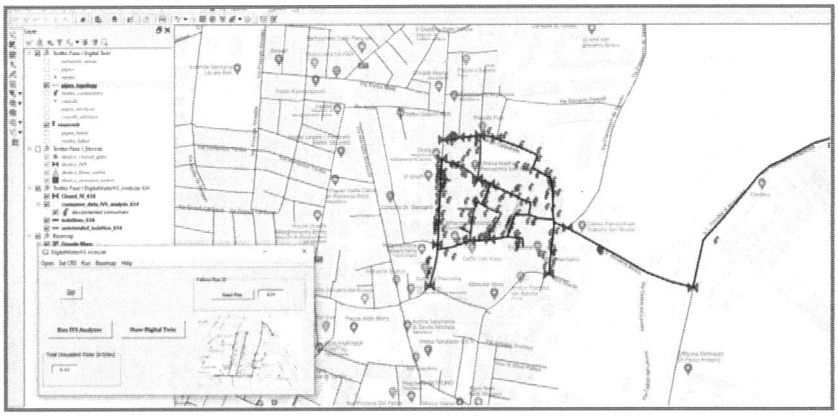

Figure 7. Visualization of the results of the *DigitalWaterIVS_Analyzer* service.

3.4. Asset management

Asset management is conceived as tactical activities (i.e., DMA design and pressure control) and strategic planning (i.e., pipeline replacement) – Graphical User Interfaces are presented in figure 8.

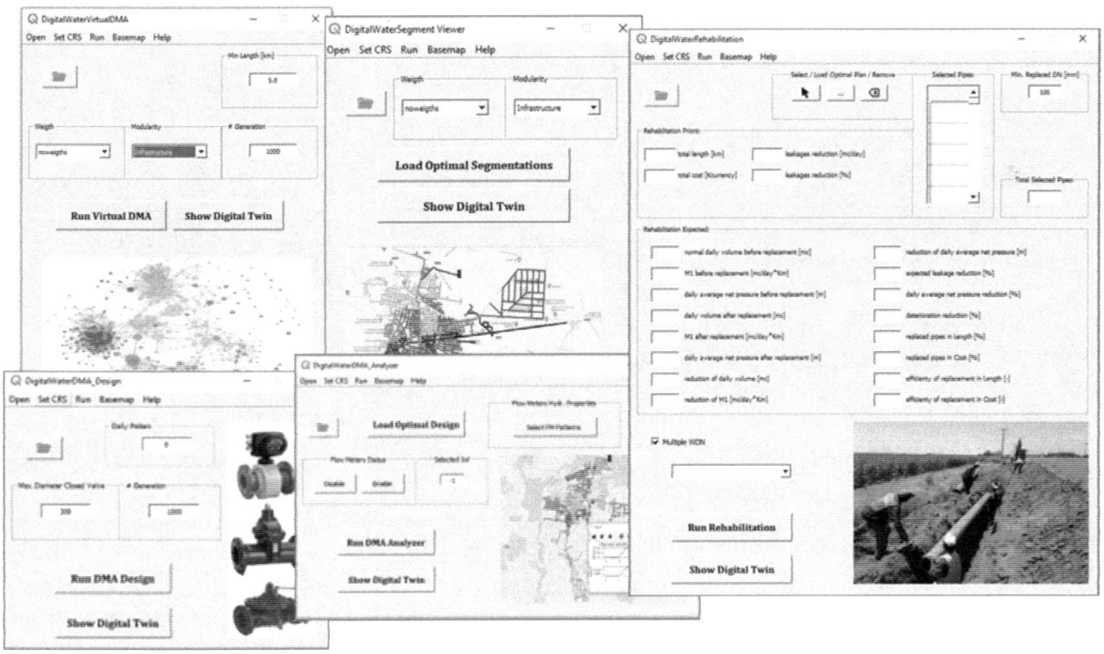

Figure 8. Graphical User Interfaces (GUIs) of *DigitalWaterVirtualDMA*, *DigitalWaterDMA_Design*, *DigitalWaterSegmentViewer*, *DigitalWaterDMA_Analyzer* and *DigitalWaterRehabilitation*.

The DWSs support the strategy for DMAs design proposed by [26], as a two-stages process.

(i) The *DigitalWaterVirtualDMA_Design* service performs the topological segmentation of the network, formulated as a bi-objective optimization that maximizes the WDNs-tailored infrastructure modularity index [27] and minimizes the number of "conceptual cuts". The user can assign the minimum length of each segment and the weight of the modularity index; Pareto front of optimal nested solutions of the segmentation, are visualized as layers. The *DigitalWaterSegment_Viewer* service allows visualizing previously generated optimal segmentation solutions.

(ii) The *DigitalWaterDMA_Design* performs the hydraulic districtualization, which solves a multi-objective optimization to minimize the number of flow meters at the edge of DMAs and optimizing the closed sectioning valves for flow paths reconfiguration and leakages reduction, while delivering sufficient water to consumers. The optimization also includes optimal setting of pressure control valves for each DMA configuration. Each DMAs configuration is represented by the DMAs layer, which shown the different districts along the network, the closed gates and the flow meters layers.
In addition, users can customize one or more solutions provided by the *DigitalWaterDMA_Design* service through the *DigitalWaterDMA_Analyzer*, which allows disabling some specific water meters of the solution, according to the technical-economical constraints.

The *DigitalWaterRehabilitation* (figure 9) supports the evaluation of pipes replacement plans: users can manually select pipes to rehabilitate or load an external optimal plan, generated by the WDNetXL Design module, and then perform the hydraulic simulation which turns out the evaluation of the expected rehabilitation performances in terms of technical requirements and economic constraints, based on the advanced hydraulic solver embedded in the [19] platform.

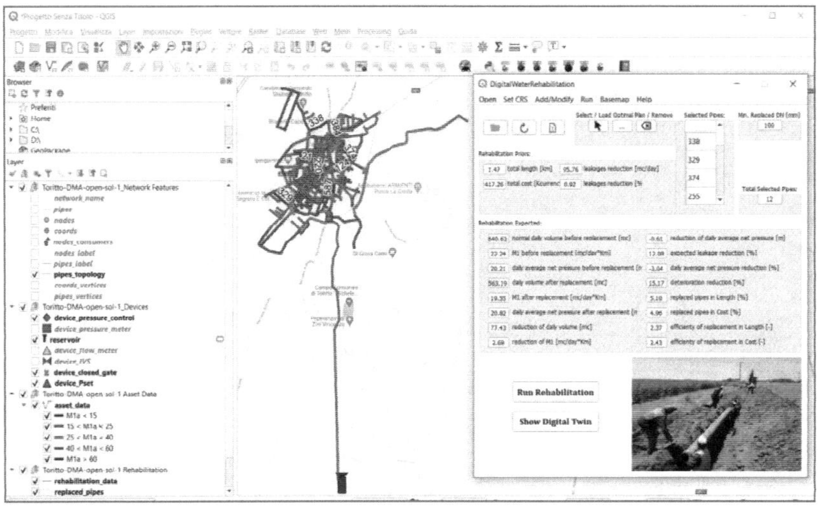

Figure 9. Visualization of the results of the *DigitalWaterRehabiliation* service.

4. Conclusions

Technicians and researchers are still in the process of assessing paradigms and developing strategies within a shared vision of the digital transition in WDNs.

In particular, the efforts are aiming at the definition of standard and robust methodologies for developing DT for WDNs, aimed at matching the peculiarities of the networks, as well as at developing a framework for coupling the physically-based hydraulic modelling with the potentialities from advanced data-driven analysis, like Artificial Intelligence, Machine Learning and Deep Learning.

In this work, several DWS developed under a recent [15] paradigm for the digital transformation of WDNs have been presented, which highlight the potentiality to provide user-friendly tools for supporting each single task of the WDNs management, planning and design workflow.

The same paradigm could provide standardized procedures for continuous update of the DT infrastructures, by acquiring the enhancement in ICT/IoT and the improvement of computing calculation, such as Quantum Computing, for enabling a more sophisticated analysis of the complexities and the interactions of WDNs.

Acknowledgments
This paper was partly financed by European Union - PON Ricerca e Innovazione 2014-2020 E FSC, call D.D. 13 July 2017 n. 1735, Project "Energidrica - Efficienza energetica nella gestione delle reti idriche"(code ARS01_00625). Data for figures of DWSs were provided by IDEA-RT (www.idea-rt.com).

References
[1] Maskuriy R, Selamat A, Maresova P, Krejcar O and David OO 2019 Industry 4.0 for the construction industry: Review of management perspective *Economies* **7**(3) 68 https://doi.org/10.3390/economies7030068
[2] Vial G 2019 Understanding digital transformation: A review and a research agenda *J. Strategic Inf. Syst.* **28**(2) 118–144 https://doi.org/10.1016/j.jsis.2019.01.003
[3] Conejos Fuertes P, Martínez Alzamora F, Hervás Carot M and Alonso Campos JC 2020 Building and exploiting a Digital Twin for the management of drinking water distribution networks *Urban Water J.* **17**(8) 704–713 https://doi.org/10.1080/1573062X.2020.1771382
[4] Baheti R and Gill H 2011 Cyber-physical systems *The Impact of Control Technology* **12**(1) 161–166
[5] Grieves M and Vickers J 2017 Digital twin: Mitigating unpredictable, undesirable emergent behavior in complex systems *Transdisciplinary Perspectives on Complex Systems* ed Kahlen J, Flumerfelt S and Alves A (Springer Cham) pp 85–113 https://doi.org/10.1007/978-3-319-38756-7_4
[6] Shafto M, Conroy M, Doyle R, Glaessgen E, Kemp C, LeMoigne J and Wang L 2012 Modeling, simulation, information technology & processing roadmap *National Aeronautics and Space Administration* **32** 1–38
[7] Wang Z, Song H, Watkins DW, Ong KG, Xue P, Yang Q and Shi X 2015 Cyber-physical systems for water sustainability: challenges and opportunities *IEEE Communications Magazine* **53**(5) 216–222 doi: 10.1109/MCOM.2015.7105668
[8] ARERA 2017 *Regolazione della qualità tecnica del servizio idrico integrato ovvero di ciascuno dei singoli servizi che lo compongono* Delibera 917/2017/R/IDR (in Italian)
[9] Taormina R, Galelli S, Tippenhauer NO, Salomons E and Ostfeld A 2017 Characterizing cyber-physical attacks on water distribution systems *J. Water Resour. Plan. Manage.* **143**(5) 04017009 https://doi.org/10.1061/(ASCE)WR.1943-5452.0000749
[10] Todini E and Pilati S 1988 A gradient algorithm for the analysis of pipe networks *Computer Applications in Water Supply* vol **1** *Systems Analysis and Simulation* ed Coulbeck B and Chun-Hou O (Taunton, UK: Research Studies Press Ltd.) pp 1–20
[11] Giustolisi O, Savic D and Kapelan Z 2008 Pressure-driven demand and leakage simulation for water distribution networks *J. Hydraul. Eng.* **134**(5) 626–635 https://doi.org/10.1061/(ASCE)0733-9429(2008)134:5(626)
[12] Van Zyl JE, Savic DA and Walters GA 2004 Operational optimization of water distribution systems using a hybrid genetic algorithm *J. Water Resour. Plan. Manage.* **130**(2) 160–170 https://doi.org/10.1061/(ASCE)0733-9496(2004)130:2(160)
[13] Berardi L, Giustolisi O, Kapelan Z and Savic DA 2008 Development of pipe deterioration models for water distribution systems using EPR *J. Hydroinform.* **10**(2) 113–126 https://doi.org/10.2166/hydro.2008.012
[14] Yazdani A and Jeffrey P 2011 Complex network analysis of water distribution systems *Chaos: An Interdisciplinary Journal of Nonlinear Science* **21**(1) 016111 https://doi.org/10.1063/1.3540339
[15] Ciliberti FG, Berardi L, Laucelli DB and Giustolisi O 2021 Digital Transformation paradigm for asset management in water distribution networks *2021 10th Int. Conf. Energy and Environment (CIEM)* Bucharest pp 1–5 https://doi.org/10.1109/CIEM52821.2021.9614864

[16] Kritzinger M, Karner M, Traar G, Henjes J and Sihn W 2018 Digital Twin in manufacturing: A categorical literature review and classification *IFAC-PapersOnLin*e **51**(11) 1016–1022 https://doi.org/10.1016/j.ifacol.2018.08.474
[17] Negri E, Fumagalli L and Macchi M 2017 A review of the roles of digital twin in CPS-based production systems *Procedia Manufacturing* **11** 939–948 https://doi.org/10.1016/j.promfg.2017.07.198
[18] Tao F, Zhang M and Nee AYC 2019 *Digital Twin Driven Smart Manufacturing* (Academic Press)
[19] *WDNetXL-WDNetGIS* 2022 http://www.idea-rt.com/digital-water-latest-innovation
[20] Rossman LA 2000 *EPANET 2: Users Manual* EPA/600/R-00/057 (Cincinnati, OH: US Environmental Protection Agency)
[21] Berardi L, Simone A, Laucelli D and Giustolisi O 2017 Feasibility of mass balance approach to water distribution network model calibration *Procedia Engineering* **186** 551–558 https://doi.org/10.1016/j.proeng.2017.03.269
[22] Giustolisi O, Simone A and Ridolfi L 2017 Network structure classification and features of water distribution systems *Water Resour. Res.* **53**(4) 3407–23 https://doi.org/10.1002/2016WR020071
[23] Giustolisi O, Ridolfi L and Simone A 2020 Embedding the intrinsic relevance of vertices in network analysis: the case of centrality metrics *Scientific Reports* **10**(1) 1–11 https://doi.org/10.1038/s41598-020-60151-x
[24] Giustolisi O 2020 Water distribution network reliability assessment and isolation valve system *J. Water Resour. Plan. Manage.* **146**(1) 04019064 https://doi.org/10.1061/(ASCE)WR.1943-5452.0001128
[25] Giustolisi O, Ciliberti FG, Berardi L and Laucelli DB 2022 A novel approach to analyze the isolation valve system based on the Complex Network Theory *Water Resour. Res.* **58**(4) e2021WR031304 https://doi.org/10.1029/2021WR031304
[26] Laucelli D, Simone A, Berardi L and Giustolisi O 2017 Optimal design of district metering areas for the reduction of leakages *J. Water Resour. Plan. Manage.* **143**(6) 04017017-1-12 https://doi.org/10.1061/(ASCE)WR.1943-5452.0000768
[27] Giustolisi O and Ridolfi L 2014 A novel infrastructure modularity index for the segmentation of water distribution networks *Water Resour. Res.* **50**(10) 7648–7661 https://doi.org/10.1002/2014WR016067

Developing water monitoring solutions with FIWARE defining, creating, populating, and extending digital twins

Gareth Lewis[1], Brett Snider[1], Lydia Vamvakeridou-Lyroudia[1,2], Albert S Chen[1], Slobodan Djordjević[1] and Dragan A Savić[1,2]

[1] Centre for Water Systems, University of Exeter, Harrison Building, North Park Road, Exeter, EX4 4QF, UK
[2] KWR Water Research Institute, Groningenhaven 7, P.O. Box 1072, 3430 BB Nieuwegein, the Netherlands

g.lewis2@exeter.ac.uk

Abstract. The water industry is often and unfairly criticised for being relatively slow to adopt IT as part of its operating processes. To address this, the FIWARE and Fiware4Water projects were tasked with developing open-source data definition, management, and processing, and applying it to the water sector. In this work, we are reviewing our experiences with developing FIWARE-based solutions for the aqua3s[1] and Fiware4Water[2] projects across the whole project lifecycle, from our initial of installing and learning to work with FIWARE context brokers[3], through building and maintaining synthetic digital twins for load and edge-case testing development and debugging and finally looking at longer-term data management and archiving. Through our work on these projects, we have started to develop a wealth of best-practice, strategic choices, and approaches to be avoided for successful and long-term context broker usage.

1 Introduction

A digital twin is usually defined as some form of virtual representation that serves as the real-time digital counterpart of a physical object or process. For the Fiware4Water project [2], the Centre for Water Systems partnered with South West Water on a 'smart metering and citizen engagement' project, to develop a digital twin representation of a smart metering network installed in Devon, UK. For the aqua3s project [1], multiple digital twins were constructed to model water delivery and monitoring networks across multiple European pilots.

For both projects, the digital twin goals were similar, in that both projects required real-world data to be defined, modelled, and collected for digital twin representations. Once collected, data could be stored long-term and analysed and processed in manners that were suitable for the projects.

From a technical architecture perspective, these projects can be considered through the model-view-controller paradigm, in that data is collected in the model perspective, processed in the controller, and presented to users through the view. Traditionally, the model perspective would be realised through a relational database, either an SQL or no-SQL. However, for the smart metering and water sensing projects, this approach isn't ideal given the scope for dissimilar data formats, e.g., different types of sensing devices recording different types and formats of data, and the limited need for 'heavy' database functionality, e.g., relationships between different tables.

IOP Conf. Series: Earth and Environmental Science 1136 (2023) 012003 doi:10.1088/1755-1315/1136/1/012003

2 FIWARE as a technology for digital twins

FIWARE [4] describes itself as a curated framework of open-source platform components, based around several NGSI specifications [5] and a set of smart data models [6], with a context broker forming the core of a digital twin architecture.

Figure 1 details a typical FIWARE-based digital twin and highlights the three key context processes. Initially, contexts are generated with an IoT agent that collects data from devices, e.g., sensors, smart meters etc, and stores it on the context broker. Context transformation describes processes that use existing contexts to create new contexts or modify existing contexts, whilst context consumption describes processes that use existing contexts as inputs to present information to users.

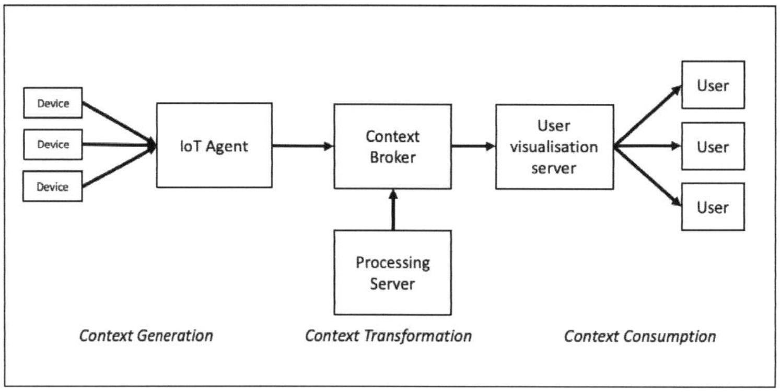

Figure 1. Typical FIWARE broker installation.

2.1 FIWARE for aqua3s

For the aqua3s project, a digital twin was developed through the periodic collection of sensor data in water distribution networks. The digital twin's users were SCADA staff who could use processed context data to assess the state of the water network and support decision making for operational issues. As per the FIWARE context broker architecture, sensor data was collected through the IoT Agent to generate appropriate context data using a pre-existing 'device' smart data model [7].
These device contexts were used as inputs to an alert process which created alert state contexts which were then used to determine the state of the system. Both the device and alert context were then consumed into user visualisations, Figure 2, with the left-hand image showing sensors visualised in a geographic context with alert context used to colour-code the sensors (green showing no current alert) along with related quantitative data. The right-hand image showing historic quantitative data from the sensor contexts.

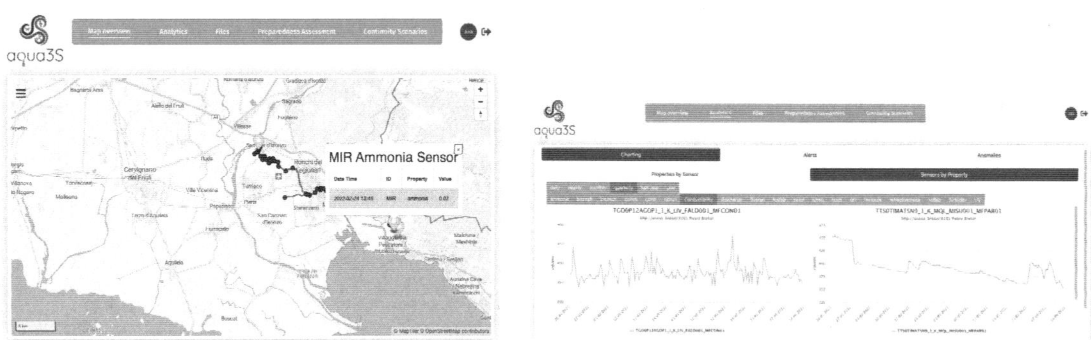

Figure 2. aqua3s user visualisations from FIWARE data. Current sensor data visualised with maplibre [8] (left), historic sensor data visualised with highcharts.js [9] (right).

2.2 FIWARE for Fiware4Water

For Fiware4Water, a digital twin was developed through the daily collection of customer water consumption through smart meters. The digital twin had two distinct users: staff at South West Water who would use the data to determine customer-side leaks through a leak reporting service and customers who could monitor their personal consumption through a mobile app. As per the FIWARE context broker architecture, smart meter data was collected daily and then processed to give daily individual and group consumption figures. This data was then accessed through the two visualisation applications on an ad-hoc basis, ***Figure 3***. The top image shows the staff leak reporting application, ranking meters by number of recent leak alarms and showing recent historic customer consumption. The bottom image shows the customer app, showing daily and historic consumption in a textual form, along with customer consumption ranking in their local group and a longer-term graphical consumption graph.

For this project, a 'WaterConsumptionObservered' model was developed and made available to the smart data models repository [10].

(a)

(b) (c)

Figure 3. Fiware4Water user visualisation from FIWARE data. Leak reporting application (a), customer consumption application (b)-(c).

3 Results

3.1 What worked well?
In general, FIWARE has been a good solution for both projects. FIWARE has a general model of 'ease of use' which comes across in most aspects of its operations. Providing context brokers as Docker images [11], [12] makes the installation and operation of brokers straightforward, notwithstanding docker issues. The context broker http request interface effectively contains just 4 commands (POST, PATCH, GET and DELETE) and the ngsi-ld/v1 spec demonstrates these operations clearly, far more so than the broker documentation, which can be hard to find.

The key benefit of FIWARE has been the use of contexts and smart models. Moving away from SQL and explicit table definitions to the JSON-LD format allows smart models to be authored in an extremely flexible manner using key-value pairs and is particularly useful in situations where a device may have multiple properties that it records. This was a huge benefit for both research projects as it enabled researchers to spend their time solving important research problems and not determining the most suitable table specification to capture arbitrary data. It should be noted that moving away from direct SQL interaction is something of a trend in web development, with Django's object relationship mapping [13] and SQLAlchemy [14].

3.2 What didn't work so well?
The first challenge of working with FIWARE is determining which broker to use as there is a wide range of solutions available which all appear to offer differing levels of functionality / adherence to the ngsi-ld specification, organisation, and requirements. For Fiware4Water, Stellio [15] was used which was presented as a single component broker which worked well for the project but did not support multiple-tenant (a core requirement for aqua3s) and tended to use a lot of resources, making it difficult to run on lightweight virtual servers for development and testing. Conversely, aqua3s used Orion-LD [12] for 'short-term' data storage and Cygnus [16] for longer-term storage, though approach was problematic to make work reliably and led to significant simplification of the device model that was used.

As expected, brokers had performance limitations, concerning sequential http requests, with a typical use case being the retrieval of entities. Whilst the ngsi-ld/v1 supports the entities command, it is limited to the number of entities it will return to keep the size of data packets down, so the normal approach is to query the broker for the number of instances of a given entity and then retrieve entities individually. This can be addressed to some degree with threading requests, but it does demonstrate the nature of http requests and the need to develop algorithms that are sympathetic to data collection.

One approach to mitigate this is the use of FIWARE's notification model, in that the broker can be programmed to send a request to a remote device upon the receipt of a particular request itself. The use case here is that when the IoT Agent sends a PATCH request to the context broker, the context broker can then send a notification to another http request server to perform some function based on the entity that has been patched. This was problematic for both the aqua3s and Fiware4Water projects, in that for aqua3s Orion-LD notifications appeared to be unreliable and for aqua3s, the operation Stellio broker could not resolve to local development machines for testing. Whilst this would clearly work in an operation context, we were reluctant to put untested functions straight into an operational environment.

From our experiences on both projects, it became apparent that broker functionality was geared around operations rather than development / testing activities. This became particularly apparent when brokers were loaded with synthetic data for testing. Typically, this would involve creating a new broker and a set of contexts using the POST request for each instance of context data. For Fiware4Water, we had a population of 100 smart meters with daily collections, giving a yearly backlog of around 36,500 data points. Aqua3s presented more of an issue with a pilot consisting of up to 20 devices and device data collected every 15 mins, giving around 100 data points per day per device, so 730,000 contexts as the total backlog to post, which was unrealistic to upload to the broker.

IOP Conf. Series: Earth and Environmental Science 1136 (2023) 012003 doi:10.1088/1755-1315/1136/1/012003

4 Developing a bespoke brokerage

During the development of the aqua3s project, development of context transformation and consumption functionality was delayed due to issues with setting up Cygnus to reliably manage historic data. Initially, delays were met with creating lightweight historic broker substitutes, but as the delays continued, it became apparent that little extra work was required to convert the 'lightweight historic broker' into a generic context broker. This work wasn't undertaken lightly as we would have preferred to use Stellio as an off-the-shelf solution, given that it was already being used, and working, on the Fiware4Water project. However, Stellio's resource requirements made it unusable on the lightweight development server we were running and Stellio's lack of multi-tenant support coupled with our lack of understanding of multiple container Dockerisation made it a non-starter.

The bespoke brokerage was built in Python using Flask as the http request frontend and SQLite as a backend database. Protocols were implemented using a mixture of the ETSI standards [5] and black box testing on Stellio by sending commands to the broker and comparing the Stellio results with the bespoke results. The broker was not designed to be anything like a full implementation, just an implementation of the features that were being used for development.

4.1 Benefits of the bespoke brokerage

The most obvious benefit of the bespoke brokerage was the development of an alternate patching strategy for building testing backlogs. An additional request was implemented that took an array of patch data for a given entity, rather than relying on multiple PATCH requests for each data point. This made it possible to build large backlogs in realistic times. For example, creating a broker with 3 tenants, each with 5 devices and 3 properties per device and 10 days of data with a 15-minute data point interval takes around 270 seconds on a broker using a 'normal' patching approach, of one patch per update, or 28 seconds using a patch array containing all the patch data in one request. Whilst this is clearly not a normal operational activity for a context broker, it's a key development / testing activity.

Having complete control over the broker opened a lot of opportunities based around status and logging. Working across multiple locations and time zones, it was always difficult to know what the status of a shared broker was, in terms of what tenants had been set up and what entities were in those tenants. Adding an unexe-broker/v1/version command returned the current state of the broker across all tenants, detailing all entity types and their instances.

Another area for improvement was broker logging. Whilst Stellio provides a lot of logging information through the Docker console, all the useful request and response information tends to get lost amongst internal component messages, typically Kafka and PostgreSQL. Whilst this is useful for Stellio developers, it didn't help our work. We realised that the logging interface could then be extended into a transaction log, which could prove invaluable for recording and re-running problematic broker requests, particularly during client testing sessions. This could also be coupled with an archival approach such that a copy of the broker can be dumped to disk prior to testing and any issues that occur would be stored in a transaction log. After testing, developers could then reload the initial archive and re-run the transaction log to determine the cause of the issue.

4.2 Issues of the bespoke brokerage

The most pressing issue of the bespoke brokerage is that it not FIWARE compliant and it doesn't completely follow the ngsi-ld specs. It follows the specs to the degree that we are using the specs in our projects, but it would be completely unrealistic to drop the broker into another project and expect it to work. However, the broker was not designed for such open-ended use cases, so it's not too much of a concern.

There is also the possibility that applications developed with the bespoke brokerage could 'drift' from FIWARE/ngsi specifications and end up using requests that are not supported by FIWARE brokers. This is a real risk, in that FIWARE brokers tend to implement interpretations of the ngsi specification and broker updates have included changes to request arguments. This is the nature of working with off-the-shelf software that is in development. From our side, we have developed a suite of black box tests for working with brokers, so issues can be raised and, hopefully, addressed.

The broker was developed as a Python/SQLite [17] application, so developer-time efficiency trumped run-time efficiency, therefore, the broker is likely to run into performance issues sooner or later. These are likely to concern the broker's ability to process multiple simultaneous clients, process database calls and so on. Again, the broker wasn't designed for such open-ended use cases, and it currently meets its performance goals.

5 Conclusions

The fundamental question to ask at this point is 'does it make sense to use FIWARE?' and, of course, the answer is 'yes, absolutely'. Like all technologies and frameworks, even the good ones are a bit rough around the edges and the key thing with them is that their value massively outweighs their issues. In general, the LD-JSON based context broker is a huge win for developers as it enables them to spend their time concentrating on solving real-world problems and not re-solving SQL representation issues. Additionally, ETSI's ngsi-ld/v1 specification makes a lot of sense for the use cases that developers will generally encounter with broker development.

For us, the bespoke brokerage has worked very well to support our development work on both projects. For aqua3s, it was a useful solution to continue with meaningful development and testing work whilst the 'proper' solution was in development. For the Fiware4Water project, although we had a working operations broker from day one of the project, the bespoke brokerage gave us a shared broker to use for development and testing work given our reluctance to mix development and operations environments.

The broker has allowed us to create development and test functionality that may be of little to no use to the operational role of FIWARE brokers but has made a huge impact on the work we have undertaken.

Developing a broker has given us a lot of insight into how brokers can work and the relationship between smart data models and underlying broker functionality. At the start of these projects, there was a conventional wisdom that smart data models should be small, in much the same way that classes are 'small', or at least of a single responsibility. The use of 'observedAt' in smart models effectively breaks a single model up into parts, given that the observedAt element of the model is stored in a time-based manner. From that, it makes sense to create fewer smart models with more observedAt elements as the model can be retrieved with a single http request. However, this may be problematic with Cygnus.

This paper isn't a stick to hit FIWARE with, it's just a report on the issues we ran into with doing development and testing work with FIWARE and what we did to address them. As researchers, we can 'travel light and break things' with our research and highlight findings that we think may be of interest to the broader development community.

Acknowledgements

The work presented in this paper was funded by the ongoing EC H2020 aqua3S (GA 832876), Fiware4Water (GA 821036) and LOTUS (GA 820881) projects.

References

[1] Aqua3s project homepage https://aqua3s.eu/ (accessed Jul. 12, 2021)
[2] Fware4Water homepage https://www.fiware4water.eu/ (accessed Jul. 12, 2021)
[3] FIWARE Components https://www.fiware.org/developers/catalogue/ (accessed Apr. 01, 2022)
[4] Fiware https://www.fiware.org/ (accessed Jul. 12, 2021)
[5] ETSI Context Information Management (CIM); NGSI-LD API
[6] FIWARE Smart Data Models https://www.fiware.org/smart-data-models/ (accessed Apr. 01, 2022)
[7] FIWARE DataModel.Device https://github.com/smart-data-

IOP Conf. Series: Earth and Environmental Science 1136 (2023) 012003 doi:10.1088/1755-1315/1136/1/012003

models/dataModel.Device/blob/master/Device/README.md (accessed Apr. 01, 2022).
[8] MapLibre Homepage https://maplibre.org/ (accessed Apr. 01, 2022)
[9] Highcharts Homepage https://www.highcharts.com/ (accessed Apr. 01, 2022)
[10] FIWARE DataModel.WaterConsumption https://github.com/smart-data-models/dataModel.WaterConsumption/tree/master (accessed Apr. 01, 2022)
[11] Stellio Docker Image https://github.com/stellio-hub/stellio-context-broker (accessed Apr. 01, 2022)
[12] FIWARE Orion-LD Docker Image https://github.com/FIWARE/context.Orion-LD (accessed Apr. 01, 2022)
[13] An introduction to the Django ORM https://opensource.com/article/17/11/django-orm (accessed Apr. 01, 2022)
[14] The Python SQL Toolkit and Object Relational Mapper
[15] Stellio Context Broker https://github.com/stellio-hub/stellio-context-broker/blob/develop/API_Quick_Start.md (accessed Jul. 12, 2021)
[16] Connecting Orion Context Broker And Cygnus https://fiware-cygnus.readthedocs.io/en/0.13.0/user_and_programmer_guide/connecting_orion/index.html (accessed Apr. 01, 2022)
[17] SQLite Homepage https://www.sqlite.org/index.html (accessed Apr. 01, 2022).

IOP Conf. Series: Earth and Environmental Science 1136 (2023) 012004 doi:10.1088/1755-1315/1136/1/012004

An artificial intelligence approach for managing water demand in water supply systems

Ariele Zanfei[1], Andrea Menapace[1], Maurizio Righetti[1]

[1] Faculty of Science and Technology, Free University of Bozen-Bolzano, Piazza Università 5, Bolzano, Italy.

corresponding author email: ariele.zanfei@natec.unibz.it

Abstract. Water demand management is essential for water utilities, which have the critical task of supplying drinking water from water sources to end-users through the distribution network. Therefore, the water utilities have to make decisions for the current and future functioning of the water distribution system. In this context, the artificial intelligence approach with data-driven methods can be used to develop powerful tools to improve overall water management. In fact, data-driven methods can model water demands for plenty of tasks and applications such as demand forecasting or anomaly detection. In this work, we propose and discuss a practical application of an artificial neural network to model the urban water demand of a water supply system. The flexibility of the proposed method allows the prediction of water demand on different horizons. Moreover, this developed model can effectively support water utilities on different operational schedules and decision tasks.

1. Introduction

The water resource is asking for more sustainable and efficient management in the future. In the world of water management, the constant increase of water demands from agriculture, energy production and humans, due to climate change and socioeconomic factors is forcing researchers to deal with the challenging request to optimize as much as possible the management of the available water resource [1]. In this context, the era of Big Data and Machine Learning provides innovative ways to accomplish a more sustainable future.

Keywords such as systems optimization, smart systems, and digital water are just some of the research challenges for sustainable urban water management [2]. In the latest years, hydro informatics research played a crucial role in developing new techniques and methodologies for supporting the sustainable management of urban water, dealing with this discipline that "*can be thought of as a continuous process of developing and using water data, models and tools, to understand the environment, to engage all stakeholders, and help make decisions that improve society*" [3,4]. In the latest decades, artificial intelligence and machine learning have been the object of increased research activity in the hydraulic field [5–8].

The aqueducts infrastructures, along with the energy grids, have undertaken an important renewal process in the last years that aims at transforming the current distribution networks into smart grids [9,10]. The reason for these demanding transitions into green and smart systems lies in the climate change that is facing our planet, which will keep increasing the global water demand [11]. A combination of both aqueduct and water demand management strategies is necessary in order to adapt to a varying environment and the associated uncertainties. As result, new challenges related to water-energy nexus and energy recovery are rising in this new era and are demanding sustainable and eco-design solutions, due to restrictions imposed by climate change, requiring more flexible and new adaptive answers [12].

Water demand modeling is key for better managing the water resource. In fact, adopting data-driven algorithms that can learn from the demand data is the basis for many applications that are fundamental for the proper water management [13]. For instance, developing a short-term model able to accurately predict the following hours of demand can help for the anomaly detection task [14]. Differently, a model that aims to predict the following 24 or even more hours of demand, can provide a useful indication regarding the operational management of the systems. Furthermore, a model that forecasts longer horizons can provide a useful indication for the planning and management of the system. This paper discusses how a conventional neural network model can provide these indications.

2. Materials and methods

To provide a discussion on the capability of modeling water demand with nowadays methodologies, it is proposed to train a conventional multilayer perceptron model (MLP) for three prediction tasks. In fact, modeling the demand and predicting the demand are two concepts strictly similar. In both cases, a model has to learn the demand. When predicting, the model has to provide the values in a time horizon that are the results of the learning task, meaning the prediction is accurate as the model was able to understand the characteristics of the demand. In this study, it is proposed to assess an MLP model for modeling the demand both with hourly and daily time resolutions.

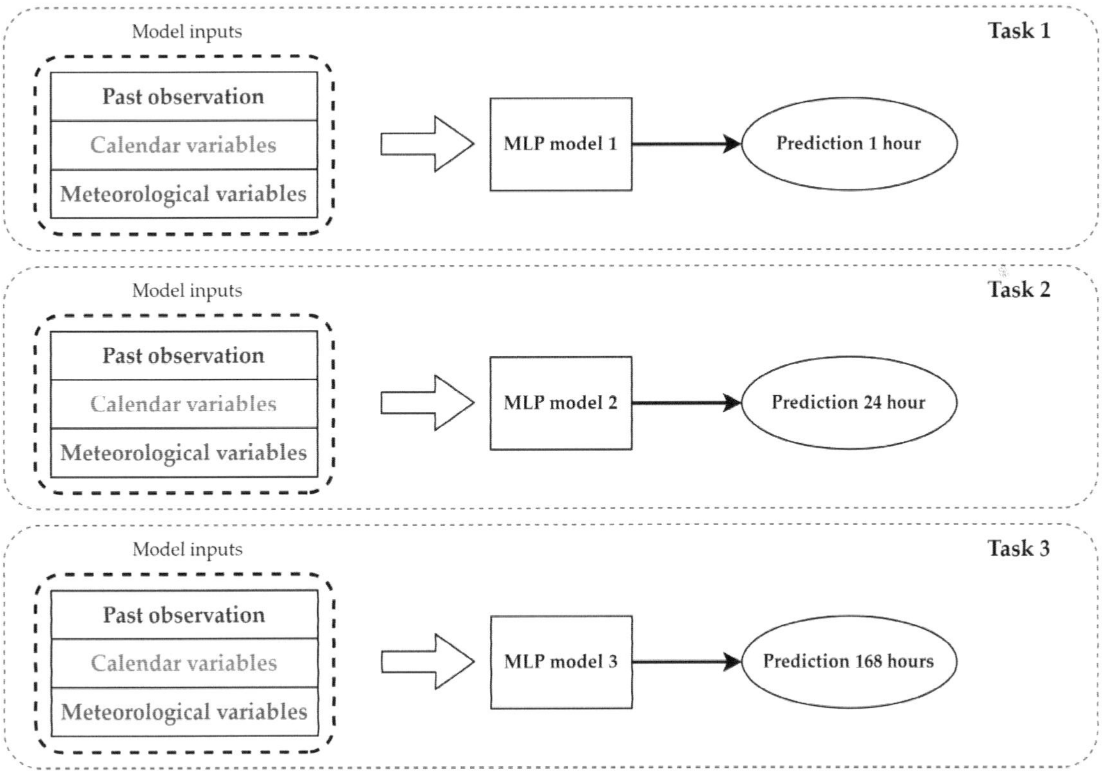

Figure 1. Proposed prediction tasks.

Therefore, it is proposed to perform three prediction tasks: predicting the following 1-hourly value (named Task 1), predicting the following 24-hourly values (named Task 1), and predicting the following 168-hourly values (named Task 3). Figure 1 summarizes the procedure.

To evaluate this problem, the MLP models are trained on a real water demand time series. The inputs for these models vary depending on the Tasks. In addition, for all the models, the input adopted comprises past observations of the demand, calendar variables (i.e., day type, month, hour of the day, day of the month), and meteorological variables (i.e., temperature, radiation, humidity and rainfall). The models are trained over 6 years of data and tested on almost 1 year of data.

2.1. Model tuning and metrics

In order to build a reliable and robust model, the hyperparameters of the neural networks model have to be tuned [7]. At this aim, the three MLP model are tuned through a conventional grid search procedure, and the results are summarized in Table 1.

Table 1. Summary of the hyperparameters.

Parameters	MLP1	MLP2	MLP3
Layers	3	3	2
Nodes	18 – 32 - 18	32 – 48 - 32	48 - 72
Activation function	ReLu	ReLu	ReLu
Optimizer	Adam	Adam	Adam
Batch size	24	1	1
Number of epochs	100	150	180
Learning rate	0.001	0.001	0.001

In addition, in order to evaluate the model performances, the following metrics of the mean absolute percentage error (MAPE), the determination coefficient (R^2) and the mean absolute error (MAE) are used:

$$MAPE = \frac{100}{N}\sum_{i=1}^{N}\left|\frac{Y_i - \hat{Y}_i}{Y_i}\right| \quad (1)$$

$$R^2 = 1 - \frac{\sum_{i=1}^{N}\left(Y_i - \hat{Y}_i\right)^2}{\sum_{i=1}^{N}(Y_i - \bar{Y}_i)^2} \quad (2)$$

$$MAE = \frac{1}{N}\sum_{i=1}^{N}\left|Y_i - \hat{Y}_i\right| \quad (3)$$

where N is the number of predicted values, while $\bar{Y}_i$ represents the average of the observed values, Y_i and $\hat{Y}_i$ are the observed and forecasted values at the i_{th} hourly time step, respectively.

3. Results and discussion

To evaluate the performance of the models in each task, it is proposed to calculate the well-known metrics of the mean absolute error (MAE), the mean absolute percentage error (MAPE), and the determination coefficient (R^2). These indicators are calculated by comparing the output of the models with the real values. The results are reported in Table 2.

Table 2. Description of the table content.

Metrics	Task 1	Task 2	Task 3
MAE (l/s)	1.14	1.52	1.73
R^2 (-)	0.88	0.79	0.73
MAPE (%)	9.47	13.07	14.49

As expected, the metrics highlight superior performances for the models that reproduced the demand for fewer hours of horizon. It is clear that, the higher the horizon, the higher the uncertainty. This led to a more complex task for the data driven model, which results in a lower ability of the model to accurately reproduce the water demand time series. However, it is also worth to mention that the models supply different information from the utility point of view. For instance, the higher error made in Task 3 is associated to a prediction that covers a whole week of operation, which make the higher uncertainty more acceptable.

Furthermore, Figure 2 highlights 3 weeks of prediction of the models for the 3 proposed tasks.

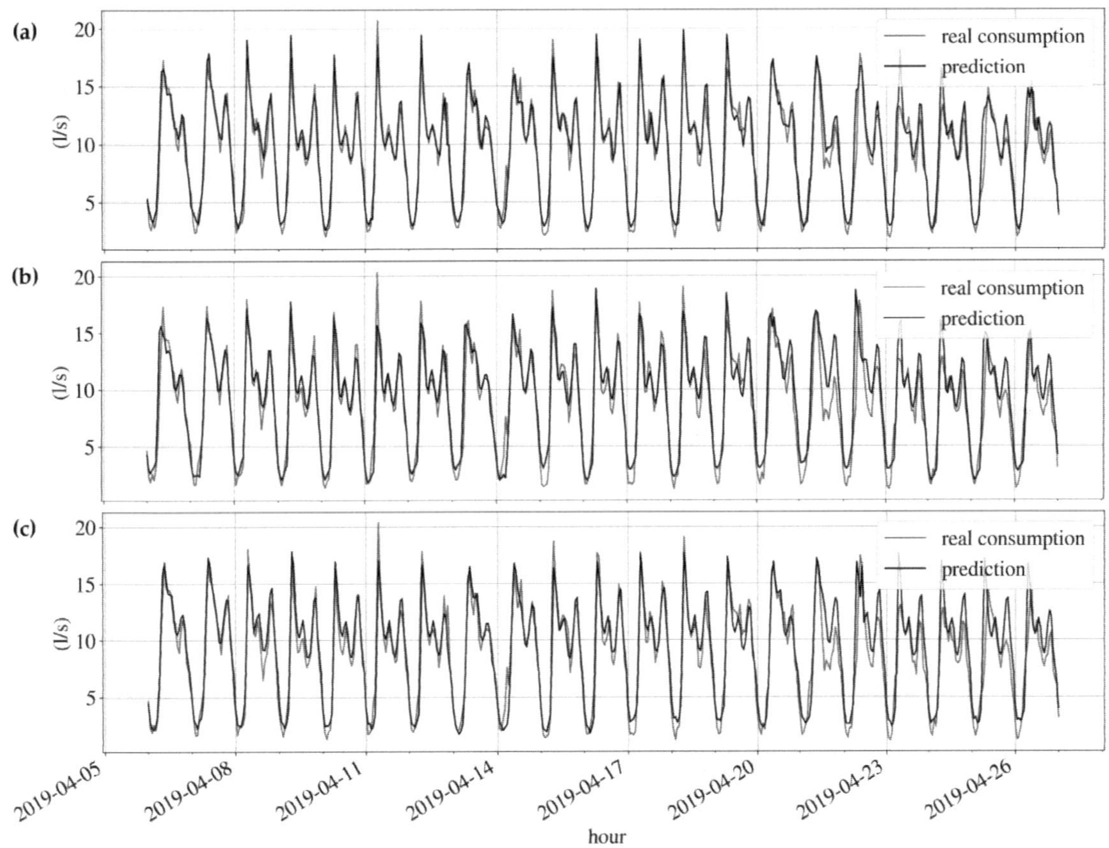

Figure 2. Results of the proposed prediction tasks. Panel (a) shows task 1, panel (b) task 2, and panel (c) the task 3.

It is important to note that all the models are capable to well model the demand and providing an accurate prediction. However, these models are consistently different. The model of task 1 is the one performing better, but is also the one with the easier task, predicting just 1 value. In this case, accuracy is crucial because such a model is typically used for anomaly detection algorithms. Therefore, its accuracy is needed for detecting fast possible anomalies like leaks, intrusions and bursts. The model of task 2 is still accurate, but clearly less than the one of task one since it is simultaneously predicting 24 values. In this case, the model could have a different aim, like managing tanks and pumps for operational activities. Finally, the model of task 3 is less accurate since it is predicting 168 values simultaneously. As already briefly discussed, this model has not to be accurate as the others, since the aim is to support the weekly management of the system.

4. Conclusion

Water demand prediction is an important task for the sustainable and optimal management of water distribution systems. In fact, the need to optimize and improve as much as possible the use of the water resources has become every year more important, due to climate change that is modifying the water availability around the globe. In the latest decades, the improvement and the novel creation of models for this task has been the target of plenty contributions. Furthermore, due to the central role achieved by machine learning and deep learning algorithms, methods like neural networks have become the state-of-art in this field.

This study discussed the importance of models that can accurately learn and predict water demand. In this case, it is proposed a series of short-term models with different aims, and these latter are discussed. It is shown also the different grade of complexity that is hidden behind these tasks, and also that the performance decrease with the increase of the horizon. It is also discussed that this decrease of performance is acceptable due to the different scopes of the involved models.

Future works will address more the role of meteorological inputs in the context of short-term water demand forecasting.

References

[1] A. Zanfei, A. Menapace, F. Granata, R. Gargano, M. Frisinghelli, M. Righetti, 2022 An Ensemble Neural Network Model to Forecast Drinking Water Consumption, *J. Water Resour. Plan. Manag.* ***148*** *04022014*. https://doi.org/10.1061/(ASCE)WR.1943-5452.0001540.

[2] D. Savic, 2019 Artificial Intelligence-How can water planning and management benefit from?.

[3] Makropoulos, Savić, 2019 Urban Hydroinformatics: Past, Present and Future, *Water.* ***11*** *1959*. https://doi.org/10.3390/w11101959.

[4] A. Menapace, D. Avesani, 2019 Global Gradient Algorithm Extension to Distributed Pressure Driven Pipe Demand Model, *Water Resour. Manag. 1–20.*

[5] A. Menapace, D. Avesani, M. Righetti, A. Bellin, G. Pisaturo, 2018 Uniformly Distributed Demand EPANET Extension, *Water Resour. Manag.* ***32*** *2165–2180.* https://doi.org/10.1007/s11269-018-1924-6.

[6] A. Menapace, A. Zanfei, M. Felicetti, D. Avesani, M. Righetti, R. Gargano, 2020 Burst Detection in Water Distribution Systems: The Issue of Dataset Collection, *Appl. Sci.* ***10*** *8219*. https://doi.org/10.3390/app10228219.

[7] A. Menapace, A. Zanfei, M. Righetti, 2021 Tuning ANN Hyperparameters for Forecasting Drinking Water Demand, *Appl. Sci.* ***11*** *4290*. https://doi.org/10.3390/app11094290.

[8] M. Righetti, C. Bort, M. Bottazzi, A. Menapace, A. Zanfei, 2019 Optimal Selection and Monitoring of Nodes Aimed at Supporting Leakages Identification in WDS, *Water.* ***11*** *629*. https://doi.org/10.3390/w11030629.

[9] S.W. Lee, S. Sarp, D.J. Jeon, J.H. Kim, 2015 Smart water grid: the future water management platform, *Desalination Water Treat.* ***55*** *339–346.* https://doi.org/10.1080/19443994.2014.917887.

[10] H. Ramos, A. McNabola, P. López-Jiménez, M. Pérez-Sánchez, 2019 Smart Water Management towards Future Water Sustainable Networks, *Water.* ***12*** *58.* https://doi.org/10.3390/w12010058.

[11] X. Wang, J. Zhang, S. Shahid, E. Guan, Y. Wu, J. Gao, R. He, 2016 Adaptation to climate

change impacts on water demand, *Mitig. Adapt. Strateg. Glob. Change.* ***21*** *81–99.* https://doi.org/10.1007/s11027-014-9571-6.

[12] H.M. Ramos, A. Carravetta, A.M. Nabola, 2020 New Challenges in Water Systems, *Water.* ***12*** *2340.* https://doi.org/10.3390/w12092340.

[13] A. Zanfei, B.M. Brentan, A. Menapace, M. Righetti, M. Herrera, 2022 Graph Convolutional Recurrent Neural Networks for Water Demand Forecasting, *Water Resour. Res. 58.* https://doi.org/10.1029/2022WR032299.

[14] A. Zanfei, A. Menapace, B.M. Brentan, M. Righetti, M. Herrera, 2022 Novel approach for burst detection in water distribution systems based on graph neural networks, *Sustain. Cities Soc. 104090.* https://doi.org/10.1016/j.scs.2022.104090.

IOP Conf. Series: Earth and Environmental Science 1136 (2023) 012005 doi:10.1088/1755-1315/1136/1/012005

Profiling urban water consumption using autoencoders and time-series clustering techniques

Danillo Lange[1], Marc Ribalta[1], Lluís Echeverria[1] and Joshua Pocock[2]

[1] Eurecat Technology Centre, Unit of Applied Artificial Intelligence, Parc Científic i Tecnològic de Lleida, 25003, Lleida, Spain

[2] South West Water Limited Registered office: Peninsula House Rydon Lane Exeter Devon, England EX2 7HR

danillo.lange@eurecat.org

Abstract. Nowadays, water utilities face the rising challenge of ensuring water availability amidst a rapidly growing society and a shifting climate. Our research aims to develop a household clustering solution based on water consumption behaviour in Southwest England, to enable utilities to identify different profiles and enhance customized control of household consumption, resulting in improved resource management. The solution is composed of three modules. The first one is based on a K-Means clustering model, designed to group domestic water use behaviours. This module uses the Dynamic Time Wrapping algorithm as a similarity mechanism to process the high-resolution water meter data. In parallel, the second module processes the market segmentation data through an Autoencoder, a specific Neural Network architecture used to reduce the high dimensionality of such data to a low dimension dataset by extracting its latent encoded space. Finally, to assemble the final household water use profiles, a blending K-Means algorithm is used to merge previous modules outputs, based on the Euclidean distance. The solution provides insightful information to water companies to better understand consumer behaviour, habits, and routines.

1. Introduction

Water is considered a key element to the growth and development of communities. In this context, many factors influence water demand, such as climate features, social and economic conditions of households, seasonal characteristics, temperature, etc. Understanding such consumption patterns becomes critical to water supply companies and identifying and characterizing this kind of consumer behaviour may result in improved comprehension of social necessities and better water resource management.

Characterizing water consumption is of great importance to water companies and has been an intensive study subject for many years. As previously stated by Laspidou *et al.* [1] in their preliminary analysis, urban water consumption can be useful to distinguish between households according to a clustering solution, even if the data is not real-time. Moreover, Leitao *et al.* [2] stated by its novel framework of water consumption patterns detection, based on time-series analysis, that some conclusions can be extracted from such clustering, to improve water management and resource saving. Finally, although being in a different area of study, Chatterjee *et al.* [3] concluded that a clustering ensemble solution can provide better performance when compared to similar approaches. Based on the context provided by those previous works, this work is proposed.

The presented work aims at providing water supply companies with an insight into identified behaviour across urban households by analysing historical values of consumption and other factors such as economic, social, and weather parameters. We propose an Artificial Intelligence (AI) data-driven ensemble approach that combines multiple clustering algorithms with time-series data analytics and advanced dimensionality reduction techniques. As a result, a predictive model is obtained, which can provide a real-time classification, among different behaviour groups, based on any given customer's short-mid and long-term historical and static information to validate it. We demonstrate the results in a specific case in the Southwest England area.

This document is structured as follows: Section 2 details the problem and introduces the available data set used for the study. Section 3 explains the employed methodology. Section 4 covers the results and discussions that it apports, and, finally, Section 5 is composed of conclusions of the study.

2. Problem and Data Set understanding

The Southwest England area has numerous households with varying sizes, incomes, and additional socio-economic factors, which results in a very diverse water consumption behaviour. This variety makes the water management task difficult for water utilities. With this problem in mind, there is an increasing need for a way to identify different consumption behaviours and cluster similar households together.

Table 1. Household consumption features explanation.

Feature	Description	Unit	N. of registers
Sensor type 1 readings	Water consumption registered from the sensor with the highest definition, in a 15-minute interval.	Litres/15 minutes	42,414,797
Sensor type 2 readings	Water consumption registered from sensor 2 in a 30-minute interval.	Litres/day	5,630,035
Sensor type 3 readings	Daily water consumption in litres registered from sensor 3.	Litres/30 minutes	820,313
Rain information	Rain historic information and location of rainfall stations	mm/string	2,047,668

Table 2. Household characteristics features explanation.

Feature	Description	Unit	N. of registers
Financial category	Represents the individual power level, providing financial services segmentation.	Categorical	4,918
Socio-economic category	Individual classification of consumers based on their net worth, combines income, family composition and financial assets.	Categorical	4,918
Neighbourhood characteristics	Demographics and neighbourhood characteristics.	Categorical	4,592
Garden Size	The floor size of the property's garden.	m^2	5,090
Building Size	The floor size of the property's building.	m^2	5,090
Household Size	Estimated number of persons that resides on the property.	Integer	5,111
Children at Home	Estimated number of children residing on the property	Integer	5,111

The data set used to demonstrate our developments, shown in Table 1, is composed of a high-resolution time-series water consumption collected from 3 different types of smart sensors scattered across the area, with dates in the range from 2016 to 2021 at an interval between 15-minutes and a day.

Moreover, it also includes a rainfall dataset containing the height, in millimetres, with irregular intervals, and lastly, a data set containing demographic characteristics and financial level segmentation of said households, presented in Table 2.

3. Material and methods

Our solution is composed of three modules. The first module covers a household consumption clustering [4] based on time-series data similarity using the Dynamic Time Warping (DTW) [5] distance metric, allowing time series of different lengths and shapes to be compared. In parallel, the second module processes the static sampled properties characteristics and meteorological features aiming to reduce its dimensionality via its latent representation identification. This step is conducted using a Deep Neural Network (DNN) implementation called Autoencoder [6,7].

Finally, the outputs of modules one and two are combined and clustered to generate groups representing the high-resolution input data set and suitable to help water supplies identify profiles and understand households' consumption behaviour in the Southwest England region.

The general strategy followed in this study is composed of: (a) Data preparation; (b) Cluster Assessment; (c) Dimensionality Reduction; (d) Final clustering; and (e) test and validation. This methodology is presented in Table 3.

Table 3. Solution methodology.

Solution Methodology
1) Data Preparation a) General data treatment b) Assemble daily time series joined data set c) Assemble the high-dimension data set 2) Cluster Assessment (K-Means + DTW) a) Season identification b) Evaluation Metrics 3) Dimensionality Reduction and latent space identification (Autoencoder) a) Neural Network structure and Autoencoder for dimensionality reduction 4) Clustering Ensemble (K-Means + Euclidean distance) a) Data joining b) K-Means algorithm + Euclidean distance metric

Following, in Figure 1, the solution's architecture is represented in flowchart format, each section or model is composed of the input data model and the respective algorithm.

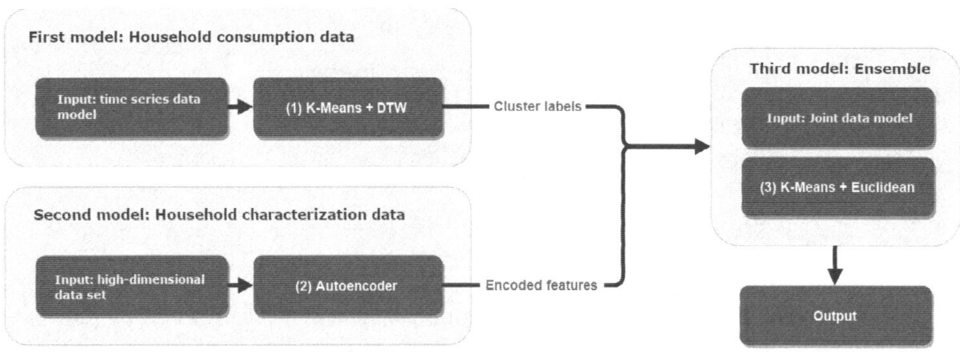

Figure 1. Methodology's representation flowchart.

3.1. Data preparation

3.1.1. General data treatment. While working with time-series data, some pre-processing steps are necessary because some error may occur during data collection, or some specific behaviour registered in the data can negatively affect the development of ML models.

The tasks performed in the earlier steps of the data analysis were the analysis of outliers and consecutive zeroes in the historic water demand data. To deal with outliers, we have decided to compare every value of the historic data set against a defined threshold at a local (property) scope. That The values follow this criterion, seen in equation (1), are then replaced by their local median value; this process was also applied to household characteristics data fields as well.

$$Value > (\overline{x} + 3 \times \sigma) \tag{1}$$

where $\overline{x}$ = mean value and $\boldsymbol{\sigma}$ = standard deviation, both calculated locally.

We have come to an understanding that 30 or fewer days with consecutive zeroes in readings are normally considered vacations of some kind, and that is why the team has decided to cut from the time series any period with more zeroes than that.

Moreover, missing, and negative values were treated as well. In this case, such values were replaced by the average value of the corresponding data field.

3.1.2. Assemble daily consumption time series joined data set. The strategy at this point was to employ DTW as a distance metric in the K-Means algorithm. As described by Huang et al. [8], it is possible to efficiently employ DTW into a K-Means to cluster time-series of different shapes and lengths, enabling a dynamic and more flexible comparison.

Combining the time series of each household (including all different smart sensors) resulted in a training dataset where each register corresponded to a consumption time series of a given household. The frequency of the readings was adjusted to be at a daily level, to make use of all sensors and achieve records with a reasonable number of registers. Table 4 shows the data model of the joined data set at this point, and later, it will be adapted to the accepted shape of the clustering algorithm.

Table 4. The data model of joined time series for the first model.

Data field	Description	Unit
Readings	Water consumption value	Litres/day
Date	Date of registered event	Date
Reference	The household identification	String
Season	Which season the register belongs to	String

3.1.3. Assembly of the household characterization data set. The purpose of the household characterization data set is to represent all quantitative and qualitative static (non-time-series) data that is available for each household.

Almost all available data fields were considered resulting in a high-dimensional data set, including for each register (household): a) Average consumption of weekdays weekend days, and weekly values for each season; b) Properties characteristics such as building and garden size; c) Rainfall data; d) Occupancy, social and financial categories (which were converted from categorical variables into binary fields); the data set achieved in the end around 147 different features. Some of the latter data fields are highlighted in Figure 2.

Furthermore, this data set is composed of both numeric features and categorical data fields, as can be observed in Table 5, the categorical features were later converted to indicator variables.

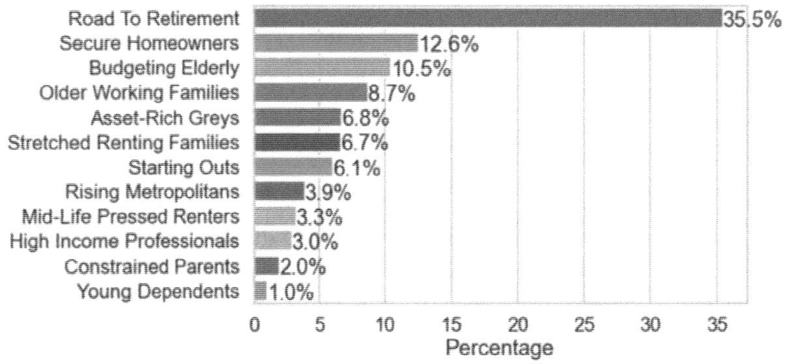

Figure 2. Percentage spread of socio-economic categories of households.

Table 5. The household characterization dataset.

Data field	Description	Unit
Seasonal weekday average	Water consumption average value on weekdays (for each season).	Litres/day
Seasonal weekend average	Water consumption average value on weekend days (for each season).	Litres/weekend
Seasonal weekly average	Weekly water consumption average value (for each season).	Litres/week
Daily rainfall average	Daily rainfall average calculated for each household, based on the closest rainfall station.	mm/day
Max value	Maximum consumption value.	Litres
Mean value	Mean consumption value.	Litres
Percentiles (25, 75)	Percentile values of consumption (25% and 75%)	Litres
Property characteristics	Characteristics include Garden and building size, and the number of children at home.	m^2/Integer
Socio-economic categories	General demographic and neighbourhood characteristics (6 categories and 49 sub-categories).	Categorical
Financial categories	Individual power's value, based on a research survey (10 categories and 50 sub-categories).	Categorical

3.2. Cluster assessment (K-Means + DTW)

3.2.1. Season identification. As identified by Ibrahim et al.[9], different seasons can influence water consumption behaviour, because of that, daily consumption time-series data was segmented into seasons, and individual ML models were generated for each period. The split of seasons was executed following United Kingdoms' season definition as follows: a) Spring was the months of March, April, and May; b) Summer, June, July, and August; c) Autumn, September, October, and November; and lastly, d) Winter was composed of December, January, and February.

Additionally, the time series data were also separated by year. We have decided to work exclusively with the years 2017 and 2018, mostly because 2019 had some gaps on it during the crucial season of summer, and 2020 was thought to be not a representative year due to the behaviour change that the sanitary crisis bought [10].

3.2.2. Evaluation metrics. K-Means is an unsupervised learning algorithm widely used for data clustering [4]. K-Means works by randomly assigning k cluster points in the hypervolume of the data set, then it assigns each data register to the closest cluster centre and recomputes the cluster centres based on current memberships. This process is repeated until it converges on the defined criterion.

To find the ideal number of clusters, the following scoring metrics were used:

- **Elbow method (EM):** Heuristic used in determining the number of clusters, consists of plotting the *explained variation* (also known as the Sum of Squares of the distance between data points, SSE) against the number of clusters, where it is visible that after the inflexion curve (resembling an elbow) the model fits better.
- **Silhouette Coefficient (SC):** Resembles how similar a sample is within its cluster when compared with other clusters, takes values between -1 and 1, with values closer to 1 meaning well-separated clusters.

Additionally, the K-Means model was trained using random initialization of centroids.

3.3. Dimensionality reduction and latent space identification (Autoencoder)

3.3.1. Deep Learning AutoEncoder. We propose the use of an AutoEncoder (AE), a specific DNN architecture, for the dimensionality reduction task. An AE is widely used for noise reduction and image compression. It is composed of three main segments which are the encoder, responsible for shrinking the original data into compressed data into a second segment known as the bottleneck, and a third one called the decoder, which does the reverse work of the encoder, transforming the low-dimensional data into the original shape. After this process, the output is then compared to the input to see if the DNN managed to reconstruct the data based on its compressed state reasonably well. In this schema, the bottleneck layer is forced to identify the latent space of the input information and can be used for dimensionality reduction purposes with non-linear capabilities.

The strategy is to extract the low-dimensional data from the bottleneck, the latent space, to use as an encoded feature that represents every property. Moreover, this extraction was only made after the AE evaluation (based on the reconstruction of data) was acceptable.

3.3.2. AutoEncoder architecture. The layers defined for our DNN were the following: a) Encoder, composed of Linear and ReLU layers with the input of 147 features and output of 3 features; b) Bottleneck data; c) Decoder set of layers including Linear and ReLU with the input of 3 features and output of 147 features. This structure is normally adopted for dimensionality reduction. Aside from the structure, some hyperparameters were optimized for this NN:

- **Optimizer:** Adam, an extension of stochastic gradient descent that is broadly used for deep learning applications.
- **Learning Rate:** 1e-2, parameter tuned to achieve an acceptable performance after 1000 epochs, values considered on hyperparameters optimization: 1e-2, 1e-3 and 0,1.
- **Weight decay:** 1e-5, a parameter used to penalize complexity, is a way to contour problems of underfitting or overfitting in NN. Values considered on optimization: 1e-5, 1e-4, and 1e-3.
- **Criterion (Evaluation):** MSE Loss, measures the loss as *Mean Squared Error* between each element in input x and target y.

3.4. Clustering Ensemble

3.4.1. Data joining. As described by Zhou et al. [11], ensemble techniques applied to supervised and unsupervised learning algorithms can result in a better performance overall. More specifically, combining algorithm outputs for the execution of a K-Means can provide clusters with a higher similarity and quality.

Joining the data is an important step of the study, and that is accomplished by converting the cluster labels from the first model to indicator variables (binary) and appending the encoded features extracted from the latent space of the AE. The joining is made at a household level, meaning that each register that represents a property, have both outputs combined and used as training data for the next K-Means model. The respective data model can be observed in Table 6.

Table 6. The data model of joined training data set.

Data field	Description	Unit
Encoded feature 1	First AE encoded feature	Float
Encoded feature 2	Second AE encoded feature	Float
Encoded feature 3	Third AE encoded feature	Float
Label 2017	Cluster label that was assigned to the 2017 model (K-Means + DTW)	Integer
Label 2018	Cluster label that was assigned to the 2018 model (K-Means + DTW)	Integer

The two latter variables were converted into binary indicator variables, so, finally, the data set would be composed of 18 features. Once again, one data set per season was made to train the individual models (one for each season).

3.4.2. K-Means algorithm + Euclidean distance metric. Like the first model, this K-Means implementation adopted the random initialization of centroids and selection of value k made by analysing the EM (SSE), SC, and CH metrics. Moreover, this model adopts the Euclidean method as a distance metric, widely adopted by default in the industry.

Another different step in this model was the necessity of scaling the data set beforehand due to the different nature and numeric scale of joined outputs which composes the training data set.

4. Results and discussions

4.1. First model (K-Means + DTW)

As defined earlier, we generated four different K-Means models, each one with its own data set (separated by season). The choice of the value *k* came based on the analysis of the values of the EM (SSE) and the SC. The value can be observed in Table 7.

Table 7. Evaluation metrics for the first model.

Season	SSE for k = 6	SSE for k = 7	SSE for k = 8	SC for k = 6	SC for k = 7	SC for k = 8
Spring	1.3342	1.2845	**1.2760**	0.0535	0.0650	**0.0476**
Summer	1.3383	1.2841	**1.2868**	0.0573	0.0536	**0.0456**
Autumn	1.3534	1.2821	**1.2797**	0.0567	0.0587	**0.0609**
Winter	1.3228	1.3233	**1.2874**	0.0505	0.0458	**0.0471**

The values marked in bold were the choices for the number *k* of clusters, based on internal discussions and the metrics obtained, the chosen *k* value for all models was 8.

4.2. Second model (Autoencoder)

The data that resides in the bottleneck is the one that will be used as a low-dimension version of the data, given that the decoder made a good job reconstructing it. The Autoencoder also allows for use of non-linear data.

The AE was evaluated by measuring the loss metric through all the epochs of training, as defined earlier. Moreover, an additional measuring during the validation phase was implemented. Those values can be observed in Table 8. Additionally, 70% of the data set was used to train the model, while the other 30% was used to evaluate it.

Table 8. Training and validation losses of the autoencoder, from epochs 1 to 1000.

Epoch	Train loss (MSE)	Validation loss (MSE)
1	0.9893	0.4063
200	0.4872	0.3547
600	0.4071	0.3285
1000	0.3889	0.3131

4.3. *Third model (K-Means + Euclidean distance)*

To evaluate the ensemble model, three different metrics are employed: the previously discussed EM and SC, but also the Calinski Harabasz index (CH), which is a metric best suited for Euclidean distance-based clustering.

Calinski Harabasz Index (CH) [12], also known as Variance Ratio Criterion (VRC), represents the ratio between the within-cluster dispersion and the between-cluster dispersion, higher values mean well-separated clusters.

As can be observed in Figure 3, the SC values for each model differ, but, overall, they represent well-defined clusters while maintaining a reasonable low value for *k*. Moreover, Figure 3 also shows the EM values for the ensemble model:

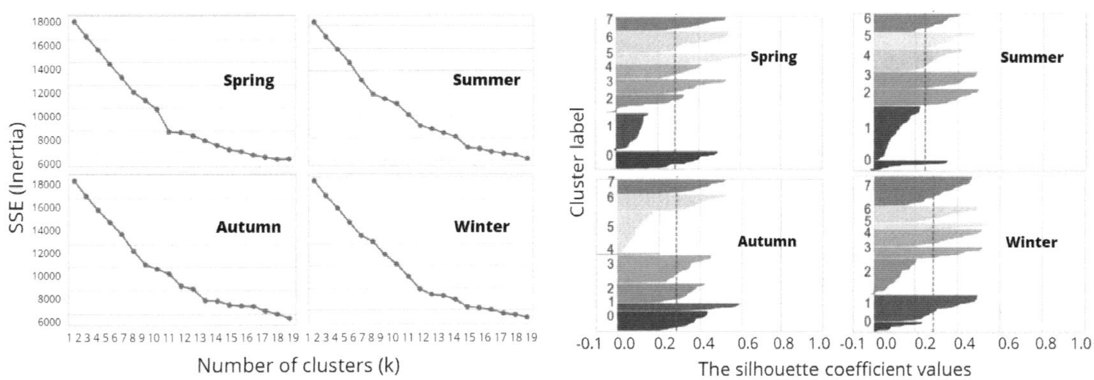

Figure 3. Left) Elbow method of the ensemble models; Right) Silhouette coefficient values for the ensemble models.

Next, in Table 9 a summary of the evaluation metrics for each ensemble model is represented, based on that, we can conclude that the spring and autumn models achieved the highest metrics and the summer model, had the worst performance overall.

Table 9. Evaluation metrics of the ensemble model.

Season	N. of clusters	SSE	SC	CH
Spring	8	9855.568	0.28443	112.6678
Summer	7	11484.491	0.25049	106.3353
Autumn	8	9871.417	0.28479	111.5321
Winter	8	10120.917	0.28838	109.7568

On the other hand, one chosen form of validation to compare the first clustering result (K-Means + DTW) with the ensemble model, and to identify and observe the impact that the *encoded features* finally had on the final set of clusters, is the Sankey Diagram (SD). The SD is useful to observe continuity between distinct sources, and it is represented in Figure 4.

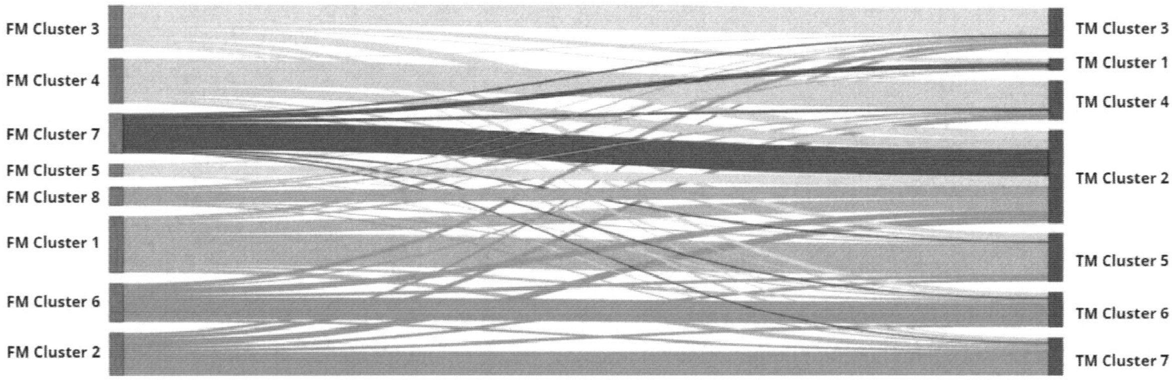

Figure 4. Sankey Diagram of continuity between First Model (FM) clusters and the Third model (TM) clusters, considering the summer season.

It is clear the creation of a new cluster, cluster 1 in the TM, that is composed of properties that were assigned before to other clusters because this cluster did not exist in the previous clustering iteration. Moreover, we can conclude from the SD that the encoded features had an impact on the new assignation of clusters, but without changing most of the clusters defined by the FM, resulting in an overall improved clustering result.

It is important to note that, this proposed solution can be easily expanded by adding more years of data, which can result in a better clustering result.

5. Conclusions

The interactions with the water utility enabled the identification of the suitable number of generated clusters by comparing the results of all models to the real distribution behaviour. More importantly, the ensemble solution succeeded in producing a well-defined clustering strategy of similar properties. By separating the data models and algorithms into seasons, it was easier to achieve a more representative categorization of properties consumption due to the differences in behaviour that each season provokes. This study has proved a useful methodology for joining time-series data with other data types successfully, resulting in the ensemble model.

Moreover, aside from meeting the demand of supplying a reasonably low number of defined *k* clusters for easier management, the fact that the ensemble is composed of different technologies and data types was enough to justify the employment of this strategy, since the objective was to make use of all available property data and turn that into useful insights for water supply companies. Also, based on the assigned labels of each property, it will be possible in future works to produce a more in-depth characterisation of consumption patterns.

In the end, the solution provided the operators with a different point of view by having the clusters assigned, the task of defining consumption trends and profiles will be easier and more accurate, which is a great benefit for them, as well to understand and define the different consumption behaviour that each season of the year represents and to correctly fit newly arrived households as well.

Acknowledgements

These developments are part of the FIWARE4Water project that has received funding from the European Union's Horizon 2020 research and innovation programme under grant agreement No. 821036.

References

[1] Laspidou C, Papageorgiou E, Kokkinos K, Sahu S, Gupta A and Tassiulas L 2015 Exploring patterns in water consumption by clustering *Procedia Engineering* **119** 1439–46

[2] Leitão J, Simões N, Marques J, Gil P, Ribeiro B and Cardoso A 2019 Detecting urban water consumption patterns: a time-series clustering approach *Water Supply* **19**(8) 2323–29

[3] Chatterjee S and Mukhopadhyay A 2013 Clustering ensemble: A multiobjective genetic algorithm based approach *CIMTA* **10**(1) 443–449

[4] Jain AK, Murty MN and Flynn PJ 1999 Data clustering: a review *ACM Comput. Surv.* **31**(3) 264–323

[5] Aghabozorgi S, Shirkhorshidi AS and Wah TY 2015 Time-series clustering – A decade review *Information Systems* **53** 16–38

[6] Becker M, Lippel J, Stuhlsatz A and Zielke T 2020 Robust dimensionality reduction for data visualization with deep neural networks *Graphical Models* **108** 101060

[7] Wang Y, Hongxun Y and Zhao S 2016 Auto-encoder based dimensionality reduction *Neurocomputing* **184** 232–242

[8] Huang X, Ye Y, Xiong L, Lau RY, Jiang N and Wang S 2016 Time series K-Means: A new K-Means type smooth subspace clustering for time series data *Information Sciences* **367-368** 1–13

[9] Ibrahim AS, Memon AF and Butler D 2021 Seasonal variation of rainy and dry season per capita water consumption in Freetown city Sierra Leone *Water* **13**(4) 499

[10] Kalbusch A, Henning E, Brikalski MP, Vieira de Luca F and Konrath AC 2020 Impact of coronavirus (COVID-19) spread-prevention actions on urban water consumption *Resources, Conservation and Recycling* **163** 105098

[11] Zhou Z-H and Tang W 2006 Clusterer ensemble *Knowledge-Based Systems* **19**(1) 77–83

[12] Calinski T and Harabasz J 1974 A dendrite method for cluster analysis *Communications in Statistics* **3**(1) 1–27.

IOP Conf. Series: Earth and Environmental Science 1136 (2023) 012006 doi:10.1088/1755-1315/1136/1/012006

From simulation to dissemination: automation of data and metadata management

Mohammad Shafi Arif[1], Peter Schade[1], Rainer Lehfeldt[1], Vikram Notay[2], Georg Carstens[3] and Frank Kösters[1]

[1] Department of Coastal Engineering, Federal Waterways Engineering and Research Institute (BAW), 22559 Hamburg, Germany
[2] Department of Central Services, Federal Waterways Engineering and Research Institute (BAW), 76187 Karlsruhe, Germany
[3] Department of Hydraulic Engineering in Inland Areas, Federal Waterways Engineering and Research Institute (BAW), 76187 Karlsruhe, Germany

mohammad-shafi.arif@baw.de

Abstract. Working with Computational Fluid Dynamics (CFD) can produce numerous datasets, which contain different physical parameters and study settings. Management and dissemination of such data benefits greatly from a comprehensive data description (metadata), ideally a specialized Metadata Information System (MIS), and adequate long-term storage (data warehouse). To avoid the manual creation of rival metadata, an automation method has been developed, which adds metadata automatically by the simulation and post-processing programs. The automation method described here is an example for data curation suitable for a professional work environment. This method begins with the collection and creation of metadata and ends with the dissemination and publication of the data. This procedure reduces the challenging amount of tedious and error-prone workload, avoids redundancy, enhances efficiency, and is thus a means of quality assurance. Basis for the automation process is an open-source metadata information system (MIS) which has been adapted to the techno-scientific demand of simulation metadata. It stores the metadata in an SQL database and provides Open Geospatial Consortium (OGC) compliant services and communication interface. A hierarchical metadata management concept was initiated for efficient management of numerous datasets. Subsequently, the metadata validation and dissemination are automated via a middleware, which compiles metadata in an XML file, imports the metadata into the MIS and transfers the data to a long-term repository via a Representational State Transfer (REST) interface. Finally, the data and metadata are interlinked and published.

1. Introduction

Computational Fluid Dynamics (CFD) are a mean of studying hydrodynamics in open-ocean, coastal and inland waters. In coastal engineering, a typical application of numerical modelling is the assessment of climate change or anthropogenic impacts on hydro- and morphodynamics of coastal waterways [1, 2]. In order to obtain reliable results, a large number of numerical simulations are usually required, e.g. when calibrating the model. Here, individual simulations may only differ in few details of the required model parametrizations [3]. On the other hand, numerical simulations for large model set-ups have high computational costs. This has been fostered in recent years by the exponential growth of computing power and use of high-performance computing (HPC) [4]. Long-term simulations with high spatial and temporal resolutions for large areas have become more feasible [5, 6] using HPC. The use of HPC has also enabled researchers to investigate a wider variety of scenarios covering natural variability that affect the hydro- and morpho-dynamics. These data are assigned the 3V characteristics of Big Data due to their Volume, Velocity and Variety [7]. Challenges such as data reproducibility, FAIR data documentation for effective use of open data and specifically the workflow automation of data curation are associated with Big Data must be addressed.

Conducting multiple CFD simulations for system analysis can yield rather similar output files. Recording the data history, i.e., processing phases and workflow is important to identify the model particulars, trace the workflow, understand and reproduce the data. Lack of custodians after project completion or data growth aggravates the loss of information on the data. However, a descriptive metadata catalogue capturing sufficient contextual information can assemble the workflow and history, and promote the data reproducibility [8]. A comprehensive data description will ensure the permanent traceability and reproducibility of the data and is a prerequisite to data FAIRness (findable, accessible, interoperable, and reusable) [9]. The metadata standards of the International Organization for Standardization (ISO-19100 series) and the Infrastructure for Spatial Information in Europe (INSPIRE) comprise an extensive set of metadata for general documentation of spatial data. An appropriate number of administrative, descriptive, and structural information can be selected from these standards to enable discovery and facilitate reproduction of the data and metadata. Extensions for both the CFD description and the quality characterization of geodata [10] can be accommodated where the ISO does not suffice.

Scientific data are understandably complex and costly [11]. Government agencies are seeking greater transparency and accountability in data management and are promoting data accessibility and exploitability beyond organizations. Taking Germany as an example, an open data initiative [12] has created the legal framework to increase free data accessibility. In addition, long-term reusability of data, transfer of know-how, and collaboration between technical and research institutions will be promoted through an appropriate data sharing approach [13]. Hence, the distribution, discovery, and retrieval approach via the web for geodata are much more evolved and standardized, whereas dissemination workflows have not yet been put on an adequate path. Manual assembly and management of data and metadata in large quantity is a tedious and error-prone job and a major challenge for data curators. An automation technique can produce better-quality metadata efficiently and cheaper than manual creation. Most importantly, it is semantically consistent by using standardized vocabulary and syntax [14]. Recently, methods of metadata extraction from CFD output and log files [15] and metadata collection for model data sharing [16] were presented. This leads to a discussion of the importance of metadata for a robust CFD infrastructure [17]. However, these studies focus on reusable data applying the Dublin core profile. They omit the interoperability and reproducibility as well as dissemination workflow automation. Moreover, an automated validation approach for the metadata schema and syntax has not yet been integrated in CFD data management workflows by the modelling community [18]. Automating the workflow for CFD data curation is especially important for capturing metadata and curating and disseminating data via a web portal.

This study focuses on the automation of the workflow for data curation with engineering applications of CFD as an example. We illustrate the workflow developed here with examples from our technical implementation which can be transferred to other software products and allow for user-specific customization. Data curation is described as data management throughout the data life cycle [19], which pertains to data and metadata processing, production, long-term preservation, and dissemination of complex scientific data. In this context, the hierarchical metadata management (HMM) [20] method based on persistent data identification (UUID) is examined and a metadata profile for computer-based studies is developed based on existing metadata standards. The importance of metadata for the data reproducibility toward a robust infrastructure is systematically examined. A suitable MIS is integrated into the infrastructure, together with a long-term storage system (LTS) for storage and preservation of CFD files. A FAIR data management of CFD data satisfying the desire for long term access to open data is illustrated. Both securing the valuable data against loss and quality measures are strongly considered in the automation steps of the new middleware.

2. Methods

A discipline-oriented approach for data curation and management of research and scientific data is technically implemented. At first, a data life cycle for the management of scientific and research data is identified (Figure 1), which emphasizes the six stages of data assembly and collection to archiving, dissemination, and reuse for research projects and downstream users [21].

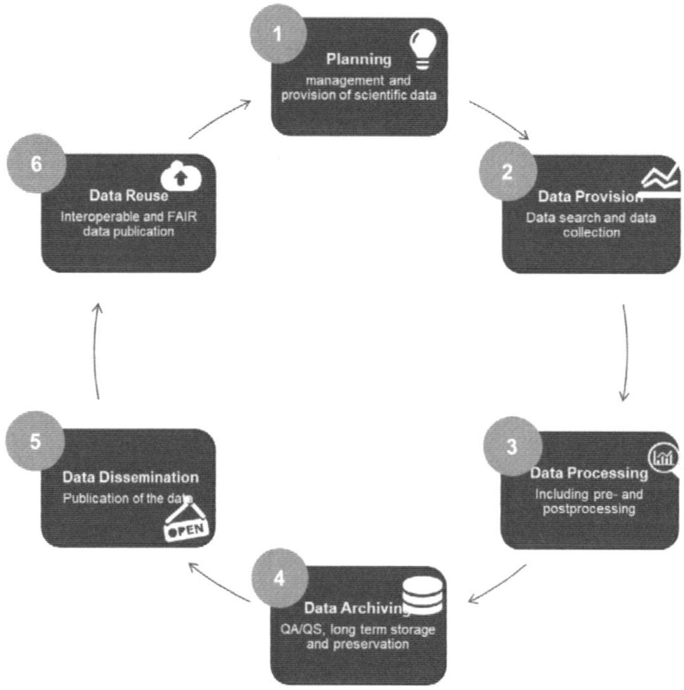

Figure 1. The data life cycle for the management of research and scientific data, adapted from [21].

The pertaining purpose-built infrastructure includes components for seamless job completion. The components are a metadata assembly mechanism for CFD data, a metadata generation component for metadata packages as Extensible Markup Language (XML) files, a data and metadata transferring application programming interface (API) to move the data to long-term storage, a metadata information system for management and administration of metadata, a data archive and a portal component for dissemination and publication of data. Data processing (3), data archiving (4) and data dissemination (5) are potential automation components of the data life cycle.

3. Results

3.1. Conceptual workflow

As a first step the conceptual workflow for data and metadata ranging from generation to dissemination was translated to a technical realization. A middleware for seamless execution of these components in the workflow was developed and integrated into the data management ecosystem (Figure 2).

The data curation workflow begins with invoking the middleware to start metadata generation in either XML or human-readable text format for the selected simulation data. Before exporting the metadata, a metadata validation is performed, during which pop-up windows can be used to check the entries for any missing elements and to add them. After successful metadata import into the MIS, the data proper is also moved to LTS. Successful completion of the data transfer automatically establishes the link between data on the LTS and the earlier imported metadata. Finally, the data publication via the geoportal is triggered.

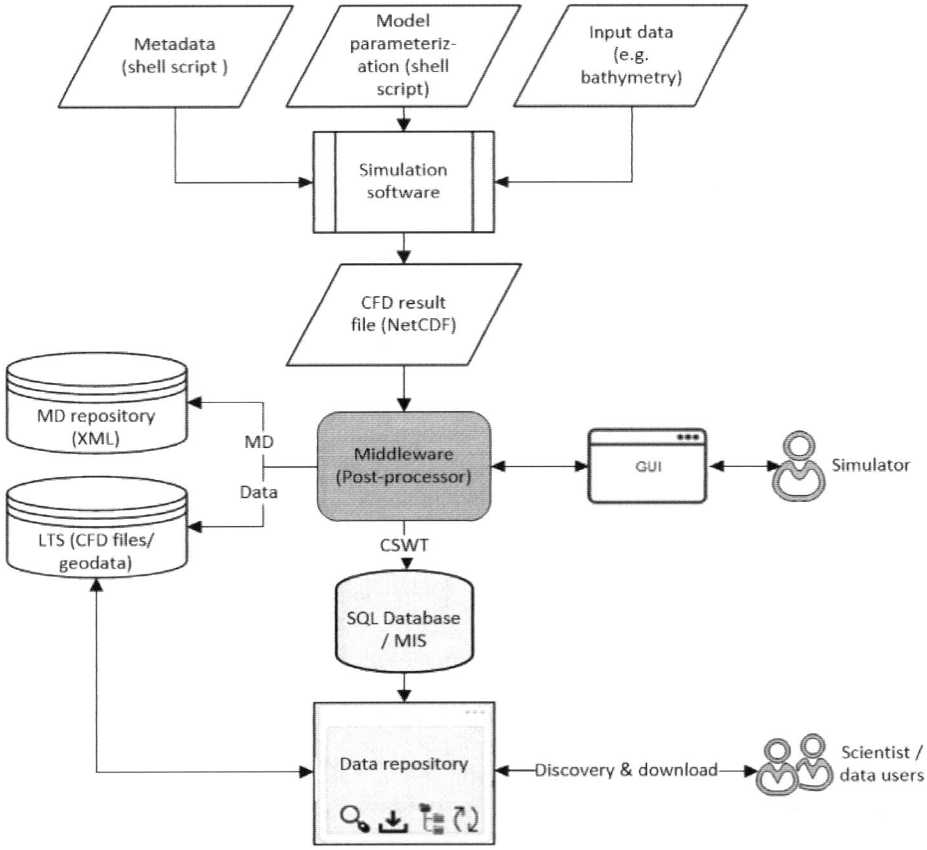

Figure 2. Workflow for data and metadata from generation to dissemination.

3.1.1. Metadata Profile (MDP). The metadata is essential for discovery, interoperability, and reuse of data in general. Metadata are developed to describe resources and facilitate the data discoverability. They give semantics to the data which they describe. Computer-based scientific studies require complex descriptions with controlled vocabularies for meaningful metadata collection.

The metadata profile MDP-GDI-BAW V.1.4 [21] has been developed at the German Federal Waterways Engineering and Research Institute (BAW) for an appropriate and uniform description of CFD simulations including investigation variations of relevant parameters in different scenarios. It can be extended to describe other types of scientific data. This new MDP extensively covers the data quality, lineage, history, and particularly the simulated physical variables. This profile emphasizes syntactical and semantical consistency of the metadata. It adheres to the ISO-19100-series defined for geographic information, INSPIRE for Europe wide interoperability, GDI-DE, for national access and cataloging of the data. This conformity supports cross platform interoperability of the metadata catalogs and enables effective harvesting for a broader user community and other institutions, e.g., thematic, national, and international portals and beyond [22].

The metadata profile tailored to computer-based studies contains additional lineage elements to document the input and configuration files used in CFD. Such a dataset history facilitates the repetition of simulations with the same parameterizations and thus the reproduction of result data. The data quality elements are also adapted to document the scope and units of physical variables in each dataset that characterize the different scenarios studied. Similarly, the data hierarchy level information - as structural metadata - is a critical class of this metadata profile, through which a hierarchy of simulation data is easily traceable, and it allows the user to identify and issue a data and project affiliation.

A comprehensive description of these complex CFD simulations needs additional metadata elements which have in part been formulated in other organizational namespaces, e.g., SeaDataNet (www.seadatanet.org), and EMODnet (https://emodnet.ec.europa.eu), or otherwise could be described as keywords, e.g., from the Leibniz Institute for Baltic Sea Research – IOW (https://iowmeta.io-warnemuende.de/).

3.1.2. Metadata collection. Metadata collection is exemplarily integrated into the simulation tool UnTRIM^2 [23] and corresponding post-processing and analyzing software. The software implementation follows the paradigm of least burden for the user through highly automated workflows. For this purpose, step-by-step metadata collection and storage is put into practice in a Network Common Data Form (NetCDF) data file, which are commonly used for scientific datasets to store and retrieve de facto multidimensional data [24]. An extension of climate and forecast (CF, 1.6) metadata [25] and the unstructured grid metadata conventions (UGrid-NetCDF) [26] integrates user-defined metadata elements and variables.

Since the CFD ecosystem is operated via shell scripts, most metadata elements are compiled from the input and configuration files relevant to a particular study scenario. Only a few administrative and project related details can be entered manually at the beginning of a study series, e.g. project name, project ID and author information. Further information can be extracted from result data such as, e.g., the spatial bounding box, the simulation parameters including their units and spatial and temporal resolution. Similarly, the post-processing software log their lineage and configured setup information along with their result data in the same output NetCDF file so that the workflow can be traced from the beginning of the simulation job. In this way, the study setup, software parametrization and results achieved are all stored in a single file. This metadata collection ensures preservation of all information required for reproducing the data.

3.1.3. Metadata generation. Many studies are conducted for each project. In its final phase, a project includes multiple datasets with a variety of parametrizations and environmental scenarios. By initiating the HMM [20] procedure, parent-child relationships of datasets are established through metadata that facilitates the understanding of data relationships.

To generate metadata packages for the data and the relevant hierarchy of simulation data, a metadata generation tool (middleware) is developed. This tool extracts the attributes and parameters existing in a NetCDF file to the metadata packages as an XML file. In addition, another overview of the generated metadata is integrated into a text document that allows to preview the given information in a human-readable format. This tool adds some unrecorded information regarding the metadata itself, e.g., language, date and time of generation, hierarchy information, the MDP version used, and metadata contact information. The compiled metadata comprehensively describe each dataset, including administrative, descriptive and structural metadata based on the specified metadata profile. The middleware is programmed to produce reproducible metadata records to ensure traceability and reproducibility of the metadata.

3.1.4. Data and metadata storage. An API in the middleware automatically stores interim result metadata files (XMLs) in a metadata repository and transfers the final version to the MIS. The Transactional Catalog Service Web (CSWT) manager interface of OGC is used to insert, update or delete the catalog records via the transaction operation of the MIS [27].

A cluster system is used for centralized and LTS, high-performance query and retrieval of the multidimensional array data, whose file size on disk can range from megabytes to terabytes. The cluster system is called via a REST interface for the import and storage of the data. This ensures secure storage of the published data from manipulation. The REST architecture is found to be a robust API for management and data transactions [28], since the CFD files transferred are mostly large datasets and interlinked [29]. Additionally, the middleware is further developed to automate the data workflow via the REST API to the LTS and grant the user access to the data via the web. The data move from a local location to the LTS is triggered after confirmation of successful metadata insertion or update and the process result is examined at the end. The workflow for the frontend user requires no other steps than switching on the data transfer to LTS option.

3.1.5. Data and metadata dissemination. Big Data technology provides hardware and software solutions for large-scale data curation in research and development projects [30][6]. The selected open-source metadata information system MIS stores the metadata in a SQL database, provides an OGC compliant interface with services of searching metadata (CSW), visualizing maps (WMS), and ATOM download of the data and includes a graphical user interface (GUI) for manual metadata entry, editing and administration.

Once the data have been moved to the LTS and the metadata have been imported into the MIS by the middleware, the data and metadata are then linked to each other and published. Here, a user can search, find, and download the data, and access and explore further options through standardized interfaces provided.

3.2. Technical realization

The presented robust spatial data infrastructure (SDI) for long-term data and metadata management and dissemination is integrated in an existing CFD environment. All simulation and postprocessing software components include methods for creating and handling metadata to automatically track the lineage of data products such as simulation results and related analyses. Automation begins with the collection of metadata, including all data history and quality information, and continues with the generation of metadata files, importing the metadata into a database and MIS, storing the data in LTS, and disseminating the data. This method achieves a reasonable workflow that integrates both the simulation process and the management of simulation data. For most scientists who are not data managers, this represents a small burden.

The applied metadata profile forms the basis for harmonized long-term metadata collections and facilitates interoperability and FAIR use of data. Keywords and terminology are selected in accordance with open data and domain ontology recommendations to enable better data discovery. The published data is optionally assigned a Digital Object Identifier (DOI) as part of the metadata profile. Data DOIs are state-of-the-art and science-fostering methods for long-term accessibility of data and linking a published article to its data [31].

Metadata assembly and collection are among the largest concerns of data scientists and domain data curators. In the developed method, for instance, the spatial and temporal resolution or geographic extent are estimated, keywords are set in code lists and physical parameters estimated are obtained from the data variables of the NetCDF file. In this way, the user is not distracted by a tedious workload of metadata collection. The automation workflow makes it easy for the user to generate metadata and publish the data with a few clicks, which should not take more than a few minutes. The quality measures for metadata are characterized by two aspects in the considered workflow, i.e., automation of metadata collection and validation during XML file creation. The validation procedure for metadata is automated and integrated into the middleware which generates the metadata file. This constitutes an additional quality assurance measure of validity via syntactical and semantical examination of metadata.

Similarly, the data validity is observed prior to publication of metadata via an inhouse software which skims the values and estimates the bounds, to identify unexpected values or erroneous computed data. Each performed CFD scenario is documented by metadata in order to allow for reproduction of their simulation results. All of these data and metadata remain in the Data Archiving component (Figure 1, component (4)) for long term storage and possible further studies. Only those studies meant for publication enter the Data Dissemination component (Figure 1, component (5)) after review, expert assessment and quality assurance. The implementation of the HMM method shows dependencies of datasets, which is especially convenient for larger research projects with a variety of physical systems to be investigated. This allows for searching of datasets in the catalog view of a project (see Figure 3).

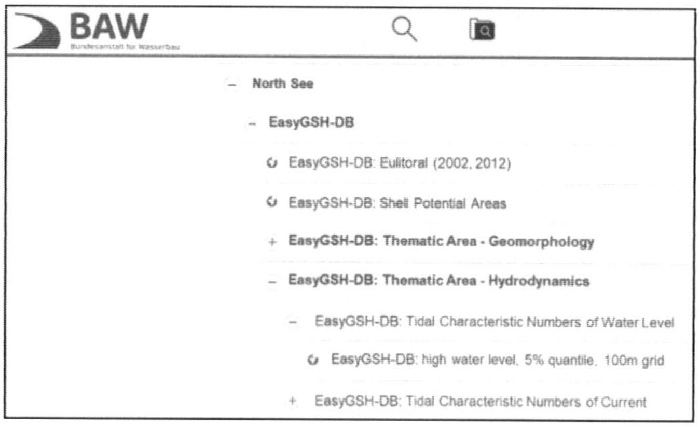

Figure 3. Hierarchy of metadata in a catalogue view for a project.

The inclusion of customization files for pre- and postprocessing tools and the recording of operations performed on the data ensure both workflow traceability and data reproducibility. The identical data and metadata generation avoids reimport and consequently republication of data and metadata. The LTS is designed to store and secure data for read-only access over the Internet. The implemented REST communication interface to write data in the LTS has proven to be an efficient communication platform. The storage system for data on the LTS strictly follows the hierarchical metadata management system in the MIS. The applied middleware is a combination of tools into a single software which is illustrated in Figure 4.

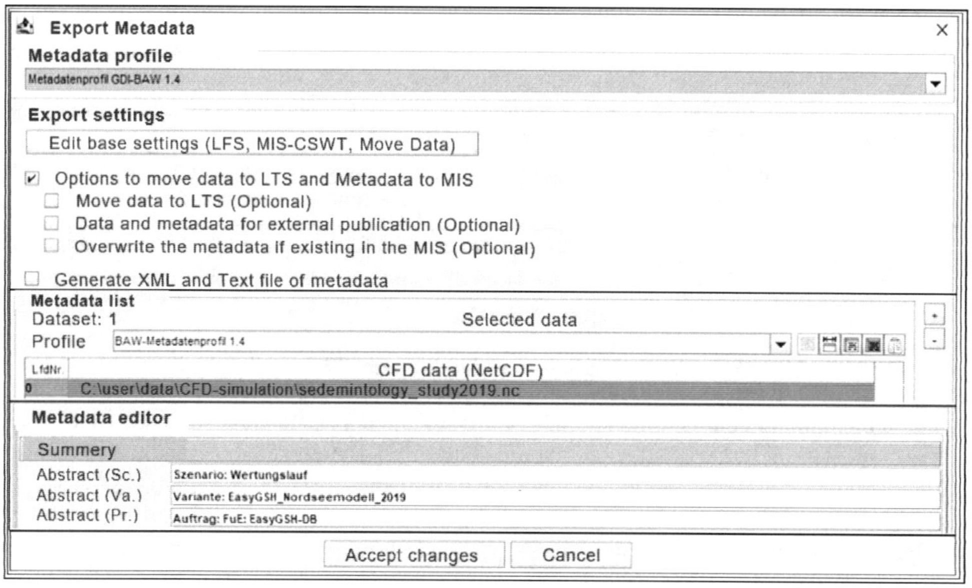

Figure 4. The middleware for executing the metadata and data dissemination job.

It consists of the API for data transaction to the LTS, the API to generate metadata sets in XML, the API to import metadata into the MIS, and a GUI. The advantages of this combination and integration are that only one piece of software needs to be invoked, the software communicates better during the execution of each job and provides feedback for the start of the next job, and the user has a single point of contact for data management tasks to control and execute the task.

The publication of scientific studies with public access to reports and related data is realized with the portal component of the MIS with a metadata catalog and search and ATOM download possibilities (Figure 5). The MIS is also used for the administration and manual generation of metadata instances as well. A prototype implementation of the data dissemination in the BAW context is depicted on the institutional portal https://datenrepository.baw.de/.

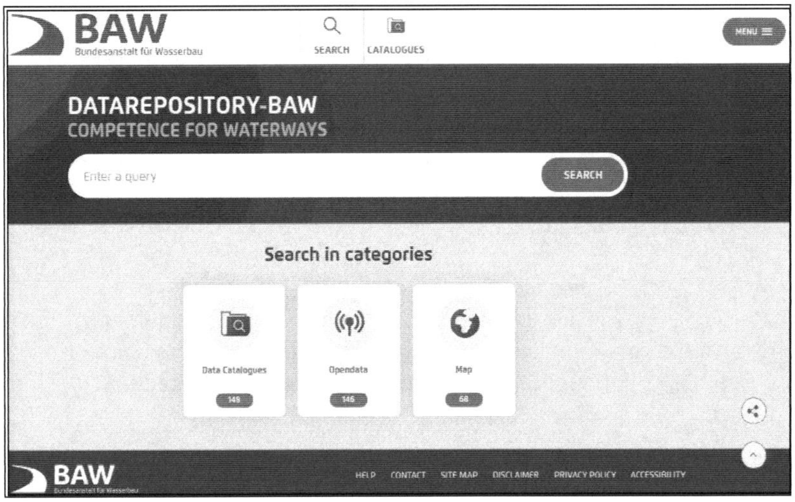

Figure 5. The data repository (geoportal) for public access to the scientific data.

The automation scheme for data curation and dissemination has attracted the interest of scientists because it reduces the effort required to create metadata for data from CFD studies and related publications. High quality metadata as a means of quality assurance in combination with reliable long-term archiving of simulation data increases the interest in the sustainable management of their data. The functionality of the new middleware introduced at BAW avoids data value decay and facilitates long-term data reusability.

4. Discussion & Conclusions

This study introduces an automation method for both generation of metadata and dissemination of data and associated metadata, which shows that metadata capturing and data management can be highly automated. The implemented automation process has significantly optimized both the workflow for scientists and the data and metadata pipeline from generation to dissemination and long-term archiving via a one-stop-shop middleware. The automation approach avoids the erroneous capture of metadata, controls the FAIRness and completeness of the data and metadata via a validation scheme, centralizes the compilation and publication workflow for many datasets in a distributed and collaborative project environment and avoids multiple publication of identical data. This approach significantly increases quality and efficiency by reducing the burden of data curation for scientists. Moreover, the descriptive and standardized metadata profile and the generated validated metadata ensure data discoverability, interoperability, and reusability.

An ISO-compliant metadata profile for computer-based studies was developed that records the entire data history and sequence to ensure the discoverability and reproducibility of archived data. The goal of publishing scientific, research and proprietary data via a geoportal has been achieved. This will promote transparency and discourse on open data and pave the way for the use of data beyond an institution itself since conceptual metadata is available. In addition, the use of LTS proves to be the key to uniform and centralized access to the data via the web browser and via a REST API, as well as to long-term data archiving. Finally, using the HMM in this study to search for relevant data, to create a project-specific metadata overview and to establish data relationships has proven to be very effective in a CFD ecosystem.

The illustrated automation approach for CFD simulation metadata compilation and generation is coupled to the NetCDF conventions for on demand recording and storing of user defined variables. With minor software adaptations for writing the attributes and variables, this approach can be implemented in other simulation software, which produces NetCDF output data of any dimension, e.g., 1D, 2D and 3D, such as the D-Flow FM [32]. Once the metadata are recorded in the output file, the implemented data curation workflow facilitates the data and metadata dissemination. The automation methods for metadata handling are available on demand according to licensing schemes. The amount of work required for implementing this method in a new model depends on the modularity of the simulation software, which allows the integration of script inputs or the computation of additional values required by the metadata profile. The proposed metadata profile documents the key parameters of each CFD study performed within the current BAW simulation ecosystem. Future refined CFD procedures may require additional metadata elements that can be included in the metadata profile according to the conventions of the ISO 19119 standard.

Acknowledgments
We thank Dr. Günther Lang and his team for the integration and implementation of the metadata workflow in CFD software environments at BAW. We would also like to acknowledge and thank the DMQS team for their contributions to the development and implementation of methods and tools required in this project.

References
[1] Rasquin C Seiffert R Wachler B Winkel N 2020 The significance of coastal bathymetry representation for modelling the tidal response to mean sea level rise in the German Bight. *Ocean Sci.* **16**(1) 31–44 https://doi.org/10.5194/os-16-31-2020.
[2] Jacob B Stanev E 2021 Understanding the impact of bathymetric changes in the German Bight on coastal dynamics: One step towards realistic morphodynamic modeling https://doi.org/10.5194/egusphere-egu21-15453.
[3] Chau KW 2014 Modelling for Coastal Hydraulics and Engineering. 1st (London: CRC Press).
[4] Lautenschlager M, Adamidis P, Kuhn M 2015 Big Data Research at DKRZ – Climate Model Data Production Workflow URL: ebooks.iospress.nl.
[5] Grandinetti L Joubert GR Kunze M Pascucci V, editors 2015 Big data and high performance computing (Amsterdam: IOS Press).
[6] Biookaghazadeh S Xu Y Zhou S Zhao M 2015 Enabling scientific data storage and processing on big-data systems. 2015 *International Conference on Big Data*. NJ: IEEE; 1978–84.
[7] Baumann P Mazzetti P Ungar J Barbera R Barboni D Beccati A Bigagli L Boldrini E Bruno R Calanducci A, et al. 2015 Big Data Analytics for Earth Sciences: The EarthServer approach. *IJDE* **9**(1) 3–29 https://doi.org/10.1080/17538947.2014.1003106.
[8] Greenwald M Fredian T Schissel D Stillerman J 2012 A metadata catalog for organization and systemization of fusion simulation data. *Fusion Engineering and Design* ***87***(12) 2205–8 https://doi.org/10.1016/j.fusengdes.2012.02.128.
[9] Wilkinson MD Dumontier M Aalbersberg IJ Appleton G Axton M Baak A Blomberg N Boiten J-W da Silva Santos LB Bourne PE, et al. 2016 The FAIR Guiding Principles for scientific data management and stewardship. *Sci Data* **3**(1) https://doi.org/10.1038/sdata.2016.18.
[10] Feistel S Kleeberg U Kohlus J Lehfeldt R Schirnick C Tamm S 2018 Automatic assessment of metadata quality in ISO 191xx. *International Conference on Marine Data and Information Systems*. In: Giorgetti A, Fichaut M, Tosello V, editors. Spain; 67–8.
[11] Preuss N Staudter G Weber M Anderl R Pelz PF 2018 Methods and Technologies for Research- and Metadata Management in Collaborative Experimental Research. *AMM* 885 170–83 https://doi.org/10.4028/www.scientific.net/AMM.885.170.
[12] Bundesministerium des Innern 2021 Open-Data-Strategie der Bundesregierung (Germany).

[13] van Noorden R 2013 Data-sharing: Everything on display. *Nature* **500**(7461) 243–5 https://doi.org/10.1038/nj7461-243a.
[14] Liu J 2007 Metadata and its applications in the digital library: Approaches and practices (Westport, Conn.: Libraries Unlimited).
[15] Schembera B Iglezakis D 2020 EngMeta: metadata for computational engineering. *IJMSO* **14**(1) 26 https://doi.org/10.1504/IJMSO.2020.107792.
[16] Morsy MM Goodall JL Castronova AM Dash P Merwade V Sadler JM Rajib MA Horsburgh JS Tarboton DG 2017 Design of a metadata framework for environmental models with an example hydrologic application in HydroShare. *Environmental Modelling & Software* **93** p.13–28 https://doi.org/10.1016/j.envsoft.2017.02.028.
[17] Leipzig J Nüst D Hoyt CT Ram K Greenberg J 2021 The role of metadata in reproducible computational research. *Patterns* **2**(9) https://doi.org/10.1016/j.patter.2021.100322.
[18] Keeley R Woodruff S Pouliquen S Conkright-Gregg M Reed G 2010 The development of the data system and growth in data sharing. *Proceedings of OceanObs'09: Sustained Ocean Observations and Information for Society*. In: Hall J, Harrison DE, Stammer D, editors. ESA.
[19] Albani M Maggio I Data Stewardship Interest Group C 2020 Long-Term Data Preservation Data Lifecycle, Standardisation Process, Implementation and Lessons Learned. *IJDC* **15**(1) 10 https://doi.org/10.2218/ijdc.v15i1.715.
[20] Arif M-S Schade P Kösters F 2021 Metadata Hierarchy for Enhanced Management of Hydro-Numerical Simulation Data. *International Conference on Marine Data and Information Systems*. In: Aalberts N, Barale V, Belov S, Bruin T de, editors 245–6.
[21] Damrau T Carstens G Binder K Diermayer E Notay V Schade P Arif M-S 2021 Datenmanagement und Qualitätssicherung im Verkehrswasserbau (DMQS) 2.0: FuE-Abschlussbericht (B3954.07.04.70006) (Germany).
[22] Notay V Carstens G Lehfeldt R 2017 A Metadata Standard for Hydroinformatic Data Conforming to International Standards. In: *Geophysical Research Abstracts*. EGU2017-13004.
[23] Casulli V 2009 A high-resolution wetting and drying algorithm for free-surface hydrodynamics. *Int. J. Numer. Meth. Fluids* **60**(4) 391–408 https://doi.org/10.1002/fld.1896.
[24] Rew R Davis G 1990 NetCDF: An interface for scientific data access. *IEEE Comput. Grap. Appl.* **10**(4) 76–82 https://doi.org/10.1109/38.56302.
[25] Eaton B, Gregory J, Drach B, Taylor K, Hankin S, Blower J, Caron J, Signell R, Bentley P, Rappa G, et al. 2021 NetCDF Climate and Forecast (CF) Metadata Conventions URL: cfconventions.org.
[26] 2018 UGRID Conventions: Unstructured Grid Metadata Conventions Home Page URL: ugrid-conventions.github.io.
[27] Open Geospatial Consortium 2007 OpenGIS® Catalogue Services Specification 2.0.2 - ISO Metadata Application Profile 2007 Jul 19 URL: www.ogc.org [cited 2022 January 24]
[28] Rodríguez C Baez M Daniel F Casati F Trabucco JC Canali L Percannella G 2016 REST APIs: A Large-Scale Analysis of Compliance with Principles and Best Practices. In: Bozzon A, editor. *Web engineering*: 16th International Conference (Heidelberg) Springer 21–39.
[29] Charest G Rogers M Planning S Leader EA Sevier R Thomas M Clevenger J Tikofsky B 2020 Data exchange methods and considerations (Cambridge) Technical report.
[30] La Beaujardière J de 2010 Ocean and Coastal Data Management. *OceanObs'09: Sustained Ocean Observations and Information for Society*. In: European Space Agency 226–36.
[31] Bloom T Ganley E Winker M 2014 Data Access for the Open Access Literature: PLOS's Data Policy. PLoS Biol **12**(2) e1001797 https://doi.org/10.1371/journal.pbio.1001797.
[32] Kernkamp HWJ van Dam A Stelling GS Goede ED de 2011 Efficient scheme for the shallow water equations on unstructured grids with application to the Continental Shelf. *Ocean Dynamics* **61**(8) 1175–88 https://doi.org/10.1007/s10236-011-0423-6.

Web-based GIS for simulation, and visualisation: EPANET and data-rich shapefiles

Gareth Lewis[1], Brett Snider[1], Lydia Vamvakeridou-Lyroudia[1,2], Albert S Chen[1], Slobodan Djordjević[1] and Dragan A Savić[1,2]

[1] Centre for Water Systems, University of Exeter, Harrison Building, North Park Road, Exeter, EX4 4QF, UK
[2] KWR Water Research Institute, Groningenhaven 7, P.O. Box 1072, 3430 BB Nieuwegein, the Netherlands

g.lewis2@exeter.ac.uk

Abstract. The management of water resources often involves with spatially and temporally varied information collected from various providers with different data formats. Although certain software or applications such as QGIS[1] and EPANET[2] have been developed to support the analyses and decision making, these solutions do have some critical weaknesses: 1) it can take users considerable time to become proficient in using a given application, 2) applications generally assume a level of domain-specific knowledge, 3) applications may require customisation through plug-ins to provide suitable information and 4) applications are often tied to specific hardware / operating system configurations. To address these issues, the aqua3s and Fiware4Water research programmes were developed as 'cloud first' projects, using the cloud/web to deliver functionality. In this work, we developed approaches for integrating and visualising information to support water management, specifically developing a web-based EPANET simulation and visualisation for large water networks (c30,000 EPANET nodes and links), and a web-based visualisation of regional flood data shapefiles for Trieste and surrounding regions. In both cases, data was processed and balanced between client and server to minimise client loading and maximise responsiveness.

1. Introduction

For those working with water systems, QGIS [1] and EPANET [2] provide fundamental visualisation and simulation platforms. QGIS allows users to visualise complex environments using a collection of layers built on top of vector or XYZ tile sets to view data graphically with map and/or satellite image contexts. Likewise, the EPANET application enables users to interactively build and simulate water networks using the EPANET hydrostatic solver to provide accurate simulation results.

Whilst these applications are incredibly useful, they can be criticised for having steep learning curves and, in the case of EPANET, only providing limited platform (Windows) support. For developers of water management systems, it is desirable to take core functionality from QGIS (layered map and data visualisation) and EPANET (water network visualisation and simulation) and present it to users in a way that can: 1) be used without the need for bespoke computers and 2) be used with little training or support.

The aqua3s project had requirements to take typical QGIS and EPANET visualisations and present them within a browser context, requiring visualisation through a suitable web-based tile-map visualisation solution.

2. Web-based tile-map visualisation

Web-based tile-maps describe the browser-based implementation of the vector and XYZ tile approaches of QGIS and describe an approach where geographic maps are split into collections of tiles allowing users to scroll over a map giving the impression of an endless map surface. Map tiles are subdivided into increasing levels of detail, to create the effect of zooming into or out of the map and this is provided for the web with services such as Google Maps, Microsoft Bing and Apple Maps.

For developers of tile-map visualisations there is an array of solutions available, all of which tend to deliver similar functionalities but often different business models. For our initial work, Google Maps[3], Mapbox[4], leaflet.js[5] and Maplibre[6] were considered, and programming was undertaken with all excluding Google Maps, given the credit card requirement of its commercial terms.

The remaining three frameworks were similar in their conceptualisations, which is not surprising given that Maplibre is an open-source drop-in for Mapbox and Leaflet.js also appears to have had some involvement with Mapbox. The frameworks work on the concept of a view being a collection of layers, which are either provided by the map tile server (ground, buildings, roads, labels etc) or provided by users, with five distinct components, Figure 1, which form the basis for user-defined visualisations.

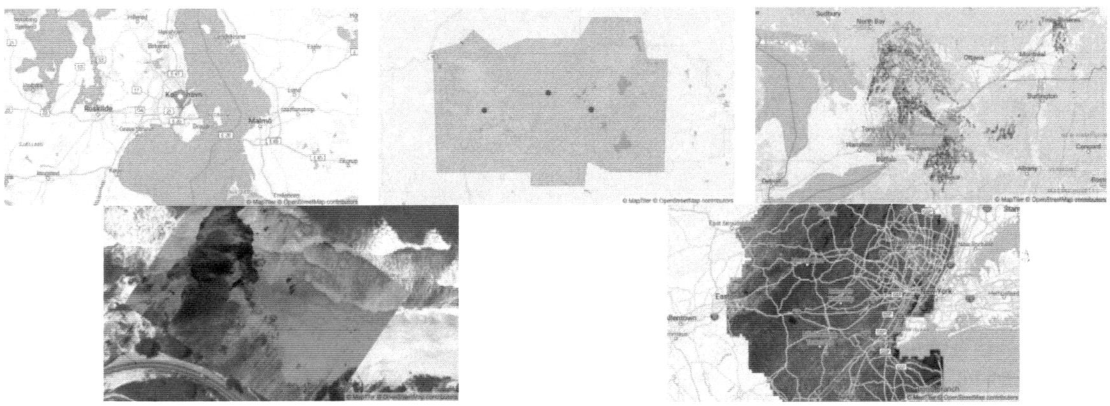

Figure 1: Typical user-defined visualisations: default marker[7], GeoJSON geometries[8], animated images[9], video[10], WMS[11].

The default marker visualisation type can be thought of as the de facto visualisation/interaction style for web-based tile-maps and consists of a 'pin' in the map at a given location. Typically, the pin can be clicked by a user to reveal pin-specific information.

The GeoJSON geometry visualisation type is a more open-ended approach to visualisation/interaction and can be considered as a data transport into the tile-map solution. GeoJSON[12] supports a range of geometry types (points, lines, and polygons) giving users the opportunity to create pin-like interactions with point and line types and/or region-based interaction with the polygon.

The animated images visualisation type can be used in 'weather map' style visualisations, where animated visualisation data is overlayed onto a map. Care should be taken that the pixel size of animated data may not work well with different map zooms, the Maplibre example[9] is a good case in point. The image of 640x456 pixels is projected onto an area of about 600 km x 600 km, giving a resolution of one pixel per km, which results in a good presentation from distance, but detail is lost when the user zooms into the map.

The video visualisation type provides a similar solution to animated images, but the use of a video format rather than a collection of images allows for more frames of animation for a given data footprint, which may be useful for managing collections of files for fluid animations. Like the animated images, video visualisation is sensitive to map zooming, the Maplibre example[10] applying a lot of detail to a relatively small area (the inverse of the animated images example).

The WMS (Web map service) visualisation types provides an interface for visualising external web maps and requires an external web map server to host content.

For our use cases of visualising shapefiles and visualising EPANET simulations, it would appear that GeoJSON is likely to be the most suitable solution, given our need to work with large amounts of geographic data and our need for 'close up' detail.

3. Visualising shapefiles

For the aqua3s project, pilot sites were using shapefiles to store a wide range of visual and quantitative data relating to the operation of their water networks, Figure 2, typically regional flood maps, areas of commercial importance, sites of wells, regional boundaries and so on.

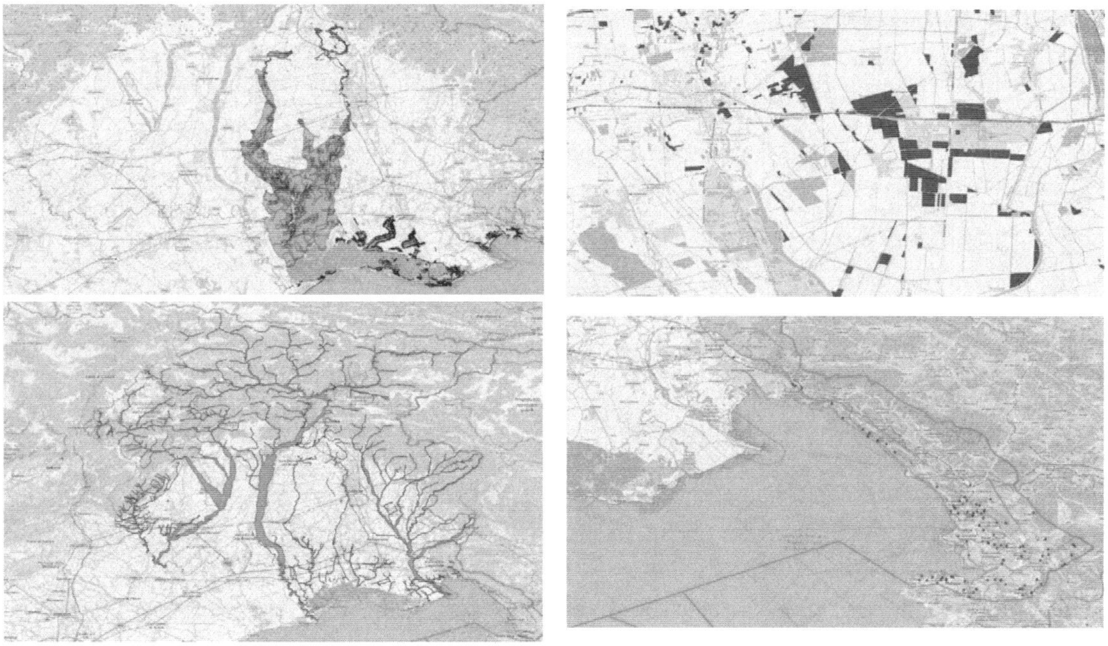

Figure 2: Typical shapefiles. Flood maps (top left), commercial locations (top right), river flows (bottom left) and wells (bottom right).

The shapefiles presented for visualisation tended to be large and complex files, typically with file sizes in the order of 100-200mb and geometry and metadata for up to 150,000 features and associated metadata per file. The shapefiles covered up to an area of 60km x 80km with a resolution of 3m x 3m. Typical operation of the application would see all the shapefiles downloaded to the client, with users having the ability to show and hide different shapefile layers, and to reveal metadata associated with shapefile features.

As stated previously, shapefiles are not a 'concrete' data type for MapLibre, so they would need to be converted into GeoJSON, a process which could be undertaken on the client or server, depending on processing requirements. It was expected that the larger shapefiles, c150,000 features, were likely to have a significant conversion load and were likely to create substantial GeoJSON files which may not work well with MapLibre, particularly giving the noticeable delays when working with the shapefiles in QGIS.

4. Approach

Initially, shapefiles were downloaded to the map-tile browser client and converted to GeoJSON in-situ, though this tended to be a time-consuming process and made for a poor user experience. It also proved to be beyond the visualisation capacity of Leaflet.js, resulting in a move to first Mapbox and then Maplibre given its open-source credentials.

An offline approach was taken to convert shapefile data into GeoJSON using geopandas[13] and to transform the projection of the resultant geometry into wgs84 format (the default for Mapbox). This was a very slow process and undertaken infrequently, with the resulting GeoJSON files stored on a file server. Once downloaded to the client, GeoJSON data was loaded into the framework using map.addSource[14] and map.addLayer[15], as per the Maplibre sample.

This resulted in a working 'proof of concept', i.e., it was possible to convert shapefiles into GeoJSON and have large GeoJSON files visualised in MapLibre with metadata available through mouse enter/exit events.

However, the large file size of the GeoJSON files (and number of files downloaded to the client) resulted in a long wait before all the GeoJSON layers were available to the user.

Visual inspection of the GeoJSON files revealed that around half the files was feature metadata with the remainder being feature geometry. The entire metadata was not required, as it was only referenced when the user's mouse entered a region and a pop-up dialog displayed it. Therefore, the metadata could be downloaded on demand.

Figure 3: Shapefiles from Figure 2 visualised in aqua3s. Flood maps (top left), commercial locations (top right), river flows (bottom left) and wells (bottom right).

To further reduce the footprint of the GeoJSON data, the resolution of coordinate data was reduced. Geopandas appeared to default to 15dp for WGS84 coordinate data which accounts for most of the resulting file size. Through trial and error, it was found that resolution could be reduced to 6dp without 'noticeable' issues, though this is entirely dependent on the resolution of data that was being used, the 3m x 3m resolution of flood maps. It should also be noted that WGS84 measures in degrees, so the length of a degree is not a consistent measure, therefore, our acceptable results and 34 north may not be valid for other latitudes.

Finally, the resulting files were zipped, given a footprint reduction from c200mb to around c10Mb. Whilst this data needs to be unpacked on the browser, it's a significantly less intensive operation than building the GeoJSON file on the browser.

5. Results

Figure 3 details the shapefiles from Figure 2 visualised as GeoJSON layers in the aqau3s client application. In general, MapLibre has proved to be a very capable tile-map platform, easily dealing with all the data that was downloaded onto it. One significant advantage in moving from Mapbox to Maplibre was the availability of source code which enabled us to solve a couple of troublesome issues with layer re-initialisation 'style.load' call-back[16].

6. Visualising EPANET simulations

For EPANET simulation, clients of aqua3s wanted to be able to visualise water networks modelled with EPANET and view the results of hydraulic and quality simulations in a manner like the existing EPANET application[17], Figure 4.

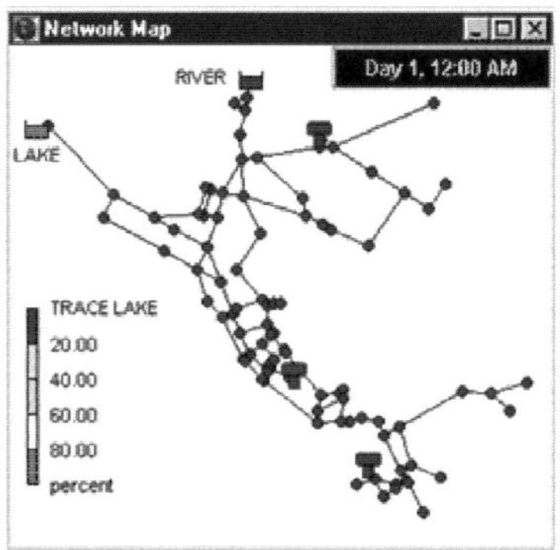

Figure 4: EPANET application water network visualisation.

6.1. Visualising Topology

On first inspection, EPANET networks appeared to be similar to shapefiles, in that they contain rich sets of geometry data that could be described as features. However, EPANET simulations give the network a dynamic quality, in the sense that the values of the network will change over time. The EPANET application displays this through colours and provides users with numeric data on demand. So, whilst the topology of the network doesn't change over time, its representation does. Maplibre provides functionality to update layers in the setData() function, so an animated solution will need to build layer data (as GeoJSON), update layer data as the result of simulation data and finally set the layer data to reflect the changes.

Constructing a GeoJSON representation of an EPANET network was a straightforward task, notwithstanding the need to spatially convert elements into WGS84, as required for MapLibre visualisation. The biggest challenge was in determining a suitable structure for the collection of EPANET components which resulted in building a set of Maplibre layers to represent the pipes, pumps, valves, junctions, reservoirs, and tanks. Whilst the Maplibre 'add multiple geometries' sample[8] does show a single GeoJSON with multiple layers, we found that separating the data made it conceptually simpler to work with, particularly when dealing with simulation data and mouse enter/exit calls.

Having worked with large shapefiles of over 100,000 features, the fairly complex (by EPANET standards) water networks, Figure 5, appeared fairly small in comparison and given the relatively small payloads (c1Mb) there seemed to be little need to compress geometry data in the way that we had for shapefiles.

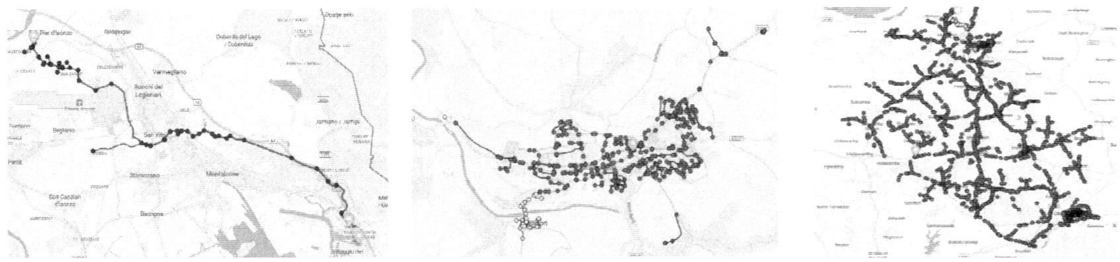

Junctions	60	Junctions	1000	Junctions	12000
Pipes	72	Pipes	780	Pipes	9500
Valves	2	Valves	250	Valves	2800
Payload	5.5k	Payload	95k	Payload	c1Mb

Figure 5: EPANET water networks of differing complexity.

6.2. Simulation Data

Within the Centre for Water Systems, EPANET simulations are typically run over 7 days with a 15-minute resolution, creating around 700 data points per component per simulation. Simulations can be run in the EPANET app and saved as binary files[18] and/or run through the EPANET libraries, either to completion as per the app, or in a stepwise form. For our use cases, it was decided to run the simulation once and store the results for visualisation. Whilst the simulation was not massively computationally expensive, typically taking less than 2 minutes in the biggest case, there seemed to be little value in re-running the simulation if users were unable to change simulation parameters.

For our large case, the resulting simulation output file was 100Mb. Whilst this could be loaded into memory using Python and converted into Python data types, it was a slow purpose and echoed back to the shapefile metadata, in that not all users were going to require all of the data all the time, so an approach was developed where the binary file was loaded, and data was unpacked on demand using Python's struct.unpack_from[19].

6.3. Simulation Process

EPANET divides components into one of two types, nodes, or links, for simulation and organises its colourisation visualisation around four modes for nodes (supply, head, pressure, and quality) and eight modes for links (flow, velocity, headloss, quality, status, setting, reaction rate and friction). Therefore, the simulation visualisation only needs to deal with a small subset of simulation data at any one time. However, unlike shapefile visualisation where only user selected metadata is downloaded, simulation visualisation requires all the node and link data to be downloaded for the given simulation frame and relevant modes. However, displaying this information visually, rather than quantitatively, requires colour information to be sent to the client which occupies a smaller data footprint, even more so when the visualisation uses EPANET's five colour lookup, Figure 4, effectively just requiring a map of indices to colour values and a colour index for each component. On the client, the simulation data was parsed and used to update the colour value of each feature in each layer of the model. On completion, setData was called and the visualisation is updated, Figure 6.

7. Conclusions

In general, using MapLibre as a tile-map for the aqua3s project has worked exceptionally well. The library is generally easy to use and has a good level of visualisation performance which has provide suitable for the size and complexity of data we have been working with for both the EPANET and shapefile domains. The open-source nature of the library enabled us to look into the underlying code when things were not working quite as expected and this did allow us to solve a particularly annoying issue with refreshing visualisations after style.load.

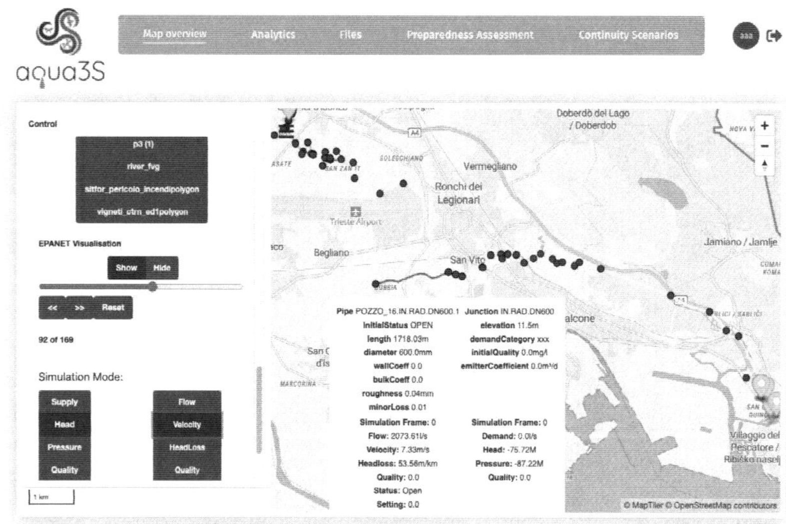

Figure 6: EPANET simulation visualization in aqua3s, showing colourised water network and simulation details.

To a large degree, choosing a suitable technology stack for the project was a bit of a beauty pageant, in that it is a time and resource-limited activity, so there is a tendency to go with whatever works, rather than the absolute best possible solution. There is also the scope to re-engineer existing technologies to meet performance needs, particularly in the case of dismantling GeoJSON data to make it 'fit' our needs and although GeoJSON felt like a 'fat' format, it compresses well and keeps readability, which moving to a binary format would not allow so easily.

Currently, the biggest issue with the overall project is the time taken for the Python-based conversion of Shapefiles to GeoJSON, though this may be addressed by moving a different conversion library.

Acknowledgements
The work presented in this paper was funded by the ongoing EC H2020 aqua3S (GA 832876), Fiware4Water (GA 821036) and LOTUS (GA 820881) projects.

References
[1] QGIS Home Page https://qgis.org/en/site/ (accessed Jul. 12, 2021)
[2] Aqua3s project homepage https://aqua3s.eu/ (accessed Jul. 12, 2021)
[3] Google maps homepage https://cloud.google.com/maps-platform/pricing/sheet (accessed Jul. 12, 2021)
[4] Mapbox Homepage https://www.mapbox.com/ (accessed Jul. 12, 2021)
[5] Leaflet.js Homepage https://leafletjs.com/ (accessed Jul. 12, 2021)
[6] MapLibre Homepage https://maplibre.org/ (accessed Apr. 01, 2022)
[7] Add a default marker https://maplibre.org/maplibre-gl-js-docs/example/add-a-marker/ (accessed Apr. 01, 2022)
[8] Add multiple geometries from one GeoJSON source https://maplibre.org/maplibre-gl-js-docs/example/multiple-geometries/ (accessed Apr. 01, 2022)
[9] Animate a series of images https://maplibre.org/maplibre-gl-js-docs/example/animate-images/ (accessed Apr. 01, 2022)

[10] Add a video https://maplibre.org/maplibre-gl-js-docs/example/video-on-a-map/ (accessed Apr. 01, 2022)
[11] Add a WMS Source https://maplibre.org/maplibre-gl-js-docs/example/wms/ (accessed Apr. 01, 2022)
[12] GeoJSON Homepage https://geojson.org/ (accessed Apr. 01, 2022)
[13] Geopandas Homepage https://geopandas.org/en/stable/ (accessed Apr. 01, 2022)
[14] MapLibre.AddSource https://maplibre.org/maplibre-gl-js-docs/api/map/#map#addsource (accessed Apr. 01, 2022)
[15] Maplibre Map.AddLayer https://maplibre.org/maplibre-gl-js-docs/api/map/#map#addlayer (accessed Apr. 01, 2022)
[16] MapLibre.map.on https://maplibre.org/maplibre-gl-js-docs/api/map/#map#on (accessed Apr. 01, 2022)
[17] EPANET Application https://www.epa.gov/water-research/epanet (accessed Apr. 01, 2022)
[18] OWA EPANET outfile format http://wateranalytics.org/EPANET/_out_file.html (accessed Apr. 01, 2022)
[19] Python struct https://docs.python.org/3/library/struct.html (accessed Apr. 01, 2022).

IOP Conf. Series: Earth and Environmental Science 1136 (2023) 012008 doi:10.1088/1755-1315/1136/1/012008

Monitoring planning for urban drainage networks

Antonietta Simone[1], Alessandra Cesaro[1], Cristiana Di Cristo[1], Oreste Fecarotta[1] and Maria Cristina Morani[1]

[1] Department of Civil, Building and Environmental Engineering (DICEA), University Federico II, Naples, 80125, Italy

antonietta.simone@unina.it

Abstract. Urban drainage network (UDN) monitoring is an important task whose planning can be related to various purposes, as for example contaminant detection and epidemiological studies. This paper proposes two different strategies for the identification of a monitoring system for UDNs. The optimal solution, in terms of location and number of sensors, is firstly addressed using a deterministic approach. A new mathematical model is developed and a global optimization solver is employed to perform the optimization procedure. Secondly, the position of devices is also investigated using a new strategy based on the complex network theory (CNT) tools. The comparison between the results achieved by both the strategies is finally presented with reference to a benchmark network.

1. Introduction

Urban drainage networks (UDNs) are complex systems [1,2] aimed at the collection and transport of wastewater and rainwater up to the treatment plant.

In recent decades, the need to propose new monitoring strategies for such complex systems has become increasingly urgent, mainly for factors related to the identification of illicit intrusions, to the control of specific contaminants [3,4] and to the limitation of the potential environmental impacts [5].

Solving this problem with optimization algorithms has been the focus of attention of many researchers, mainly relying on heuristic approaches [3,4,6,7].

These approaches are very effective but also very cumbersome, both for the hydraulic simulations they require and for the expensive calculation times. Furthermore, it could also happen, particularly with very large and complex systems, that the lack of information (e.g., flow) makes such analyzes difficult or even unreliable. In this context, the Complex Networks Theory (CNT) is proposed as a useful approach for the analysis of complex real systems [8].

The aim of this paper is to model the sensor placement problem, i.e., determining the candidate positions to host water quality sensors useful for detecting the presence of contaminants or pathogens in the system, using both an optimization procedure and complex network theory approach.

The first strategy is based on an optimization procedure aiming at searching for the best location of a fixed number of water quality sensors, in order to maximize the reliability of the whole system in case of contamination. This strategy relies on the development of a mathematical model, which is solved by the use of a global optimization solver.

The second one uses a CNT centrality metric by integrating both information on the connectivity structure and on the intrinsic relevance of nodes of the system. The strategy considers the different role of nodes (e.g., inlet nodes, connection nodes, outfall nodes, etc.) embedding the information about their intrinsic relevance [12] as lateral inflow and the presence of spatial constraints (e.g., slope) derived from the flow direction of the system. The goal is to apply the In-Relevance-based harmonic centrality to the direct graph of the UDN to support the monitoring system planning.

The paper is organized as follows. The next section describes the two methodologies, the third section presents the results of the analyses applied to a benchmark UDN and their comparison. Concluding remarks are drawn in the last section.

2. Material and Methods

Determining the ability of a node to detect the contaminant in UDNs is essential for planning an efficient monitoring system. This paper presents two modeling approaches for optimal sensor placement aimed at detecting contaminants and pathogens in UDNs.

The first methodology proposes an optimization problem based on one single objective function, that is, the maximization of the network reliability when a spill of contaminant occur within the system.

The second methodology investigates the sensor monitoring problem from a more topological perspective, using a metric proposed by the complex network theory (CNT) and adapted to infrastructural systems like UDNs.

2.1. The optimization method

The first strategy is a deterministic optimization, relying on the analytical properties of the problem to generate a sequence of points converging to a global optimum or an approximately global optimum. Deterministic methods include: linear programming (LP), non-linear programming (NLP), mixed-integer linear programming (MILP) and mixed-integer non-linear programming (MINLP) [13,14]. Most of the available MILP/MINLP solvers can achieve global optima only in convex problems [15]. The only solvers managing to find global optima in both convex and non-convex problems are the global optimization solvers.

In this study, the optimization procedure is performed by SCIP (Solving Constraint Integer Programs) [16] solver, which is a global optimization solver implementing a spatial branch and bound and various heuristics.

2.2. The optimization procedure

Given a network consisting of L links and N nodes, the aim of the optimization procedure is to find the best location of a fixed number of sensors in order to maximize the network reliability (R_N) in case of pollutant introduction. In this study, the optimization procedure has been decoupled by the hydraulic and quality modelling of the network, thus flow velocity (V) and pollutant concentration (C) are preliminary computed by means of an external software (i.e. SWMM) [17].

The network can be modelled as a directed graph, so that each link l ($l = 1....L$) has a proper direction and the discharge flows from the initial node (i_i) to the final node (i_f) of the generic link l. In this study, each node of the network is a possible candidate for the installation of a sensor and it has been modelled by means of a binary variable (I_n), which is equal to 1 if the device is installed, 0 otherwise.

According to the proposed procedure, the sensors should be located in order to maximize the reliability of the network, which means maximizing the number of detected points when a pollutant is inserted within the network. In this study, the network reliability is assessed over several scenarios, each one differing in the node assumed as inlet point of the pollutant.

A further variable of the optimization is represented by the pollutant detection ($\delta_{i,s}$), which is defined for each node i ($i = 1...N$) and for each scenario s ($s = 1...\sigma$). Such a binary variable has been properly modelled by means of a set of constraints, written for each link l ($l = 1...L$) of the network:

$$\delta_{i_i,s} = I_{i_i} B_{i_i}^{s} + I_{i_f} B_{i_i}^{s} + \left(1 - I_{i_f}\right)\left(1 - I_{i_i}\right)\left(\delta_{i_f,s} B_{i_i}^{s}\right) \quad (1)$$

where $B_{i_i}^s$ is a parameter equal to 1 if the concentration at the i_i-th node ($C_{i,s}$) is greater than a threshold concentration (C_0), 0 otherwise. The parameter $B_{i_i}^s$ results from the concentration values obtained by the quality model performed by the software SWMM.

It is worth underlining that the outfall (ω) is not constrained by means of Equation (1), since it cannot be the initial node of any links. In order to properly define the pollutant detection at the outfall node, the following constraint has been defined:

$$\delta_{\omega,s} = I_\omega B_\omega^s \tag{2}$$

According to Equation (2), the pollutant at the outfall is detected only in case of device installation (i.e., I_ω equal to 1) and concentration value greater than the threshold value (i.e., B_ω^s equal to 1).

Finally, a further constraint has been considered in order to set the total number of sensors (N_s) installed within the network:

$$\sum_{i=1}^{n} I_n = N_s \tag{3}$$

As objective function of the optimization, the system reliability has been considered and defined as follows:

$$maximize\left(R_N = \frac{\sum_{s=1}^{\sigma}\sum_{i=1}^{n}\delta_{i,s}}{\sum_{s=1}^{\sigma}\sum_{i=1}^{n}B_{i,s}}\right) \tag{4}$$

2.3. In-Relevance-Harmonic centrality for monitoring planning of UDNs

Centrality is one of the most studied concepts in CNT. Several centrality metrics have been proposed [8,9,12,18] to evaluate the most central element in real systems with respect to different physical phenomena and about the way information flow in a network [14]. The proposed approach recalls the concept of Harmonic centrality [19] and investigates the possibility of using this metric to efficiently plan a quality monitoring system in UDNs.

The Harmonic centrality is a metric that well detects the ability of each node to disseminate and receive information in the system based on the concept of proximity. The original formulation of the Harmonic centrality, $H^C{}_i$, is mathematically expressed as:

$$H_i^C = \sum_{j=1}^{N}\frac{1}{d_{ij}} \tag{5}$$

where d_{ij} is the distance from node i to node j in the network.

The standard metric is extended to embed the information about the intrinsic relevance of the nodes (outfall, connection nodes, etc.) through the function $f(R_i, R_j)$. The Relevance-harmonic centrality [12] is mathematically expressed as:

$$H_i^C = R_i\sum_{j=1}^{N}\frac{1}{d_{ij}} \tag{6}$$

The intrinsic relevance R_i (i=1, …, N) is an information depending on the type of the network and the analysis to perform. The proposed strategy is performed using the function $f(R_i, R_j) = (R_i+R_j)/2$, corresponding to the mean value of the intrinsic relevance of each pair of nodes i and j because it highlights the role of the intrinsic relevance in the connections between nodes [12].

It is important noting that the harmonic centrality assumes that information through the network move only along the shortest possible paths, just as happens in UDNs, for which the shortest paths between the various pairs of nodes are uniquely determined by the slope of the system.

In the present case, the In-Relevance-Harmonic centrality, is evaluated for each node to determine which nodes can detect the greatest amount of information disseminated on the network, and therefore, which are best suited for positioning sensors.

3. Case study

The benchmark UDN reported in Figure 1 [20] is considered to perform the two approaches. The UDN is composed of 77 nodes (manholes), 79 edges (sewer pipes) and 1 outfall. The network is direct according to the flow directions imposed by the slope of the system.

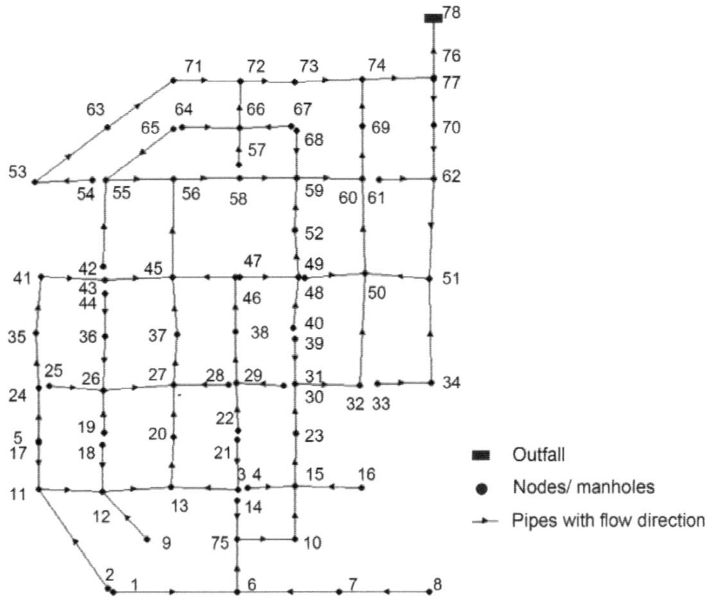

Figure 1. Network layout with flow direction.

3.1. *Application of the optimization procedure to the case study*

For the analysed case study, each node of the network has been assumed as possible inlet of the pollutant. Therefore, the total number of analysed scenarios (σ) is equal to the total number of nodes (N). Regarding the inserted pollutant, the inflow concentration has been assumed equal to 1 mg/l. The pollutant has been considered to decay with a first order kinetic, with the decay coefficient set as 0.2.

Once a threshold concentration value (C_0) and a number of installed sensors (N_s) are fixed, the optimization process is accomplished by the solver SCIP in less than 1 second. The achievement of the global optimum is extremely fast since SCIP is a very high-performance solver for MILP problems.

It is worth underlining that, by increasing the number of installed sensors, the value of the objective function (i.e., the network reliability) progressively increases up to the maximum value (i.e., 1). In particular, Figure 2 reports the value of the objective function as function of the number of sensors for the case $C_0 = 0.008$ mg/l, showing that four sensors are sufficient for reaching the maximum value.

3.2. *In-Relevance-Harmonic centrality for the sensor system planning*

In this paragraph the In-Relevance-Harmonic centrality is applied to the case study to understand if it can draw useful information about the optimal sensor location. The intrinsic relevance is set equal to the lateral inflow for each node, which is proportional to the concentration of contaminant introduced into each node as mg/l.

According to the performed analysis, the maximum value occurs for node 78, i.e., the system outfall. This result indicates that the outfall is the most suitable node to detect the information of other nodes and thus, it can be seen as a good candidate for sensor placement, consistently with the system operation.

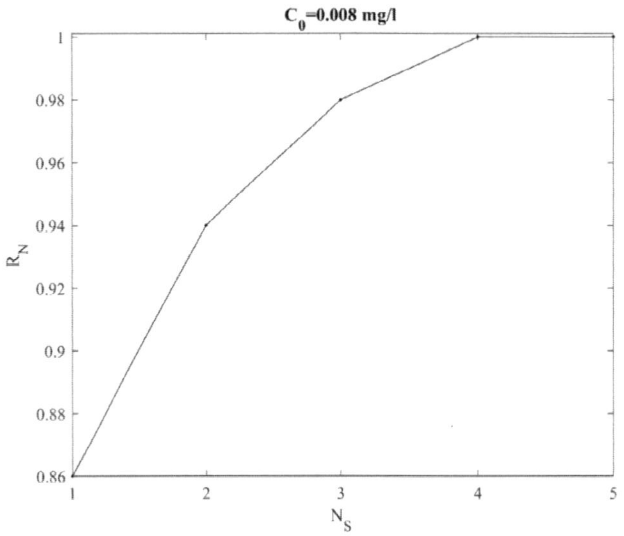

Figure 2. Variability of the objective function depending on the number of installed sensors for C_0 = 0.008 mg/l.

To increase the number of sensors to be installed in the network, it is possible to refer to the metric values and select the nodes immediately less central than the outfall. It indicates that nodes 76 and 59 as good information receptors. It is evident that node 76 is alternative to node 78 since they are connected to each other. The analysis suggests that node 76 receives a lot of information about the system, being close to the outfall, and therefore represents an excellent position to host a sensor, but at the same time its proximity to the receptor makes it an alternative, to be considered, perhaps, as a sensor backup and for step-testing during monitoring campaigns.

Therefore, the analysis suggests placing two sensors in nodes 78 and 59. The ranking of the metric values allows, as the number of sensors to be positioned increases, to immediately identify the most suitable position to host them. With such an interpretation of the metric, nodes 15, 27 and 45 follow as candidate nodes for hosting sensors.

The analysis provides null values of the metric for all extremal nodes because no information arrives in these nodes. This result implies that these points are not candidates for hosting measurement points.

The results of the analysis show that the In-Relevance-Harmonic centrality is effective in evaluating the ability of the nodes to receive contaminant and therefore it could be effective in the study related to the spread of contaminants and the planning of monitoring systems.

3.3. Optimization procedure vs In-Relevance-Harmonic centrality

The present paper proposed two modeling approaches for optimal sensor placement in UDNs. The two methodologies, despite being based on different concepts and approaches, provided quite comparable results, as shown in Figure 3. Indeed, both the strategies identify the outfall node as the optimal point for sensor location. Moreover, according to the results achieved by the In-Relevance-Harmonic centrality, the selected points are very close to the nodes selected by the optimization method. For example, Figure 4 shows the comparison between the In-Relevance-Harmonic centrality and the optimization procedure considering C_0 = 0.008 mg/l, and in both cases the nodes 78, 59 and 50 are indicated as possible candidates for the positioning of sensors.

4. Conclusions

The present paper faced the sensor placement problem using two different approaches based on an optimization procedure and a CNT centrality metric. The analyses produced comparable results for the two strategies in determining the candidate positions to host water quality sensors.

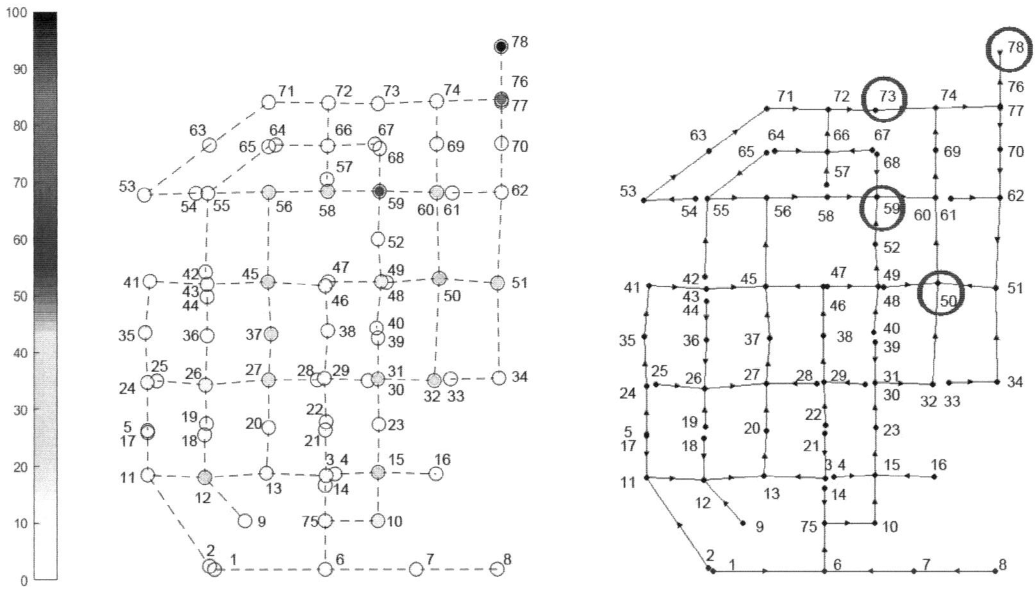

Figure 3. In-Relevance-Harmonic centrality values (left) and position of installed sensors for C_0=0.008 mg/l using the optimization procedure (right) for the benchmark UDN.

Furthermore, considering the good results obtained, this study suggested the idea of integrating the two approaches to achieve a more performing, rational and effective strategy for the planning of an advanced monitoring system, in order to envisage both the connective structure and the hydraulic behaviour of the system in the sensor placement problem. A first topological phase that identifies the best candidates to host the sensors using the In-Relevance-Harmonic centrality, and a second hydraulic phase, which starting from the results provided by the first phase, provides the real design of the monitoring system, in which optimization has as the goal to define the actual optimal positioning of the sensors.

Acknowledgments
Authors thank the University of Naples Federico II that supported this research under the Project Program Finanziamento della Ricerca di Ateneo 2020.

References

[1] Yazdani A and Jeffrey P 2011 Complex network analysis of water distribution systems *Chaos* **21**(1) 016111 https://doi.org/10.1063/1.3540339
[2] Giustolisi O, Simone A and Ridolfi L 2017 Network structure classification and features of water distribution systems *Water Resources Research* **53**(4) 3407–3423 https://doi.org/10.1002/2016WR020071
[3] Banik BK, Di Cristo C, Leopardi A and de Marinis G 2017 Greedy algorithms for sensor location in sewer systems *Water* **9**(11) 856 https://doi.org/10.3390/w9110856
[4] Sambito M, Di Cristo C, Freni G and Leopardi A 2020 Optimal water quality sensor positioning in urban drainage system for illicit intrusion identification *Journal of Hydroinformatics* **22**(1) 46–60 https://doi.org/10.2166/hydro.2019.036
[5] Gromaire MC, Garnaud S, Saad M and Chebbo G 2001 Contribution of different sources to the pollution of wet weather flows in combined sewers *Water Research* **35**(2) 521–533 https://doi.org/10.1016/S0043-1354(00)00261-X
[6] Banik BK, Alfonso L, Di Cristo C, Leopardi A and Mynett A 2017 Evaluation of different

formulations to optimally locate sensors in sewer systems *Journal of Water Resources Planning and Management* **143**(7) https://doi.org/10.1061/(ASCE)WR.1943-5452.0000778

[7] Sambito M and Freni G 2021 Optimal positioning of water quality sensors in sewer during dry and wet weather *AIP Conf. Proc.* **2343** 110004 https://doi.org/10.1063/5.0048105

[8] Newman M 2010 *Networks: An Introduction* (Oxford, UK: Oxford University Press) https://doi.org/10.1093/acprof:oso/9780199206650.001.0001

[9] Borgatti SP 2005 Centrality and network flow *Social Networks* **27**(1) 55–71 https://doi.org/10.1016/j.socnet.2004.11.008

[10] Reyes-Silva JD, Zischg J, Klinkhamer C, Julian D, Klinkhamer C, Rao PSC, Sitzenfrei R and Krebs P 2020 Centrality and shortest path length measures for the functional analysis of urban drainage networks *Appl. Netw. Sci.* **5**(1) https://doi.org/10.1007/s41109-019-0247-8

[11] Zuluaga VZ, García JM, Mesa MJ and Acosta-Minoli CA 2020 Use of network theory to model water quality parameters in a hydrological network *Journal of Physics Conference Series* **1448**(1) 012006 doi: 10.1088/1742-6596/1448/1/012006

[12] Giustolisi O, Ridolfi L and Simone A 2020 Embedding the intrinsic relevance of vertices in network analysis: the case of centrality metrics *Sci. Rep.* **10** 3297 https://doi.org/10.1038/s41598-020-60151-x

[13] Belotti P, Kirches C, Leyffer S, Linderoth J, Luedtke J and Mahajan A 2013 Mixed-integer nonlinear optimization *Acta Numerica* **22** 1–131 doi:10.1017/S0962492913000032

[14] Morani MC, Carravetta A, D'Ambrosio C and Fecarotta O 2021 A new mixed integer non-linear programming model for optimal PAT and PRV location in water distribution networks *Urban Water Journal* **18**(6) 1–15 https://doi.org/10.1080/1573062X.2021.1893359

[15] Belotti P, Lee J, Liberti L, Margot F and Wächter A 2009 Branching and bounds tighteningtechniques for non-convex MINLP *Optimization Methods and Software* **24**(4-5) 597–634 https://doi.org/10.1080/10556780903087124

[16] Vigerske S and Gleixner A 2018 SCIP: global optimization of mixed-integer nonlinear programs in a branch-and-cut framework *Optimization Methods and Software* **33**(3) 1–31 https://doi.org/10.1080/10556788.2017.1335312

[17] Rossman LA 2010 *Storm Water Management Model User's Manual Version 5.0* EPA/600/R-05/040 (Cincinnati, OH: U.S. Environmental Protection Agency)

[18] Freeman L 1978 Centrality in social networks: Conceptual clarification *Social Networks* **1**(3) 215–239 https://doi.org/10.1016/0378-8733(78)90021-7

[19] Marchiori M and Latora V 2000 Harmony in the small-world *Physica A: Statistical Mechanics and its Applications* **285**(3-4) 539–546 https://doi.org/10.1016/S0378-4371(00)00311-3

[20] Li G and Matthew RGS 1990 New approach for optimization of urban drainage systems *Journal of Environmental Engineering* **116** 927–944 https://doi.org/10.1061/(ASCE)0733-9372(1990)116:5(927)

IOP Conf. Series: Earth and Environmental Science 1136 (2023) 012009 doi:10.1088/1755-1315/1136/1/012009

Using artificial intelligence models to support water quality prediction in water distribution networks

Laura Enriquez[1], Juan Saldarriaga[1], Luigi Berardi[2], Daniele Laucelli[3] and Orazio Giustolisi[3]

[1] Water Distribution and Sewerage Systems Research Center (CIACUA), University of the Andes, Bogotá, 111711, Colombia
[2] Department of Engineering and Geology, University "G. D'Annunzio" of Chieti Pescara, Pescara, 65127, Italy
[3] Department of Civil, Environmental, Land, Building Engineering and Chemistry, Technical University of Bari, Bari, 70126, Italy

Corresponding author: lv.enriquez@uniandes.edu.co

Abstract. Water chlorination is the most used disinfection method in water distribution networks (WDNs). Nonetheless, water quality parameters, including chlorine concentration, are not available at every point of the WDN, although such information is of direct relevance to drive the operation at water treatment plants to keep the correct chlorine residual through the system. This work proposes the use of data-driven models, i.e., Artificial Neural Networks and Evolutionary Polynomial Regression, to predict the water quality parameters in most areas of a WDN, using water quality data measures at few sampling points. The study is demonstrated on the case studies of the trunk network of Bogota's water distribution system.

1. Introduction

The most common disinfection method for water distribution networks (WDNs) is chlorination [1], which generates disinfection by-products that mainly consist of trihalomethanes (THMs). On the one hand, the chlorine dose at water treatment plants (WTPs) must be such that at least a minimum concentration is achieved at the system's extremities [2]. On the other hand, the regular consumption of THMs has been related to cancer formation [3] and congenital formations [4], thus it is desirable to minimize the chlorine dose at WTPs to avoid the excessive formation of THMs. Therefore, chlorine dose must be carefully monitored and controlled within water distribution networks.

There is a high uncertainty level regarding the water quality parameters throughout a network since sensors are usually located only at some points of the system and the measurements are not always available on real time. For this reason, water utilities tend to create water quality models for management and planning purposes. There are two types of models: data-based, in which there is not an explicit relation between the data and it is only useful for management applications, and process-based, that relates variables through equations and can be used both for planning and management [5]. According to [6], the data-based models are only recommended when the parameter estimation for process-based models is "imprecise or difficult to obtain".

Regarding free chlorine concentration, the most common data-based method to predict this parameter in WDNs are artificial neural networks (ANNs). The inputs that have been used in ANNs include temperature, chlorine concentration at the sources, flow [6], pH and turbidity [7]. Although some authors have implemented input removal by analysing highly correlated inputs [8] or using the Akaike information criterion [9], the time lags between the inputs are usually estimated by trial and error [9]. Furthermore, the ANN structure is commonly defined by trial and error [8]. In general, the application of ANNs to predict free chlorine has proved having a high level of accuracy using input data obtained by a modelling process or direct measurements. Nevertheless, it has only been tested in networks that have up to 92 nodes [8] with a number of output nodes that ranges from one [6,9] to 26 nodes [8] using the data of the sources [6,8,9] and/or local condition parameters [7].

Regarding THMs formation, there are two approaches that have been taken to create data-based models. On the one hand, according to [10], linear procedures such as simple linear regression [4] and multilinear regression analysis (MLR) [4,11] have been explored by several authors. On the other hand, according to [10], non-linear methods such as support vector machine regression (SVR) and multi-layer perceptrons [11], a type of neural networks, have been tested in different researches. In this sense, it has been found that linear models like MLR have a better extrapolation ability than data-driven models such as ANNs [11], but the latter outperforms the former for high THMs concentrations [12]. The THMs models usually include multiple independent variables such as free residual chlorine, temperature, pH, Total Organic Carbon (TOC), UV_{254} and residence time [4,11,13].

Hence, the development of ANNs for the prediction of water quality in WDNs has been limited to relatively small networks, which is why this investigation extends it to a bigger network: the trunk network of Bogotá, that has approximately 550 km of pipes with 4316 junctions. A comprehensive process-based water quality model has not been developed in this system due to the scarcity of measured data. In fact, the chlorine concentration is monitored at the exit of WTPs every two hours, and at some tanks and nodes inside the system every 3 to 8 days. Similarly, trihalomethanes are only measured six-monthly in around 70 points of the network. For this reason, this research implements data-based models such as ANNs, trained with information obtained by a modelling procedure, as an alternative to process-based models for the prediction of chlorine and TTHMs concentrations.

Furthermore, the literature shows that ANNs creation may be improved by optimizing the inputs time lags and the ANN structure within a single algorithm. Consequently, this investigation applies ANN-MOGA to support the prediction of water quality in WDNs since it minimizes the input dimension considering time lags and simultaneously optimizes the number of hidden neurons based on a fitness function using a multi-objective genetic algorithm (MOGA).

In addition, given that ANNs have been tested in small-sized WDNs, water sources have been mostly used as inputs to forecast water quality. Conversely, this study proposes a two-step sampling design methodology to select the most significant input nodes for the water quality prediction at each output node applying TEVA-SPOT and the Evolutionary Polynomial Regression data-modelling strategy using MOGA optimization (EPR-MOGA). In this sense, given the large size of the case study, this research explores the relation between chlorine and THMs concentration throughout the network and other hydraulic variables, such as pressure. In this way, this study aims to test the accuracy of EPR-MOGA and ANN-MOGA to predict chlorine and THMs in the trunk network of Bogota's water distribution system and to assess if the information of a few points can be used to forecast these variables throughout the network.

2. Case study

The case study of this research is the trunk network of Bogota's water distribution system, that serves a population of nine million inhabitants, who demand an average daily flow of 16.5 m^3/s for 2019 [14]. The hydraulic model of the network, that was provided by the Bogota's water utility, consists of 66 tanks, 4362 pipes, 4316 junctions, 73 pumps and 145 valves, including flow control, throttle control, pressure breaker and pressure reducing valves. The model has approximately 550 km of pipes with diameters that range from 38.1 mm to 3500 mm and an absolute roughness that varies from 0.0015 mm to 0.6 mm with a predominant use of Concrete Cylinder Pipes (CCP) [14]. The model is updated up to November of 2014, but the accuracy of its parameters is unknown, and its calibration is beyond the objective of this research.

The hydraulic model includes three WTPs as reservoirs: Tibitoc, Francisco Wiesner and El Dorado, as shown in figure 1. In general, the WTPs implement a conventional treatment process that includes aeration, coagulation, flocculation, sedimentation, filtration, and disinfection, except for Wiesner, that does not perform flocculation nor sedimentation because its water source contains a low sediment load.

Figure 1. Location of WTPs in the trunk network of Bogotá.

In addition, the hydraulic model incorporates a low-demand scenario with different 48-hour demand patterns that are assigned to multiple junctions. The duration of the hydraulic simulation was increased from 48 hours to 100 hours to achieve periodic patterns in the concentrations of the chemical species. Moreover, some of the parameters of the model, such as rule-based controls and the initial level of the tanks, were modified to avoid negative pressures when the simulation was extended.

3. Methodology

3.1. *Water quality modelling*

The data to train ANN-MOGA was generated using the MATLAB EPANET Toolkit developed by [15], that incorporates EPANET 2.2 and EPANET-MSX capabilities, given the scarcity of measured data. The multi-species model adopted the first-order models to represent bulk and wall chlorine decay, as shown in equation (1), while TTHMs formation was simulated with a linear model that depends on the chlorine consumption [16], as displayed in equation (2). In addition, the k_b and Y coefficients were computed on a pipe basis as a weighted average considering the concentration of conservative tracers assigned to Tibitoc and El Dorado WTPs, as suggested by [17]. This is presented in equation (3) and equation (4), respectively.

$$\left(\frac{\mathrm{d}C_{Cl_i}}{\mathrm{d}t}\right)_{total} = -k_{b_i} C_{Cl_i} - \frac{4}{D_i}\frac{k_{w_i} \cdot k_{f_i}}{\left(k_{w_i} + k_{f_i}\right)} C_{Cl_i} \tag{1}$$

$$\frac{\mathrm{d}C_{TTHM_i}}{\mathrm{d}t} = Y_i k_{b_i} C_{Cl_i} + \frac{4}{D_i}\frac{Y_i \cdot k_{w_i} \cdot k_{f_i}}{\left(k_{w_i} + k_{f_i}\right)} C_{Cl_i} \tag{2}$$

$$k_{b_i} = TT_i \cdot k_{b_t} + \left(1 - TT_i - TD_i\right) \cdot k_{b_w} + TD_i \cdot k_{b_d} \tag{3}$$

$$Y_i = TT_i \cdot Y_t + \left(1 - TT_i - TD_i\right) \cdot Y_w + TD_i \cdot Y_d \tag{4}$$

where C_{Cli} is the residual bulk chlorine concentration for the pipe i; k_{bi} is the first-order bulk coefficient for the pipe i; D_i is the diameter of the pipe i; k_{wi} is the first-order wall decay coefficient for the pipe i ;k_{fi} is the mass transfer coefficient for the pipe i ; $TTHM_i$ is the bulk total trihalomethanes concentration for the pipe; Y_i is the TTHMs yield coefficient; TT_i and TD_i are the concentrations of the Tibitoc and El Dorado conservative tracers for the pipe i; and the subscripts t, w and d stand for the coefficients for the Tibitoc, Wiesner and El Dorado WTPs, respectively.

The coefficients for the chlorine-decay models were assumed based on a literature review, whereas the parameters for the TTHMs model were found through a calibration process using Monte Carlo simulations. The assumed values for the first-order bulk coefficients were 0.203 days^{-1}, 0.163 days^{-1} and 0.436 days^{-1} for the Tibitoc, Wiesner and El Dorado WTPs, respectively, based on the results of previous research [18]. Regarding the wall coefficient, the pipes were grouped in 23 sets based on the diameter, the absolute roughness, and the mean flow velocity. The coefficient was assigned to each set with a linear function that depended on the mean pipe diameter considering the values described by [19]. Finally, regarding the TTHMs formation, the results of the yield coefficients were 33.32 μg/mg, 32.12 μg/mg, and 36.82 μg/mg for the Wiesner, Tibitoc and El Dorado WTPs, respectively.

The three WTPs of the model were set as the chlorine sources with different 24-hour patterns using a mean concentration of 1.56 mg/L, 1.47 mg/L and 1.66 mg/L for Tibitoc, Wiesner and El Dorado WTPs, respectively. The initial concentrations of chlorine and TTHMs at each node were set through an iterative process. Moreover, a 45-hour warm-up period was used to achieve approximate periodical patterns in the behaviour of the chemical species. Therefore, the water quality data was sampled from 45 hours to 100 hours with a 30-minute time step.

3.2. Sampling design

3.2.1. TEVA-SPOT. This investigation applied TEVA-SPOT to pre-select the input and output points throughout the network. This software stands for Threat Ensemble Vulnerability Assessment-Sensor Placement Optimization Tool, and it was created by the EPA with collaborators from the Sandia National Laboratories, Argonne National Laboratory and the University of Cincinnati [20]. The software finds the optimal location of water quality sensors based on the impacts obtained from the simulation of contamination events [20].

The contamination scenario for this research replicated the settings used by [20]. The simulation consisted in the injection of a chemical contaminant at a rate of 800 mg/min from 0 hours to 2 hours at every non-zero demand node, assuming no decay or loss with a concentration zero threshold equal to 1E-5 mg/L. Next, the sensor design was performed with GRASP, only considering non-zero demand nodes as feasible locations, that corresponds to 322 nodes for the case study. In this sense, the sources were dismissed as inputs for the ANNs. The sensor set sizes were 0, 20, 30, 40, 50, 60 and 70. The objective function was the mean time to detection, that is the time from the injection start time until contamination is first detected at each node plus a response time [21]. The selected response time was 15 minutes, and a 0.003 mg/L detection limit was considered.

Subsequently, the 20-sensor design was selected as the sampling design for the system to restrict the number of possible inputs for each output. Regarding the output nodes, where the water quality would be predicted, they were chosen from the 30-, 40- and 50-sensor designs limiting its number to 15 nodes and guaranteeing that at least one output node was located at each zone defined by the water utility.

3.2.2. EPR-MOGA. The selection of the most sensitive inputs for each output variable was done using the EPR-MOGA software. EPR consists in performing a symbolic regression as data-model of observed data. Furthermore, EPR-MOGA integrates EPR with a multi-objective genetic algorithm named OPTIMOGA, that stands for Optimized Multi-Objective Genetic Algorithm, which includes three objective functions: maximizing the model accuracy while minimizing the number of polynomial coefficients and the number of inputs [22,23]. This software performs a two-step optimization process to define: 1) the set of input combinations and their exponents within candidate user-defined bounds and

2) the constants for each term in the polynomial regression expression using the linear least-squares method [23]. EPR-MOGA can dismiss some inputs when zero is included as a candidate exponent, which allowed to reduce the number of inputs prior to the implementation of ANNs.

On the one hand, the candidate input variables included pressure, chlorine and TTHMs, which resulted in 60 total candidate inputs. On the other hand, the variables to predict at each output point were chlorine and TTHMs concentrations, that resulted in 30 output variables. The input removal was performed using all the data for training in a two-step process because the automatic generation of plots within the software caused an out of memory error. First, each output was tested using three different sets of inputs, each of 20 variables. Second, a further input selection was done to achieve a suitable prediction performance. The results for every test were analyzed and a single model was chosen based on maximizing the accuracy while keeping a low number of inputs. Next, the inputs of each test were joined in a single file for every output variable, and an individual expression was selected considering a limit of 9 input variables.

The expression structure selected for the EPR-MOGA analysis is shown in equation (5); an exponential function was chosen due to its similarity with the chlorine first-order bulk decay model. The input and output data were scaled between 0 and 1. The number of terms was limited to 5, while the number of past input values was restricted to 3 and no past output values were considered as inputs. The candidate exponents were [-2, -1.5, -1, -0.5, 0, 0.5, 1, 1.5, 2]; that included zero to dismiss some of the inputs. Finally, two strategies were implemented to avoid overfitting: the coefficients were constrained to only have positive values and they were eliminated when its value was not sufficiently greater than zero with a confidence level of 99%.

$$Y = \exp\left(a_0 + \sum_{j=1}^{m} a_j \cdot \left(\mathbf{X}_1\right)^{\mathbf{ES}(j,1)} \cdot \ldots \cdot \left(\mathbf{X}_k\right)^{\mathbf{ES}(j,k)} \right) \tag{5}$$

where Y is the least squares estimate of the output value, a_j is a constant parameter of the jth term, a_0 is an optional bias, m is the number of terms in the expression, $\mathbf{X}_k$ is the kth vector of inputs, k is the number of input variables and **ES** is a matrix of exponents $m \times k$.

3.3. *Artificial Neural Networks (ANNs)*

The software ANN-MOGA, written in MATLAB and implemented in MS-Excel, was applied for this research [24]. This software performs a multi-objective optimization of single-hidden-layer ANNs considering a Multiple Input Single Output (MISO) model. The optimization is carried out with the OPTIMOGA software including three objective functions: the coefficient of determination computed on the validation data, the input dimension, and the number of hidden neurons. The training of the ANNs is done though the Levenberg Marquardt algorithm and adaptive search direction [24]. In this way, the software presents several ANN models that lie in the Pareto front with the following decision variables: number of inputs and hidden neurons, weights, and kernel function parameters. Therefore, besides optimizing the ANNs structure, ANN-MOGA further reduces the number of inputs of the model.

The water quality data to develop ANNs was separated in two sets: the training and validation set, that corresponded to 70% of the records, and the test set, with 30% of the records. Subsequently, the ANNs designs were done using four kernel functions, as shown in table 1. The tests 2-4 were only done for 10 of the output variables, that did not fulfill a test coefficient of determination (R^2) of at least 0.7. The test performance of the models was only assessed using the 1-step ahead prediction. The autoregressive with exogenous input (ARX) model was selected limiting the number of past input values to 2 and the number of hidden neurons to 5. The number of generations was set to 1000.

IOP Conf. Series: Earth and Environmental Science 1136 (2023) 012009 doi:10.1088/1755-1315/1136/1/012009

Table 1. Characteristics of the test performed with ANN-MOGA.

Test	Kernel function	Output variables
1	Hyperbolic tangent	30
2	Sigmoid	10
3	Quadratic	10
4	Linear	10

4. Results and discussion

4.1. EPR-MOGA

The comparison between the fitness results of the EPR and ANN models in terms of the global coefficient of determination for TTHMs and chlorine are shown in table 2 and table 3, respectively. The name of each output has the initial of the variable (T for TTHMs and C for chlorine), followed by a hyphen and the node ID. In general, the EPR models outperformed ANNs, except for T-1005, T-2375 and C-1295, in which ANNs dominated with a difference below 10%. The greatest increase in the coefficient of determination of EPR models compared with ANNs arose for T-190, T-920 and C-1191, in which the rise was above 99%. However, if the training and the test R^2 are analyzed separately, the mean training fitness of ANNs is better than that of the EPR models, while the mean test accuracy has the opposite behaviour. Therefore, EPR models exhibit a slight better generalization performance than ANNs. Nevertheless, it is important to consider that the EPR models were trained using all the available records, which means that they cannot be directly compared to ANNs.

Table 2. Comparison of fitness results -TTHMs.

Output	EPR - R^2	ANN - R^2
T-1005	0.6981	0.7006
T-10745	0.9860	0.9792
T-1191	0.9681	0.8215
T-1257	0.9949	0.9781
T-1295	0.7350	0.6350
T-1351	0.9409	0.9192
T-1497	0.9169	0.8682
T-1562	0.9717	0.9164
T-1715	0.9925	0.8062
T-190	0.7033	0.3315
T-2312	0.7659	0.4780
T-2375	0.9749	0.9786
T-1028	0.9639	0.9633
T-3213	0.9711	0.9423
T-920	0.9911	0.1416

Table 3. Comparison of fitness results - chlorine.

Output	EPR - R^2	ANN - R^2
C-1005	0.6130	0.5058
C-10745	0.9883	0.9851
C-1191	0.9781	0.4902
C-1257	0.9939	0.9691
C-1295	0.6568	0.7296
C-1351	0.9627	0.9589
C-1497	0.8429	0.7576
C-1562	0.9078	0.8245
C-1715	0.9857	0.9722
C-190	0.8027	0.7158
C-2312	0.7290	0.6070
C-2375	0.9737	0.8198
C-1028	0.9296	0.9156
C-3213	0.9713	0.9495
C-920	0.9722	0.9275

In addition, the descriptive statistics of the number of inputs for each type of output, achieved using the EPR-MOGA strategy, are shown in table 4. The number of inputs does not include the defined time lag, which means that each type of input variable is only counted once at every input location. The most selected type of input variables to predict TTHMs and chlorine were TTHMs and chlorine, respectively. In addition, it was found that the pressure and the TTHMs concentrations at other nodes have a low level of influence in chlorine prediction. Conversely, the pressure inputs proved having more impact in the TTHMs concentrations than chlorine inputs.

Table 4. Statistics of the number of inputs for each type of output.

Output	No. of Pressure Inputs			No. of TTHMs Inputs			No. of Chlorine Inputs		
	Min	Mean	Max	Min	Mean	Max	Min	Mean	Max
TTHMs	0	2.13	5	1	3.60	6	0	1.27	5
Chlorine	0	1.00	4	0	1.40	4	2	4.07	8

4.2. ANN-MOGA

The attributes of the best ANN models for the prediction of TTHMs and chlorine are displayed in table 5 and table 6, where the # inputs column does not include the time lag, while the #inputs with lag column does include it; and the #input EPR column refers to the number of inputs of the EPR-MOGA models without incorporating the time lag. In addition, #hidden is the number of hidden neurons, #weights refers to the number of weights and # Test is the test number from which the best ANN was obtained.

The number of inputs of the ANNs were lower or equal than the number of inputs obtained in the EPR-MOGA models, with a maximum reduction of 83% for T-3213 and C-920, that achieved a training and a test R^2 greater than 0.9. Additionally, when the ANNs included the same number of inputs than the EPR models, the latter had a better general performance, which may indicate that the significant inputs to predict the output variables depend on the structure of the mathematical structure that is implemented. Furthermore, the results show that a greater number of inputs is related to lower training R^2 and/or test R^2 values.

Regarding the structure and parameters of the ANNs, the number of hidden neurons ranged from 1 to 4, which allowed for achieving sufficient fitness values in most of the cases. Therefore, a single hidden layer is enough to accomplish a suitable prediction accuracy. Moreover, the test that most frequently led to the best accuracy results was number 1, that was executed using the hyperbolic tangent as a transfer function, that demonstrates the suitability of this function for these types of models.

Table 5. Best ANNs models for TTHMs outputs.

Model	# inputs	# inputs with lag	# inputs EPR	# hidden	# weights	Training R^2	Test R^2	# Test
T-1005	6	11	6	3	37	0.8859	-1.2053	1
T-10745	5	5	6	2	13	0.9868	0.9535	1
T-1191	5	10	6	1	12	0.7320	0.7223	3
T-1257	3	4	7	2	12	0.9776	0.9696	1
T-1295	7	9	9	2	20	0.7014	0.3933	3
T-1351	3	4	6	1	6	0.9367	0.8759	1
T-1497	6	8	8	1	10	0.9023	0.3233	3
T-1562	6	8	8	2	20	0.9237	0.8185	1
T-1715	2	2	5	4	16	0.7633	0.8783	1
T-190	6	9	9	2	21	0.7069	-0.4300	3
T-2312	5	6	9	1	8	0.5189	0.2929	2
T-2375	3	4	6	2	11	0.9799	0.9467	1
T-1028	2	2	6	3	10	0.9683	0.9334	1
T-3213	1	2	6	1	4	0.9559	0.8586	1
T-920	5	6	8	2	17	0.1849	-0.5000	1

Regarding the training R^2, 90% of the models accomplished values above 0.7. On the other hand, a large group of models had a test R^2 below 0.7, including six TTHMs models and seven chlorine models. The causes of a poor test performance include overfitting due to a high number of inputs or the variation of the tendency between the training and test dataset. Moreover, three output variables had low training

R^2 and test R^2, namely T-2312, T-920 and C-2312. This may be caused by an insufficient number of training data, a lack of relevant inputs to describe the behaviour of the outputs, such as data from the sources, or the variation of the determining input variables over time.

Table 6. Best ANNs models for chlorine outputs.

Model	# inputs	# inputs with lag	# inputs EPR	# hidden	# weights	Training R^2	Test R^2	# Test
C-1005	7	7	7	3	25	0.8794	-1.0575	1
C-10745	3	3	5	4	17	0.9905	0.9648	1
C-1191	2	2	6	3	12	0.7394	-2.6555	1
C-1257	4	6	6	3	24	0.9704	0.9357	1
C-1295	8	14	9	2	31	0.8468	0.3661	2
C-1351	6	7	7	2	17	0.9765	0.9306	1
C-1497	3	3	6	2	9	0.7847	0.5138	1
C-1562	2	2	4	1	3	0.8157	0.6907	4
C-1715	3	3	7	1	5	0.9771	0.9186	1
C-190	7	9	7	3	34	0.9071	0.2002	1
C-2312	6	8	8	1	10	0.6058	0.5318	3
C-2375	1	2	4	4	16	0.7653	0.7189	1
C-1028	4	5	8	2	13	0.9478	0.7135	1
C-3213	3	4	7	3	16	0.9758	0.8576	1
C-920	1	2	6	3	13	0.9172	0.9486	1

On the one hand, the comparison between the observed and the predicted data for C-10745, that was the output with the best performance, is shown in figure 2. This output variable had a low number of inputs and a consistent trend over the time, for which a suitable performance was expected. On the contrary, figure 3 displays the behaviour for T-920, that has the worst fitness results. This output variable does not show a different behaviour when comparing the training and the test dataset. Hence, the poor test performance may be caused by an insufficient amount of training data or the absence of determining inputs to describe this variable.

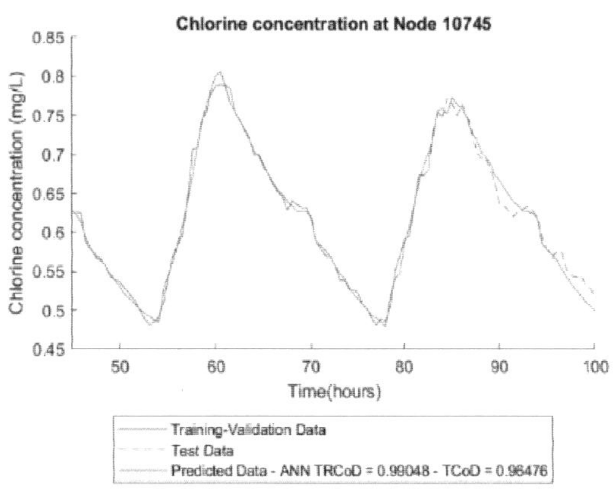

Figure 2. Chlorine concentration at Node 10745.

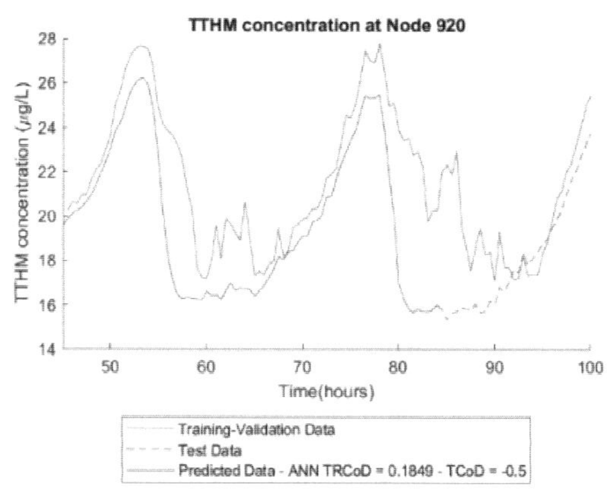

Figure 3. TTHM concentration at Node 920.

5. Conclusions

This research extended the use of ANNs for the prediction of water quality to a larger water distribution network: the trunk network of Bogotá. The design of the ANNs was optimized using the ANN-MOGA strategy, while the inputs for the models were selected applying TEVA-SPOT and EPR-MOGA.

Although this investigation mainly used EPR-MOGA to perform a sampling design, its prediction accuracy was also evaluated. In this sense, it was found that EPR-MOGA generally outperforms ANNs. However, since all the data was used for training the EPR models and the test set was not presented separately, both methods are not directly comparable and future research should further analyse the accuracy of EPR models for the prediction of water quality.

Regarding the forecast performance of the ANNs models, suitable values of the coefficient of determination were found for most of the output variables. Nevertheless, a few models did not achieve a satisfactory training R^2, and several models had poor test fitness values. The accuracy of the ANNs models may be improved by 1) increasing the number of training records; 2) including other hydraulic and water quality variables as inputs, such as data from the sources or changes in the operation scenario; and 3) using a lower number of inputs to avoid overfitting.

Hence, even though the prediction fitness of the ANNs and the EPR models needs to be further analysed and improved, the results show that data-driven approaches have a suitable generalization capacity and that they are a promising modelling alternative. Furthermore, this investigation demonstrated that the information of a few points is sufficient to forecast water quality throughout the trunk network of Bogotá.

Acknowledgments
The authors wish to acknowledge the help and the data of the trunk network of Bogotá provided by Dr. Ivonne Navarro, Dr. Mauricio Jiménez and the Water Supply and Sewerage Company of Bogotá.

References

[1] WHO 2011 Guidelines for Drinking-water Quality FOURTH EDITION WHO Library Cataloguing-in-Publication Data [Internet] *World Heal. Organ.*

[2] Fisher I, Kastl G and Sathasivan A 2011 Evaluation of suitable chlorine bulk-decay models for water distribution systems *Water Research* Elsevier Ltd **45**(16) pp 4896–908 https://doi.org/10.1016/j.watres.2011.06.032

[3] Mazhar M A, Khan N A, Ahmed S, Khan A H, Hussain A, Rahisuddin et al. 2020 Chlorination disinfection by-products in municipal drinking water – A review *J. Clean. Prod.* https://doi.org/10.1016/j.jclepro.2020.123159

[4] Espigares M, Lardelli P and Ortega P 2003 Evaluating Trihalomethane Content in Drinking Water on the Basis of Common Monitoring Parameters: Regression Models *Journal of Environmental Health* **66**(3)

[5] Fisher I, Kastl G, Sathasivan A and Jegatheesan V 2011 Suitability of chlorine bulk decay models for planning and management of water distribution systems *Critical Reviews in Environmental Science and Technology* Taylor & Francis Group **41**(20) pp 1843–82 https://doi.org/10.1080/10643389.2010.495639

[6] Rodriguez M J, West J R, Powell J and Sérodes J B 1997 Application of two approaches to model chlorine residuals in Severn Trent Water LTD (STW) distribution systems *Water Science and Technology* https://doi.org/10.1016/S0273-1223(97)00489-7

[7] Cuesta Cordoba G A, Tuhovčák L and Tauš M 2014 Using artificial neural network models to assess water quality in water distribution networks *Procedia Engineering* https://doi.org/10.1016/j.proeng.2014.02.045

[8] D'Souza C D and Kumar M S M 2010 Comparison of ANN models for predicting water quality in distribution systems *Journal / American Water Works Association* https://doi.org/10.1002/j.1551-8833.2010.tb10152.x

[9] Wu W, Dandy G C and Maier H R 2011 Application of artificial neural networks to forecasting water quality in a chloraminated water distribution system *MODSIM 2011 - 19th International Congress on Modelling and Simulation - Sustaining Our Future: Understanding and Living with Uncertainty* https://doi.org/10.36334/modsim.2011.c1.wu

[10] Platikanov S, Martín J and Tauler R 2012 Linear and non-linear chemometric modeling of THM

IOP Conf. Series: Earth and Environmental Science 1136 (2023) 012009 doi:10.1088/1755-1315/1136/1/012009

formation in Barcelona's water treatment plant *Science of the Total Environment* **432** https://doi.org/10.1016/j.scitotenv.2012.05.097

[11] Godo-Pla L, Emiliano P, Poch M, Valero F and Monclús H 2021 Benchmarking empirical models for THMs formation in drinking water systems: An application for decision support in Barcelona, Spain *Science of the Total Environment* **763** https://doi.org/10.1016/j.scitotenv.2020.144197

[12] Milot J, Rodriguez M J and Sérodes J B 2002 Contribution of Neural Networks for Modeling Trihalomethanes Occurrence in Drinking Water *Journal of Water Resources Planning and Management* https://doi.org/10.1061/(asce)0733-9496(2002)128:5(370)

[13] Kumari M and Gupta S K 2015 Modeling of trihalomethanes (THMs) in drinking water supplies: a case study of eastern part of India *Environmental Science and Pollution Research* **22**(16) https://doi.org/10.1007/s11356-015-4553-0

[14] Empresa de Acueducto y Alcantarillado de Bogotá -E.A.A.B 2018 Manual de operación del sistema matriz acueducto Bogotá

[15] Eliades D G, Kyriakou M, Vrachimis S G and Polycarpou M M 2016 EPANET-MATLAB Toolkit: An Open-Source Software for Interfacing EPANET with MATLAB *Computing and Control for the Water Industry CCWI 2016*

[16] Boccelli D L, Tryby M E, Uber J G and Summers R S 2003 A reactive species model for chlorine decay and THM formation under rechlorination conditions *Water Research* https://doi.org/10.1016/S0043-1354(03)00067-8

[17] Shang F, Uber J and Rossman L 2011 EPANET Multispecies Extension - User's Manual U.S. Environmental Protection Agency, Cincinnati

[18] Saldarriaga J, Díaz O R and Bohórquez J 2016 Water Quality Model Calibration in Potable Water Distribution Systems [Internet] *World Environ. Water Resour. Congr. 2016* p. 542–52 https://doi.org/doi:10.1061/9780784479865.057

[19] Carrico B and Singer P C 2009 Impact of Booster Chlorination on Chlorine Decay and THM Production: Simulated Analysis *Journal of Environmental Engi*neering **135**(10) https://doi.org/10.1061/(asce)0733-9372(2009)135:10(928)

[20] Schal S, Lothes A, Bryson L S and Ormsbee L 2013 Water Quality Sensor Placement Guidance Using TEVA-SPOT [Internet] *World Environ. Water Resour. Congr. 2013* p. 1022–32 https://doi.org/doi:10.1061/9780784412947.099

[21] U.S EPA National Homeland Security Research Center and Sandia National Laboratories 2015 Water Security Toolkit User Manual: Version 1.3

[22] Giustolisi O and Savic D A 2006 A symbolic data-driven technique based on evolutionary polynomial regression *Journal of Hydroinformatics* **8**(3) https://doi.org/10.2166/hydro.2006.020b

[23] Giustolisi O and Savic D A 2009 Advances in data-driven analyses and modelling using EPR-MOGA *Journal of Hydroinformatics* **11**(3–4) https://doi.org/10.2166/hydro.2009.017

[24] Giustolisi O and Simeone V 2006 Optimal design of artificial neural networks by a multi-objective strategy: Groundwater level predictions *Hydrological Sciences Journal* https://doi.org/10.1623/hysj.51.3.502

IOP Conf. Series: Earth and Environmental Science 1136 (2023) 012010 doi:10.1088/1755-1315/1136/1/012010

Water management issues in the context of music festivals

Corina Maria Giurgea[1], Carmen-Anca Safta[2], Ciprian Lupu[3], Mihaela Ordean[4] and Dan Opruța[1]

[1] Faculty of Automotive, Mechatronics and Mechanical Engineering, Technical University of Cluj-Napoca, Cluj-Napoca, 400114, Romania
[2] Faculty of Energy Engineering, University POLITEHNICA of Bucharest, 060042, Romania
[3] Faculty of Automatic Control and Computer Science, University POLITEHNICA of Bucharest, 060042, Romania
[4] Faculty of Automation and Computer Science, Technical University of Cluj-Napoca, Cluj-Napoca, 400114, Romania

corina.giurgea@termo.utcluj.ro

Abstract In recent years, festivals, especially music festivals have become important vectors for promoting their host cities. Festivals are complex phenomena with high economic and socio-cultural impact. High CO_2 emissions and huge resource consumption are commonly related to the festivals' negative environmental impact. Ensuring the suitable amount of water for a large-scale event is essential for both the audience experience and the city residents, especially when their access to water is hindered by excessive water usage for the festival. Therefore, an accurate estimation of water consumption during festivals is a prerequisite for rigorous water planning. In the Covid-19 world, due to the worsening crisis of resources and to the increasing pressure exerted by audiences asking for more eco-conscious events, festival organizers are called to get together with other stakeholders and re-examine the water management issues. Water companies have already started to embrace digital solutions. Their further implementation will result in a large amount of environmental data that could contribute to a paradigm shift concerning festival sustainability. The present paper was conceived around this topic, with the goal of increasing the awareness of all the actors involved in festival operational processes. Outcomes of a desk research and a case study focused on the UNTOLD and Electric Castle music festivals are presented.

1. Introduction

Today's image of Romania on the worldwide stage could not be conceived without the essential support of some artistic and cultural events that are held in the country. During the last years, the major Romanian cities have been playing host for prominent Festivals like: George Enescu International Music Festival, Sibiu International Theater Festival, TIFF (Transilvania International Film Festival), Untold, Electric Castle, Neversea etc. Gathering people from all over the world, these Festivals become not only public or cultural, but also economic development instruments, through tourism. As almost all of them take place regularly in the summertime between mid-May until end-September, the policy makers might think about a "Festivals' season in Romania" as a true cultural and tourism country brand [1].

The city of Cluj Napoca is considered to be „the champion-city" of the festivals' based re-branding process, an exercise that the city had already taken over in an particular way in 2012 with a blow-up in 2015 [2]. In 2015 the city of Cluj Napoca was named European Youth Capital (EYC) as part of a program run by the Council of Europe. In the same year Untold – a city center-based electronic dance music festival was launched as the EYC's core event with the aim of raising the awareness about Cluj-Napoca and boosting the local tourism. But the narrative behind the syntagm "Cluj Napoca – the city of festivals" is thought to have originated in 2002 with the first edition of TIFF, a festival that grew into one of the biggest film festivals in Eastern Europe.

In June 2013, Electric Castle (EC) – a rurally organized multi-genre music festival came on the stage. The festival is held within the mystic grounds of a medieval castle – the Banffy Castle, in Bontida, a commune situated 30 km far from Cluj Napoca. Since their first edition, all these festivals have been attracted a lot of people from all over the world to Cluj. The role of the culture as engine for the city's urban regeneration had already been confirmed in the Cluj Napoca's Development Strategy 2014-2020 that stipulated the resolution of the local administration to support the emergent creative industries, to encourage the enlargement of festival industry as a conduit for internalization and wellbeing [3,4]. In 2021 Cluj Napoca was named Smart City [5] and in the development strategy of the city, the cultural paradigm switched places with the digital one [6].

After an almost two-year break imposed by the Covid-19 pandemic, the events industry struggles to resume its activity and to adapt to the new normality. This could be a true moment for reset. In light of recent resolutions of COP26 (2021 Glasgow, United Nations Climate Change Conference) the event industry's negative impact on emissions has to become a priority for change [7]. The time has come to undertake some self-evaluation, looking for ways to create memorable events by considering the demands for sustainable ones. Organizing sustainable large-scale festivals should not be done only for the credentials or image and financial benefit. The cultural and operational managers, the creditors and local authorities should understand that sustainability concerns all the stages of such an event: planning, development and deliveries, and requires their combined forces and responsibility [8].

The present paper owes its inception to (informal) talks with a series of actors involved in organizing and/or financing cultural events taking place in Cluj: from technical staff of Untold Festival, EC or TIFF to cultural managers and representatives of local authorities (e.g. the Public Events Service of the Mayor's Office). Each of these is aware of the fact that, given the worsening crisis of resources such as water and energy, the Covid-19 pandemic, and their respective consequences at all levels of life, a paradigm shift is imminent as regards the cultural activities promoted, and the way in which financial support will be made available from public or private funds.

In this sense, one can mention the telling example of the European Commission supplementing by 50% the budget of the cultural and creative sectors, by means of the „Creative Europe 2021-2027" program. The events and project that will apply for financial support will have to present clearly the innovative and novel elements which will contribute to the recovery of these sectors, reinforcing their efforts to become more digital, greener, more resilient, and more inclusive [9].

The present paper aims to sensitize those involved in organizing cultural events with respect to the importance of resource consumption, water-energy-waste, but also with respect to costs when large-scale events with tens of thousands of participants daily are organized. The present study is a desk-research and should be seen as a first step of an intended future analysis of the water infrastructure in Cluj County and the ways to manage it in a sustainable way in conditions of excessive water usage due to festivals. After presenting the motivation for the paper and the context of the topic, in section two an analysis is undertaken of the very few works available in the academic literature which are related to the specific topic of water use in large scale festivals. The third section of the paper presents the water and wastewater infrastructure of the city of Cluj Napoca, and that of the village of Bontida, the host of the Electric Castle Festival. In the same section the benefits of digital transformation for the utilities' sector (water, wastewater, energy, communications, and transportation) are also briefly presented with a focus on water digitalization in the context of the optimization of water and energy consumption. The fourth section is devoted to a case study. Based on data coming from the organizers,

IOP Conf. Series: Earth and Environmental Science 1136 (2023) 012010 doi:10.1088/1755-1315/1136/1/012010

media and Somes Water Company, some aspects regarding the water use at both Untold and Electric Castle Festivals are presented and discussed. The conclusion and the references can be found at the end of the paper.

2. Literature review

In the last two decades, festival growth in number and popularity contributed not only to the development of a sector in tourism industry – festival tourism – but also to the founding of a new field of academic research – festival studies.

Chronologically one can identify, according to the themes on which these studies focused, the way in which the area evolved. Thus, in the early 2000s, most studies discussed the economic impact of festivals, especially of the positive effects, including community cohesion, place branding, or pure financial profit. At the start of the second decade, when climate change and its effect determined an increased concern with durable development, the idea of sustainability entered not only the agenda of policy makers of all areas, but also "in a specific segment of enquiry within the tourism and events research, namely the festival sector" [10].

In his article, Zifkos draws attention to the fact that the sustainability of a music festival must be considered beyond „embedding green" or "eco-friendly" practices into the festival management", requiring a holistic approach of the concept. He qualifies as short-sighted the limitation of "sustainable festival practice" concerning irreversible resource consumption and environmental degradation to mere "technological solutions to a technological problem, embedded in a rhetoric of natural engineering". After highlighting the fact that the negative impact of mega-festivals is substantial, Zifkos indicates that academic interest for the topic is very recent [at the time of writing], quoting Getz with his 2010 article. Indeed, in this reference piece for festival research and festival tourism concerning their negative impact, Getz [11] had highlighted the paucity of academic literature on the negative environmental impact of events and festivals.

Stew [13], who not only worked for 8 years as Festival and Event Site Producer for the Glastonbury Festival (one of the biggest Music Festival in the World) but who also worked as Festival Water Analyst and undertook research in the field of Festival Sustainability, had already claimed the very little available literature focused on sustainable water use for festivals. Before referring to blogs or media pages as a source of information in this field, Stew mentions an essential contribution by Meegan Jones in his work Sustainable Event Management. The work contains a breakdown by category, labeled by colors, of the water used or „produced" during a festival. In a festival's water management, especially a festival taking place in a location without permanent infrastructure, each of these categories produces its own problems that need solutions in logistics, sources and quantities required. Disregarding these aspects in the planning phase can lead to a series of problems during the run of the event [14]

Although this labeling is encountered in practical guise in the literature for professionals in the field [15] as far as we know, it has not been implemented yet by the operational teams of the Romanian festivals we are focused on. We consider it useful to preface the case study in section 4 with the following observations regarding clear (drinking) and blue water, gleaned from practical guides and professional blogs.

The clear water of good quality is essential for the festival goers' hydration and comfort, especially during the summer festivals. Hence, a ready supply of free water should always be available. Tankers along with onsite installation should provide the appropriate number of drinking station, calculated based on the number of participants in conformity with the water regulations related to temporary events. To keep all the supplies flowing, the suitable water pressure should be assured by means of the temporary water infrastructure. For sustainability reasons, supplies of drinking water in plastic bottles should be discouraged [14]. "Drastic on plastic" seems to become the imperative logo for the festivals [16]. Thus, they are either encouraging participants to bring with them or providing themselves the re-usable branded bottles made of metal that can be refilled at drinking stations.

The blue water is the clean water not suitable for drinking, but for washing, showering and "other human contact activities". Following the authors' discussions with the operational management of both Untold and Electric Castle, it looks like the flushing toilets are the largest water consumer in the context of water use for festivals. On the other hand, the toilet facilities are a crucial part of the festival experience for the festival goers and are considered as one of the crucial success factors for festival organizers [17]. We might quote here the statement coming from production manager for the greatest Sweden Rock festival: "the toilet system is more important than the sound system".

3. Water infrastructure: current status in Cluj Napoca and facts versus water digitalization

Located in Cluj County, Romania, Cluj-Napoca city is extended in Someşul Mic River Valley on an area of 179.52 km^2. The city is surrounded by hills with heights between 500 m and 700 m, so that the land configuration is irregular with valleys and slopes (see Figure 1). Thus, in the northern part of the city there are hills, in the eastern part there is Someş plateau, in the southern part there is the upper terrace of Feleac Hill with its northern slope, and finally, in the western part of the city there are the hills of Calvaria and Cetăţuia.

Figure 1. Cluj-Napoca City – metropolitan area and peri-urban area.

The water supply and sewage systems in Cluj County are managed by Someş Water Company www.casomes.ro. The Company buys water from The National Administration Romanian Water www.rowater.ro and has the water supply and sewage infrastructures to provide the necessary water to all the types of consumers in Cluj County (see Figure 2).

The water supply system of Cluj County is very complex, with 95 water supply sub-systems (from which 21 are under construction), with 107 intake water systems from sources of water surface and groundwater, and 118 water treatment plants. The total length of the intake pipes, discharge pipes and pipes distribution network are about 3,500 km. The sewage system in Cluj County consists of 67 wastewater treatment plants and the total length of the sewage system pipes is about 2,500 km.

The whole city of Cluj-Napoca with its metropolitan and peri-urban areas is supplied with raw water from Tarniţa Lake, as main water surface source. Someşul Cald and Gilău Lakes, as well as the undergroundwater in Floreşti are backup water sources. The raw water from Tarniţa Lake is treated in Gilău Water Treatment Plant (WTP) that has a capacity of 2,650 L/s. The intake water in Gilău WTP is gravitational. After the treatment the water is transformed into drinking water and is stored into three reservoirs of 1,600 m^3 each. From these reservoirs, Cluj-Napoca city is gravitationally supplied with drinking water through two pipes of 1000 mm and 1400 mm diameters, and the length of 16 km. From the input node in water distribution network of Cluj-Napoca City, the length of pipes distribution network is 91,472 km and the diameters vary between 400 mm and 1400 mm.

Gilău Water Treatment Plant was technologically modernized, and a SCADA (Supervisory Control and Data Acquisition) system was implemented to control and monitor the technological process of drinking water. The whole controlling and monitoring system needs to be improved with new sensors of pressure and flow rate to provide real-time data regarding the input of raw water flow rate in WTP and the output drinking water flow rate from the water treatment plant.

The topography of the city is uneven, with valleys and hills, which determines a water supply network difficult to optimize. Thus, around the city area 20 drinking water tanks are placed, out of which 40% are controlled and monitored with SCADA system. Also, 17 pumping stations, 26 pumping stations with hydrophore and other more hydrophore pumping stations located in the basements of the blocks are needed to supply water to the consumers of Cluj-Napoca city.

The sewerage infrastructure of Cluj-Napoca has 662,779 km length of sewer pipes with average diameters between 200 mm and 3000 mm. From the wastewater collection network, 558,279 km are only in the metropolitan part of the city and 104.5 km in the surrounding area. Of the total length of the sewer pipes in the metropolitan area, only about 51% are less than 20 years old.

The collection of wastewaters from the city of Cluj-Napoca and their transport to the treatment plant is done using 54 pumping stations with flow rates and pumping heights ranging between 7.2 m^3/h and 720 m^3/h, respectively 4 mwc and 45 mwc.

The Cluj-Napoca wastewater treatment plant was put into operation in 1977 and was designed for an estimated capacity of 414,022 population. The station was rehabilitated and modernized in 2013. The station has a SCADA system for monitoring and controlling the technological process of water purification.

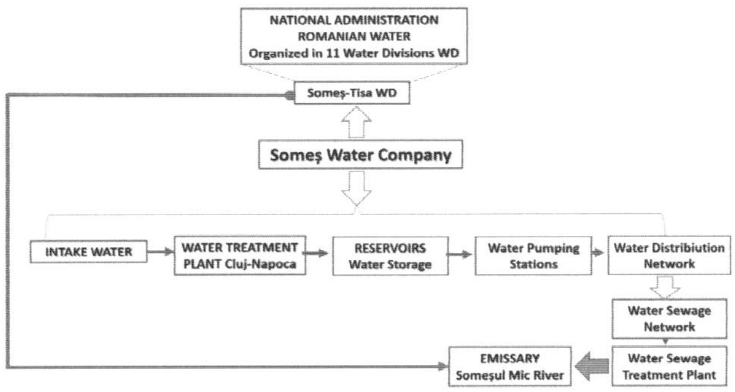

Figure 2. Flow chart of water circuit technology.

In the same context of water infrastructure, Bontida is a commune located 30 km northeast of Cluj-Napoca. Beginning with 2013, the first festival of music was organized here, which combines various musical areas with technology, street art, alternative art, and culture. In Bonţida commune (Figure 1), the distribution network has a length of 9.3 km, and the diameters of the pipes vary between 50 and 160 mm.

The description of the water supply system of Cluj-Napoca, and the wastewater collection and treatment system shows the complexity of the two systems on one hand, but also the need to optimize these systems in order to have a sustainable and synchronous development with the city development and transformation into a ”Smart City”.

In this sense, the optimization of such systems is considerably facilitated by the new technologies for monitoring the parameters of raw water transformation into drinking water and its distribution operation. The monitoring operation is also completed by the development of data storing and processing methodology by identifying and analyzing the existing data (data mining concept); the data refer not only to the technological process, but also to the consumption habits. The data from water utilities' billing systems are also identified and analyzed. The phrase "Smart" as "Smart City", "Smart Water", "Smart Agriculture" is increasingly used and is organically related to the process of water digitization. Figure 3 shows the architecture of a "Smart Water" system.

The “Smart City” infrastructure on the water component requires a very precise and detailed measurement of the parameters specific for the technological processes of capturing, treating, and distributing drinking water, respectively collecting, and treating wastewater. The real-time measurement and parameter monitoring process requires a specific hardware and software architecture SCADA, Figure 4. Treatment drinking water plants and wastewater plants are geographically positioned outside the city or in an area adjacent to the city so that the interaction with urban infrastructure (roads, surface and underground transport, electricity distribution networks, gas, heat, telecommunications) is often limited.

The same cannot be said about water distribution and wastewater collection networks, which have many “intersections” and constraints with other networks. Very often, in cities with a historical past, the development of a new infrastructure for water distribution and collection networks is severely limited or even impossible. However, this does not prevent the necessary measures to be taken to speed up the process of digital transformation of distribution and water collection networks and the installation of sensors to monitor technological parameters or smart flow meters for billing purpose water consumption. All monitored data are stored in (historical) databases [19,20].

Stored data on human behavior in relation to water consumption and usage are also important in stored historical data. Based on histories, current and future decisions can be made on the optimal operation and development of these networks [21,22]. Unfortunately, there are few localities where these aspects are developed correctly and coherently [22,24]. The problem of the presence of sensors and monitoring data transfer and storage systems is incipient in most localities in Romania. The most "difficult" problems are encountered in large cities where there are old pipes, the replacement of which is very difficult and expensive.

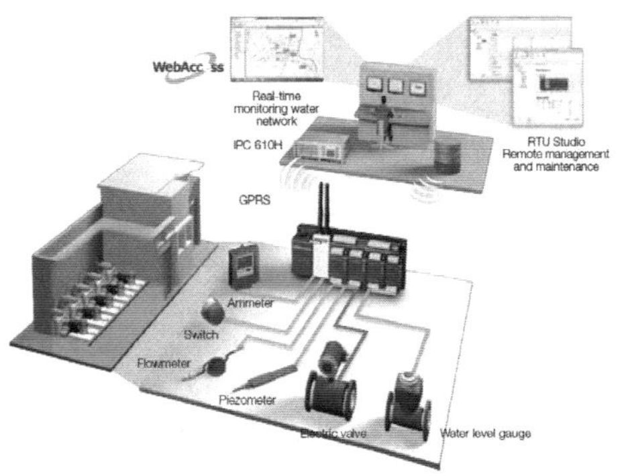

Figure 3. "Smart Water" architecture [18].

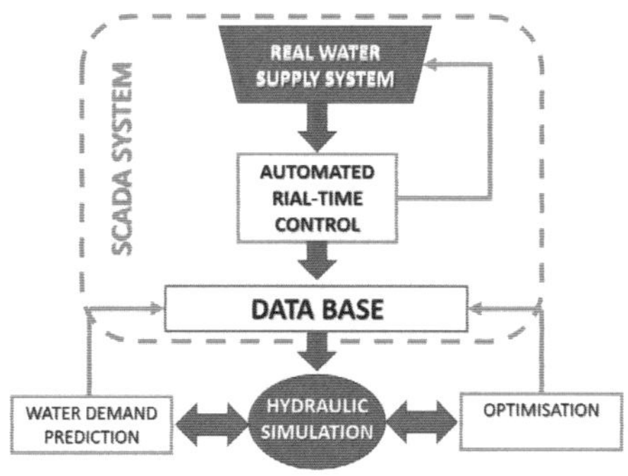

Figure 4. SCADA system and optimal operation of water distribution network.

A modern distribution network requires each component segment to include a minimum package of pressure, flow, and temperature sensors. This "package" is connected through a wired or wireless network, installed, or even included in the "structure" of pipes. Acquired, primary and/or processed data may also provide information on leak detection or water loss [25]. Digital transformation in water industry improves even the maintenance management of water infrastructure [26].

Any water distribution system must be supplied with pressurized water. The supply can be made gravitationally or using pumping and booster pump stations, as in the case of the city of Cluj-Napoca. The operation of the pumping stations complies with the flow and pressure requirements of the system with the optimization of energy consumption which implies the use of variable speed pumps and the optimal coupling of the pumps for their operation in parallel [27,28]. Automating the operation of variable speed pumps requires power inverters, and the parallel operation of the pumps requires laborious coupling/decoupling procedures [24].

The performance increase of the pumping systems, drinking water distribution and wastewater collection systems is achieved using high-performance control algorithms. These performances are obtained from the adaptability or "intelligence" of these algorithms. Thus, the control architectures based on machine learning elements, multi-model systems [29], etc. give great results.

4. Case study

Although different in the musical genres approached, the two festivals, Untold and Electric Castle, brought economic benefits, and contributed to the international promotion of the image of "City of Festivals", turning it into a tourist destination (event tourism). In 2015, after its 1st edition, Untold - the largest festival in Romania and one of Europe's biggest and best dance electronic music festival - received the award for the "Best Major Festival" at the European Festivals Award and was designated as the Brand of the Year by International Advertising Association Romania. Untold is also the first European festival to establish a partnership with the Environment, Public Health, and Food Safety Committee of the European Parliament with the objective of becoming 100% sustainable and green festival [30].

Electric Castle - a green field camp festival that takes place in a village - was designated the Best Medium Sized by European Festivals Award in 2019 after having been nominated and shortlisted for this award since its first edition in 2013.

Untold is the only festival in Romania that has managed to gather almost 100,000 participants in one day [1]. This happened at the 2019 edition, which was a record in terms of the total number of participants, as can be seen in Figure 5; the data in this figure are based on the information provided by the organizers. It should be noted that the upward trend in the number of participants was interrupted at the 2021 edition. This decrease in the number of participants is explained by the fact that this edition - rescheduled from 2020, when the festival did not take place – was a special one. On one hand, it was the only edition that took place in September, the usual period of the festival being the first weekend in August. On the other hand, the security measures before the fourth wave of the COVID 19 pandemic imposed the restriction of the number of participants.

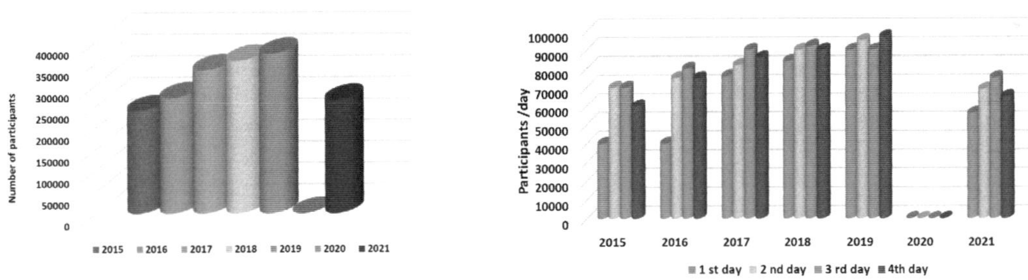

a) b)

Figure 5. Untold total number of participants: a) each edition; b) each day of each edition.

A peculiarity of Untold is that it is the only large-scale music festival that is held in the actual center of a city with a large population. The festival takes place in the central area of the city, within a closed perimeter of about 230,000 m^2. Every year, two weeks before the event's opening, more than two-thirds of the Central Park, the Cluj Arena, the BT Arena (Polyvalent Hall) as well as the areas and streets in their immediate vicinity are surrounded by fences and are closed to people who do not attend the festival. During the entire event, the gates of the Untold territory are opened daily at 2 p.m and closed at 6 a.m, the area becoming crowded after 8 p.m with a peak within 10 p.m and 1 a.m.

Not only the smaller stages, the food court, the vendor & leisure areas from the Central Park and Parking lots, but also the temporary water supply, sanitation, and wastewater infrastructures to service this part of the Untold territory are built from scratch every year.

IOP Conf. Series: Earth and Environmental Science 1136 (2023) 012010 doi:10.1088/1755-1315/1136/1/012010

The temporary water supply is taken from three existing connections of the municipal public water line, the consumption being registered by means of the temporary water meters they are equipped with. It should be mentioned here that Untold participants are prohibited from bringing with them water even if it is in sealed bottles. During the entire festival the several permanent drinking fountains from the Central Park are closed due to fear of contamination and of people queueing. The drinking water is available only in sealed bottles sold at numerous kiosks.

According to the information received from the organizers, the consumption of bottled water was 210,000 bottles of 0.5L in 2018 and 203,000 bottles in 2019, which means approximately an average consumption of 100 m^3, almost insignificant compared with the total water consumption from the respective editions, as can be seen below.

Thus, the temporary overground water supply arrangement provides only the water for the temporary sanitation infrastructure (290 flushable portable toilets and 10 portable hand washing stations each having 10 taps) as well as the water used for cooking/washing in the foot court. To avoid the risk of a water shortage (it had already happened during the 2019 edition) due to high water demand, especially for the toilets, during the peak hours a tank of 30 m^3 filled with water twice daily was used as a buffer during the 2021 edition. However, the food court area was depleted of water at the peak hours. In search for solution, the organizers and the Somes Water Company managed to identify another possibility to connect to the main water line by using a connection pipe larger in size than the current one having a nominal diameter of 32 mm. As far as we know, the Somes Water Company is going to make the necessary changes to their system to ensure an appropriate water supply for the 2022 edition.

The wastewater from the temporary sanitation facilities is collected through the temporary over ground sewer service connection pipelines and pumped into the municipal sewage system by means of nine Sanimaster G pumps working at a rate of 20 m^3/h.

Apart from the temporary clean, respectively wastewater infrastructures that serve the Central Park area of the festival, Untold time-shares the whole permanent water infrastructure and the sanitary facilities of the Cluj Arena - the host of the Main Stage (249 flushing toilets, 235 urinals, 61 showers and 254 hand wash sinks) and a large part of those of the BT Arena (52 flushing toilets, 28 urinals and 56 hand wash sinks).

The history of the clean water consumption based on the clean water meter readings took by the public water company at the beginning and at the end of the Untold territory arrangement respectively refurbishment is depicted in Figure 6.a. As can be seen, every year the water usage was well balanced between the three infrastructures (Cluj-Arena, BT Arena, Park Central), whatever the yearly total registered consumption was.

However, the results of the study focused on the perception of the participants in the 2019 edition - the top edition of the festival [31] shows, rather, a modest performance in terms of availability of sanitary facilities, toilets (note 6.03/10) and cleanliness on the festival site (note 7.27 out of 10). Given the importance of toilets in the public's perception of the festival experience [17] we can advance the idea that the deficiencies related to the temporary infrastructure in the park area have contributed to the dissatisfaction of a part of the public.

In Figure 6.b it is shown the historical yearly total registered water consumption during Untold festival. The data used to plot this graph have been taken from the public water company's invoices.

Figures 5 and 6 show that in the first five editions (2015-2019) an increase in the number of spectators corresponds to an increase in water consumption (even if the proportional increases are not the same). Although the number of spectators at the 2021 edition was 10.4% higher than in the 2015 edition, the billed water consumption decreased slightly, by 5%. An explanation for this "reversal" can be found in the fact that, as we have shown, the 2021 edition was the only one that took place in mid-September when the temperatures in Cluj, especially in the evening and at night are much lower than in early August. This observation can be considered as illustrative for the influence of the weather conditions on the water consumption in a festival in which the duration, location, water infrastructure remained the same from one edition to another. On the other hand, the observation according to which there was a decrease of water consumption despite the increase of the number of participants in 2021 compared to 2015 can be considered a counterargument to the organizers' practice of "scaling" the increase of water consumption and number of participants with the same multiplication factor.

A comparison between the water consumption during the Untold festival and that in other large-scale festivals might not be done with the available data. Both scholars [32] and professionals [13,15] relate the water use in festivals to "per capita per day figures" and compare them with the national per capita consumption average [32].

Nevertheless, Stew [13] found it unrealistic to compare festivals from the point of view of water use by means of "per capita per day figures" because these figures are strongly influenced by a series of factors like the audience demographic, festival size, weather conditions and infrastructure.

The assessment of the "per capita per day" water consumption was not possible in our case study, anyway. For this, it would have been necessary to know, for each edition, not only the total number of participants in the festival, respectively the number of participants in each day, but it would have been necessary to "break them down" according to the frequency of participation (e.g. read spectator unique, read for 2 days, etc).

However, the authors consider it inappropriate for some festival organizers to predict water consumption at the festival edition in preparation for the consumption/per capita value calculated as the ratio between the total consumption recorded at the last festival edition and the total number of participants. Moreover, the authors suggest that the daily consumption read on meters must be considered, even if there are no meters with information storage. This practice would be more appropriate because it is a well-known fact that in the festival industry a music show held in weekends and having heavy names in the line-up can cause sudden and unforeseen increases in water consumption with an impact on the participants and community, too.

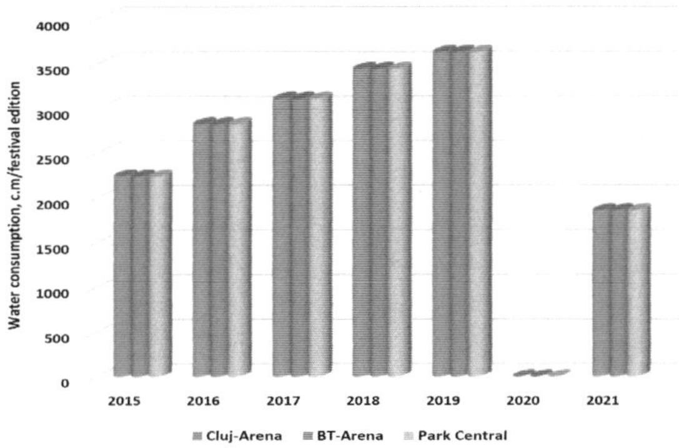

a)

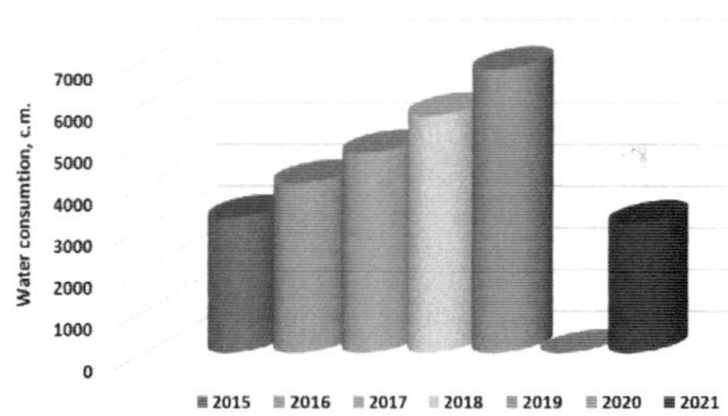

b)

Figure 6. Untold festival water consumption: a) in different locations; b) total water consumption.

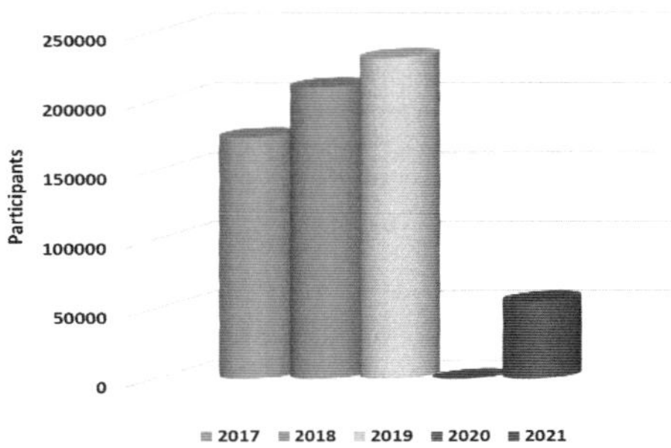

Figure 7. Bonțida Electric Castle festival, number of participants.

Illustrative in this sense is the case of the Electric Castle (EC) festival, which in the edition of 2018 crossed an episode of "shortage of water". Starting from the premise of maintaining the same duration and the same infrastructure, the estimated water consumption for the 2018 edition has been approximated as being almost the same with that of the 2017 edition. But the spectacular increase in the number of the first day participants – as can be seen in Figure 7, there were 50% more participants on the first day in 2018 than on the first day of 2017, as well as the increase in the number of accommodation places offered by the villagers in Bontida and the outflow of participants in the weekend resulted in a sharp increase in water consumption. Thus, the festival left without water on Saturday - the day of maximum audience. At the request of Bontida local authorities and festival organizers, and to prevent a new shortage at the 2019 edition, Somes Water Company modified the water infrastructure by introducing a pipe (diameter 110 mm and length 1 km) which would supply water from the main line of the village and would serve EC.

This change in infrastructure combined with the use of three 50 m^3 buffer bags apparently solved the festival's water supply problem even if the traffic within the festival increased in weekend (approx. 50,000 participants). However, the number of festivalgoers accommodated in the village also increased, equalling that of the campers (approximately 10,000 people), which meant that, this time, part of the village remained without water. The organizers forecast for the next three years a 20% increase in traffic inside the festival, forecasting a 15-20% higher consumption based on a scale with the same percentage of health infrastructure and outbuildings. Based on these approximations, to eliminate the risk of water shortage, the solution proposed and approved by Somes Water Company to be put into practice before the 2022 edition is that of a new pipe (with nominal diameter of 160 mm) to be supplied during the festival from the supply pipe Cluj-Gherla upstream of Bontida village, and to make the junction with the village pipe having the same internal diameter, closing a ring to serve the village during the festival.The lack of data on the history of water consumption at EC compared to that of UNTOLD explains the focus of the case study on Untold.

5. Conclusions

The literature consulted and the case studies analysed evidence various aspects, positive and negative, useful for approaching a coherent and sustainable management plan regarding the water requirements of a music festival. Thus, identification of consumption peaks according to the number of participants, location (in or outside the city), the season the event takes place, the technical condition of the existing water infrastructure, the quality of the water distributed to participants are just a few elements requiring monitoring both during the festival and from one edition to the next.

Using current means of digital monitoring the water and energy consumption has the advantage of providing datasets analysable according to mathematical models and through AI techniques that can estimate possible resource consumption. Practically, this analysis must the benchmark to start from in any plan of resource management for a festival. In this respect, the organisers should cooperate closely with the contractors to elaborate a detailed plan for the way water and wastewater should be managed. The plan should include a description of the water system, a map of the water (including wastewater) infrastructure as well as consistent and clear description of the procedure to be used in measuring and monitoring processes. More particular, the procedures should address the way of collecting data (the number and the type of water meters respectively clamp on flow meters for the wastewater to be fit especially in the critical zones, the frequency of checking the reading with the aim of finding the peaks of use), the monitoring tools (types and ways of tracking and analysing the collected data), solution for the water conservation.

In the light of the above, and taking into consideration the outlook for festival growth in large cities in Romania [1], cities aspiring to the smart city status as well, the authors propose to festival organizers, to local authorities and policy makers, to providers of services who have already embraced digital solutions [33] to reflect together on how going digitally could contribute to the paradigm shift concerning the sustainability for the anticipated new wave or new age in music festivals' industry as well as in urbanism: smart events for smart cities. „The time is right"!

Acknowledgments
The authors thank Calin Chiver from Untold festival and Andrei Cojocariu from Electric Castle festival as well as the technical staff from the Somes Water Company for their support in gathering the information required for the study. The study did not receive any funding.

References
[1] Tanasescu A 2022 *Tinerii din întreaga lume ar avea un motiv în plus să descopere România prin intermediul tuturor festivalurilor – Edy Chereji, director de comunicare Untold & Neversea* (in Romanian) https://culturaladuba.ro/tinerii-din-intreaga-lume-ar-avea-un-motiv-in-plus-sa-descopere-romania-prin-intermediul-tuturor-festivalurilor-edy-chereji-director-de-comunicare-untold-neversea/
[2] Ţichindeleanu O 2019 O re-educatie estetica: Orasul creativ si ruinele capitaliste *Hipsteri, bobos si clase creative* (in Romanian) ed State C and Gutu D (Chisinau: Cartier) pp 189–205
[3] Dunford J 2020 10 of the best things to do in Cluj-Napoca, Romania: a local's guide *The Guardian* 27 January https://www.theguardian.com/travel/2020/jan/27/cluj-napoca-romania-city-break-locals-guide
[4] Trandafoiu R 2021 Cluj Napoca, Romania: Electronic Dance Music and local policy (2015-2020) *Electronic Cities. Music, Policies and Space in the 21st Century* ed S Darchchen, D Charrieras, J Willsteed and P Macmillan Chapter 13 pp 207–222 https://www.researchgate.net/publication/350986576_ClujNapoca_Romania_Electronic_Dance_Music_and_Local_Policy_2015-2020
[5] Radu C 2021 *Cluj-Napoca, the most "business friendly" city in Romania* https://economedia.ro/cluj-napoca-the-most-business-friendly-city-in-romania Retrieved 27 March 2022
[6] Ghilas O 2020 Strategia Clujului pentru 30 de ani: paradigma digitală schimbă hiba culturală *Ziar de Cluj* 22 January, http://www.ziardecluj.ro/strategia-clujului-pentru-30-de-ani-paradigma-digitala-schimba-hiba-culturala
[7] https://www.consilium.europa.eu/ro/meetings/international-summit/2021/11/01/
[8] Shakila R 2020 *Can music festivals dance to a sustainable tune?* https://www.soundoflife.com/blogs/experiences/music-festivals-sustainability-going-green
[9] European Commission, Directorate-General for Education, Youth, Sport and Culture 2021 *Creative Europe 2021-2027: push boundaries* https://data.europa.eu/doi/10.2766/794740
[10] Zifkos G 2015 Sustainability Everywhere: Problematising the "Sustainable Festival" Phenomenon *Journal of Tourism Planning and Development* **12**(1) pp 6-19 https://doi.org/10.1080/21568316.2014.960600
[11] Getz D 2009 Policy for Sustainable and Responsible Festivals and Events: Institutionalization of a New Paradigm *Journal of Policy Research in Tourism, Leisure & Events,* **1**, pp 61–78 https://doi.org/10.1080/19407960802703524
[12] Collins A and Cooper C 2017 Measuring and managing the environmental impact of festivals: the contribution of the Ecological Footprint *Journal of Sustainable Tourism* **25**(1), pp 148–162 doi: 10.1080/09669582.2016.1189922
[13] Stew D. (not dated) *Sustainable Water Management for Music Festivals. The Basics* https://www.agreenerfestival.com/wp-content/uploads/pdfs/StewDennyWater.pdf
[14] Robertson A 2021 *Essential water supplies for music festivals* https://www.festivalpro.com/festival-management/1429/news/2021/6/18/Essential-Water-Supplies-for-Music-Festivals.html
[15] *Julie's Bicycle Practical Guide: Water Management at Outdoor Events* https://juliesbicycle.com/resource/water-management-at-outdoor-events-guide/
[16] Association of Independent Festivals 2018 *Initiatives Drastic on plastic* http://aiforg.com/initiatives/drastic-on-plastic/#:~:text=A%20digital%20campaign%20launched%20to,awareness%20of%20single

%20use%20plastic.

[17] Jets™ Magazine 2019 *A cleaner, greener festival experience* https://jetsgroup.com/jets-magazine/article/en/a-cleaner-greener-festival-experience

[18] Rqptech.com. 2022 *Smart Water Grid and Pump Station Monitoring* RQP Technology AB. [online] Available at: http://www.rqptech.com/smart-water-gridand-pump-station-monitoring/ [Accessed Aprl. 2022]

[19] Da Costa Coelho B 2016 *Energy efficiency of water supply systems using optimisation techniques and micro-hydroturbines* (PhD Thesis, Universidade de Aveiro, Portugal) http://hdl.handle.net/10773/21255

[20] Guyer J 2013 *An Introduction to Pumping Stations for Water Supply Systems* (Createspace Independent Pub: Scotts Valley CA, USA)

[21] Abunada M, Trifunović N, Kennedy M and Babel M 2014 Optimization and reliability assessment of water distribution networks incorporating demand balancing tanks *Procedia Engineering* **70** pp 4–13 https://doi.org/10.1016/j.proeng.2014.02.002

[22] Predescu A, Truică C-O, Apostol E-S, Mocanu M and Lupu C 2020 An advanced learning-based multiple model control supervisor for pumping stations in a smart water distribution system *Mathematics* **8**(6) 887 https://doi.org/10.3390/math8060887

[23] Vrachimis SG, Eliades DG and Polycarpou MM 2014 Enhanced adaptive control of water quality in water distribution networks by incorporating abrupt hydraulic changes *Procedia Engineering* **89** pp 239–246, doi:10.1016/j.proeng.2014.11.183

[24] Castro-Gama M, Pan Q, Lanfranchi E A, Jonoski A and Solomatine D P 2017 Pump scheduling for a large water distribution network *Procedia Engineering* **186** pp 436–443, doi:10.1016/j.proeng.2017.03.249

[25] Lupu C, Predescu A, Lupu M and Gliga L 2017 Technical procedure for model identification and for leak/fault detection in fluid transport infrastructures *Proc. 21st Int. Conf. on System Theory, Control and Computing (ICSTCC)*, pp 535–539, doi: 10.1109/ICSTCC.2017.8107090.

[26] Di Nardo A, Boccelli L D, Herrera M, Creaco E, Cominola A, Sitzenfrei R and Taormina R 2021 Smart Urban water networks: Solutions, trends and challenges, *Water*, **13**(4) 501 https://doi.org/10.3390/w13040501

[27] Nowak D, Krieg H, Bortz M, Geil C, Knapp A, Roclawski H and Böhle M 2018 Decision support for the design and operation of variable speed pumps in water supply systems *Water* **10**(6) 734 doi:10.3390/w10060734

[28] Menke R, Abraham E, Parpas P and Stoianov I 2017 Extending the envelope of demand response provision though variable speed pumps *Procedia Engineering* **186**, 584–591, doi:10.1016/j.proeng.2017.03.274.

[29] Lupu C, Popescu D, Ciubotaru B, Catalin-Dumitru P and Florea G 2006 Switching solution for multiple-models control systems *Proc. 14th Mediterranean Conf. on Control and Automation,* (Ancona, Italy, 16–19 June) pp 1–6, doi:10.1109/MED.2006.328827

[30] Moisescu O I, Gica O A, Coros M M and Yallop A C 2019 The UNTOLD Story Event tourism's negative impact on residents'community life and well-being *Worldwide Hospitality and Tourism Themes* **11**(5) 492–505, https://doi.org/10.1108/WHATT-06-2019-0036

[31] Moisescu O I 2021 *Analiza cheltuielilor participantilor si a experientei festivaliere la editile din 2019 ale Untold si Electric Castle* (in Romanian) Technical report DOI: 10.13140/RG.2.2.18198.06724 https://www.researchgate.net/publication/354684141

[32] Browne A, Jack T and Hitchings R 2019 Already existing' sustainability experiments: Lessons on water demand, cleanliness practices and climate adaptation from the UK camping music festival *Geoforum* **103** pp 16–25 https://doi.org/10.1016/j-geoforum.2019.01.021

[33] Leverton J 2020 *Essential digital investments for the water industry. The future of infrastructure* https://infrastructure.aecom.com/2020/essential-digital-investments-for-the-water-industry

Added value of deterministic models in Decision Support Systems

Philippe Gourbesville[1]

[1] Polytech Lab, Université Côte d'Azur, Polytech Nice Sophia, Sophia Antipolis, 06903, France

gourbesv@unice.fr

Abstract. Management of water uses requests to harmonize demands and needs which are getting more and more complex and sophisticated especially with the growing urbanization. Modern cities request a larger number of services for their inhabitants and expect, at the same time, to limit investments to constrain the tax pressure. The need of optimization appears at various levels and request the wide spread of monitoring strategies. At the same time, urban growth mobilizes last available spaces that are frequently under the thread of natural hazards like inundations. Decisions Supports Systems that may integrated various components both for real-time monitoring and forecast through model, appear as one of the most relevant answers to the urban environment management's expectations. The model's integration is a challenging task that requests to build a global vision that ensures both technical feasibility and sustainability. As demonstrated with the AquaVar approach implemented in several catchments on the French Riviera, several deterministic models can be orchestrated within a single environment that can address the diversity of the water related issues handled by local technical services. The model's selection must integrate the evolution of the tools and the possibility to integrate gradually new approaches and methods that are more data oriented and using the results produced from the implemented deterministic tools. The recent extreme events recorded in October 2020 during the ALEX storm and the disaster within the Vésubie valley have underlined the interest of the deterministic tools. In fact, those models are the only solutions that can produce a meaningful estimation of the extreme hydrological processes and deliver the expected results to decision makers.

1. Context and challenges

Within the current context of competition among water uses, the management requests to harmonize demands and needs. The growing urbanization exacerbates the situation and requests a more and more complex and sophisticated approach. The environment of modern cities is characterized by numerous services that are provided to the demanding inhabitants against a reasonable tax pressure. Municipalities need to optimize through monitoring their strategies on technical investment and maintenance. Simultaneously, urban developments are taking place in remaining free spaces that are frequently associated to a higher vulnerability regarding natural hazards. Development and massive production of sensors available at limited cost have multiplied the possibilities to measure and quantify the various hydrological processes and, thanks to the availability of the communication networks, to interface those field data with managing environments frequently deployed within metropolis, municipalities, and catchment authorities. However, for ensuring an optimized efficiency and answering to the demands of the field operators, the simple observations must be enriched with forecasts mostly delivered in real-time. Over the last decade, many attempts and systems have been elaborated and implemented in various contexts. In most of the cases, the developed environments had to conciliate several constraints such as accurate and reliable field observations based on trusted sensors, broadband for data transmission, computational power for running models in real-time and useful results for decision makers. Solving this equation where uncertainties are numerous remains a challenging task and the question on which models to implement for real-time operation is still open.

The management of urban environments and related services is currently conducted through numerous monitoring systems that are expending rapidly. To ensure efficiency, the collected data, associated to modelling tools should be integrated within a planned Information System (IS). The numerous initiatives taken by various institutions (from cities to catchment management entities) have generated a multiplicity of local ecosystems that are currently evolving to a more mature environment where unified protocols (such as IoT) and standards are flourishing. The urbanization of the Information Systems within the municipal environments has been initiated since more than a decade and the process is still at an early stage with the request to move from a static based approach to a flow-based dynamic one [1]. Obviously, the water sector doesn't lead the evolution and the energy sector has been more proactive within the integration process. Complexity of water related issues requests to design and to implement systems that can address diversity of processes, numerous interactions, and timely forecasts. For real-time monitoring actions and models-based forecasts, the Decision Supports Systems (DSSs) are identified as one of the most efficient approaches for answering to the expectations of urban environment management. To achieve the integration of the models, a global vision including architecture and development road map should be established at the early stage.

Due to the complexity of challenges - water supply security issues from groundwater resources, inundation risk and water resources management under the perspective of climate change – the difficulty to handle them, the diversity of actors and the vulnerability of the area regarding water related hazards, the need for a Decision Support System (DSS) has gradually emerged since the late 90s [2]. The objective for such system is first to provide an overview of the current situation regarding the various water processes and second, to produce in real-time a forecast of those processes to ensure an efficient support to technical teams in charge of management and public safety. In most of the cases, even if the need is clearly identified due to a complex and sensitive situation, both availability of data and technical tools (from communication protocols to modelling tools) have not permitted to engage the development of a real DSS that is integrated within the IS of the local managing entities.

2. The AquaVar approach

2.1. Background & expectations

Nice Côte d'Azur Metropolis, along the Mediterranean coast in the south of France, concentrates 540,000 inhabitants. Within the territory of the Metropolis, the city of Nice federates 350,000 inhabitants and is the 5th largest city in France. The territory of the Metropolis concentrates a dense population and is associated to a high vulnerability due to multiple inter-connected hazards related to tectonic, geomorphic, and climatic context that is characterized by frequent extreme rainfall events. The city of Nice is developing and expending mainly on the left bank of the Var River, in a densely urbanized floodplain of 20 km long.

To manage this complex environment that concentrates multiple water competing activities, the decision to elaborate a specific Decision Support System (DSS) was taken in 2014 with AquaVar project. The aim for the platform is to display hydrological conditions within the Var catchment and to generate, on an hourly base, forecasts for the next 72 hours to support the technical teams and the first responders in case of emergency. Data and modelling tools availability have been major constraints for the development of the DSS despite the early identification of its potential interest.

Systematic data collection on climate, hydrology and topography over the last 15 years has allowed to build a meaningful knowledge about the various water related processes over the Var catchment and the Nice Côte d'Azur Metropolis territory. The AquaVar development was engaged in 2014 with the objective to address issues from flood emergency response to global water resources management at the catchment scale [3]. The Decision Support System is also used as reference tool for the consensus building process among the various stakeholders committed to the water management.

The Var catchment is regularly affected by rapid and severe floods that reach a discharge of 3,500 m^3/s in less than 12 hours. Such type of event was observed in 1994 and in 2020. This Mediterranean characteristic is explained by the convective and stationary rainfall events that can affect the partly or totally the catchment. The flood management strategy should integrate the fast response of the catchment and develop the proper adaption capacity. The recent event observed in October 2020 during the Alex storm has demonstrated the limited efficiency of the classical monitoring devices that were all destroyed during the event. The improvement of the preparedness request that the technical teams and the first responders can obtain a clear forecast of the current flood process on an hourly base over the full Var catchment. The request for the DSS is then to deliver an hourly set of forecasts covering the next 72 hours based on the rainfall forecasts produced by Météo France, the French meteorological services.

2.2. Selected modelling architecture

The AquaVar concept (Fig.1) [3, 4, 5] is defined as a services environment based on platform a service bus dedicated that is collecting and gathering in real time field data and forecasts produced by various models to address hydrologic processes and to support water uses and to mitigate potential water related hazards. The field and produced data are globally addressed with standard procedures that are defined as Key Performance Indicators (KPIs). Those KPIs cover transfers of data within the system but also defined thresholds used for alert protocols and mitigation directives [6]. AquaVar system is based on three distributed physically based modelling systems that provide an accurate simulation of the hydrologic conditions at the catchment scale; The system delivers forecasts that are used by technical teams to optimize the management strategies defined and monitored by the Operation Center. To ensure a good communication towards all the decision makers, AquaVar is offering a Web interface that is allowing to visualize the current situation and the various simulation results that are presenting the hydrologic dynamic of the catchment.

The AquaVar platform includes three physically based models that can simulate the hydrologic conditions of the catchment: 2D surface runoff, 2Dsurface flows within the river network and 3D groundwater situation with the coupling of the 2D and 3D models. The system includes pollutants exchanges. The three deterministic models are included within a hypervisor environment. The decision to use deterministic distributed models was made due to their capacity to simulate extreme hydrologic conditions that have never been recorded such as intense rainfall events. The deterministic tools provide a very good performance for such cases because the different variables remain the same and are operating within a range that is linked to a physical reality. In the opposite way, the non-distributed, data driven, and conceptual tools are not able to perform at a similar level for the extreme situations because they don't find the relevant references within the knowledge training data that were collected from the field observations. The different models have been selected and linked to produce the needed simulations and forecasts:

- MIKESHE model [7] is used to generate hydrologic data that are applied as boundary conditions for the groundwater model (FEFLOW) and the hydraulic model (Mike21).
- Mike21 hydraulic model [7] addresses the surface flow conditions in 2D within the low Var valley and is linked with the groundwater 3D model built with FEFLOW to simulate the interactions between surface and aquifer. The morphological changes of the riverbed occurring during major events are also integrated within the simulations.
- FEFLOW 3D hydrogeological model [7] is used to represent the aquifer conditions. As mentioned, this module is connected to the hydraulic model to represent properly the interactions between surface waters and aquifer that are taking place within the riverbed.

The three models have been validated over the period of four years: from 10th of September 2009 to 26th of February 2013. Six piezometers among the 24 piezometers used for daily monitoring of the

IOP Conf. Series: Earth and Environmental Science 1136 (2023) 012011 doi:10.1088/1755-1315/1136/1/012011

aquifer along the low valley have been used for the validation of the hydrogeologic model. The locations of the selected piezometers provide a global overview of the low valley condition.

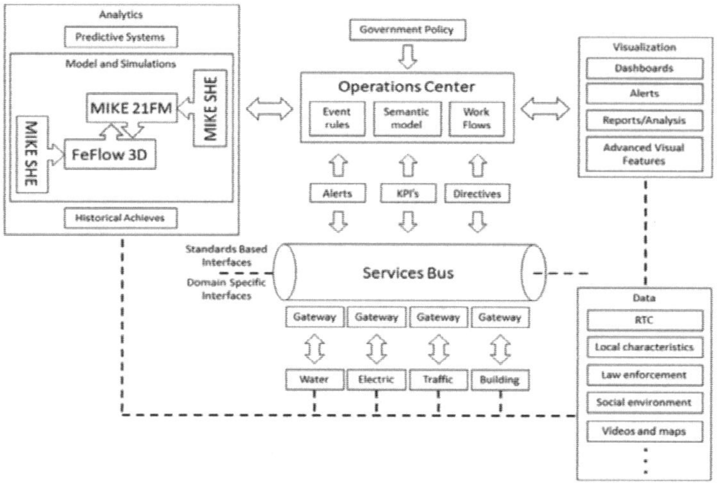

Figure 1. AquaVar concept and architecture.

The obtained outcomes from the model show that the model can simulate efficiently the complex processes like river-aquifer exchange, aquifer recharge as well as the groundwater extraction [7]. Based on those results, the model can be mobilized within the aquifer management tasks and implemented within the AquaVar platform. For the free surface flows and the morphological evolution, the same approach has been implemented for le model build with Mike 21 FM. The validation has been performed with the 1994 and 2011 flood events that are major recorded situations. The morphological evolution was achieved with the Sand Transport module that estimates the sediment transport capacity, the bed level changes' initial rates and the morphological changes for non-cohesive sediment due to currents. Within the model, the hydrodynamics conditions and the sediment properties determine the sediment transport. In addition, several meshes were generated to simulate the flood event recorded from the 3rd of October to the 6th of October 2015 to optimize the hydraulic model. The final model with a 10 meters resolution combines quadrangular and triangular elements. The efficiency of this approach has been already demonstrated and allows to match with recorded observations. The various hydraulic structures have been integrated with the model with high resolution mesh to properly simulate effects. A similar procedure has been carried out for the hydrologic modelling at the catchment scale with Mike SHE. The procedure of validation was developed over 3 hydrologic years after optimization of the computational grid resolution. The model is performing very well and provides input data that are used by the hydraulic and the hydrogeologic models. The three models are orchestrated and embedded with the AquaVar platform that is deployed with the Information System of Nice Côte d'Azur Metropolis.

2.3. Computational efficiency
The main objective of AquaVar is to deliver hourly forecasts covering the next 72 hours. This target requests that the three models can run in less than one hour. The objective was jointly defined by field operators and users of the DSS. The computational architecture is particularly important and should be based on the relevant resources that can ensure the performance. Several architectures based on CPU and GPU resources have been tested with the three models to assess the various performances.

The three deterministic models have been tested over a period of seven days representing the hourly update planned for the real-time operation mode. The selected period is from 1st to 8th of November 2011 and corresponds to a major flood event associated with inundation. The period was selected as many cells were active due to the inundation process and the interactions between surface flows and aquifers. The MIKE Zero 2017 version from DHI and running under Microsoft Windows Server 2012 was used for all the tests.

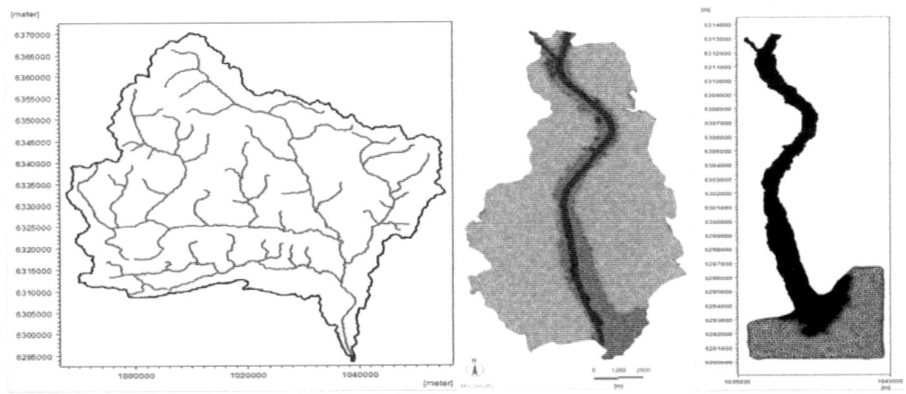

Figure 2. AquaVar DSS models: hydrologic (left), hydrogeological (centre) and hydraulic (right).

Hydrologic model – The hydrologic model is built with Mike She modelling systems. The model uses a standard computation architecture based on CPU resources. The simulation period is extended to seven days integrating four days of past observations and three days for forecasting period.

Table 1. Hydrologic simulation results based on CPU architecture.

Model	Cells	Start / End	Cores	Virtual CPUs	RAM (Go)	Threads	Computation (minutes)	Usage %
Model A	560,000	01/11/2011 - 08:00 to 08/11/2011 - 08:00	19	76	304	76	57	100

The deployed solution allows to deliver the needed simulation in less than 60 minutes. The obtained results can be converted and displayed within the web mapping services of the AquaVar system. The technical solution remains reasonable and can be easily deployed within the AquaVar environment.

Hydraulic model – the MIKE 21 FM code was used to develop the hydraulic model. The modeling system can mobilize GPU resources and parallel computing approach. In a similar way as for the hydrologic model, the simulation period is extended to seven days integrating four days of past observations and three days for forecasting period. Three different models were developed with different resolution and type of elements to assess the performance of GPU architecture:

- model A is the reference model and combines quadrangular elements for the riverbed and triangular for the floodplain.
- model B is obtained from model A with a reduction of elements that may generate instabilities.
- model C is designed with constant surface triangular elements (100 m^2). The aim of this approach is to investigate the impact of higher resolution on computation time and accuracy of results.

The GPU architecture provided a meaningful solution as a 7 days' period simulation was performed in less than 60 minutes. The performance of model A and B is very similar, and the optimized computation mesh don't affect the computation time.

The model C with a higher resolution - C is 2.6 times larger than A and B – demands a computation effort that is not compliant with real-time forecast production. The GPU architecture with 1 and 4 units still needs from 12 to 8 hours to achieve a simulation. The outcomes results underline the fact that GPU performance is not linear. The use of 4 GPU units reduces the computation time for A and B but not in a significative way that can support the use of the more powerful architecture. The structure of the code structures and the mesh characteristics of the mesh used for the model can explain this situation. At the same time, the CPU resources are not fully mobilized. The GPU architecture used for running the model A and B can deliver the needed computation in almost 60 minutes. The obtained results can be converted and displayed within the web mapping services of the AquaVar system. The technical requirements are reasonable as for the previous server and they can be deployed within the AquaVar system.

Table 2. A, B & C models, and simulation conditions.

Model	Types	Nodes	Elements	Start / End	Timestep	Number of timesteps
Model A	Quadrangular + Triangular	197,549	325,437	01/11/2011 - 08:00 to 08/11/2011 - 08:00	2 s	300,600
Model B	Quadrangular + Triangular	197,531	325,401			
Model C	Triangular	437,738	869,554			

Table 3. Simulation performances for A, B & C models with various GPU architectures.

Model	Cores	Virtual CPU	GPU device	Nb GPU	Subdomains	RAM (Go)	Threads	Simulation (minutes)	Usage %
Model A	16	32	V100-NVIDIA	1	1	256	32	68	47
Model A				4	4			50	32
Model B				1	1			77	50
Model B				4	4			51	35
Model C				1	1			765	71
Model C				4	4			478	41

Hydrogeologic model – The hydrogeologic model developed with the FEFLOW code is using a CPU architecture. However, as the number of timesteps is limited, the principle was to use the same CPU server already used for the Mike21 model. In a similar way as for the precedent tools, the simulation period is extended seven days integrating four days of past observations and three days for forecasting period. Outcomes show that the approach is delivering a very good performance for the hydrogeologic model that can run a seven day's simulation in less than 30 minutes with the server equipped with a single GPU resource. The quality of the source code of FEFLOW explains this outcome as computing efficiency has been optimized recently. The needed simulation can be produced in less than 30 minutes with the used architecture using only a single GPU resource. The obtained results are quickly converted and displayed within the web mapping services. AS for the previous model, the current technical architecture remains reasonable and can be integrated within a DSS or a monitoring platform. As the hydrogeologic model don't request a massive computational power, it can be easily hosted on the same GPU unit.

An operational architecture can be defined and implemented for real-time operations based on the tests' outcomes and the constraint to deliver an hourly update for all models. To achieve this aim, a cloud-based approach should be privileged for hosting the three deterministic models. The global architecture can be designed around two servers, one with a CPU architecture and one with a GPU architecture. The hydrologic model is hosted on the first one and the two other models are running on the GPU unit. The two computing servers are linked to additional servers dedicated to the mapping services (GIS) that are used to display the simulation results within the AquaVar environment.

3. The challenge of extreme situations

The extreme hydrological situations like major floods are a major challenge for the communities and request tremendous efforts for the development and the implementation of mitigation measures. Over the last five years, numerous extreme flood events have been recorded in various regions of the world. The magnitude of recorded damages has demonstrated the vulnerability of our modern urbanized environments under such conditions. For most of the observed situations, the extreme floods and their combined inundations are initiated by specific meteorological conditions combining wet and warm air masses in convective cells. The possibility to face such for the development of convective and intense events is increasing within the climate change conditions and the gradual increase the sea temperature especially in close seas like the Mediterranean.

Table 4. Model A and simulation conditions.

Model	Types	Elements	Nodes	Number of timesteps	Timestep	Start date / End date
Model A	Triangular	392,161	198,954	1,670	360 s	01/11/2011 - 08:00 08/11/2011 - 08:00

Table 5. Simulation performances with a GPU architectures Results for the hydrogeologic model simulations for 7 days.

Model	Cores	Virtual CPUs	GPU device	Nb GPU	Subdomains	RAM (Go)	Threads	Simulation (minutes)	CPU usage %
Model A	16	32	V100-NVIDIA	1	1	256	32	27	22

Several extreme events have been recorded over the last two decades along the coastline of the Mediterranean Sea. France (2015, 2019, 2020), Italy (2011, 2014, 2018), Spain (2000, 2005, 2021) and Algeria (2001, 2020, 2021) are some of the locations that have been deeply affected. During those extreme events, the losses have frequently exceeded € 1 billion for the direct damages. The managing authorities facing this type of situation should have the possibility to assess the potential impacts and to define mitigation strategies that could improve resilience. The experience has demonstrated that the traditional structural measures such as levees, dikes, embankments, hydraulic structures, … are frequently insufficient. Those measures cannot cope with the magnitude of the hydrologic phenomena. In a surprising way, the field data collection regarding discharges and flow conditions remains very uncertain and the measurement quality is questionable despite the engaged technical efforts.

The multiplication of recent cases around the Mediterranean Sea has underlined the vulnerability of modern societies and the difficulty for the urban environments to cope with sudden and massive runoffs. At the same time, the need for a better understanding of the events for an efficient anticipation has been formulated by many communities with the ambition to reduce damages and impacts on people and goods. This request leads to formulate resilience strategies based on modelling tools that can offer meaningful simulation of extreme rainfall events and their associated runoffs.

The challenge for modelling systems is then to have the capacity to represent properly processes that have never been observed and obviously not used during the instantiation phase of the models. The AquaVar DSS must address the extreme flood risk and the extreme event of October 2020 was a real size test for the system. The analysis of extreme Mediterranean rainfall events over the past decades highlights an intensification of heavy rainfall (+22% on the annual maxima of daily accumulations during the period of 1961 to 2015) and a frequency's increase for the heaviest Mediterranean events that are exceeding the threshold of 200 mm in 24 hours. Extreme analyses applied to regional climate projections indicate an increase in the intensity of intense precipitation over the northern part of the Mediterranean Basin. The 6th IPCC report [8] published during the summer of 2021, confirms the expected intensification of these episodes of heavy precipitation if global warming continues to intensify and exceeds 2°C.

IOP Conf. Series: Earth and Environmental Science 1136 (2023) 012011 doi:10.1088/1755-1315/1136/1/012011

3.1. The storm ALEX

Storm ALEX hit France on the night of Thursday October 1st to Friday October 2nd, 2020. The storm had an extremely original trajectory: it stagnated on the English Channel during the day of October 2, still producing heavy rains in the west of the country, before plunging towards the Atlantic coasts on the 3rd to finally head towards the center of the country. On the 4th, the storm shifted towards the south of England, before evacuating to the North Sea the following day. Storm ALEX triggered an unprecedented Mediterranean episode. The low-pressure conditions' maintenance over the northwest of the country, combined with cold air aloft plunging towards Iberian Peninsula, established a massive and fast southerly flow over the Mediterranean Sea on the 2nd of October. It is ahead of its cold front that this phenomenon occurs. Caught in a convergence, thunderstorms associated with the front dump very large amounts of rain into the hinterland. The valleys of Estéron, Tinée, and especially Vésubie and Roya are particularly affected. The hot and humid air stumbled on the alpine relief, causing a historic Mediterranean episode over the department of Alpes Maritimes. Exceptional rain accumulations were recorded within the department and over the north-east part of the neighbour department of Var. The torrential rains, locally reaching 500 litters per square meter, caused devastating floods in the Var catchment and tributary rivers. 500 mm of rain were measured by Météo France at the station in Saint-Martin-Vésubie in 24 hours, a new record for all months combined in 24, 48 or 72 hours at the departmental level.

The Roya and Vésubie valleys in the hinterland of Nice were impacted by torrential rains with rapid floods and extremely destructive torrential lava. To the east of the Alpes-Maritimes, the municipalities are particularly affected by torrential floods, four of which are extreme: Breil-sur-Roya, Roquebillière, Saint-Martin-Vésubie and Tende. Human toll established in January 2021 (judicial source): 10 dead and 8 missing. The damages are enormous and the destructions multiple. The damage estimation reaches about € 2 billions for public infrastructures and about € 600 million for private insured goods.

3.2. The storm ALEX simulation

The rainfall data collected by Météo France during the storm ALEX on October 2nd and 3rd at the 14 rainfall stations located within the Var catchment have been introduced within the AquaVar system to generate flood hydrographs within the various tributaries and to estimate discharges at key locations where serious damages have been observed. The following figures – Fig.3 and Fig.4 – show the results obtained with AquaVar for key locations: the Vésubie valley that was the most affected by the rainfall event and the downstream Var valley where most of the economic activities and assets are concentrated. The AquaVar system allows to extract the hydrographs from the computation grid that is covering the full catchment. The produced hydrographs can be compared with field observations and the various discharge estimations as well.

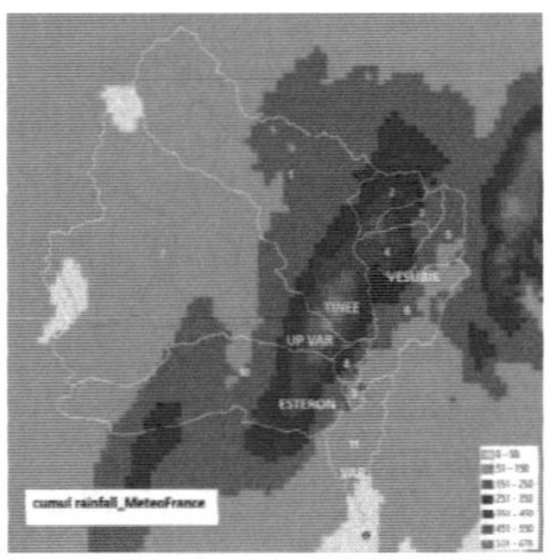

Figure 3. Cumulative rainfall (in mm) in 24 hours, from October 2nd, 2020, at 06:00 UTC to October 3rd, 2020, at 06:00 UTC (Météo France) over the Var catchment and main tributaries.

The hydrographs obtained for the upstream area of the Vésubie valley demonstrate the extreme hydrological behaviour of the catchment during the rainfall event. The two peaks obtained for the Vésubie and the Boréon are very consistent with the local observation and the estimated value above

400 m^3/s after the confluence in St Martin Vésubie is perfectly matching the hydraulic estimation provided by state services. The dual peaks reflect the rainfall distribution in time. In a logical way, the contribution of the upstream part of the Vésubie catchment has been massive. The hydrographs produced at points allow to understand the dynamic within the catchment. An estimation of the runoff for the flood event has been computed and a runoff coefficient of 79% has been obtained for the upstream surface of 393 km^2. This value underlines the extreme intensity of the processes and explains why an unusual quantity of materials was mobilized during the event. The intensity and the weight of water (rainfall) has initiated the morphological processes that have driven to a complete and major evolution of the slopes, flood plains and riverbeds.

The hydrograph obtained at Napoléon III bridge (Figure 5) located at the mouth of the Var River is estimating the discharges at 3,000 m^3/s at the time of the second peak. This value is consistent with field observations. The reported observed values are extracted from the online VigieCrues database (https://www.vigicrues.gouv.fr) and are the only available data for that extreme event as all monitoring were destroyed.

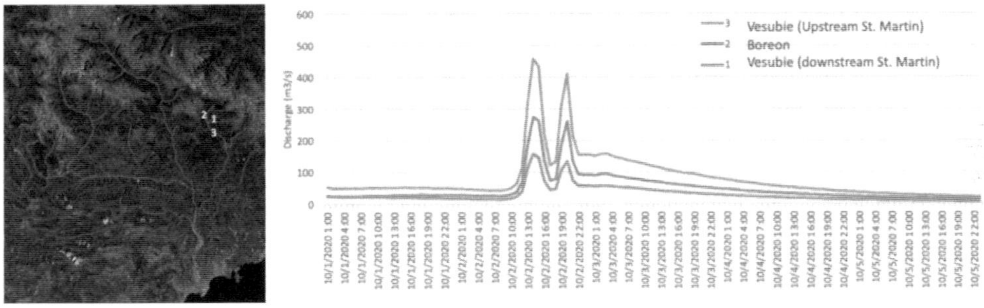

Figure 4. Hydrographs for the upstream area of the Vésubie catchment generated by AquaVar.

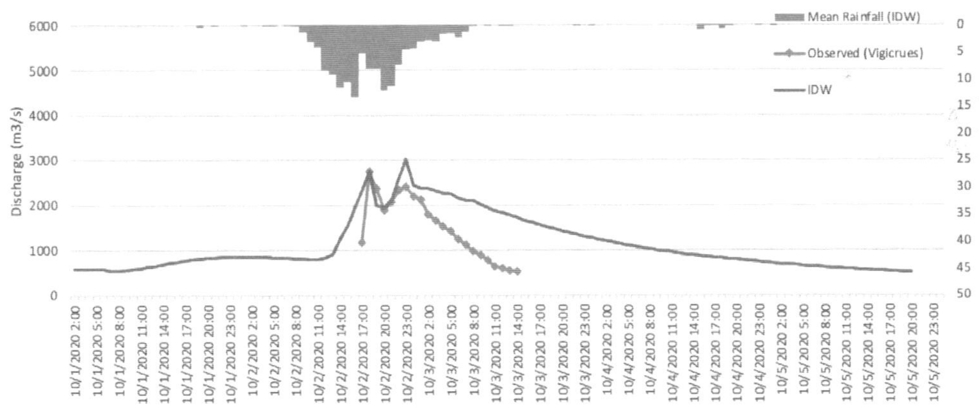

Figure 5. Hydrograph at Napoleon III bridge station: observed data and simulation from AquaVar. Observed data are subject to bias due to temporary civil engineering works".

The station at Napoléon III bridge was not fully operational due to works taking place with the riverbed since September 2020. The flow was diverted to the arches of the right bank and the station was able to operate only when the discharge was strong enough to remove the peddle accumulations what were used to divert the flows. The VigieCrues data are extrapolated from the rating curve that has been built for this station only up to 600 m^3/s. So, the quality of observed values is highly questionable. A hydraulic estimation based on the geometry of the weir located below the bridge and on the maximum water level observed confirms the discharge of 3,000 m^3/s. The estimation of the runoff coefficient at Napoléon III bridge is 61% for a total surface of 2,821 km^2.

4. Conclusions

The AquaVar DSS combines three deterministic models that are used for producing forecasts on surface runoff, flood propagation and groundwater evolution. For the specific context of the Nice Côte d'Azur Metropolis and the associated valleys, the deterministic models offer a relevant approach due to their robustness and their capacity to produce results within a physically based range. The implementation of these models within DSSs requests to develop a strategy for their orchestration and their exploitation in real-time. The constraints expressed by the operational entities represent boundary conditions that must be integrated within the DSS design. The necessity to obtain simulation results in one hour time is a major point in the case of the Var low valley. The various tests conducted – simulation of 7 days with 4 days from the past and 3 days ahead based on meteorological forecasts - with the different technical configurations have demonstrated that an operational architecture based on two computational servers using CPU and GPU resources can deliver the expected performance: a forecast for the next 3 days delivered less than 1 hours.

The architecture of the system can also be updated with the implementation of data driven tools that can be combined with the deterministic approach. The addition of the data driven models can reduce the computation time and permit faster forecast production. The 2020 storm ALEX is a good test to demonstrate that the deterministic modelling system was delivering efficient estimation of the discharges within the catchment based on the meteorological data collected on the field. The very limited efficiency of data driven, and conceptual models is clearly underlined with the extreme rainfall as those tools cannot produce relevant results as such type of situation is not integrated within their initial knowledge base. The AquaVar concept underlines the performance of the deterministic distributed models for extreme events and their possibility to generate accurate hydrographs. The current results are used by the field operators and decisions makers who can adjust their strategies, implement mitigation measures, and improve resilience of exposed populations.

Acknowledgments

AquaVar is supported by Nice Côte d'Azur Metropolis, Université Côte d'Azur with the support of the Rhône Méditerranée Water Agency. The author would like to thank all institutions that have actively supported the development with data collection and field expertise.

References

[1] Bertin E and Crespi N 2014 Urbanization of information systems: an outdated method? *Digital Enterprise Design & Management* (Springer, Cham) pp 83–91

[2] Nohara D, Gourbesville P and Ma Q 2018 Towards development of effective decision support systems for integrated water resources planning and management *Disaster Prevention Research Institute (DPRI) Annuals* **61**(B) 702–710 https://www.dpri.kyoto-u.ac.jp/nenpo/no61/ronbunB/a61b0p47.pdf

[3] Gourbesville P 2020 Which models for Decision Support Systems? Proposal for a methodology. *Advances in Hydroinformatics* (Singapore: Springer) pp 3–17

[4] Gourbesville P, Du M, Zavattero E and Ma Q 2016 DSS architecture for water uses management *Procedia Engineering* **154** 928–935 https://doi.org/10.1016/j.proeng.2016.07.512

[5] Ma Q, Gourbesville P and Gaetano M 2020 Aquavar: Decision Support System for surface and groundwater management at the catchment scale *Advances in Hydroinformatics* (Singapore: Springer) pp 19–28

[6] MIKE Zero by DHI https://www.mikepoweredbydhi.com/products

[7] Du M, Zavattero E, Ma Q, Delestre O, Gourbesville P and Fouché O 2018 3D modeling of a complex alluvial aquifer for efficient management – application to the lower valley of Var river, France *La Houille Blanche* February (1) 60–69 https://doi.org/10.1051/lhb/2018009

[8] Legg S 2021 Climate Change 2021: The Physical Science Basis (IPCC, 2021) *Interaction* **49**(4) 44–45.

Analysis of planetary boundary layer feedbacks on the multiple equilibria of the continental water cycle

Abhinav Dengri[1] **and Tomohito Yamada**[2]

[1] Graduate School of Engineering, Field Engineering for the Environment, Hokkaido University, Sapporo, 060-8628, Japan
[2] Faculty of Engineering, Hokkaido University, Sapporo, 060-8628, Japan

dengriabhinav@gmail.com

Abstract. A simple soil-atmosphere interaction model is developed to study the bimodal distribution of soil moisture in regions where there is feedback between precipitation and soil moisture. Our study includes planetary boundary layer (PBL) feedback in the coupled analytical soil-atmosphere interaction model with a varying PBL height and entrainment from the free atmosphere. This study aims to reveal the dominant PBL feedbacks that impact the occurrence of soil moisture bimodality by comparing the evolution of prognostic variables for a detailed PBL scheme (with varying PBL height and entrainment of heat and moisture from the free atmosphere) with the simple PBL scheme (with fixed PBL height and no entrainment). This study has shown the existence of multiple equilibria in the continental water balance for both detailed PBL and simple PBL schemes when the model is integrated for different initial soil moisture up to equilibria. Although both with and without PBL feedback, the results are in qualitative agreement; the soil moisture initialization to realize multiple equilibria of a detailed PBL scheme is larger than the simple PBL scheme. Although the direct quantitative comparison with observation is clearly limited in the simple model, it also demonstrates how the non-linear dynamics of the coupled soil-atmosphere system may be expected to occur in observations.

1. Introduction

The water stored on the land as soil moisture is the key component in the water and energy cycle. It is the source of moisture to the atmosphere through evapotranspiration and also controls latent and sensible heat flux, thus influencing the stability of the boundary layer, the temperature of the air, and precipitation [1]. There are numerous feedbacks between the soil layer and the surrounding atmosphere, of which soil moisture–precipitation coupling has gained much attention since the early 1980s [2].

The soil moisture-precipitation coupling can be described as the precipitation response to changes in soil moisture. The Global Land-Atmosphere Coupling Experiment (GLACE) identified hotspots of soil moisture-precipitation interaction to be located in the transition region, i.e. between wet and dry climates [3]. In these hotspots, evapotranspiration was not only sensitive to soil moisture but also large enough to affect precipitation. Moreover, global climate modeling and observational studies have been shown to improve subseasonal predictions of temperature and precipitation over hotspots by accounting for accurate soil moisture initialization [4,5].

One of such regions with statistically significant lagged correlations between soil moisture anomalies and successive summer rainfall is Illinois, United States [5]. Furthermore, the distribution of long-term summer soil moisture in Illinois was bimodal (soil moisture bimodality), that is, having two distinct peaks, with one peak representing the dry soil moisture state and another wet soil moisture state [6]. It has been argued that the soil moisture-precipitation coupling drives bimodal distribution of soil moisture, where dry pre-summer soil moisture led to lower summer precipitation, thus further drying the soil moisture and precipitation. On the other hand, wet presummer soil moisture led to increased precipitation, thus sustaining the wet soil moisture. Not only Illinois, but also the bimodal distribution of summer soil moisture was detected over soil moisture–precipitation hotspots using AMSR-E satellite data [7]. These bimodal distributions are crucial to explaining continental mid-latitude summer heat waves [8].

The simplified bucket models have been the pioneers in understanding the effect of soil moisture dynamics on the climate system and are still attractive as they shed light on primary mechanisms. The land-atmosphere interaction model used for this study provides a simple physical representation of the planetary boundary layer's thermodynamics, convection and precipitation[8]. In addition, the model includes a representation of planetary boundary layer growth and entrainment processes. The height of the planetary boundary layer (PBL) is one of the key characteristics in both atmospheric models and observations, as it is related to the land surface conditions and determines the vertical extent to mix heat, moisture, and momentum. For example, the height of the planetary boundary layer increases for rapidly drying soils due to an increased Bowen ratio[9,10]. Similarly, the entrainment of warm and dry air from the free atmosphere affects the PBL temperature and humidity, and thus impacts evapotranspiration and surface fluxes. In short, there is a feedback loop that potentially influences evapotranspiration and PBL growth: the sensible heat flux controls the PBL growth and rate of entrainment of drier and warmer air, which modifies saturation deficit within the PBL; this saturation deficit then exerts control on evapotranspiration rate, which is related to surface energy budget, and thus, sensible heat flux. Observational studies based on soil moisture and the flux station over the southern great plains have found the evolution of the PBL to be related to surface soil moisture and atmospheric stability [11].

This study using the continental-scale soil layer-PBL interaction model with the representation of the PBL growth and entrainment processes studies the bimodal distribution of surface soil moisture. We found that the initial soil moisture and the entrainment of water vapor from the free atmosphere are crucial factors in creating the soil moisture bimodality.

2. Methodology

2.1. Box Model

The water and energy interaction between the soil layer and the planetary boundary layer is simulated using the idealized box model. Figure 1 shows the schematic of the model and the various PBL feedbacks of the soil layer. The depth of the soil layer is fixed at 0.5 m, and the depth of PBL is estimated empirically (section 2.4) using soil moisture and atmospheric stability. There are four prognostic variables, temperature and humidity, for both the PBL and the soil layer, whose evolution is given by equations (1)-(4). The key model parameters and their values are given in Table 1.

The PBL temperature (θ_a) increases from sensible heat flux (Q_s), absorption of longwave radiation from the ground ($\epsilon_s \sigma T_s^4$), and entrainment of warm air (H_i) from the free troposphere and decreases through convective drying as given by equation (1). Convective drying is present only when convection occurs, i.e., when PBL is unstable. The details of convection parameterization are described in Sections 2.2 and 2.3, respectively. Next, the PBL humidity (q_a) is moistened by evapotranspiration (E) from the soil layer and incoming atmospheric moisture driven (F_q) by large-scale synoptic flows. Incoming atmospheric moisture is an important parameter in the study, whose impact is studied in detail. On the other hand, PBL humidity is reduced by entrainment of dry air into the PBL and by convective drying as represented by Equation (2). The soil temperature (T_s) is driven by incoming net radiation (F_{rad}) and outgoing sensible heat flux, latent heat flux (L_e), and longwave radiations ($\epsilon_s \sigma T_s^4$). The incoming radiation is fixed throughout the computation to represent a mid-latitude summer day. Finally, the soil moisture (q_s) is the relative soil moisture with respect to saturation. Volumetric soil moisture is obtained by multiplying q_s by the porosity (n). Soil moisture is calculated as the balance of precipitation, evapotranspiration, and Leakage ($L(q_s)$ as given in equation (4). All soil layer parameters have been obtained for sandy loam [12].

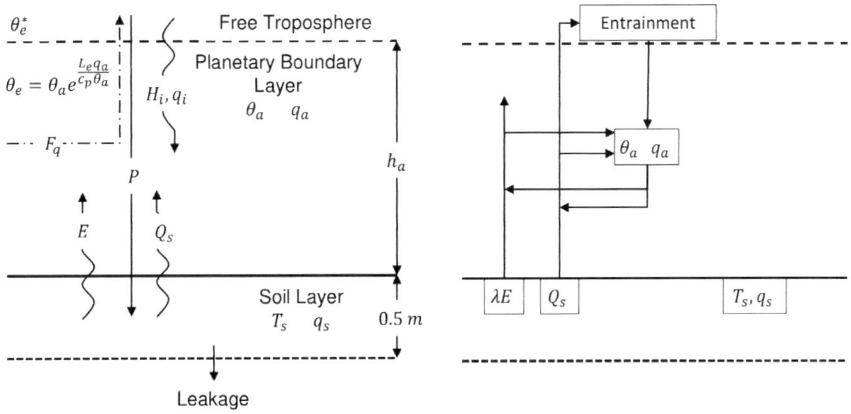

Figure 1: Schematic of the soil layer-PBL interaction box model (left). The depth of soil layer is 0.5 meters, and the height of the PBL (h_a) is calculated dynamically based on atmospheric stability and soil moisture state. The lateral moisture flux (F_q) brings the atmospheric moisture into the model. The heat and water fluxes are represented by the curve arrows. The formulation to computer equivalent potential temperature of the PBL is given on the left corner. The Planetary boundary layer and soil layer feedbacks (right). Surface fluxes modify the entrainment from free atmosphere and thus PBL condition which further modifies surface fluxes.

$$\rho c_p h_a \frac{\partial \theta_a}{\partial t} = Q_s + \varepsilon_a \varepsilon_s \sigma T_s^4 - \rho c_p h_a \frac{\partial \Delta\tilde{\theta}_a}{\partial t} + \frac{\theta_a^* - \theta_a}{\tau} + H_i \tag{1}$$

$$\rho h_a \frac{\partial q_a}{\partial t} = E - \rho h_a \frac{\partial \Delta\tilde{q}_a}{\partial t} + F_q - q_i \tag{2}$$

$$\rho_s c_{ps} h_s \frac{\partial T_s}{\partial t} = F_{rad} - Q_s - L_e(E) - \varepsilon_s \sigma T_s^4 \tag{3}$$

$$w_o h_s \frac{\partial q_s}{\partial t} = P - E - L(q_s) \tag{4}$$

Table 1. Key model parameters with their values [8] used for the model computation.

Symbol	Meaning	Value	Units
F_{rad}	Net radiation at the surface	500	W m^{-2}
L_e	Specific latent heat of evaporation	2.5×10^6	J kg^{-1}
R	Ideal gas constant	287	J kg K^{-1}
c_{pa}	Air-specific heat	1000	J kg^{-1} K^{-1}
c_{ps}	Soil specific heat	1000	J kg^{-1} K^{-1}
h_a	Thickness of the atmospheric boundary layer	1000	m
h_s	Depth of the active soil layer	0.5	m

w_o	Soil water holding capacity	1500	kg m^{-3}
ϵ_a	Blackbody absorptivity of the PBL	0.3	-
ϵ_s	Blackbody absorptivity of the Earth	0.8	-
ρ	Air density	1	kg m^{-3}
ρ_s	Soil density	1800	kg m^{-3}
$\overline{u_s}$	Mean wind velocity	6	m s^{-1}
C_D	Bulk aerodynamic drag coefficient	0.008	-
q_{fc}	Field capacity	0.56	-
q^*	Maximum plant efficiency point	0.46	-
q_w	Wilting point	0.18	-
q_h	Hygroscopic point	0.14	-
n	Porosity	0.43	-

2.2. Convective Parameterization

The simple convection parameterization is assumed to present the bulk effect of the convectively active period in a large area rather than in an individual storm. Convection occurs only when the PBL is unstable. The stability of the PBL is decided as the difference between the equivalent potential temperature of the free atmosphere (θ_e^*) and the PBL (θ_e). The parameter, θ_e^* is assumed to be fixed at 300K, however the results are insensitive to reasonable change in the parameter value. Only when $\theta_e - \theta_e^* > 0$, PBL is unstable, leading to spontaneous convective cooling and drying of the PBL as represented in equations(1)-(2). The convective cooling and drying are represented by the equations (5) and (6).

$$\Delta\tilde{\theta}_a = \frac{\theta_e - \theta_e^*}{1 + \frac{L_e}{c_p} q_{rel} \delta q_{sat}} \tag{5}$$

$$\Delta\tilde{q}_a = q_{rel} \delta q_{sat} \Delta\tilde{\theta}_a \tag{6}$$

2.3. Precipitation

The amount of precipitation is modeled as a fraction of the amount of moisture uplifted into the free atmosphere during convection. The amount of precipitation depends on the intensity of the convection updraft, where higher convection updrafts lead to greater precipitation and vice versa. Precipitation rate is expressed as $f\rho h_a \frac{\widetilde{\Delta q_a}}{\delta t}$, where f is the precipitation efficiency of the convection and $\rho h_a \frac{\widetilde{\Delta q_a}}{\delta t}$ is the convection updraft. If the convection updraft is greater than 3mm/day, precipitation efficiency is 0.9, and if the convection updraft is less than 1mm/day, the precipitation efficiency is 0.2.

2.4. Evapotranspiration

Evapotranspiration is modeled as a function of the saturation deficit of the atmosphere and the average soil moisture. Evapotranspiration vanishes for the average soil moisture below hygroscopic point (q_h). Above the hygroscopic point, the evapotranspiration of the soil grows linearly, with a higher rate of growth beyond wilting point (q_w). Beyond the maximum plant efficiency point (q^*), evapotranspiration is constant to its maximum value, E_{max}. The formula for evapotranspiration is given by Equations (7) and (8).

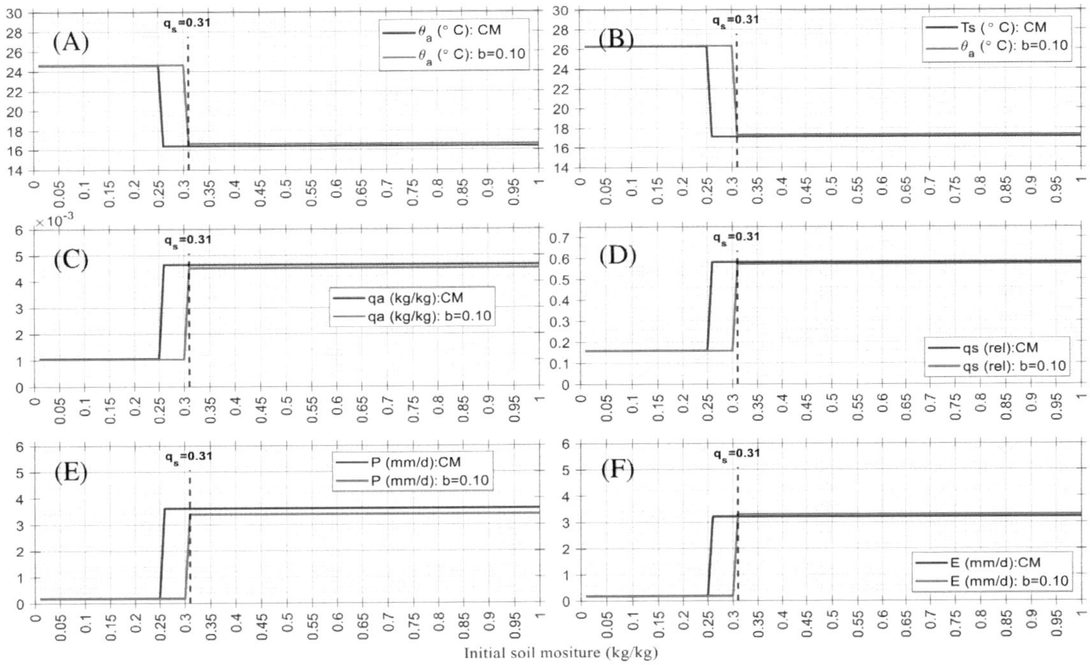

Figure 2. Equibrium states of the control model (fixed PBL height, and no entrainment fluxes), blue line, and dynamic PBL model, red line as a function of initial soil moisture. Starting for the given initial soil moisture, the model is integrated till the equilibrium is reached. The equilibrium state for the dynamic PBL model is obtained when the fraction of entrained moisture to evapotranspiration, b is 0.10. Black dashed line shows the initial soil moisture threshold for the dynamic PBL model. (A) Air temperature, (B) Soil Temperature, (C) air humidity, (D) soil moisture, (E) precipitation and (F) evapotranspiration.

$$E = E' \frac{q_{sat} - q_a}{q_{sat}} \tag{7}$$

$$E' = \begin{cases} 0 & q_s \le q_h \\ \dfrac{q_s - q_h}{q_w - q_h} E_w & q_h \le q_s \le q_w \\ E_w + \dfrac{q_s - q_w}{q^* - q_w}(E_{\max} - E_w) & q_w \le q_s \le q^* \\ E_{\max} & q_s \ge q^* \end{cases} \tag{8}$$

2.5. *Planetary Boundary Layer height and Entrainment*

The control model (CM) has a fixed PBL height of 1000m. This study incorporates the variable PBL height (h_a) and entrainment of moisture (q_i) and heat (H_i) from the free troposphere. The height of the PBL is simulated as a function of the volumetric soil moisture (q_{sv}) and atmospheric stability ($\gamma_v = 2\times10^{-3}$) as given by equation (9). The second-order polynomial empirical equation is obtained from the daily observation at a site in the southern Great Plains[11]. The equation parameters are calculated by [13] for the site with sandy loam soil are $p_o = 4.23\times10^3$, $p_1 = -5.24\times10^5$, $p_2 = -9.30\times10^1$, $p_3 = 2.55\times10^8$, $p_4 = -7.72\times10^4$ and $p_5 = 3.81\times10^0$.

$$h_a(q_{sv},\gamma_v) = p_o + p_1\gamma_v + p_2 q_{sv} + p_3\gamma_v^2 + p_4 q_{sv}\gamma_v + p_5 q_{sv}^{\ 2} \tag{9}$$

Equation (10) gives the heat entrainment from the free atmosphere is parameterized as the function of the surface heat flux. The heat entrainment rate parameter, A_r is assumed to be fixed at 0.2; however, the model is not sensitive to the chosen value. We adopted the simple parameterization for water vapor entrainment (equation 11), where water vapor flux at the top of PBL is proportional to surface evapotranspiration [13]. The moisture entrainment rate parameter, b represents the fraction of entrained air to total evapotranspiration.

$$H_i = A_r Q_s \tag{10}$$

$$q_i = bE \tag{11}$$

3. Results

3.1. Multiple Equilibria of water cycle in coupled soil-planetary boundary model

For the realistic values of the parameters and for given initial soil moisture, the model is integrated until the equilibrium is reached. For the fixed incoming lateral moisture flux ($F_q = 0.8$ mm/day), the control model has two stable equilibria, i.e., dry stable state and wet stable state (Figure 2). The x-axis shows the initial soil moisture conditions and y axis shows the final equilibrium state of the model variables. The dry stable equilibria are reached when the initial soil moisture is less than 0.26, and is associated with higher air and soil temperatures and lower air and soil humidity. These conditions classify mid-latitude summer as hot and dry. On the other hand, soil moisture initialization greater than 0.26 leads to wet stable equilibria, which has lower air and soil temperature and higher air and soil humidity, symbolizing cool and wet mid latitude summer. A detailed study of the control model configurations and in initial conditions suggests that the model is most sensitive to initial soil moisture while insensitive to initial values of other prognostic variables.

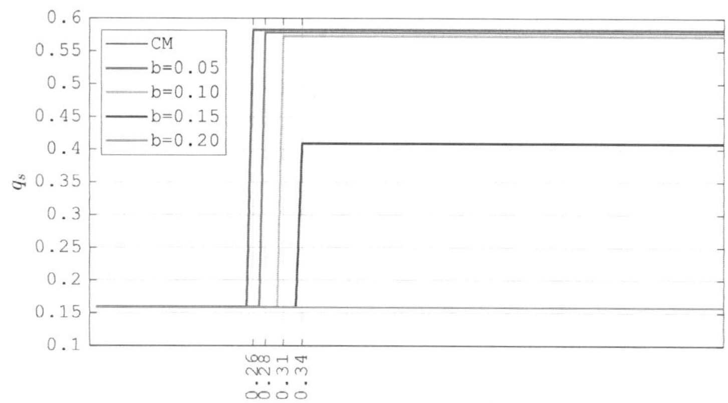

Figure 3. Soil moisture equilibrium states as a function of soil moisture initialization. The moisture entrainment parameter, *b*, is varied from 0.05 to 0.20. For the control model there are two equilibrium states, i.e., dry and wet equilibria. For the dynamic PBL model, as the parameter, *b* increases, the initial soil mositure initialization to attain wet equilibria increases, until *b* = 0.20.

The control model is modified to include PBL growth, heat and moisture entrainment (section 2.5). To determine the impact of the dynamic PBL model on multiple equilibria of water cycle, it is compared with the control model. For fixed incoming lateral moisture flux ($F_q = 0.8$ mm/day), heat entrainment rate, $A_r = 0.20$ and moisture entrainment rate, $b = 0.10$, the model is in qualitative agreement with the control model, i.e., it has two stable equilibria. Unlike the control model, the dynamic PBL model requires higher soil moisture initialization, i.e, $q_s = 0.31$, to reach wet stable equilibria.

The influence of moisture entrainment parameter, b, on the soil moisture initialization is shown in figure 3. The x-axis is the soil moisture initialization, and y-axis is the equilibrium state of soil moisture. For higher values of the moisture entrainment parameter b, the initial soil moisture threshold to distinguish between dry and wet stable states increases. When $b = 0.15$ the equilibrium state of soil moisture decreases, representing a relatively less cool and wet summer. On the other hand, a large influx of dry air into the PBL, $b = 0.20$ leads to dry stable state regardless of higher initial soil moisture initializations.

3.2. Coupled Soil-Planetary Boundary Model Sensitivity to Synoptic Flow.

The important control parameter of the coupled soil-planetary boundary model is the lateral moisture convergence flux (F_q) which represents the synoptic flows. The value of lateral moisture convergence over mid-latitude summer varies from negative (divergence) to positive (convergence) in few millimeters per day. For the mid-latitude regions, the mean F_q is 0 mm/day, with standard deviations as large as 3mm/day [8].

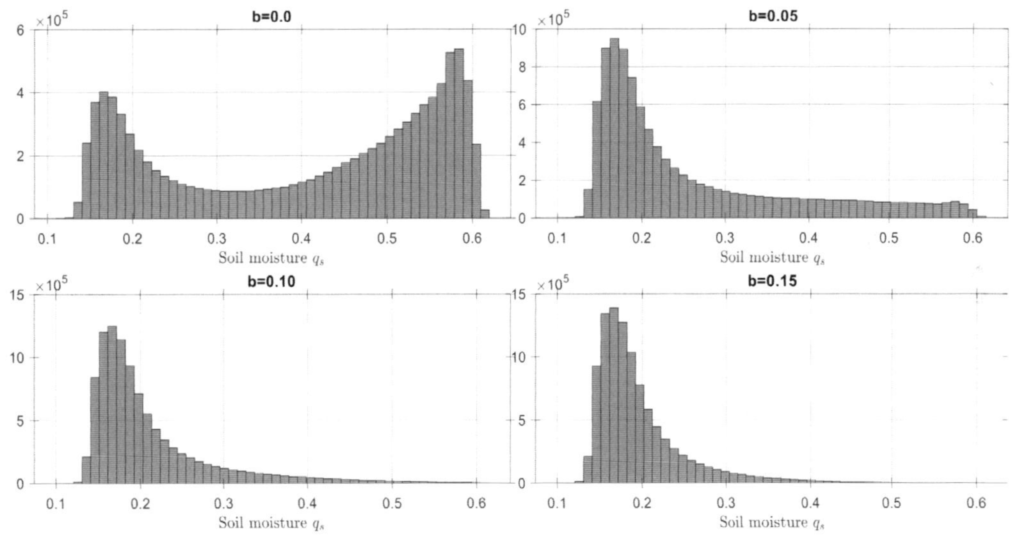

Figure 4. Time evolution histogram of soil moisture (q_s) in case of variable synoptic forcing. The dynamic PBL model is evaluated for different values for the moisture entrainment parameter, b. For lower values of b, the model attains two equilibrium states, however with increase in entrainment of dry air from free atmosphere the model attains unimodal soil moisture distribution.

In order to simulate the impact of synoptic flow variability, F_q is stochastically varied. For the given initial soil moisture, $q_s = 0.20$, model is integrated for 10^8days during which F_q is randomly selected every ten days from a uniform distribution of -0.8mm to 2.6mm/day. With the described forcing, the control model ($b = 0$) produces the bimodal distribution of soil where soil moisture oscillates between dry and wet states. Figure 4 shows the histogram of the soil moisture for the different values of the moisture entrainment parameter, b, when model is forced to account for synoptic variability. Bimodal distribution gradually changes to unimodal distribution with a dominant peak on the drier soil moisture as the moisture entrainment parameter increases. These results highlight the significant role of moisture entrainment in influencing the bimodal distributions.

4. Summary

This study adds PBL growth and entrainment schemes to the simple continental-scale soil layer – PBL interaction model. This study is an extension of [8], where the authors assumed that the PBL was a closed box of fixed height. In our approach, the height of the PBL is parameterized as a function of volumetric stability and atmospheric stability. In addition, the entrainment of warm and dry air is taken into account. Heat and moisture entrainment is parameterized as a fraction of sensible heat flux and evapotranspiration.

It is shown that the model is sensitive to moisture entrainment parameters in addition to initial soil moisture. If the entrainment parameter increases beyond 0.20, no multiple equilibria are obtained. Although we have employed a simple set of equations to represent entrainments, it is crucial to emphasize the role of entrainment in influencing the soil layer-PBL interactions.

Acknowledgments
This study is supported by ArCS II (Arctic Challenge for Sustainability), JSPS KAKENHI Grant Number 22H01594 and Hokkaido University DX Doctoral Fellowship Grant Number JPMJSP2119.

References

[1] Seneviratne SI, Corti T, Davin EL, Hirschi M, Jaeger EB, Lehner I et al 2010 Investigating soil moisture–climate interactions in a changing climate: A review *Earth-Sci Rev.* **99**(3–4) 125–61
[2] Schär C, Lüthi D, Beyerle U and Heise E 1999 The soil–precipitation feedback: A process study with a regional climate model *J. Clim.* **12**(3) 722–41
[3] Koster RD, Dirmeyer PA, Guo Z, Bonan G, Chan E, Cox P et al 2004 Regions of strong coupling between soil moisture and precipitation *Science* **305**(5687) 1138–40
[4] Yamada TJ and Pokhrel Y 2019 Effect of human-induced land disturbance on subseasonal predictability of near-surface variables using an atmospheric general circulation model *Atmosphere* **10**(11) 725 https://doi.org/10.3390/atmos10110725
[5] Findell KL and Eltahir EAB 1997 An analysis of the soil moisture-rainfall feedback, based on direct observations from Illinois *Water Resour. Res.* **33**(4) 725–35
[6] D'Odorico P and Porporato A 2004 Preferential states in soil moisture and climate dynamics *Proc. Natl. Acad. Sci.* **101**(24) 8848–51 https://doi.org/10.1073/pnas.0401428101
[7] Vilasa L, Miralles DG, de Jeu RAM and Dolman AJ 2017 Global soil moisture bimodality in satellite observations and climate models: Global soil moisture bimodality *J. Geophys. Res. Atmospheres* **122**(8) 4299–311 https://doi.org/10.1002/2016JD026099
[8] D'Andrea F, Provenzale A, Vautard R and De Noblet-Decoudré N 2006 Hot and cool summers: Multiple equilibria of the continental water cycle *Geophys. Res. Lett.* **33**(24) L24807 https://doi.org/10.1029/2006GL027972
[9] Entekhabi D, Rodriguez-Iturbe I and Castelli F 1996 Mutual interaction of soil moisture state and atmospheric processes *J. Hydrol.* **184** 3–17 https://doi.org/10.1016/0022-1694(95)02965-6
[10] Entekhabi D and Brubaker KL 1995 An analytic approach to modeling land-atmosphere interaction: 2. Stochastic formulation *Water Resour. Res.* **31**(3) 633–43
[11] Santanello JA, Friedl MA and Kustas WP 2005 An empirical investigation of convective planetary boundary layer evolution and its relationship with the land surface *J. Appl. Meteorol. Climatol.* **44**(6) 917–32
[12] Laio F, Porporato A, Ridolfi L and Rodriguez-Iturbe I 2001 Plants in water-controlled ecosystems: active role in hydrologic processes and response to water stress *Adv. Water Resour.* **24**(7) 707–23
[13] De Bruin HAR 1983 A model for the Priestley-Taylor parameter α *J. Appl. Meteorol. Climatol.* **22**(4) 572–578.

Implementation of water related Sustainable Development Goals in Romania: overview of current and future challenges

Camelia Teau[1,2], Ioana Popescu[3,2], Constantin Florescu[2], Albert Constantin[2], Cristian Moisescu Ciocan[4,2] and Valentin Vlaicu[5,2]

[1] Mures Water Administration, Romanian Waters, Tg. Mures, 540057, Romania
[2] Faculty of Civil Engineering, Politehnica University of Timisoara, 300223, Romania
[3] Department of Hydroinformatics and sociotechnical innovations, IHE Delft Institute for Water Education, Delft, 2611 AX, The Netherlands
[4] Banat Water Administration, Romanian Waters, Timisoara, 300173, Romania
[5] Aquatim, Timisoara, 300081, Romania

Corresponding author: i.popescu@un-ihe.org

Abstract. The collection of information on the achievement of the water related Sustainable Development Goals (SDGs), as well as their analysis through research, is defined as being difficult especially in the absence of data and tools that can model these performances. An analysis of SDG indicators at the level of 2015, for 41 countries where the population has high and medium incomes shows that a special level of complexity in monitoring is given by the fact that each country measures certain indicators specifically, and these cannot be compared. It is recommended to formulate methods of measurement and monitoring that can be compared. All these consistent methods between countries must be subject to data protection rules, both in terms of access to them and in terms of their management. Present paper shows the main challenges in monitoring water related SDGs (mainly 6, 11 and 13) in Romania, what is the status so far and what are the challenges for the future.

1. Introduction

The development of the society in the 20th century led to the need to diversify the water requirements for different purposes: water supply for house and industry consumption; capitalizing on the hydropower potential of watercourses; protecting the population and socio-economic objectives by mitigating the maximum flows on the main watercourses in the river basins.

Water needs are constantly increasing as the water resource is declining, consequently there is the need for a sustainable management of environmental resources. Practicing sustainable development entails defining policies for environmental resources exploitation that are transformed into services for population.

There is a general consensus worldwide that the 1992 development goals (the Millennium Development Goals - MDGs), of the UN Agenda 21, have not been well implemented ([1]).

After 1992, the implementation of the MDGs was analyzed every 10 years, and the unsatisfactory results of the implementation led to the definition of a number of 17 Sustainable Development Goals (SDGs), which retains part of the millennium goals. For this reason, the new development objectives are defined as necessary to be sustainable ([2]).

The implementation of the SDGs can be possible only if governance policies offers tools able to facilitate their implementation. Such tools are: monitoring the decision-making aspects; monitoring and evaluation of the performance indices defined for each SDG; and transparency in implementation. The principles of any governance involve decision-making at the local, national, regional and global levels to ensure coherence between the SDGs between these levels.

The collection of information on the achievement of the water related SDGs, as well as their analysis through research, is defined as being difficult especially in the absence of data and tools that can model these performances ([3])

Therefore, it is recommended to formulate methods of measurement and monitoring that can be compared. Present paper shows the main challenges in monitoring water related SDGs (mainly 6, 11 and 13) in Romania, what is the status so far and what are the challenges for the future.

2. Overview of SDGs implementation in Romania

In 2018 the government of Romania issues the strategy for Romania's Sustainable Development for the year 2030, in which the emphasize is on goal 17 (Partnerships for the Goals), especially because of the six months European presidency that the country had at the beginning of 2019. The defined strategy sets goals in two stages, one almost immediate, for 2020, and one for a longer term, for 2030. The main highlights of the document are the shortcomings of implementing the SDGs in Romania, complemented by an analysis of what are the specific areas in which additional efforts and resources are needed. The ultimate aim is to have the level of the SDGs the same as the ones of the European Union. ([4])

Goals are proposed to be achieved through indicators, defined at length in chapter two of the strategy. In total there are 100 indicators for the 17 SDGs, out of which 42 are defined to monitor more than one single SDG. Implementation and monitoring is defined in a qualitative way in the last chapter of the report. ([5]).

Romania's performance overall, at the level of year 2021 is 39 out of 165 countries monitoring their SDGs. Figure 1 shows the overall performance as well as performance for each SDG in Romania. This overall position is a good one, however at European level, Romania is on quite low position, which shows the need for intense monitoring and tools that will ensure achievement of SDG goals by 2030.

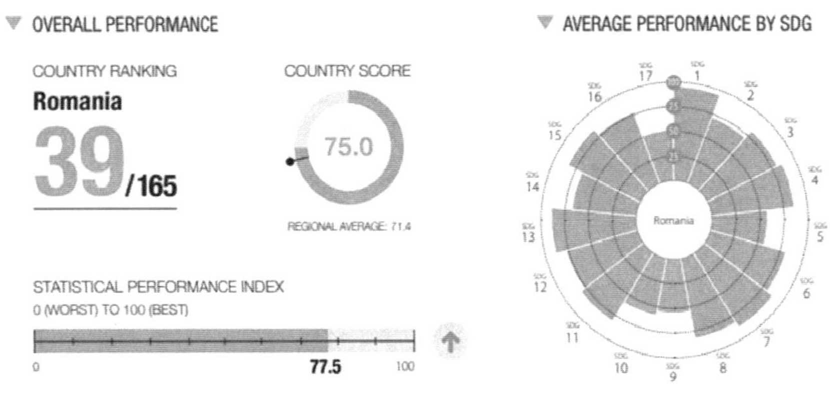

Figure 1. Romania's SDG overall performance (source: dashboards.sdgindex.org/rankings).

The set of indicators that measures each SDG level are achieved differently by each of the 165 countries that are monitored. The aim is to eventually have all them achieved at same level. Worldwide special databases collect the indicator statuses for each country. These are available in form of reports. The latest year when this data is available is 2021 in [6], while Romania's data in national opensource databases is available only up to the year 2019. Updates are not done regularly. This paper shows in Figure 1 the 2021 status for Romania, based on international overview. Table 1 gives an overview on the indicator status in 2019, as presented by [4], based on their own calculation at the level of year 2019.

IOP Conf. Series: Earth and Environmental Science 1136 (2023) 012013 doi:10.1088/1755-1315/1136/1/012013

Table 1. Romania's SDG status in 2017.

SDG	Romania SDG indicators achievement (%)	Status	Challenges
SDG 1 – No Poverty	50	On track	Moderate
SDG 2 – Zero Hunger	60	Improving	Significant
SDG 3 – Good Health and Well-Being	44,4	Improving	Significant
SDG 4 – Quality Education	0	Decreasing	Moderate
SDG 5 – Gender Equality	12,5	Improving	Significant
SDG 6 – Clean Water and Sanitation	33,3	Improving	Significant
SDG 7 – Affordable and Clean Energy	71,4	Improving	Moderate
SDG 8 – Decent Work and Economic Growth	11,1	Improving	Moderate
SDG 9 – Industry, Innovation and Infrastructure	26,8	Improving	Significant
SDG 10 – Reduced Inequalities	12,5	No info	Major
SDG 11 – Sustainable Cities and Communities	50	Stagnant	Significant
SDG 12 – Responsible Consumption and Production	37,5	No info	No info
SDG 13 – Climate Action	50	Stagnant	Significant
SDG 14 – Life Below Water	100	Decreasing	Major
SDG 15 – Life on Land	33,3	Stagnant	Significant
SDG 16 – Pace, Justice and Strong Institutions	50	Improving	Significant
SDG 17 – Partnerships for the Goals	75	Improving	Significant

Methodologies of assessing these indicators based on statistical available data in [7] and [8] are proposed, not only by [4] but also by other authors. Notably is a method based on Earth Observation proposed by [9], which gives a good visualization of the SDG status in Romania at the level of 2019. The approach given in [9] is very detailed and uses a lot of publicly available data to compute and store results, however it is a one-time off effort, which is not maintained as data base, nor could be accessed outside of the open access paper.

The data presented in [4] is available as clear numeric values analysis over a period of 10 years (2007 – 2017), without detailing the approach taken to compute the values that are composite of several indicators. What it is noticed in the analysis of [4] and [9] is that though each SDG is computed as an indicator, a lot of attention in analysis is, rightfully, given to poverty, education and health, while the ones related to water are not looked at.

According to [10] SDG implementation is important for any country. This is seen in reports on SDG implementation in Romania, including the government official reports, which emphasize their importance for the country society, therefore their monitoring and implementation has become fundamental for the development.

3. Water related SDGs and their achievement in Romania

There are mainly three SDGs that directly relate to water, 6, 11 and 13. There are SDGs that can be related to water, such as SDG 7, in which clean energy of hydropower can be counted as a water SDG, however the main focus in SDG 7 is on energy overall and not directly on water.

The sixth SDG, of clean water and management of it looks at the access right of population to basic water quantity and quality services. Though overall Romania is well connected to drinking water supply networks in the urban areas, comparable with the rest of Europe, significant challenges still remain for rural areas. Moreover, access to bath, shower or indoor toilet, is also a problem, in Romania 22% of the population do not have access to it. Figure 2 shows that overall Europe less than 2% of the population is not having access to basic sanitation services, while the worst situation after Romania is with a less than 10% no access. One of the reasons that Romania is so much behind is also the high number of populations in the rural area, while in many European countries less population lives in the countryside. Population in rural areas is difficult to be connected to a water supply system. However, the government made a lot of investments since 2015, when 45% of population was not having the access.

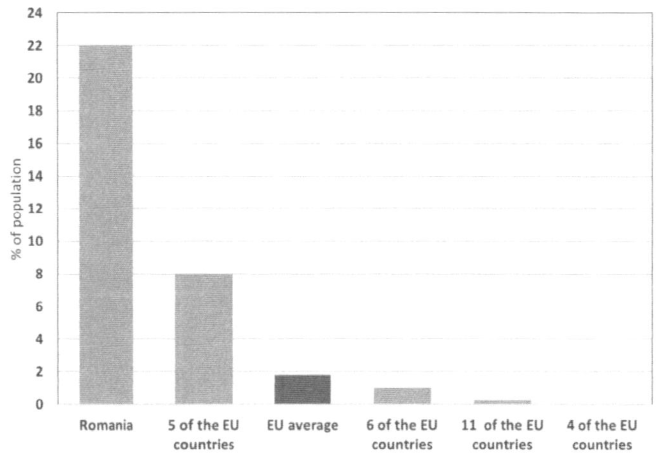

Figure 2. Romania % of population with no access to bath water, in 2019.

Access to basic water supply indicator is strongly linked with SDG 1, 3 and 4, where vulnerable population, such as poor, old or uneducated persons will not always have the opportunity to have access to clean water, even if the government makes investments and bring the water mains in front of their houses. Vulnerable population will not invest connecting to these mains. Apart from investment, government need to have strong educational campaign and financial support for convincing them to connect to basic water services. Monitoring of number of households that are not yet connected, in a form of a database, as proposed by [4] would help.

SDG 6 is not only about services to drinking water and sanitation, but also about management of water resources as a whole, looking at water quantity and avoiding risks that come due to pollution of surface and ground water due to waste disposals. Closely connected is also the risk of flooding, which is a common indicator with SDG 13. In order to avoid flooding Plans for the Prevention of, Protection from and Reduction of the Effects of Flooding and the Flood Risk Management Plans, were defined for the 11 drainage basins of Romania. ([11]). Monitoring how these plans are working in the context of climate change is important. Moreover, the whole area of Romania, is within the Danube basin, which is shared between 16 countries. Therefore, Romania contributes to the development of the Danube River Basin Management Plan, with the main aim avoid environmental decline, enhance the quality of life of communities in the basin. This is done through smart technologies.

Strongly linked with SDG 6 is SDG 11, which aims to make cities and human settlements inclusive, safe, resilient and sustainable. Apart from transport and monitoring environment impact, sanitation and waste reduction, will greatly improve impact on water pollution. Waste water treatment plants are important in this context. In 2019 Romania reports that 50% of population is not connected to a waste water treatment plant of some sort, as opposed to Europe average which is 20%. This is mainly due to scattered rural areas, with small communities, far from a centre where wastewater could be treated.

Actual climate change trends show that Romania will face more and more droughts in the years to come, which will generate water shortages. In order to overcome these shortages, as well as possible outcomes due to flood hazards, Romania needs to look at SDG13 and meet its targets.

4. Conclusions and recommendations

All data presented herein are referring to water related SDGs, and to publicly available data up to 2019. Though indicators are improving throughout the years and the 2030 target seem easy to be achieved, data updates are not done regularly, hence the two years lagging behind in the analysis. Covid situation may be a reason for slow updates in receiving the information, however there is no transparency in why the data is not up to date. Recommendations would be that any data base with quantitative values on SDG indicators should have specific dates when updates are done (once a month, once a semester, etc). In case some information is not available at the scheduled date, a specific remark needs to be noted. This will ensure all NGOs and public institutions monitoring indicators to have access to most recent data.

Policies and water related law are a key element in achieving SDGs, as could be for example compulsory connection to existing sewer network. This should come not only as policy but also with proper economic support for vulnerable communities, i.e reduced connection fees for the first time connection.

Use of prediction models for flood and droughts, testing potential measures to reduce effects of climate change and minimize the risk of population for such hazards is important. This can be achieved by having training of specialists to use such models, disseminating results in a comprehensible manner for all citizens and involving citizens in observing the environment and providing data to specialists.

It is advisable that the newly created department of Sustainable Development makes transparent all their data on the SDG indicators, such that their effort to improve monitoring in Romania is easy accessible. Initiation of workshops and other programs to disseminate knowledge on the use, adoption, and benefits of different services, including e-services, which would increase the pace of achieving the proposed SDG targets are advisable.

References

[1] Lomazzi, M, Borisch, B, Laaser, U 2014 The Millennium Development Goals: experiences, achievements and what's next, *Global Health Action*, **7** (1), pp 1-9, https://www.tandfonline.com/doi/full/10.3402/gha.v7.23695

[2] United Nations. Sustainable Development Goals: 17 Goals to Transform Our World http://www.un.org/sustainabledevelopment/sustainable-development-goals/(UN, 2015).Swamee P K and Sharma A K 2008 *Design of Water Supply Pipe Networks* (Hoboken, New Jersey: John Wiley & Sons) p 45, (accessed on 10 April 2022)

[3] Xu, Z., Chau, S.N., Chen, X., Zhang, J., Li, Y, Dietz, T., Wang, J, Winkler,J.A., Fan, F, Huang, B., Li, S., Wu, S., Herzberger, A., Tang, Y, Hong, D., Li, Y., Liu, Y 2020 Assessing progress towards sustainable development over space and time. *Nature* **577**, 74–78 https://doi.org/10.1038/s41586-019-1846-3

[4] Firoiu, D.; Ionescu, G.H.; Băndoi, A.; Florea, N.M.; Jianu, E. 2019 Achieving Sustainable Development Goals (SDG): Implementation of the 2030 Agenda in Romania. *Sustainability* **11**, 2156. https://doi.org/10.3390/su11072156

[5] Romanian Government. National Sustainable Development Strategy Romania 2013–2020–2030. 2008.https://sustainabledevelopment.un.org (accessed on 10 March 2022)

[6] Sachs, J., Kroll, C., Lafortune, G., Fuller, G., Woelm, F. 2021. The Decade of Action for the Sustainable Development Goals: Sustainable Development Report 2021. Cambridge: Cambridge University Press. Available online: https://dashboards.sdgindex.org/rankings (accessed on 10 March 2022).

[7] European Commission. Standard Eurobarometer 89—Spring 2018 "Public Opinion in the European Union—First Results". 2018. Available online: http://ec.europa.eu/commfrontoffice/publicopinion/ (accessed on 10 March 2022).

[8] National Institute for Statistics Romania. 2019. Available online: http://statistici.insse.ro:8077/tempoonline/#/pages/tables/insse-table (accessed on 10 March 2022).

[9] Benedek, J., Ivan, K, Török, I., Temerdek, A., Holobâcă, I., H. 2021 Indicator-based assessment

of local and regional progress toward the Sustainable Development Goals (SDGs): An integrated approach from Romania, *Sustainable Development* **29** (5) pp 860-875, doi: 10.1002/sd.2180

[10] Diaz-Sarachaga, J.M., Jato-Espino, D, Castro-Fresno, D 2018 Is the Sustainable Development Goals (SDG) index an adequate framework to measure the progress of the 2030 Agenda? *Sustainable Development* **26** (6) pp 663-671, doi: 10.1002/sd.1735

[11] Romanian Waters Planul de management al riscului la inundatii, 2016 Available on line at https://rowater.ro/despre-noi/descrierea-activitatii/managementul-european-integrat-resurse-de-apa/planurile-de-management-ale-bazinelor-hidrografice/planuri-de-management-ale-bazinelor-spatiilor-hidrografice/planuri-de-management-ale-bazinelor-spatiilor-hidrografice-2016-2021/ (accessed on 10 April 2022).

IOP Conf. Series: Earth and Environmental Science 1136 (2023) 012014 doi:10.1088/1755-1315/1136/1/012014

Smart water chain: Immutable, distributed and decentralized water transaction ledgers

Talat Kemal Satilmisoglu[1] and Huzur Keskin[1]

[1] Department of Earth System Science; Graduate School of Natural and Applied Science; Middle East Technical University, Ankara, 06800, Türkiye

talat.satilmisoglu@metu.edu.tr

Abstract. Blockchain is a transactional data storage system where data can be stored reliably without the need for a central database or trusted authority. The data can be anything like financial transactions, supply chain processes or medical records. It is similar to a classical database but uses a decentralized ledger and allowing each participant in the network to have their own copy of the ledger and be able to see all transactions. Data stored in the distributed ledger can only be read or written, not deleted or updated unlike traditional central database. Reliable data is essential for the water industry for information about the current status of any water system, to build trust between stakeholders at all scales, and for effective forecasting and future scenarios by reducing uncertainty. The aim of this study is to examine the potential of using blockchain-based algorithms (smart contracts, chain codes, decentralised identifiers etc.) for the water industry on the edge of digital transformation, and to propose water-related data processing system architectures for different water quantity-quality based scenarios.

1. Introduction

In the first quarter of the 21st century, while many sectors such as energy, economy, health services, transportation, banking, water management, etc., took a step towards digitalization, blockchain technology has become a cornerstone because of its main features; immutability, decentralisation, enhanced security, distributed ledgers, consensus, and faster settlement. The basic results of these features are data reliability and trust built between stakeholders [1]. Smart contracts or chain codes are programs stored on a blockchain that run when prespecified conditions are met. They are used to automatize the execution of an agreement so that all stakeholders can be simultaneously certain of the result without the mediation process [2].

The qualitative and quantitative sustainability of water depends on the resilience and efficiency of water systems from source to user. In that case, reliable data is essential for system efficiency, real-time monitoring and forecasting in terms of sustainable water management. This essentialness requires reliable storage and processing of water-related data at all scales [3]. Although an autonomous water system works entirely on blockchain and smart contracts has not been implemented yet, there are theoretical or experimental studies such as pollution monitoring systems, intelligent water management systems [4-5].

The aim of this study is to propose four blockchain-based decentralised autonomous system architecture for measuring, monitoring and storing water-related data for different cases. Firstly, a blockchain enabled water market is introduced by focusing the water market and water economics concepts. Later on, three cases are introduced to underline how blockchain-based system architectures have huge potential to create trustless systems, that the stakeholders do not need to know or trust each other or any third party for the system operations, between water users at all scales. The purposed system architectures aim to ensure: i) enable the usual benefits of blockchain technology to be utilized in the digitalization of the water markets such as; trust, security, immutability, transparency, coordination. ii) reliability of urban water systems for consumers. iii) reliability between riparian countries discharge treated wastewater into the same inland sea or located in the same transboundary river basin. iv) reliability between stakeholders of circular water economy.

2. Material and Methods

2.1. Water Markets

Basically, a market is a platform for transactions in which something of value may be exchanged. Free market economy is defined as where products are exchanged by a willing buyer or seller. The water market concept extends this idea to water [6].

Although the free market economy leaves only the producers that innovate and creates products consumers want and at the same time, driven by profit, incentivise producers to increase the efficiency of production, in that case the main beneficial points are to avoid monopolies in water access and distribution, defining property rights clearly and well-defining the economic value of water to maximize welfare. By revealing the true value of water, water markets can provide effective and urgent solutions to water scarcity. It can also achieve ecosystem restoration and sustainable water management by pricing unit change in water quality or valuing unit discharge to water resources. Water markets are based on rights to the quantity or quality of tradable unit water that allows water to be transferred from one user to another, temporarily or permanently. A well-managed water market, together with the determination of property rights, seriously encourages water savings and brings all stakeholders to the table to find the water supply demand balance. Strategy to solve water-related problems by creating a free market economy is a common method by irrigators to assist agricultural water management [7].

However, water also has a socio-cultural value. In case of income inequality, water usage rights may be scattered in a way that harms social justice and socio-cultural values of water. It is very likely that people or stakeholders using water will make very bad financial decisions and act irrationally. The lack of knowledge about the market of the buyer or seller wants to trade in water may be the reason. Unconscious market management may ignore the need for environmental water and cause excessive use of water in the excessive market. According to studies on Australia - Murray Darling Basin, which is also one of the most popular water markets, some people who need water very little may stockpile their water rights for purely economic reasons, ignoring the environmental consequences, creating agricultural inefficiency by preventing the use of water by those who need it more. These and similar reasons can create socio-economic, environmental and ecological problems. Traditional water markets can promise weeks or even months of bureaucratic processes for the transaction to take place. [8-12].

The digital system architectures based on blockchain and smart contracts have the potential to maximize the advantages of water markets while minimizing the risks. First of all, it creates a peer-to-peer market for water trade, and automates trade agreements with smart contracts, considering environmental, socio-cultural and ecological rules. Moreover, all data required for an effective water management and market operations is immutable and reliably accessible to economic, environmental policy makers and water traders.

2.2. Trust in Water Systems

Sociologically, trust is the most basic social construct. Every time people interact with each other, a certain level of trust or distrust begins to form. People have a natural tendency to trust and judge the reliability of other individuals or groups of people and things, social and physical systems they interact. Natural processes, on the other hand, have gained trust of humankind because they occur as a result of precise and fixed physics rules and can be tested by experiment-observation. When the air temperature drops below a certain level, the water freezes or as a result of rainfall surface runoff occurs. Thousands of years of hydrological observations and infinite number of test results make this cause and effect relationship reliable. Water systems are hybrid systems that need to be managed together with the natural cycle of water and the sociocultural perception of people. No matter how a person perceives himself, water always flows according to certain physical laws. People, on the other hand, want to survive by manipulating this flow wherever and whenever the water starts to flow [13].

For a well-managed and resilient water system at all scales, there should not be a trust problem between stakeholders, policy makers and the financial and environmental system operators. For customers in urban water systems, it is the most natural right to be informed about all the storage, transmission or treatment processes of the water and the quantity and quality of the water they pay for. All supply chain stages of water from source to tap should be transparent in order to sustain trust in socio-hydrological system.

Similarly, transboundary river systems and inland seas involve stakeholders with unlimited demand who share a water system with a limited capacity. Stakeholders (countries) located in the same river basin may not only provide their agricultural, domestic and industrial water use from the same basin, but also discharge used or treated water to the same basin. If this complexity is combined with a trust issue, conflict between riparian countries is deepened. The transboundary river basin faces a severe drought or the inland sea is teste with a pollution load well above its absorptive capacity as a result of unconscious discharges. In both cases, ecological and hydrological restoration is expensive and very arduous [14-16]. Sustainable and integrated water resources management is a complex process that requires huge effort, but unless socio-hydrological trust in any water system is established, whole optimizations, uncertainty analyses or technical assessments cannot go beyond just theoretical studies.

A simple distributed digital architecture based on blockchain and smart contracts is able to ensure socio-hydrological trust straightforwardly and data security. Recorded data in an immutable manner can help resolve the uncertainty of hydrologically complex multi-stakeholder systems.

2.3. *Water Footprint and Virtual Water*

The water footprint is an indicator of freshwater use by taking the water used directly and indirectly together. The water footprint has three components: green, blue and grey. A green water footprint is water obtained directly from precipitation. It is a concept especially applicable to the agricultural sector and is related to the water stored in the root zone of the plant and lost by evapotranspiration. The blue water footprint is the water obtained and used from surface and ground water sources. It is valid for all water uses, especially irrigated agriculture, industrial products, domestic use. The grey water footprint is the amount of fresh water required to absorb pollutants to meet certain water quality standards. The grey water footprint considers point source pollution discharged directly into a freshwater source via a pipe or indirectly through runoff or seepage from soil, impermeable surfaces or other diffuse sources [17]. Virtual water is the total water consumed in the production process of an agricultural, industrial product or service. If a country or a region imports or exports a product, it can also virtually import/export water [18].

The water footprint draws attention to the embedded water consumed in the global supply chain of any product that is conceptually a local product. In other words, it underlines the fact that in a world where international trade is valid, water is never a local and domestic source, but rather a global natural resource. This means that one country's consumption habits may be directly related to another country's water scarcity. This point shows that economic policies and water policies should be integrated. Even a country that produces and consumes its own coffee beans is considered to have imported water if it uses imported products or services (energy, production technology, etc.) in any of the production stages. Water footprint can also be included in the policy in the form of water footprint caps and permits. A water footprint cap can be institutionalized at the basin or country scale; this can then translate into maximum water footprint permissions to be granted. On a smaller scale, water footprints can be created for processes and products that will enable companies to formulate water footprint reduction targets for their operations and supply chains. Water footprints indicate that consumption patterns based on raw materials from water-scarce regions may need to change. This encourages companies, investors, governments and consumers to explore how they can make their supply chains more sustainable in terms of water use. After all, relying on an unsustainable supply chain involves many risks and cannot be permanent [19].

The supply chain of products or services in terms of water footprint can be recorded and certified on the blockchain. Countries or companies have the chance to be aware of the investment they pay for or invest in virtual water, especially for water intensive products. Immutable and distributed data about virtual water and water footprint not only provide rationality in investment decisions in agriculture and industry, but also enable producers, consumers and policy makers to take water wise actions.

3. Results and Discussion

Digitally transformed water markets with a blockchain-based architecture offer new advantages by eliminating the disadvantages of traditional water markets. The Smart Water Chain (SWC) architecture provides a transparent and immutable market data availability. It prevents the asymmetric information problem that may occur in traditional water markets (Figure 1).

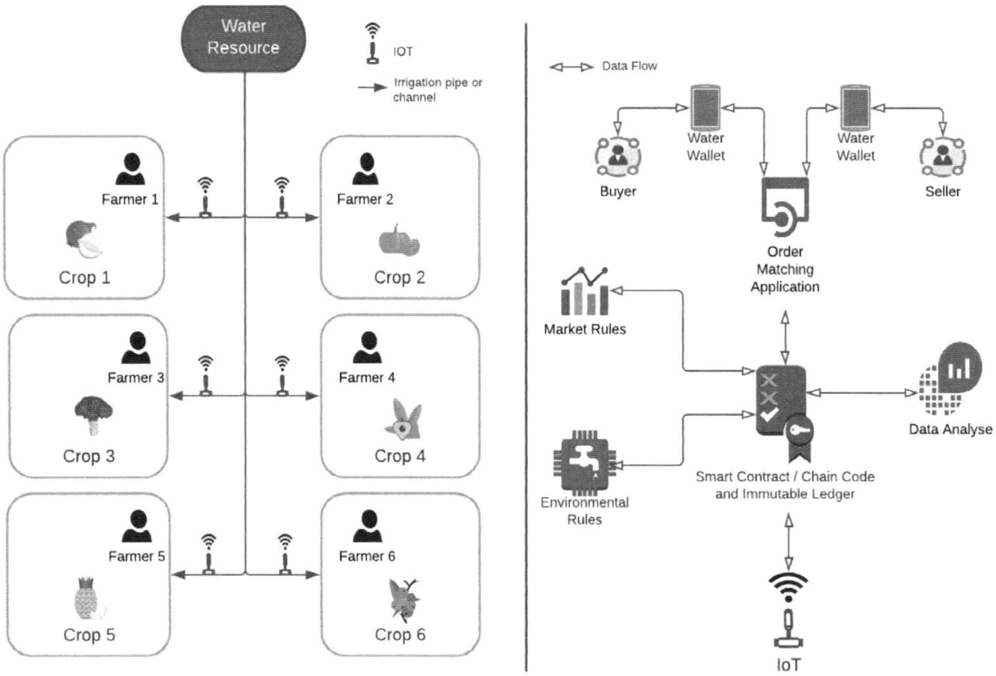

Figure 1. Illustration of an irrigation system and SWC architecture design for blockchain-based water market.

Market actors do not have to trust or distrust each other because every transaction is transparent. Farmers in the same basin system can grow crops that are subsidized at different rates by the agricultural policy maker. They can be sure that all other stakeholders comply with market or environmental rules that are strictly defined by smart contracts or chain codes, and they have maximum knowledge of the market. Farmers can easily access their water wallets, being aware of all market information and willing to trade as buyers or sellers. Water wallets can be thought of as any electronic system that allows a person to make water transactions. This can be any decentralized web or mobile application. SWC architecture conceptually follows these processes:

- Checks the compatibility of the information in the water wallets of the buyer and the seller with the IoT data and reports possible incompatibilities.

IOP Conf. Series: Earth and Environmental Science 1136 (2023) 012014 doi:10.1088/1755-1315/1136/1/012014

- Allows buyer and seller to view offers and match them
- Checks the compliance of the transaction agreed by the buyer and the seller with the environmental and market rules
- If all the requirements are met, the finalized transaction and all related data is recorded on immutable ledger.

The crucial point here is that market data must be transparent (not real identities). It is essential that all stakeholders have access to read the recorded data but not change or modify. The transparency and reliability of the data should not be even a matter of discussion, rather than the system's efforts to create trust between stakeholders. For this reason, the genesis block or the network to be selected is very important. Networks that have adopted the revolutionary differences brought by blockchain technology and that have proven themselves for several natural resource management applications should be selected. Distributed water transaction ledgers may be part of any public blockchain that provide opportunity to work compatible smart contracts and water-related data type. Distributed Ledger Technology (DLT) platforms such as; Ethereum, IOTA and Hyperledger are capable to process real-time water-related data [20].

Although a blockchain-based water trading infrastructure offers uniqueness, the trust and transparency provided by decentralized and distributed architectures also provides resilience for water systems at different scales. Figure 2 is the illustration of a simple urban water system.

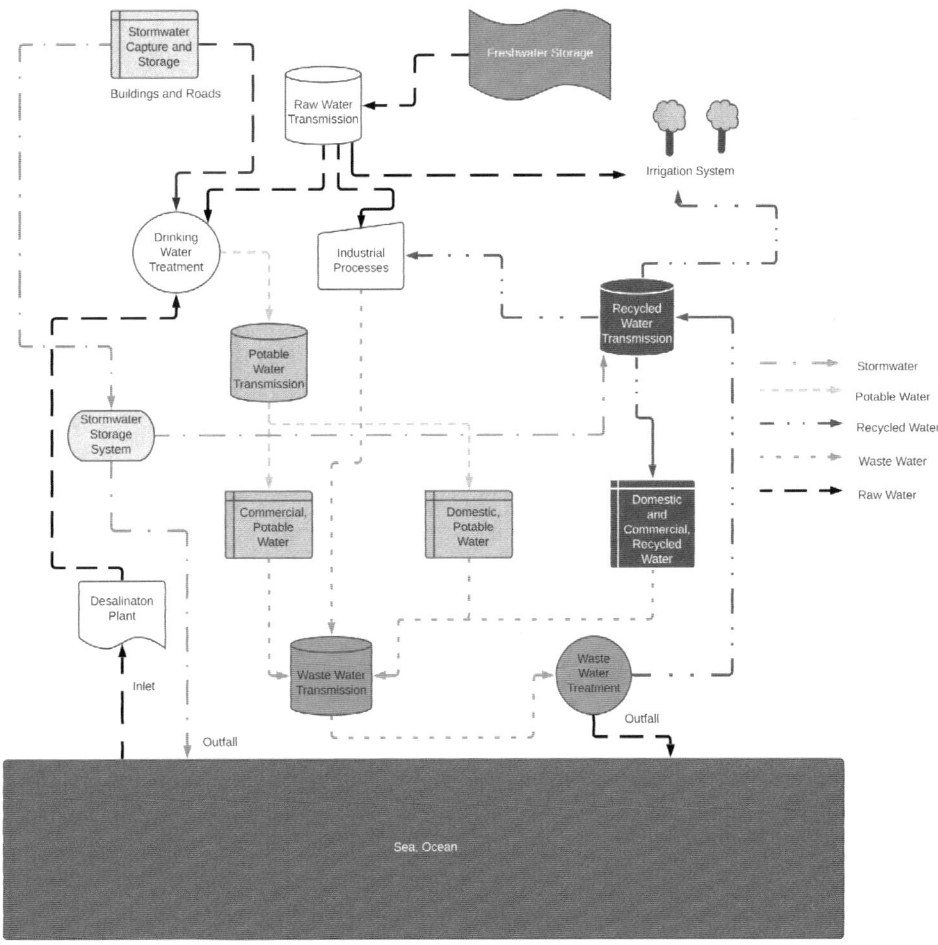

Figure 2. Illustration of simple urban water system.

Each unit where water is physically transmitted, stored or treated also functions as digital twin and measuring, monitoring and sending real-time data about the quantity and quality of water to the distributed water transaction ledger.

In this way, the consumers not only aware of the quantity and quality of consumed potable or industrial water, but also all the water-related data in the system becomes trusted and transparent. Through smart contracts, consumers pay water bills only if the consumed potable or industrial water meets quality standards and whole system meets environmental criteria. All inefficiencies in the system, measures to be taken, climatic trends, and leaks can be reliably analyzed and prevented thanks to immutable data.

Similarly, Figure 3.a and 3.b (may be considered as the Baltic Sea and the Nile Basin) indicated the basic transboundary river basin water cycle and inland sea waste water discharges. Resilience of transboundary river basins or inland sea systems highly depends on trust and coordination between stakeholders and reliable, immutable and transparent environmental data [21]. It is clear that, transboundary water system can be transformed digitally by the agency of blockchain-based algorithms to ensure essentialness for more resilient international water systems.

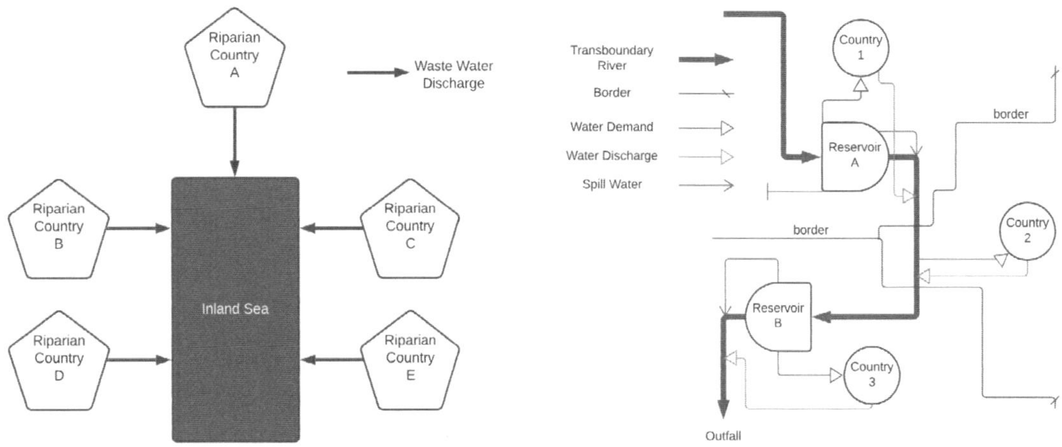

(a) (b)

Figure 3. Illustration of a (a) wastewater-based polluted inland sea system (b) transboundary river system.

Proposed system architecture is demonstrated in Figure 4.a. All riparian countries can be easily aware of the fact that the transboundary water system or inland they are part of has a layer of trust in terms of water quantity and quality, and that welfare maximization is aimed. Although it is not essential for resilience of every water system, a free market economy may also be applied to these systems in which any water demand and water right, or unit permission to pollution load by discharge can be opened to trade. All regulative red and yellow lines can be drawn with the smart contracts (Figure 4.b).

It is widely recognized that reducing the water footprint should be part of the sustainable water management strategy like carbon footprint. In production systems where process water is one of the main inputs, especially in the textile and beverage sector, the money spent for water footprint and virtual water is vital. The management of the water used in production and the amount and quality of the wastewater generated are correlated with financial sustainability [22]. Likewise, the conscious consumer wants to know the water used and the wastewater generated throughout the entire supply chain of the product. Collecting, combining and presenting water-related data, which is valid for the entire supply chain of a product reaching the consumer, with traditional database architecture, requires a very complex process. With a blockchain-based solution, data in all processes can be collected and made into open access for all consumers in a reliable and transparent way. This is also a way of challenge for brands and companies and can play an encouraging role to prove their environmental sensitiveness. Similarly, the resulting integrated and transparent data helps farmers reduce their water footprint. In addition, water labeling, certification and benchmarking plans can be collaborated on and reports generated on supply chain water footprints and associated impacts of their products.

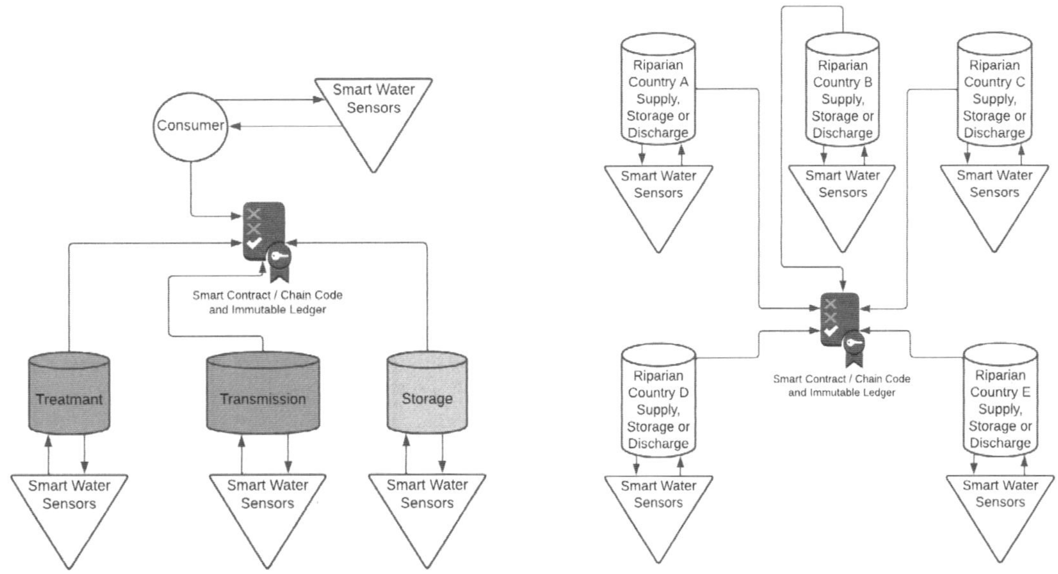

(a) (b)

Figure 4. SWC architecture design (a) urban water system (b) transboundary river system / inland sea system.

Countries with high virtual water trade volumes may need more transparent background information on the water impact of the products they import. Thanks to a reliable, real-time and immutable dataset, import/export policies can be reviewed and more rational steps can be taken. Governments can adjust their development cooperation policies according to their desire to promote the consumption and trade of sustainable products. They can analyze their current import/export trends and consider water footprint and economic importance in new policies. Special precautions can be taken for products and services that are considered to be of critical importance in this regard, and the regulation can be modified. All these steps can be taken theoretically with traditional methods. However the bureaucratic and digital collaboration required takes excessive effort [23-24]. A blockchain-based architecture can provide this instantly. As an example, a simple water footprint-oriented supply chain of cup of coffee and SWC architecture can be seen in Figure 5.

An important point to be noted here is that the IoT infrastructure, which plays a key role in data transmission, should be strong. The relevant electronic and digital calibration should be done with precision. It should be known that although it has the potential to provide the highest level of system efficiency, DLT is a set of autonomous data processing and storage architectures. The SWC architecture mentioned here cannot measure data. It uniquely combines the data processing, recording and management system with the physical system. It's a great hydroinformatics tool, but not a flow meter and water quality sensor. Flow meters and water quality sensors are components of the proposed architecture.

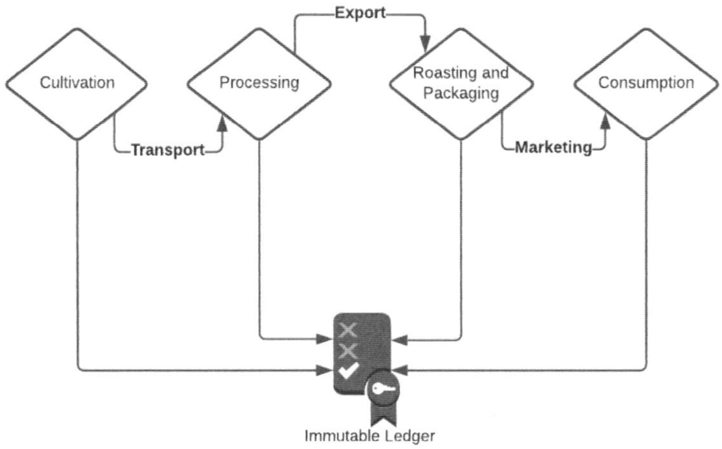

Figure 5. SWC architecture design for water footprint-oriented supply chain.

On the other hand, smart contracts can only be as smart as the professionals who design and code them. The environmental and market rules to be included in the smart contract should be carefully prepared and programmed. Due to the general programming principle and gas fees, it should have maximum application potential with as few lines of code as possible. Once a blockchain-based water market begins to be traded in real time, detected structural errors are nearly impossible to fix.

Another issue discussed during the study is whether architectures should be consisting of public or private blockchains. A private blockchain architecture that is accessible only to a limited number of stakeholders within the basin, city or transboundary water system has been evaluated. However, it is realized that digital water system architectures designed with private blockchain will eventually converge to monopolization, centralization and opaqueness. This goes against the revolutionary potential of blockchain technology and the sociocultural value of water.

4. Conclusion

Water is a commodity that is traded, that is paid to use. Water is the part of socio-cultural life of humankind. Water is also a natural resource, but unlike the usual, it is a two-dimensional natural resource. For efficient use, it should not only be in sufficient quantity, but also be in sufficient quality. Moreover, water does not act as a natural resource that you can leave alone and manage when the time comes. It should be managed continuously in a dynamic, sustainable, rational way. This situation necessitates being aware of the real value of water as a natural resource and having a transparent and reliable management perspective due to its socio-cultural value.

In the 21st century, while every field such as industries, cities, transportation, communication methods, entertainment and sports etc. are rapidly digitalizing, it is impossible for the water industry to act conservatively on this transformation or to limit digital transformation to decision support systems or hydrological models. As a result, the sine qua non of a next generation water management system, directly overlapping with the advantages of blockchain-based distributed algorithms; trust, reliability, immutability, transparency, integration and coordination.

SWC, which combines, smart contracts and decentralized water transaction ledgers, ensures that the inputs and outputs of all components of the urban water system meet environmental standards. The consumers trust the quantity and quality of the water they pay for. Similarly, incentives or penalties in cases where environmental standards are met or not can be regulated by smart contracts. SWC can resolve the issue of trust between all stakeholders who are dependent on the same water resource. SWC can certificate water footprint of any product or service peerlessly. An immutable database of real-time water data provides golden opportunity for data-driven modelling and forecasting.

References

[1] Varfolomeev AA, Alfarhani LH and Oleiwi ZC 2021 Secure-reliable smart contract applications based blockchain technology in smart cities environment *14th Int. Symp. Intelligent Systems* https://doi.org/10.1016/j.procs.2021.04.188

[2] Buterin V 2015 *A Next-Generation Smart Contract and Decentralized Application Platform* https://ethereum.org/whitepaper

[3] Angelakis AN, Vuorinen HS, Nikolaidis C, Juuti PS, Katko TS, Juuti RP, Zhang J and Samonis G 2021 Water quality and life expectancy: Parallel courses in time *Water* **13** 752 https://doi.org/10.3390/ w13060752

[4] Niya SR, Jha SS, Bocek T and Stiller B 2018 *Design and Implementation of an Automated and Decentralized Pollution Monitoring System with Blockchains, Smart Contracts, and LoRaWAN* Zurich Open Repository and Archive, University of Zurich

[5] Dogo EM, Salami AF, Nwulu NI and Aigbavboa CO 2019 Blockchain and Internet of Things-based technologies for intelligent water management system *Artificial Intelligence in IoT. Transactions on Computational Science and Computational Intelligence* https://doi.org/10.1007/978-3-030-04110-6_7

[6] Garrick D, Lane-Miller C and McCoy AL 2011 Institutional innovations to govern environmental water in the western United States: lessons for Australia's Murray-Darling Basin *Economic Papers* **30**(2) 167–184

[7] Wheeler S, Loch A, Zuo A and Bjornlund H 2014 Reviewing the adoption and impact of water markets in the Murray-Darling Basin, Australia *Journal of Hydrology* **518**(A) 28–41

[8] Musgrave W 2008 Historical development of water resources in Australia. Irrigation policy in the Murray-Darling Basin *Water Policy in Australia* ed Crase L (New York: Routledge) chapter 3 pp 28–43

[9] National Water Commission (NWC) 2009 *Australian water markets report* 2008–09, Canberra

[10] National Water Commission (NWC) 2012 *Impacts of water trading in the southern MDB between 2006/7 and 2010/11*, Report, Canberra

[11] Quiggin J 2001 Environmental economics and the Murray-Darling river system *Aust. J. Agric. Resour. Econ.* **45**(1) 67–94

[12] The Economist 2003 *Survey: Liquid assets* 368, 13 Available at http://www.economist.com/node/1906938

[13] Luhmann N 1995 *Social Systems* (California: Stanford University Press) p 127

[14] Stockholm International Water Institute 2013 *Independent Evaluation of the Nile Basin Trust Fund, Final Report* Avaliable at https://documents1.worldbank.org/curated/en/461521468202171071/pdf/892770WP0P09230on0of0NBTF0June02013.pdf

[15] Susskind L 2012 *Water Diplomacy: Creating Value and Building Trust in Transboundary Water Negotiations – Israel and Jordan, From War to Water Sharing* Harvard Law School Daily Blog

[16] Grupper MA, Sorice MG, Stern MJ and Schreiber ME 2021 Evaluating determinants of social trust in water utilities: implications for building resilient water systems *Ecology and Society* **26**(4) 41 https://doi.org/10.5751/ES-12833-260441

[17] Chapagain AK and Hoekstra AY 2004 *Water Footprints of Nations, Volume 1: Main Reports, Value of Water Research Report Series No: 16, UNESCO-IHE* Delft, The Netherland p 80

[18] Schubert H 2011 The virtual water and the water footprint concepts *Acatech Materialien* **14**

[19] Hoekstra AY and Hung PQ 2002 *Virtual Water Trade A Quantification of Virtual Water Flows Between Nations in Relation to Crop Trade* IHE Delft

[20] Pincheira M, Vecchio M, Giaffreda R and Kanhere SS 2002 Cost-effective IoT devices as trustworthy data sources for a blockchain-based water management system in precision agriculture *Computers and Electronics in Agriculture* **180** 105889 https://doi.org/10.1016/j.compag.2020.105889

[21] United Nations Economic Comission for Europe International Network of Basin Organizations

2015 *Water and Climate Change Adaptation in Transboundary Basins: Lessons Learned and Good Practices* p 86

[22] Dong H, Zhang L, Geng Y, Li P and Yu C 2021 New insights from grey water footprint assessment: An industrial park level *Journal of Cleaner Production* **285** doi: 10.1016/j.jclepro.2020.124915

[23] Yu Y, Hubacek K, Feng K, and Guan D 2010 Assessing regional and global water footprints for the UK *Ecological Economics* **69**(5) 1140–1147 doi: 10.1016/j.ecolecon.2009.12.008

[24] Aivazidou E, Tsolakis N, Vlachos D and Iakovou E 2018 A water footprint management framework for supply chains under green market behaviour *Journal of Cleaner Production* **197** 592–606 doi: 10.1016/j.jclepro.2018.06.171

Hydroinformatics tools used to quantify the impact of urban expansion on water bodies, in the context of the intensification of extreme phenomena (Bacău urbanization case study)

Nicolae Marcoie[1], Daniel Toma[1], Valentin Boboc[1], Tomi Alexăndrel Hrăniciuc[1] and Cătălin Dumitrel Balan[2,3]

[1] Faculty of Hydrotechnical Engineering, Geodesy and Environmental Engineering; "Gheorghe Asachi" Technical University of Iasi, 700050, Iasi, Romania
[2] Faculty of Chemical Engineering and Environmental Protection; "Gheorghe Asachi" Technical University of Iasi, 700050 Iasi, Romania
[3] Author to whom any correspondence should be addressed

cbalan@ch.tuiasi.ro

Abstract. The present study presents the results of the use of hydraulic and hydrological modeling in order to maintain the quantitative and qualitative parameters of the groundwater Capture Front Gherăești, Bacau County. With the help of specialized hydro-computer programs such as: Visual MODFLOW for groundwater dynamics, and Mike 11 for hydraulic modeling, the impact of urban development projects on surface and groundwater bodies can be simulated and anticipated. The results obtained are used in the Project to assess the impact of the project on water bodies (SEICA). SEICA contributes significantly to maintaining the quality of water bodies by making reasoned decisions that ensure the sustainability of urban development projects.

1. Introduction and study objectives

Population growth and its concentration around large cities lead to a horizontal and vertical expansion of the urban area with a significant impact on groundwater and surface water bodies. Whether we are talking about new industrial parks, residential neighborhoods, leisure areas or shopping malls, these are a challenge for integrated water resources management. By integrated management in this context, we understand both the provision of water and the maintenance of its quality parameters.

This paper uses a recent method of impact assessment SEICA (2019) which through the impact quantification algorithm significantly reduces the subjectivity of the team of evaluators.

The objective of the study is to maintain the quality of water bodies (Gherăești Capture Front) and to protect the urban development project (Recreation Area Project) against floods and landslides due to extreme weather phenomena (Tudose et al 2011). The global warming felt in recent years by the intensification of extreme weather phenomena such as strong winds, large amounts of precipitation in short periods of time, and floods have also been reported in the Bacau study area.

Solving these challenges (maintaining water quality and protection against extreme weather events) must be done from the design phase of projects taking into account impact assessment studies (Neamtu et al 2021, SEICA 2019). The analyzed project is located in the northeast of Bacau on the left bank of the river Bistrita, as can be seen in Figure 1.

Figure 1. Study area Project Bacau Leisure Center.

The investment made by the beneficiary contributes to solving a very current problem, namely the expansion of the urban area (M.D.R.A.P., 2017) ensuring a high degree of safety (against floods) and comfort (through the created recreational area), in a city full development such as Bacău. The protection dam and the recreational lake lead to an increase in land use. The realization of the project aims in the first phase (fencing with protection dam and lake) the establishment of an area protected from floods, thus creating the possibility of using the land for the recreational base.

By making the investment, the following benefits will be created in the studied area:

- protection, development of the land;
- beautification of the current location;
- road arrangement.
- in the lateral part towards the left bank of the Bistrița river, the protection dam will be built (protection against floods and any other dangerous hydro-meteorological phenomena), L = 2257.38m, H variable.

Simultaneously with the works for the proposed protection dam, works will be carried out for the establishment of the recreational lake with an area S = 3.18 ha (SEICA 2019).

Currently, the perimeter under study is a former orchard of fruit trees, currently partially deforested, located in the city of Bacău, Șerbănești neighborhood. The reason for the deforestation was the fact that the normal operating life of the orchard had expired, as it was aged and unproductive.

The purpose of the impact assessment is to identify and quantify the pressures and impacts of the protection dam, the lake and the operation of the recreational base on the groundwater and surface water body in accordance with EU Directive 2014/52 of the European Parliament and of the Council of 16.04.2014, amending the EU Directive 2011/92 on the elaboration of the effects of certain public and private projects on the environment (Marcoie et al 2019, Comanita et al 2018, SEICA 2019). In order to remove the land from the effect of the floods in order to place some objectives of socio-economic interest, the following were used:

- Detailed topographic studies to determine the flood curves and the location of future investment objectives;
- Hydrological studies prepared by ABA Siret Bacău through which it was possible to determine the elevations of the flood curve and the elevations of the protective dam.

IOP Conf. Series: Earth and Environmental Science 1136 (2023) 012015 doi:10.1088/1755-1315/1136/1/012015

2. Methodology

The SEICA (Impact Assessment Study on the Water Body) project (SEICA 2019) was used as working tools. which, in addition to the classical procedure imposed, used in the study the results obtained by the adequate assessment study software Mike 11 hydraulic modeling program for surface water (Hraniciuc 2016), the Visual MODFLOW program developed by U.S. Geological Survey Ground-Water software for the underground body and The adequate assessment study.

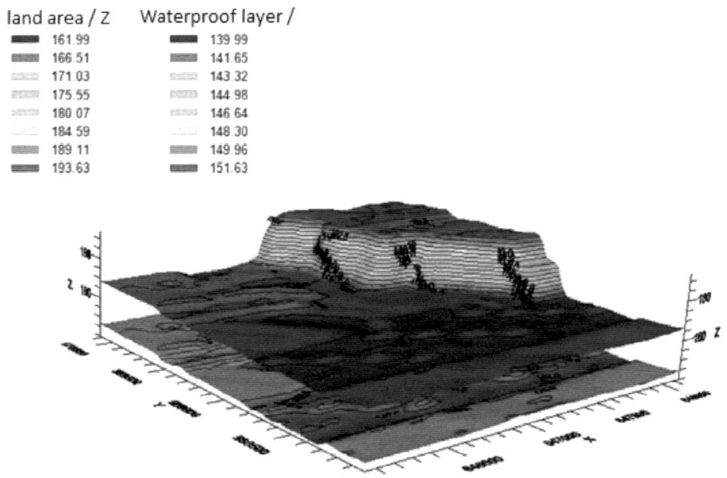

Figure 2. 3D representation of the studied surface in the Visual MODFLOW program software.

An image of the results obtained after running the software Mike 11 hydraulic modeling program for surface waters can be seen in Figures 2 and 3. Figure 2 is an example of a graphical representation of the terrain under study using the Visual MODFLOW software program.

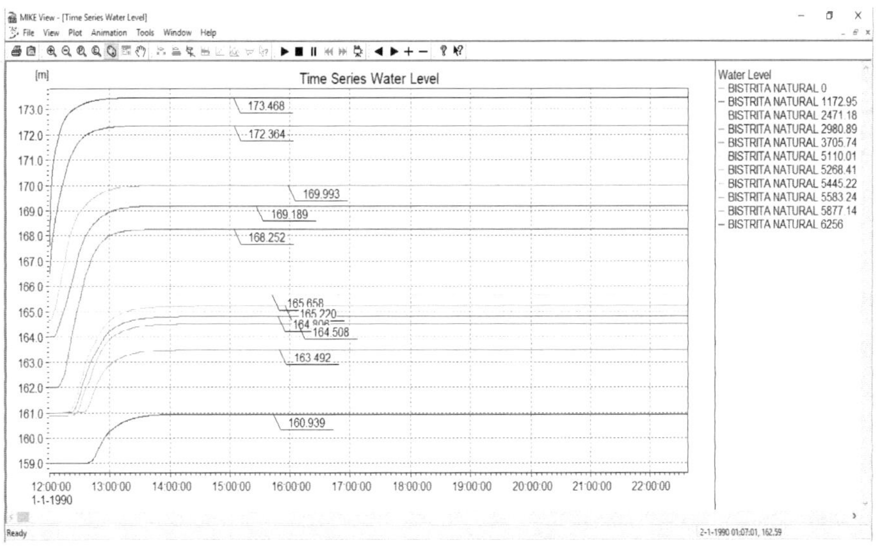

Figure 3. The values of the maximum water levels in the sections studied from P8 to P17 at a flow rate of Q = 1900 m^3/s.

The software used simulates the parameters of the surface and groundwater body in different stress situations (in the presence of the project, at maximum and minimum flows etc).

3. Case study impact assessment Bacau Recreation Area Project

The Ministry of Waters and Forests issues on July 4, 2019 the SEICA Framework Content - Water Impact Assessment Study, which is a recent procedure for assessing the impact on surface or groundwater bodies. With publication in the Official Gazette no. 615bis of July 25, 2019 becomes indispensable in assessing the impact of projects on water bodies (Marcoie et al 2019, Robu et al 2008).

The assessment of the impact of the project on the groundwater body uses a series of multidisciplinary information that ultimately leads to the quantification and determination of the level of impact, including the cumulative impact, its duration and whether it leads to the deterioration of the water body (Zechariah 2003).

The elaboration of the impact study took into account the following:

- Potential impact on the groundwater body – Impact on the hydrostatic level in the boreholes on the right bank of the Bistrița River (Gherăești catchment front), Influences on the groundwater flow by creating the recreational lake, which takes over some of the groundwater.
- Potential impact on the body of surface water – Narrowing of the drainage section of the Bistrița riverbed, through the construction of the protection dam, The potential impact induced by the reduction of the drainage section of the Bistrița river was analyzed, namely the possibility of flooding the areas of interest (Hraniciuc et al 2016).
- Potential impact on biodiversity – By conducting the Adequate Evaluation Study on the Bacau Recreation Base project.

In this study it was followed if the structural and functional relations that maintain the integrity of the site ROSPA0063 Buhuși-Bacău-Berești accumulation lakes are affected (Gache 2005, Feneru 2000).

4. Results and conclusions

The main potential ways of affecting the quality of the groundwater from which the water supply of the Gheraiesti Capture Front is made, caused by the construction of the recreational lake and the perimeter dam on the left bank of Bistrita, were analyzed. The results of the evaluation are summarized in Table 1.

Table 1. Evaluation of potential ways to affect groundwater quality.

No.	Potential pressure exerted	Potential impact and quantification	Preventive measures	Residual impact	Reduction measures
1	On the groundwater flow by making the agreement lake (Gheraes Capture Front)	Above the hydrostatic level in the wells drilled on the right bank. Minor negative	Application of the operating and maintenance regulations	Not	Not necessary
2	By narrowing the drainage section of the Bistrița riverbed, by building the protection dam	Exceeding the right dike canopy at a probability of exceeding 0.5% Moderately negative	Raising the right dike with minimum guard heights	Not	Not necessary
3	Impaired site ROSPA0063 Buhuși-Bacău-Berești accumulation	The project in the analyzed perimeter will not affect the structural and functional relations that maintain the integrity	The execution of construction works for the creation of a lake and the perimeter dam,	Not	Not necessary

lakes	of the site ROSPA0063 accumulation lakes Buhuşi-Bacău-Bereşti Moderately negative	should not take place during the nesting and rearing periods of the chicks, i.e. between May 15 and August 15	

1. Even if the impact is manifested, it is in all cases minor, local, temporary and reversible, so that no significant depreciation is foreseen on the quantity and quality of the groundwater from which the water system is supplied.

2. In conclusion, the elevation level of the proposed dam on the left bank is not exceeded by the maximum water level for the considered flow ($Q = 1900\ m^3/s$). Regarding the elevation of the existing dam on the right bank, for the flow taken into account, in some sections, it is exceeded by the maximum water level, and in the others it is at the limit in danger, very close to the canopy. Practically with or without the construction of the project, the right bank presents the same dangers regarding the overcoming of the canopy at a probability of 0.5% (Hraniciuc et al 2019, Marcoie et al 2019).

3. As a result of the presented aspects, we consider that the project from the perimeter subject to analysis will not affect the structural and functional relations that maintain the integrity of the site ROSPA0063 Buhuşi-Bacău-Bereşti accumulation lakes. In order to avoid any impact on bird species, we recommend that the construction of recreational lake and perimeter dam construction works not take place during the nesting and rearing periods of the chicks of the two species that could become vulnerable due to possible presence on the site, i.e. in May 15 - August 15 (Doniţa et al 2015, Neamtu et al 2021, Comanita et al 2017).

Based on the data obtained by simulating the existence of the project in the proposed area with the help of Visual MODFLOW and Mike 11 software for hydraulic modeling, respecting the steps required by the SEICA procedure (SEICA 2019) impact assessment we have reached the following results:

- The project does not present the risk of effects, respectively the risk of deterioration of the water body condition, at the level of quality and quantity element.
- The assessment of the cumulative impact of the project with the projects on water or in connection with the authorized / under authorization / approved / under approval / planned waters on which the project will belocated on the Gherăeşti Capture Front is insignificant.
- The project adds value, safety and aesthetics to the site through the landscaping works and facilities created, contributing to the sustainable modernization of the Bacau leisure area.

References

[1] Brînduşa Robu, Laura Bulgariu, M Macoveanu 2008 The impact in environment by presence in soil of chrome ions, *University of Agricultural Sciences and Veterinary Medicine of Iaşi, Agronomie Journal of Scientific Papers - Agronomy Series*, **51**, 185-190

[2] Carmen Gache 2005 The monitoring of the human impact on the avifauna's evolution in the important birds' areas from the Prut River basin, *Scientific Annals of the University „Al. I. Cuza "Iaşi, s. Animal Biology*, 195-203

[3] Comanita E.-D., Simion I.M., Cozma P., Hlihor R.M., Campean T., Gavrilescu M. 2017 Application of cost-benefit analysis for an eco-product manufactured from production waste, *Proceedings of Int. Chemical, Biological and Environmental Engineering*, **101**, 91-98

[4] Comanita E.-D., Cozma P., Simion I.M., Rosca M., Gavrilescu M. 2018 Evaluation of eco-efficiency by multicriteria decision analysis. Case study of eco-innovated and eco-designed products from recyclable waste, *Environmental Engineering and Management Journal*, **17**, 1791-1804

[5] Doniţa, N., Popescu, A., Pauca-Comanescu, M., Mihailescu, S., Biris, I.A. 2005 Habitats of

Romania, *Silvica Technical Publishing House. Bucharest*

[6] Feneru Fl., 2000 - Population's dynamics of some aquatic bird species in the lake Lilieci natural reserve - 1996-2000, Scientific Studies and Research vol.5, pp. 245-248, *Alma Mater Publishing House, University of Bacău*

[7] Hraniciuc T.A., Nicolae Marcoie, Cătălin Dumitrel Balan 2019 Determination of the flood defence embankment elevation crown, using advanced hydrodynamic modelling, *Int. Conf. On Environmental Engineering and Management, ICEEM10,* Iasi, Romania

[8] Hraniciuc Tomi Alexandrel, Petru Cercel 2016 Coupled Hydrological and Hydraulic Modeling of Dangerous Meteorological Phenomena, *16th International Multidisciplinary Scientific Geoconference SGEM*, Albena, Bulgary, **I**, 211-218

[9] Marcoie N., Toaca D., Hraniciuc T.A., Balan C.D., Andriescu P., Boboc V., Toma D., Pavel V.L. 2019 The impact of the embakment and the recreational lake on capturing front with drilling wells, *Int. Conf. on Environmental Engineering and Management, ICEEM10,* Iasi, Romania

[10] Ministry of Waters and Forests – Framework Content of July 4, 2019 of SEICA Water Impact Assessment Study, Official Gazette no. 615bis of July 25

[11] Ministry of Regional Development and Public Administration, National Housing Strategy, 2017

[12] Water Law, Law no. 107 of September 25, 1996, Official Gazette no. 244 of October 8, 1996

[13] Neamtu R, B Sluser, O Plavan, C Teodosiu 2021 Environmental monitoring and impact assessment of Prut River cross-border pollution, *Environmental Monitoring and Assessment* **193**(6), 1-16

[14] Robu B., Bulgariu L., Bulgariu D., Macoveanu M. 2008 Quantification of impact and risk induced in surface water by heavy metals: case study Bahlui river Iasi, *Environmental Engineering and Management Journal*, **7**, 263-267

[15] ROSPA0063 Buhusi-Bacau-Beresti accumulation lakes, Management Plan of the Special Avifauna Protection Area

[16] Zaharia, C. 2003 Legislation for environmental protection, Alexandru Ioan Cuza University Publishing House, Iaşi

[17] Tudose O.G., Cazacu M.M., Timofte A., Balin I. 2011 ESYRO Lidar system developments for troposphere monitoring of aerosols and clouds properties, *Conf. of Remote Sensing of Clouds and the Atmosphere*, vol. **XVI**, 8177

[18] Directive 2011/92 / EU of the European Parliament and of the Council of 13 December 2011 on the assessment of the effects of certain public and private projects on the environment

[19] Visual MODFLOW Pro, U.S. Geological Survey Modular Ground-water model 2007

[20] Danish Hydraulic Institute 2014 MIKE 21 Flow Model FM_Hydrodynamic Module_User Guide

[21] Danish Hydraulic Institute 2014 MIKE 21 & Mike 3 Flow Model FM_Hydrodynamic and Transport Module_Scientific Documentation.

Spatiotemporal correlation and multivariate analysis between vegetation health, terrestrial water storage and precipitation

Sithmini Wijekoon[1], Izni Zahidi[1], Badronnisa Yusuf[2] and Helmi Zulhaidi Mohd Shafri[2]

[1] Civil Engineering Discipline, Monash University Malaysia, Bandar Sunway, 46150, Malaysia
[2] Department of Civil Engineering, University Putra Malaysia, Seri Kembangan, 43400, Malaysia

izni.mohdzahidi@monash.edu

Abstract. Vegetation health is an essential indicator in the global hydrologic cycle as it is interrelated with the hydrological components. In tropical areas where vegetation dominates, analysing their correlation at a regional scale helps forecast the hydrologic cycle and understand vegetation's response to climate change. However, the interactions between vegetation, terrestrial water storage and climate factors such as precipitation remain poorly understood in this region. Therefore, using Landsat and Gravity Recovery and Climate Experiment (GRACE) remote sensing and observed precipitation data, this study analysed the spatiotemporal correlation of Normalized Difference Vegetation Index (NDVI), Terrestrial Water Storage Anomaly (TWSA) and precipitation for the whole Peninsular Malaysia. The correlation coefficient (R) was used to assess the temporal variability of NDVI with TWSA and precipitation separately. Furthermore, a Geographically Weighted Regression (GWR) model was constructed to evaluate the spatial non-stationarity and heterogeneity relationships between the multi variables. The findings revealed complex interactions between the variables, where the strength of the correlations varied depending on the localised region and study period. The results suggest that downscaled GRACE-derived TWSA data would be helpful for detailed vegetation modelling and water resources management.

1. Introduction

As an essential component of the terrestrial ecosystem, vegetation plays a crucial role in the global hydrologic cycle, carbon cycle and climate change [1-3]. For more than half of the global ecosystems, vegetation growth is limited by water availability [4]. Vegetation can significantly affect terrestrial water storage by regulating the water balance at regional, basin and global scales [5-7]. Consequently, hydrological components such as evapotranspiration (ET), precipitation (P), runoff and soil moisture are both directly and indirectly affected by changes to vegetation [8-9]. Thus, it is vital to understand the interactions between vegetation and terrestrial water to forecast the hydrologic cycle, carbon cycle and its response to global climate change [4, 10].

With the population growth and agriculture driven economic growth, the forest land cover of Malaysia has been greatly impacted by development activities over the past few decades [11]. Despite rainforest vegetation currently covering up to two-fifths of Peninsular Malaysia, the country has lost an average of 0.43% of forest cover between 1990 to 2010 [12]. Subsequent regional studies also provide evidence of such land degradation. For example, a study based on Pahang state by [13] showed degradation of 8.9% of the forest cover in the region, while [14] reported that dense vegetation has decreased from 78.57% to 65.44%, while area without vegetation has increased from 3.55% to 7.25% in the city of Seremban, Negeri Sembilan. This further implies that the tropical rainforest vegetation cover has been subjected to continuous degradation and defragmentation over the years [13]. It is of great importance to study vegetation changes in Malaysia to understand better the consequences and the response of the environment to such changes.

Many studies have investigated the response of vegetation to water conditions using conventional hydrological indicators such as precipitation and soil moisture [5, 15, 7]. Donhue et al [5] stated that in water limiting environments, factors that affect water availability, such as precipitation, govern vegetation's response. These studies have concluded that precipitation is one of the most influencing factors for vegetation growth [16]. Thus, it is widely used as an indicator to assess the impact of water conditions on vegetation [7]. However, the relationship between precipitation and vegetation is complex as precipitation only provides indirect information on water availability [17]. Vegetation rather depends on the "effective precipitation", which is determined by factors such as precipitation intensity, soil properties, vegetation type, density, and terrain [17, 18].

Conversely, soil moisture is the water stored in the unsaturated zone of soil above the groundwater level, and studies have shown soil moisture to be useful in analysing vegetation dynamics and ecological processes [19, 20]. However, soil moisture is widely affected by changes in groundwater and deep percolation, further complicating the relationship between vegetation and precipitation [17]. Moreover, the limitations in acquiring accurate soil moisture monitoring data [15, 20] that reflects root-zone moisture conditions has led to gaps in the understanding of vegetation dynamics and its response to hydrological components [17].

In recent years, remotely sensed data such as Gravity Recovery and Climate Experiment (GRACE) have been widely used to analyse spatiotemporal variations in terrestrial water storage changes [21, 22]. Terrestrial Water Storage (TWS) is a summation of all types of water storage, including surface waters such as rivers, lakes and wetlands, groundwater aquifers, soil moisture, snow, ice and water stored in vegetation [23]. Changes in soil moisture contribute the most to terrestrial water storage variations, followed by groundwater and intermediate zone storage [24]. However, TWS variability due to changes in soil moisture is often seen on a seasonal basis [23], while groundwater storage changes are often reflected on interannual or decadal time scales [24, 25]. Terrestrial Water Storage Anomaly (TWSA) derived from GRACE is a good measure of climate change [21] and has been used to investigate and forecast extreme hydrological events such as floods and droughts [26, 27].

Normalised Difference Vegetation Index (NDVI) is one of the most frequently used indicators of environmental resources [28]. Vegetation indices, such as NDVI, derived from space-borne data, provide information on terrestrial vegetation cover, vigour, and growth dynamics [29]. Many studies have utilised NDVI to assess its spatiotemporal variations [30, 31] and vegetation feedback to climate change [32, 33, 10]. For example, using mainland Australia, [17] found interannual and seasonal trends between TWSA and surface greenness measured using NDVI. Velicogna et al [34], through a study based in Eurasia, showed that TWSA alters the response of vegetation to temperature changes, in addition to directly impacting vegetation growth. Xie et al [7] provide a global view on this aspect and derived significant positive relationships between NDVI and TWSA for arid, semi-arid and sub-humid regions, with significant negative relationships for humid regions, high northern latitudes and tropical rainforests.

Subsequently, regression methods such as Ordinary Least Square (OLS) and Geographically Weighted Regression (GWR) have been utilised before to analyse NDVI change in response to climate factors [30, 10]. However, studies have shown that the relationship between NDVI and climate factors display significant spatial non-stationarity and heterogeneity [35, 36, 31]. Non-stationarity and spatial autocorrelation of model residuals makes the statistical analysis less reliable. Thus, GWR, a local form of regression analysis that models spatially varying relationships, has effectively addressed these issues. Studies have proven GWR to perform and explain the complex relationships of NDVI with environmental and climate factors better [35, 10] whilst showing reduced spatial autocorrelation of model residuals [36].

In summary, space-borne data products such as GRACE-derived TWSA overcome the limitations in traditional methods and thus, are widely used to analyse the effects of climate change regionally and globally [37, 38, 10]. Existing studies have focused on the interactions between NDVI and hydrological factors such as terrestrial water storage and precipitation which have provided valuable observations in analysing vegetation response [33, 7]. However, the interactions between vegetation, terrestrial water conditions and climate factors in tropical regions such as Malaysia remain poorly understood. This study addresses this gap by analysing the spatiotemporal correlations between vegetation health in terms of NDVI, TWSA by GRACE and in-situ monthly precipitation for Peninsular Malaysia based on the hypothesis that the three variables are closely interrelated. The two main objectives of the study are: (1) to analyse spatiotemporal variations of NDVI in relation to TWSA and precipitation and (2) to assess spatial heterogeneity between NDVI, TWSA and precipitation using GWR.

2. Materials and method

2.1. Study area

Peninsular Malaysia is located in the tropical latitudes of 1°N and 7°N and longitudes of 100°E and 104°E. Peninsular Malaysia experiences an equatorial climate where the climate within the region is vastly influenced by its topography of central mountainous cores with coastal lowlands and alternating monsoonal winds. The rainfall climatology consists of two monsoons, Northeast Monsoon (NEM) from November to February and Southwest Monsoon (SWM) from May to September. However, the onset and retreat dates of the monsoonal winds are not adequately defined. The temperature is uniformly high throughout the year with the daily temperature varying between 25.1° C in January and 26.8°C in May [39]. The average temperature in most lowlands is 27°. On Eastern coasts, it varies between 23 to 32°C, while the highlands tend to be colder and wetter. In the present study, analysis on temporal variations of NDVI with respect to TWSA and precipitation and regression analysis using GWR were carried out based on 500 random points generated across Peninsular Malaysia to better understand the impact of climate change in Peninsular Malaysia (figure 1).

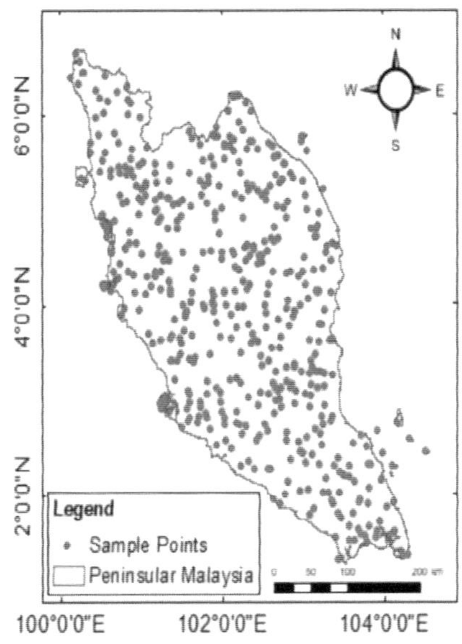

Figure 1. Study area and sample points.

2.2. Data

2.2.1. NDVI data. In this study, Landsat 8 imagery was used to derive NDVI, which is freely available via Earth Explorer by the United States Geological Survey at https://earthexplorer.usgs.gov/. Landsat 8 collects images of the Earth with a 16-days cycle and the Operational Land Imager (OLI) sensor onboard Landsat 8 provides multispectral images (Bands 1 – 9) with a spatial resolution of 30 m. Level 1 data products, corrected radiometrically and geometrically, were downloaded for Peninsular Malaysia for 2016 (January, March, June, and July) and 2017 (April and June). Images with minimal cloud cover were considered such that the effect of cloud on the calculated NDVI is negligible considering the large study area. The data gaps in GRACE data, from which TWSA is derived, were also considered when selecting the study duration [41] defines NDVI as the ratio between the maximum radiation absorption in the red spectral band by vegetation and the maximum reflection of radiation in the near-infrared band by vegetation. In Landsat 8 multispectral image, Band 4 represents red radiation (R), and Band 5 represents the near-infrared radiation (NIR). Dense and healthy vegetation produce the largest NDVI values while clouds, barren soil and surface water produce the lowest NDVI values [28]. NDVI function in ArcGIS, which computes a single band dataset representing NDVI from the given multispectral image, is used in this study.

2.2.2. TWSA from GRACE. The TWSA data for Peninsular Malaysia were obtained from GRACE Level 3 gridded monthly data products produced by the University of Texas, Centre for Space Research at Austin (UTCSR). These monthly data products have a spatial resolution of 1° and were obtained from the RL06 time-variable gravity field model, which is available at https://podaac-tools.jpl.nasa.gov/drive/files/allData/grace/L3/land_mass/RL06/v03/CSR.

GRACE maps the spatiotemporal variations of the Earth's gravity field using two identical satellites that obit 220 km apart at a low altitude of 450-500 km. Monthly analysis of the measurements provides temporal variations of the gravity field caused by redistribution of masses [23]. Once the static masses such as the mass of the solid Earth is removed from the monthly solutions, the time-varying component is due to atmospheric and oceanic circulations and redistribution of terrestrial water due to the hydrological cycle [23]. During data processing of GRACE, tidal and non-tidal gravity effects are removed such that non-atmospheric and non-oceanic mass variations representing the water mass storage in the continental region are obtained [42]. In Level 3 processing, several successive filters are applied (de-striping filter, geo-centre correction, Glacial Isostatic Adjustments, spatial smoothening, and spatial leakage correction) so that the provided monthly gridded solutions are TWSA relative to 2004 - 2009 time-mean baseline. TWSA measurements are represented in centimetres of Equivalent Water Thickness, the vertical extent of concentrated mass changes near the Earth's surface.

2.2.3. Precipitation data. Monthly precipitation data from 50 rain gauge stations across Peninsular Malaysia were acquired from the Department of Irrigation and Drainage Malaysia for 2016 - 2017. Ly et al [43] reported that for monthly rainfall, geostatistical interpolation methods such as kriging are preferable to deterministic methods such as Inverse Distance Weighting (IDW). Therefore, spatial kriging interpolation was used to create a continuous raster based on the acquired gauge data.

2.3. Correlation analysis

In this study, the overall interactions of NDVI with TWSA and monthly precipitation are first assessed separately for Peninsular Malaysia to determine whether a significant association exists between the variables. For this purpose, 1° spatial resolution of coarse GRACE data was considered. Thus, the maximum NDVI for each cell was derived from the corresponding random sample points. The maximum NDVI in each cell represents the climax productivity of vegetation in each cell. Similarly, the monthly precipitation was averaged to a spatial resolution of 1° using the corresponding random sample points. Correlation coefficient (R) was adopted to assess the degree of correlation between different data sets. A positive correlation indicates an increasing trend between the two data sets, whereas a coefficient closer to 0 indicates weak to no correlation.

2.4. Geographically Weighted Regression (GWR)

GWR analysis was carried out with NDVI as the dependent variable, while TWSA and monthly precipitation were independent variables in the GWR model. GWR accounts for spatially varying relationships between the three variables by generating local regression coefficients varying with spatial location. GWR is an extension of the traditional regression equation which allows estimation of local parameters rather than global parameters and can be defined as follows [44]:

$$y_i = a_0(u_i, v_i) + \sum_k a_k(u_i, v_i) x_{ik} + \varepsilon_i \tag{1}$$

where y_i is the dependent variable, x_{ik} is the k-th independent variable, (u_i, v_i) are the coordinates of the i-th point in space, ε_i represents the error term at point i and $a_k(u_i,v_i)$ is the continuous function at point i, accounting for spatial non-stationarity at different locations.

3. Results and discussion

3.1. Interactions of NDVI with TWSA and precipitation

Figures 2(a) and 2(b) illustrate the scatter plots of maximum NDVI with TWSA and monthly precipitation, respectively. It can be observed that a significant association could not be derived for either of the two data sets for Peninsular Malaysia. The R^2 for the relationships were as low as 0.0051 and 0.0009. The weak correlation is probably due to the spatial heterogeneity of the relationships, where the variables are most likely to be closely correlated in some regions more than the others.

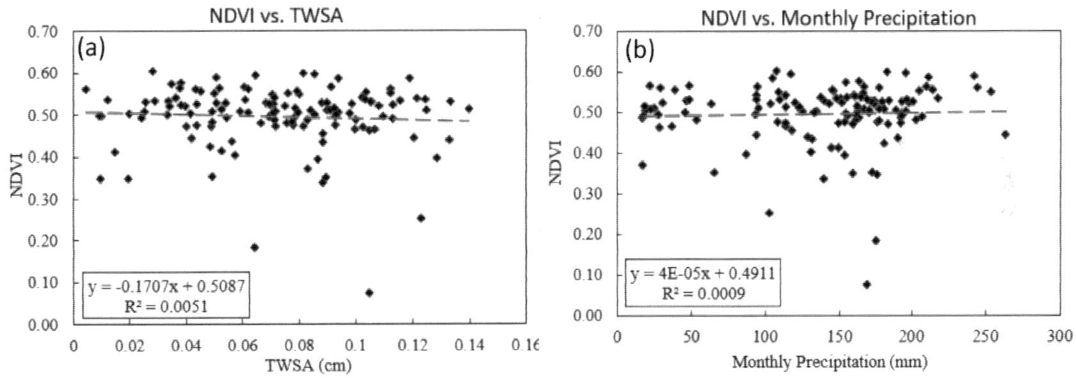

Figure 2. Scatter plots of maximum NDVI (a) with TWSA (b) with monthly precipitation.

Table 1 shows the correlation coefficients for maximum NDVI with TWSA and monthly precipitation corresponding to each GRACE cell for the study duration. The R values at 1° spatial resolution capture the spatially varying correlation between the variables well, with both positive and negative correlations. The correlation between maximum NDVI and precipitation shows varying strengths, as shown in table 1. The values of R vary between -0.77 and 0.90. Strong positive correlations are observed in some northern regions (G2 and G11), while strong negative correlations are observed in some southern regions (G19 and G20) of Peninsular Malaysia. However, most cells exhibit either negative or positive weak correlations (G3, G12 and G13).

The distributions of the correlation coefficients in a spatial resolution of 1° of Peninsular Malaysia with the corresponding comparisons between maximum NDVI with TWSA and monthly precipitation are shown in figure 3 and figure 4, respectively.

Table 1. Correlation coefficients.

Cell	Correlation Coefficient (R)		Cell	Correlation Coefficient (R)	
	NDVI and TWSA	NDVI and Monthly precipitation		NDVI and TWSA	NDVI and Monthly precipitation
G1	0.54	-0.16	G12	-0.38	-0.01
G2	-0.17	0.67	G13	-0.78	-0.08
G3	-0.06	0.02	G14	-0.63	-0.53
G4	-0.47	0.27	G15	-0.25	-0.18
G6	-0.76	0.52	G16	0.16	0.24
G7	-0.26	0.39	G17	-0.29	0.35
G8	-0.26	0.66	G18	-0.90	-0.31
G9	-0.59	0.34	G19	0.23	-0.77
G10	-0.77	0.22	G20	0.04	-0.70
G11	-0.28	0.90	G21	0.79	-0.47

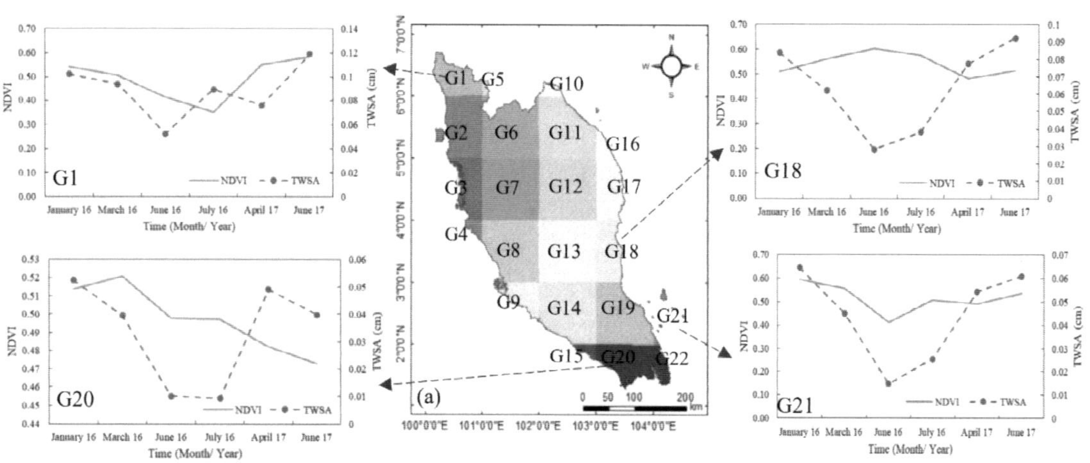

Figure 3. Distribution of correlation coefficients between maximum NDVI and TWSA in a spatial resolution of 1° with the corresponding comparison for the grid cells.

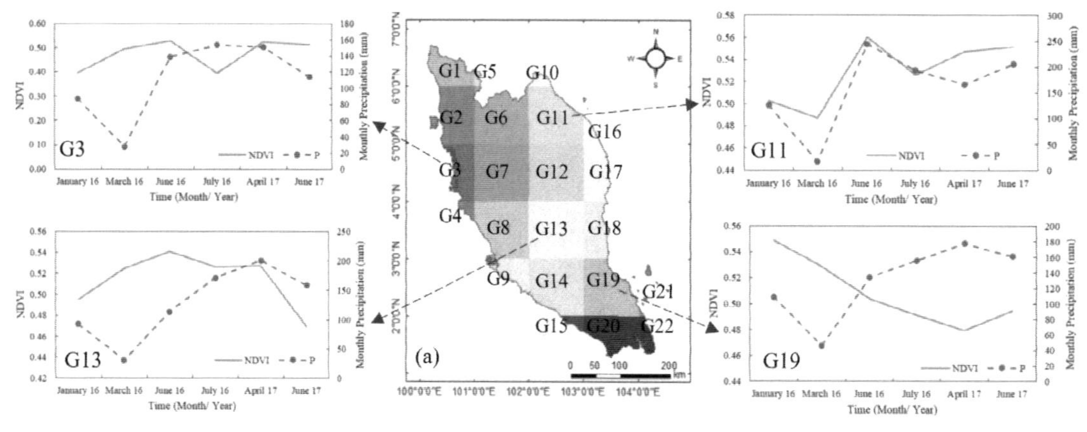

Figure 4. Distribution of correlation coefficients between maximum NDVI and monthly precipitation in a spatial resolution of 1° with the corresponding comparison for the grid cells.

When considering maximum NDVI and TWSA, negative correlations are primarily observed for inland regions, while some coastal regions show positive correlations. This observation is roughly consistent with that suggested by [7] as the inlands of Peninsular Malaysia mainly consist of forest vegetation cover. Peninsular Malaysia does not consist of significant arid and sub-arid regions, and it experiences a relatively high humid climate, which further explains the negative correlation observed in most regions. Positive correlations of varying strengths are observed in some coastal regions, with G21 having the highest positive correlation coefficient. The R values vary between -0.90 and 0.79, with some cells showing very weak correlations between maximum NDVI and TWSA (G3, G16 and G20).

The diverse strengths in correlations suggest that the response of vegetation to precipitation is most likely to be influenced by many factors such as vegetation types, soil properties, temperature, and time of the year. The overall time-lag effect of precipitation on vegetation greenness [33, 7] could also affect the correlations' strengths [45] further reported that increased precipitation results in reduced sunshine hours, possibly affecting the vegetation health as photosynthesis is not facilitated [37]. Thus, abundant precipitation might not necessarily result in higher NDVI and, therefore, a stronger correlation as one would expect.

3.2. Temporal heterogeneity of NDVI in relation to TWSA and precipitation

In early 2016, Malaysia experienced a drought caused by the lingering El Nino phenomenon [46]. Although this mainly affected Sabah and Sarawak of East Malaysia, the temperature within Peninsular Malaysia was increased, affecting the vegetable, paddy and rubber cultivation [47]. Effects of such phenomenon are observed with the gradual decrease in TWSA and monthly precipitation in almost all regions from January to March of 2016. Similarly, NDVI shows a reasonably consistent relationship, as a decrease in NDVI is detected from January to March of 2016 for specific regions.

When considering the trends in TWSA, an overall increase in TWSA is observed for most regions from June to July of 2016 and April to June of 2017. This increase is likely caused by the SWM, which starts in May and continues until late September. It is also interesting to note that TWSA is relatively higher in 2017 than in 2016, probably indicating its recovery from the drought experienced earlier in 2016. A consistent relationship is also exhibited by precipitation, as a sudden increase can be seen in June of 2016, following the El Nino phenomenon from January to March (figure 5). However, such an increase cannot be seen in June 2017 as precipitation decreases from April to June 2017 (figure 6).

Moreover, [48] reported that NDVI increases during the rainy season and decreases towards the end of the rainy season. Additionally, [37] reported that insignificant negative correlations are probable if the increased precipitation during the rainy season is stored in soils and is available for vegetation growth even after the rainy season. Thus, NDVI values are most likely to be affected by monsoonal and inter-monsoonal seasons. However, this study did not consider continuous monthly data due to cloud cover limitations. As a result, a consistent response of vegetation that can be used to derive a functional relationship was not obtained.

Furthermore, NDVI is highly sensitive to background brightness, including soil background effects [29]. Therefore, NDVI is likely to produce inaccurate results in the early stages of crop growth when the green leaf area is relatively small [49]. Models that are established to improve the sensitivity of NDVI, such as Enhanced Vegetation Index (EVI), Soil Adjusted Vegetation Index (SAVI) and Atmospherically Resistant Vegetation Index (ARVI), are expected to provide a better representation of vegetation health. Therefore, more obvious relationships reflective of seasonal and interannual trends are expected in vegetation response to TWSA and precipitation with continuous monthly data and improved vegetation index models.

3.3. Spatial heterogeneity of NDVI in relation to TWSA and precipitation

As shown in figure 5, spatial heterogeneity between NDVI, TWSA and monthly precipitation is evident. Variations in local R^2 showed that the model performed moderately well for some areas and poorly for others. The overall R^2 and the adjusted R^2 are low for the study period, ranging from 0.03 to 0.18 and 0.02 to 0.16, respectively. The highest overall R^2 and adjusted R^2 are obtained for June 2016 (figure 5). Figure 5(a) shows that central and north-western regions mainly indicate negative coefficients for TWSA, while north-eastern and eastern coasts indicate positive coefficients. As previously stated, the central region of Peninsular Malaysia is mainly covered with forest vegetation, and the obtained results are consistent with the findings by [7]. However, such apparent relationships are only evident for June 2016 and not other months.

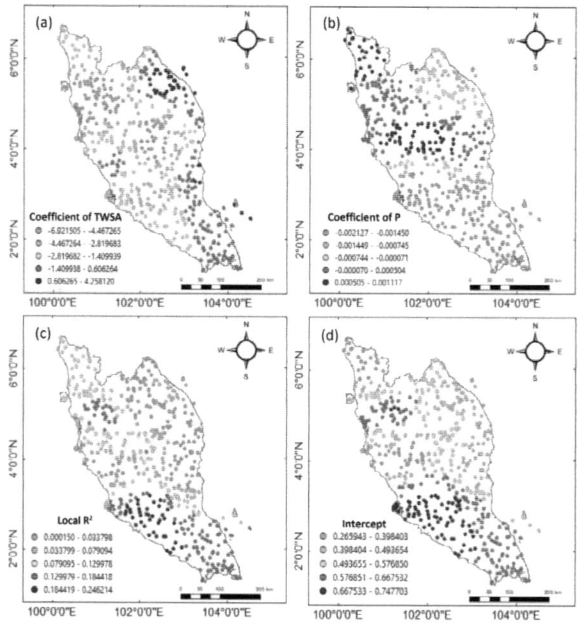

Figure 5. GWR model outputs in June 2016 for (a) Coefficient of TWSA, (b) Coefficient of P, (c) Local R^2 and (d) Intercept.

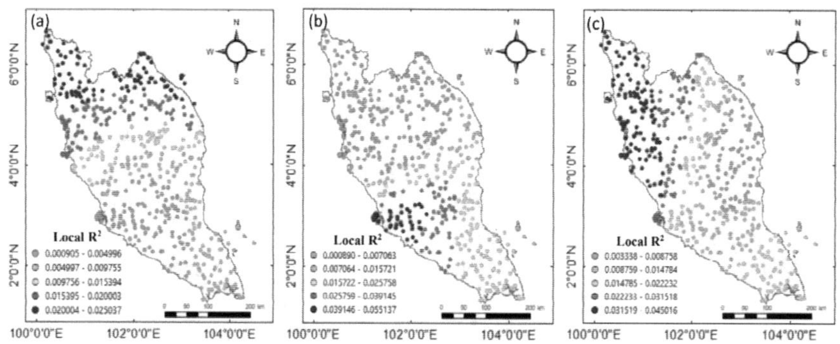

Figure 6. Local R^2 in (a) January 2016, (b) April 2017 and (c) June 2017.

The model performance varies across the regions and with the months considered during the study period (figure 6), suggesting that model performance depends on spatiotemporal variations. Generally, the model performs relatively better for northern and north-western regions and poor for southern and south-eastern regions, except for June 2016 and April 2017 (figure 5(d) and figure 6). For June 2016 and April 2017, an opposite relationship was seen.

A consistent relationship was not obtained between the variables by considering coefficients of the independent variables. The relationship between NDVI with TWSA and NDVI with monthly precipitation varies across the region, with some regions being negatively correlated. Interestingly, the relationship obtained for TWSA and precipitation with NDVI is opposite to each other for specific regions; i.e., in areas where TWSA is positively correlated, monthly precipitation is negatively correlated (figures 5(a) and 5(b)). This phenomenon is consistent with [37], indicating that increased precipitation during the rainy season is stored in soils and is available for vegetation even after, resulting in increased vegetation growth following the rainy season even if the precipitation is reduced.

In addition, limitations were encountered in deriving functional spatial relationships between NDVI and TWSA due to the coarse spatial resolution of GRACE-derived TWSA data. Thus, statistical and dynamic downscaling approaches that improve the spatial resolution of GRACE derived products is recommended [50]. Downscaled GRACE-derived TWSA is also expected to improve the correlations as the temporal analyses of NDVI with TWSA and precipitation can be carried out comprehensively at a higher resolution.

4. Conclusions

In this study, correlation coefficients and geographically weighted regression were used to analyse the temporal variation of NDVI in relation to TWSA and monthly precipitation, and the spatial heterogeneity among the multivariable for Peninsular Malaysia. It was evident that vegetation response to terrestrial water storage changes and precipitation is complex, and apparent spatial heterogeneity existed. Due to intermittent data periods used, seasonal and interannual vegetation variations and the time-lag effect of precipitation on vegetation could not be detected. Thus, it is recommended to use continuous monthly data to capture more accurate temporal trends of NDVI in relation to TWSA and precipitation. Improved sensitivity models such as EVI, SAVI and ARVI are also recommended to measure vegetation health. Additionally, the coarse resolution of GRACE data resulted in limitations in deriving functional correlations and more accurate trends between the variables, despite the high resolution of NDVI and precipitation data. Thus, downscaling of GRACE data is recommended to derive more useful relationships. Regional downscaling of GRACE data would allow understanding the response of vegetation health to climate factors at a higher resolution and identification of regional flood and drought-prone areas, which can then be incorporated into local water resource planning.

Acknowledgments

The authors would like to acknowledge the Ministry of Higher Education Malaysia (MOHE) for the financial support of this research under the Fundamental Research Grant Scheme (FRGS) with project code: FRGS/1/2019/TK01/MUSM/03/2.

References

[1] Seddon AW, Macias-Fauria M, Long PR, Benz D and Willis KJ 2016 *Nature* **531** 229-232 https://doi.org/10.1038/nature16986
[2] Lemordant L, Gentine P, Swann AS, Cook BI and Scheff J 2018 *P. Natl. Acad. Sci. USA* **115**(16) 4093-8 https://doi.org/10.1073/pnas.1720712115
[3] Richardson AD, Hufkens K, Milliman T, Aubrecht DM, Furze ME, Seyednasrollah B, Krassovski MB, Latimer JM, Nettles WR, Heiderman RR and Warren JM 2018 *Nature* **560** 368-371 https://doi.org/10.1038/s41586-018-0399-1
[4] Heimann M and Reichstein M 2008 *Nature* **451** 289-292 https://doi.org/10.1038/nature06591
[5] Donhue RJ, McVicar TR and Roderick ML 2009 *Glob. Change Biol.* **15**(4) 1025-39 https://doi.org/10.1111/j.1365-2486.2008.01746.x
[6] Koirala S, Jung M, Reichstein M, de Graaf IEM, Camps-Valls G, Ichii K, Papale D, Ráduly B, Schwalm CR, Tramontana G and Carvalhais N 2017 *Geophys. Res. Lett.* **44**(9) 4134-42 https://doi.org/10.1002/2017GL072885
[7] Xie X He B Guo L Miao C Zhang Y 2019 *Remote Sens. Environ.* **231** 111259

https://doi.org/10.1016/j.rse.2019.111259
[8] Ellison D N, Futter M and Bishop K 2012 *Glob. Change Biol.* **18**(3) 806-820 https://doi.org/10.1111/j.1365-2486.2011.02589.x
[9] Zeng Z, Piao S, Li LZ, Wang T, Ciais P, Lian X, Yang Y, Mao J, Shi X and Myneni RB 2018 *J. Climate* **31**(7) 2633-50 https://doi.org/10.1175/JCLI-D-17-0236.1
[10] Wang Y and Gao W 2020 *IOP Conf. Ser.: Earth Environ. Sci.* **568** 012012 https://doi.org/10.1088/1755-1315/568/1/012012
[11] Al-Amin AQ and Filho WL 2011 *Int. J. Global Warm.* **3**(4) 390-402 doi: 10.1504/IJGW.2011.044402
[12] FAO 2010 Global Forest Resources Assessment 2010 – Main Report *FAO Forestry Paper* **163** https://www.fao.org/3/i1757e/i1757e.pdf
[13] Othman MA, Ash'aari ZH, Aris AZ and Ramli MF 2018 *IOP Conf. Ser.: Earth Environ. Sci.* **169** 012047 https://doi.org/10.1088/1755-1315/169/1/012047
[14] Aburas MM, Abdullah SH, Ramli MF and Ash'aari ZH 2015 *Procedia Environ. Sci.* **30** 238-243 https://doi.org/10.1016/j.proenv.2015.10.043
[15] Chen Y, Yang K, Qin J, Zhao L, Tang W and Han M 2013 *J. Geophys. Res.-Atmos.* **118**(10) 4466-75 https://doi.org/10.1002/jgrd.50301
[16] Liu L, Zhang Y, Wu S, Li S and Qin D 2018 *Sci. Rep.* **8** 2962 https://doi.org/10.1038/s41598-018-21339-4
[17] Yang Y, Long D, Guan H, Scanlon BR, Simmons CT, Jiang L and Xu X 2014 *J. Geophys. Res.-Biogeo.* **119**(12) 2245-60 https://doi.org/10.1002/2014JG002670
[18] Ding M, Zhang Y, Liu L, Zhang W, Wang Z and Bai W 2007 *J. Geogr. Sci.* **17**(3) 259-268 https://doi.org/10.1007/s11442-007-0259-7
[19] Chen T, de Jeu RAM, Liu YY, van der Werf GR and Dolman AJ 2014 *Remote Sens. Environ.* **140** 330-338 https://doi.org/10.1016/j.rse.2013.08.022
[20] Wang H, He B, Zhang Y, Huang L, Chen Z and Liu J 2018 *Sci. Total Environ.* **612** 347-357 https://doi.org/10.1016/j.scitotenv.2017.08.212
[21] Lenk O 2013 *J. Geodyn.* **67** 106-110 https://doi.org/10.1016/j.jog.2012.04.010
[22] Yang P, Xia J, Zhan C, Qiao Y and Wang Y 2017 *Sci. Total Environ.* **595** 218-228 https://doi.org/10.1016/j.scitotenv.2017.03.268
[23] Girotto M and Rodell M 2019 Chapter Two - Terrestrial water storage *Extreme Hydroclimatic Events and Multivariate Hazards in a Changing Environment: A Remote Sensing Approach* ed V Maggioni and C Massari (Elsevier) pp 41-64 https://doi.org/10.1016/B978-0-12-814899-0.00002-X
[24] Rodell M and Famiglietti JS 2001 *Water Resour. Res.* **37**(5) 1327-39 https://doi.org/10.1029/2000WR900306
[25] Li B, Rodell M, Zaitchik BF, Reichle RH, Koster RD and van Dame TM 2012 *J. Hydrol.* **446-447** 103-115 https://doi.org/10.1016/j.jhydrol.2012.04.035
[26] Reager JT and Famiglietti JS 2009 *Geophys. Res. Lett.* **36**(23) https://doi.org/10.1029/2009GL040826
[27] Long D, Shen Y, Sun A, Hong Y, Longuevergne L, Yang Y, Li B and Chen L 2014 *Remote Sens. Environ.* **155** 145-160 https://doi.org/10.1016/j.rse.2014.08.006
[28] Kogan FN 1995 *Adv. Space Res.* **15**(11) 91-100 https://doi.org/10.1016/0273-1177(95)00079-T
[29] Xue J and Su B 2017 *J. Sensors* **2017** 1353691 https://doi.org/10.1155/2017/1353691
[30] Guo L, Zuo L, Gao J, Jiang Y, Zhang Y, Ma S, Zou Y and Wu S 2020 *Remote Sens.-Basel* **12**(8) 1332 https://doi.org/10.3390/rs12081332
[31] Yang L, Guan Q, Lin J, Tian J, Tan Z and Li H 2021 *Remote Sens. Environ.* **254** 112247 https://doi.org/10.1016/j.rse.2020.112247
[32] Piao S, Mohammat A, Fang J, Cai Q and Feng J 2006 *Global Environ. Change* **16**(4) 340-8 https://doi.org/10.1016/j.gloenvcha.2006.02.002
[33] Wu D, Zhao X, Liang S, Zhou T, Huang K, Tang B and Zhao W 2015 *Glob. Change Biol.* **21**(9)

3520-31 https://doi.org/10.1111/gcb.12945
[34] Velicogna I, Kimball JS and Kim Y 2015 *Environ. Res. Lett.* **10**(12) 124024 https://doi.org/10.1088/1748-9326/10/12/124024
[35] Foody GM 2003 *Remote Sens. Environ.* **88**(3) 283-293 https://doi.org/10.1016/j.rse.2003.08.004
[36] Kupfer JA and Farris CA 2007 *Landscape Ecol.* **22**(6) 837-852 https://doi.org/10.1007/s10980-006-9058-2
[37] Pang G, Wang X and Yang M 2017 *Quatern. Int.* **444** 87-96 https://doi.org/10.1016/j.quaint.2016.08.038
[38] Ghebrezgabher MG, Yang T, Yang X and Sereke TE 2020 *Egypt. J. Remote Sens. Space Sci.* **23**(3) 249-61 https://doi.org/10.1016/j.ejrs.2020.08.003
[39] Pour SH, Wahab AKA and Shahid S 2020 *Atmos. Res.* **233** 104720 https://doi.org/10.1016/j.atmosres.2019.104720
[40] Bee OJ, Lockard CA, Ahmad ZB and Leinbach TR "Malaysia" *Encyclopedia Britannica*, https://www.britannica.com/place/Malaysia (accessed 2 December 2022)
[41] Tucker CJ 1979 *Remote Sens. Environ.* **8**(2) 127-150 https://doi.org/10.1016/0034-4257(79)90013-0
[42] Jiang D, Wang J, Huang Y, Zhou K, Ding X and Fu J 2014 *Adv. Meteorol.* **2014** 725131 https://doi.org/10.1155/2014/725131
[43] Ly S, Charles C and Degré A 2013 *Biotechnol. Agron. Soc. Environ.* **17**(2) 392-406 https://popups.uliege.be/1780-4507/index.php?id=10003
[44] Fotheringham AS, Charlton ME and Brunsdon C 1998 *Environ. Plann.* A **30**(11) 1905-27 doi: 10.1068/a301905
[45] You Q, Kang S, Flügel WA, Sanchez-Lorenzo A, Yan Y, Huang J and Martin-Vide J 2010 *Theor. Appl. Climatol.* **101**(3) 445-457 https://doi.org/10.1007/s00704-009-0231-9
[46] ECHO Emergency Response Coordination Centre: ECHO Flash, 'Malaysia – Drought', European Civil Protection and Humanitarian Aid Operations https://erccportal.jrc.ec.europa.eu/ECHO-Flash#/echo-flash-items/9628 (accessed 2 December 2022)
[47] Staff S 2016 Drought beyond India: Malaysia faces a massive water crisis as South East Asia swelters *Meltdown* **2016** https://scroll.in/article/806812/drought-beyond-india-malaysia-faces-a-massive-water-crisis-as-south-east-asia-swelters
[48] du Plessis WP 1999 *J. Arid Environ.* **42**(4) 235-60 https://doi.org/10.1006/jare.1999.0505
[49] EOS Data Analytics 2019 NDVI FAQ: All you need to know about Index https://eos.com/blog/ndvi-faq-all-you-need-to-know-about-ndvi/
[50] Yin W, Hu L, Zhang M, Wang J and Han SC 2018 *J. Geophys. Res.-Atmos.* **123**(11) 5973-87 https://doi.org/10.1029/2017JD027468

IOP Conf. Series: Earth and Environmental Science 1136 (2023) 012017 doi:10.1088/1755-1315/1136/1/012017

Which models for extreme flood events in Mediterranean catchments?

Philippe Gourbesville[1] **and Masoud Ghulami**[1]

[1] Polytech Lab, Université Côte d'Azur, Polytech Nice Sophia, Sophia Antipolis, 06480, France

gourbesv@unice.fr

Abstract. Over the last years, most of the Mediterranean countries have been affected by catastrophic flood events generated by never observed before extreme rainfalls. Those events are frequently convective and stationary for several hours. The recorded intensities have frequently values exceeding 100 mm/hour and can last for several hours. Under those specific conditions as during the storm ALEX of October 2020, the runoff processes are particularly intense and can generate massive floods. Mitigation strategies should be developed and based on modeling that can support the understanding of the processes and the forecast the magnitude of the flood events and the associated impacts. The choice of the relevant hydrological modeling tools is a key element. The selected ones should be able to provide the relevant information, on time and with the relevant accuracy for the decision makers. The hydrological deterministic distributed models have reached a sufficient maturity to provide an accurate representation of the processes. The storm ALEX was generating more than 500 mm in less than 6 hours and producing a devastating flood within the Vésubie valley (France) in October 2020. The analysis of the event has demonstrated the difficulty to obtain consistent field observations and the limitations of many stochastic hydrological tools. The application of the AquaVar approach, combining several distributed deterministic models, has demonstrated its capacity to generate meaningful hydrographs and to quantify process with the catchment. The proposed solution is currently implemented within a real-time decision support system and can be used in real-time.

1. Context and challenges

Over the last years, most of the Mediterranean countries have been affected by extreme flood events generated by never observed before intense rainfalls. Those situations are frequently convective and stationary for several hours. The recorded intensities have frequently values exceeding 100 mm/hour and can last for several hours. Under those specific conditions as during the storm ALEX of October 2020, the runoff processes are particularly intense and can generate massive floods. The specific topography of Mediterranean catchments frequently characterized by steep slopes and large urban developments within the downstream areas and along the shoreline reinforces the propagation process and amplifies the flood wave. Within this complex situation and environment, mitigation strategies should be developed and based on modeling that can support the understanding of the processes and the forecast the magnitude of the flood events and the associated impacts.

The selection of an efficient hydrologic model tools is a key point within the methodology. The identified model ones should deliver the needed results, on time and with the proper accurateness the field managers. As the hydrologic situation is expected to be continuously known, the modelling systems should run continuously and provide updated forecasts that can allow the decision makers and field teams to initiate mitigation actions. In most of the cases, due to the limited size of the Mediterranean catchments, the reaction time is frequently below one hour and doesn't exceed 12 hours. The challenge is then to setup models with sufficient resolution that can produce meaningful forecasts within a computation time compliant with the field operations (frequently below 1 hour). Very frequently, the real-time modelling strategy is oriented towards conceptual models that can deliver results in few minutes with a limited computational effort. However, those tools have a serious limitation regarding the extreme events they cannot address properly because those events were not used during their calibration. The interest of deterministic models is then to overcome this limitation and to produce meaningful results even for extreme situations. Of course, the computational effort is more significant than for the conceptual tools and has to remain compliant with the expectations of the field operators. The following paragraphs present the application of this approach to the storm ALEX that has generated a major catastrophic flood event in October 2020 within the Vésubie valley, France.

2. The storm ALEX

2.1. Meteorological situation

Storm ALEX hit France on the night of Thursday October 1st to Friday October 2nd, 2020. The storm had an extremely original trajectory: it stagnated on the English Channel during the day of October 2, still producing heavy rains in the west of the country, before plunging towards the Atlantic coasts on the 3rd to finally head towards the center of the country. On the 4th, the storm shifted towards the south of England, before evacuating to the North Sea the following day. ALEX comes from the very rapid digging of a depression on the Atlantic, about 600 km from Brittany. Such a deepening was made possible by a strong dynamic at altitude, and in particular by the "jet stream".

Storm ALEX triggered an unprecedented Mediterranean episode. On Friday October 2nd, the maintenance of low-pressure conditions over the northwest of the country, combined with cold air aloft plunging towards the Iberian Peninsula, established a rapid and massive southerly flow over the Mediterranean. It is ahead of its cold front that this phenomenon occurs. Caught in a convergence, thunderstorms associated with the front dump very large amounts of rain into the hinterland. The valleys of Estéron, Tinée, and especially Vésubie and Roya are particularly affected. The hot and humid air stumbled on the alpine relief, causing a historic Mediterranean episode in the Alpes-Maritimes department. Exceptional accumulations of rain were recorded in the interior of the Alpes-Maritimes and the extreme north-east of the Var. The torrential rains, locally reaching 500 litters per square meter, caused devastating floods in the Var and tributary rivers. 500 mm of rain were measured by Météo France at the station in Saint-Martin-Vésubie in 24 hours, a new record for all months combined in 24, 48 or 72 hours at the departmental level. Some of the extreme values recorded are the following:

- 380 mm in Andon, the equivalent of two and a half months of rain in October;
- 343 mm in Tende;
- 336 mm at Coursegoules, i.e. 2 months of rain;
- 319 mm at the Mas, i.e. 2 months of rain;
- 271 mm in Breil-sur-Roya, i.e. 2 months of rain;
- 199 mm in Sospel, i.e. 1 month and a half of rain;
- 178 mm in Caussols, i.e. 1 month of rain.

The analysis of extreme Mediterranean rainfall events over the past decades highlights an intensification of heavy rainfall (+22% on the annual maxima of daily accumulations between 1961 and 2015) and an increase in the frequency of the strongest Mediterranean episodes, especially those exceeding the threshold of 200 mm in 24 hours.

Extreme analyses applied to regional climate projections indicate an increase in the intensity of intense precipitation over the northern part of the Mediterranean Basin. The 6th IPCC report [1] published during the summer of 2021, confirms the expected intensification of these episodes of heavy precipitation if global warming continues to intensify and exceeds 2°C.

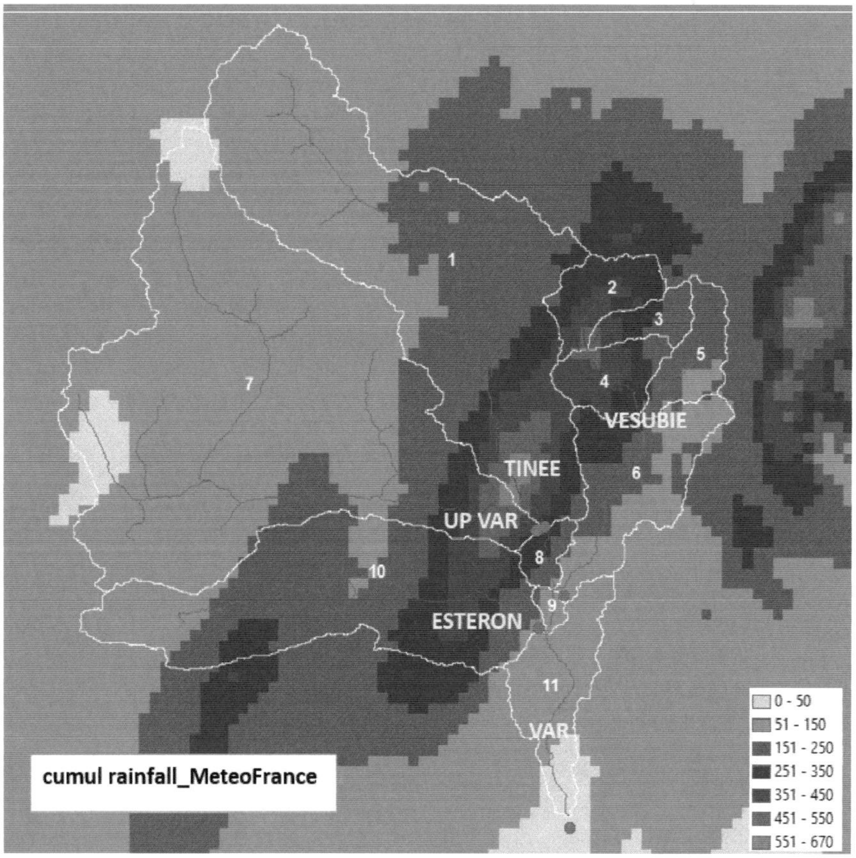

Figure 1. Cumulative rainfall (in mm) in 24 hours, from October 2nd, 2020 at 06:00 UTC to October 3rd, 2020 at 06:00 UTC (Météo France) over the Var catchment and main tributaries.

2.2. *Impacts and damages*

The Roya and Vésubie valleys in the hinterland of Nice were impacted by torrential rains with rapid floods and extremely destructive torrential lava. To the east of the Alpes-Maritimes, the municipalities are particularly affected by torrential floods, four of which are extreme: Breil-sur-Roya, Roquebillière, Saint-Martin-Vésubie and Tende. Human toll established in January 2021 (judicial source): 10 dead and 8 missing. The damages are enormous and the destructions multiple. The following destructions can be mentioned: hundred houses destroyed or uninhabitable, the gendarmerie in Saint-Martin-de-Vésubie, one EDF plant (hydropower plant), several wastewater treatment plants, 200 km of roads and 6 bridges, Tende cemetery, village hall in Puget-Theniers, etc. In total, the damage estimation reaches about € 2 billion for public infrastructures and about € 600 million for private insured goods.

2.3. *Hydrological synthesis*

The assessment of the flood processes by the state technical services [2] are reporting the following main points for the Var basin:

- the upstream Var basin was relatively spared with annual flood levels;
- the Tinée sub-basin was the first to react. Between 13:00 and 14:00, the discharge of this watercourse increased from less than 100 m^3/s to more than 900 with an increase of 350 m^3/s in 30 minutes between 14:00 and 14:30;
- discharge at Tournefort can be described as centennial. However, only the downstream part of the Tinée was really affect by flood. The occurrence of the phenomenon is therefore much greater than 100 years on this downstream section;

- the Vésubie reached from 18:00, a discharge between 650 and 700 m^3/s, with a hundred-year recurrence, with a rise slightly slower than the Tinée but with a valley with more challenges. This discharge was maintained until the morning of October 3. The result is a catastrophic situation, worse than the historical events of 1994 and 1997, in particular in the communes of St-Martin-Vésubie and Roquebilière;
- the Estéron, the most downstream tributary of the Var river, experienced a flood with a peak discharge of 470 m^3/s deemed fifty-year return period;
- the downstream Var reached a discharge rate at the Nice-Napoleon III station of 2,940 m^3/s with a rapid rise from 1,300 m^3/s at the maximum value in 60 minutes between 17:15 and 10:15 It approaches that of the historic flood of November 5, 1994 (3,200 m^3/s in Nice).

Most of the discharges' values presented have been evaluated from direct observations and field recognitions that took place during and immediately after the flood. All gauging stations with the Var catchment and tributaries have been destroyed during the event. Obviously, the quality of those estimations is highly questionable and should be understood as indicative of the magnitude of the processes. A similar situation was observed during the flood of 1994 and it has to be underlined that no real improvement has been achieved so far on the flood monitoring at the catchment scale.

3. The AquaVar approach

The storm ALEX represents a major challenge for local communities and requests the implementation of mitigation measures and the development of a resilience strategy. Within this specific context, the understanding of the processes is a major issue, and an accurate assessment of the flood event should be produced. The modeling strategy is potentially an efficient approach. However, the choice among various modeling strategies remains open. The selected approach is use distributed hydrological deterministic models that have the capacity to represent the hydrological processes within a catchment. Today those modeling system are sufficiently mature now to deliver a performant simulation of the hydrologic conditions for various conditions including the extreme events. The objective is then to see if they can be used also for extreme events that were not used for the calibration and validation phases.

Nice Côte d'Azur Metropolis, along the Mediterranean coast in the south of France, concentrates 540,000 inhabitants. Within the territory of the Metropolis, the city of Nice federates 350,000 inhabitants and is the 5th largest city in France. The territory of the Metropolis concentrates a dense population and is associated to a high vulnerability due to multiple inter-connected hazards related to tectonic, geomorphic, and climatic context that is characterized by frequent extreme rainfall events. The city of Nice is developing and expending mainly on the left bank of the Var River, in a densely urbanized floodplain of 20 km long. To manage this complex environment that concentrates multiple water competing activities, the decision to elaborate a specific Decision Support System (DSS) was taken in 2014 with AquaVar project.

The AquaVar system (Fig.1) [3, 4, 5] is defined as a services environment based on platform a service bus dedicated that is collecting and gathering in real time field data and forecasts produced by various models to address hydrologic processes and to support water uses and to mitigate potential water related hazards. The field and produced data are globally addressed with standard procedures that are defined as Key Performance Indicators (KPIs). Those KPIs cover transfers of data within the system but also defined thresholds used for alert protocols and mitigation directives [6]. AquaVar system is based on three distributed physically based modeling systems that provide an accurate simulation of the hydrologic conditions at the catchment scale; The system delivers forecasts that are used by technical teams to optimize the management strategies defined and monitored by the Operation Center. To ensure a good communication towards all the decision makers, AquaVar is offering a Web interface that is allowing to visualize the current situation and the various simulation results that are presenting the hydrologic dynamic ofteh catchment.

The AquaVar platform includes three physically based models that can simulate the hydrologic conditions of the catchment: 2D surface runoff, 2Dsurface flows within the river network and 3D groundwater situation with the coupling of the 2D and 3D models. The system includes pollutants exchanges. The three deterministic models are included within a hypervisor environment. The decision to use deterministic distributed models was made due to their capacity to simulate extreme hydrologic conditions that have never been recorded such as intense rainfall events. The deterministic tools provide a very good performance for such cases because the different variables remain the same and are operating within a range that is linked to a physical reality. In the opposite way, the non-distributed, data driven and conceptual tools are not able to perform at a similar level for the extreme situations because they don't find the relevant references within the knowledge training data that were collected from the field observations. The different models have been selected and linked to produce the needed simulations and forecasts:

- MIKESHE model [7] is used to generate hydrologic data that are applied as boundary conditions for the groundwater model (FEFLOW) and the hydraulic model (Mike21).
- Mike21 hydraulic model [7] addresses the surface flow conditions in 2D within the low Var valley and is linked with the groundwater 3D model built with FEFLOW to simulate the interactions between surface and aquifer. The morphological changes of the riverbed occurring during major events are also integrated within the simulations.
- FEFLOW 3D hydrogeological model [7] is used to represent the aquifer conditions. As mentioned, this module is connected to the hydraulic model in order to represent properly the interactions between surface waters and aquifer that are taking place within the riverbed.

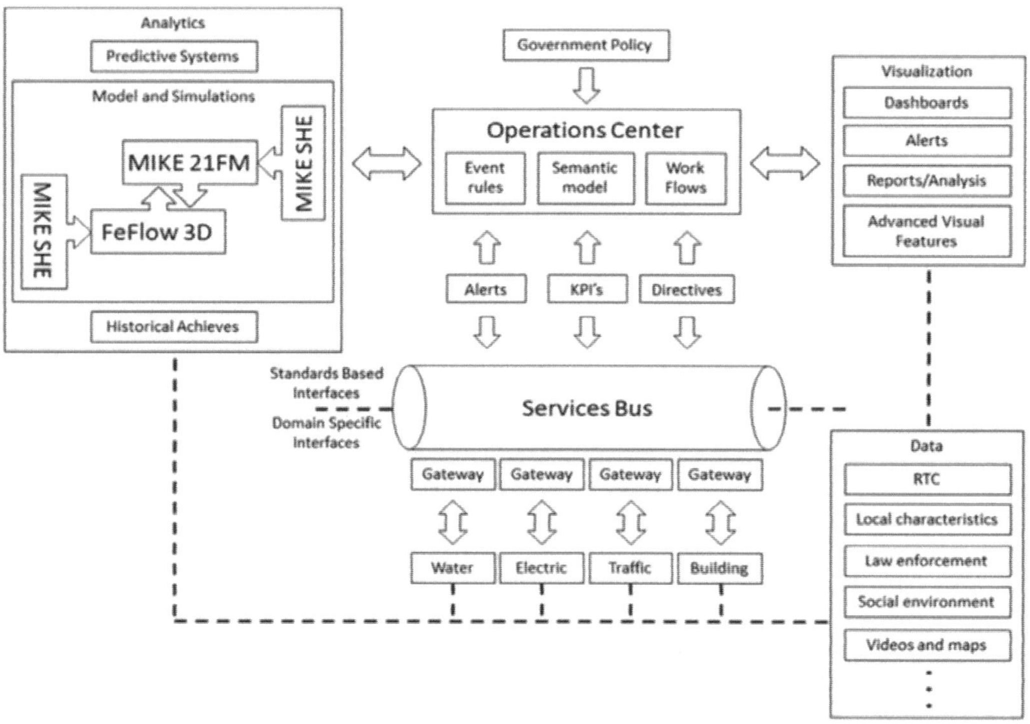

Figure 2. AquaVar concept and architecture.

The AquaVar DSS has developed an innovative approach that mobilizes modeling tools as non-interactives services. In the classical approach, the simulation engines like FEFLOW, MikeSHE and Mike21 are used independently and in an interactive mode with the user to benefit from the numerous visualization tools. Within the AquaVar approach, the modelling tools are mobilized in batch mode and are integrated within a modeling procedure defined by an orchestrator.

The various models are automated with the AquaVar DSS (Fig.3) as well as the data flows that are requested to run the simulations. The DSS gathers several components that are globally orchestrated to deliver the expected simulations:

- Data acquisition: the needed input data are extracted from a data storage component that is gathering the field data transmitted by the various sensors. A standard interface is used to easily connect the various data sources.
- Configuration: this module is used to setup automatically the various parameters needed for the model simulations and to update the format of the data that may request a conversion.
- Scheduler: the module is used to launch the simulation tools according to the defined time intervals. A table following the Unix crontab template is used and can be updated by the user of the system.
- Simulation engines: the modelling tool FEFLOW, Mike21 and MikeSHE are wrapped within a simulation engine. The wrapping approach allows to easily add a new simulation engine with no changes within the global architecture of the system.
- Data delivery: this last component transfers the results obtained from the modelling tools into a database. As for the data acquisition module, a standard and shared interface is used to deliver the results in various format such as plain file system, database, etc.
- Orchestrator: the module is dedicated to the management of the various components. web-based user interface allows the user to parameterize the orchestrator: he can define all the simulation parameters such as the timesteps for the various models.

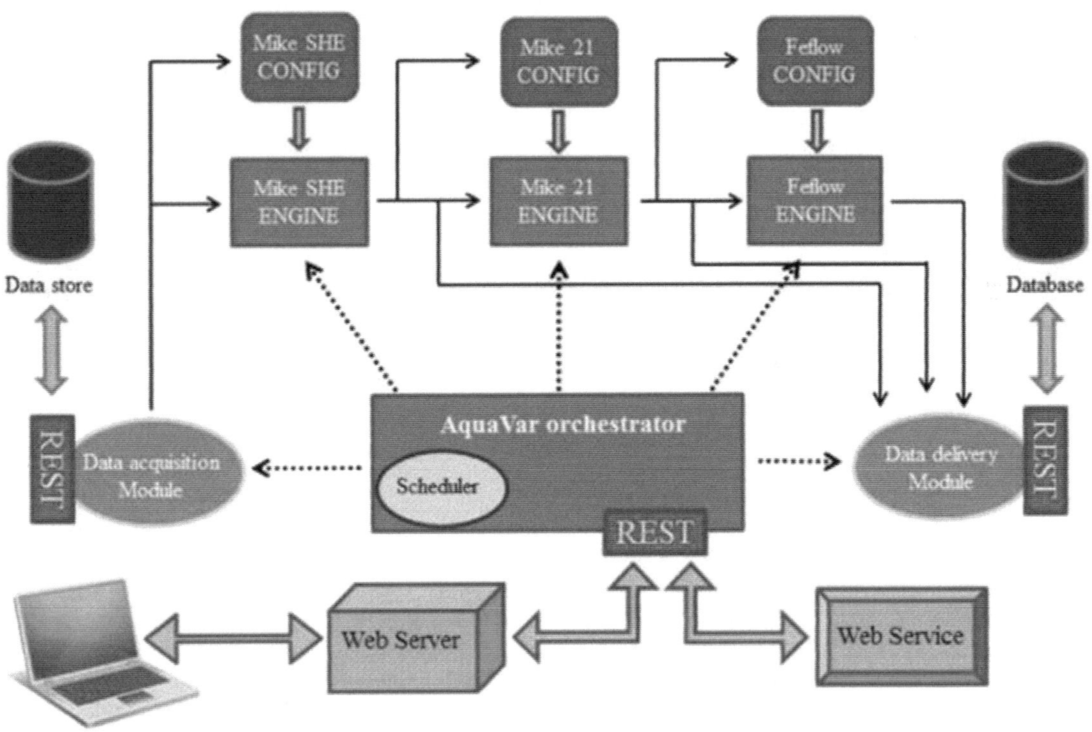

Figure 3. Orchestration of the various components within the DSS.

4. Storm ALEX simulation

The Vésubie valley within the Var catchment has been affected by an exceptional rainfall event and the generate flood has massively transformed the morphology of the stream network and the riverbed of the river. This unprecedent event has generated major damages with tremendous costs that request a very significant effort for communities to recover. The assessment of this catastrophic event is a priority for the field managers and for the local communities that are willing to implement efficient mitigation measures.

Within the specific context of this extreme situation and due to the difficulty to collect meaningful field data during the flood event because all data collection stations were destroyed, the only alternative approach is to use a modelling approach that can produce a meaningful and trustable estimation of the processes. Obviously, various options can be selected for developing a modelling solution and many kinds of models should, be on the theoretical point of view, mobilized. However, as the system should be mobilized for never observed situations as in October 2020, the data-oriented solutions are less relevant and cannot provide the needed results. The deterministic hydrological models offer a better alternative as they can potentially represent any type of flood and can be also deployed within DSSs that are implemented real-time management based on produced forecasts. Within the AquaVar system, the modelling tools are running continuously and reproduce the hydrologic conditions of the catchment. The requested inputs, as the models are linked, are limited to variables – observations and forecasts - related to rainfall, temperature, evapotranspiration and water extractions related to various uses of ground and surface water.

The rainfall data collected by Météo France during the storm ALEX on October 2nd and 3rd at the 14 rainfall stations located within the Var catchment have been introduced within the AquaVar system to generate flood hydrographs within the various tributaries and to estimate discharges at key locations where serious damages have been identified. Theresults obtained with AquaVar for key locations are presented on the figures 4, 5 and 6: the Vésubie valley that was the most affected by the rainfall event and the downstream Var valley where most of the economic activities and assets are concentrated. The AquaVar system allows to extract the hydrographs from the computation grid that is covering the full catchment. The produced hydrographs can be compared with field observations and the various discharge estimations as well.

The hydrographs obtained for the upstream area of the Vésubie valley demonstrate the extreme hydrological behavior of the catchment during the rainfall event. The two peaks obtained for the Vésubie and the Boréon are very consistent with the local observation and the estimated value above 400 m^3/s after the confluence in St Martin Vésubie is perfectly matching the hydraulic estimation provided by state services. The dual peaks reflect the rainfall distribution in time. Obviously, the runoff over the upper part of the catchment was massive. Hydrographs produced at points 4, 5 and 6 allow to understand the dynamic with the catchment.

Contributions from various sub-catchments like the Gordolasque were very limited as the rainfall was not directly affecting them with the same intensity as for the upstream part. An estimation of the runoff for the flood event has been computed and a coefficient of 79% has been obtained at point 6 (Utelle) for a surface of 393 km^2. This value underlines the extreme intensity of the processes and explains why an unusual quantity of materials was mobilized during the event. The intensity and the weight of water (rainfall) has initiated the morphological processes that have driven to a complete and major evolution of the slopes, flood plains and riverbeds.

The hydrograph obtained at the last bridge located in the estuary of the Var River is estimating the discharges at 3,000 m^3/s at the time of the second peak. This value is consistent with field observations. The reported observed values are extracted from the online VigieCrues database (https://www.vigicrues.gouv.fr).

The station at Napoléon III bridge was not fully operational due to works taking place with the riverbed since September 2020. The discharge was diverted to the arches of the right bank and the station was able to operate only when the discharge was strong enough to remove the peddle accumulations what were used to divert the stream. The VigieCrues data are extrapolated from the rating curve that has been built for this station only up to 600 m^3/s. So, the quality of observed values is highly questionable. A hydraulic estimation based on the geometry of the weir located below the bridge and on the maximum water level observed confirms the discharge of 3,000 m^3/s. The estimation of the runoff coefficient at Napoléon bridge is 61% for a total surface of 2,821 km^2.

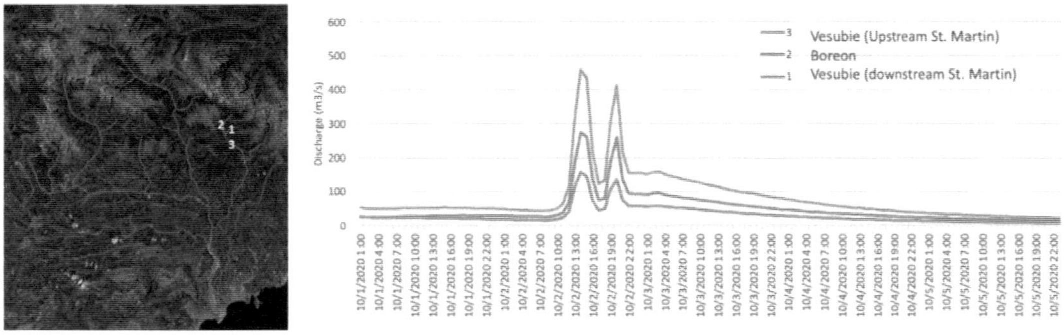

Figure 4. Hydrographs for the upstream area of the Vésubie catchment generated by AquaVar.

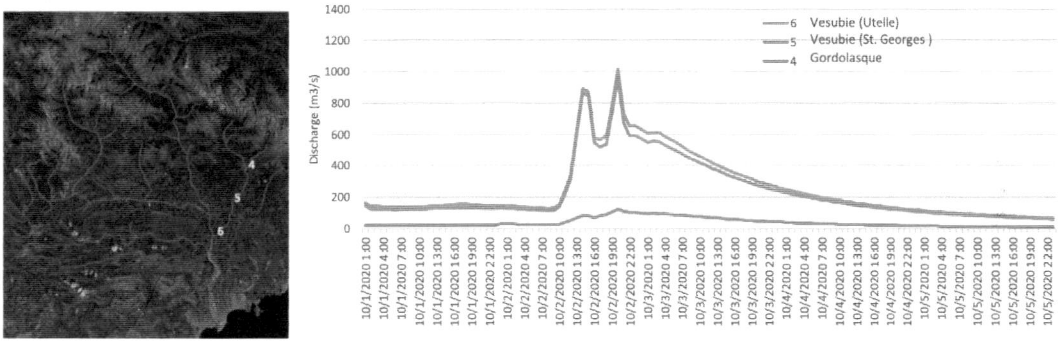

Figure 5. Hydrographs for the downstream area of the Vésubie catchment generated by AquaVar.

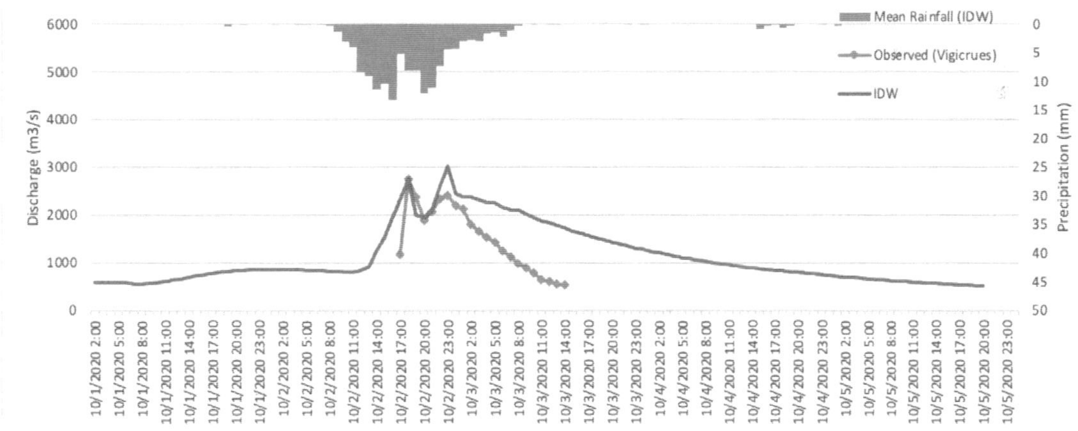

Figure 6. Hydrograph at Napoleon III bridge station: observed data and simulation from AquaVar.

5. Conclusions

The application of the AquaVar approach demonstrates its added value for the simulation of extreme rainfall events and the production of meaningful hydrographs all over the affected catchment. The obtained results can be easily used by the technical services and stakeholders in charge of mitigation measures design and resilience improvement of the exposed populations.

The storm ALEX storm of 2020 is a perfect example where the AquaVar system was able to produce a consistent estimation of the flood discharges within the Var catchment from the Météo France

IOP Conf. Series: Earth and Environmental Science 1136 (2023) 012017 doi:10.1088/1755-1315/1136/1/012017

observation data. The weakness of stochastic and conceptual tools has been highlighted with this extreme event because those models were not able to produce meaningful results as the magnitude was exceeding the defined range of the models' application.

The AquaVar DSS combines three deterministic models that are used for producing forecasts on surface runoff, flood propagation and groundwater evolution. For the specific context of the Nice Côte d'Azur Metropolis and the associated valleys, the deterministic models offer a relevant approach due to their robustness and their capacity to produce results within a physically based range. The implementation of these models within DSSs requests to develop a strategy for their orchestration and their exploitation in real-time. The constraints expressed by the operational entities represent boundary conditions that must be integrated within the DSS design. The necessity to obtain simulation results in one hour time is a major point in the case of the Var low valley. The various tests conducted – simulation of 7 days with 4 days from the past and 3 days ahead based on meteorological forecasts - with the different technical configurations have demonstrated that an operational architecture based on two computational servers using CPU and GPU resources can deliver the expected performance: a forecast for the next 3 days delivered less than 1 hours.

Table 1. Performance for hydrologic model.

Model	Start / End	Elements	Cores	Virtual CPUs	CPU Usage %	RAM (Go)	Threads	Simulation time (min)
Hydrologic model	01/11/2011 - 08:00 08/11/2011 - 08:00	560,000	19	76	100	304	76	57

Table 2. Performances for hydraulic and hydrogeologic models.

Model	Start / End	Elements	Cores	Virtual CPUs	GPU device	Nb GPU	RAM (Go)	Threads	Simula-tion (min)	Usage %
Hydraulic model	01/11/2011 - 08:00 08/11/2011 - 08:00	325,437	16	32	V100-NVIDIA	4	256	32	50	32
Hydrogeolo-gic model	01/11/2011 - 08:00 08/11/2011 - 08:00	392,161	16	32	V100-NVIDIA	1	256	32	27	22

The architecture of the system can also be updated with the implementation of data driven tools that can be combined with the deterministic approach. The addition of the data driven models can reduce the computation time and permit faster forecast production. The current approach is currently applied and implemented within the AquaVar system used by Nice Côte d'Azur Metropolis.

The efficiency of the AquaVar approach implemented for the full Var catchment has been established for the operational management of standard daily conditions and extreme situations as well. From the initial application over the Var catchment, Nice Côte d'Azur Metropolis has decided to expand the modelling tools over the Paillons (east area of Nice – 450 km^2) and Cagne (west area of Nice – 90 km^2). The modelling environment is based on the similar tools and covers the flow conditions for surface runoff, propagation, and aquifers. The forecasts are produced for a period of 72 hours based on the previous 4 days observed. The results are available within a web-based environment that is available for all the managing teams in charge of water related issues and emergency responses.

IOP Conf. Series: Earth and Environmental Science 1136 (2023) 012017 doi:10.1088/1755-1315/1136/1/012017

Acknowledgments
The AquaVar development is supported by Nice Côte d'Azur Metropolis, Université Côte d'Azur and the Rhone Méditerranée Water Agency. Authors would like to express their recognition to all technical advices and field data received for the managing institutions.

References

[1] Legg S 2021 IPCC, 2021: Climate Change 2021 – the Physical Science basis *Interaction* **49**(4) 44–45
[2] Direction Départementale des Territoires et de la Mer Alpes Maritimes (DDTM 06) 6 octobre 2020 *Note de synthèse – Événements du 2 et 3 octobre 2020 dans les Alpes-Maritimes*
[3] Gourbesville P 2020 Which models for Decision Support Systems? Proposal for a methodology *Advances in Hydroinformatics* (Springer, Singapore) pp 3–17
[4] Gourbesville P, Du M, Zavattero E and Ma Q 2016 DSS architecture for water uses management *Procedia Engineering* **154** 928–935
[5] Ma Q, Gourbesville P and Gaetano M 2020 AquaVar: Decision Support System for surface and groundwater management at the catchment scale *Advances in Hydroinformatics* (Springer, Singapore) pp 19–28
[6] Nohara D, Gourbesville P and Ma Q 2018 Towards development of effective decision support systems for integrated water resources planning and management *Disaster Prevention Research Institute (DPRI) Annuals* **61**(B) 702–710 https://www.dpri.kyoto-u.ac.jp/nenpo/no61/ronbunB/a61b0p47.pdf
[7] MIKE Zero by DHI https://www.mikepoweredbydhi.com/products

IOP Conf. Series: Earth and Environmental Science 1136 (2023) 012018 doi:10.1088/1755-1315/1136/1/012018

Assimilation of SAR-derived flood extent observations for improving fluvial flood forecast – A proof-of-concept

Thanh Huy Nguyen[1], Sophie Ricci[1], Andrea Piacentini[1], Christophe Fatras[2], Peter Kettig[3], Gwendoline Blanchet[3], Santiago Peña Luque[3] and Simon Baillarin[3]

[1] CECI, CERFACS/CNRS UMR 5318, Toulouse, 31057, France
[2] Collecte Localisation Satellites (CLS), 31520, Ramonville Saint-Agne, France
[3] Centre National d'Études Spatiales (CNES), Toulouse, 31401, France

thnguyen@cerfacs.fr

Abstract. As the severity and occurrence of flood events tend to intensify with climate change, the need for flood forecasting capability increases. In this regard, the Flood Detection, Alert and rapid Mapping (FloodDAM) project, funded by Space for Climate Observatory initiatives, was set out to develop pre-operational tools dedicated to enabling quick responses in flood-prone areas, and to improve the reactivity of decision support systems. This work focuses on the assimilation of 2D flood extent data (expressed in terms of wet surface ratios) and in-situ water level data to improve the representation of the flood plain dynamics with a Telemac-2D model and an Ensemble Kalman Filter (EnKF). The EnKF control vector was composed friction coefficients and corrective parameter to the input forcing. It is then augmented with the water level state averaged over several floodplain zones. This work was conducted in the context of Observing System Simulation Experiments (OSSE) based on a real flood event occurred in January-February 2021 on the Garonne Marmandaise catchment. This allows to validate the observation operator associated to the wet surface ratio observations as well as the dual state-parameter sequential correction implemented in this work. The merits of assimilating SAR-derived flood plain data complementary to in-situ water level observations are shown in the control parameter and observation spaces with 1D and 2D assessment metrics. It was also shown that the correction of the hydraulic state significantly improved the flood dynamics, especially during the recession. This proof-of-concept study paves the way towards near-real-time flood forecast, making the most of remote sensing-derived flood observations.

1. Introduction

Floods account for more than 40% of all the globally recorded disasters between 1998 and 2017, according to the United Nations Office for Disaster Risk Reduction (https://www.prevention-web.net/knowledgebase). They are the most common and also the costliest natural disasters worldwide. Yet, quality estimate, assessment, and forecasting of flood hazards remain insufficient in many regions of the world. In order to mitigate flood impacts, hydrodynamic numerical models providing simulations of flood events in analysis and forecast modes are of paramount essential. However, these models remain imperfect due to the uncertainties inherently existing within the models and the inputs, namely friction and boundary conditions (BC), which would be translated into uncertainties in the model outputs.

Flood reanalysis and forecast capability have been greatly improved, owing to the advancements in data assimilation (DA). These methods aim at combining time-series measurements with numerical models to correct the hydraulic states and sequentially reduce the uncertainties in the model parameters

[1]. In particular, EnKF relies on stochastically computing the forecast error covariance matrices using a number of member simulations. However, this approach relies strongly on the frequency, density and statistics of errors of the observing network [2]. In reality, limnimetric in-situ gauge stations are only available at a few locations within a catchment [3], usually exclusive in the river bed. This situation can be mitigated by using other data sources such as remote-sensing flood observations which, despite low observation frequency, offer valuable information about the flow dynamics through a 2D representation. In recent years, SAR data has established as a major data source in operational flood management, due to its reliable ability to collect all-day images regardless of weather conditions. Water surfaces and flooded areas typically exhibit low SAR backscatter, as most of the incident radar pulses are specularly reflected away upon arrival at the surfaces. Indeed, flood extent delineation is relatively straightforward on SAR images, with several exceptions, namely built environment and vegetation areas. In our previous works [4, 5], flood extent maps were derived from Sentinel-1 (S1) SAR images by applying a Random Forest classifier. Being used as independent validation observations, they allowed to improve the DA results in both re-analysis and forecast modes.

Taking further advantage of such observations, the assimilation of 2D flood extent maps, represented by wet surface ratios (WSR) provided by the count of wet pixels observed on SAR data, is investigated to comprehensively reduce the inherent uncertainties, and to improve the overall re-analysis and forecast performance. This article presents a novel approach to take into account these WSR observations, jointly with in-situ water-level measurements, within an EnKF framework on top of a 2D hydraulic model. As such, we carry out a dual state-parameter estimation strategy to reduce the uncertainties, namely in friction coefficients, upstream inflow discharge, and hydraulic state in several floodplain areas. This research work is conducted in an OSSE framework, where a reference simulation with pre-determined settings, which is considered as the *truth,* is used to generate synthetical observations. Resulting synthetical in-situ water level and WSR observations are then assimilated into an EnKF with a priori (or *background*) settings which are different from the truth's settings. Such a strategy is quite common in DA studies in order to assess a DA method. It aims at validating whether the proposed method can yield an analyzed control vector that is closer to the truth's settings, as well as providing the resulting model state closer to the observations compared to the a priori information.

2. Study Area, Data, Model

2.1. Shallow Water Equations (SWEs) in T2D

The hydrodynamic numerical model Telemac-2D (T2D, www.opentelemac.org) is used to simulate and predict water surface elevation and velocity. T2D solves the Shallow Water equations (SWE) derived from depth-averaged Navier-Stokes equations with an explicit first-order time-integration scheme, a finite-element scheme, and an iterative conjugate gradient method [6]. The model results, namely water depth and velocity averaged over the azimuth axis, are provided at each node of the mesh that represents the catchment topo-bathymetry. The non-conservative form of SWEs is written in terms of water level h [m] and velocity $\boldsymbol{u} = (u, v)$ (with horizontal components u and v [m.s^{-1}]). Using similar mathematical notations to [4, 7], the SWEs read:

$$\frac{\partial h}{\partial t} + \frac{\partial}{\partial x}(hu) + \frac{\partial}{\partial y}(hv) = 0 \tag{1}$$

$$\frac{\partial u}{\partial t} + \boldsymbol{u} \cdot \nabla u = -g\frac{\partial z}{\partial x} + F_x + \frac{1}{h}\nabla \cdot (h\nu_e \nabla u) \tag{2}$$

$$\frac{\partial v}{\partial t} + \boldsymbol{u} \cdot \nabla v = -g\frac{\partial z}{\partial y} + F_y + \frac{1}{h}\nabla \cdot (h\nu_e \nabla v) \tag{3}$$

where z [m NGF69] is the water surface elevation ($h = z - z_b$ with z_b [m NGF69] being the river bottom elevation) and ν_e [m^2.s^{-1}] is the water diffusion coefficient and g [m.s^{-2}] is the gravitational acceleration constant. F_x and F_y [m.s^{-2}] are the horizontal components of external forces (i.e. friction, wind and atmospheric forces):

$$\begin{cases} F_x = -\dfrac{g}{K_s^2}\dfrac{u|\boldsymbol{u}|}{h^{4/3}} - \dfrac{1}{\rho_w}\dfrac{\partial P_{atm}}{\partial x} + \dfrac{1}{h}\dfrac{\rho_{air}}{\rho_w}C_d U_{w,x}\sqrt{U_{w,x}^2 + U_{w,y}^2} \\ F_y = -\dfrac{g}{K_s^2}\dfrac{v|\boldsymbol{u}|}{h^{4/3}} - \dfrac{1}{\rho_w}\dfrac{\partial P_{atm}}{\partial y} + \dfrac{1}{h}\dfrac{\rho_{air}}{\rho_w}C_d U_{w,y}\sqrt{U_{w,x}^2 + U_{w,y}^2} \end{cases} \quad (4)$$

where K_s [m$^{1/3}$.s^{-1}] is the river bed and floodplain friction coefficient according to the Strickler formulation [8], ρ_w and ρ_{air} [kg.m^{-3}] are the water and air density, P_{atm} [Pa] is the atmospheric pressure, C_d [-] is the wind drag coefficient that relates the free surface wind to the shear stress, and lastly $U_{w,x}$ and $U_{w,y}$ [m.s^{-1}] are the horizontal wind velocity components. In order to solve Eq. (1)-(3), the initial conditions $\{h(x,y,t=0)=h_0(x,y);\ u(x,y,t=0)=u_0(x,y);\ v(x,y,t=0)=v_0(x,y)\}$ are required, whereas the upstream BC is described with a time-dependent hydrograph, and a rating curve at the downstream BC.

2.2. *Study area and description of the uncertainties*

The study area, illustrated in Figure 1, is the Garonne Marmandaise catchment which extends over a 50-km reach between Tonneins and La Réole of the Garonne River (southwest France). Three observing stations indicated as black circles are located at Tonneins, Marmande, and La Réole, operated by the VigiCrue network (https://www.vigicrues.gouv.fr/). The white circle indicates a diagnostic location in the floodplain near Marmande; denoted as FPM in the following. A T2D model was developed and calibrated over this catchment [9]. In order to translate the observed water levels into a discharge time-series, the local rating curve at Tonneins is used. This discharge is then applied over the whole upstream interface, covering both river bed and floodplain boundary cells (implemented by EDF R&D [9]), to allow a cold start of the model with any inflow value. However, it prompts a temporary over-flooding of the upstream first meander, until the water returns to the river bed. The downstream BC at La Réole is described in the model with a local rating curve.

Over the simulation domain, the friction coefficient is defined over seven zones, K_{s_1} to K_{s_6} (respective calibrated values: 45, 38, 38, 40, 40 and 40 [m$^{1/3}$.s^{-1}]) for the river bed and K_{s_0} (calibrated value 17 [m$^{1/3}$.s^{-1}]) for the entire floodplain, as illustrated in Figure 1 with solid colored segments of the river bed. The initial design of this model by EDF consisted in only three friction segments within the river bed. Based on our previous study [4, 5], a finer zoning definition was favored, which led to the former second and third segments (covering, respectively, the middle and downstream parts of the river bed) were subdivided into two and three smaller segments, respectively $K_{s_{2,3}}$ and $K_{s_{4,5,6}}$. Due to the absence of in-situ data for all of these zones, their a priori values are set based on the calibration process, yet they will allow for a fine adjustment in the DA algorithm. Such a description of the friction coefficients is highly uncertainty-prone due to zoning assumption and calibration procedure. In the following, these coefficients are considered as random variables normally distributed with mean and standard deviation estimated from the calibration process. In addition, the limited number of in-situ observations also yields errors in the formulation of the rating curve that is typically used to translate the observed water-level into discharge, especially for high flow. Therefore, a multiplicative factor a applied on the discharge time-series is considered as a random variable with a normal Probability Density Function (PDF) centered at 1.

Lastly, in order to account for the rainfall, evapotranspiration, and ground infiltration processes that are unavailable in the T2D Garonne model, a correction of hydraulic state in the floodplain has been implemented. Five subdomains (hereafter called *zones*) delineated in the floodplain beyond the dykes, involving a uniform water level correction over each zone, are added to the control vector. These state corrections $\delta H_{[1:5]}$ are considered as zero-mean normal random vector. These zones were determined based on the storage areas, as a result of the dyke system of the catchment [9]. It is worth-noting that the first storage area of the model, at the first meander near Tonneins, is excluded in this study because of the aforementioned artificial temporary over-flooding effect. In addition, several storage areas near the downstream area are not considered, because these areas are not fully observed by S1, and spurious dynamics may be caused by the errors in topography [9]. Over each of the five zones, the wet surface ratio (WSR) between the area of observed wet surfaces and the total area of the zone is measured.

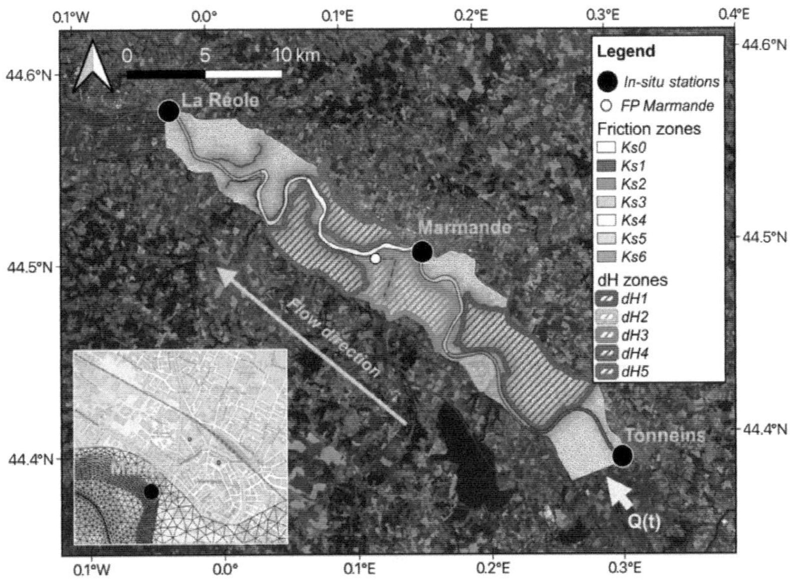

Figure 1. T2D Garonne Marmandaise model and control vector. The inset figure shows the impacted urban area near Marmande (Note: from [11]).

The proof-of-concept of the DA was carried out for a substantial flood event that occurred on the Garonne catchment in late January and February 2021. This event exceeded the highest threshold level for flood risks, set out by the French national flood forecasting center (SCHAPI) in collaboration with the departmental prefect. The flood peak was reached on 2021-02-04. In this work, we examine an extended length of this event between 2021-01-16 and 2021-02-10, during which the catchment was observed nine times by S1 images.

3. Data Assimilation

3.1. Description of the control vector

The DA algorithm consists in a cycled stochastic EnKF, where the control vector $\mathbf{x}$ is composed of the seven friction coefficients $K_{s_{[0:6]}}$, one multiplicative parameter a to modify the time-varying upstream BC, and five state corrective variables $\delta H_{[1:5]}$ over the floodplain zones. These $n = 13$ parameters are assumed to be constant within a DA cycle, and allowed by the DA to evolve from one cycle to another. Each cycle k covers a time window, denoted by $W_k = [t_{start}, t_{end}]$ of 18-hour length over which N_{obs} observations are assimilated. The cycling consists in sliding the windows by $T_{shift} = 6$ hours so that two consecutive cycles W_k and W_{k+1} overlap for 12 hours. This EnKF algorithm relies on the propagation of N_e members with perturbed values of $\mathbf{x}$, i.e. the forecast values denoted by $\mathbf{x}_k^{f,i}$ with $i \in [1, N_e]$ represents the ensemble member index.

3.2. Forecast step

The forecast step stands in the propagation in time of the control and model state vectors, whereas the analysis step (section 3.3) stands in the update of these vectors.

However, the EnKF is here applied to model parameters (in the control vector $\mathbf{x}$) that do not evolve during each W_k. Artificial dispersion is further introduced when adding perturbations $\boldsymbol{\theta}_k^i$ to the mean of the analysis from the previous cycle $\overline{\mathbf{x}_{k-1}^a}$ to avoid ensemble collapse. The forecast step thus reads:

$$\mathbf{x}_k^{f,i} = \begin{cases} \mathbf{x}_0 + \boldsymbol{\theta}_1^i & \text{if } k = 1 \\ \overline{\mathbf{x}_{k-1}^a} + \boldsymbol{\theta}_k^i & \text{if } k > 1 \end{cases} \tag{5}$$

with $\overline{\mathbf{x}_{k-1}^a} = \left(\sum_{i=1}^{N_e} \mathbf{x}_{k-1}^{a,i}\right)/N_e \in \mathbb{R}^n$ and $\boldsymbol{\theta}_k^i \sim \mathcal{N}(\mathbf{0}, \sigma_{i,k}^2)$ where

$$\sigma_{i,k} = \begin{cases} \sigma_{\mathbf{x}} & \text{if } k = 1 \\ \lambda \sqrt{\frac{1}{N_e} \sum_{i=1}^{N_e} \left(\mathbf{x}_{k-1}^{a,i} - \overline{\mathbf{x}_{k-1}^a}\right)^2} + (1-\lambda)\sigma_{\mathbf{x}} & \text{if } k > 1 \end{cases} \tag{6}$$

The controlled variables issued from $\overline{\mathbf{x}_{k-1}^a}$ is further dispersed with perturbations $\boldsymbol{\theta}_k^i$ drawn from the zero-mean normal distribution with the standard deviation obtained from the linear combination (weighted by $\lambda = 0.3$) of the $\sigma_{\mathbf{x}}$ and the standard error of the analysis at previous cycle. This technique is an alternative for anomaly inflation in order to avoid the ensemble from collapsing, while preserving part of the information from the background statistical characteristics. The corrections to the hydraulic state variable $\delta H_{[1:5]}$ are drawn within zero-mean normal distributions, thus fully described by perturbations $\boldsymbol{\theta}_k^i$ (i.e. without the mean of analysis from previous cycle).

The background hydraulic state associated with each ensemble member, denoted by $\mathbf{s}_k^{f,i}$, results from the integration of the hydrodynamic model $\mathcal{M}_k: \mathbb{R}^n \to \mathbb{R}^m$ from the control space to the model state space over W_k: $\mathbf{s}_k^{f,i} = \mathcal{M}_k\left(\mathbf{s}_{k-1}^{a,i}, \mathbf{x}_k^{f,i}\right)$. For the very first cycle, the initial condition for $\mathcal{M}_k$ at t_{start} is provided by a restart file defined by user. Then, for the following cycles k, it requires in a full restart from $\mathbf{s}_{k-1}^{a,i}$, saved from the analysis run of the previous cycle $\mathbf{s}_{k-1}^{a,i} = \mathcal{M}_{k-1}\left(\mathbf{s}_{k-2}^{a,i}, \mathbf{x}_{k-1}^{a,i}\right)$ at the beginning of each W_k. In order to ensure the consistencies between the hydraulic state and the analyzed parameters at t_{start}, a short (3-hour preceding t_{start}) spin-up integration can be carried out.

The control vector equivalent in the observation space for each member, noted $\mathbf{y}_k^{f,i}$, stems from:

$$\mathbf{y}_k^{f,i} = \mathcal{H}_k\left(\mathbf{s}_k^{f,i}\right) - \mathbf{y}_{bias} \tag{7}$$

where $\mathcal{H}_k: \mathbb{R}^m \to \mathbb{R}^{N_{obs}}$ is the observation operator from the model state space to the observation space that selects, extracts and eventually interpolates model outputs (if necessary) at times and locations of the observation vector $\mathbf{y}_k^o$ over W_k. In this work, the observation operator also includes a correction to account for a systematic model-observation bias, denoted by $\mathbf{y}_{bias}$. Such a bias was estimated based on the 24-hour quasi-stationary non-overflowing period of 2021-01-15, prior to the real flood event in January-February 2021 [4].

3.3. *Analysis step*

When applying a stochastic EnKF [10], the observation vector $\mathbf{y}^{o,i}$ is perturbed, thus an ensemble of observations $\mathbf{y}_k^{o,i}$ $(i \in [1, N_e])$ is generated:

$$\mathbf{y}_k^{o,i} = \mathbf{y}_k^o + \boldsymbol{\epsilon}_k \text{ with } \boldsymbol{\epsilon}_k \sim \mathcal{N}(0, \mathbf{R}_k) \tag{8}$$

where $\mathbf{R}_k = {\sigma_{obs}}^2 \mathbf{I}_{N_{obs}}$ is the observation error covariance matrix, assumed to be diagonal, of standard deviation σ_{obs} (and $\mathbf{I}_{N_{obs}}$ is the $N_{obs} \times N_{obs}$ identity matrix). For the in-situ water level observation, the standard deviation is proportional to the observations $\sigma_{obs,k} = \tau y_k^o$, whereas the σ_{obs} takes fixed values for the WSR observations. The innovation vector (over W_k) computes the difference between the perturbed observation vector $\mathbf{y}_k^{o,i}$ (from Eq. (7)) and the model equivalent $\mathbf{y}_k^{f,i}$ (from Eq. (8)). Such vector is weighted by the Kalman gain matrix $\mathbf{K}_k$ and then added as a correction to the background control vector $\mathbf{x}_k^{f,i}$, so that the analysis control vector $\mathbf{x}_k^{a,i}$ is as follows,

$$\mathbf{x}_k^{a,i} = \mathbf{x}_k^{f,i} + \mathbf{K}_k\left(\mathbf{y}_k^{o,i} - \mathbf{y}_k^{f,i}\right) \quad (9)$$

The Kalman gain reads: $\mathbf{K}_k = \mathbf{P}_k^{x,y}\left[\mathbf{P}_k^{y,y} + \mathbf{R}_k\right]^{-1}$ with $\mathbf{P}_k^{y,y}$ being the covariance matrix of the errors in the background state equivalent in the observation space $\mathbf{y}_k^f$ and $\mathbf{P}_k^{x,y}$ the covariance matrix between the errors in the control vector and the error in $\mathbf{y}_k^f$. They are stochastically estimated within the ensemble:

$$\mathbf{P}_k^{x,y} = \frac{1}{N_e}\mathbf{X}_k^T\,\mathbf{Y}_k \in \mathbb{R}^{n\times N_{obs}} \quad \text{and} \quad \mathbf{P}_k^{y,y} = \frac{1}{N_e}\mathbf{Y}_k^T\,\mathbf{Y}_k \in \mathbb{R}^{N_{obs}\times N_{obs}} \quad (10)$$

with $\mathbf{X}_k = \left[\mathbf{x}_k^{f,1} - \overline{\mathbf{x}_k^f}, \cdots, \mathbf{x}_k^{f,N_e} - \overline{\mathbf{x}_k^f}\right] \in \mathbb{R}^{n\times N_e}$ and $\mathbf{Y}_k = \left[\mathbf{y}_k^{f,1} - \overline{\mathbf{y}_k^f}, \cdots, \mathbf{y}_k^{f,N_e} - \overline{\mathbf{y}_k^f}\right] \in \mathbb{R}^{N_{obs}\times N_e}$ where $\overline{\mathbf{x}_k^f} = \frac{1}{N_e}\sum_{i=1}^{N_e}\mathbf{x}_k^{f,i} \in \mathbb{R}^n$ and $\overline{\mathbf{y}_k^f} = \frac{1}{N_e}\sum_{i=1}^{N_e}\mathbf{y}_k^{f,i} \in \mathbb{R}^{N_{obs}}$.

The analyzed hydrodynamic state $\mathbf{s}_k^{a,i}$, associated with each analyzed control vector $\mathbf{x}_k^{a,i}$ is denoted by $\mathbf{s}_k^{a,i}$. It results from the integration of the hydrodynamic model $\mathcal{M}_k$ with the updated controlled parameters over W_k, starting from the same initial condition (for the first cycle), then from each background simulation within the ensemble: $\mathbf{s}_k^{a,i} = \mathcal{M}_k\left(\mathbf{s}_{k-1}^{a,i}, \mathbf{x}_k^{a,i}\right)$.

4. Observing System Simulation Experiment (OSSE)

4.1. Truth and observation generation

The framework of an OSSE consists in a deterministic simulation with a chosen set of parameters, which are time-varying in the present case. This reference simulation is considered the *truth*. In the present work, the so-called true friction parameters and inflow correction are set from two previous DA experiments on the real 2021 flood event that assimilate in-situ observations and WSR observations. The state correction true values were set up with negative cosine curves at the three first groups of S1 observations, and then with a constant negative water of -18 cm at the recession period. The *synthetic observations* are generated based on the reference simulation. They are then assimilated in a DA experiment, with initial calibrated/default values that are different from the truth's settings. The upstream time-varying BC at Tonneins is also provided by the real flood event in January-February 2021. However, only synthetical observations are assimilated.

It is worth-noting that the corrections of the hydraulic state in the floodplain ($\delta H_{[1:5]}$) are only performed at the assimilation window in which WSR observations are present. A full study including real flood events can be found in [11].

4.2. Experimental setup

One free run and three DA experiments were carried out (Table 1) with different types of observations that are assimilated and the different components of the control vector. Two types of observations are concerned: *(i)* in-situ water-level observations at three observing stations Tonneins, Marmande, and La Réole, *(ii)* WSR measurements on the five floodplain zones (associated with $\delta H_{[1:5]}$). For the DA experiments, the value of τ for in situ data is fixed to 15%, meaning that σ_{obs} amounts to 15% of the observation value, whereas the value of σ_{obs} associated with WSR data varies from 0.2 down to 0.1 according to when the observation time is located within the 18-hour assimilation window.

Table 1. Summary of the experiment settings.

Exp. name	*FR or DA?*	*Assimilated observations*	N_e	*Control vector*
FR	FR	-	1	-
IDA (in-situ DA)	DA	In-situ water level	75	$K_{s_0}, K_{s_{[1:6]}}, a$
IWDA	DA	In-situ water level & WSR	75	$K_{s_0}, K_{s_{[1:6]}}, a$
IHDA	DA	In-situ water level & WSR	75	$K_{s_0}, K_{s_{[1:6]}}, a, \delta H_{[1:5]}$

5. Results and Discussions

Quantitative performance is assessed in the control and observation spaces. They involve the comparisons between *(i)* the analyzed parameters and the truth's parameters, *(ii)* the analyzed water-level time-series and synthetical in-situ observations, *(iii)* the analyzed WSR and synthetical WSR observations in the floodplain, and lastly the evaluation of *(iv)* the overall Critical Success Index (CSI) computed over the analyzed and the truth's flood extent maps.

5.1. Results in the control space

Figure 2 shows the analyzed parameters from the DA experiments, with IDA represented by the blue curves, IWDA by green, and IHDA by red ones.

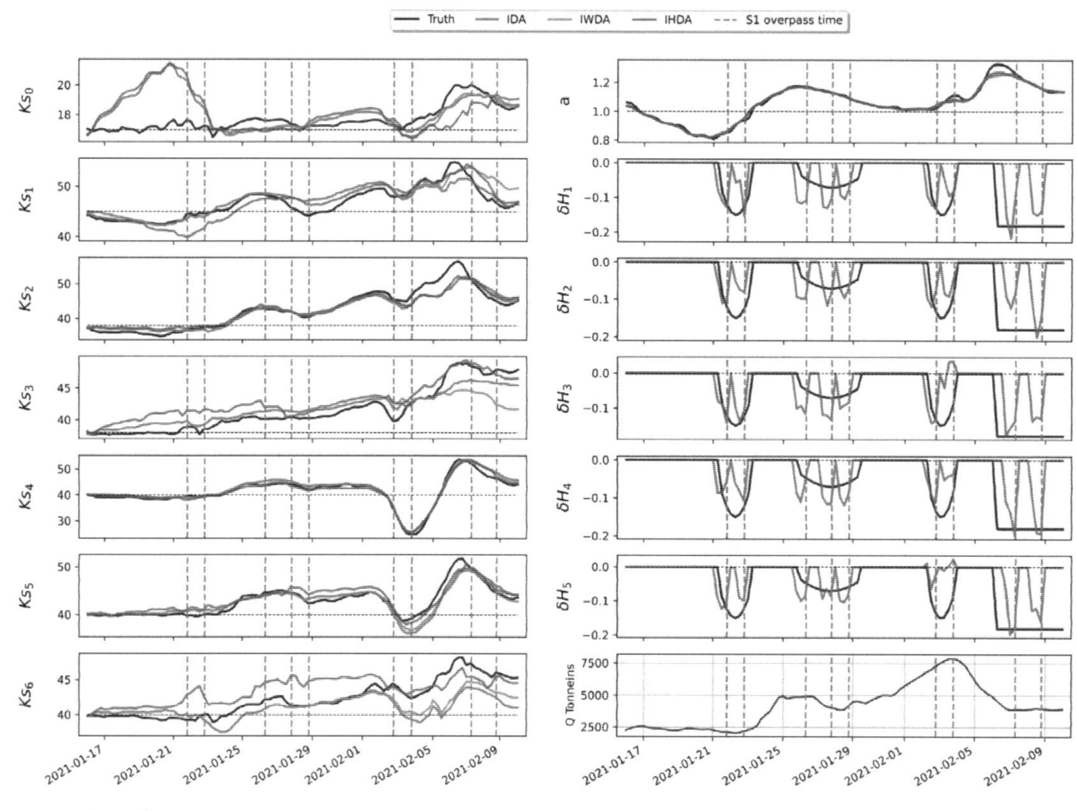

Figure 2. Evolution of controlled parameters for friction, multiplicative correction applied on the inflow, water level correction in the floodplain zones. Last panel depicts the forcing inflow at Tonneins.

In each panel, the *truth* parameters are depicted in black, whereas the calibrated and default values x_0 are shown by the horizontal dashed lines. The overpass times of S1 in this event are depicted by the vertical dashed lines. It appears that for most of riverbed friction zones (except K_{s_6}), the three DA analyses are able to retrieve the true value of the friction coefficients. Among them, IHDA (red curves) with the extended control vector provides the closest estimation of the *truth*. In the floodplain, the correction of the friction K_{s_0} exhibits some difficulty for all DA experiments and it takes several days to converge to the true parameter, despite the assimilation of WSR observations. It should be noted that in spite of slow convergence or equifinality issues, all analyzed friction coefficients remain within physically feasible ranges.

All DA experiments provide excellent results for the inflow multiplicative factor α, except near the flood peak. Since the analysis for the state correction in the floodplain $\delta H_{[1:5]}$ is only activated when WSR observations are present over the assimilation window, starting from 18 hours before each observation, hence the state correction is null-valued for most of the time, including in between S1 overpass times (minimum 24 hours between two observations). Yet, the correction reaches values that are close to the true values.

5.2. *Results in the observation space: Water levels at observing stations*

The dynamic of the river bed and the floodplain is assessed with respect to in situ data at Tonneins, Marmande, La Réole, and FPM location (not assimilated) respectively, by the Root-Mean-Square Errors (RMSE) computed between the reference water level time-series and the FR/DA experiments. For each column, the lowest RMSEs are boldfaced. Table 2 shows that all DA experiments succeed in significantly reducing the water level errors, compared to that of FR. The DA experiments RMSE at the three observing stations, where water lever synthetical data are assimilated remains under 5.3 cm. This level of precision can be expected in OSSE, as it validates the performance of the EnKF algorithm even though the ensemble size is limited (with $N_e = 75$ members). The RMSE reduction with respect to FR amounts to 80%, 89%, and 91%, respectively, at Tonneins, Marmande, and La Réole, while is reaches 34.4% (for IDA and IWDA) or 61.1% (for IHDA) at FPM where no data is assimilated. The merits of correcting the water level state are shown as IHDA outperforms the other two DA experiments, especially at FPM.

Table 2. Water level RMSE w.r.t. in-situ water levels at VigiCrue observing stations and FPM.

			RMSE [m]	
Exp. name	*Tonneins*	*Marmande*	*La Réole*	*FPM (only validation)*
FR	0.259	0.393	0.572	0.483
IDA	**0.051**	**0.043**	0.051	0.317
IWDA	0.053	0.044	0.050	0.317
IHDA	**0.051**	**0.043**	**0.048**	**0.188**

5.3. *Results in the observation space: WSR in the floodplain*

The WSR in the five floodplain zones for FR and DA are compared to the reference values, as shown in Figure 3 with the same color code as in Figure 2. The difference between the WSR simulated by FR and DA experiments and the reference *true* are shown. Up to the water level rise (around 2021-02-01), the impact of DA is not significant as the floodplain is not active. At the flood peak, the FR (orange curves) over-predicts the flooding in zone 1, while under-predicting highly the flooding in zones 2, 3, and 4. Same results are obtained with IDA and IWDA in zone 1 at the flood peak, yet both of them allow for an improvement in zones 2-4 around the flood peak.

The correction of the water level state in IHDA allows for a significant improvement, especially during the flood recession when the T2D model alone fails to evacuate the water. It appears that the augmented control vector together with the assimilation of WSR data is an efficient strategy that allows to account for model error and leads to the best results.

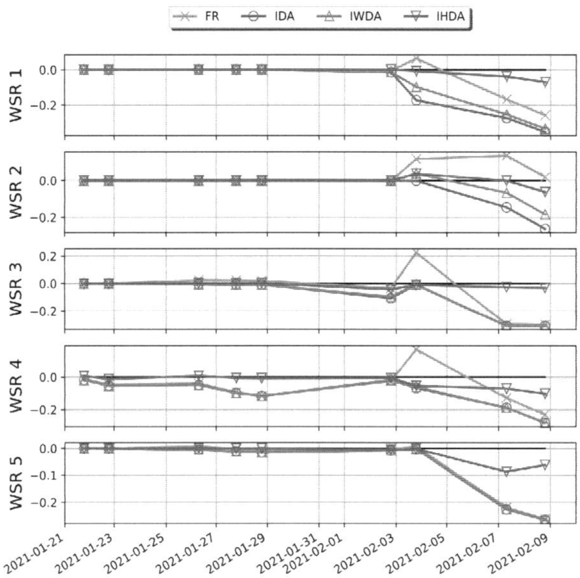

Figure 3. Errors between truth's WSR values and simulated WSR values in the 5 floodplain zones.

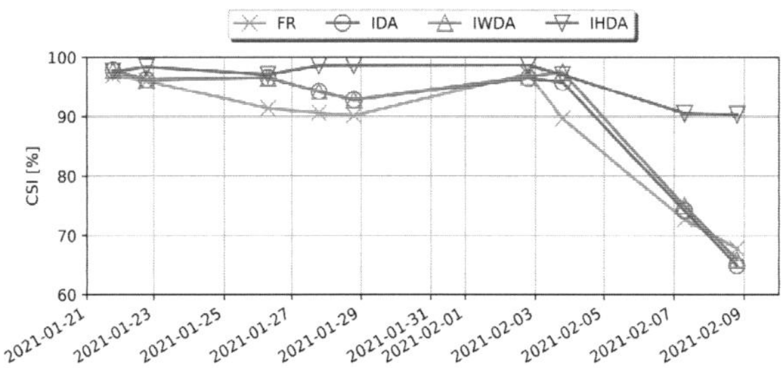

Figure 4. CSI scores between the realized experiments. 2D validation with Critical Success Index (CSI).

The CSI is computed to validate the agreement between the flood extent maps predicted by the FR (and DA) experiments with respect to those based on the *truth* simulation (i.e., the reference flood extent maps). This metric relies on the pixel counts combining the four possible outcomes: the number of pixels correctly predicted as flooded and non-flooded, the number of non-flooded pixels incorrectly predicted as flooded (or *over-prediction*), and the number of the flooded pixels being missed by the prediction (or *under-prediction*). Figure 4 summarizes the CSI yielded by all four performed experiments at every S1 overpass times. It should be noted that all the resulting flood extent maps from DA experiments are in relative agreements with the reference flood maps. As such, all three DA experiments present a significant improvement near the flood peak (2021-02-03 19:00) compared to FR. In particular, IHDA results in CSI above 90% at every S1 overpass times (and above 96% before recession period). The performance is shown to be better at various moments during the whole flood event, especially during the water recession period where the other two DA experiments IDA and IWDA yield a low CSI (65-75%).

6. Conclusions and Perspectives

This work studies the assimilation of flood extent maps, represented by WSR, to improve the hydrodynamic simulation results based on numerical T2D models with EnKF. The study was carried out over the Garonne Marmandaise catchment, focusing on the a synthetical flood event. Four experiments were realized, namely one in free run mode and three in DA mode in an OSSE framework. All of the DA experiments were implemented by a cycled EnKF with an 18-hour assimilation window sliding with 12-hour overlapping. This study confirms the assertion that a densification of the observing network allows to mitigate the equifinality issues and improve the flood forecast capability. These first two DA experiments (IDA and IWDA) focus on sequentially correcting friction coefficients and inflow discharge (through a multiplicative factor). However, IDA assimilates only in-situ water-level observations whereas IWDA involves jointly in-situ water level and WSR observations derived from SAR flood extent maps. They are shown to be capable of retrieving the true controlled parameters, thus providing relevant results. Nevertheless, this article highlights particularly the merits of IHDA experiment, which also assimilates both types of observations. In addition, it deals with friction coefficients and inflow discharge as well as the hydraulic state variable in five determined floodplain zones, representing rainfall, evapotranspiration and/or ground infiltration processes, based on a dual state-parameter estimation within the EnKF. This approach is able to result in simulated water levels and WSRs that are very close to the synthetic observations, as well as yielding better estimates of true friction and discharge parameters, compared to IDA and IWDA. In terms of 1D assessment, IHDA presents similar RMSE to IDA and IWDA (if not slightly better) on the three observing stations. The largest error on both three DA experiments is only 5.3 cm, which confirms a very high precision for the performed cycled DA approach. Nevertheless, IHDA results in a water level RMSE on the floodplain which is reduced 40.7%, compared to the RMSEs provided by IDA and IWDA. The 2D assessment also shows the advantage of the IHDA over the other two DA experiments, in particular before the flood peak and during the recession period, highlighted by a CSI always above 90% (and above 96% before the recession).

Moreover, this article has validated the developed observation operator dedicated to assimilating WSR observations, as well as the appropriate augmented control vector. These observations were able to be accommodated into the EnKF paradigm without conflicts with in-situ observations. Yet, it should be stressed that the selection of floodplain zones to be corrected with δH is quite a challenging task, which involved crucial understanding of the T2D model. In addition, within this research work, we have also carried out several different approaches concerning the localization within EnKF, meaning the isolation of the influences of WSR observations onto friction coefficients and inflow discharge, as well as the impact of in-situ observation on the δH variables. However, since the resulting effects are quite insignificant for a sufficiently large ensemble, this article only presents the experiments and results with 75 members and without any localization.

A particular perspective of this work concerns the non-gaussianity of these bounded WSR observations. This can lead to a suboptimality of the EnKF algorithm, since it requires the condition that observational errors follow normal distributions. Several existing studies have been investigated on a variable change, widely known as Gaussian anamorphosis, in order to realize the analysis step in a Gaussian space. Furthermore, we aim to tackle the subject of the assimilation of SAR-derived flood extent maps in near real time, expressed similarly in WSR measurements. In addition, the proposed dual state-parameter estimation approach can also be applied to correct the highly-uncertain topography over the downstream areas. Lastly, we are also interested in expanding the observing network from space with other types of data, namely a higher resolution SAR dataset (TerraSAR-X) and optical imagery dataset (Sentinel-2).

Acknowledgement
This work was supported by CNES, CERFACS and SCO-France. The authors gratefully thank the EDF R&D for providing the Telemac-2D model for the Garonne River, and the SCHAPI, the SPC Garonne-Tarn-Lot and the SPC Gironde-Adour-Dordogne for providing the in-situ data.

References

[1] Neal, J. C., Atkinson, P. M., and Hutton, C. W. 2007 Flood inundation model updating using an ensemble Kalman Filter and spatially distributed measurements. *Journal of Hydrology*, **336**(3-

4), 401-415. https://doi.org/10.1016/j.jhydrol.2007.01.012

[2] Mirouze, I., Ricci, S., and Goutal, N. 2019 The impact of observation spatial and temporal densification in an ensemble Kalman Filter. *Proc. 26th TELEMAC-MASCARET User Conf.*, October 15-17, 2019, Toulouse. https://doi.org/10.5281/zenodo.3549572

[3] Mason, D. C., Schumann, G. P., Neal, J. C., Garcia-Pintado, J., and Bates, P. D. 2012 Automatic near real-time selection of flood water levels from high resolution synthetic aperture radar images for assimilation into hydraulic models: A case study. *Remote Sensing of Environment*, **124**, 705-716. https://doi.org/10.1016/j.rse.2012.06.017

[4] Nguyen, T. H., Ricci, S., Fatras, C., Piacentini, A., Delmotte, A., Lavergne, E., and Kettig, P. 2022 Improvement of flood extent representation with remote sensing data and data assimilation. *IEEE Transactions on Geoscience and Remote Sensing*, **60**, 1-22, 2022, Art no. 4206022, https://doi.org/10.1109/TGRS.2022.3147429

[5] Nguyen, T. H., Delmotte, A., Fatras, C., Kettig, P., Piacentini, A., and Ricci, S. 2021 Validation and improvement of data assimilation for flood hydrodynamic modelling using SAR imagery data. *Proc. 27th TELEMAC-MASCARET User Conference 2020,* October 2021 (pp. 100-108). https://hdl.handle.net/20.500.11970/108298

[6] Hervouet, J. M. 2007 *Hydrodynamics of Free Surface Flows: Modelling with the Finite Element Method* **360** (Wiley Online Library) https://doi.org/10.1002/9780470319628

[7] Laborie, V., Ricci, S., De Lozzo, M., Goutal, N., Audouin, Y., and Sergent, P. 2020 Quantifying forcing uncertainties in the hydrodynamics of the Gironde estuary. *Computational Geosciences*, **24**(1), 181-202. https://doi.org/10.1007/s10596-019-09907-7

[8] Gauckler, P. 1867 Études théoriques et pratiques sur l'écoulement et le mouvement des eaux. *Compte Rendus de l'Académie des Sciences Paris*, **64**, 818–822.

[9] Besnard, A., and Goutal, N. 2011 Comparison between 1D and 2D models for hydraulic modeling of a floodplain: Case of Garonne River. *Houille Blanche-Revue Internationale De L'Eau*, **97**(3), 42-47. https://doi.org/10.1051/lhb/2011031

[10] Asch, M., Bocquet, M., and Nodet, M. 2016 *Data Assimilation: Methods, Algorithms, and Applications* (Society for Industrial and Applied Mathematics) https://doi.org/10.1137/1.9781611974546

[11] Nguyen, T. H., Ricci, S., Piacentini, A., Fatras, C., Kettig, P., Blanchet, G., Peña Luque, S., and Baillarin, S. 2022 Dual state-parameter assimilation of SAR-derived wet surface ratio for improving fluvial flood reanalysis *Water Resources Research* **58** e2022WR033155 https://doi.org/10.1029/2022WR033155

IOP Conf. Series: Earth and Environmental Science 1136 (2023) 012019 doi:10.1088/1755-1315/1136/1/012019

Introduction of integrated decision support system for flood disaster management

Hao-Ming Hsu[1] **and Philippe Gourbesville**[1]

[1] Polytech Nice Sophia, University of Côte d'Azur, Nice, 06103, France

hao-ming.hsu@etu.univ-cotedazur.fr

Abstract. Heavy precipitation events and fluvial flood can cause serious damage to property and fatality, and thus flood disaster management is essential to lower the risk, to prevent and to mitigate the crisis. Effective and efficient flood disaster management requires on-site information, forecast and appropriate corresponding response strategies and measures. Therefore, a concept of integrated decision support system (DSS) for flood disaster management is proposed and discussed in this paper. The DSS integrates observation, rainfall forecast, fast flood simulation, historical events and crisis scenarios, and response strategies and measures, by means of information and communication technology (ICT) and internet of things (IoT). The DSS can provide relevant information and offer suggestions of action before and during the crisis. The DSS can also be utilised for disaster drill, education, and demonstration with the help of virtual reality (VR), augmented reality (AR) or Metaverse technology. The aims of this DSS are to improve existing response strategies and emergency services, to enhance the community-based disaster risk management, and to raise public crisis awareness.

1. Introduction

Heavy precipitation events and fluvial flood can cause serious damage to property and fatality, and thus flood disaster management is essential to lower the risk, to prevent and to mitigate the crisis. Effective and efficient flood disaster management requires on-site information, forecast and appropriate corresponding response strategies and measures. In order to interpret the complex data to responders and decision makers, decision support systems (DSSs) are good tools for providing useful and meaningful information. For instance, Gourbesville et al. [1] established a DSS combining several hydrological and hydraulic models and providing real-time assessment and forecasts for water management in south France; Yang et al. [2,3] developed a DSS integrating monitoring sensors, rainfall forecast, model simulation and response strategies against disasters for a science park in central Taiwan.

In this paper, we expand those former applications and propose a concept of integrated DSS structure for flood disaster management. The integrated DSS integrates real-time observation, rainfall forecast, fast flood simulation, historical events and crisis scenarios, and response strategies and measures, by means of information and communication technology (ICT) and internet of things (IoT). The proposed concept of the DSS can provide relevant information and offer suggestions of action before and during crisis. The DSS can also be utilised for disaster drill, education, and demonstration. The aims of the concept are to improve existing response strategies and emergency services, to enhance the community-based disaster risk management, and to raise public crisis awareness.

IOP Conf. Series: Earth and Environmental Science 1136 (2023) 012019 doi:10.1088/1755-1315/1136/1/012019

2. Proposed Framework

The integrated DSS consists of 6 different components: risk assessment, response strategies, database, smart water systems, decision support systems and its services, as shown in Figure 1. The detailed information of each component will be elaborated in the following sections from 2.1 to 2.6.

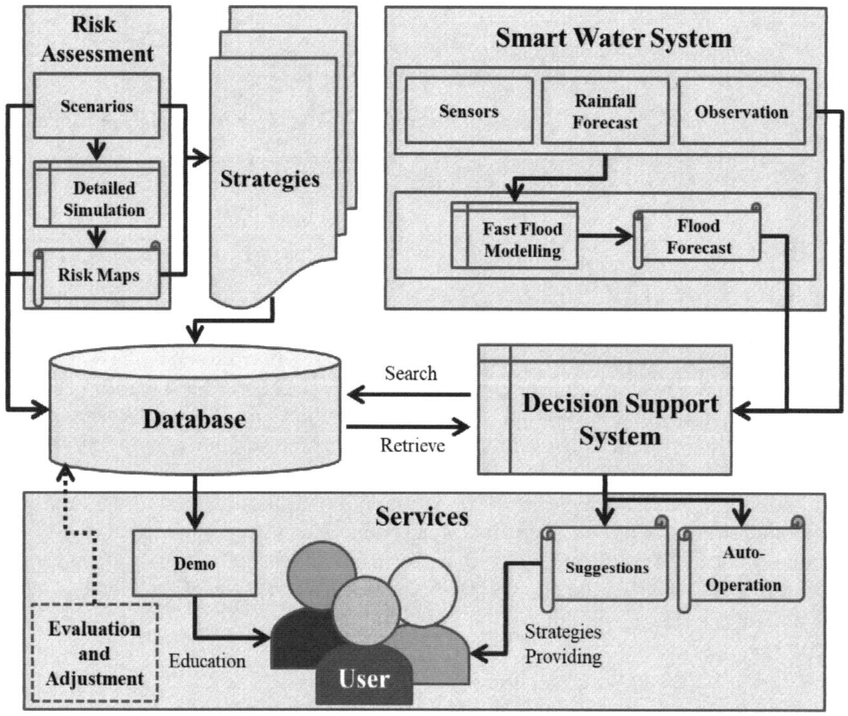

Figure 1. The structure of the decision support system.

2.1. Risk assessment

First of all, every possible scenario and meaningful historical disaster event should be determined, such as specific design return periods of rainfall and flood, possible operations of disaster prevention facilities and potential malfunction of the facilities, response strategies against disasters, deployment of disaster prevention devices and personnel, possible compound disasters et cetera. Then, based on the scenarios, we can perform corresponding detailed inundation simulations with high-resolution topographic data and with high-order numerical method. With the simulations, we can delineate potential inundated areas in detail and generate flood maps.

The flood maps can serve as references to evaluate the impact on socio-economic aspects. Other than hydrological and hydraulic modelling, we can also conduct simulations of interaction between human behaviours and the environment with the help of agent-based models (ABMs). For instance, antecedent research utilised ABM to analyse evacuation behaviours considering traffic interruption and traffic congestion during flood events [4] and tsunami, [5] and to analyse manpower for disaster management. [6] Finally we can even generate risk maps on various aspects apart from flood maps. The information of the location, the extent and the magnitude of the impacts should all be stored with the help of geographic information system (GIS). For instance, the point shapefile format for storing water level information, locations of historical flood events (damage, water level, discharge), locations of key facilities, and so on; the polyline shapefile format for representing sewer systems, drainage systems, dykes, and so on; the polygon shapefile format for historical flood maps, flood hazard maps, risk maps and so on.

2.2. Strategies

Response strategies should be developed based on the scenarios and the risk maps in the risk assessment phase. The corresponding response strategies can cover all phases of disaster, including disaster preparedness, disaster prevention, disaster mitigation, crisis responses, recovery after disaster. The strategies can include maintenance and operation of structural measures such as sewer systems, sluice gates, deployment and operation of pumps, and non-structural measures such as warning and alarm systems, food forecast, mobilisation, traffic diversion, evacuation, participation of volunteers, business continuity plan [7] and so on. Several different response strategies may be adopted in one scenario, and in the same vein, one strategy may satisfy various scenarios. Thus, the triggers of the all the strategies should be well-defined. The information of the strategies should be stored in tables indicating the relations, and therefore the relation between the scenarios and the strategies can be utilised for information retrieval in the DSS with the help of database management. Afterwards, the DSS can provide the relevant information and offer suggestions of action before, during and after crisis.

2.3. Database

A database is vital for storing and organising the data collected in the risk assessment phase and the strategies phase. The data should be well-categorised according to the connection between the scenarios, the risk maps and the strategies; therefore, the data can be retrieved accordingly. There are several different kinds of database model, such as hierarchical database, network database, relational database and so on. In this paper, a relational database structure is recommended due to its easy data fetching process and rapid response. Although a hierarchical structure can be intuitively adopted through the process of risk assessment and strategy development, i.e., scenario-flood-risk maps-strategies, it is not ideal for complex systems in which many factors are interconnected. Database management systems, such as MySQL, Microsoft Access, Oracle database and so on, can be employed to manage the data, enabling users to easily retrieve relevant data, to maintain the database, and to update the database. The illustration of the proposed database structure is shown as Fig. 2. The information of the historical events and the simulation output is stored in the Scenarios table and every scenario has a specific ID. Every scenario has one flood map and many risk maps on various aspects, and the relation is connected through the MapID. The outcomes of the scenarios are stored in the Phenomena table, containing the information about the scenarios. Many phenomena may result from the same scenario, and in other words, there are many phenomena have the same scenariosID in the table. Every phenomenon has its corresponding response strategies, and in this case, the same PhenomenaID may appear in different strategies. The related responders are connected with the strategies via ContactID, and therefore, the system can find the contact method of the responders.

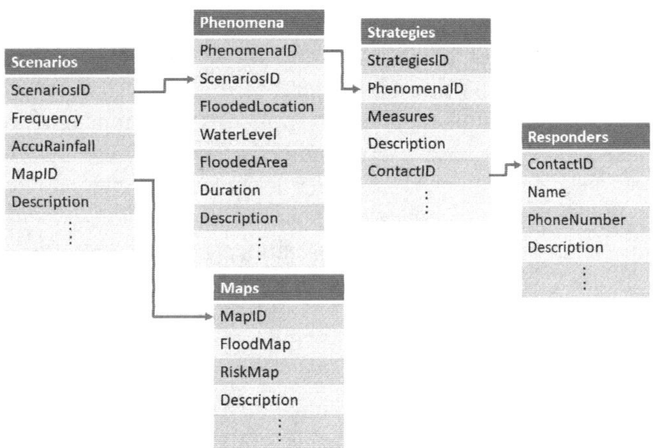

Figure 2. The illustration of the database structure.

2.4. Smart water system
This paper applies a concept of cyber-physical system, which uses available real-time observations and rainfall forecast to produce flood forecasts. First, the system collects observation from on-site sensors and monitors, such as rain gauges, water gauges in rivers and in sewer systems, for analysing. The observation data can be imported as initial condition and boundary condition into hydrological and hydraulic models. Then the system should collect rainfall forecast information for hydrological and hydraulic simulation. In this series of process, information and communication technology (ICT) and internet of things (IoT) technology are crucial to data collection and transmission, which should be well-maintained especially during crisis. [8] In order to provide flood forecasts in a short period of time for emergency response, a fast flood model is needed. There are various ways to shorten the computational time: reducing the complexity of 2D-1D (terrain-sewer) model to 1D-1D (street-sewer) model [2,3], running 2D model and 1D model separately instead of 2D-1D coupled model [9], using cellular automata for 2D modelling [10] et cetera.

2.5. Decision support system
The decision support system (DSS) should automatically collect all the relevant information, analyse the information, and then provide useful suggestions of actions. First, the DSS should be designed to automatically collect all the data from smart water system: on-site information, rainfall forecast and flood forecast. Then the DSS should automatically search from the database and retrieve flood map and risk maps and provide suggestions of actions from the designed response strategies according to the relation between all the strategies and the scenarios. The maps and the suggestions can be displayed on a specific website, and users can check the information on-line manually. The webpage should update regularly to provide the latest real-time information for end-users. Besides, the contact information of every responder and authority can also be retrieved from the database by the DSS, and therefore, the DSS can send the suggestions to them directly via mobile applications, email, text messages and so on.

2.6. Services
The database and the DSS can serve disaster management in all phases. In prevention and mitigation phase, the scenarios and the detailed risk maps can be utilised for education, and demonstration with the help of virtual reality (VR), augmented reality (AR) or Metaverse technology to facilitate the training of emergency personnel and to educate the public. For instance, water level from the risk maps and real-time data can be visualised to perceive the potential dangers through AR [11]; VR can also be used to visualise the flood scenarios for educating the public and to survey their behaviour when facing disaster. [12,13] In addition, all the scenarios and the response strategies can also be introduced into serious games and into a virtual world, such as Metaverse environment, for disaster education.

Previous studies show the benefit of serious games [14] and immerse experience in a virtual environment [15] for disaster education. In disaster preparedness phase, response phase and recovery phase, DSS can automatically provide meaningful suggestions of actions retrieved from the database. Decision makers and responders can react accordingly. Furthermore, the DSS can also be connected with disaster response-related facilities, such as water gates, pumps, traffic lights and so on; then the system can directly operate the facilities, under responders' supervision, during emergency situation. After crisis, DSS can offer recovery strategies. Information of the latest crisis can also be added into the database as new scenarios, and then the risk maps and the strategies can be updated as well after evaluation of the event.

3. Discussion
This paper proposes a concept of integrated decision support system for flood disaster management. The aims of this DSS are to improve existing response strategies and emergency services, to enhance the community-based disaster risk management, and to raise public crisis awareness. The system is

designed to provide suggestions of strategies and to be utilised for disaster drill, education and demonstration. Decision makers, first responders and the public can benefit from the system. Despite the advantages in various fields, there are still many gaps inside the structure to be filled. The DSS contains several different components and each one of these has been developed separately over a long period of time. However, these components are not usually well-interconnected. The most important issue about the operation of the DSS is the way of format standardisation, communication and interoperability between the various existing components.

The DSS can help to improve existing response strategies and emergency services. The users can save time and improve performance of reaction to the crisis under emergency condition. We can perform detailed simulations to assess the impacts of disaster and then develop corresponding strategies based on the scenarios. Afterwards, we can carry out further simulations to check the strategies and to assess the effect on various aspects. Moreover, the hydrodynamic simulation and agent simulation can also be combined together, and thus the autonomous agents (i.e., first responders, local residences, rescuers et cetera) can interact with the other agents and with the local environment (i.e., water depth, flow speed, traffic networks et cetera) in the models. We can refine the strategies through these iterations. At last, all the data should be well categorised into a database with well-organised relation between the data, and thus the DSS can automatically provide information and the optimum suggestions of strategies to users according to the real-time situation and flood forecast. the DSS can also be connected with disaster response-related facilities and hence the system can directly operate the facilities during emergency situation. Decision makers and first responders do not need to take a long time to analyse complex data for taking actions.

The DSS can also provide real-time observation information and flood forecasts. The system should first collect information from on-site sensors and monitors and then conduct fast flood simulation. For this service, ICT and IoT are the main concern. Every sensor and monitor should be well-connected and transmit information stably; however, the transmission commonly has problems during crisis. People should have back-up equipment to maintain the service during crisis. By the same token, we should also consider possibly invalid observation data and adopt substitute algorithm to generate flood forecast and to retrieve relevant responding strategies.

Additionally, we can also utilise the DSS to enhance the community-based disaster risk management and to raise public crisis awareness. The enormous data in the database, such as historical events and model predictions of predefined scenarios can serve for education purpose through AR, VR, Metaverse, serious games and so on. Previous reports emphases the importance of disaster prevention education and of self-protection and self-rescue [16]. Without crisis awareness and risk culture, even with well-designed response strategies, people may still have improper behaviours. Many individuals may not be able to understand the importance of self-protection and self-rescue and to prepare for risk of disaster unless they experience the disaster. Studies shows that virtual environment can help to increase motivation to evacuation and buy flood insurance [13], to increase investments for disaster prevention. [17] and to raise people's interest in disaster prevention and to offer information efficiently. Besides, serious games, as educational interventions, can offer positive experience to the users and increase their interest. [18] Users can experience, learn, and improve those they have learnt.

4. Conclusions

In this paper, we propose a concept of integrated decision support system for flood disaster management. The aims of this system are to improve existing response strategies and emergency services, to enhance the community-based disaster risk management, and to raise public crisis awareness. The system is designed to provide suggestions of strategies and to be utilised for disaster drill, education and demonstration. Decision makers, first responders and the public can benefit from the system. Despite the advantages in various fields, there are still many gaps inside the structure to be filled.

References

[1] Gourbesville P, Du M, Zavattero E, Ma Q and Gaetano M 2018 Decision support system architecture for real-time water management. *Advances in Hydroinformatics* pp 259–272 (Springer, Singapore) https://doi.org/10.1007/978-981-10-7218-5_17

[2] Yang TH, Yang SC, Kao HM, Wu MC and Hsu HM 2018 Cyber-physical-system-based smart water system to prevent flood hazards *Smart Water* **3**(1) 1–13

[3] Yang TH, Hsu HM and Kao HM 2020 Integrations of an early warning system and business continuity plan for disaster management in a Science Park *Advances in Hydroinformatics* pp 171–179 (Springer, Singapore) https://doi.org/10.1007/978-981-15-5436-0_14

[4] Dawson RJ, Peppe R and Wang M 2011 An agent-based model for risk-based flood incident management *Natural Hazards* **59**(1) 167–189. https://doi.org/10.1007/s11069-011-9745-4

[5] Wang H, Mostafizi A, Cramer LA, Cox D and Park H 2016 An agent-based model of a multimodal near-field tsunami evacuation: Decision-making and life safety *Transport. Res. C–Emer.* **64** 86–100 https://doi.org/10.1016/j.trc.2015.11.010

[6] Dressler G, Müller B, Frank K and Kuhlicke C 2016 Towards thresholds of disaster management performance under demographic change: exploring functional relationships using agent-based modeling *Natural Hazards and Earth System Sciences* **16**(10) 2287–2301

[7] International Standards Organisation 2012 *Societal security – business continuity management systems – Requirements* ISO 22301:2012 (London, British Standards Institution)

[8] Chang LC, Chang FJ, Yang SN, Kao IF, Ku YY, Kuo CL and Amin IMZBM 2018 Building an intelligent hydroinformatics integration platform for regional flood inundation warning systems *Water* **11**(1) 9

[9] Hsu MH, Chen SH and Chang TJ 2000 Inundation simulation for urban drainage basin with storm sewer system *Journal of Hydrology* **234**(1-2) 21–37 https://doi.org/10.1016/S0022-1694(00)00237-7

[10] Ghimire B, Chen AS, Guidolin M, Keedwell EC, Djordjević S and Savić DA 2013 Formulation of a fast 2D urban pluvial flood model using a cellular automata approach *Journal of Hydroinformatics* **15**(3), 676–686 https://doi.org/10.2166/hydro.2012.245

[11] Haynes P, Hehl-Lange S and Lange E 2018 Mobile augmented reality for flood visualisation *Environ. Modell. Softw.* **109** 380–389 https://doi.org/10.1016/j.envsoft.2018.05.012

[12] Fujimi T and Fujimura K 2020 Testing public interventions for flash flood evacuation through environmental and social cues: The merit of virtual reality experiments *International Journal of Disaster Risk Reduction* **50** 101690 https://doi.org/10.1016/j.ijdrr.2020.101690

[13] Zaalberg R and Midden CJ 2013 Living behind dikes: mimicking flooding experiences *Risk Analysis* **33**(5) 866-876 https://doi.org/10.1111/j.1539-6924.2012.01868.x

[14] Sermet Y, Demir I and Muste M 2020 A serious gaming framework for decision support on hydrological hazards. *Science of the Total Environment* **728** 138895 https://doi.org/10.1016/j.scitotenv.2020.138895

[15] Bredl K, Groß A, Hünniger J and Fleischer J 2012 The Avatar as a Knowledge Worker? How Immersive 3D Virtual Environments may Foster Knowledge Acquisition *Electronic Journal of Knowledge Management* **10**(1) 15–25

[16] Groupe URD 2020 *Évaluation en temps réel de la réponse à la tempête Alex dans les Alpes-Maritimes - 15 et 16 octobre 2020* [Real-time assessment of the response to storm Alex in the Alpes-Maritimes] Retrieved from: https://en.calameo.com/read/00368008687178aa8fdc5

[17] Mol J M, Botzen WJW and Blasch JE 2022 After the virtual flood: Risk perceptions and flood preparedness after virtual reality risk communication *Judgm. Decis .Mak.* **17**(1) 189

[18] Skinner C 2020 Flash Flood!: a SeriousGeoGames activity combining science festivals, video games, and virtual reality with research data for communicating flood risk and geomorphology *Geoscience Communication* **3** 1–17 https://doi.org/10.5194/gc-3-1-2020

Flood forecasting with Machine Learning in a scarce data layout

Théo Defontaine[1], Sophie Ricci[2], Corentin Lapeyre[2], Arthur Marchandise[3] and Etienne Le Pape[4]

[1] CERFACS, INP Toulouse, Toulouse, 31057 Cedex 1, France
[2] CERFACS-CECI, UMR 5318, Toulouse, 31057 Cedex 1, France
[3] Direction des Risques Naturels, DREAL Occitanie, Toulouse, 31074 Cedex 9, France
[4] Service Central d'Hydrométéorologie et d'Appui à la Prévision des Inondations, Ministère de la transition écologique, Toulouse, 31057 Cedex 1, France

Corresponding authors: defontaine@cerfacs.fr, ricci@cerfacs.fr, lapeyre@cerfacs.fr

Abstract. Flooding is one of the major natural disasters occurring in the world, with climate change increasing their occurrence and severity. Reliable flood forecasting models are needed to have better insurance in the emergency services' actions. This work reflects the capacities of Machine Learning models to improve discharge prediction results from empirical lag and route models based on hourly measured water level at gauge stations on the Garonne River. With scarce flooding data (30000 points) over the last 15 years, several learning algorithms have been implemented to predict floods in Toulouse at a 6-hour lead time from upstream stations providing hourly observations. A Linear Regression, a Gradient Boosting Regressor (Machine Learning) and a MultiLayer Perceptron (Neural Network, Deep Learning) are compared, using various strategies for learning, validating and predicting. Preliminary results show that the various strategies score as well as the empirical lag and route model. Further improvements are being investigated regarding the constitution of learning and validation data bases. This paper highlights how AI algorithms allow to improve the reliability of flood forecasts and how the layout of the limited volume of data influences their performance.

1. Introduction

As the effects of climate change become ever more significant, a surge of extremely large floods in European rivers has occurred in recent years. Accurate forecasts with long lead-times are crucial to give emergency services the opportunity to better react to these events and reduce their harmful effects. A good knowledge of various data on the geometry of the catchment, the properties of the land surface cover and the dynamics of the flow as in-situ observations of water level for instance is necessary to properly forecast flooding. The access to a large enough of good quality data remains a challenge.

This study was carried over the Garonne upstream of Toulouse, in south-western France. Full bathymetry and topography are not well known for this catchment as the river beds geometry strongly varies in the Pyrenees foothills and as bathymetry campaigns are still necessary. Fine hydrodynamics model solving Shallow Water Equations (SWE) such as MASCARET-TELEMAC [1, 2, 3, 4], is demanding and simpler approaches should be investigated. Rainfall-runoff model could be good candidate as rain data are available on the foothills but were not implemented so far for operational purposes over this catchment [5]. Hydrology modeling in Toulouse is currently based on the use of upstream measurements at a large number of hydrometric stations on this catchment.

The local Flood Forecasting Service (*SPC*, Service de Prévision des Crues) in Toulouse has implemented an empirical model based on upstream timeseries in order to issue short-term forecasts at Toulouse station. This upstream part of the Garonne River is landlocked by surrounding mountains and the flow is linear enough to be predicted empirically at short lead times. Yet, the extension of the forecast lead time of such model is a key challenge as security and economics are at stake. In that perspective, the use of more advanced models based on learning from observed data time series is investigated here, for instance with Machine Learning (ML) approaches [6, 7]. It should be noted that the limited knowledge of the topography/bathymetry as well as scarce data layout for significant flood events over the catchments are constraints that heavily weigh on the learning strategy. The objective of this paper highlights how AI algorithms allow to improve the reliability of flood forecasts and how the layout of the limited volume of data influences their performance.

In the literature, several ML algorithms have been explored and preliminary conclusions have been drawn. For instance, the use of Recurrent Neural Networks (RNNs) as well as Long Short-Term Memory neural networks (LSTMs) are too demanding in terms of volume of data [8]. The use of Bayesian Linear algorithm [9] could be investigated. In the present work, classical ML algorithms were considered as preliminary study. Here, a Linear Regression (LR), Gradient Boosting Regressor (GBR) [10] and a MultiLayer Perceptron (MLP) are investigated in a scarce data layout. The performance and the robustness of the algorithms were assessed with multiple criteria, with a focus on the impact of the data layout, including outliers, within the learning and validation databases.

The remainder of this article is organized as follows. Section 2 described the catchment and its hydrodynamics, the data that are available for the study and how they could be used in quasi-operational mode. The ML models and their implementation are described in Section 3, along with assessment criteria. The results for water level and discharge estimation and forecast are detailed in Section 4. Conclusions and perspectives are finally given.

2. Study case

2.1. Study area and forecast model

The Garonne catchment in Toulouse shown in Figure 1, is equipped with a hydrometric station named *Toulouse Pont Neuf*, where observations are made hourly. This station represents a catchment of 10 133.95 km^2. There are several water height measurements (also hourly) stations upstream of Toulouse, each situated at a certain propagation distance and associated lead time from Toulouse. The stations of *Marquefave* (watershed area of 5 237.98 km^2 on the Garonne main branch) and *Auterive* (3 471.79 km^2 on the Ariege tributary) are close to Toulouse, and located on a reach that is characterized by a quasi)-linear dynamics of the flow. The propagation times to Toulouse are estimated by the Toulouse SPC to about 4h. The effect of the rain between these stations was shown to be negligible. For these reasons a lag and route model adapted to a 4h forecast was successfully calibrated at SPC Toulouse.

When longer lead time are considered, five stations should be considered upstream of Toulouse as shown in Figure 1: *Mancioux* (catchment area of 2 810.96 km^2, Garonne main branch), *Roquefort sur Garonne* (1 574.95 km^2, Salat tributary), *Mas d'Azil* (220.21 km^2, Arize tributary), *Saverdun* (1 813.83 km^2, Ariege tributary) and *Mazères* (1 376.84 km^2, Grand Hers tributary). The *SPC* has set an empirical lag and route flow forecasting model between those 5 stations and Toulouse. It has been calibrated specifically for 6h forecasts during floods. As the dynamics on this reach is less linear than that of reaches that are closer to Toulouse and as the number of tributaries is large, the calibrated model shows his reach shows fewer satisfying results. It should also be noted that the flow dynamics is now significantly influenced by rain fall, thus advocating for a more advanced modelling strategy for extended lead times and larger catchments.

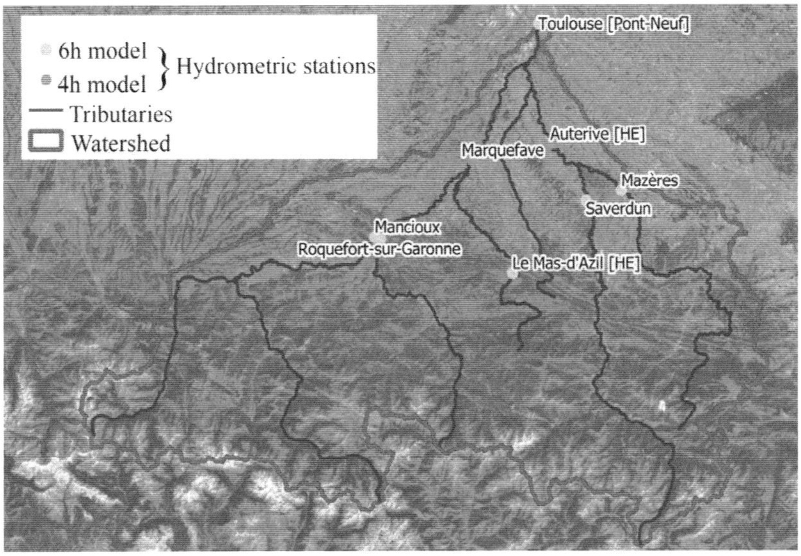

Figure 1. Garonne Catchment (in red) and its tributaries (in blue). Red dots are for the two hydrometric stations 4h upstream of Toulouse, yellow dots are for the five hydrometrics stations 6h upstream of Toulouse. Toulouse takes part in both models' outputs.

2.2. The database

The hydrometric stations used in this case provide hourly water level measurements throughout the year. Each of them has an experimental setting curve estimate by SPC from a limited number of discharge and height measurements on field, rarely in high flow. The rating curve is usually extrapolated for high flows, leading to a significant uncertainty in discharge estimates from water level measurement during flooding events.

All models that are considered in the present paper are calibrated for floods; the data sets are thus only composed of data measured during flooding events. Since 2005 (time at which the stations were consistently measuring), there has been about 40 flood events, half of them not rising the *SPC* vigilance level. The selection criterion for these events is based on a water level gauge in *Toulouse* which alerts the *SPC* when the threshold is exceeded. A preliminary study with 20 of those events from 2005 to 2015 showed promising results [11] for learning algorithms. Here, the database was extended so as to consider 30 flood events between 2007 and 2018. The data for these events were rearranged to fit those of a rainfall-runoff model calibration set with the Plathynes platform [12] at SPC for research purposes. These 30 flood events amount to 30000 data points available to train, validate and test the AI algorithms as each event lasts around 100 hours.

It should be noted that these events differ in various ways (severity, length, multiple/single peak, date, origin, ...) For instance, rain does not always fall on the same location of the catchment, with the same intensity, the input flow varies with the month of the year and consequently with different impact of the snow melt process from the Pyrenees. The database is thus heterogeneous and scarce, it is assumed here that it is representative of the possible flood event that may occur over the catchment (for less than centennial flooding).

2.3. The SPC model: a baseline

The model discussed here is the *SPC*'s **6h-model** (similar to the 4h model); it was calibrated at SPC with gauge measurements (including some minor events).

SPC has set an empirical lag and route model (similarly to [13]) based on the hypothesis that the upstream stations are all situated at a 6h propagation time from Toulouse. For each time step, they use a 6h fixed flow lag (water levels are converted into discharge through local rating curves) for each upstream station before summing them as shown in equation (1):

$$\underset{t\to t+6}{\Delta} Q_{\mathrm{Tls}} = Q_{Tls}(t+6) - Q_{Tls}(t) = a_{\mathrm{Mn}} \underset{t-6\to t}{\Delta} Q_{\mathrm{Mn}} + a_{\mathrm{Rq}} \underset{t-6\to t}{\Delta} Q_{\mathrm{Rq}} + a_{\mathrm{Ms}} \underset{t-6\to t}{\Delta} Q_{\mathrm{Ms}} + a_{\mathrm{Sv}} \underset{t-6\to t}{\Delta} Q_{\mathrm{Sv}} + a_{\mathrm{Mz}} \underset{t-6\to t}{\Delta} Q_{\mathrm{Mz}} \quad (1)$$

with $a_{\mathrm{Mn}} = 0.91$, $a_{Rq} = 0.70$, $a_{Ms} = 3.00$, $a_{Sv} = 0.40$ and $a_{\mathrm{Mz}} = 0.65$.

SPC manually calibrated this model with 20 events in the 2005-2015 temporal window, selecting events for their features. Some of those events featured only one tributary contribution thus allowing the SPC to calibrate the corresponding tributary static coefficient. The five static coefficients shown in equation (1) are calibrated one by one with events that are mainly influenced by one. These coefficients compensate for the fact that rain input data are not accounted for.

2.4. Experimental settings

As seen in subsections 2.2 and 2.3, the *SPC* model is fed with flows in input, through setting curves. It also brings some uncertainties. To see the extent of their reliability, an experiment was made for the AI algorithms without the setting curves, named E_h in table 1. The experiment without the curves is named E_Q in the following table 1.

The learning experiments also differ from the data base setting as shown in Figure 2. In the following, for each learning algorithm, the data was split into a training set (~80% of the whole dataset) and a testing set (~20%). When the events in the learning data base are selected chronologically, meaning that whatever happens in the future is caused by what happened in the past, the experiment is called E_c (in table 1). In this case, the test set is composed of the 6 last events. When the events in the learning data base are taken arbitrarily, the experiment is called E_a. Figure 2 shows the repartition of the events in the sets. In this latter case, the validation set is similar to that in [12].

Table 1. Specific designation of the modeling experiments.

Experiment	E_Q	E_h
E_c	**dqc**	**dhc**
E_a	**dqa**	**dha**

The experiments here are related to two modifications of the setup: one being the choice of the testing set (E_c and E_a in table 1), the other being the choice of the kind of quantity learned (E_Q and E_h in table 1). Each experiment has thus two variations (one for each experiment related to the other modification).

Figure 2. Location of test set events for E_c and E_a. Orange coloured dots stands for training events, green for test events.

It should be noted that the input data are considered as instantaneous values. The treatment of the temporality of the data within the learning process is beyond the scope of this paper.

3. Machine Learning algorithms and validation criteria

3.1. Machine Learning Algorithms

Three different Machine Learning algorithms were implemented in python 3 using scikit-learn [14] and TensorFlow with Keras [15].

3.1.1. Linear Regression. For this scikit-learn's *linear_regression* module, basic settings were used. Here, the output y_i is supposed to be a linear combination of the inputs x_i plus random noise ϵ_i, as shows equation (2):

$$\forall i \in [\![1, n]\!], \qquad y_i = \sum_{i=1}^{n} x_i \alpha_i + \epsilon_i \quad with\ \epsilon_i \sim \mathcal{N}(0, \sigma^2) \tag{2}$$

where α_i and σ are à priori unknown. α_i being the weights to be determined. The σ are supposed to be independent of each other. This scikit-learn algorithm aims at minimizing the quadratic error norm of the weights. It is a least squares regression. As expected, this simple algorithm performs poorly when tackling complex processes.

3.1.2. Gradient Boosting Regressor (GBR). The idea behind the gradient boosting algorithm is to minimize the error of a chosen simple estimator on the observations. To do so, other simple estimators are used recursively to correct the error of the previous estimator. The recursion follows a steepest gradient descent. The whole mathematical process is described in [16].

With this more complex algorithm, ranges of operability are used for a subsample of the hyper-parameters that may be optimized as explained in Section 3.2. The scikit-learn version of this algorithm was used involving the Python class *GradientBoostingRegressor*. Two different setups are implemented, with *absolute error* metric and loss:

- The first setup takes as input the data from the upstream stations and tries to minimize the *absolute* error relative to the observations in Toulouse.
- The second setup takes as input the upstream observations together with the output of the *SPC* model and the *Linear Regression*. The main purpose of this setup is for the *GBR* to correct the outputs of the other two given as inputs. This setup makes the *GBR* act as an error correction algorithm. In the following, this algorithm will be referred to as *GBR+*.

3.1.3. MultiLayer Perceptron (MLP). This *Neural network* is the simplest multilayer Deep Learning algorithm. Its implementation is done with Keras [15, 17] (with the *TensorFlow* underlying system). It had 3 to 4 fully connected layers (1-2 hidden), with up to 100 nodes each. The last layer was a single node layer to have a scalar output.

The choice was made to convert the basic *Sequential* algorithm into an *estimator.* It is then easier to use other ML tools (especially those from scikit-learn). With this DL algorithm, ranges of operability are used for a subsample of the hyper-parameters that may optimized as explained in Section 3.2. Here, two different setups were also implemented, the same ones as explained for the *GBR*, with *absolute error* metric and loss*:*

- The first setup takes as input the data from the upstream stations and tries to minimize the *absolute* error relative to the observations in Toulouse.
- The second setup takes as input the upstream observations together with the output of the *SPC* model and the *Linear Regression*. The main purpose of this setup is for the *MLP* to correct the outputs of the other two given as inputs. This setup makes the *MLP* act as an error correction algorithm. In the following, this algorithm will be referred to as *MLP+*.

3.2. Optimization process
For GBR and MLP, an optimization process was implemented. *RandomizedSearchCV* from scikit-learn was investigated. It consists in giving each hyperparameter a range of values to be tested during training. To limit overfitting on the training set with this method, a cross validation step during training is mandatory.

3.2.1. Cross Validation. Cross validation is used during training to prevent optimized algorithms from overfitting. A grouped cross validation algorithm was used to prevent events being cut in parts. The *GroupKFold* (from scikit-learn) was splitting the training data into 5 groups that were used one by one as a validation set (one group) and training set (the remaining 4 groups). Thus, there is 5 different training plus validation splits on which we can train the algorithms.

3.2.2. Randomized Search. Following a Monte Carlo method, several algorithms are tested at once. The sampled algorithms have randomly chosen hyperparameters in the ranges given for them. With enough samples over the ranges given, it is considered that the best found hyperparameters are close to optimal ones that could have been found by searching exhaustively over the ranges [18]. This optimizer uses a Cross-Validation as explained above, it trains on one set of hyperparameters over the 5 different splits and then calculates the estimator's average metric value for this set of hyperparameters.

3.3. Performance Criteria
For the training/validation, as shown above, each algorithm had its own metric. For better comparison, the same 3 criteria were used to assess their performances on the test set and on the result of the training. In this section, every time series is considered discrete and thus numbered in the range $[\![1, n]\!]$. Each criterion was also computed in two different ways: by test/training set (a set is a list of several events) and by event.

3.3.1. Coefficient of determination (R^2 score). *This first criterion has the* SPC *model as a baseline and computes the relative squared error of the tested algorithm with respect to the reference model:*

$$R^2(\tilde{Q}, \hat{Q}) = 1 - \frac{\sum_{i=1}^{n}(\tilde{Q}_i - \hat{Q}_i)^2}{\sum_{i=1}^{n}(\tilde{Q}_i - \bar{\tilde{Q}}_i)^2} \tag{3}$$

where $\hat{Q}_i$ is the predicted value, $\tilde{Q}_i$ is the *SPC* baseline and the mean of the reference is $\bar{\tilde{Q}}_i = \frac{1}{n}\sum_{i=1}^{n}\tilde{Q}_i$.

Computing this over an event implies that n is the length of the event. When computing over a set of several events, *n* is the sum of the length of the events. When computed by set is also not equal to the mean of the corresponding events values. It variates between 1 (perfect prediction) and $-\infty$ (poorest prediction). A value of 0 means that the target has a mean value equal to the mean of the reference.

3.3.2. Nash-Sutcliffe Efficiency coefficient (NSE). This criterion formulates the relative error of an algorithm with respect to the observed flow. It reads:

$$NSE(Q, \hat{Q}) = R^2(Q, \hat{Q}) = 1 - \frac{\sum_{i=1}^{n}(Q_i - \hat{Q}_i)^2}{\sum_{i=1}^{n}(Q_i - \bar{Q}_i)^2} \tag{4}$$

where $\hat{Q}_i$ is the predicted value, Q_i is the *observed flow* and $\bar{Q}_i = \frac{1}{n}\sum_{i=1}^{n}Q_i$. The *NSE coefficient*'s equation is the exact same as the R^2 *score*, the only difference being the reference data, here it is the observed flow, whereas with the R^2 *score* it is the *SPC model*'s results. This criterion has the same behaviour as the R^2 *score*.

3.3.3. Persistence criterion. This criterion computes the temporal shift between the modelled value and the observed value, relatively to a predetermined shift as shown in equation (5):

$$Persistence(Q, \hat{Q}) = 1 - \frac{\sum_{i=\Delta j}^{n}(\hat{Q}_i - Q_i)^2}{\sum_{i=\Delta j}^{n}(Q_{i-\Delta j} - Q_i)^2} \tag{5}$$

where $\hat{Q}_i$ is the predicted value, Q_i is the *observed flow* and Δj the predetermined shift of the time series, which is 6 in our case. For this criterion, the behaviour in terms of values obtained is the same as the previous criteria. In terms of computation, the criterion is only obtained per event. It would be meaningless to compute it over a set as it is time-dependent. The set score is thus the mean of the events scores.

4. Results and discussion

4.1. Machine Learning Algorithms' performances in the different experiments

Figure 3 displays the criteria values (test and training) for all 4 experiments' configurations. The metrics on the left (a), (c) and (e) are computed by set (training in light orange/test in dark green) for each one of the 4 experiments' configurations given in table 1. The metrics on the right (b), (d) and (f) are computed by event for the same experiments. Light orange dots indicate events that were used during training and dark green dots indicate events that were used for test. R^2 scores are plotted between -1 and 1. The other metrics are plotted between 0 and 1. It was observed that all three ML algorithms behave in the same way for the different experiments. For clarity purposes, only results for GRB+ are shown in the following, with loss of generality in the conclusions.

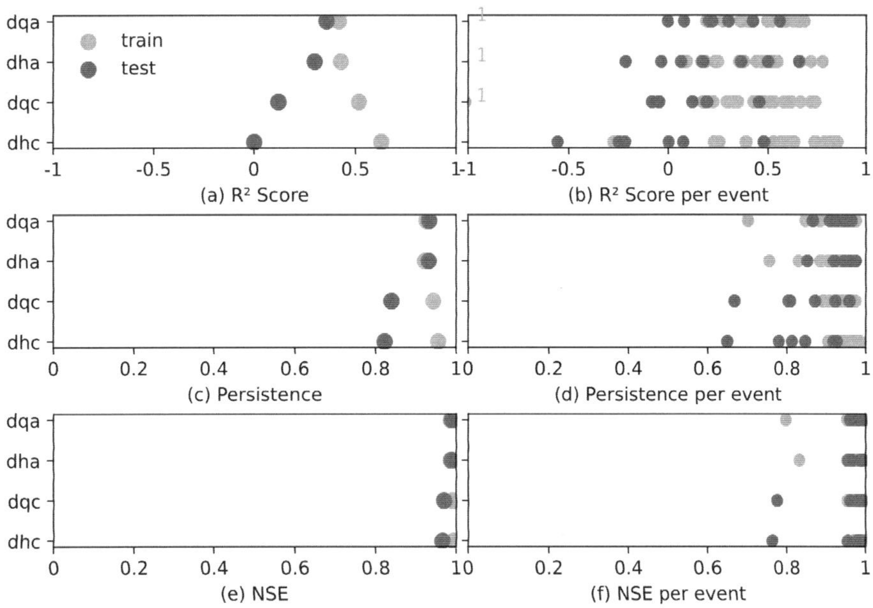

Figure 3. *GBR+ Algorithm.* Comparison of the different experiments' performances with respect to the various criteria. In (a), (c) and (e) each experiment is represented on a line (dqa, dha, dqc and dhc), test score in green, training score in orange. The orange line shows that training scores better than test. In (b), (d) and (f) the same scores are computed by event and grouped in colours by test/training set.

Figure 3 shows slightly better results for E_a with even better results in test than in training. This can be explained by the fact that more recent events (since June 2018) are less similar to the older ones, thus testing on them while training on older ones deteriorates measured performances (chronologic configuration). An outlier also seems to be present in the causal experiment E_c test set, visible on figure 2 (d) and (f). Arbitrarily chosen test events also appear to be among the most favorable ones in the dataset when it comes to scoring. They seem to be amongst the typical events on which the algorithms are well trained because they are well represented in the database.

The second part of these results show that there are higher uncertainties when learning directly from heights (dha, dhc). It confirms the *SPC*'s opinion that their rating curves are calibrated well enough and do not lead to significant errors. Plus, learning with flow variables is safer as it is derived from the mass balance equation. In this case, that equation is not closed because rain inputs are not taken into account and only balanced in the coefficients of the algorithms, thus uncertainties still remain.

4.2. Comparing algorithms performances for the same experiment

The most favourable strategy *dqa* is further investigated here, considering all available algorithms. Table 2 shows that the *GBR+* is the best scoring version of our algorithms, and the most consistent in that configuration. This result is not surprising as Gradient Boosting algorithms are easier to calibrate and have better manoeuvrability than MLPs in a scarce data layout.

The algorithms are in a favourable configuration as most test scores are above training scores in table 2. This is unusual for Machine learning algorithms. This can be explained by the low outlier that shows up in the training set. It is visible in figure 3 (b) where the orange number 1 shows that one event has performances worse than -1. The persistence and NSE scores of *dqa* in figure 3 (d) and (f) also shows a training event lower than others.

Another important feature shown in table 2 is that our algorithms all seem to outperform the *SPC model*. This was a targeted feature, our algorithms being more complex, it is not that surprising to see them perform better. It also shows the robustness of their implementation as they outperform SPC in all 3 criteria.

Table 2. Performances of the various algorithms in the *Arbitrary, flow experiment* (**dqa**)

Model	R^2 Score		Persistence		NSE	
	Train	Test	Train	Test	Train	Test
SPC	0	0	0.867	0.902	0.976	0.983
LR	0.237	0.325	0.898	0.929	0.982	0.989
GBR	0.344	0.274	0.916	0.926	0.984	0.988
GBR+	0.417	0.354	0.926	0.932	0.986	0.989
MLP	0.255	0.279	0.904	0.927	0.982	0.988
MLP+	0.251	0.272	0.904	0.926	0.982	0.988

4.3. Scarce data effects

Figure 4 gives a focus on a chosen test event with the configuration *dqc* (see table 1). It appears that all model fails to properly forecast this event (observations are plotted in black), similarly to the SPC model forecast represented in red. As the inputs upstream of Toulouse show limited flow rise, the flood peak is either due to some upstream influence that is not accounted for in the list of selected stations or to influence of rainfall in-between observing stations. This type of event and its influence in the learning and test sets should be further investigated. Indeed, the models outperforms with significant errors: this event has a mean absolute error (MAE) of 20cm and a maximum error of 109cm for the SPC model (the worst one) and a MAE of 19.69 cm, maximum error of 96cm for the *extended GBR*. These are way above average (less than 10 cm for MAE, less than 50cm for maximum error).

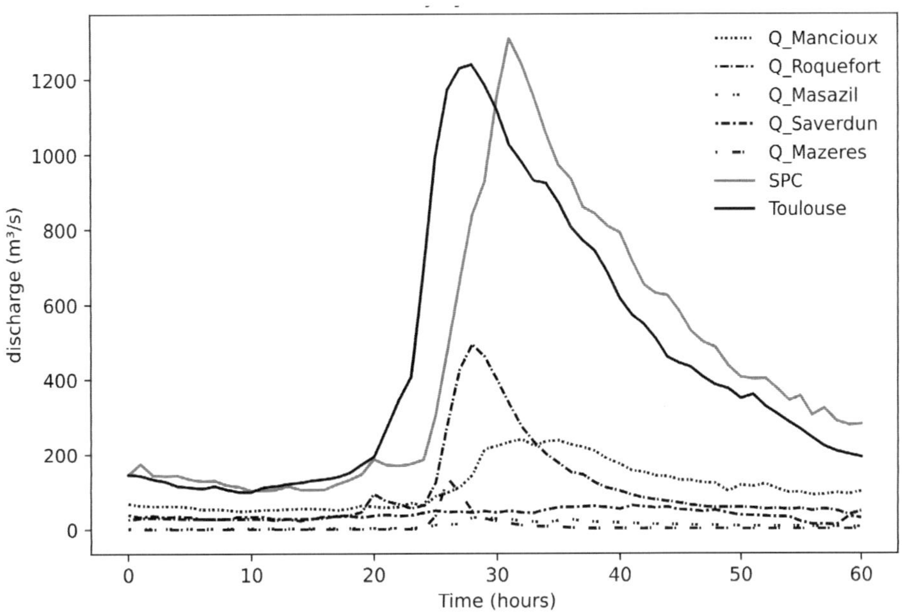

Figure 4. 15 July 2018 event. Observed flow inputs with a 6h-lag and flow output are shown (with the *SPC model*'s result for comparison).

5. Conclusion

Even though the data is scarce in this problem, good performances can be achieved with these models, with even better results than the operational model of the *SPC*. Attention must be put on the robustness of the algorithms used when the dataset is scarce as it is the case here. The experiments show that working with variables derived from a closed equation has better results, as the information then gathered by the algorithms is more robust. The scarcity also makes the problem less stable. The experiments conducted here also shows that there is a real influence of the choices made to separate the data into sets. Thus, this must be done with the most specific attention. There still lies some outliers that require to be better treated. They must be identified and the results treated with this knowledge as their influence is non negligeable. It can be wise to gather non global results to tackle these two points.

Although the dataset is pretty restricted, convincing results with simple algorithms are obtained. The scarcity of the events is the major problem here, making the algorithms less stable. Indeed, when complexity is added, most of the algorithms are bond to become divergent. Thus, any complexity add-on in a scarce layout must be done carefully. Complexity and scarce data do not fit well together.

To improve results, the authors aim to input more information, such as rain data or temporal window for data use. In both cases, the learning task becomes more complex and the stability and robustness of the algorithms becomes more challenging. The preliminary results obtained here are promising, with room for improvement.

References

[1] Tiberi-Wadier A L Goutal N Ricci S Sergent P Monteil C 2019 Sensitivity Analysis of the Mascaret Model on the Odet River *XXVIth Telemac & Mascaret User Club*

[2] Broich K Pflugbeil T Disse M Nguyen H 2019 Using Telemac-2D for Hydrodynamic Modeling Rainfall-runoff *XXVIth Telemac & Mascaret User Club*

[3] Barthelemy S Ricci S Rochoux M C Le Pape E Thual O 2017 Ensemble-based data assimilation for operational flood forecasting – On the merits of state estimation for 1D hydrodynamic forecasting through the example of the "Adour Maritime" river *Journal of Hydrology 552* pp 210-224

[4] Barthelemy S Ricci S Morel T Goutal N Le Pape E Zaoui F 2018 On operational flood forecasting system involving 1D/2D coupled hydraulic model and data assimilation *Journal of Hydrology* pp 623-634
[5] Todini E 1988 Rainfall-runoff modeling – Past, present and Future *Journal of Hydrology* pp 341-352
[6] Dawson C D Wilby R 1998 An artificial Neural Network approach to rainfall-runoff modeling *Hydrological Sciences Journal* pp 47-66
[7] Toukourou M Johannet A Dreyfus G Ayral P-A 2011 Rainfall-runoff modeling of flash floods in the absence of rainfall forecasts: the case of "Cévenol flash floods" *Applied Intelligence, Springer Verlag (Germany)* pp 178-189.
[8] Kratzert F Klotz D Shalev G Klambauer G Hochreiter S Nearing G 2018 Rainfall-runoff modeling using Long Short-Term Memory (LSTM) networks *Hydrology and Earth System Sciences* pp 6005-6022.
[9] Noymanee J Theeramunkong T 2019 Flood Forecasting with Machine Learning Technique on Hydrological Modeling *Procedia Computer Science* pp 377-386
[10] Sanders W Dongfeng L Wenzhao L Zheng N F 2022 Data-driven Flood Alert System (FAS) using Extreme Gradient Boosting (XGBoost) to forecast flood stages *Water 14 number 5: 747* https://doi.org/10.3390/w14050747
[11] Nony B Lapeyre J C Ricci S 2019 Prédiction et reconstruction de données hydrauliques par apprentissage machine *unpublished*
[12] Demazels L Le Pape E 2021 Modélisation Hydrologique de la Garonne à Toulouse *unpublished*
[13] Vidal J J Dupouyet J P Murillo T Deltheil T Boignard J P 1998 SOPHIE : Système Ouvert de Prévisions Hydrologiques Informatisé avec Expertise *Journées de l'Hydraulique* pp 389-398
[14] Pedregosa *et al.* 2011 Scikit-learn: Machine Learning in Python, *JMLR 12* pp 2825-2830.
[15] Chollet *et al.* 2015 Keras https://keras.io
[16] Friedman J H 2001 Greedy function approximation: a gradient boosting machine. *Annals of statistics* pp 1189-1232
[17] Gull A Pal S 2017 Deep Learning with Keras *Packt Publishing Ltd.*
[18] Bergstra J Bengio Y 2012 Random search for hyper-parameter optimization *The Journal of Machine Learning Research*

IOP Conf. Series: Earth and Environmental Science 1136 (2023) 012021 doi:10.1088/1755-1315/1136/1/012021

Integrated modelling approach for flood forecasting in small Mediterranean catchment – Application to the Cagne Catchment, France

Mingyan Wang[1], Paguédame Game[1] and Philippe Gourbesville[1]

[1] Université Côte d'Azur, Polytech'Lab, Sophia Antipolis, 06410, France

mingyan.wang@etu.univ-cotedazur.fr

Abstract. The French Mediterranean region is vulnerable to flooding caused by extreme rainstorm events. To better predict floods and anticipate consequences, the technical managing services operating on the field need to analyze the hydrological situation, receive accurate forecasts and issue alerts to the exposed population. Within this context, the various needed modelling tools are integrated within a common framework of a Decision Support System (DSS). This approach has been developed and implemented in the Cagne catchment located in the French Riviera. The Cagne catchment covers 96 km^2 area and it is densely urbanized with high risks of inundation in the downstream area. The hydrological model MIKE SHE covers 96 km^2 with a 20 m resolution. The river network is represented in a 1D MIKE 11 model by 18 branches and 1899 cross sections. The downstream area is covered with a high-resolution hydraulic model built with MIKE 21 FM. The model receives inputs from the hydrological model and can produce detailed inundation maps and forecasts. A major rainfall event observed from October to December 2020 is currently used as validation situation and the built modelling system has demonstrated its efficiency to reproduce the observed processes and to deliver the forecasts on due time.

1. Introduction

The French Mediterranean region is characterized by increased vulnerability to flooding caused by extreme rainstorm event due to global climate change and by the intense urbanization that generate an increase of the surface runoff. Recently the French Riviera suffered from 2 sever flood disasters in 2015 and 2020, which caused serious economic losses and casualties [1].

To efficiently manage this kind of flood hazard and effectively reduce the damage costs, a Decision Support System (DSS) has been requested to produce a comprehensive view of current situation in real time and further provide accurate forecasts as fast as possible for supporting the decision makers and the field operators. The DSS for flood forecast has been a hot topic in water management research since numerous decades [2]. Ford DT. [3] has made a flood-warning decision support system including the HEC programs as the analytic tools in Sacramento County, California in 2001. In 2005, Ahmad S, et al. [4] developed a DSS called DESMOF (Decision Support for Management of Floods) in Red River Basin in Manitoba, Canada. it consists of one-dimensional (HEC-RAS) and two-dimensional (MIKE 21) hydrodynamic modelling tools to simulate the runoff process. The artificial neural network is also introduced in flood forecast. In 2016, a GIS based DSS is developed by Zhang W et al. [5] for emergency evacuation against dike-break flood in Jinjiang, China. MIKE 21 model is used for two-dimensional surface flow simulation. In 2018, Şensoy A et al. [6] established a DSS integrated by HEC-HMS hydrological model, HEC-RAS hydraulic model and HEC-ResSim model for a critical dam reservoir in Turkey. This study will therefore focus on the modelling approach that can be integrated in the DSS for flood forecast.

The study area, the Cagne catchment, is a small catchment (96 km^2) lies in the French Mediterranean region. The Cagne River originates from the Coursegoules at an altitude of 1,535 meters, it flows through the city Cagnes-sur-Mer and into the sea. The hydrological conditions of the Cagne catchment are typical Mediterranean regime with rainfall floods, strongly affected by autumn precipitations, leading to an increased discharge from October to January [7]. Over the past 10 years, a large amount of topography, climate and hydrological data has been collected in this catchment, which provides the possibility to build hydrological and hydrodynamic models and to use them for flood prediction as the vulnerability of the urbanized downstream area has increased. The objective is then defining the flood simulation models that best suit to this specific catchment size and hydrological conditions and then to integrate them within a DSS.

2. Needs and Selected Methods

Water information technology, especially modelling tools (hydrological and hydraulic modelling), has been widely used in the integrated water management filed to analyze the historical hydrological events and to make forecast of potential flood events. To provide an efficient support to the field decision makers, these models should be accurate and then will play an important role in the Decision Support System.

In this study, the distributed hydrological model MIKE SHE [8] and the hydraulic model MIKE 21 FM [9] (both developed by DHI) were selected for the local Decision Support System operated by the local teams.

There are mainly two types of hydrological models - lumped hydrological model and distributed deterministic hydrological model – that can be integrated within a DSS. Lumped model homogeneously simplifies the distributed characteristics of the catchment in the simulation. Even though the parameters in this type of model may have physical meaning, they were most often set through model calibration rather than establishing their values from catchment physical characteristics. As a result, the lumped model is unable to accurately categorize hydrological processes and could only analyze catchment responses at the outflow without accounting for particular sub-basin responses. In the other hand, the distributed model represents the catchment by breaking down the whole research region into several smaller sections. This type of model keeps physical information with a specific level of precision and considers the spread nature of hydrological variables including soil, land use, and slopes.

The biggest challenge of the distributed models is the shortage of high-quality input data. However, with the development of data collection technology, systematic data collection on topography (including high resolution data produced with LIDAR solutions), climate and hydrological variables have permitted to gather a significant knowledge on the main hydrological processes within the Cagne catchment and to engage the hydrological analysis with a distributed approach.

1 m resolution DEM data provided by Métropole Nice Côte d'Azur is used to build a one-dimensional free-surface hydraulic MIKE11 model, the cross-sections are extracted from the DEM with MIKE HYDRO tool. The river network is made up of 18 branches and 1889 cross-sections with an interval of 50 m (as shown in Figure 1.). Then it is coupled with MIKE SHE to increase the accuracy. The 100 m resolution land use map collected from Métropole Nice Côte d'Azur is used to estimate the value of the Strickler coefficient. Soil type data based on satellite images is produced by European Soil Center with 500m resolution, it is used to estimate the related parameters and the initial soil depth. One distributed data base called SAFRAN [10] produced by Météo-France including daily meteorological information such as rainfall, air temperature and evapotranspiration with 8000 m by 8000 m resolution is also used as input data for MIKE SHE models. Last but not least, the rainfall data is collected from rain gauge station St. Jeannet with a 6-minutes interval, covering a period from 1st October 2020 to 31st December 2020.

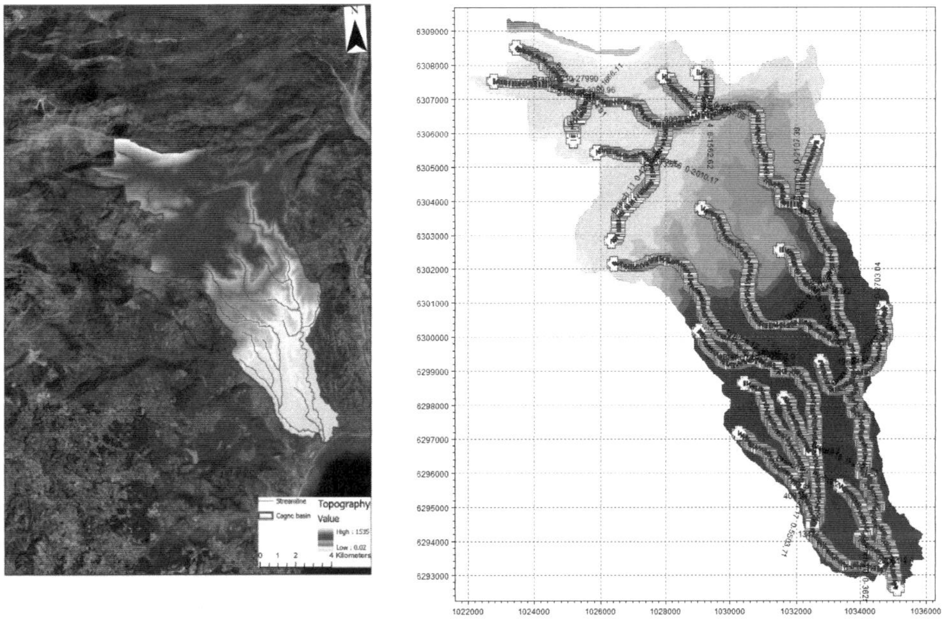

Figure 1. Cagne catchment and MIKE 11 river network.

The hydrologic model will allow to generate the flood hydrographs and the flooding process located within the downstream area will be simulated with a 2D deterministic high resolution hydraulic model. The hydraulic model can represent the discharges within the riverbed and also simulate the extension of the inundation processes. To obtain a good representation of the complex hydraulic processes, a MIKE 21 FM hydrodynamic two-dimensional flexible mesh model was selected. The model can represent a wide range of physical and chemical processes, including hydrodynamic behavior, pollutant transport, and morphological changes [11]. MIKE 21 FM is based on a flexible mesh approach. The nonorthogonal triangular construction of the mesh enables a flexibility in the resolution across the model area compared to the strict raster grid utilized in the MIKE 21 Classic approach. This allows the user to focus their attention on certain sections of the model. Special care may be required, for example, to better represent a small region with complicated elevation conditions or to provide a realistic representation of buildings that are important to the hydraulic situation. In contrast, the user's flexibility allows them to pay less attention and lower the resolution in less important regions. A lower resolution will result in fewer mesh components and, as a result, a faster processing time. In other words, the flexible mesh enables the user to divert computing power from homogenous low-importance areas to heterogeneous high-importance areas [12].

In this study, the MIKE 21 FM model is built to simulate the surface runoff and to generate the inundation maps in the downstream part of the Cagne catchment (8.1 km2). It is selected to do the hydraulic simulation because upstream part is mostly forest and agricultural land, downstream part is dense urban area Cagnes sur Mer city center, which is more prone to the inundation. An extreme rainfall event of 23rd October 2020 to 25th October 2020 that covers a peak rainfall is selected to do the simulation. 1 m topography data from Métropole Nice Côte d'Azur covering the whole study area is used as topography input. To make the simulation more accurate and efficient, the domain is divided into 3 parts: riverbed, floodplain and sea area. Different resolutions are applied to different areas, 10 m for riverbed, 30 m for floodplain, 100 m for sea area (as shown in Figure 2.). The discharge result from MIKE SHE model is used as upstream boundary condition, the sea area boundary condition is set as free outflow, and rest of the land boundary is set as zero normal velocity. The rainfall data for the period of 23rd October 2020 to 25th October 2020 from rain gauge station St. Jeannet. Within the

IOP Conf. Series: Earth and Environmental Science 1136 (2023) 012021 doi:10.1088/1755-1315/1136/1/012021

downstream area, the part of the Cagne River is fully covered. This sector called Cagne Couverture, it is 250 meters long with a rectangular profile of 3 meters high and 8 meters wide. This culvert is simulated by using the long culvert function in MIKE 21 FM as shown in Figure 3.

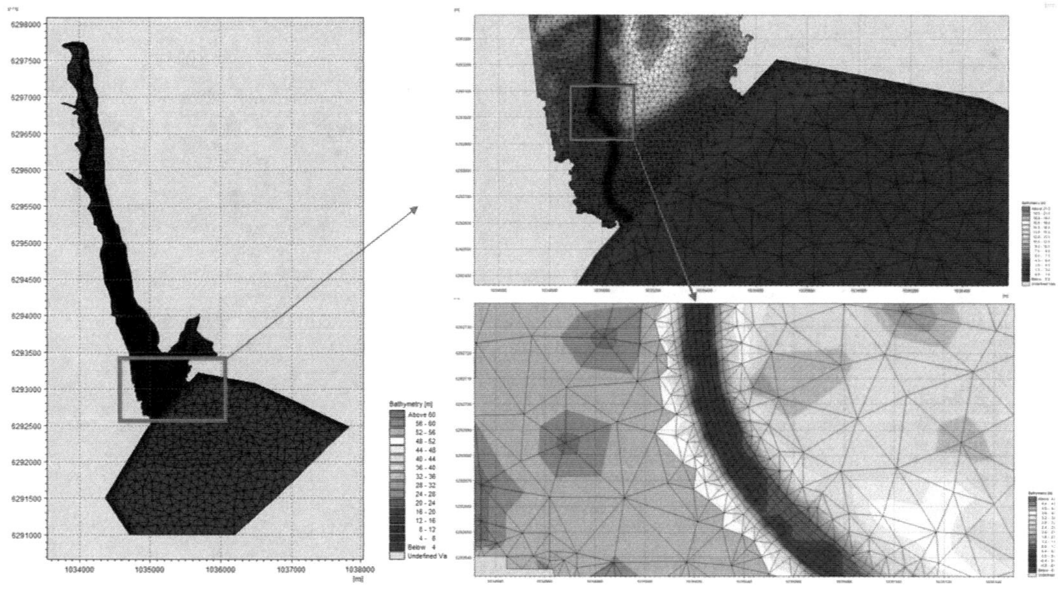

Figure 2. MIKE 21 FM mesh.

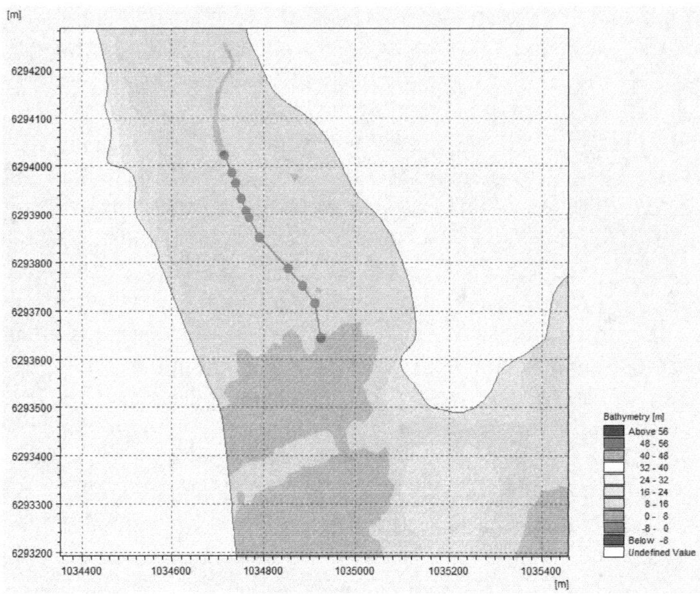

Figure 3. MIKE 21 FM culvert.

3. Results

The modeling procedure begins with a simulation of the catchment hydrological cycle by MIKE SHE. The MKE SHE application in the Cagne catchment is able to provide precise descriptions of most major hydrological and hydrogeological characteristics at any location in the catchment, and to support the applications of MIKE 21FM to have more detail simulation of the Cagnes sur Mer urban area, based on high resolution simulation within daily and hourly time intervals, respectively.

Based on the available data and the constructed model, the rainfall event from 1st October 2020 to 31st December 2020 was simulated. The result is analyzed and compared to data collected from the Cagne Aval station, which is located near the Cagne River's sea exit. Figure 4 shows a comparison of observed water depth data with simulated water depth data at two different resolutions. It shows that the model can track the trend with a topographic resolution of 30 m, however there is an abrupt drop after every peak, and certain areas don't match the real data even if uncertainties are attached to those data. The simulated data with a resolution of 20 m are more accurate. Except for the initial high, which is due to inadequate simulation time at the start, the peak and trend are tightly matched. For the MIKE SHE model, it has been validated with an one year event and all Mike SHE components have been integrated.

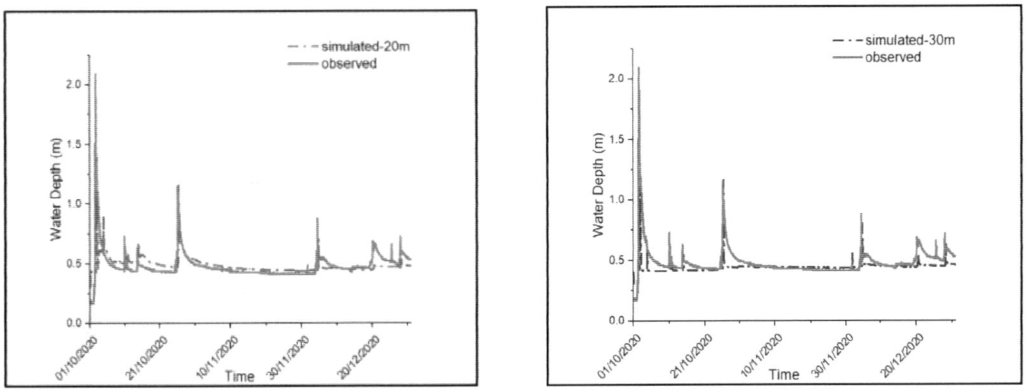

Figure 4. MIKE SHE water depth observed data against simulated data in Aval station.

A 3-days extreme rainfall event is simulated with the hydraulic model with 10 m resolution of riverbed bathymetry, 30 m in the floodplain and 100 m in the sea. Figure 5 shows the observed and simulated water depth data at Cagne Aval station. The simulated result matches the observed water depth well. The inundation map was generated as shown in Figure 6. It can be seen that the underground part is well represented, and it shows no overflow in the river floodplain, which is true as many observers have not reported overflows or inundation during the event. In summary, this model can reproduce the hydrodynamics behaviour of the lower Cagne River.

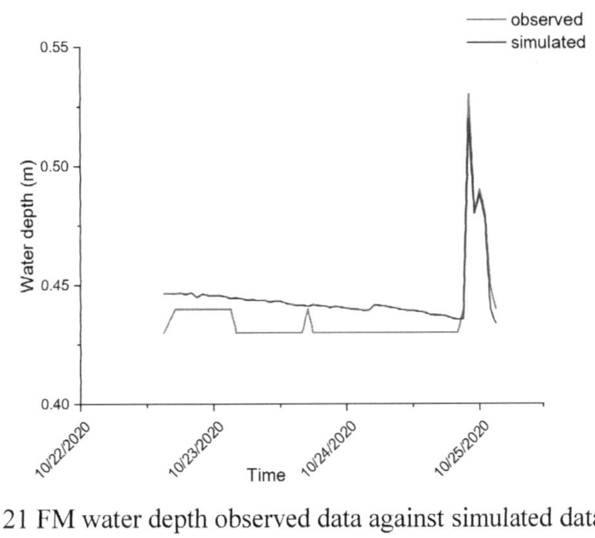

Figure 5. MIKE 21 FM water depth observed data against simulated data in Aval Station.

Figure 6. Inundation map.

4. Conclusions and perspectives

The defined Decision Support System can produce correct and adequate information for improving the decision-making process and maximizing the efficiency of the management process. Obviously, the use of deterministic models represents a major asset within the approach. The two developed models can properly simulate the hydrological process of Cagne catchment and hydrodynamics behaviour of the Cagne River during extreme events. These two models are combined in a modelling chain within the DSS system that should be integrated within the Information System of the local managing technical services. Considering both accuracy and efficiency, the hydrological model is set up with a 20 m resolution and the 2D hydraulic model is set up with a high resolution (10 m) in targeted area. A recorded rainfall event was simulated, the result of water depth matches the observed data and an inundation map was generated. The lack of field measurements and recorded observations has been overcome with the good performance of the two deterministic models that can be now used within the DSS. Considering that the complete water cycle also includes groundwater, the DSS will be extended with a 3D groundwater model based on FEFLOW [13]. This complete DSS system for integrated water management can be applied to other catchments.

Acknowledgments
The work benefited from the data provided by the Métropole Nice Côte d'Azur, Conseil Départemental 06 and Météo France. This work is funded by China Scholarship Council.

References

[1] Chochon R, Martin N, Lebourg T and Vidal M 2021 Analysis of extreme precipitation during the mediterranean event associated with the alex storm in the alpes-maritimes: atmospheric mechanisms and resulting rainfall. *SimHydro 2021*: *Models for complex and global water issues*

[2] Nohara D, Gourbesville P and Ma Q 2018 Towards development of effective decision support systems for integrated water resources planning and management. 京都大学防災研究所年報. B= Disaster Prevention Research Institute Annuals **61**(B) 702–10

[3] Ford DT 2001 Flood-warning decision-support system for Sacramento, California. *Journal of Water Resources Planning and Management*. **127**(4) 254–60

[4] Ahmad S and Simonovic SP 2006 An intelligent decision support system for management of floods. *Water Resources Management*. **20**(3) 391–410

[5] Zhang W, Zhou J, Liu Y, Chen X and Wang C 2016 Emergency evacuation planning against dike-break flood: a GIS-based DSS for flood detention basin of Jingjiang in central China *Natural Hazards* **81**(2) 1283-301

[6] Şensoy A, Uysal G and Şorman AA 2018 Developing a decision support framework for real-time flood management using integrated models *J. Flood Risk Manag.* **11** S866-83

[7] Folton N 2014 *Modélisation hydrologique du bassin versant de la Cagne pour la confrontation des estimations DOE et différentes situations hydrologiques, selon des évolutions des prélèvements et selon des scénarios d'évolutions climatiques* (PhD Thesis, IRSTEA, France) https://hal.archives-ouvertes.fr/hal-02601232/

[8] DHI 2012 *Mike SHE User's Manual* DHI Software (DHI Water & Environment)

[9] DHI 2017 *Mike 21 Flow Model User Guide* DHI Software (DHI Water & Environment)

[10] Durand Y, Brun E, Merindol L, Guyomarc'h G, Lesaffre B and Martin E 1993 A meteorological estimation of relevant parameters for snow models. *Annals of Glaciology* **18** 65–71

[11] Zavattero E. *Integration of surface water modeling in a decision support system: application to the Lower Var Valley, France* (PhD Thesis, Université Côte d'Azur, France) https://theses.hal.science/tel-02442008

[12] Salmonsson A. MIKE 21 FM in Urban Flood Risk Analysis: A comparative study relating to the MIKE 21 Classic model.

[13] Diersch HJ 1999 *User Manual FEFLOW, Version 4.8* (Berlin: WASY GmbH).

IOP Conf. Series: Earth and Environmental Science 1136 (2023) 012022 doi:10.1088/1755-1315/1136/1/012022

Rainfall variability assessment in the hydrological catchment of Addis Ababa

Giulia Failla[1], Marcella Lo Bianco[1], Gabriele Freni[2], Gaetano Beninati[1]

[1] Department of hydraulic design, Sering Ingegneria, Palermo, 90139, Italy
[2] Faculty of Engineering and Architecture, University of Enna, Enna, 94100, Italy

gr.failla@gmail.com, gabriele.freni@unikore.it

Abstract. The Intensity-Duration-Frequency (IDF) relationship is a key input for defining the hydrograph and peak flow discharge required for the hydraulic design. The IDF curves describe the relationship between rainfall intensity, rainfall duration, and return period. The objective of this study was to develop an accurate rainfall analysis in the Addis Ababa catchment. To compute reliable IDF curves, long rainfall data at fine time step scale and uniformly distributed are necessary. In Ethiopia, there is limited availability of data at sub-hourly and at spatial scales. To cope with these issues, (1) stochastic disaggregation techniques have been employed to disaggregate daily observed data into sub-hourly data and (2) to guarantee homogeneity in the project area, the merge of satellite data and observed gauged data have been considered. In this study, IDF curves have been computed for the entire hydrological catchment of Addis Ababa (~1430 km^2). The results of the study include tabular and graphical presentation of IDF curves under changed climatic conditions, for return periods of 2, 5, 10, 25, 50 and 100 years for the durations of 15,30 min and 1, 3, 6, 12 and 24 hours.

1. Introduction

Uncontrolled rainfall-runoff process can cause a multitude of problems, such as degradation of public resources and flooding that endangers life and properties. Increased stormwater flows, due to urbanization or change in climatic conditions, expand floodplains, bringing flooding to locations where it did not occur before and worsening flood problems in areas already prone to flood.

Hydrologic design of storm sewers, culverts, retention/detention basins and other components of storm water management systems are typically performed based on specified design storms derived from the rainfall intensity-duration-frequency estimates and an assumed temporal distribution of rainfall.

The present study is part of a larger project for the development of the Drainage Master Plan (DMP) for Addis Ababa city. The DMP is aimed at individuating the main intervention for the improvement of the storm drainage system of the city and defining hydraulic design criteria as reference for new developments. Rainfall Analysis is the prime input of the rainfall-runoff process, and it has the aim of deriving IDF curves. The IDFs curves describe the relationship between rainfall intensity, rainfall duration, and return period. IDFs provide a graphical representation of the rainfall distribution for a specific area, which can be subsequently used for drainage design and are the scope of this project.

2. Material and methods

The study was conducted in the hydrological catchment of Addis Ababa. The Project area is geographically located between 8°46' to 9°13' North Latitude and 38° 35' to 39°00' East Longitudes. The total area is of about 1,430 km^2. An accurate measurement of precipitation across space and time remains essential to assess the hydrological conditions of an area and precipitation gauge should be uniformly distributed. The methodology to assess the IDF curves includes the following components: (1) rainfall data collection and data pre-processing, (2) disaggregation of daily precipitation times series to sub- hourly scale based on the Bartlett-Lewis model; (3) spatial calibration of the satellite data through the comparison of satellite data and rain gauge observations; (4) calibration of the sub-hourly data of the disaggregated and satellite time series using the recorded 15min time series of Bole station; (5) statistical analysis of rainfall for various durations and development of IDF curves under changed climatic conditions. Here below, in the scheme, the summary of the rainfall activities is summarized while in the next paragraphs, the rainfall study process is given.

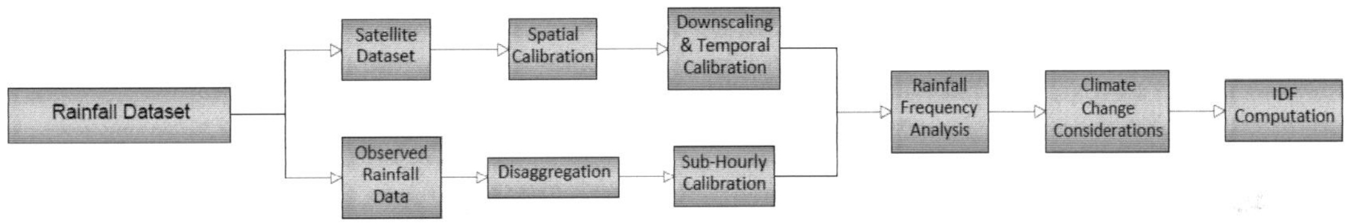

Figure 1: Scheme rainfall assessment methodology

2.1. Rainfall Data Collection and data pre-processing

The rainfall data set includes:

- Daily rain gauges data, collected from 11 existing rainfall stations;
- 15 min observed rainfall data from one station (Bole Station) times series;
- Satellite rainfall data: the GPM Multi-Satellite Precipitation with gauge Calibration, with spatial resolution of 0.1° x 0.1° and a time step of half-hour has been used. The gridded precipitation datasets have been acquired in NetCDF format. Each square represents a rain gauge station influence area while its gravity centre localizes the virtual station position. In this study, 13 virtual rainfall station have been individuated.

The pre-processing analysis was necessary to define the goodness of the data collected and to define the time series to be considered in our study. The quality control pre-processing includes: detection of Gaps in the data, detection of values above set thresholds (outliers value), definition of homogeneous time series, computation of basic statistics. At the end of this process, out of the 11 existing rainfall stations only 9 have been considered to be robust enough and reliable.

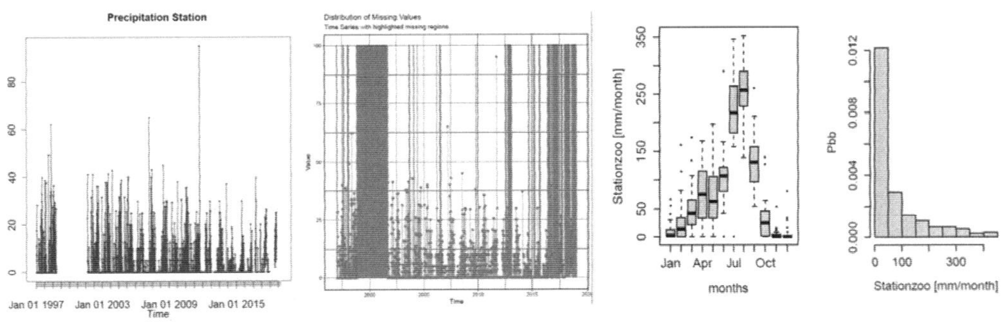

Figure 2: Sample of quality control processing

IOP Conf. Series: Earth and Environmental Science 1136 (2023) 012022 doi:10.1088/1755-1315/1136/1/012022

In addition, the daily and sub-hourly data of the Bole station have been compared. The sub-hourly recorded data of Bole station have been aggregated and compared with the daily data recorded in the same station. Although the two distributions are linearly related, the rainfall depth recorded in a day do not perfectly correspond. This underlines some difference in the accuracy of the rainfall measurement. For this reason, the two sources of data have been used with two different scopes: the daily data, characterized by a longer timeseries, have been used for the rainfall frequency analysis and for the definition of the IDFs curves while the 15-min rainfall data were used for the calibration of rainfall intensity in the sub-daily and more precisely sub-hourly time step.

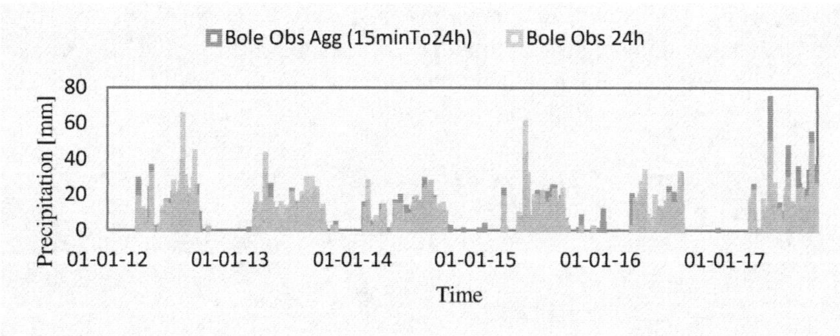

Figure 3: Comparison of recorded rainfall data in Bole Station (sub-hourly and daily data)

The area of influence for each rain gauge stations has been calculated by the Thiessen Polygon methodology. The Thiessen method is based on the assumption that measured amounts at any station can be applied halfway to the next station in any direction, which means that for any point rainfall is equal to the observed rainfall at the closest gauge. The weights of the rain gauges are computed by their relative areas, which are estimated with the Thiessen polygon network. The area of each polygon is used to weight the rainfall amount of the station in the center of the polygon. In the figure below, the Thiessen polygon and the rainfall network in the project area is showed in the figure below.

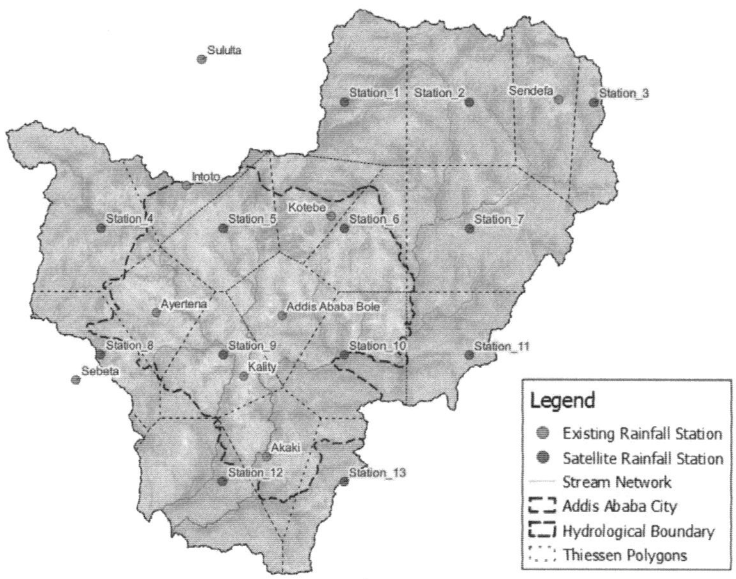

Figure 4: Rainfall Network and Thiessen Polygon

2.2. Disaggregation of observed daily rainfall data

Hourly and sub-hourly data for hydrological application are indispensable to understand short duration events, which are characterized by high intensity. The Hyetos Minute R tool based on the Bartlett-Lewis Rectangular Pulse (BLRP) model has been selected to disaggregate the daily rainfall to sub-hourly data (time step of 15 minutes) using a proportional adjusting procedure. More precisely, the sub-hourly data recorded from Bole Station have been used to calibrate the BLRP model and to reproduce rainfall event with short duration. The BLRP model, depends on 6 parameters: λ, β, γ, η, μX, σx that have been computed for each rainfall station. After fixing the parameters of BLRP model, the 15min precipitation has been estimated using the DigaSimul function of Hyetos R package. The BLRP model was fitted on a year basis. For a sequence of wet days, the model runs several times and generates the sequence that best matches with the observed daily data. To achieve a better reproduction of extremes, a distribution with a heavier tale is required for cell intensity; for this reason, the Gamma Distribution for the cell Intensity was applied. The 15 min disaggregated and observed data comparison is illustrated below.

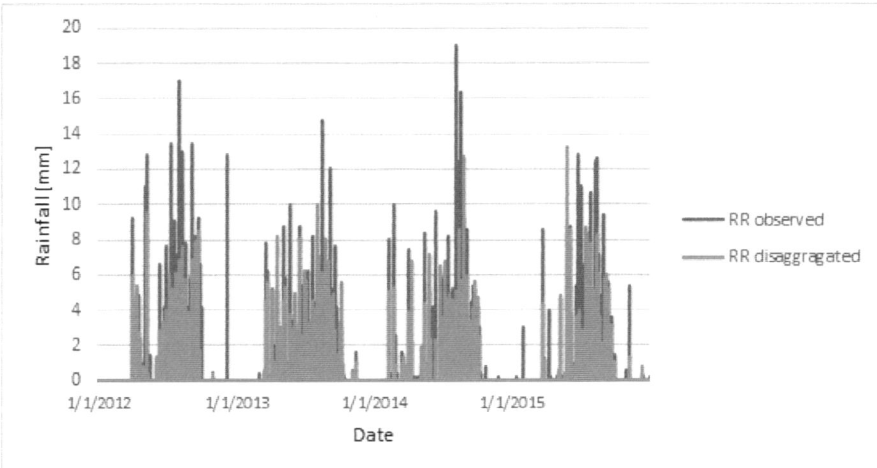

Figure 5: Comparison between disaggregated and observed data in Bole station

2.3. Spatial calibration of satellite data

The satellite rainfall data have been compared with the nearest observed daily rainfall data; for each pair of daily data a QQ-plot with the regression equation has been defined and a climate factor was identified. The calibration process was carried out by dividing the satellite data by the corresponding climatic factor.

Table 1. Climate Spatial Factor

Rainfall Station Relationship	Climate Spatial Factor
Sendefa-Satellite station 3	1.8
Intoto-Satellite station 5	2.0
Kotebe-Satellite station 6	2.1
Ayertena-Satellite station 9	2.0
Bole-Satellite station 9	2.2
Sebeta-Satellite station 8	2.3
Akaki-Satellite station 12	2.1

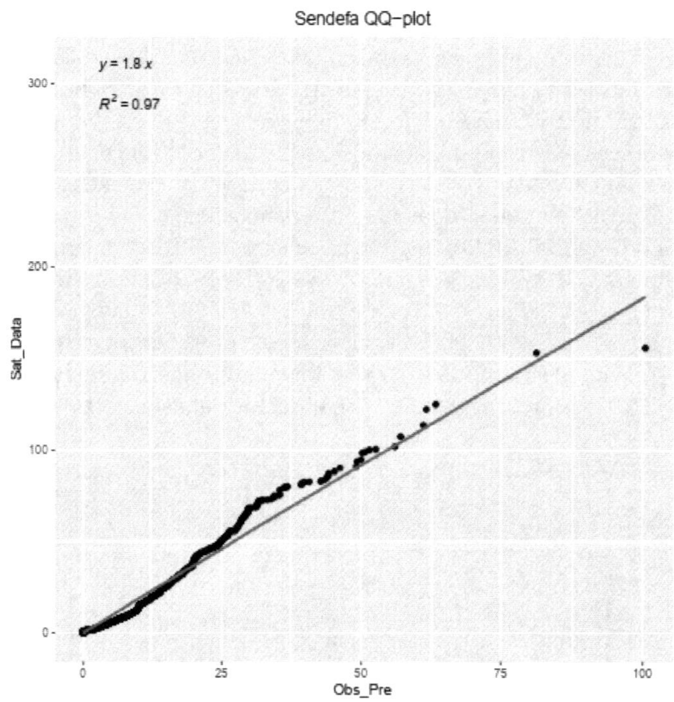

Figure 6. Sample of QQ plot

2.4. Temporal calibration and downscaling of satellite data
In order to calibrate the rainfall behaviour for sub-daily data, the maximum intensity has been compared with the observed recorded sub-hourly data of Bole station: the rainfall maximum depth for the duration of 30min, 1,3,6,12,24 hours of the satellite rainfall data and the observed rainfall data have been compared in a Q-Q plot. The coefficient of the regression line has been used to calibrate the sub-hourly satellite data. To implement the frequency analysis to a time step of 15 minutes, the satellite data were downscaled according to the observed sub hourly data recorded in the Bole station.

Table 2. Climate Temporal factor.

Time Scale	Temporal Climate Factor
30 min	0.39
1h	0.56
3h	0.91
6h	0.98
12h	0.97
24h	0.97
30minto15min	1.50

2.5. Climate Change consideration
According to several studies, an increasing in rainfall is expected over the city. The Climate Change Knowledge Portal of the World Bank has predicted in 2050 an increase of 11% of the Projected Average Largest 1-Day Precipitation in the worst predicted scenario: SSP5 RCP8.5. The Average largest 1-day precipitation is an important parameter for the extreme rainfall event analysis. In this study, to consider the effect of the climate change in the extreme rainfall event, the maximum precipitation value of each station and for each return period has been precautionary increased by 11%.

2.6. Frequency analysis

At the end, the rainfall frequency analysis has been developed for each of the stations by calculating the parameters for Addis Ababa city. For each gauge station, a time series of 15-20 years has been analysed to define the IDF curves for different return periods. The frequency analysis has been carried out on the maximum annual series, for all active stations and with sufficient data recorded, using the Gumbel distribution. For each given return period a three-parameters equation has been fitted between the maximum intensity and duration.

$$i = \frac{c}{(D + a)^b} \quad (1)$$

where: i=intensity of rainfall (mm/hr), D= duration (hrs), c, a and b are rainfall coefficients.

3. Results and discussion

The Barlett-Lewis model and the disaggregation process have provided a reliable reproduction of the statistical properties of the observed precipitation. The distribution of the rainy days was clearly reproduced by the model which gives high difference between variance and skewness values. The results allow for a high precision description of the season variability. In addition, the quality of the results was assessed by comparing the main statistics of observed and disaggregated data. This comparison has shown that the statistic of the disaggregated precipitation values is comparable with the observed one. However, the disaggregation process attenuates the peak of the intensity for the time step of 15min and 30 min. In order to mitigate this error, the maximum value of the intensity for short duration have been adjusted according to the observed and available 15min time series.

Satellite data have been calibrated using the observed data ("the true" station) in order to adjust the precipitation value recorded. To define the robustness of the rain gauge network and the uniformity of the data, error difference matrixes have been estimated, comparing the rainfall intensity (for each duration and return period) of the satellite data with the nearest observed rainfall data.

$$error : \frac{satellite\ value - observed\ value}{observed\ value} \quad (2)$$

Furthermore, the estimated error of each satellite station and then of the entire rainfall network was calculated by applying the arithmetic method. The error found was of 14.3 %. This value proves a reliable and robust precipitation homogeneity and the goodness of the rainfall frequency analysis. In fact, being the hydrological data characterized by high variability from spatial and temporal point of view, an error of 14.3 % is considered acceptable, also in light of the different dataset and complexity of the methods used.

4. Conclusions

A homogeneous spatial distribution of rainfall data is of paramount importance for rainfall assessment and subsequent water resources management planning. The runoff analysis is determined by the rainfall distribution. In this study, to define a homogeneous rainfall network, observed and satellite rainfall data have been merged. A total of 22 rainfall stations (physical and virtual) have been identified within the project area and analysed. Generally, satellite data are non-sensitive to topographic influences on precipitation. On the other hand, the gauge observed data are commonly not adequate to completely quantify spatial variations due to topographic location influences. That is why, the integration of the two sources of data, after calibration, is a valid tool for hydrological assessment in gauged and ungauged areas. The IDF curves have been generated for 22 rainfall station, for the return periods of 2, 5, 10, 25, 50 and 100 years and for durations of 15,30 min and 1, 3, 6, 12 and 24 hours. The IDF curve have been updated according to the future climate scenario taken into account the time horizon of 30 years (2050s).

Acknowledgments
We would like to express our gratitude to the Addis Ababa City Road Authority and to the National Meteorological Agency of Addis Ababa for providing support for data acquisition (http://www.ethiomet.gov.et/)

References

[1] Kossiers P., Makropoulos C., Onof C. and Koutsoyiannis D., 2018, A rainfall disaggregation scheme for sub-hourly time scales: Coupling a Bartlett-Lewis based model with adjusting procedures, *Journal of Hydrology,* **Volume 556,** pp 980-992
[2] Villani V., Di Serafino D. Rianna G. and Mercogliano P.,2015, Stochastic Models for the Disaggregation of Precipitation Time Series on Sub-Daily Scale: Identification of Parameters by Global Optimization, CMCC Research Paper, No. **RP0256**
[3] Engida A. N. and Esteves M., 2011, Characterization and disaggregation of daily rainfall in the Upper Blue Nile Basin in Ethiopia, Journal of Hydrology, Volume 399, Issues 3–4, pp 226-234
[4] Abeba T., Birhane H. and Mekin M., 2018, Rainfall Intensity-Duration-Frequency Development for Soil and Water Conservation Interventions Under Changing Climate in Southern Tigray, Ethiopia, *Journal of Natural Sciences Research*, **Volume 8**, No 19
[5] Hu Q., Yang H, Meng, X., Wang, Y. and Deng, P., 2015, Satellite and gauge rainfall merging using geographically weighted regression, *Proceedings of the International Association of Hydrological Sciences*. **Volume 368**, pp 132-137.

Points of consideration on identification of the atmospheric fronts depicted on weather charts

Maki Miyamoto[1] **and Tomohito J. Yamada**[2]

[1] Graduate School of Engineering, Hokkaido University, Sapporo, 060-8628, Japan
[2] Faculty of Engineering, Hokkaido University, Sapporo, 060-8628, Japan

miyamoto@elms.hokudai.ac.jp

Abstract. Atmospheric fronts play an important role on the day-to-day variation of rainfall in Japan, North-western Pacific. They are defined as a discrepancy of air density and have been depicted on surface weather charts for more than 100 years by Japan Meteorological Agency (JMA). This study develops an approach to detect fronts from the charts published from 1978 to 2020 and investigates how detected fronts could be utilized. The fronts which are depicted on the charts might have a diverse physical characteristic in each age. One of factors is a difference in how to represent fronts. The difference also prevents from detecting fronts under a certain accuracy. Previous study demonstrates detecting fronts based on the color available from the charts. However, this approach is applicable for weather charts published after 2000 by JMA. This paper describes consideration points for a unified detection of fronts from the charts for 43 years in detail. Furthermore, we compare fronts detected by several methods and discuss a spatial scale in which their differences could be neglected.

1. Introduction

Atmospheric fronts are defined as a discontinuity of density and is one of disturbances which usually occurs in the mid-latitudes. It is considered that fronts are located in the region which has a large difference in temperature as for an analysis of weather charts.

In Japan, weather charts have been made by Japan Meteorological Agency (JMA) for over 100 years [1]. Thus, a position of fronts is recorded and utilized for understanding an annual change and decadal change of the climate in Japan [2, 3]. A method for making a gridded dataset of fronts depicted on weather charts is proposed [4]. However, it could be applied to weather charts published after 2000 because the format of weather charts and representation of fronts have changed with the development of forecasting techniques. It is useful to reveal the construction and developing process of atmospheric fronts. In this study, we detect fronts from weather charts published after 1978 and compare them.

2. Methodology

Surface weather charts have been made in an image or document format. There are two types of charts. One's target area is Asia-Pacific region, and the other is Japan as shown in Figure 1. Both are presented as a result of temporal analysis and reanalysis is conducted after a few days. In this study, we use surface weather charts for the Asia-Pacific region between 1978 to 2001 and for Japan up to 2020, as published by JMA.

As shown in Figure 1, the format of weather charts has differences with ages and fronts are represented as a line which has triangles and semicircles.

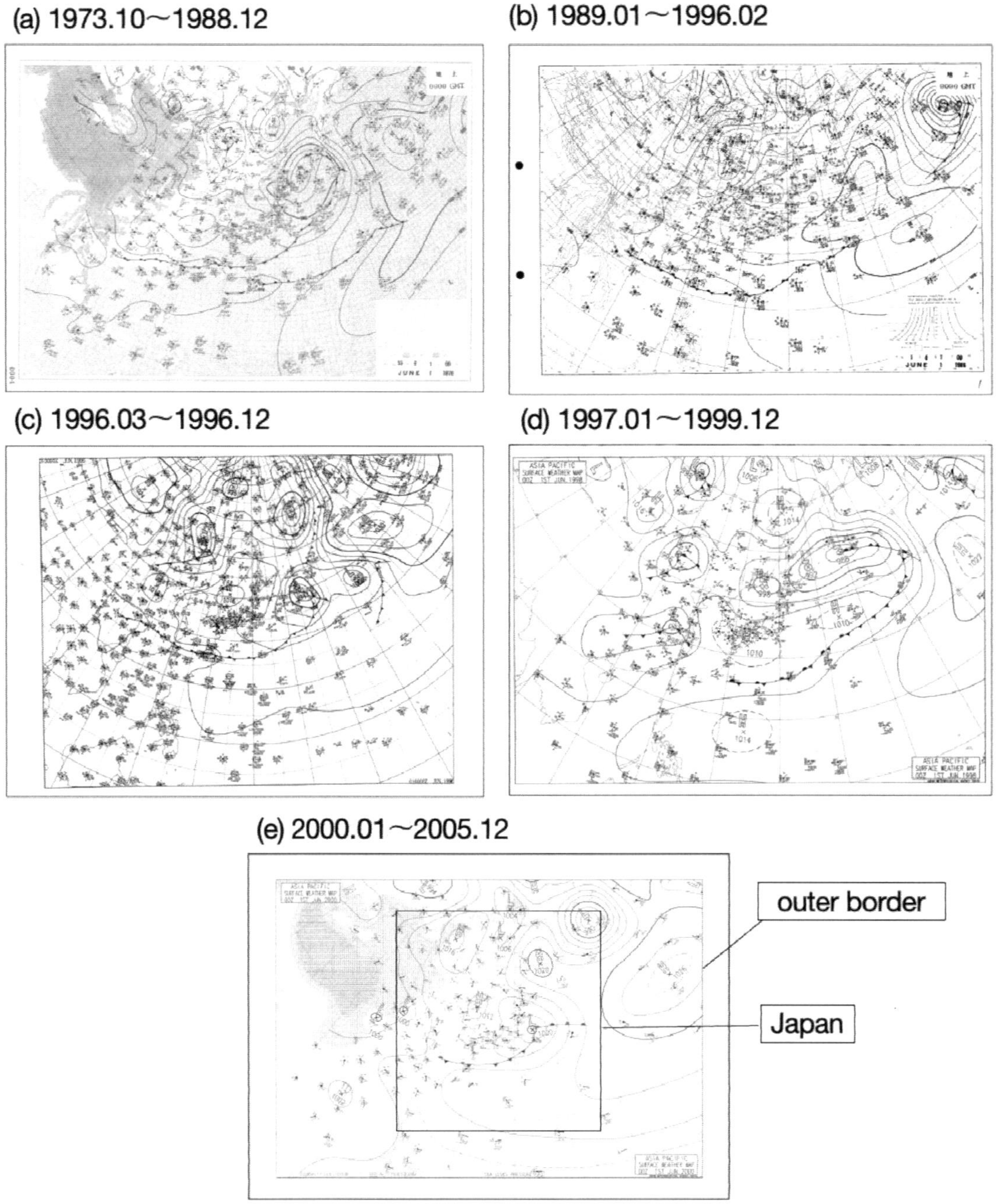

Figure 1. The charts of surface weather analysis in Asia-Pacific from (a) October 1973 to December 1988, (b) January 1989 to February 1996, (c) March 1996 to December 1996, (d) January 1997 to December 1999, and (e) January 2000 to December 2005. JMA also provides the charts for Japan indicated with the black box in (e). The outer border of weather charts is different for each image.

A procedure of detecting fronts from weather charts is presented in Figure 2. As mentioned in section 1, surface weather charts are provided as images to convert image elements (hereafter, pixels) to grids. First, on Step 1 shown in Figure 2, we find an outer border where a column and row have the most pixels with black colour (or brown in case of weather charts before 1988) and crop the image to fit the border, which is almost vertical and horizontal, respectively.

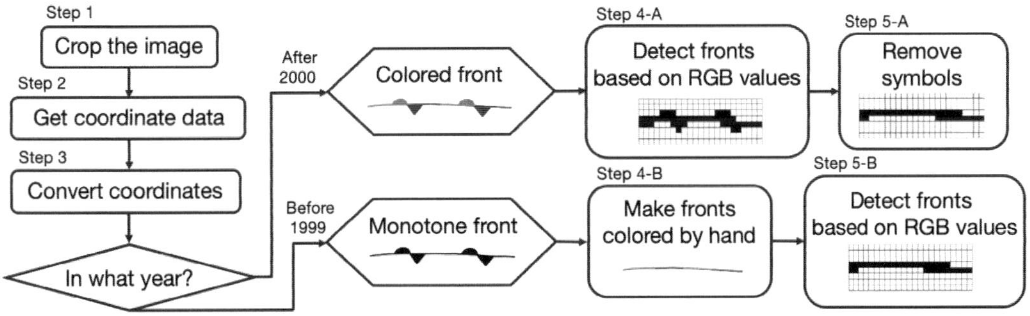

Figure 2. A Flow chart of converting images to gridded datasets as in fronts.

Notice that the border is slightly tilted from vertical and horizontal direction and cropping isn't completely along the border because weather charts published before 1996 are made of paper and converted paper to image by scanning [5]. Then, cropped images are resized to a certain size. Second, we introduce Step 2 indicated in Figure 2. We get coordinates for every 10 degrees of longitude and latitude. In this study, we assign coordinates to corresponding points in advance (described in detail in Section 3) and use linear interpolation and extrapolation. Step 3, shown in Figure 2, is converting the lambert coordinates to the Mercator projection. Atmospheric fronts are extracted based on the RGB value of each pixel from images and expressed with a grid composed of columns and rows. The size of the grid is 0.1degrees of latitude and 0.125 degrees of longitude, corresponding to 100 km^2, over the range of latitude from 22.4 degrees to 47.6 degrees and longitude from 20 degrees to 50 degrees. Monotone fronts are depicted on weather charts published before 1999. Therefore, we make fronts coloured to detect them using the RGB value. We remove symbols from detected fronts by identifying the shape, such as a triangle and semicircle as in after 2000. It is needed to identify objects and make them connect with the neighbourhood because a part of front lines also may be removed in this process. Therefore, gridded dataset of fronts expressed with only a line are converted from weather charts for every year.

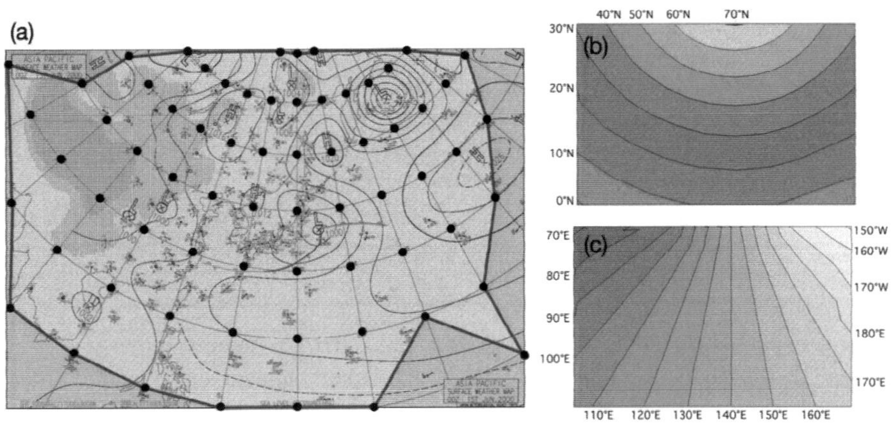

Figure 3. An assignment of coordinates of latitude and longitude to image elements (a) with an interpolation and extrapolation by giving the coordinates every 10 degrees of latitude and longitude (black dots). (b) The diagram of contour for providing the coordinates of latitude. Contours are drawn every 10 degrees. (c) Same as (b) but for longitude.

3. Results

To obtain the coordinates, we assigned it every 10 degrees of latitude and longitude to some pixels as shown with black dots in Figure 3 after cropping the weather charts. It is presented in Figure 3(b and c) that the distribution of latitude and longitude on pixels is given by applying linear interpolation and extrapolation to the inside and outside of the black solid line in Figure 3(a), respectively. Even though we don't give the coordinates to pixels at all, it can be obtained by finding the intersection of lines of latitude and longitude in each image. It is noteworthy that we couldn't obtain the coordinates of the intersection, which are covered with a symbol representing the condition of weather. This is done so estimated coordinates can be different from what we read from weather charts.

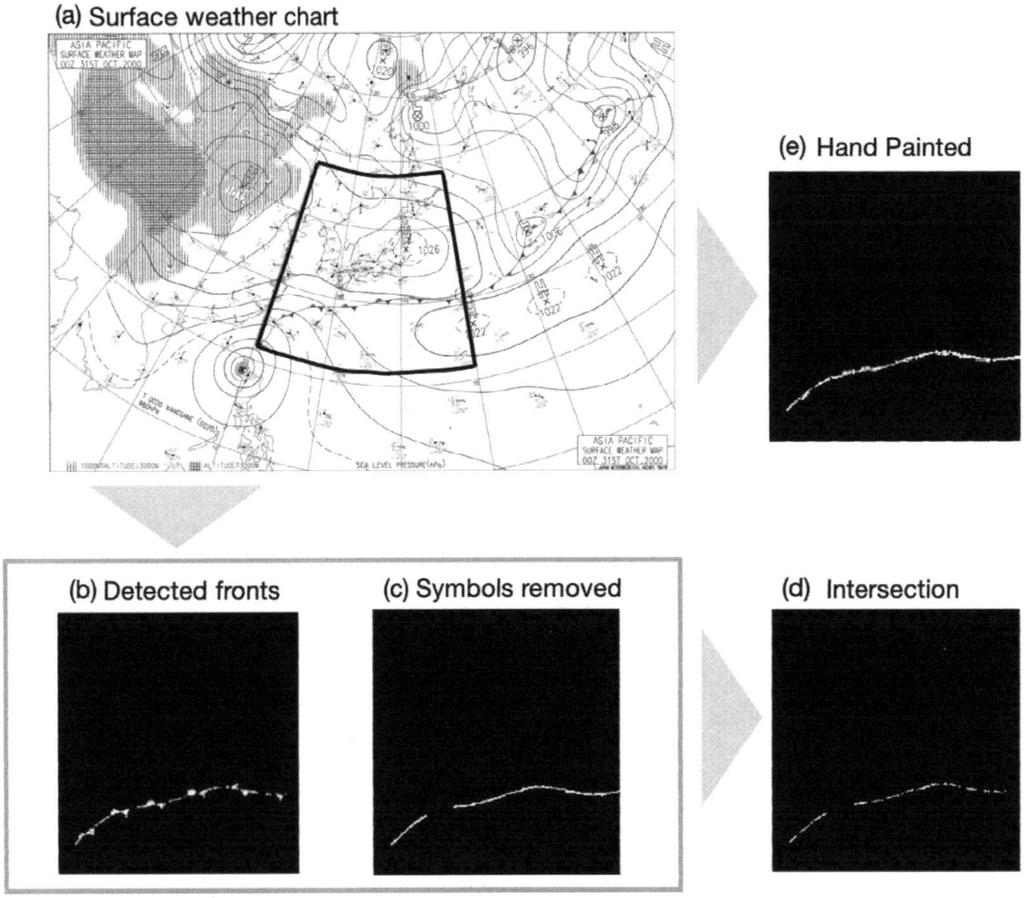

Figure 4. A front detected based on the RGB values from (a) surface weather charts which have (b) an originally coloured front, (e) manually coloured front. (c) Same as (b), but with removed symbols, triangles and semicircles from (b). (d) The intersection of (b) and (c). A black box in (a) indicates the target area. White grids in (b)-(e) represent detected fronts.

According to Figure 1(e), fronts depicted on weather maps published after 2000 are coloured in red/blue and detected by getting the RGB value in each pixel as seen in Figure 4(a). The types of fronts are distinguished by colours. For example, the red represents warm fronts, and the blue represents cold fronts. A warm front is a boundary between cold and warm air mass, but the warm is superior to the cold. In contrast, the warm is inferior to the cold around a cold front. Fronts coloured with pink are occluded fronts, which are formed because cold fronts catch up on warm fronts. If the line is alternately coloured red and blue at short intervals, it represents stationary fronts, where cold and warm air masses are equally pushing each other. Then, we removed symbols such as triangles and semicircles from detected fronts, shown in Figure 4(b), and combined them with a continuous line (Figure 4(c)). This often contains a front which doesn't exist in the weather map in case there is a bonding point between the occluded front and others. Figure 4(d) shows an intersection of fronts, which are just detected from weather charts and removed symbols.

It is difficult to distinguish fronts from others in weather charts based on RGB value because they are coloured with black before 1999. Additionally, observation data at each point is described on weather charts. Hence, fronts are hardly distinguished from such characters through object recognition. We made fronts on the weather chart coloured with red, blue, pink, and green by hand-painting. Each colour is corresponding to warm, cold, occluded, and stationary fronts, respectively. All fronts have no symbols but lines. Thus, we easily obtained fronts without symbols as shown in Figure 4(d). Fronts detected as same as each one in Figure 4 are referred to as front A, B, C, and D.

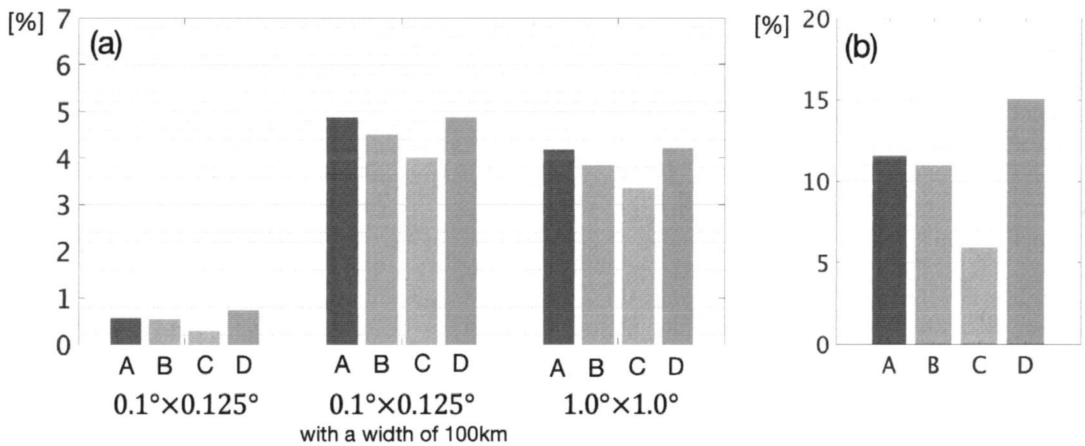

Figure 5. The ratio of frontal grids to (a) all grids, (b) grids of front D with a width of 100 km in spatial resolution of about 10 km. Each colour corresponds to four types of fronts shown in Figure 4, indicated as alphabets below the graph. (a) Left side of the group are fronts in grids of 0.1° × 0.125°; The centre group is same as the left but fronts with a width of 100 km; Right side of the group in grids of 1.0° × 1.0°.

To compare those four types of fronts, we made a gridded dataset from weather charts published in 2000. It's a reasonable choice because of fronts depicted on the charts with some colours. Thus, all types of fronts can be detected. As for a spatial resolution of gridded data, three types of datasets are available. As described in Section 2, grid intervals are 0.1 degrees in latitude and 0.125 degrees in longitude. It is known that a front has a width of about 100 km [6]. Therefore, it is assumed that fronts exist in 11×11 grids cantered on each grid where a front exists. Fronts are also represented with a grid which has a size of 1.0 degrees in latitude and longitude. Figure 5(a) shows that a ratio of the grids where fronts exist to all grids in each dataset and type of fronts. If we take the width of fronts into consideration, the number of frontal grids is about 8 or 9 times more than the otherwise.

A ratio of actual frontal grids to the virtual ones corresponding to a width of front D is presented in Figure 5(b). We compare fronts which are detected based on the RGB value from surface weather charts (front A) with those that are composed of no symbols with hand-coloured fronts (front D). This is like comparing the fronts detected from weather charts published before 1999 by Step 5-B of the process as shown in Figure 2, to after 2000 by Step 4-A. The number of front A is about 1.3 times greater than that of front D and an intersection of those occupies about sixty percent of front A and fifty percent of front D. It is assumed that grids where only front A exists represent symbols embedded with fronts and where only front D exists are caused by a bold line with hand painted.

The front A is compared to fronts with removed symbols by distinction of the shape from front A (front B). According to Figure 2, it indicates the comparison between fronts detected by Steps 4-A and 5-A of the procedure. The number of them is almost the same and the intersection is less than half of each front. In the process for the front B, symbols are removed from fronts and frontal objects are combined with each other. This may cause the lack of fronts. In particular, it is likely that the whole of occluded fronts is missing. Fronts are also detected at grids with no fronts, which cause grids where only front B exists.

To neglect symbols embedded with front A and a part of front B detected by mistake, we made front C, an intersection between front A and B, composed of lines without symbols even though a part of fronts on weather charts may be lost. The front C covered with no front D is smaller than twenty percent of the whole front C. It seems that front C coincides with front D and the number of front D exceeds that of their intersection because of the wider width of fronts. Considering fronts have a width of 100 km, the number of front D is about 1.2 times greater than that of front C.

4. Summary

The atmospheric front is a discrepancy of density between air masses, which have been estimated and recorded on surface weather charts by JMA since the late 19th century. This study explored an approach to detect the fronts from the charts and examined a difference in a specified spatial resolution for representing fronts in summertime. The method for detecting fronts, based on the RGB value, is applied to the charts for 43 years, which are different in the format of the charts. The fronts before 1999 are drawn with black on the charts and needed to be coloured with the other objects. About half of them are covered with those detected from the charts published after 2000. By removing the symbols from the fronts after 2000, the positions of the fronts were generally consistent, and the manually coloured fronts were about 2.5 times more systematically common. This difference is thought to be due to the width of the fronts. The results were 1.2 to 1.3 times greater when fronts are assumed to have a width of 100 km.

Hand-painted fronts are detected somewhat accurately from surface weather charts. However, fronts detected by only a use of image analysis from the charts may lack some fronts that the hand-painted ones have. Therefore, both of them can be regarded as the fronts in the environmental field of synoptic scale, with a horizontal length scale on the order of 1000 km. As for non-widened fronts, a front width is generally different for 10 km between those detected by a combination of image analysis and hand-painted fronts, and only using image analysis. They would be suitable for the meso-α-scale (200 – 2,000 km).

It would be different in a physical characteristic of fronts depicted on surface weather charts over time. One of the reasons for this is a change of how fronts are represented, and another is that understanding of fronts has deepened. It is also due to forecasts that estimate the position of fronts [7]. The items mentioned above should be taken into account for a use of dataset of detected fronts. However, we discussed a representation of fronts and format of surface weather charts because they are crucial for detecting the fronts.

Acknowledgments

This research is supported by Research Field of Hokkaido Weather Forecast and Technology Development (endowed by Hokkaido Weather Technology Center Co. Ltd.), Science and Technology Research Partnership for Sustainable Development (SATREPS) in collaboration between Japan Science and Technology Agency (JST, JPMJSA2009) and Japan International Cooperation Agency (JICA), MEXT-Program for the advanced studies of climate change projection (SENTAN), and The integrated Research Program for Advancing Climate Models (TOUGOU) Grant Number JPMXD0717935561 and JPMXD0717935498 from the Ministry of Education, Culture, Sports, Science and Technology (MEXT), Japan, the establishment of university fellowships towards the creation of science technology innovation. Surface weather charts is provided by Japan Meteorological Agency (JMA).

References

[1] Japan Meteorological Agency 2018 A method of analysis for surface weather charts of Asia-Pacific *Techniques for Quantitively Forecasting* **23** 84–103 (in Japanese)

[2] Sato N and Takahashi M 2001 Long-term variations of the Baiu Frontal Zone and Midsummer Weather in Japan *Journal of the Meteorological Society of Japan. Ser. II* **79**(3) 759–770 https://doi.org/10.2151/jmsj.79.759

[3] Senoo Y and Kato K 2008 Climatological study on large-scale situation associated with cool summer around the North-eastern Japan in 1990s (in comparison with that in 1970s) *Earth Science Reports* **15**(1) 67–77 (in Japanese)

[4] Utsumi N, Kim H, Seto S, Kanae S and Oki T 2014 Climatological characteristics of fronts in the western North Pacific based on surface weather charts *Journal of Geophysical Research* **119**(15) 9400–9418 https://doi.org/10.1002/2014JD021734

[5] Japan Meteorological Business Support Center 2016 *Surface Weather Charts Provided as DVD by JMA* JMBSC October 10, 2016 http://www.jmbsc.or.jp/jp/offline/dvd0010.html (accessed on May 6, 2022) (in Japanese)

[6] Sanders F 1955 An investigation of the structure and dynamics of an intense surface frontal zone *Journal of the Atmospheric Sciences* **12**(6) 542–552 https://doi.org/10.1175/1520-0469(1955)012<0542:AIOTSA>2.0.CO;2

[7] Renard JR and Clarke CL 1965 Experiments in numerical objective frontal analysis *Monthly Weather Review* **93**(9) 547–556.

IOP Conf. Series: Earth and Environmental Science 1136 (2023) 012024 doi:10.1088/1755-1315/1136/1/012024

Statistical evaluation on hydrologic performance of bioretention design parameters under different rainfall conditions

Husnain Tansar[1,3], **Huan-Feng Duan**[1,3], **Ole Mark**[2]

[1] Department of Civil and Environmental Engineering, The Hong Kong Polytechnic University, Hung Hom, Kowloon, Hong Kong, SAR, China
[2] Innovation Urban Drainage, Kruger A/S, Gladsaxevej 363, Denmark
[3] Author to whom any correspondence should be addressed

husnain.tansar@connect.polyu.hk; hf.duan@polyu.edu.hk

Abstract. Effectiveness of a bioretention cell (BC) in mitigating stormwater peak flow load to an urban drainage system is highly dependent on the design configuration and retrofitting scale. The selection of appropriate BC design parameters has always been critical to achieving its maximum benefits. The BC modeling in the stormwater management model (SWMM) relies on numerous design parameters (in total of 18) with wide selection ranges, making parameter selection challenging. This study investigated the effect of hydrologic dynamics of BC design parameters under different rainfall conditions (i.e., intensity, duration, time-to-peak) on surface infiltration, surface outflow, and storage. Firstly, the seven influential BC design parameters (i.e., conductivity, berm height, vegetation volume, suction head, porosity, wilting point, and soil thickness) were selected based on one-factor-at-a-time (OAT) method, and 1000 random samples of each parameter were generated across the factor space by following respective uniform distributions. The simulations were performed with a python wrapper of SWMM (PySWMM) that used random samples for one parameter while leaving the others constant each time. In general, the statistical results of each parameter represent a significant variation in the volume distribution under various design rainfall conditions. Comparatively, the parameters of conductivity, berm height, and vegetation volume are categorized as the most influential compared to other BC design parameters for surface infiltration, storage, and surface outflow, respectively.

1. Introduction

Bioretention Cell (BC) is widely recognized as one of the most promising low impact developments (LID) facilities employed for quantitative and qualitative control of urban stormwater [1]. Research on comparative performance evaluation of different LIDs ranked BC as one of key and effective components place among all LIDs because of its larger infiltration and storage capacities, and cost-effectiveness [2]. Majority of previous research studies focused on evaluating BC's performance on surface runoff reduction, peak flow reduction and delay, flood reduction, and water quality improvement at the catchment scale, and proved its major contribution towards significant improvement in stormwater control and management. A handful of studies including [3,4] conducted on evaluation and optimization of structural design parameters of BC for alterations of its hydrological functions (e.g., infiltration volume, storage volume and outflow volume) under a single design rainfall scenario. However, accurate BC design configurations are critical for achieving various pre-defined design goals, and effective designing and implementation of BC can significantly enhance its efficiency for better utilization of urban impervious lands. This study evaluated effect of key design parameters on hydrological dynamics of BC under different design rainfall scenarios (i.e., return period, storm duration, time-to-peak). The main objective of this research is to (1) distinguish influential and non-influential parameters used for designing of BC in SWMM; (2) evaluate how BC's hydrological behavior toward different outputs changed under different rainfall conditions; and (3) evaluate variations of parameters' responses to different outputs within its design space.

2. Methodology

A simple and computationally efficient sensitivity analysis methodology was adopted for understanding hydrological dynamics of influential parameters of BC towards different outputs (e.g., surface infiltration, surface outflow, and storage volume) under different design rainfall conditions (e.g., return period, storm duration, time-to-peak). First, the one-factor-at-a-time (OAT) method was adopted to eliminate non-influential parameters among total parameters (18), and seven parameters including conductivity, berm height, vegetation volume, suction head, porosity, wilting point, and soil thickness were selected for further assessment. Following this, 1000 random samples of each influential parameter, which may provide acceptable probability of engineering accuracy [5], were uniformly generated across the factor space for equal representation of each part of parameter' space, as shown in Figure 1a. Factor space of design parameters is decided based on BC design guidelines from Storm Water Management Model (SWMM) reference manual [6], and Technical Guide of Sponge City Construction of Low Impact Development [7], and can be found in Appendix A. The samples of BC design parameters are simulated by Python wrapper of SWMM (PySWMM) on a hypothetical catchment of 1 km^2 (Figure 1b), with 5% treatment of the impervious area. The step-by-step methodology is represented in Figure 2. Furthermore, four design rainfalls of each scenario (i.e., intensity, storm duration, time-to-peak) were considered, with return period of (e.g., 3, 5, 10 & 20-yr), storm duration (e.g., 1, 2, 3 & 4hr) and time-to-peak (e.g., 0.2, 0.4, 0.6 & 0.8r), as shown in Figure 3. Synthetic rainfall was designed based on the Chicago design storm method [8] together with Shanghai rainstorm IDF formula, which was developed and employed for the design of local drainage network of Shanghai [9]. The alternating block method is employed for the formulation of synthetic rainfall hyetographs. The 3-year design rainfall was employed for evaluation of storm duration and time-to-peak scenarios.

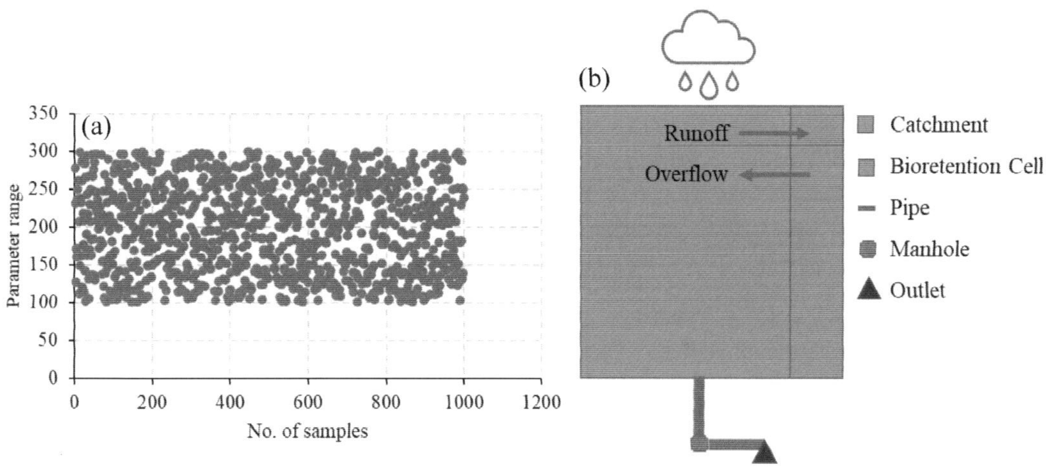

Figure 1. (a) Uniform random distribution of berm height parameter; (b) Test catchment.

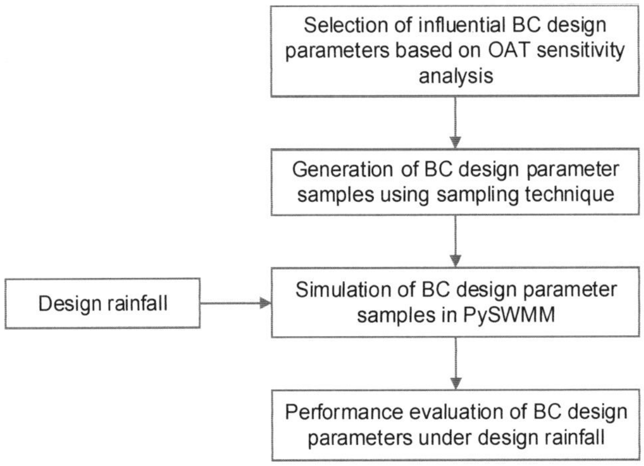

Figure 2. Methodological framework of sensitivity analysis.

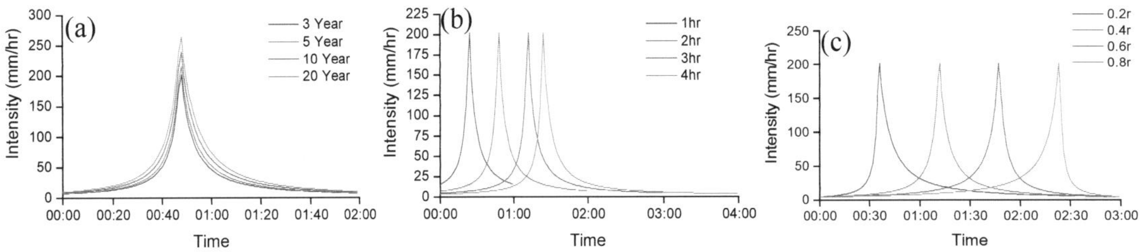

Figure 3. Design rainfall for (a) return period, (b) storm duration and (c) time-to-peak.

3. Results and Discussion

3.1 Surface Infiltration Volume

Generally, each BC design parameter has its unique hydrological dynamics under different design rainfall conditions, as represented in figure 4. Taking conductivity as an example, the surface infiltration volume represented a gradual increment in its median and interquartile range (IQR) from lower to higher and shorter to longer storm events, however, a reversal effect was observed for time-to-peak scenarios. These results illustrate higher return periods and longer storm events are more likely to be influenced by variation in conductivity parameter compared to lower intensity and shorter rainfall events. However, the earlier peak storm events are more sensitive to conductivity variation compared to later peak scenarios. Besides rainfall effect, as the first quarter of parameter space exhibited larger potential for variation of surface infiltration volume (as the first quartile (Q1) has larger spread characteristics), however, the remaining part of parameter design space denoted less variation in infiltration volume (as upper quartiles' results (Q2-Q4) are in concentrated form), showing dynamic parameter behavior within its design space. In case of other design parameters, berm height, porosity, and wilting point represented limited variation in their median and IQR for return period scenarios; however, the gradual reduction in their median infiltration volume from shorter to longer durations and earlier to later peaks highlights their maximum efficiency, because maximum water can be infiltrated and stored due to availability of larger storage capacity of soil layer (i.e., empty void spaces). Furthermore, these three parameters showed approximately equal representation of parameter's design space towards infiltration volume. Suction head showed very limited variation in its median and IQR and equal representation of parameter's design space. Vegetation volume has effect only on first quartile of infiltration, however a very insignificant contribution to change surface infiltration volume was observed for upper three quartiles (Q2-Q4). Soil thickness represented a gradual increase in infiltration volume from lower to higher return periods, however, the opposite effect was observed for shorter to longer and earlier to later peak storm events. It is because larger rainfall volume is infiltrated during higher storms events compared to relatively lower storms. On the other hand, the maximum infiltration capacity is utilized for shorter and earlier storms events compared to longer and later peak storm events because the infiltration capacity of soil gradually reduced with time leading to a reduction in infiltration volume. Soil thickness represented no significant variation in infiltration volume by changing depth within its design range.

3.2 Storage Volume

Sensitivity of BC design parameters substantially varied with unique characteristics for storage volume under different rainfall scenarios, as shown in Figure 5. For example, in case of conductivity, the gradual increase in median and decrease in IQR of storage volume from lower to higher return periods explain more shrinkage of storage volume towards higher parameters design space values. However, the expansion of storage volume around its median from shorter to longer storms showed an increase in storage volume by the extent of variation of higher design space values. Moreover, a gradual reduction in median and IQR of storage volume from earlier to later peak storms showed a reduction in storage capacity of BC for earlier to later peaks. Overall, conductivity showed a larger variation on storage volume within its design space by representing more concentrated storage and larger variation for higher and lower design space values, respectively. In case of berm height, a limited variation in storage volume for return period scenarios and gradual reduction in median and IQR from shorter to longer and earlier to later peak storms were observed. Vegetation volume showed very insignificant variations in storage volume for all rainfall scenarios. Furthermore, both parameters (berm height and vegetation volume) showed equal potential of each part of parameter design space towards storage volume. Suction head showed a gradual increment in storage volume from lower to higher return periods and a significant reduction from shorter to longer and earlier to later peak storm events. Effect of porosity and wilting point on storage volume was similar with a gradual increase in storage from lower to higher return periods and it substantially reduced from shorter to longer and earlier to later peak storm events. Further, overall, both parameters' design space outputs are bounded to limited ranges of storage volume. In case of soil thickness, it can be clearly seen a gradual increase in infiltration volume from lower to higher storm events, and a decrease from shorter to longer and earlier to later peak storm events (like its results of surface infiltration volume). However, results represent soil depth has an equal effect on storage volume within its design space.

3.3 Surface Outflow Volume

Overall, results illustrate significant effect on outflow volume by conductivity, berm height, and vegetation volume, and insignificant variation by suction head, porosity, and wilting point, as shown in Figure 6. Conductivity showed significant increase in outflow volume from lower to higher return periods, and limited growth in outflow volume from shorter to longer and earlier to later peak storm events. Contrary to surface infiltration and storage volume, maximum volume outflowed by lower range of parameter design space (Q1-Q3) and larger variations observed by upper part of design space (Q4). Based on design parameter results, it can be evidently inferred that the performance of BC is significantly reduced for intense storms (due to exploitation of maximum storage capacity of soil and storage layers) and gradually reduced for longer and later peak storms and leads to maximum runoff overflow towards drainage system. In case of berm height and vegetation volume, both parameters represented a significant increase in outflow volume from lower to higher and shorter to longer storm events, and no variation was noticed for time-to-peak scenarios. Berm height showed larger variations in outflow volume within its design space, and vegetation volume has relatively smaller compared to berm height. Other three parameters including suction head, porosity, and wilting point showed a similar pattern of results having considerable incremental variation during return periods and storm duration scenarios and no substantial change has been observed for time-to-peak scenarios. Moreover, these three design parameters showed no significant variation on surface outflow volume in the overall design parameter space. Higher and longer storms overflow more runoff compared to lower and shorter storms with a variation of soil thickness parameter, however, no countable effects were measured by variation of peak of the storm on outflow volume. The OAT sensitivity analysis results and ranking of parameters based on the median values of 3-year return storm are represented in Appendix A.

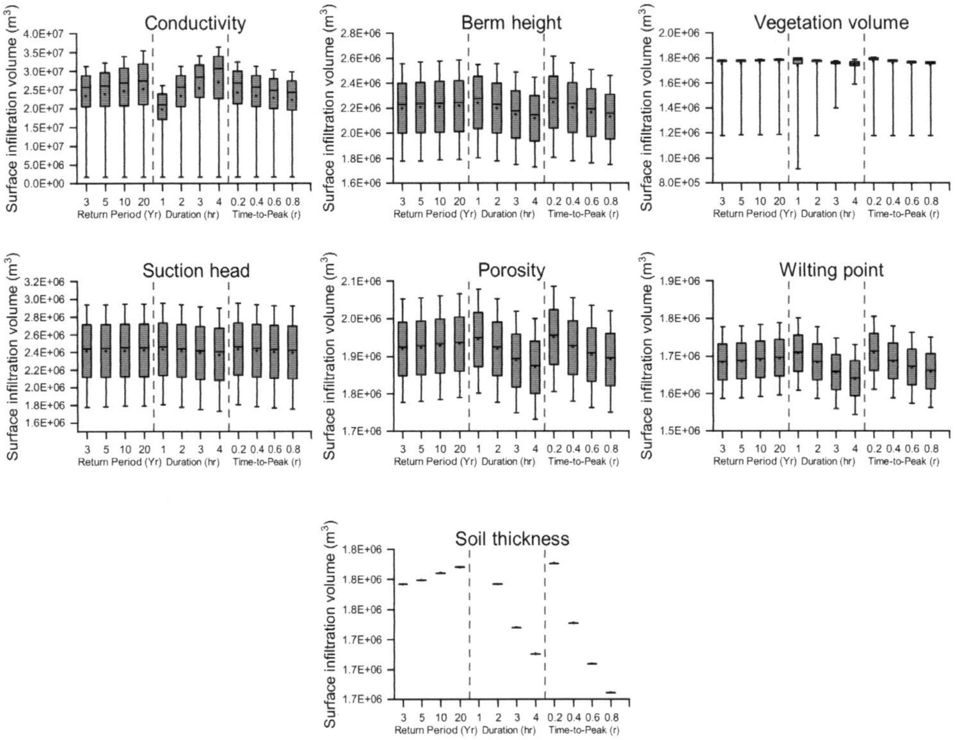

Figure 4. Hydrological response of different BC design parameters on surface infiltration volume under return period, storm duration, and time-to-peak scenarios.

Based on above-mentioned findings, it can be clearly identified that BC design parameters can be classified as influential and non-influential parameters. Each influential design parameter has its own unique performance when it comes to various outputs and storm conditions. Majority of the past research studies and government development programs on planning and designing of bioretention cell employed design parameters with guidance from SWMM reference manual [6] or with minor modifications based on local design guides. However, BC design cannot be generalized at a larger scale (global) as each region or catchment has its unique climatic dynamics, soil characteristics, groundwater table conditions, local design guidelines, and target outputs.

Results conclude that understanding the dynamical behavior of design parameters within their design space and under different rainfall conditions provides substantial guidance about their accurate selection of design values during bioretention cell planning, modeling, optimization, and implementation stages. Furthermore, the findings of study facilitate the accurate selection of design variables for bioretention cell implementation, including their soil characteristics and design elements (e.g., berm height), which can significantly improve the collective performance of bioretention cell, particularly at the catchment scale. On the other hand, inaccurate design parameters lead to erroneous BC design unit and its replication at catchment scale may result in a significant decline in overall performance.

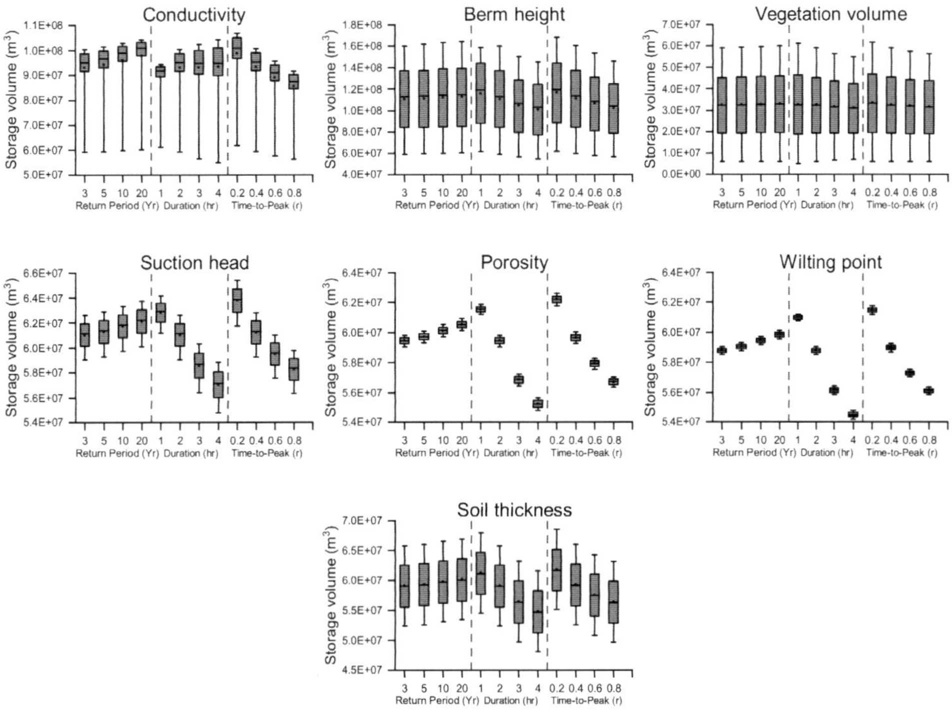

Figure 5. Hydrological response of different BC design parameters on storage volume under return period, storm duration, and time-to-peak scenarios.

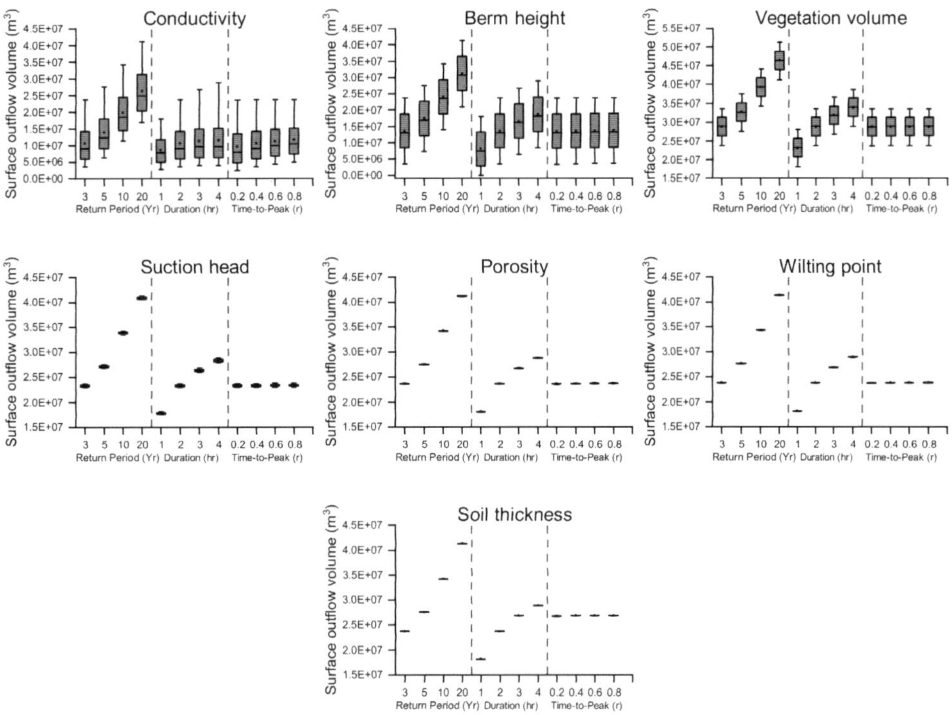

Figure 6. Hydrological response of different BC design parameters on surface outflow volume under return period, duration, and time-to-peak scenarios.

4. Conclusion

An effective and efficient statistical performance evaluation approach is employed for assessing hydrological response of different design parameters of BC in SWMM under different design rainfall conditions. Specifically, the conductivity, berm height, vegetation volume, suction head, porosity, wilting point and soil thickness are chosen as the most influential parameters among eighteen design parameters. However, overall, each influential BC design parameter represented its unique hydrological response towards different outputs (e.g., surface infiltration, storage capacity, and outflow volume) under various design rainfall scenarios (e.g., return period, storm duration, time-to-peak). Results highlighted significant variation in performance of BC design parameters within its design space or limited variability towards model outputs provide guidelines about ranges of most influential and less influential design parameters. Understanding design parameters response variation provides opportunities for better selection and designing of the best possible parameters that leads to improvement in BC design unit and accurate design replication at catchment scale may collectively produce substantial improvement in target design goals.

Acknowledgments
This work was supported by the Hong Kong Research Grants Council (RGC) under project no. 15200719 and the Hong Kong Polytechnic University under projects no. ZVR5 and no. ZVWM. We are thankful to Computational Hydraulics International (CHI) for providing license of PCSWMM to conduct this research.

Appendix A

Ranking of BC design parameters are presented in figures A1-A3 and table A1.

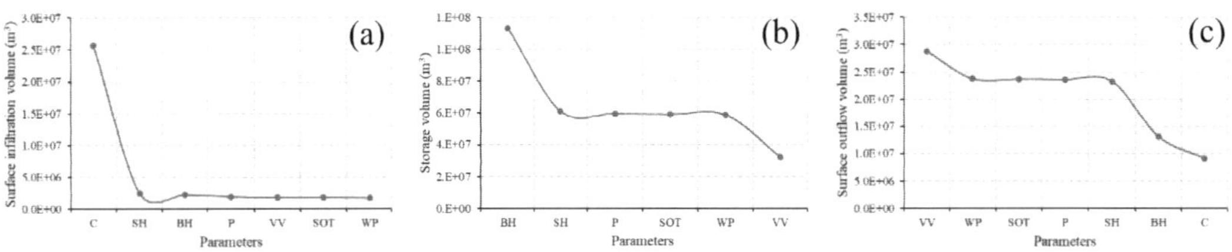

Figure A1. Ranking of BC design parameters for performance evaluation of (a) surface infiltration volume (b) storage volume and (c) surface outflow volume.

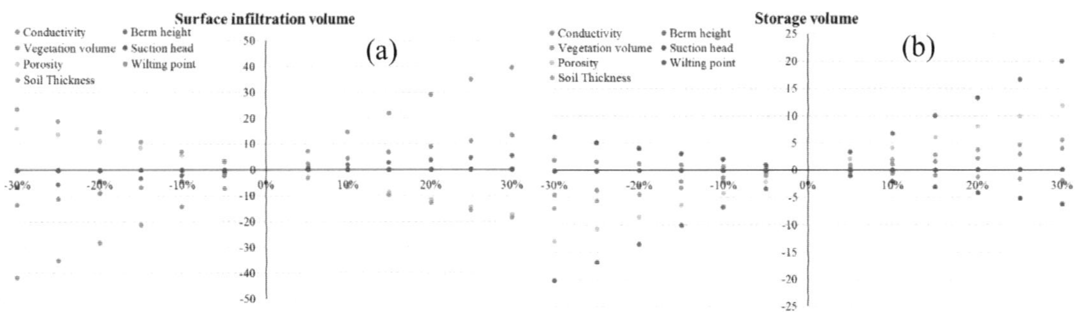

Figure A2. One-factor-at-time (OAT) sensitivity analysis of bioretention cell parameters for (a) surface infiltration volume and (b) storage volume.

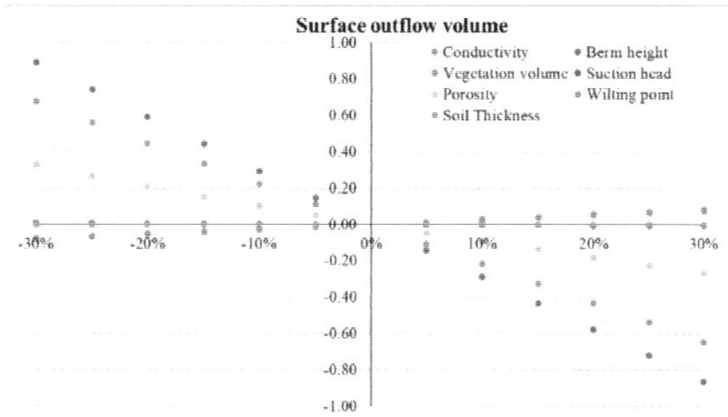

Figure A3. One-factor-at-time (OAT) sensitivity analysis of bioretention cell parameters for surface outflow volume.

Table A1. BC designed parameters with their ranges.

Layers	Parameters	Range	Short Name
Surface	Berm height (mm)	100-300	BH
	Vegetation volume (fraction)	0-1	VV
	Surface roughness (Manning's n)	0-0.25	SRM
	Surface slope (%)	0-5	SS
Soil	Thickness (mm)	450-900	SOT
	Porosity (volume fraction)	0.45-0.60	P
	Field capacity (volume fraction)	0.15-0.25	FC
	Wilting point (volume fraction)	0.05-0.14	WP
	Conductivity (mm/hr)	0.4-120	C
	Conductivity slope	30-60	CS
	Suction head (mm)	49-320	SH
Storage	Thickness (mm)	150-900	STT
	Void ratio (Voids/Solids)	0.2-0.4	VR
	Seepage rate (mm/hr)	127-380	SR
	Clogging factor	-	CF
Underdrain	Drain coefficient (mm/hr)	0-0.5	DC
	Drain exponent	0-0.5	DE
	Drain Offset height(mm)	0-305	DOH

References

[1] Yang Y and Chui TFM 2018 Integrated hydro-environmental impact assessment and alternative selection of low impact development practices in small urban catchments *J. Environ. Manage.* **223** 324–37 https://doi.org/10.1016/j.jenvman.2018.06.021

[2] Tansar H, Duan H-F and Mark O 2022 Catchment-scale and local-scale based evaluation of LID effectiveness on urban drainage system performance *Water Resour. Manag.* **36** 507-26 https://doi.org/10.1007/s11269-021-03036-6

[3] Li J, Liu Z, Jiang C, Li Y, Li H and Xia J 2021 Optimization design of key parameters for bioretention cells with mixed filter media via HYDRUS-1D model and regression analysis *Ecol. Eng.* **164**(1) 106206 https://doi.org/10.1016/j.ecoleng.2021.106206

[4] Sun YW, Wei XM and Pomeroy CA 2011 Global analysis of sensitivity of bioretention cell design elements to hydrologic performance *Water Science and Engineering* **4**(3) 246–57 https://doi.org/10.3882/j.issn.1674-2370.2011.03.002

[5] Tung Y-K, Yen B-C and Melching CS 2005 *Hydrosystems Engineering Reliability Assessment and Risk Analysis* (McGraw-Hill Civil Engineering Series)

[6] Rossman LA 2016 *Storm Water Management Model Reference Manual* vol **III** National Risk Management Research Laboratory, Office of Research and Development (Cincinnati, OH, USA: U.S. Environmental Protection Agency)
[7] MoHURD China 2014 *Technical Guide for Sponge Cities-Construction of Low Impact Development*
[8] Keifer CJ and Chu HH 1957 Synthetic storm pattern for drainage design *J. Hydr. Eng. Div.-ASCE* **83**(4) 1325–32 https://doi.org/10.1061/JYCEAJ.0000104
[9] Shanghai Municipal Engineering Design Institute 2003 *Water Supply & Drainage Design Handbook: Urban Drainage* (Beijing, China: China Architecture & Building Press) (in Chinese).

Characteristics of line-shaped rainbands regarding duration, shape, and rainfall intensity in northern Japan using the Radar/Raingauge-Analyzed Precipitation Product

Yuta Ohya[1] and Tomohito J Yamada[2]

[1] Graduate School of Engineering, Hokkaido University, Sapporo, Hokkaido, 060-8628, Japan
[2] Faculty of Engineering, Hokkaido University, Sapporo, Hokkaido, 060-8628, Japan

y_ohya@eis.hokudai.ac.jp

Abstract. Line-shaped rainbands (LRBs), including Mesoscale Convective Systems (MCSs), stagnate on river basins and cause floods and landslides. Moreover, the frequency of LRBs is increasing around Hokkaido a part of northern Japan. Due to the complexity of the three-dimensional structures found in LRBs, some research dealt with two-dimensional observation data to identify them for the sake of simplicity. In previous studies, several methods to widely identify LRBs from radar rainfall data in terms of rainfall intensity, shape of rainfall area, and duration have been proposed. However, using the nationally consistent definition, the number of LRBs occurring in this region would be evaluated as very few due to the little rainfall climatology of this area. In this study, based on the previous studies, we adjust many combinations of threshold values, such as rainfall intensity, aspect ratio, and time duration in order to verify of validity the LRBs that occurred in northern Japan. The authors visually identified the six LRBs that caused the disaster and showed thresholds at which they were mechanically extractable by a computer program. The combination of thresholds that included at least six of those disaster-scale cases was "a rainfall area defined by a closed curve with a previous 3-hour rainfall of 40 mm or more, stagnating for 2 hours or more (covering 40% or more of its area), and with a maximum aspect ratio of 2.5 or more."

1. Introduction

In Japan, about 60% of heavy rainfall events, including the direct effects of typhoons and tropical cyclones, are related to LRBs or similar meteorological disturbances [1]. Such phenomena are commonly referred to as mesoscale disturbances in the classification scheme by Orlanski [2]. Recently, the development of non-hydrostatic modeling and remote sensing has made it possible to reduce mesoscale convective processes, which has improved our understanding of the maintenance and development mechanisms of these phenomena as MCSs [3]. Among them, the rainfall areas in linear shapes are represented by squall line types, back-building types, and are characterized by the systematic development and linear formation of series of precipitation cells [4-7]. However, it was noted that the majority of such linearly shaped mesoscale disturbances in Japan are caused by back-building or both back-and-side-building [8,9]. Regarding such types there has been much discussion, beginning with Kato (1998) and others concerning their emergence, disappearance processes, and its maintenance and developmental mechanisms [10]. Additionally, early forecasting using high-resolution meteorological modeling and data assimilation technique based on advanced observational networks has been attempted in recent years [11].

The above-mentioned studies are focused on representative heavy rainfall events related to linearly shaped mesoscale disturbances, either for early prediction of their occurrence or for extensive analysis of their maintenance and development mechanisms.

On the other hand, some studies have been conducted to identify linearly shaped rainfalls from only the spatiotemporal distribution of rainfalls based on several decades of radar observation datasets and to summarize statistical characteristics, such as the frequency of occurrence and spatiotemporal distribution of such patterns. Hereafter, rainfall that is stagnant with a linear shape is described as a LRB.

Fairman (2016) extracted heavy rainfall areas greater than 500km^2 with a reflection intensity of 20 dBZ (30dBZ, 40dBZ with different thresholds) from 12 years of radar datasets in United States. Then, an elliptical approximation was applied to rainfall areas and they defined a linear shape as the ratio of the short axis to the long axis of the areas being greater than 1:3. The results showed that the extracted LRBs tended to be larger in both area and intensity than all rainfall patterns regardless of shape, while noting that the number of extracts varied by more than one order of magnitude depending on the threshold. By using this threshold, the number of LRBs showed a trend of high occurrence east of the Rocky Mountains and a maximum frequency in spring (March-May) over the same area. This is consistent with the qualitative appearance trend of squall lines [12]. Elsewhere, Ashley (2019) used an artificial intelligence (AI) method to extract LRBs over the past 22 years with the training information of several previous LRBs that were visually identified by MCS researchers [13,14]. They classified 32% of the MCS cases as LRBs which were concentrated during the spring in the area east of the Rocky Mountains. They also showed that 50% of the strong winds and hail over the same period and area were due to linearly shaped rainfall areas.

In Japan, Tsuguchi and Kato (2014) defined a LRB event as a rainfall area of 50 mm or more of precipitation in 3 hours with a short axis to long axis ratio of 1:3 or greater. On the above definition, the LRB is concentrated during the rainy season of June-July in Southern Japan while the rainfall area is wider in comparison to other shapes [15]. Hirockawa et al. (2020), who developed the above study, defined a heavy rainfall area (HRA) as a closed curve with 80 mm of rainfall per 3 hours containing a grid of 100 mm/3h, and LRBs as a case where the ratio of the short axis to the long axis is greater than 1:2.5 and stagnant for 5 hours (henceforth H20). This is also the definition of disaster prevention information on LRBs provided by the Japan Meteorological Agency (JMA).

The results showed that 80% of LRBs occur in the summer season from June to September, that 80% of LRBs have typical MCS structures, and that these are distributed in Southern Japan, mainly in Kyushu [16]. However, a study by Yamada et al. (2012) of Hokkaido and its surrounding areas indicates that even in the northern area some LRB events occurred mainly over the Sea of Japan, and that the number of such events has been increasing in recent years. Such study defined LRBs as three conditions, stagnates for over 2 hours, contains a grid of up to 20 mm/h, and is linearly shaped using radar information [17].

Because such a single index for the whole of Japan underestimates the number of LRBs extracted from northern Japan, the authors redefined the threshold over the same area, following the method of H20, and extracted LRBs. The definition is described in this paper as OY22, as opposed to H20, in detail. As a side note, another paper (under review) applied OY22 to ensemble data considering future prediction. It showed that their numbers and intensity increase with global warming scenarios, which can be characterized by the climatology of sea surface temperatures and environmental fields.

For the objective extraction of LRBs, various definitions have been proposed, depending on the area of interest and the phenomenon. In this study, to provide an indication of the suitability of the definition of LRB in this region, we evaluated the change in the recognition of LRBs when the thresholds determined univocally in OY22 were modified with increasing/decreasing parameter values, based on previous rainfalls and disasters in the corresponding area.

IOP Conf. Series: Earth and Environmental Science 1136 (2023) 012025 doi:10.1088/1755-1315/1136/1/012025

2. Methodology

In this study, only the observed data were used; and the data analysis was based on the H20 method. The dataset used and the extraction method are described in detail below.

2.1. Datasets

The observation data used in this study is the Radar/Raingauge-Analyzed Precipitation Product (R/A) [18]. R/A is gridded rainfall data based on radar rainfall observations installed and managed by JMA, which are corrected by ground observations. Since the start of data delivery in 1988 the resolution has been improved three times, beginning with a spatial resolution of 5 km with a temporal resolution of 1 hour, then to a spatial resolution of 2.5 km in 2001 with a temporal resolution of 30 minutes in 2003. It was later improved to a spatial resolution of 1 km in 2006 and onwards. First, these datasets with different resolutions were changed to 5 km through bilinear interpolation of the spatial resolution and to 1-hour intervals by reducing the time resolution from 30-minute to 1-hour intervals.

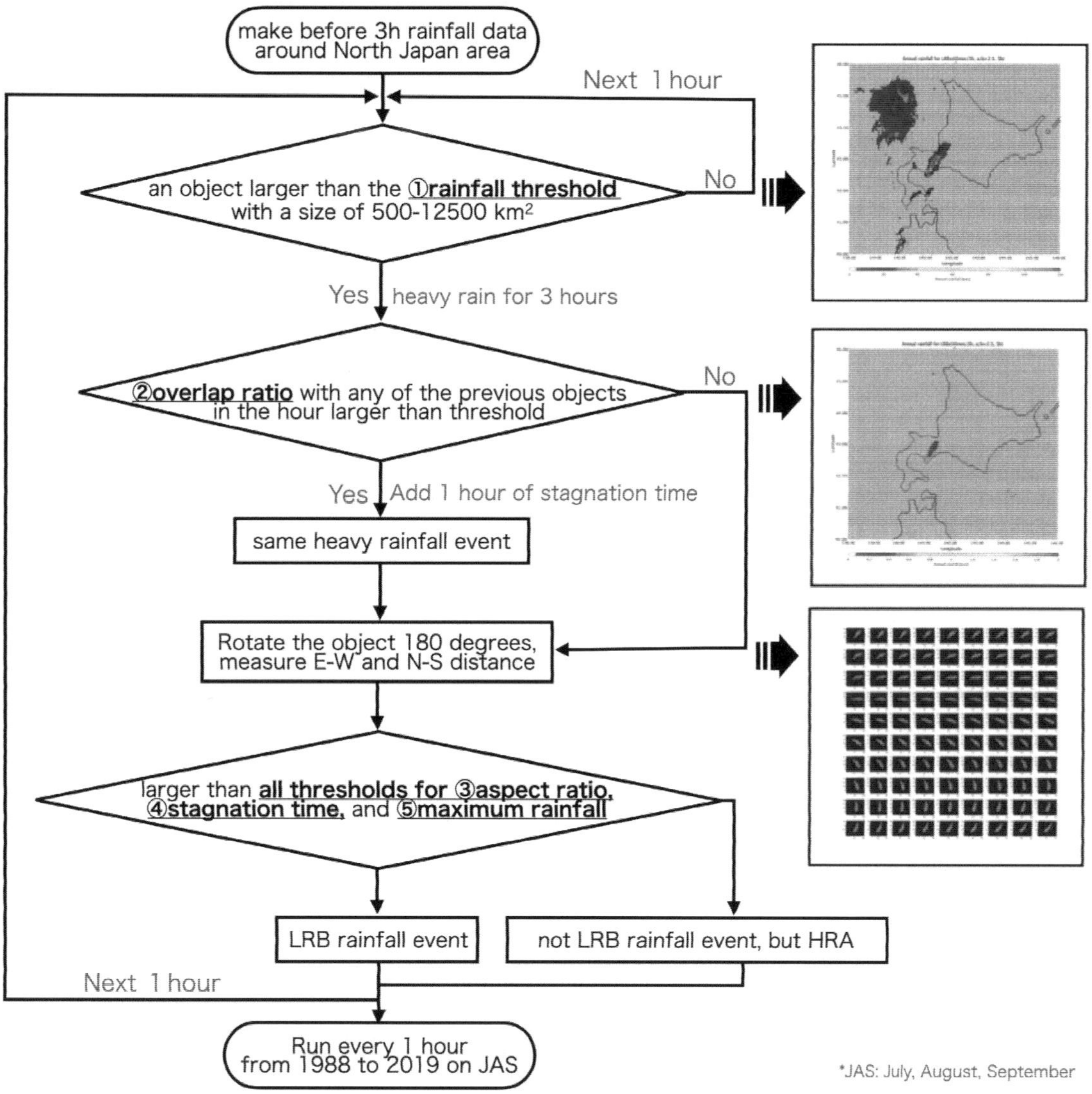

Figure 1. Flowchart of LRB extraction (the five thresholds are underlined and some processes are shown in a small window on the right side).

2.2. Extraction of LRB

A flowchart for extracting HRAs and LRBs using R/A gridded data with uniform resolution is shown in Figure 1.

Initially, the previous 3-hour rainfall was calculated, and grids with a closed curve above a certain rainfall threshold – the threshold shown on ① of Figure 1 – were extracted as a heavy rainfall object, which has an area range of 500÷12,500 km^2 (this indicates a meso-beta scale). One hour later – when the rainfall for the previous three hours satisfied the threshold – it was also extracted as a heavy rainfall object, which was later overlapped on the heavy rainfall objects of the previous hour. Then, when the ratio of the matched grids to the total grids exceeded the threshold for coverage – the threshold on ② of Figure 1 – it was defined as one heavy rainfall event. The stagnation time was defined as the time from the first radar rainfall observation to the last one used for the heavy rainfall event defined here. The ratio of the short axis to the long axis length of the HRA object contained within the stagnation time of its corresponding event extracted in this way was calculated; the maximum was recorded on the list as the aspect ratio of the HRA event. Likewise, stagnation time, amount of rainfall, area, and maximum hourly rainfall were also recorded on the list. Multiple thresholds were applied to the stagnation time, aspect ratio, and maximum 1-hour rainfall – as shown on ③, ④, and ⑤ in Figure 1, respectively. As such, a HRA event was defined as an LRB when all the thresholds were satisfied among the collected HRA events.

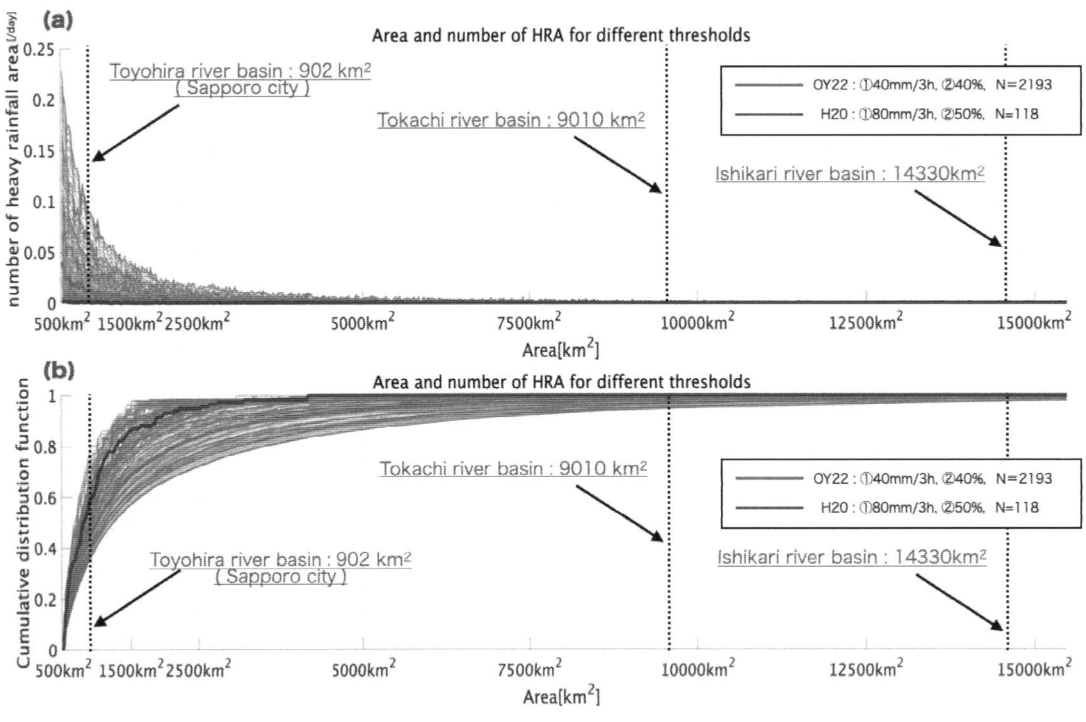

Figure 2. Distribution of the HRA occurrence by rainfall area size.

The rainfall threshold was changed from 30 mm/3h to 100 mm/3h at 5 mm/3h intervals, and the overlap ratio threshold was changed from 30% to 90% at 5% intervals, and the results for each combination are plotted. In particular, H20 is shown in blue, OY22 in red, and the range of the rainfall threshold around 10% is shown in yellow. (a) shows the probability distribution normalized to the number of occurrences per day. (b) shows the probability distribution of the normalized cumulative number of occurrences.

3. Results

Figure 2 shows the frequency of heavy rainfall events for each area with various combinations of ① rainfall intensity and ② overlap ratio thresholds. The blue lines are the results of applying the H20 thresholds (① 80 mm/3h and ② 50% of overlaps) to Hokkaido and its surrounding areas. As a result, 118 HRAs (0.04/ day) have been extracted over the past 32 years. However, these are not the only heavy rainfall events that have a risk of flooding and landslide-causing disasters in this region. Ohya and Yamada (2022) selected the following thresholds for the six cases of river disasters caused by LRBs in the past, which can be extracted using the same method (① 40 mm/3h, ② 40%, ③ 2.5, ④ 4 h, and ⑤ 35 mm/h). The red line in Figure 2 shows the number of extracted HRAs under the above definitions of ① and ②. Because the designed level rainfall of national rivers in the Kyushu area is 521mm/2days (1/80) while that of national rivers in the northern Japan area is 260mm/3days (1/150), it is understood that approximate range 30-50mm/3h is suitable for the definition of heavy rainfall in the northern Japan area. When the threshold of 40mm/3h varies by about 10% (yellow color area), the number of extractions differs from approximately half to twice as much (as shown by the red width on Figure 2). The distribution of the number of occurrences by size shows that half of the HRAs are around 1000 km^2 for any threshold.

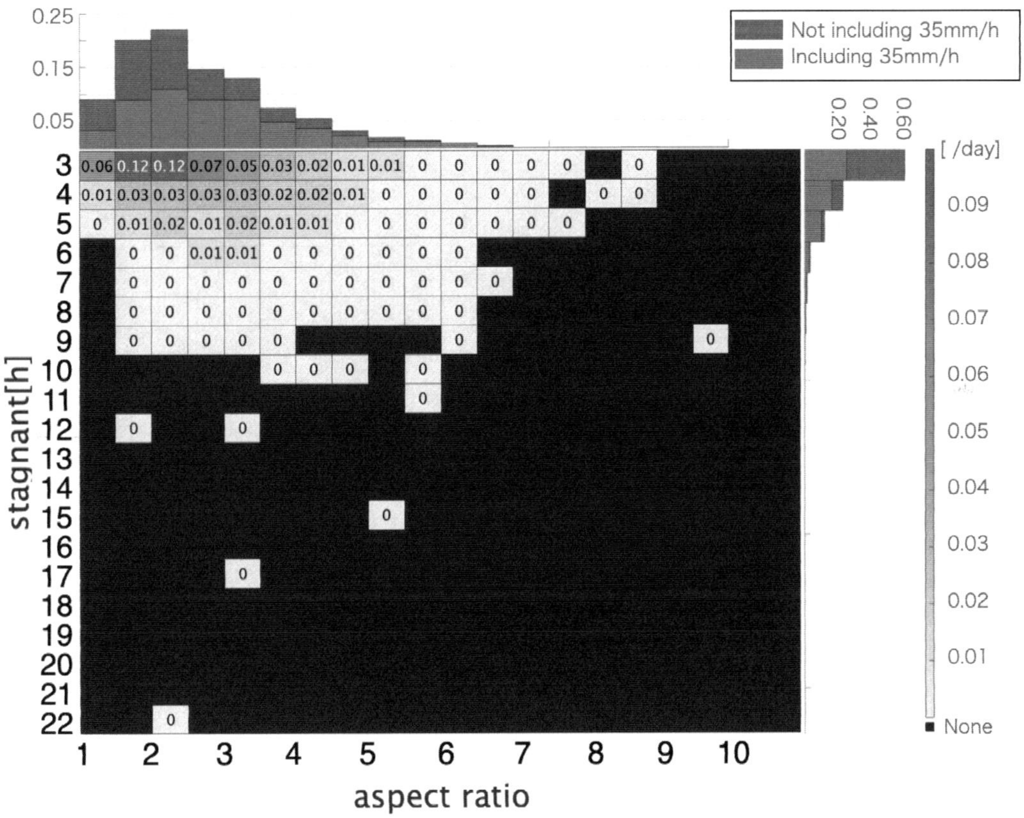

Figure 3. Distribution of occurrence of stagnant and aspect ratio in HRAs.

The threshold of rainfall is 40 mm/3h and the threshold of overlap ration is 40%. The horizontal axis is the aspect ratio and the histogram is shown at the top. The vertical axis is stagnant, and the histogram is shown on the right. The black color in the heatmap indicates that the observation does not exist, and 0 means that the observation has very few significant digits. In the histogram, the red color indicates the percentage of the grid containing 35 mm/h within the event.

Figure 3 shows the aspect ratio, stagnation time, and the inclusion of a 35 mm/h grid for the extracted heavy rainfalls based on the criteria of ① 40 mm/3 h and ② 40%. The figure shows that 60% of the heavy rainfalls are no stagnation (i.e., coverage of 40% or more with the previous 3-hour rainfall one hour before). From this data it becomes clear that many heavy rainfalls are not stagnant. Although the figure is not shown here, these features are especially observed near the Hidaka Mountains. This area is frequently characterized by topographic rainfalls associated with approaching typhoons. Additionally, the stagnation criteria in the same analysis may contribute to the elimination of such heavy rainfalls that are not caused by MCSs. On the other hand, the aspect ratio shows that the mode is from 2 to 2.5, which can be considered a tipping point for determining the existence or lack of linear shapes (the same as H20). As the aspect ratio increases, so does the tendency to have a strong rainfall grid of 35 mm/h during a heavy rainfall event. Thus, the linear shape of the rainfall may determine the presence or absence of a strong convergence band. The shape of the rainfall is directly related to the increase in disaster risk when it matches the shape of the river basin or the direction of river flow. Therefore, the area, shape, and direction of rainfall should be organized by river basin.

Table 1. Extraction result of previous disaster events.

Case	date	River basin	Stagnant [h]	Aspect [-]	Area [km^2]	Rain [mm/h]	Accum [mm]
1	25 Aug. 1988	Rumoi	17	3.4	10475	79.3	394.7
2	13 Aug. 1994	Teshio	6	3.6	7654	55.2	143.9
3	18 Aug. 2006	Mukawa, Saru	6	5.6	2875	66.3	115.8
4	24 Aug. 2010	Ichikari (Chubetsu)	5	6.6	4627	68.5	124.9
5	11 Sep. 2014	Ishikari (Chitose)	12	3.2	5325	89.9	382.5
6	2 Jul. 2018	Ishikari (Uryu)	6	5.0	11600	38.4	110.9
Definition (40h/3h, 40%)			4	2.5	500-12500	35.0	-

The characteristics (aspect ratio, stagnation time, amount of rainfall, area, etc.) of the six cases that caused the disasters described above are listed on Table 1. Case 4 is a typical LRB with a back and side-building structure. However, because of its small area, the maximum aspect ratio was also evaluated to be small. It should be mentioned that in the same case, the rainfall area spreads slowly within the stagnation time, so that a linear shape determination is not made by regarding only the shape of the accumulated rainfall. In contrast, Cases 5 and 6 have large rainfall areas and aspect ratios because the typhoons supplied large amounts of moisture to the stationary fronts on synoptic scales. Such long and narrow cases often bring heavy rainfalls in corresponding regions, but a small fluctuation of the rainfall area causes a large change in overlap ratio. Therefore, we responded by setting the lower limits of ② overlap ratio and ④ stagnation time lower than H20. As described in OY22, this method still contains many topographical features in mountainous areas.

4. Conclusions

This study discusses the differences in the results of different thresholds for the method of extracting LRB for Hokkaido and its surrounding areas. H20 threshold is mainly designed to extract heavy rainfall events in the Kyushu region during the rainy season in Japan, so the number of events extracted would be extremely small if applied to the target area of this study. Therefore, we have shown that disaster-scale rainfall events can be included by reducing the threshold of rainfall. The frequency distribution of the moderated HRA was also compared for the combinations of aspect ratio, stagnation time, and presence/absence of a strong rainfall area. As a result, it was shown that the extraction results changed significantly when the aspect ratio and stagnation time were 2.5 and 4 h. Although this study has shown the features of cases in which rivers flooded due to LRBs, there is a challenge in separating LRBs from events that do not belong to the MCS.

Acknowledgments
The authors greatly appreciate two anonymous reviewers for their constructive comments. This research is supported by Science and Technology Research Partnership for Sustainable Development (SATREPS) in collaboration between Japan Science and Technology Agency (JST, JPMJSA2009) and Japan International Cooperation Agency (JICA) and Research Field of Hokkaido Weather Forecast and Technology Development (endowed by Hokkaido Weather Technology Center Co. Ltd.), and JSPS KAKENHI Grant Number 20J21938 and 22H01594.

References
[1] Kato and Hirockawa 2021 Quasi-stationary band-shaped convective systems causing localized heavy rainfall, *Nagare*, **40**, 3–8. (in Japanese)
[2] Orlanski, I. 1975 A rational subdivision of scales for atmospheric processes. *Bulletin of the American Meteorological Society*, 527-530
[3] Houze Jr, R.A. 2018 100 years of research on mesoscale convective systems. *Meteorological Monographs*, **59**, 17-1
[4] Bluestein, H.B., and Jain, M.H. 1985 Formation of mesoscale lines of precipitation: Severe squall lines in Oklahoma during the spring. *Journal of the Atmospheric Sciences*, **42**(16), 1711-1732
[5] Houze Jr, R. et al. 1990 Mesoscale organization of springtime rainstorms in Oklahoma. *Monthly Weather Review*, **118**(3), 613-654
[6] Parker, M.D., and Johnson, R.H. 2000 Organizational modes of midlatitude mesoscale convective systems. *Monthly Weather Review*, **128**(10), 3413-3436
[7] Schumacher, R.S., and Johnson, R.H. 2005 Organization and environmental properties of extreme-rain-producing mesoscale convective systems. *Monthly Weather Review*, **133**(4), 961-976
[8] Seko H 2010 Study of the shapes and maintenance mechanisms of meso-β scale line-shaped precipitation systems in the middle-latitudes, *Journal of Meteorological Research* **62**, Nos. 1-3 (in Japanese)
[9] Kato, T 1998 Numerical simulation of bandshaped torrential rain observed over the southern Kyushu, Japan on 1 August 1993. *J. Meteor. Soc. Japan*, **76**, 97-128
[10] Moroda, Y., Tsuboki, K., Satoh, S., Nakagawa, K., Ushio, T. and Shimizu, S 2021 Structure and evolution of precipitation cores in an isolated convective storm observed by phased array weather radar. *Journal of the Meteorological Society of Japan*. Ser. II
[11] Taylor, J. et al. 2021 Predictability of the July 2020 Heavy Rainfall with the SCALE-LETKF. SOLA
[12] Fairman Jr, J.G. et al. 2016 Climatology of banded precipitation over the contiguous United States. *Monthly Weather Review*, **144**(12), 4553-4568
[13] Haberlie, A.M., and Ashley, W.S. 2019 A radar-based climatology of mesoscale convective systems in the United States. *Journal of Climate*, **32**(5), 1591-1606
[14] Ashley, W.S., Haberlie, A.M., and Strohm, J. 2019 A climatology of quasi-linear convective systems and their hazards in the United States. *Weather and Forecasting*, **34**(6), 1605-1631
[15] Tsuguchi H. and Kato T. 2014 Objective extraction of heavy rainfall events and statistical analysis on their characteristic features, *Tenki*, **61**(6), 455-469 (in Japanese)
[16] Hirockawa, Y. et al. 2020 Identification and classification of heavy rainfall areas and their characteristic features in Japan. *Journal of the Meteorological Society of Japan*. Ser. II
[17] Yamada, T. J. et al. 2012 Climatology of line-shaped rainbands over northern Japan in boreal summer between 1990 and 2010. *Atmospheric Science Letters*, **13**(2), 133-138
[18] Nagata, K. 2011 Quantitative precipitation estimation and quantitative precipitation forecasting by the Japan Meteorological Agency. *RSMC Tokyo–Typhoon Center Technical Review*, **13**, 37-50.

IOP Conf. Series: Earth and Environmental Science 1136 (2023) 012026 doi:10.1088/1755-1315/1136/1/012026

Challenges & solutions for deterministic hydraulic modelling in Mediterranean coastal catchment. Application to the lower Paillons River, Nice, France

Paguédame Game[1], Mingyang Wang[1], Philippe Audra[1] and Philippe Gourbesville[1]

[1] Polytech'Lab UPR UCA 7498, Université Côte d'Azur, Nice, 06200, France

paguedame.game@etu.univ-cotedazur.fr

Abstract. Different methods are used in hydrological and hydraulic modelling in a catchment. Deterministic modelling can be used to produce detailed information and optimize the organization of massive data collection. In the catchment of Paillons, there is a need to improve knowledge of hydrological processes through deterministic modelling. In fact, Nice Côte d'Azur metropolis faces numerous challenges in the Paillons catchment. The area provides water resources for large communities. However, it is exposed to flooding and droughts. Complex hydrological processes generate runoff over the 246 km^2 watershed. Few existing monitoring stations provide runoff data. In addition, there are limitations in the understanding of surface hydraulics. Thus, this study uses DHI Mike21FM for surface hydraulics with 5 m grid size for riverbed and 2 m for tunnels. The study area is limited to the lower Paillons River. Observed and modelled water depths vary between 0.1 and 0.5 m for a maximum discharge of 38 m^3/s in the tunnels. Flood maps created with a discharge of 1000 m^3/s, show clearly high flood risk zones and flow directions. The selected CFL condition under 0.8 is respected. The tool is suitable for modeling flooding in areas of interest within the catchment of study. The results obtained are satisfactory and demonstrate that the constructed tool makes it possible to reproduce the overall behavior of surface hydraulics.

1. Introduction

Climate change is projected to intensify extreme weather events in many locations [1]. Flooding and drought already affect cities like Nice metropolis in south of France. Existing monitoring systems are used for managing water issues. But there are gaps and errors in collected information. Integrated modelling tools are built to fill those gaps [2,3].

Several studies used deterministic modelling tools for surface hydraulic. The DHI MIKE21 FM is a two-dimensional hydrodynamic tool widely used for flood modelling. It is applicable to various environments including coastal areas. As input data, it requires topography, land use, runoff, and weather information. Modelling results are used for flood management activities. It can be coupled with hydrological and groundwater modelling tools like DHI Mike SHE and FEFLOW. Coupling allows assessments of flood extents and river-aquifer exchanges. When monitored runoff data are not available, hydrological modelling results are used as inputs to hydraulic models. Many parameters required for model set up can be retrieved from literature. Calibrated sets of parameters are obtained for areas of interest [3,4,5]. Mike21 flow model allows integration of hydraulic structures like dikes, weirs, and culverts. In some case studies, tunnels are modeled as open channels or represented by long culverts. Depending on the modelling objectives, each approach can generate useful information for management purposes. For pipes and open channels, it is possible to use DHI Mike 1D pipe flow model. An integration of comparison 1D and 2D modelling tools in decision support can be valuable. Both have advantages and disadvantages. Computing power and project objectives determine the most suitable tool. In an integrated Decision Support System (DSS), appropriate computing time is required. Therefore, it is useful to assess different approaches in the area of study. In the Aquavar project [1], MIKE21FM is implemented for the low Var River. The Var is located few kilometers from the lower Paillons river. The Paillons is smaller, but both rivers can be affected by the same extreme rainfall events. There are similarities in the geology, topography, and land use of the surrounding areas. Therefore, parameters in the Var River model can be applied to the Paillons.

The lower Paillons River drains out runoff water from the Paillons catchment. Hydrological processes are complex in the area. Tributaries of lower Paillon are steep, and floodplains are densely urbanized. Of the 246 km^2 area, 16 km^2 feeds the urban drainage network. Studies show that lower Paillon River is affected by sediment deposition. On the contrary, its steep tributaries are mostly eroded. Channelization of the riverbed is also contributing to the problem. Mainly marls and limestone compose rock outcrops. Irrigation and industry rely also on groundwater in the area [6, 7, 8, 9, 10, 11]. September to February is usually intense in extreme events [12]. In recent years, events in 2000 and 2019 caused major erosion and destructions.

The goal is to build a hydraulic model that can be used for real time DSS tool integrating hydrology, hydraulic and groundwater models. The period 2000-2021 has useful monitored runoff data. After constructing and evaluating each model component, AquaPaillon DSS platform will become operational. In this context, this analysis uses MIKE21FM to explore flow processes in the riverbed and flood risk.

2. Methods

The method focuses on hydraulic modelling of the lower Paillon River. DHI Mike21FM is a 2D surface flow modelling tool used to assess risk of flooding of the road tunnel and the capacity of closed-channel river reach outlet to drain runoff water to the sea. Calibration and validation rely on observed water depths and runoff at different locations in the riverbed. Model set up included processing mesh geometry with a 25 cm DEM data. In addition, a 1D pipe flow model based on DHI Mike 11 is applied to the road tunnel located on the banks of the river.

2.1. Case study: lower Paillon River in city of Nice

The lower Paillon River drains out runoff water from 230 km2 of Paillons watershed. It is channelized with flood protection levees and weirs. The last downstream stretch of the river is piped up to the sea. It also crosses the centre of the urban area of Nice metropolis. Figure 1 presents the layout of the river and associated structures. An important road tunnel is located to the left bank of the river. This structure is designed to be flooded. It is actually part of several tunnels in the riverbed. Six are used to drain river flow underground. And the one used as road tunnel can be flooded at different locations. For extreme events, it receives overflow from river and avoids the city centre to be affected.

There are monitoring stations (managed by local water authorities of the Nice metropolis) in the modelling domain. This stream gauges are part of a flood warning systems operated in the area. On one hand, three stations monitor water depth and discharge in the open riverbed. Stations ABA, PASS and COTY have records available at six-minutes time interval. But based on observations, data quality issues only allowed the selection of two flood events. One event occurred on February 5th, 2017, and reached a peak discharge of 40 m^3/s at station ABA. For that event, observations in the closed-channel river reach are used for validation. The second event occurred on November 23rd, 2019. An estimated discharge of 200 m^3/s caused localized overflow in the road tunnel. In fact, studies suggests that the capacity of main riverbed is decreasing due to accumulation of sediments over the years or lack of proper riverbed management activities.

The riverbed is composed of pebbles and clay. It is part of a small urbanized alluvial zone. Vegetation exists on the riverbanks and during low flow periods in the riverbed. There are periods of the year when the riverbed is dry. The low flow periods can happen in any season depending on the rainfall over the catchment. Google images in figure 2 highlight this issue. For January 2003, October 2006, August 2018 and September 2019, the riverbed has few or no water.

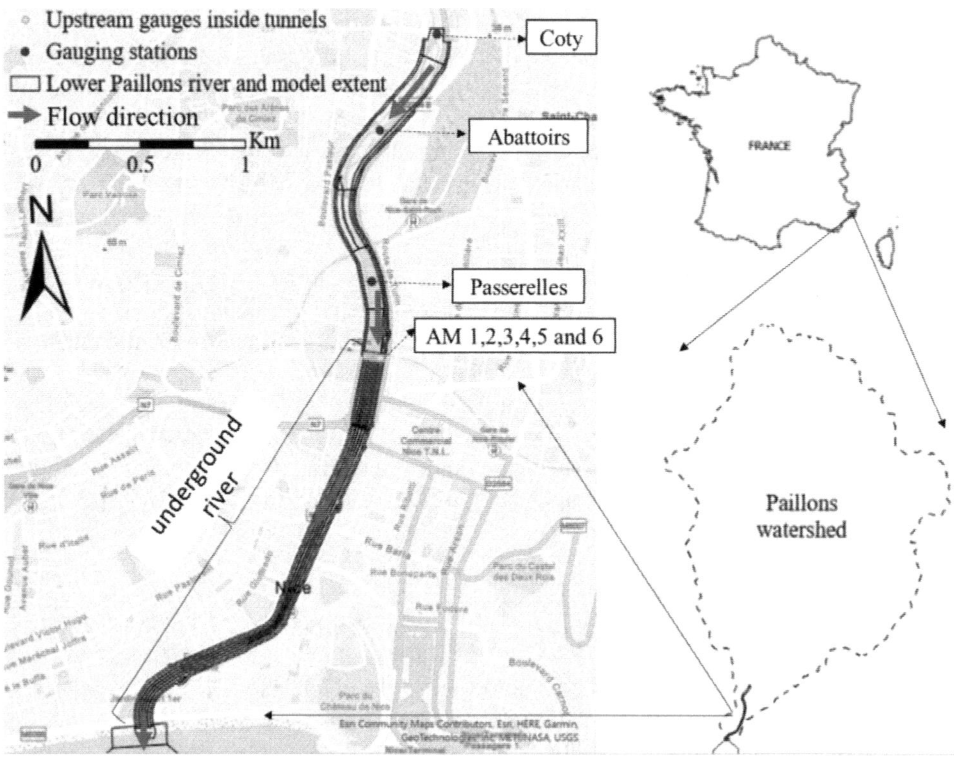

Figure 1. Study area of lower Paillons River with its associated catchment in Nice, France (ESRI background image).

Figure 2. Google images of riverbed with existing structures (Credit Images 2021 Maxar technologies).

Management activities and flow processes modify riverbed geometry. In Figure 2, a 0.5 km stretch section of the riverbed is presented. Downstream of that section, river flows underground. Riverbanks are urbanized. Construction of a bridge and road tunnel affected flows over several years. One of the seven tunnels in the riverbed is used as a road way. Wet seasons, flow processes are torrential. Extremes events move sediments and materials toward the sea. This creates obstruction in the riverbed and increases water depth at river sections. At the entrance and outlet of the tunnels, measurements gauges recorded runoff information. In figure 1, upstream gauges inside tunnels are the green dots.

2.2. Modelling strategy

DHI MIKE21 FM [4] is suitable for surface hydraulic modelling. It uses a flexible mesh approach to reproduce flow processes. The mesh is based on a cell-centred Finite Volume method. A strict stability criterion is required to solve the explicit numerical solution. The Froud-Number must be less than 1. Its hydrodynamic module solves the two-dimensional shallow water equations.

A model based on Mike21FM reconstructs the geometry of the lower Paillon River. An open channel represents each closed river stretch. This approach allows calculations of flow parameters within the tunnels. The river tunnels did not reach their maximum capacity during the past two decades. Seven tunnels upstream merged downstream into five Hydraulic structures like culverts can be used to represent tunnels. But the objective being to visualize flow parameters is possible when the tunnels are considered as open channels in Mike21FM. And the tool allows to model torrential flow and create flood maps. Another possibility is to use 1D hydrodynamic Mike 1D pipe flow model. This tool is suitable for pressurized flow applications and can map the geometry of closed channels. It is based on one-dimensional shallow water equations and applicable to flow conditions that are vertically homogenous.

This assessment focuses on using the 2D hydrodynamic modules with a mesh and dikes. Strickler coefficients describe bed resistance for each land use type. The critical Courant (CFL) number is set at 0.8. Time steps for solving shallow water equations range between 0.001 seconds and 360 seconds. Mesh of different cell sizes integrates topography and structures. An assessment of grid resolution allows to select the most suitable one for this project. Other physical processes like effects of wind, tide, infiltration, or precipitation are not considered. In this case, only flow from runoff hydrograph imposed at upstream boundary is evaluated.

Runoff information from monitoring stations is not always available. In many cases, and especially for real time forecasting, DHI MIKE-SHE hydrological model results will be used as input to hydraulic models. This paper presents analysis of flow processes in the riverbed and risk of flooding in the tunnels.

2.3. Data processing

Input data needed for assessing flow processes in the river come from monitoring stations. Six-minute time interval data is applied upstream as inflow to the domain.

2.4. Calibration and validation

An initial evaluation of flow conditions uses Strickler coefficient of 20 for riverbed and 60 for structures. Observed data are necessary for calibration and validation. Strickler coefficient and mesh size are the main variables that are adjusted to reproduce flow depths and velocities in the modelling domain. Given the topographic data, uncertainties exist in the geometry of the riverbed and flood protection structures. Model results are compared with observed data at different locations and flow maps are assessed.

2.5. Model setup

The model is set up with an initial mesh with maximum grid size 10 m in the riverbed, 2 m in the tunnels and 100 m at sea. The flood plain is represented with 25 m grid size. Figure 3 shows the mesh elements. With a surface area of 15.4 km^2, the model integrates 6.4 km of river channel. It is sea area extends 3 km wide and 2 km large.

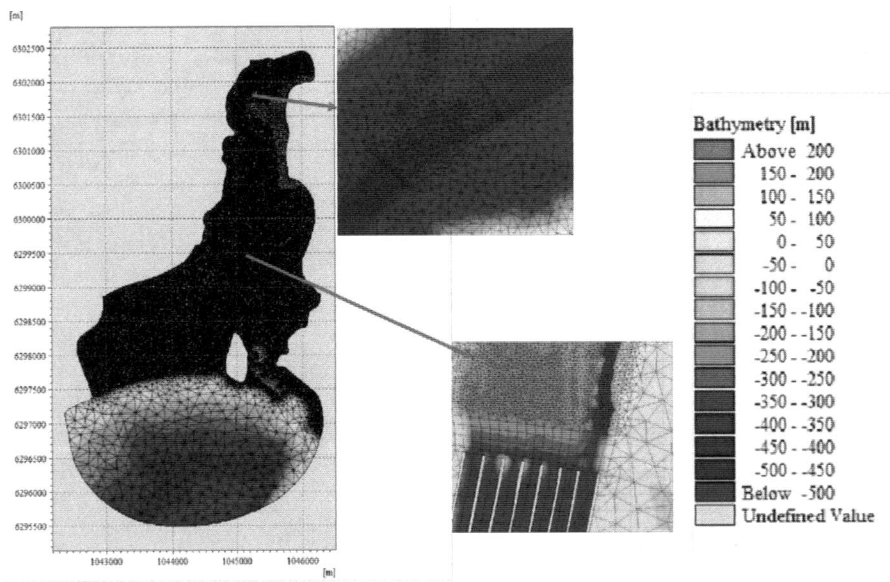

Figure 3. Example of mesh and bathymetry at the junction between open river and closed river channels.

For the entire modelling area, bathymetry varies between 0 and 35 m. Riverbanks and tunnel walls benefit of smaller grid sizes to accurately represent the edges of riverbed limits. Water moving in the riverbed with larger grid size needs less computation time. Figure 4 presents location of boundary conditions. Inflow 1 and 2 receive flow hydrographs. Sea level is fixed at 0.5 m. Strickler coefficients (K) translate flow resistance of different land use types. For the riverbed K is 20. For the flood plain it is 70. And for the sea it is 40. The closed channels are given the same coefficients as the floodplain.

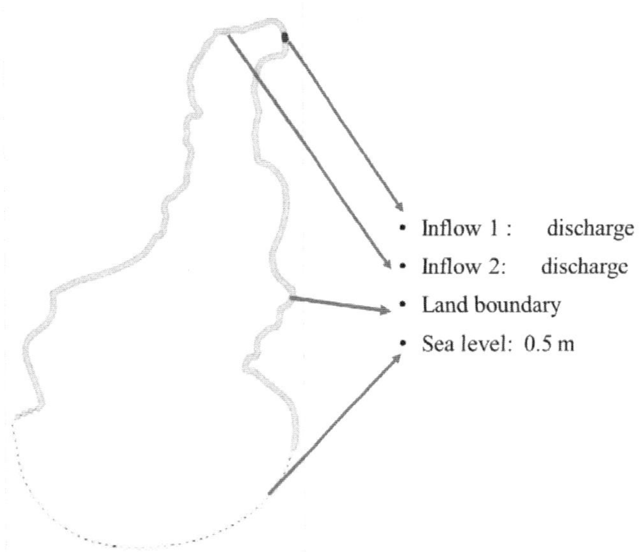

Figure 4. Type of boundary conditions imposed to the model.

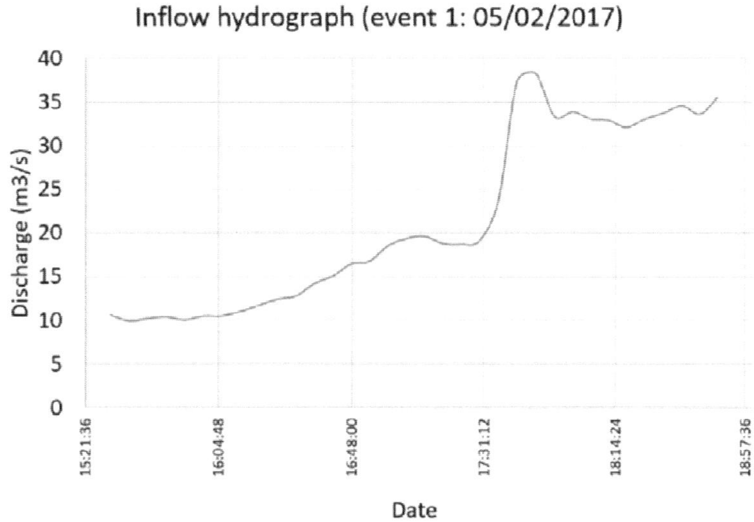

Figure 5. Inflow hydrograph for event of 05/02/2017.

In Figure 5, there is the inflow hydrograph used for evaluating flow in the riverbed. The maximum discharge for that event is 38 m^3/s. It imposed at inflow 1. Inflow to is assigned a discharge of 2 m^3/s. In addition to that hydrograph, one constant boundary of 1000 m^3/s is used to evaluate flooding (900 for inflow 1 and 100 for inflow 2). This discharge is just above the estimated 100-year return period discharge of 800 m^3/s of the associated watershed.

3. Results and discussions

Results show water depths and velocities in the riverbed. In figure 6, simulations of observed discharge less than 20 m^3/s are presented.

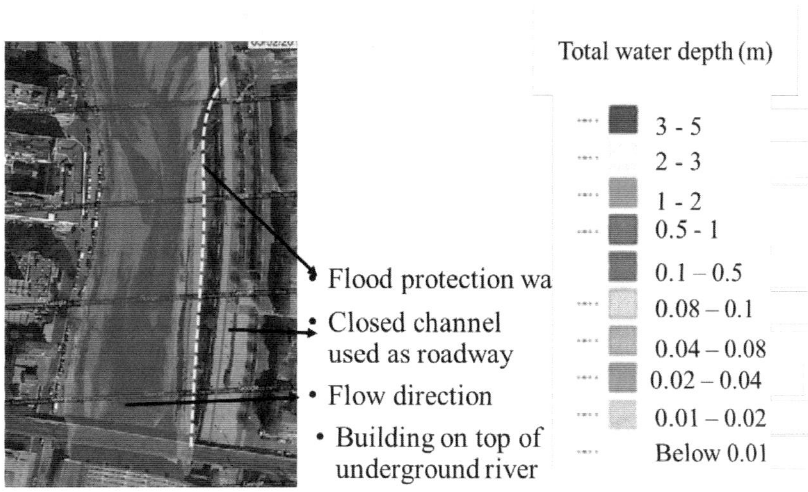

Figure 6. Total water depths from an inflow discharge of 38 m^3/s.

Water depths vary from 0 to 1 m. This is a flow map for the event of 02/05/2017. No overflow of the protection wall and no flooding of the roadway. This event is interesting because observed water depths in the underground river channels are available. In the figure 7, observed water depths are compared with model results.

IOP Conf. Series: Earth and Environmental Science 1136 (2023) 012026 doi:10.1088/1755-1315/1136/1/012026

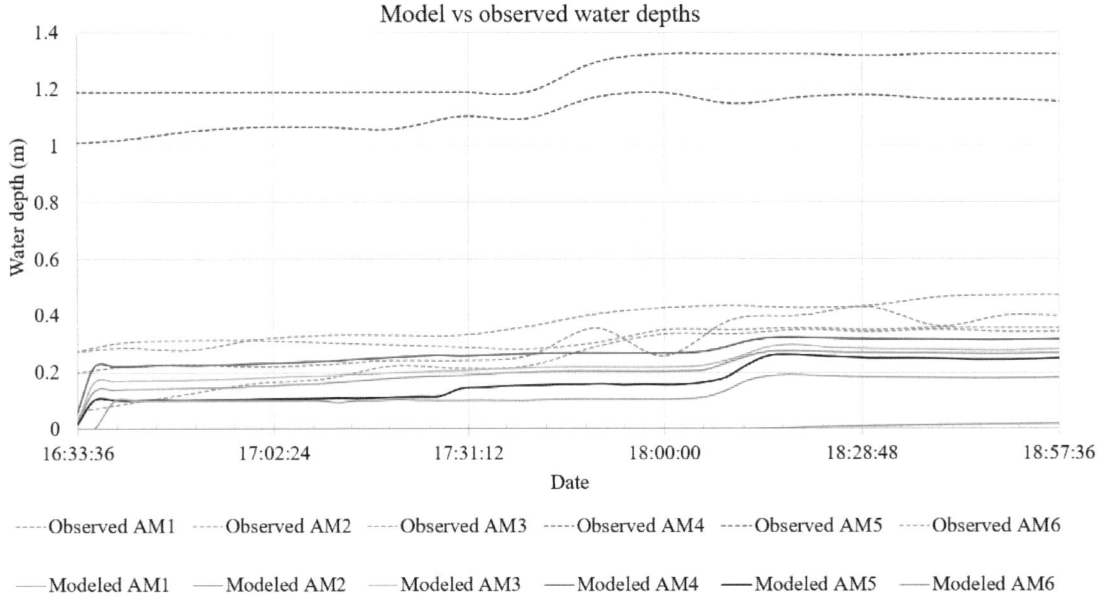

Figure 7. Observed and modeled water depths at the upstream of underground river channels.

Initial results show good agreement between observed and model results for 4 out of 6 evaluation points. Water depths are between 0.1 and 0.5 m. But for 2 channels, observations ranges from 1 to 1.4 m. There are many possible causes. It seems that the base flow in those two channels is above 1 m before the event started. Therefore, there is a need to thoroughly evaluate initial conditions.

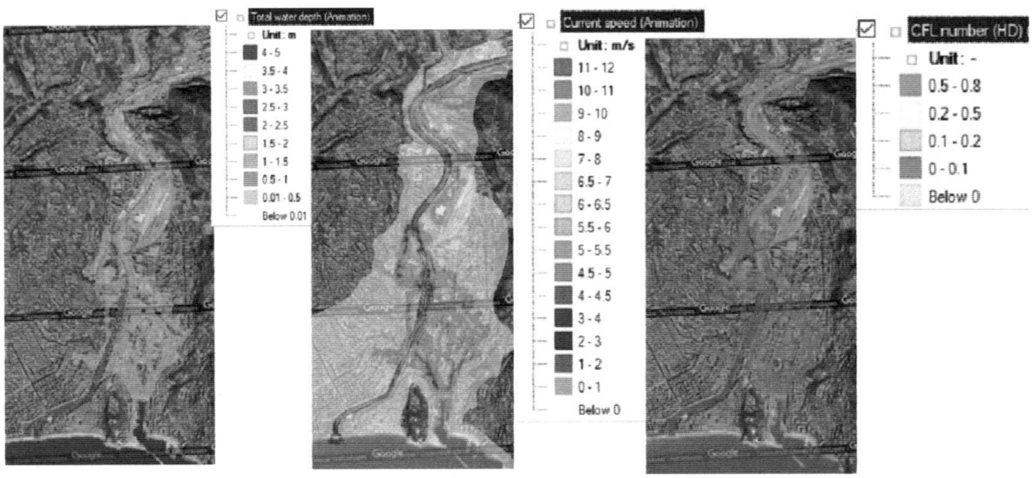

Figure 8. Modelled flood extent, velocities and Courant numbers (Q= 1000 m^3/s > Q_{100} = 800 m^3/s).

Figure 8 show a flood map associated with an inflow discharge of 1000 m^3/s. This discharge is a little more than the 100-year return period discharge. It allows an overview of flood extent and flow directions. In the study area, flood water flows toward the port. In the river bed, water depths can reach 4 to 5 m high. Similar water depths can be observed in the tunnels. Velocities vary between 0 and 12 m/s. Flow speed is higher in the riverbed than in the floodplains. Stability conditions are respected during the simulations with CFL numbers below 0.8. Not all flood risk areas are evaluated. There is a need to run simulations for longer periods.

4. Conclusions

This study aimed building a hydraulic model that can be used in a real time decision support system. Flow conditions in the lower Paillon riverbed are assessed. Results show that Mike21FM is suitable for reproducing hydrodynamic flow in a channelized river. Hydraulic structures like dikes and weirs can be implemented. Mesh size is important to represent variabilities of the topography in the river. The selected CFL condition under 0.8 is respected. The tool is suitable for modelling flooding in areas of interest within the catchment of study. The results obtained are satisfactory and demonstrate that the constructed tool makes it possible to reproduce the overall behaviour of surface hydraulics.

Additional work is needed to complete this modelling project. Several runoff events will be assessed. Flow parameters in each tunnel will be evaluated. Results from 2D model will be compared with 1D model. A real time decision support system (DSS) will also be built to integrate all models, provide flood, and resource management platform to water managers.

Acknowledgments
This work is part of a global PhD project funded by Eau d'Azur, the local water service, with the participation of Métropole Nice Côte d'Azur and the University of Nice Côte d'Azur.

References

[1] Soubeyroux J-M, Bernus S, Corre L, Drouin A, Dubuisson B, Etchevers P and Tocquer F 2020 les nouvelles projections climatiques de référence DRIAS 2020 pour la métropole France Météo-France
[2] Zavattero E 2019 Integration of surface water modeling in a decision support system: application to the Lower Var Valley France Nice Risques (Thèse de doctorat Université Côte d'Azur, France) http://www.theses.fr/2019AZUR4004
[3] Ma Q 2018 Deterministic hydrological modelling for real time decision support systems: application to the Var catchment France Nice (Thèse de doctorat Université Côte d'Azur) http://www.theses.fr/2018AZUR4015
[4] DHI 2017 Mike 21flow model user guide DHI software 2017 DHI Water & Environment (DHI)
[5] Sloan P G and Moore I D 1984 Modeling surface and subsurface storm flow on steeply-sloping forested watersheds *Water Resour. Res.* **20**(12) 1815–1822 https://doi.org/10.1029/WR020i012p01815
[6] De Saint-Seine J 1995 Monographie hydrologique et hydraulique du Paillon de Nice en vue de la gestion du risque inondation (Thèse de doctorat I.N.P.G., Grenoble, France) http://www.theses.fr/1995INPG0053
[7] Syndicat intercommunal des Paillons 2016 Etude de restauration de la continuité écologique des Paillons Phase 2 Diagnostic des obstacles Rapport 80386 Version D. Nice Syndicat intercommunal des Paillons
[8] Emily A and Tennevin G 2009 La source karstique de Sainte-Thècle Evolution des débits annuels depuis 1940 (Peillon, Alpes-Maritimes) *Ann. Mus. Hist. Nat.* **2**(XXIV fasc.) 1–14
[9] Durozoy G and Gouvernet C 1965 Les reservoirs aquiferes entre Nice et Menton Alluvions du Paillon Massifs calcaires Jurassiques Marseille DSGR65 A69
[10] Martin B 1999 Plan d de Prévention des Risques Naturels Prévisibles d'inondation du Paillon Rapport de Présentation Nice COMMUNE DE NICE
[11] Pline C 1991 Contribution à l'étude géologique, hydrogéologique et géotechnique de la basse-plaine alluviale du Paillon (Alpes-Maritimes, France) Nice (Thèse de doctorat Université Côte d'Azur, France) http://www.theses.fr/1991NICE4471
[12] Chapuis M, Elabas A, Souriguére K, Compagnon F, Mayen V, and Terrier B 2018 Quantification of the morphodynamics and ecological functionality of a Mediterranean river *E3S Web Conf.* **40** 02042 https://doi.org/10.1051/e3sconf/20184002042

IOP Conf. Series: Earth and Environmental Science 1136 (2023) 012027 doi:10.1088/1755-1315/1136/1/012027

Changes in sediment erosion rate in the Azuma River basin after the Hokkaido eastern Iburi earthquake in 2018

Kotaro Miyazaki [1], **Tsuyoshi Hoshino**[2], **Keita Shimizu**[1], **Sourabh Shrivastava**[1] **and Tomohito J Yamada**[3]

[1] Graduate School of Engineering, Hokkaido University, Sapporo, 060-8628, Japan
[2] Graduate School of Engineering, Hokkaido University, Sapporo, 060-8628, Japan (now at Watershed Environmental Engineering Research Team, Civil Engineering Research Institute for Cold Region, Sapporo, 062-8602, Japan)
[3] Faculty of Engineering, Hokkaido University, Sapporo, 060-8628, Japan

hoshino-t@ceri.go.jp

Abstract. The Hokkaido Eastern Iburi earthquake in 2018 induced the largest landslide over the Azuma River basin in Hokkaido, the north island of Japan. This study estimated the sediment erosion rate over the Azuma River basin using the universal soil loss equation (USLE) for the pre- and post-earthquake conditions. The results showed that the potential sediment erosion rate from the entire basin increased by 8–29 times more than the pre-earthquake condition. The primary cause of the increase in sediment erosion rate was deforestation due to the landslide, which generated 13% of the bare land area in the entire basin. In addition, the influence of interannual variability of rainfall on sediment erosion rate was calculated from observed rainfall data. As a result, the minimum annual sediment erosion rate after the earthquake exceeded the maximum before the earthquake, indicating that the post-earthquake sediment erosion rate is substantially higher than the pre-earthquake rate. These findings suggest that in order to assess the sediment erosion risk over the Azuma River basin, it is necessary to predict and monitor the vegetation recovery, considering the interannual variability of rainfall.

1. Introduction

An earthquake of magnitude 6.7 struck the eastern Iburi area of Hokkaido, Japan, at 3:07 a.m. JST on September 6, 2018 (Figure 1) [1]. The Japan Meteorological Agency (JMA) named the earthquake "the Hokkaido Eastern Iburi Earthquake (HEIE)". The earthquake caused 6,117 landslides around the Iburi region, covering an area of 48.4 km^2 (Figure 1a, Figure 2) [2]. The landslide area caused by the event was a record-breaking large, which is approximately four times larger than the 2nd largest landslide area in Japan (caused by the Niigata Chuetsu Earthquake (magnitude 6.8); 11.2 km^2 [3]). Although large and deep landslides occurred in the basin of the Hidaka-Horonai River, which is a tributary of the Azuma River, most of the landslides caused by the HEIE were shallow landslides (Figure 1). Several studies have discussed the details of the HEIE and landslides. Osanai et al. (2019) [4] reported the characteristics of landslide slopes based on field surveys and maps of the distribution of landslide areas. Wang et al. (2019) [5] analyzed the satellite images and showed that seismic motion, slope gradient, and soil moisture were the primary factors for landslides. In Atsuma town, six people died due to mudslides, 30 people died due to landslides, and 2,032 houses were damaged [6].

Agriculture-related damage was approximately 114.5 billion JPY [7]. In addition to the direct damage, long-term damage induced by sediment yield from the landslide slopes is anticipated. The accumulation of sediment on the riverbed and the bottom of dam reservoirs is projected to reduce the flow capacity of rivers and dam reservoirs, respectively. To consider countermeasures for such risks, it is necessary to understand the impact of landslides on the spatial distribution of sediment erosion at a sub-basin scale.

This study aimed to understand the influence of HEIE-induced landslides on sediment erosion at the sub-basin scale. Moreover, the contributions of topographic change, deforestation, and inter-annual rainfall variability over the Azuma River basin were evaluated using the universal soil loss equation (USLE).

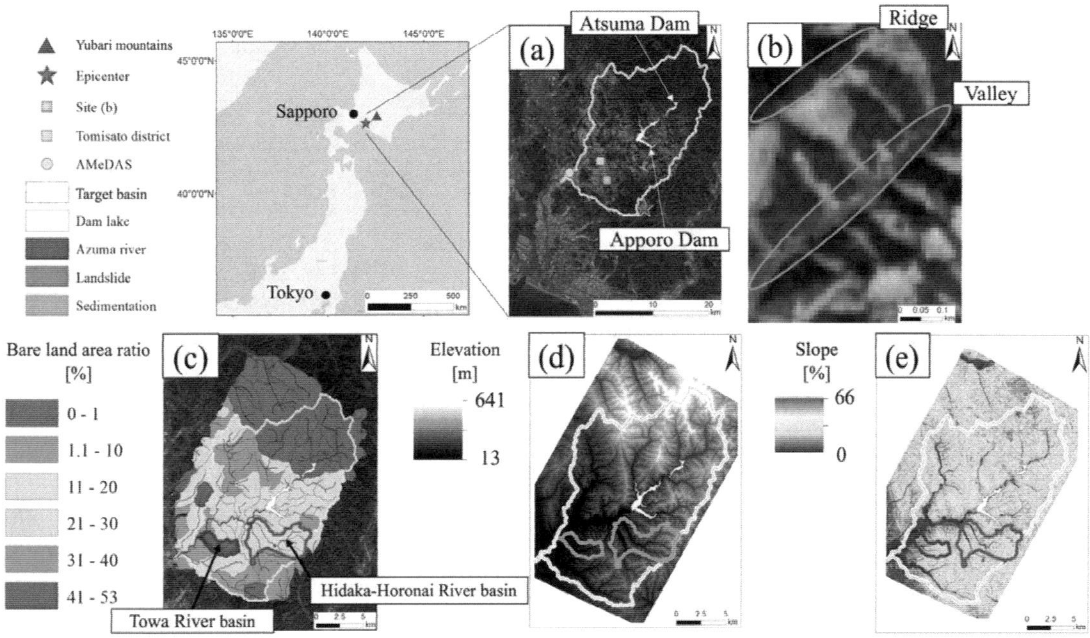

Figure 1. (a) Azuma River basin (b) Satellite image at Site (b) (the green square in (a)). The color image obtained by combining distinct visible light band images obtained from the Sentinel-2 satellite. The orange lines indicate ridge and valley. (c) The bare land area ratio in sub-basins. (Left: Towa River basin and Right: Hidaka-Horonai River basin are enclosed by a border). Pre-earthquake (d) Elevation and (e) Gradient of topography.

Figure 2. Landslides along tributaries of the Azuma River (photo captured on July 26, 2019).

2. Methodology

2.1. Target area
The Azuma River basin in southwestern Hokkaido was the target region (Figures 1a and b). The Azuma River, which flows south-westward from the Yubari mountains in central Hokkaido to the Pacific Ocean, is 52.3 km long. The basin covers an area of 382.9 km^2, with the two dams (Atsuma and Apporo Dams) located upstream (Figure 1a). The pre-earthquake average elevation of the basin was 185 m, ranging from 16 to 641 m (Figure 1d). The average slope of the basin calculated from the digital elevation model (DEM) described below was 22% (range: 0–66%) (Figure 1e). The upper reaches of the basin are mixed broad-leaved/needle-leaved forests, whereas the lower plains include paddy fields, farmlands, and urban areas. The soil type is Ta-d, *i.e.*, volcanic ash soil derived from a volcanic eruption and caldera formation in central Hokkaido, and brown forest and grey lowland soils.

The target basin of the upper-middle reaches of the Azuma River, where the landslide slopes are concentrated, was divided into 46 catchments for the following analysis (Figure 1c). The bare land area ratios for each sub-catchment were calculated by dividing the landslide area by the catchment area. The bare land area ratios of 16 sub-basins were greater than 20%. The Towa River basin had the largest bare land area ratio (41%).

2.2. Data
This study used a DEM, the landslide map (i.e., the spatial distribution of HEIE-induced landslide area), soil type map, vegetation distribution map, and precipitation data. The pre-earthquake DEM with a 10-m resolution made by the photogrammetry method was provided by the Geospatial Information Authority of Japan (GSI) [8]. The vertical accuracy of the data is less than 0.7 m. The post-earthquake DEM (taken in September 2018) with a 0.5-m resolution measured by the airborne laser scanning was provided by the Hokkaido Development Bureau of the Ministry of Land, Infrastructure, Transport and Tourism (MLIT). The landslide map with an 83-m resolution, including collapsed and accumulated areas, was provided by GSI [9]. The soil type map with a resolution of 1:200,000 was provided by the Ministry of Agriculture, Forestry, and Fisheries of Japan [10]. The vegetation distribution map with a resolution of 1:25,000 was provided by the Ministry of the Environment of Japan [11]. The 10-min interval rainfall data observed at the Automated Meteorological Data Acquisition System (AMeDAS) [12] of Atsuma observation point by JMA from April to November between 1994 and 2020 was used.

2.3. Sediment erosion rate estimation
The sediment erosion rate was estimated using the USLE. The USLE calculates the sediment erosion rate by multiplying empirical factors determined from the topography, vegetation, soil type, land use pattern of the target area, and rainfall [13], [14], [15]. The annual sediment erosion rate is expressed by

$$A = LS \cdot C \cdot R \cdot K \cdot P \qquad (1)$$

where A is the annual sediment erosion rate (t·ha^{-1}·year^{-1}), LS is the topographic factor (dimensionless), C is the cover factor (dimensionless), R is the rainfall factor (MJ·mm·ha^{-1}·h^{-1}·year^{-1}), K is the soil erodibility factor (t·h·MJ^{-1}·mm^{-1}), and P is the support practice factor (dimensionless).

LS is defined as the multiplication of the slope length factor L by the slope factor S. In this study, those factors were calculated separately. L was calculated from a DEM using a modified version of the method reported by Desmet and Govers [16]. S was calculated by the method of McCool et al. [17], which has been modified to correspond to steep slopes. According to Renard et al. [14], the USLE has limited applicability for both L and S. Therefore, the grids corresponding to flatland and river channels were excluded from the analysis target. In addition, 17% of the area was excluded from the pre- and post-earthquake conditions, respectively. To match the resolution of the pre-earthquake DEM, post-earthquake elevation data were interpolated to a 10-m resolution from a 0.5-m resolution by the cubic convolution interpolation. Note that Mondal et al. [18] pointed out that the sediment erosion rate calculated by the USLE is affected by the resolution of the DEM.

C is the ratio of sediment erosion rate compared to a bare slope. Kitahara et al. [19] summarized the data for mountainous areas in Japan, reporting that *C* ranges from 0.005 to 0.02 regardless of whether the forest is natural or artificial and that a value of 0.01 is appropriate except for no forest condition. In this study, *C* was set to 0.01 (0.005 and 0.02 for sensitivity analysis) according to the survey for the forest area. The main land-use types in the target basin were forests, bare land, urban areas, and logging sites. *C* was set to 1.0 for bare land. Urban areas were excluded from the calculation of USLE because sediment erosion does not occur in urban areas. Similarly, the logging area was excluded due to its limited size in the target river basin (2.9% of the Azuma River basin).

R represents the degree of sediment erosion associated with the detachment of surface soil layers by raindrops. *R* is obtained from the kinetic energy of raindrops for each rainfall event, the total rainfall, and the maximum 30-min rainfall intensity [13]. In this study, the kinetic energy of raindrops was calculated by the method described by Brown and Foster [20]. The annual rainfall factor R_{year} was calculated for each year to evaluate the interannual variability of sediment erosion rate from the observed 27 years of rainfall.

K is defined as the soil erodibility factor. The factor was calculated by the monograph method based on the sediment characteristics, including material, particle size, permeability, and organic matter content. *K* values summarized by Taniyama [21] for typical soil types in Japan were used in this study. *P* indicates the effect of the shape and features of the ground surface, such as ridges and tillage, on agricultural land. It corresponds to hillslope construction and slope protection by erosion control works for mountain forests. Because the focus of this study was to evaluate the change in erosion potential due to the landslides, *P* was set to 1.0 to exclude the effects of artificial conservation.

3. Results

3.1. Factors of USLE

Figure 3 shows the calculated *C* for the pre- and post-earthquake conditions and change ratio of *C* (*C* after the HEIE divided by *C* before the HEIE). The change ratios for each basin ranged from 1 to 44, with no basin having a change ratio less than that of the pre-earthquake condition. Figure 1c and Figures 3a–c show that *C* is greater for the sub-basins with a significant percentage of bare land area.

Figure 1a shows that the change rate of the upper reach of the Atsuma Dam area is lower than that of the lower reach of the target basin. The average value of *C* for all grids in the target area increased 12-fold after the earthquake, which is a much more significant contribution than the topographic change.

Figure 4 shows the annual rainfall and R_{year} for each year. R_{year} is correlated with the annual rainfall. The average value of R_{year} was 1,055 $MJ \cdot ha^{-1} \cdot mm^{-1}$. After the earthquake, the average value of R_{year} in 2019 (945 $MJ \cdot ha^{-1} \cdot mm^{-1}$) and 2020 (347 $MJ \cdot ha^{-1} \cdot mm^{-1}$) were smaller than the average value. The results indicate that if the rainfall exceeds the average value in the future, sediment erosion will be the most severe erosion in the past 27-years because of the *C* increment.

3.2. Annual sediment erosion rate

Figure 5 shows the calculated sediment erosion rate *A* with the average of R_{year} from 1994 to 2020 for the pre- and post-earthquake conditions and change ratio. The change ratios of *A* for each sub-basin ranged from 0.9 to 49.0, with a 20-fold increase in the values for 19 sub-basins. In particular, *A* was intensified for sub-basins of the lower reaches of the Atsuma Dam, which have higher bare land area ratios. The annual sediment erosion rate increased 49-fold in the Towa River basin, where the ratio of the bare land area became 41% due to the landslide. The Hidaka-Horonai River basin, where a large and deep-seated landslide occurred, also had a significant annual sediment erosion rate.

The sensitivity analysis was conducted to clarify the effect of C on the sediment erosion rate. The values of 0.005 and 0.02 were applied for C over the forest area. Table 1 shows the integrated values of the sediment erosion rate of the entire target basin under the pre- and post-earthquake conditions with different C values (Case 1: 0.01, Case 2: 0.005, Case 3: 0.02). The sediment erosion rates were calculated by applying the average of R_{year} from 1994 to 2020. The units were converted from $[t\cdot ha^{-1}\cdot year^{-1}]$ to $[t\cdot 100\ m^{-2}\cdot year^{-1}]$, *i.e.*, the annual erosion rate per grid cell (100 m^2). The annual erosion rate of the target basin under the pre- and post-earthquake conditions was 0.11×10^8 $t\cdot year^{-1}$ and 1.60×10^8 $t\cdot year^{-1}$, respectively. After the earthquake, A was 15, 8, and 29 times larger than the pre-earthquake conditions for Case 1, Case 2, and Case 3, respectively. These results indicate that the sediment erosion potential increases significantly regardless of the uncertainty of C value.

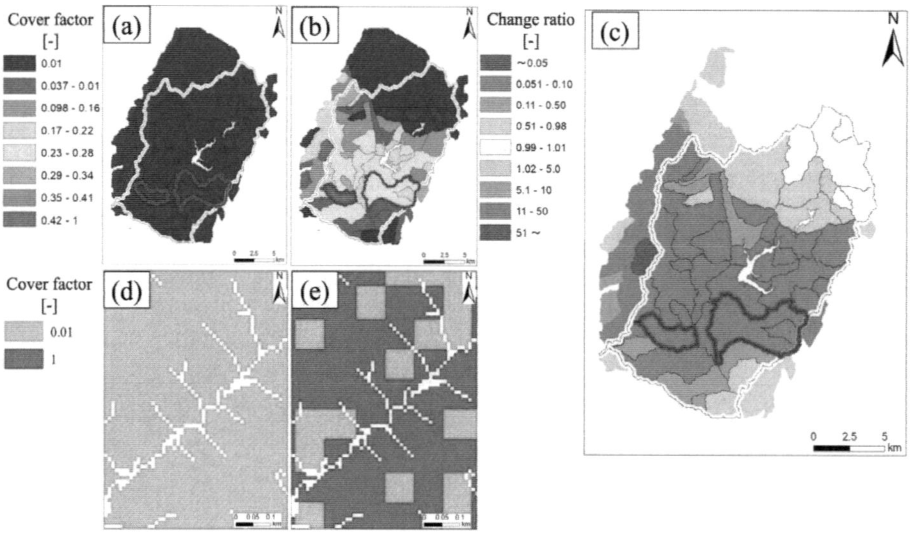

Figure 3. Distribution of cover factor (C) for sub-basins (a) before and (b) after the Hokkaido Eastern Iburi Earthquake (HEIE). (c) the ratio of change in C before and after the HEIE. (d,e) Spatial distribution of C over the same area as Figure 1b.

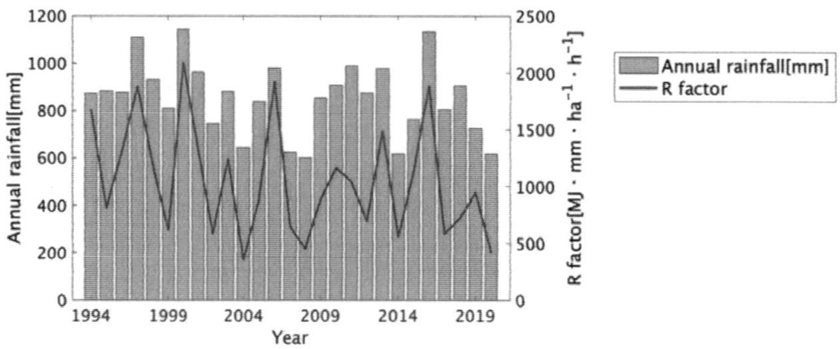

Figure 4. Correlation between annual rainfall factor, R_{year}, and annual rainfall.

To estimate the influence of the interannual variability of R_{year} on sediment erosion rates, A was calculated for each year under the pre- and post-earthquake conditions (54 matrices in total, 27 matrices each) using R_{year} from 1994 to 2020. Figure 6 shows the relative frequencies of sediment erosion rates for the target basin. The standard deviation of the annual sediment erosion rates in the pre-earthquake condition was 0.05×10^8 $t\cdot year^{-1}$; whereas in the post-earthquake condition it was

0.77×10^8 t·year^{-1}, which is 15.4 times larger. The minimum A after the earthquake exceeded the maximum A before the earthquake, indicating that the post-earthquake sediment erosion rate is substantially higher than the pre-earthquake rate.

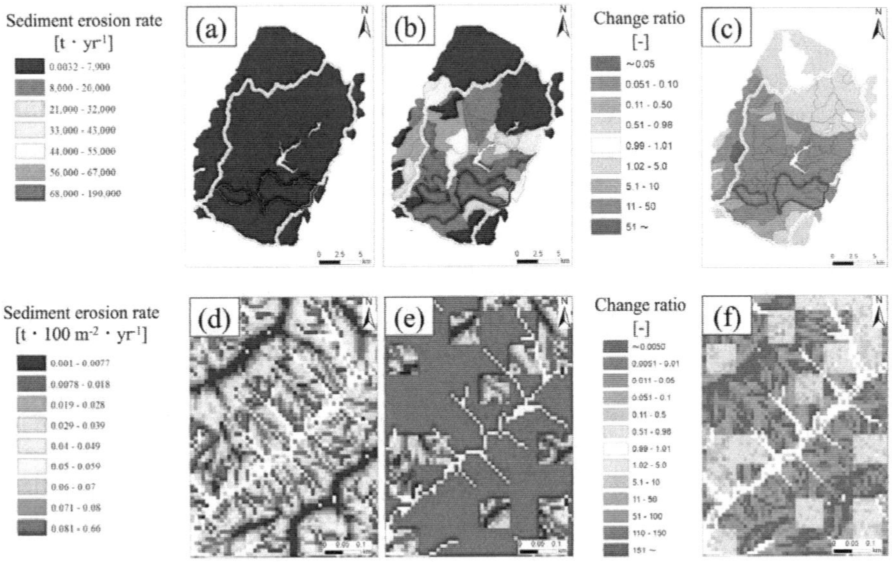

Figure 5. Distribution of integrated sediment erosion rate (A) for sub-basins (a) before and (b) after the Hokkaido Eastern Iburi Earthquake (HEIE). (c) The ratio of change in A before and after the HEIE. (d-f) Spatial distribution of A over the same area as figure 1b.

Table 1. Integrated annual sediment erosion rates from the entire basin under pre- and post-earthquake conditions with different forest cover factor (C) (general, minimum, and maximum).

Case No.	C factor [dimensionless]		Sediment erosion rate [10^8 t/yr]		
	Forest	Bare land	A_{bf}	A_{af}	Change ratio
Case 1	0.01	1.0	0.11	1.60	15
Case 2	0.005		0.05	1.56	29
Case 3	0.02		0.21	1.68	8

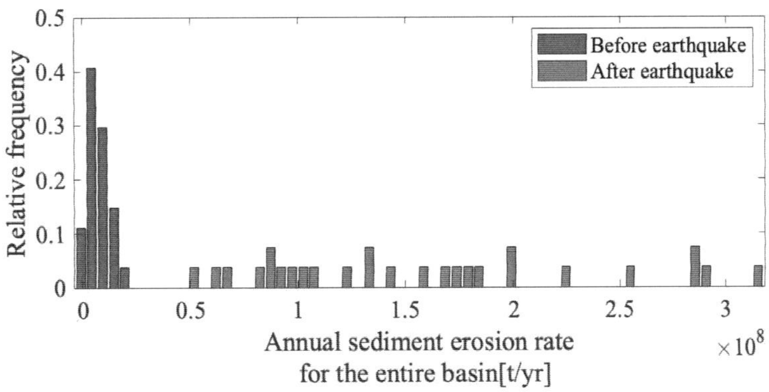

Figure 6. Relative frequency of integrated sediment erosion rate for the target basin calculated from rainfall factor (R_{year}) for each year under pre- and post-earthquake conditions.

4. Summary

This study quantified the influence of the HEIE on the sediment erosion rate using the USLE over the Azuma River basin, where severe landslides occurred. The advantage of introducing USLE is the ability to quantify the degree of contributions of deforestation and topographic change. As a result, the potential annual sediment erosion rate in the Azuma River basin increased by 8 to 29 times compared to the pre-earthquake conditions. Here, the change in the cover factor was approximately 50 times larger in the areas with large landslide areas. However, the change in the topographic factor was much smaller than that in the cover factor, indicating that deforestation was the primary cause for the change in sediment erosion rate. Moreover, the frequency distribution of the annual soil erosion rate under the pre- and post-earthquake conditions was evaluated using the interannual rainfall variability calculated from the observed rainfall data for 27 years. The minimum annual sediment erosion rate after the earthquake exceeded the maximum before the earthquake, indicating that the post-earthquake sediment erosion rate is significantly higher than the pre-earthquake rate.

The quantitative analysis of the changes in the cover and topographic factors indicates that the decrease of vegetation is the most important factor in increasing annual sediment erosion in the Azuma River basin. These results indicate the importance of vegetation in reducing the sediment erosion risk. Lin et al. [22] reported a case study of vegetation recovery in central Taiwan. They showed that approximately 90% of vegetation had recovered after the 6-years passed after the catastrophic Chi-Chi earthquake in 1999. The result indicates that appropriate monitoring of the recovery speed of vegetation is necessary to assess the risk of sediment erosion. Furthermore, the amount of heavy rainfall is projected to increase in the Hokkaido region, where the Azuma River basin is located, due to global warming [23]. On the other hand, the vegetation of the region has a possibility of becoming more revitalized with the proceeding of global warming. Therefore, further studies should be conducted to evaluate the sediment erosion risk under climate change by considering the intensification of rainfall and the degree of recovery speed of the vegetation.

Acknowledgments

The post-earthquake data used in this study was provided by the Hokkaido Development Bureau of Ministry of Land, Infrastructure, Transport, and Tourism. We would like to express our sincere gratitude to Hokkaido government and Atsuma Town for their cooperation. This study was supported by JSPS KAKENHI [grant No. 19H02241, 19H00963, 22H01594]; the River Center of Hokkaido.

References

[1] Geospatial Information Authority of Japan 2018 *The 2018 Hokkaido Eastern Iburi Earthquake: Fault Model (preliminary)* https://www.gsi.go.jp/cais/topic180912-index-e.html [Accessed June 17, 2021] (in Japanese)

[2] Kasai M and Yamada T 2019 Topographic effects on frequency-size distribution of landslides triggered by the Hokkaido Eastern Iburi Earthquake in 2018 *Earth, Planets and Space* **71**(1) 1–12 https://doi.org/10.1186/s40623-019-1069-8

[3] Sekiguchi T and Sato H 2006 Feature and distribution of landslides induced by the Mid Niigata Prefecture Earthquake in 2004, Japan *J. of the Jpn. Landslides Soc* **43** 142–154 https://doi.org/10.3313/jls.43.142

[4] Osanai N et al 2019 Characteristics of landslides caused by the 2018 Hokkaido Eastern Iburi Earthquake *Landslides* **16**(8) 1517–1528 https://doi.org/10.1007/s10346-019-01206-7

[5] Wang F, Fan X, Yunus AP, Subramanian SS, Alonso-Rodriguez A, Dai L, Xu Q and Huang R 2019 Coseismic landslides triggered by the 2018 Hokkaido, Japan (M_w 6.6), earthquake: spatial distribution, controlling factors, and possible failure mechanism *Landslides* **16**(8) 1551–66 https://doi.org/10.1007/s10346-019-01187-7

[6] Cabinet Office 2019 Damage caused by 2018 Hokkaido Iburi East Earthquake Retrieved from http://www.bousai.go.jp/updates/h30jishin_hokkaido/pdf/310128_jishin_hokkaido.pdf

[Accessed June 1, 2021] (in Japanese)
[7] Ministry of Agriculture, Forestry and Fisheries 2020 *Damage caused by 2018 Hokkaido Eastern Iburi Earthquake* http://www.pref.hokkaido.lg.jp/sm/ktk/300906/higaihou121.pdf [Accessed June 1, 2021] (in Japanese)
[8] Geospatial Information Authority of Japan 2021 *Digital Elevation Map 10 m Grid (Elevation)* https://fgd.gsi.go.jp/download/menu.php [Accessed January 25, 2020] (in Japanese)
[9] Geospatial Information Authority of Japan 2018 *Technical Report of the Geospatial Information Authority of Japan, D1-No.991, Distribution of Landslides and Sedimentation* https://www.gsi.go.jp/BOUSAI/H30-hokkaidoiburi-east-earthquake-index.html [Accessed February 16, 2020] (in Japanese)
[10] Obara H, Takada Y, Kamiyama K, Ohkura T, Maejima Y, Wakabayashi M and Kanda T 2016 *1/200,000 soil map of Japan based on the first proposal for a comprehensive soil classification* Retrieved from https://soil-inventory.dc.affrc.go.jp
[11] Ministry of Environment 2021 *1/25000 Current vegetation map* Retrieved from https://www.biodic.go.jp/kiso/vg/vg_kiso.html [Accessed December 15, 2020] (*In Japanese*)
[12] Japan Meteorological Agency 2021 *Automated Meteorological Data Acquisition System* https://www.data.jma.go.jp/obd/stats/etrn/index.php [Accessed February 18, 2021] (in Japanese)
[13] Wischmeier WH and Smith DD 1978 *Predicting Rainfall Erosion Losses: A Guide to Conservation Planning* **537** (Department of Agriculture, Science and Education Administration)
[14] Renard KG 1997 *Predicting Soil Erosion by Water: A Guide to Conservation Planning with the Revised Universal Soil Loss Equation (RUSLE)* (United States Government Printing)
[15] Alewell C, Borrelli P, Meusburger K and Panagos P 2019 Using the USLE: chances, challenges and limitations of soil erosion modelling *International Soil and Water Conservation Research* **7**(3) 203–225 https://doi.org/10.1016/j.iswcr.2019.05.004
[16] Desmet PJJ and Govers G 1996 A GIS procedure for automatically calculating the USLE LS factor on topographically complex landscape units *Journal of Soil and Water Conservation* **51**(5) 427–433
[17] McCool DK, Brown LC, Foster GR, Mutchler CK and Meyer LD 1987 Revised slope steepness factor for the universal soil loss equation *Transactions of the ASAE* **30**(5) 1387–96 https://doi.org/10.13031/2013.30576
[18] Mondal A, Khare D, Kundu S, Mukherjee S, Mukhopadhyay A, Mondal S 2017 Uncertainty of soil erosion modelling using open source high resolution and aggregated DEMs *Geoscience Frontiers* **8**(3) 425–436 https://doi.org/10.1016/j.gsf.2016.03.004.
[19] Kitahara H, Okura Y, Sammori T and Kawanami A 2000 Application of Universal Soil Loss Equation (USLE) to mountainous forests in Japan *Journal of Forest Research* **5**(4) 231–236 https://doi.org/10.1007/BF02767115
[20] Brown LC and Foster GR 1987 Storm erosivity using idealized intensity distributions *Transactions of the ASAE* **30**(2) 379–386 https://doi.org/10.13031/2013.31957
[21] Taniyama I 2003 Elucidation of soil factors related to the occurrence of Surface Water Runoff from Agricultural Land and Preparation of MI *Comprehensive research on agriculture, forestry and fisheries, trade in agricultural, forestry and fishery products, and the environment of resources* 149–152 (in Japanese)
[22] Lin WT, Lin CY and Chou WC 2006 Assessment of vegetation recovery and soil erosion at landslides caused by a catastrophic earthquake: a case study in Central Taiwan *Ecological Engineering* **28**(1) 79–89 https://doi.org/10.1016/j.ecoleng.2006.04.005
[23] Yamada TJ 2019 *Adaptation Measures for Extreme Floods Using Huge Ensemble of High-Resolution Climate Model Simulation in Japan* Retrieved from https://unfccc.int/documents/197307 [Accessed on June 15, 2021].

ONASP: A web application for groundwater data visualization in Panama

María G. Castrellón[1,2], Richard A. Pauloo[3], Ioana Popescu[1], José Fábrega[2,4]

[1] IHE Delft Institute for Water Education
[2] Centro de Investigaciones Hidráulicas e Hidrotécnicas (CIHH), Universidad Tecnológica de Panamá (UTP)
[3] Water Data Lab
[4] Sistema Nacional de Investigación (SNI), SENACYT, Panamá

g.castrellonromero@un-ihe.org

Abstract. Groundwater data in Panama is scattered and held by different institutions, which complicates water management decision-making. Integrating different data sources into a central, web-accessible location that displays near-real-time and accurate groundwater data will facilitate informed decision-making. In this paper, we present a web application to visualize the location and hydrographs associated with monitoring wells and rain gages from the La Villa Watershed, located in south-central Panama. The app was built using the R programming language and interactive elements from the shiny framework. It represents a first step towards organizing, centralizing, and making groundwater data in Panama web-accessible, to enable informed, evidence-based water management decisions.

1. Introduction

Groundwater in Panama is a critical resource for drinking water supply and agriculture, especially in rural areas of the Dry Arc Region (Figure 1). Despites its importance, groundwater resources are not properly managed across the country due to insufficient measurements and inadequate access to data, information, and knowledge. Existing groundwater data is scattered throughout different institutions in charge of overseeing groundwater use, namely the Ministry of the Environment (MIAMBIENTE), the Ministry of Health (MINSA), the Ministry of Agriculture (MIDA) and the National Institute of Aqueducts and Sewers (*Instituto de Acueductos y Alcantarillados Nacionales*, IDAAN). In order to secure water availability and enable scientifically informed decision making in the face of climate change and population growth for critical areas of the country such as the Dry Arc Region, Panama needs technical solutions that integrate, organize, and effectively communicate the country's decentralized groundwater data.

Worldwide, there are numerous examples of web-based visualization tools for groundwater data. These include the SGMA Data Viewer from California's Department of Water Resources [1]; the California Groundwater Data Observatory [2]; the Texas Water Dashboard by the USGS [3]; the OZCAR-RI ROSES Groundwater Network Observatory from France [4]; the WaterConnect Groundwater Data Portal from Australia [5]; the Groundwater Observatory from the Peru's National Water Authority [6]; and the Philippine Groundwater Outlook [7]. Nevertheless, Panama lacks an integrated solution that provides quick access to groundwater data for research and stakeholders.

The National Groundwater Observatory of Panama, or ONASP (*Observatorio Nacional de Aguas Subterráneas de Panamá*), aims to organize, centralize, and communicate groundwater data in Panama for stakeholders, authorities, and the public. Of particular importance is making these data available for water managers to enable informed, evidence-based decisions to improve groundwater resources management in the country. As a first step towards building the ONAPS, an interactive web application was created to host and share groundwater level and precipitation data from a pilot monitoring network in the La Villa Watershed located in the Dry Arc Region in south-central Panama (Figure 1). The web application is currently hosted in the shinyapps.io platform (https://mgcastre.shinyapps.io/ONASP/). This paper explains the methods for collecting and organizing the data, the implementation of the web application, and it discusses the main advantages of the tool, its intended use and future development.

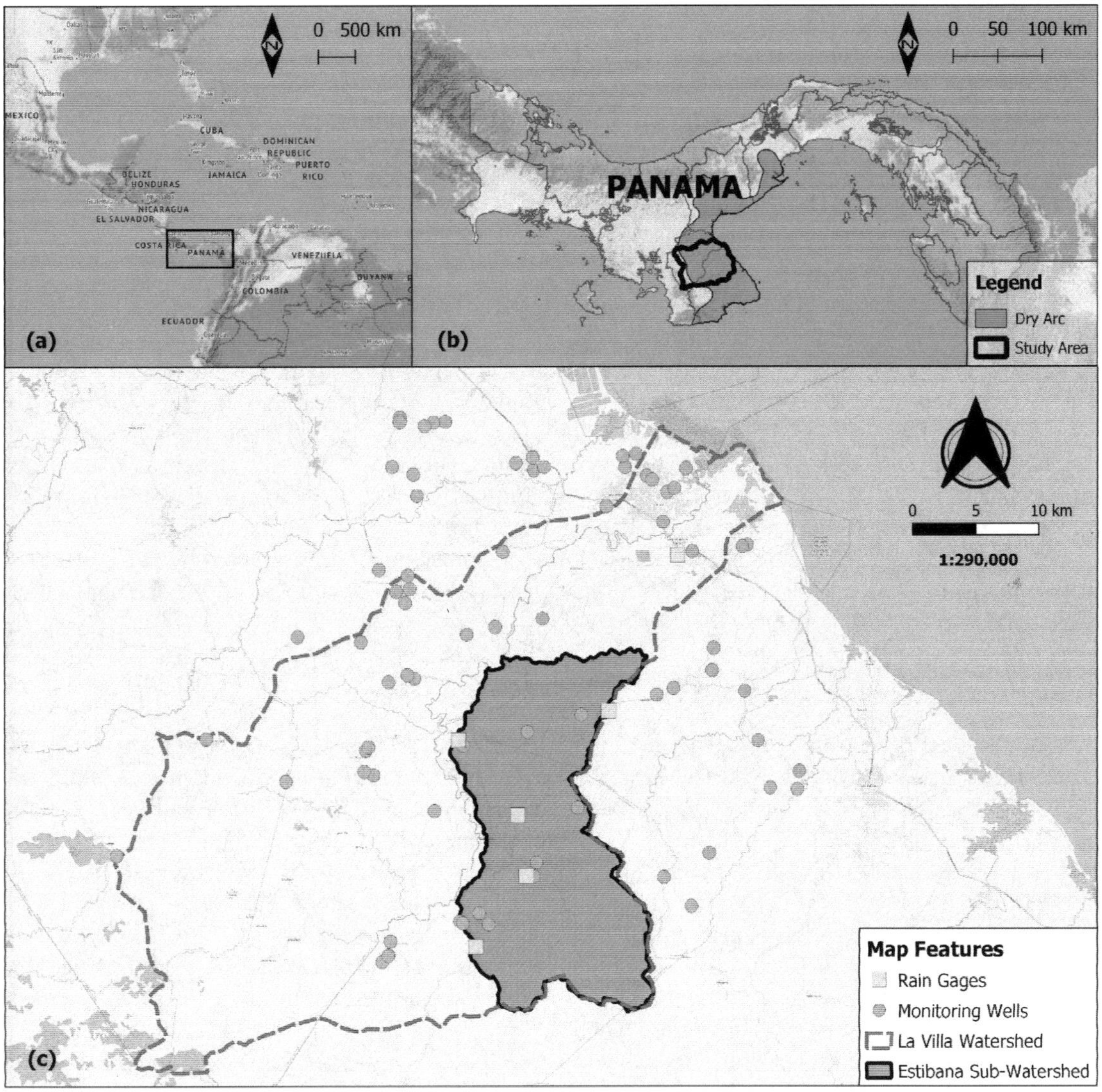

Figure 1. (a) Location of Panama; (b) location of the study area within the Dry Arc Region of Panama; (c) Monitoring wells and rain gages within the La Villa Watershed and surrounding areas.

2. Study Area

The Central Pacific Hydrologic Region of Panama also known as the Dry Arc Region (Figure 1b) has an average precipitation of 1,400 mm per year [8] but its climate is characterized as a tropical savanna with wet and dry months in which precipitation can be as high as 500 mm/month and 10 mm/month, respectively [9]. Since the year 2000, groundwater exploitation has significantly increased in this area of the country, mainly to supply water for domestic and agricultural purposes during the dry season, which can extend up to seven months in years under the effects of El Niño Southern Oscillation (ENSO) phenomenon [10]. Nevertheless, government authorities lack the necessary data and tools to design effective well drilling campaigns and make informed decisions regarding groundwater management, thus risking the depletion of this resource.

The La Villa Watershed is located in the heart of the Dry Arc Region in south-central Panama and has an area of 1,284 km^2 (Figure 1b). This area's geology is very complex with the presence of multiple tectonic features and a predominance of volcanic and consolidated sedimentary rocks. In terms of its hydrogeology, aquifers of secondary permeability and fractured rock are predominant [9]. The main economic activities in this watershed are cattle farming and agriculture with the main crops being corn, rice, and sugar cane. Understanding groundwater behaviour, especially the fluctuation of groundwater levels in response to changes in precipitation, is crucial to secure water availability for current and future generations and ensure socioeconomic development for the region.

3. Materials and Methods

This web application began as a side project during the first author's master's degree programme at the University of California, Davis. Its development continued slowly throughout the years out of pure enthusiasm until it was incorporated under the scope of a research project conducted by the CIHH-UTP. This web app was designed and built as a proof of concept to start organizing and sharing hydrogeologic data from a pilot watershed in Panama that could be extrapolated to the rest of the country. Currently, the web application contains and provides visualization tools for groundwater level data from monitoring wells (owned by IDAAN) and precipitation data from rain gages (owned by CIHH-UTP) located in the La Villa Watershed and surrounding areas (Figure 1c). These data can be visualized and downloaded for free from the web application. In the following sections, the process of data collection, processing and organization is explained, and the construction and deployment of the web application is described.

3.1. Data Collection, Organization and Storage

3.1.1. Groundwater level data. To the author's knowledge, IDAAN's first groundwater level monitoring network in Panama was established as a pilot project in the La Villa Watershed in May 2015 under the scope of the first author's undergraduate thesis project [11]. The initial network consisted of 39 inactive wells (15 to 20 cm in diameter and 30 to 100 m deep) that were drilled for production by IDAAN, MIDA or MINSA but were later abandoned due to poor water quality or because they were not productive enough to be connected to the distribution system. Groundwater levels were manually measured in these wells almost consistently (usually once a month) from May 2015 to March 2020 by personnel from IDAAN's Department of Groundwater Sources and IDAAN's Regional Offices of Herrera and Los Santos. In July 2016, private wells were added to the monitoring network but for most of them levels could not be measured more than once due to site inaccessibility, reluctant well owners and lack of funding. Also, measurements in some wells from the initial network had to be discontinued because of clogging, destruction of the well, connection of the well to the distribution system, or because the observation well was too close a production well, thus no longer representing static groundwater levels. In 2018 researchers from the CIHH-UTP started a project to investigate the groundwater behaviour of the Estibaná sub-watershed (Figure 1c), which allowed to continue groundwater level measurements in most monitoring wells in the south-eastern part of the La Villa Watershed. As of late

2019, groundwater levels were being monitored in 29 observation wells in the La Villa Watershed, 10 out of which were located within the Estibaná sub-watershed.

3.1.2. Precipitation data. In August and September 2018, six Odyssey tipping bucket rain gages were installed in the eastern part of the La Villa Watershed as part of the research project taking place in the Estibaná sub-watershed (Figure 1c). Precipitation data was retrieved during field visits by technical personnel from CIHH-UTP. Two out of the six rain gages collected data until July 2019 and then stopped due to equipment malfunction. The remaining four rain gages are still actively collecting data.

3.1.3. Database design and implementation. Once received from field technicians, the groundwater level data were initially organized and stored locally in Excel Spreadsheets. In 2017, when the development of the ONASP beta version was underway, the data was transferred to a Google Spreadsheet in which the technicians could automatically report their measurements. Once downloaded from the rain gage's loggers, the raw precipitation data were cleaned using custom R scripts to obtain daily accumulated precipitation values. This was then stored locally in CSV files. In late 2021 the aforementioned groundwater level and precipitation data were organized using a relational database implemented in SQLite together with other hydrogeologic data from the Estibaná sub-watershed project. The schema for this database was developed based on the one provided by AkvaGIS, a free and open-source module from the FREEWAT plugin for QGIS which provides tools for storing, managing, and visualizing hydrochemical and hydrogeological data [12]. The file containing the SQLite database is currently being stored in Dropbox.

3.2. Web Application Development

The ONASP web application was designed as an interactive dashboard containing reactive and interactive visualization elements and it was built using the R programming language [13]. The functionality and design of this tool was based on the California Groundwater Observatory [14] and the California Water Quality Portal [15]. The layout of the dashboard was developed using an R Markdown document and the flexdashboard library [16]. The data for ONASP, stored in an SQLite database, was read into R for further processing using the DBI [17] and RSQLite [18] packages. Reactivity was introduced to the web app using elements from the shiny web framework [19] and the shinyWidgets library [20]. Interactive maps and plots were created using the leaflet [21] and plotly [22] libraries, respectively. The knitr package [23] was used to create the HTML file which forms the basis of the ONASP dashboard. The web app is currently deployed using the shinyapps.io platform by RStudio (https://mgcastre.shinyapps.io/ONASP/). The code for this web application can be found on the first author's personal GitHub repository [24].

4. Results

The ONASP web application is organized into three sections: Map, Data and About. The Map section contains a leaflet map showing the spatial location of the monitoring wells and the rain gages, as well as other GIS layers such as the watershed delineations, the hydrologic network and the geology. In this section the user can zoom in and out, turn on or off the different map layers and change the basemap using the controls located in the bottom left corner of the map. The user can select whether to render monitoring wells or rain gages on the map using the selection menu located on the top right corner of the map. The user can click on the monitoring well or rain gage and a hydrograph showing a time series of groundwater levels or daily accumulated precipitation, respectively, will show up on in a panel on the right side of the map below the selection menu (Figure 2).

The Data section contains a sidebar for user inputs on the lefthand side and a board with two panels, Graph and Table, on the righthand side (Figures 3 and 4). The user can select to visualize data from a single, multiple or all monitoring wells or rain gages. As the user changes the inputs, the selected points show up on map in the bottom half of the sidebar. Also, the content of the Graph and Table panels updates dynamically. In the Graph panel an interactive time series plot is rendered where the user can

IOP Conf. Series: Earth and Environmental Science 1136 (2023) 012028 doi:10.1088/1755-1315/1136/1/012028

zoom in, adjust the axes and change the range of dates displayed in the plot (Figure 3). In the Table panel the user can choose to either display a table with time series data or the location information of monitoring wells or rain gages. The columns in this table can be filtered and the displayed data can be copied or downloaded either in CSV or Excel format (Figure 4). Downloading the data is free of charge but the original data sources (IDAAN and CIHH-UTP) must be attributed.

Finally, the About section contains a brief history of the development of ONASP, including a description of how the groundwater level monitoring network was setup. This section also provides information regarding the project funding and explains the importance of collecting and sharing groundwater data, especially groundwater levels.

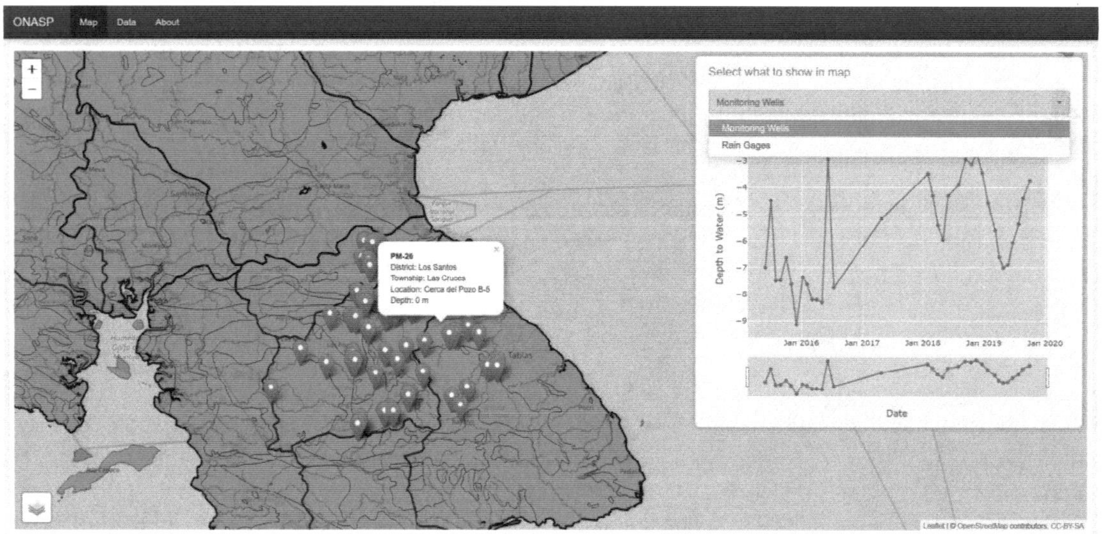

Figure 2. Screenshot of the Map section of the web application.

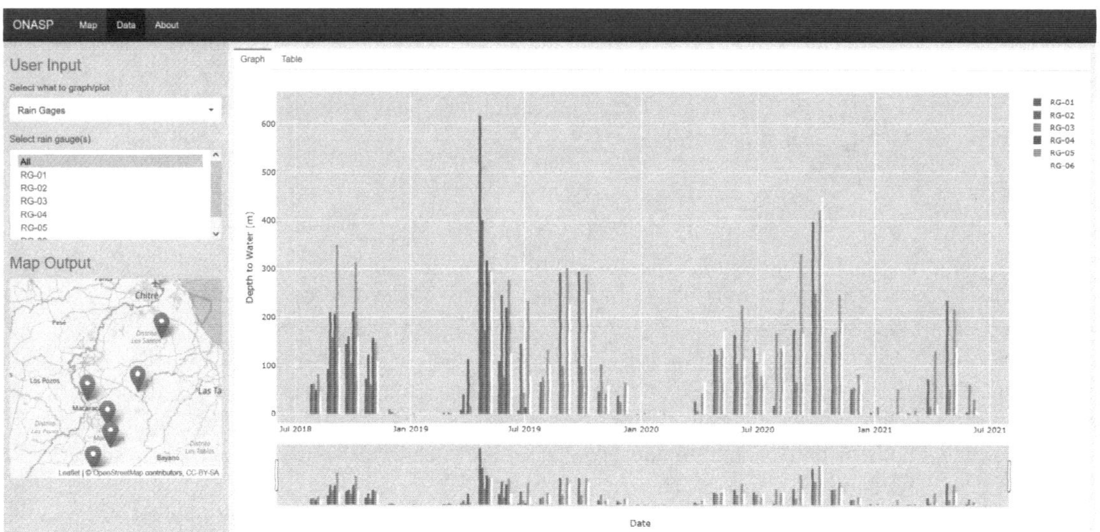

Figure 3. Screenshot of the Graph panel in the Data section of the web application.

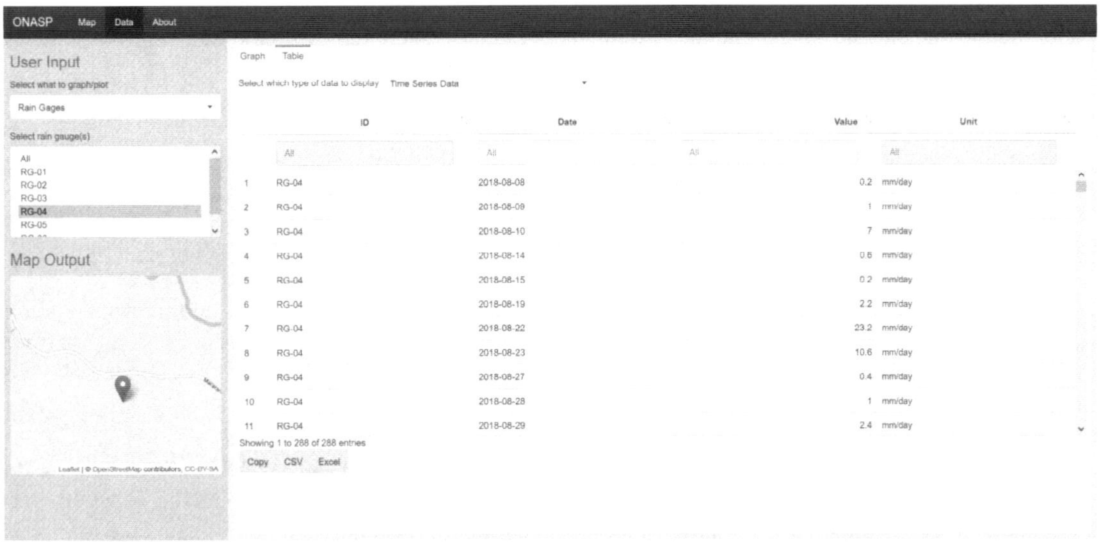

Figure 4. Screenshot of the Table panel in the Data section of the web application.

5. Discussion

The ONASP web application provides access to invaluable groundwater data, especially groundwater levels, that were not available before for authorities, stakeholders, researchers, or the general public. For researchers the availability of this data means that they can focus on analyzing data and not waste time and resources having to collect the data from different sources. For authorities and stakeholders this web can be used as a tool to make informed decisions regarding groundwater management and for the general public the web app can be a medium to educate them regarding the behavior and importance of groundwater resources.

Previous to the development of the groundwater level monitoring network that motivated the ONASP, groundwater level data was not being collected in a consistent manner in Panama except for a few punctual measurements done by consultancy projects hired by government agencies. Thanks to IDAAN, the CIHH-UTP and the ONASP development team there is now a methodology in place for collecting and organizing and sharing groundwater data in the country. Moreover, ONASP has the potential to become a platform for collecting, centralizing, organizing, and sharing hydrogeologic data and information from research projects and government agencies across the country and ensures it is findable, accessible, interoperable, and reusable (FAIR) [25].

Due to the pandemic, groundwater level monitoring ceased in March 2020 and it has not started again due to lack of funding. Given the importance of groundwater level measurements for groundwater management, it is crucial to continue measuring them in the existing monitoring network, as well as establishing new ones in other critical watersheds of the country. Nevertheless, it is sometimes difficult for regional agencies to measure groundwater levels manually once a month in a consistent manner. Calderwood et al. [2] mention that a more cost-effective way to monitor groundwater levels continuously is using pressure transducers and telemetry with a free and open-source software (FOSS) solution to clean and visualize the data. However, the elevated capital cost up front for buying and installing pressure transducers is a huge barrier for implementing this technical solution in developing countries in the global south, such as Panama. A better socioeconomical strategy might be to develop a citizen science programme where people from the community are trained to collect groundwater level data and report it using a Google Spreadsheet or a Google Form. This way, not only a sustainable long-term monitoring programme is created, but also community members get empowered and become aware and responsible of managing their water resources.

6. Conclusions and Recommendations

A web application was designed to share groundwater level and precipitation data from a pilot case study, namely the La Villa Watershed in the Dry Arc Region of Panama. The app was developed with the R programming language and contains interactive maps, plots and tables that allow the user to visualize and download the data. Previous to the development of this web app, groundwater data, especially groundwater levels, were either not available or not easily accessible for researchers, stakeholders, and the general public. The intuitive design of the app will allow all type of users to gain a deeper understanding about the behaviour of groundwater levels and precipitation patterns in the Dry Arc Region of Panama.

Improvements to the app in the near future should include gathering, storing, and adding visualization capabilities for other types of groundwater data such as lithologic records from boring logs, results from pumping tests and hydrogeochemical data for the La Villa Watershed and other critical watersheds in Panama. Also, in order to empower local communities and establish sustainable long-term groundwater level measurements, citizen science programmes should be developed.

The authors believe that findable, accessible, interoperable, and reusable (FAIR) groundwater data is key to achieving a sustainable management of this critical resource. Therefore, the ONASP web application seeks to grow into a platform for organizing, centralizing, hosting, communicating and sharing all kinds of hydrogeologic information from across the country thus becoming an essential tool for stakeholders, authorities, researchers, and the public.

Acknowledgments

We thank the National Secretariat of Science, Research and Innovation of Panama (SENACYT) for funding fieldwork campaigns, providing funds to purchase and install the rain gages, and providing financial support for the first author of this paper. We acknowledge Dr. Amvrossios C. Bagtzoglou and Dr. Gonzalo Pulido for supervising the first author's undergraduate thesis project during which the first set of wells for the groundwater level monitoring network of the La Villa Watershed were selected. We thank the Office of Undergraduate Research of the University of Connecticut for partially funding fieldwork activities during 2015 through the IDEA Grant Award. We acknowledge the institutional support received from IDAAN throughout this project, especially for collecting and sharing the groundwater level data. We especially acknowledge personnel from IDAAN's Regional Office of Herrera (Emilio Díaz, Karen Ascárraga) and Los Santos (Yolibeth Morales), and from IDAAN's Department of Groundwater Sources (Iván Cedeño, Edgardo Vergara, Abel Castillo, Aladino Naranjo, Gilberto Sánchez), for organizing and carrying out the majority of the groundwater level measurement campaigns. Also, we thank the support from the CIHH of the Technological University of Panama for collecting and sharing precipitation data from the study area. We especially acknowledge researchers Iris Arjona, Elsa Flores, Job Noel, Denisse Morales, Sidney Saavedra, Eny Serrano, Ana Franco from the Technological University of Panama for collaborating with fieldwork activities from 2017 onwards under the scope of the Estibaná sub-watershed project funded by SENACYT (Project IOMA17-006). Additionally, we would like to thank Francisco Marchena and Nichol Sánchez Kirsch for helping with the design and implementation of the SQLite database.

References

[1] DWR. SGMA Data Viewer, https://sgma.water.ca.gov/webgis/?appid=SGMADataViewer#gwlevels (accessed October 28, 2019).

[2] Calderwood AJ, Pauloo RA, Yoder AM, et al. 2020 Low-cost, open-source wireless sensor network for real-time, scalable groundwater monitoring. *Water (Switzerland)*; **12**. Epub ahead of print April 1. DOI: 10.3390/W12041066.

[3] USGS. Texas Water Dashboard, https://txpub.usgs.gov/txwaterdashboard/ (accessed October 28, 2019).

[4] ADES - Groundwater National Portal, https://ades.eaufrance.fr/Spip?p=/spip.php?action=converser&hash=373e5d2fe32c3f4a5047b0bdce76d174c04dfe07&redirect=welcome&var_lang=en (accessed October 30, 2019).
[5] WaterConnect Groundwater Data, https://www.waterconnect.sa.gov.au/Systems/GD/Pages/Default.aspx (accessed October 22, 2019).
[6] Autoridad Nacional del Agua. Observatorio de Aguas Subterráneas, http://snirh.ana.gob.pe/visorPozos/ (accessed October 22, 2019).
[7] Barkwith A, Guzman MaALG. Philippine Groundwater Outlook, https://admuwater.com/phigo (2019, accessed April 7, 2022).
[8] Comité de Alto Nivel de Seguridad Hídrica. 2016 Plan Nacional de Seguridad Hídrica 2015-2050: Agua para Todos. Panamá, República de Panamá.
[9] Castrellon Romero MG. Characterization and Modeling of a Tropical Groundwater Basin: La Villa Watershed, Panama, https://search.proquest.com/openview/5d2795a1c9f10564fbcad7c2e95b1c0a/1?pq-origsite=gscholar&cbl=18750&diss=y (2018).
[10] ANAM. Las aguas subterráneas de la región del Arco Seco y la importancia de su convervación. Panamá, República de Panamá, 2013.
[11] Castrellon Romero MG. 2016 Hydrogeological Conceptual Model of La Villa River Watershed, Republic of Panama. Undergraduate Honors Thesis, University of Connecticut, https://opencommons.uconn.edu/srhonors_theses/545/ (2016).
[12] Criollo R, Velasco V, Nardi A, et al. 2019 AkvaGIS: An open source tool for water quantity and quality management. *Computers and Geosciences* 2019; 127: 123–132.
[13] R Core Team. R: A Language and Environment for Statistical Computing, https://www.r-project.org/ (2018).
[14] Pauloo R. cosumnes_shiny Github Repository, https://github.com/richpauloo/cosumnes_shiny (2020, accessed August 9, 2021).
[15] Pauloo R. caccr.github.io Github Repository, https://github.com/caccr/caccr.github.io (2019, accessed October 14, 2019).
[16] Iannone R, Allaire JJ, Borges B. flexdashboard: R Markdown Format for Flexible Dashboards, https://CRAN.R-project.org/package=flexdashboard (2020).
[17] R Special Interest Group on Databases (R-SIG-DB), Wickham H, Müller K. DBI: R Database Interface, https://CRAN.R-project.org/package=DBI (2021).
[18] Müller K, Wickham H, James DA, et al. RSQLite: SQLite Interface for R, https://CRAN.R-project.org/package=RSQLite (2021).
[19] Chang W, Cheng J, Allaire JJ, et al. shiny: Web Application Framework for R, https://CRAN.R-project.org/package=shiny (2021).
[20] Perrier V, Meyer F, Granjon D. shinyWidgets: Custom Inputs Widgets for Shiny, https://CRAN.R-project.org/package=shinyWidgets (2022).
[21] Cheng J, Karambelkar B, Xie Y. leaflet: Create Interactive Web Maps with the JavaScript "Leaflet" Library, https://CRAN.R-project.org/package=leaflet (2021).
[22] Sievert C. Interactive Web-Based Data Visualization with R, plotly, and shiny. Chapman and Hall/CRC, https://plotly-r.com (2020).
[23] Xie Y. knitr: A General-Purpose Package for Dynamic Report Generation in R. R package version 1.36, https://yihui.org/knitr/ (2021, accessed September 10, 2021).
[24] Castrellón MG, Pauloo RA. ONASP Interactive Web Application. Epub ahead of print April 13, 2022. DOI: 10.5281/ZENODO.6564017.
[25] Wilkinson MD, Dumontier M, Aalbersberg IjJ, et al. 2016 The FAIR Guiding Principles for scientific data management and stewardship. *Scientific Data 2016 3:1* 2016; 3: 1–9.

Assessment of alluvial aquifer conditions during a construction phase for a deep excavation

Masoud Ghulami[1] **and Philippe Gourbesville**[1]

[1] Polytech Lab, University of Côte d'Azur, Sophia-Antipolis, 06903, France

masoud.ghulami@unice.fr

Abstract. The Var low valley and its unconfined alluvial aquifer is the main water resource for around 600,000 inhabitants who live in the French cities and towns near the river mouth, such as Nice and St Laurent du Var. Several commercial and residential complexes are planned to be built in the downstream floodplain areas of the Var River and close to the Nice airport. Some of these buildings may have an impact on the groundwater table and consequently on the supply of drinking water. In this study, the groundwater flow was simulated using MIKE FEFLOW numerical model to assess the impacts of new development projects by estimating the drawdown of groundwater table due to the planned pumping activities as well as tracing the trajectories of groundwater movements to make sure the salt intrusion will not happen. The study focuses on a planned building, near well line used for water supply. The building was initially designed with several underground floors as parking lots. The model is developed and validated using the available piezometric data of 2010-2012 for the study area. The model was able to simulate the hydraulic head with a good precision (the Nash Sutcliffe Efficiency coefficient of 0.7 and higher). As the local groundwater level is very high and only a few meters below the ground surface, during the construction period of the foundation, the site must be kept dry, and the water should be pumped out. It will normally last around six months. However, this construction period is the most critical stage and will have serious impacts on the groundwater table, but our simulation results suggest that even after the construction period, the groundwater table will be dropped and consequently the salt intrusion into the groundwater due to the short distance of the site to the sea will be continued. Therefore, different foundation levels were tested to find the best solution.

1. Study area

For around 600,000 people who reside in French cities and towns close to the river mouth, such as Nice and St Laurent du Var, the Var low valley and its unconfined alluvial aquifer serve as their primary water source. The building site's location and the neighbouring well lines that the Nice municipality uses to supply drinking water are depicted in Figure 1. At the building site, the ground is elevated to 6.5 M.A.S.L. (meters above sea level). The elevation of the foundation's deepest point was then planned to be 3.5 m for part one and -10 m for section two, respectively. This is called in this study as case one as it was originally designed. Later, using the trial-and-error method, an optimal solution was suggested which is called here as case two.

2. Methods

Initially, (Du et al., 2018) used the commercial software MIKE FEFLOW to simulate groundwater flow in the unconfined aquifer in the lower Var valley (Hans 2005). The same model configuration is used in this study, including the input data, simulation times, grid size, and calculation time steps. There are studies done to determine how the well line and building site would affect the groundwater table.

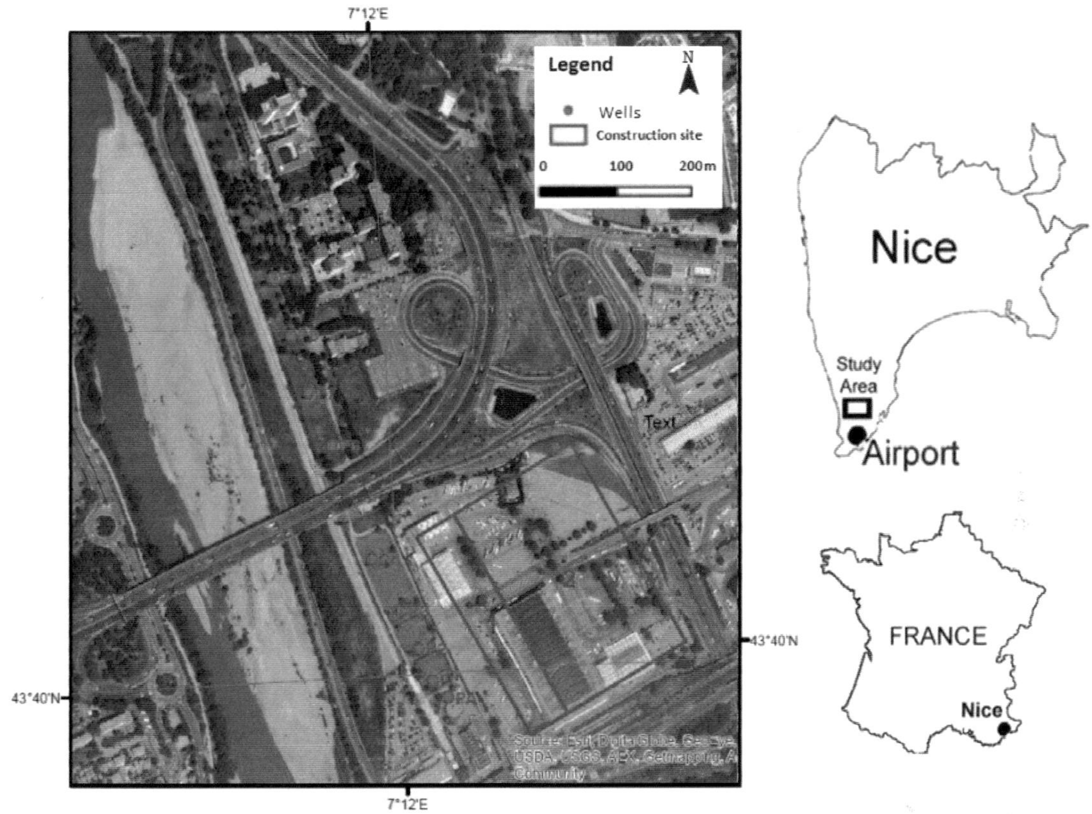

Figure 1. Map of the construction site and the nearby wells used for drinking water supply near the Nice airport.

The finite element method is used by FEFLOW to solve the 3D groundwater flow equations in porous media. The Darcy equation and the equation for the conservation of fluid mass are the controlling equations (Diersch and Kolditz, 1998):

$$S_S \cdot s(h)\frac{\partial h}{\partial t} + \varepsilon \frac{\partial s(h)}{\partial t} + \nabla q = Q \tag{1}$$

$$q = -K_r(s)K(\nabla \psi_g + \chi e) \tag{2}$$

where,

q : the Darcy flux vector (m/s);

Q : the specific mass supply (m/s);

S_S : the specific storage, $S_S = \varepsilon \gamma + (1-\varepsilon)Y$;

ε : the porosity;

γ : the fluid compressibility;

Y : the coefficient of skeleton compressibility;

ψ : the hydraulic head (m), $\psi = h + z$;

z : the elevation (m);

h : the pressure head (m);
$s(h)$: the saturation (if medium is saturated then $s = 1$);
$K_r(s)$: the relative hydraulic conductivity ($0 < K_r < 1$, $K_r = 1$ if saturated at $s = 1$);
K : the hydraulic conductivity for the saturated medium;
χ : the buoyancy coefficient including fluid density effects;
e : the gravitational unit vector.

3. Results

With a combined maximum pumping rate of 1000 L/s for all the pumps, Figure 2 displays the simulation results for the initial state and the building phase. The groundwater levels in relation to sea level are shown by the blue lines. The streamlines which are trajectories of water particles following the velocity vectors of the flow are represented by red lines.

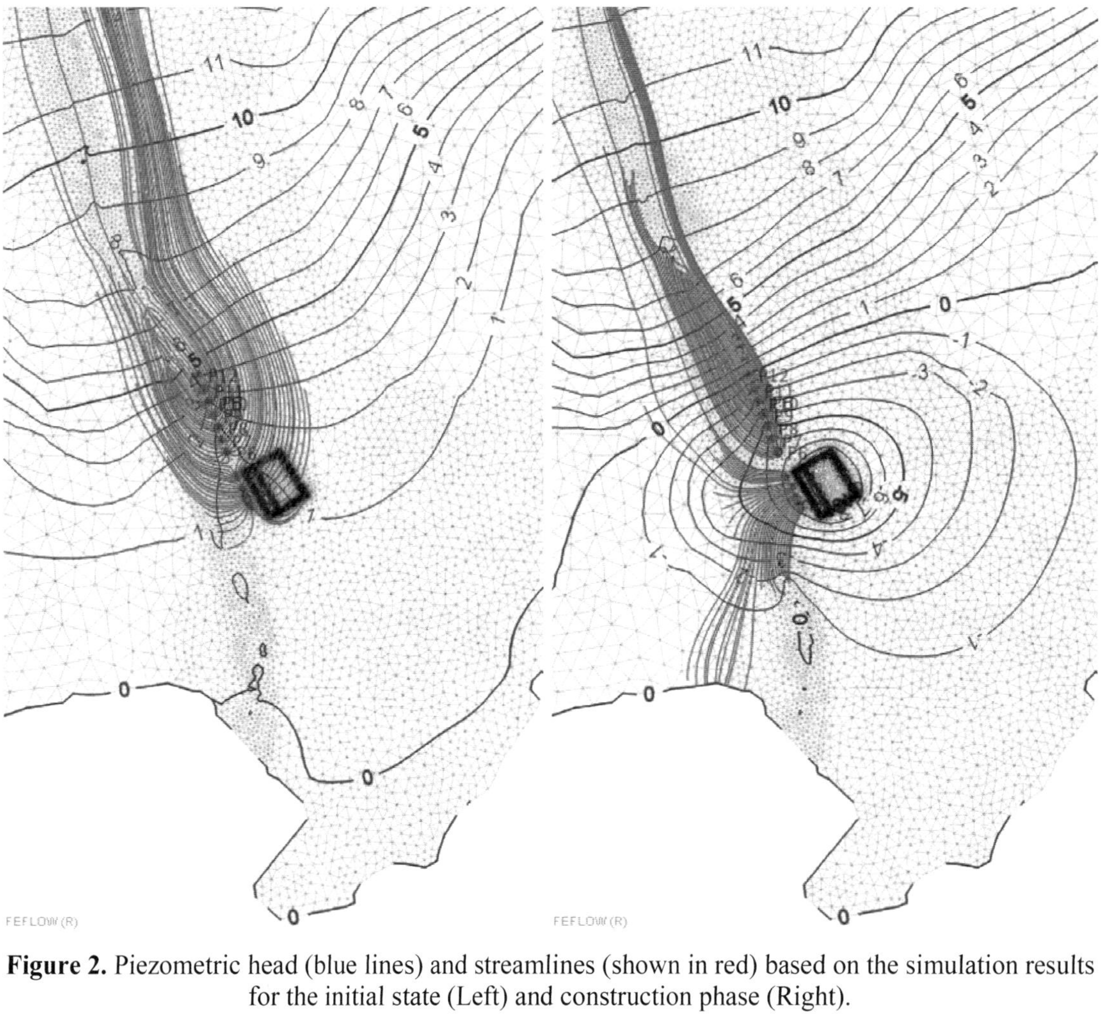

Figure 2. Piezometric head (blue lines) and streamlines (shown in red) based on the simulation results for the initial state (Left) and construction phase (Right).

Two cross-section views are given in Figure 3 to help clarify the groundwater table's decline during the building phase. Pumping the water out and creating an impermeable base for the building are required during construction. This might be finished in around six months. The groundwater level will return to normal after this time, but the building will be under intense pressure. It might be important to keep pumping the water out in some circumstances, particularly if the parking lots are situated near the base of the structure, to ensure that it is not seeping into the structure.

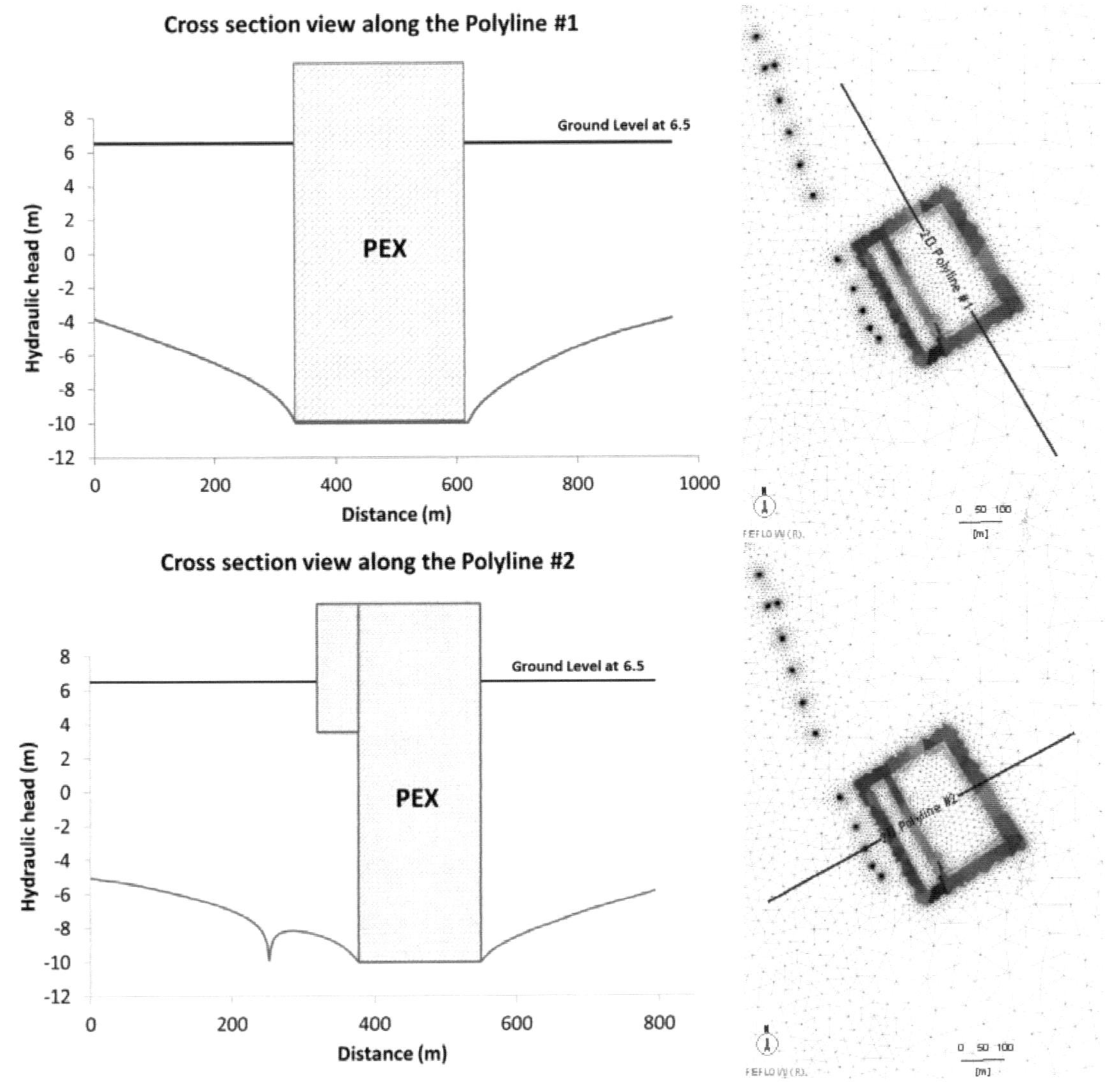

Figure 3. Cross section profiles of piezometric head during the construction period (case one: original design).

The simulation results of construction period show that the groundwater level declines of up to 10m. It consequently causes the salt intrusion into the groundwater following the streamlines. The next step was to find an optimal solution. The elevation of the deepest point for the foundation was then set to be at 3.5 m and 2 m respectively for part one and part two to avoid the salt intrusion. It is, therefore, necessary for similar projects to assess the impacts of new constructions on the groundwater table by considering the nearby related activities to see the overall impacts rather than the construction site itself to avoid challenging issues such as salt intrusion.

Another aspect which should be considered is the current elevation in which the pumps are installed inside the wells, which are used for the water supply system. If the drawdown is significant similar to the case of this study, which is already discussed, then some of the pumps which are not installed in a very deep location within the wells, will be above the groundwater level and so they cannot be used during the construction period.

Based on the trial and error, an optimal solution was suggested to find the maximum foundation depth without having the negative impacts on the groundwater level. The dimension of the optimal building will be same as the original one except for the groundwater levels. The foundation level for the part one was kept as 3.5m and for the part two the optimal elevation was calculated to be at -1m instead of -10m in the original design. Following this, the impacts on the well line will be limited and the salt intrusion will not happen.

Figure 4 shows the groundwater table, during the construction period, by drawing two cross-section views for the optimal solution named as case two.

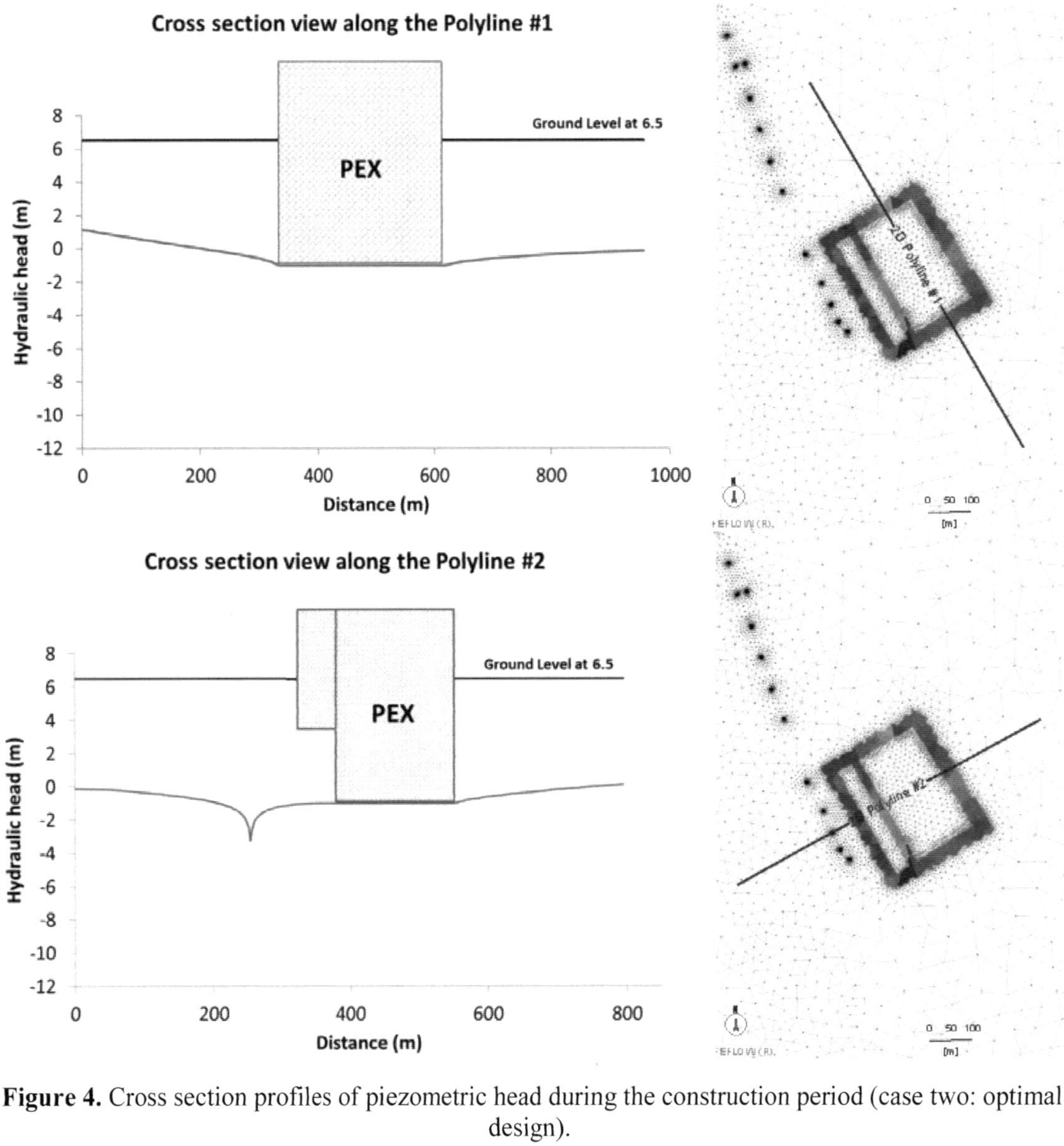

Figure 4. Cross section profiles of piezometric head during the construction period (case two: optimal design).

Implementing the case two instead of case one, is limiting the negative impacts on the well line. The drawdown will be much more limited and therefore majority of pumps (eight out of twelve) will be still under the groundwater level and will be functioning normally. This is shown and compared between two cases in Figure 5.

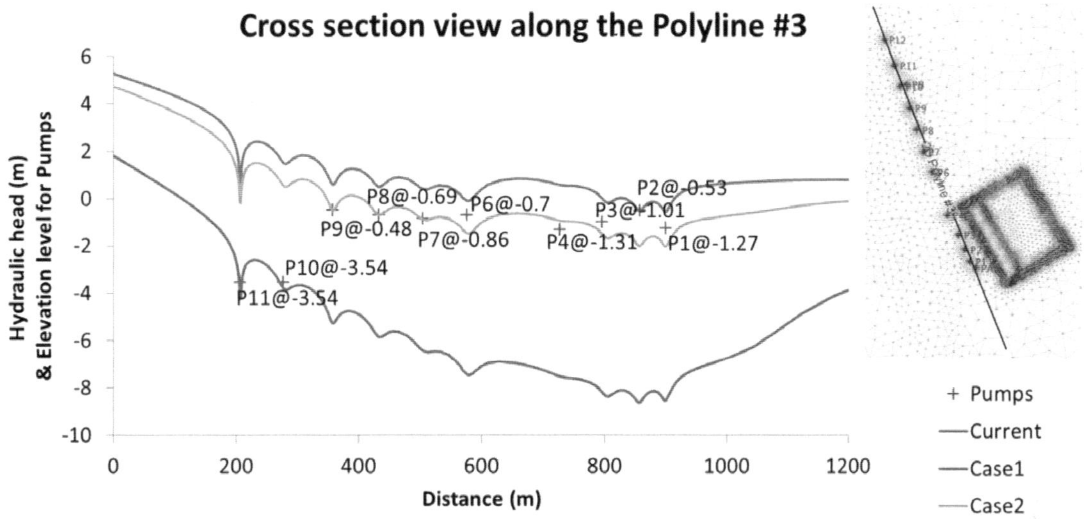

Figure 5. Cross section profiles of piezometric head along the well line during the construction period for the different scenarios. The current elevation in which the pumps are installed is also shown.

Pujades and Jurado (2021) did a literature review to highlight the progress made and the challenges remained for the groundwater-related aspects during the development of deep excavations below the water table and concluded that: the topic to which authors pay more attention is the impacts produced by dewatering tasks, among which dewatering settlements stand out.

4. Conclusion

The groundwater level decreases by up to 10 meters for case, the original design, according to the modelling findings of the building phase. As a result, the groundwater close to the well line utilized for water supply experiences salt intrusion. An optimal solution, therefore, was investigated to limit the impacts. Following this optimal solution, the drawdown of groundwater table will be limited and the problem of salt intrusion or dropping the water level under the current elevation of the installed pumps will not happen. The maintenance issues after the construction should be also considered as the groundwater pressure on the foundation makes it difficult to keep the parking lots dry without pumping out the water beneath the building.

References

[1] Diersch H J and Kolditz O 1998 Coupled groundwater flow and transport: 2. Thermohaline and 3D convection systems *Advances in Water Resources* **21**(5) 401-425 https://doi.org/10.1016/S0309-1708(97)00003-1

[2] Du M, Zavattero E, Ma Q, Gourbesville P and Delestre O 2018 Groundwater modeling for a decision support system: The Lower Var Valley, Southeastern France *Advances in Hydroinformatics* pp 273–283 https://link.springer.com/chapter/10.1007/978-981-10-7218-5_18 (Singapore: Springer)

[3] Diersch H-JG 2005 *DHI-WASY Software FEFLOW Reference Manual* (Berlin, Germany: DHI WASY GmbH)

[4] Pujades E and Jurado A 2021 Groundwater-related aspects during the development of deep excavations below the water table: A short review *Underground Space* **6**(1) 35-45 https://doi.org/10.1016/j.undsp.2019.10.002

IOP Conf. Series: Earth and Environmental Science 1136 (2023) 012030 doi:10.1088/1755-1315/1136/1/012030

Application of Delft3D computational model to estimate the influence of El Bala run-of-the-river dam on the morphological activity of Beni River, Bolivia

Adriana R Rivera[1] **and Marcelo W Heredia**[1]

[1] Hydraulics Laboratory, Universidad Mayor de San Simón, Cochabamba, Bolivia

marceloheredia.g@fcyt.umss.edu.bo

Abstract. Beni river is an important tributary in the Amazon basin. There, the Bolivian government decided the construction of a two-dam cascade hydropower system, including Chepete dam and El Bala run-of-the-river dam. The main concern is the high sediment transport in Beni river. Then, the aim of the research is the application of Delft3D model to estimate the influence of El Bala run-of-the-river dam on the sediment transport and morphological activity of Beni river. The model was calibrated for natural conditions based on ADCP data and sediment transport measurements collected from the river. The implemented model was able to reproduce hydrodynamics and sediment transport in natural conditions. Hence, several scenarios were tested. A hydrodynamic and morphological model was set to test the influence of the dam on the river for hydrodynamics, sedimentation and erosion processes in Beni river. Results show that Delft3D performed very well in modelling sedimentation and erosion processes in Beni river, the model was able to predict the expected erosion process due to the presence of Chepete dam, upstream of El Bala, which retains the sediment load of Beni river affecting the morpho-dynamics of the river. Additionally, the model was able to predict a sedimentation in the reservoir.

1. Introduction

The Bolivian government decided to increase the electricity generation for the internal and external markets. Hence, it was decided to build hydropower plants at different locations of the country. One of the regions with the highest potential for generation is Beni river, which is a part of the Amazon basin in Bolivia. The Bolivian Electricity Company (ENDE) started with the Project El Bala-Chepete, which considers the construction of a hydropower cascade system along Beni river, by the implementation of 2 dams: Chepete and El Bala, with a projected generation of 3251 MW and 425.18 MW respectively. Hence, GEODATA Engineering Company was commissioned for this Project.

The Hydraulics Laboratory of Universidad Mayor de San Simón was commissioned by GEODATA in 2018 to start a research project on the influence of the implementation of hydropower plants on the fluvio-morphological activity of Beni river. This research project included the application of numerical modelling to simulate the hydrodynamics and morphology of Beni river, since this river carries the highest suspended sediment concentrations in the Bolivian part of the Amazon basin [1,2].

The present study is aimed to the application of Delft3D computational model to estimate the influence of El Bala run-of-the-river dam on the fluvio-morphological activity of Beni river. All the data collected from the study zone was processed and implemented to setup the model, which was calibrated for hydrodynamics and sediment transport. Delft3D model was able to reproduce the observed hydrodynamic and sedimentological variables in natural conditions, allowing the implementation of hypothetical scenarios which included the implementation of El Bala dam. The results of the scenarios showed that el Bala Dam has a direct influence on the erosion and sedimentation processes of Beni river. Moreover, the computational model was able to take into account the influence of Chepete Dam, which is located upstream of El Bala Dam, by increasing the erosion rates since Chepete would reduce the sediment supply.

The paper is composed of six sections. Section 1 refers to the introduction and background of the present study. Section 2 shows the materials and methods used for the study, where all the data of the study area is described, jointly with the theoretical background of numerical modelling and the setup of the model. Section 3 presents the results and discussion, showing the results of the calibration process and the simulations of the hypothetical scenarios. Section 4 draws the conclusions of the study. Finally, Section 5 presents the bibliographic references.

2. Materials and Methods

2.1. Description of the study area

Beni river is located in the northwest of Bolivia and is part of the Amazon basin, being a tributary of Madeira river. The study area is located between 245 and 190 M.A.S.L. El Bala dam is located 10 km upstream of Rurrenabaque town, at the border of Departments of La Paz and Beni (see Figure 1).

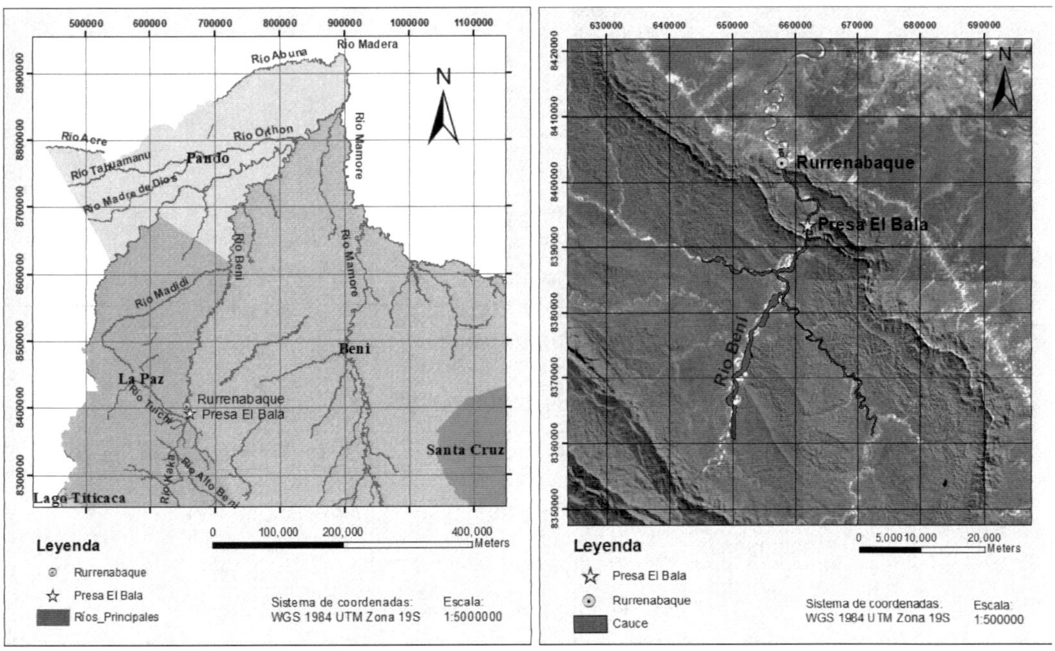

Figure 1. Location of Beni river and El Bala Dam.

The catchment of Beni river has an extension of 282,500 km^2, being one of the largest catchments of Bolivia. Two regions are clearly identified, the Andean part and the flat part of the river [3], where the Andean part is characterized by high rates of soil erosion [4]. The projected location of El Bala dam is characterized by a significative change of the river longitudinal slope, being the natural border between the Andean and the flat part of the river. The authors of article [5] mentions that the river slope decreases from 0.0002 up to 0.00007 m/m at this point, facilitating the sedimentation process from El Bala downstream.

2.2. Data of the study area

2.2.1. Bathymetry. A digital elevation model (DEM) of the study area was provided by GEODATA, the DEM has a resolution of 1x1 m (see Figure 2). However, this DEM does not cover the portion of the river located between El Bala and Rurrenabaque, this part was filled by using a DEM from a public database (i.e. Alos Palsar) with a lower resolution, 12.5x12.5 m.

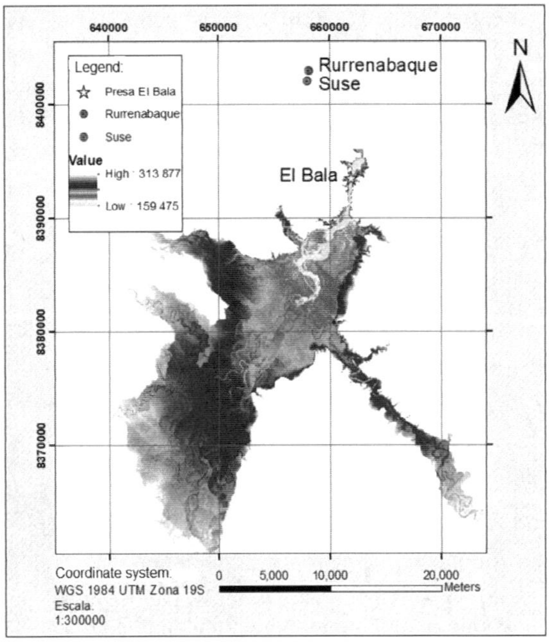

Figure 2. Digital Elevation Model of the study area.

2.2.2. Hydrometric information. The hydrometric information was collected from measuring stations located within the study area, 3 stations were used for this purpose to collect data of water levels, flow velocities and discharges, Table 1 details the location coordinates of the 3 stations:

Table 1. Location coordinates of measuring stations (WGS1984 – UTM Zone 19S).

Station	X-coordinate	Y-coordinate
El Bala	661963	8393427
Suse	658049	8402055
Rurrenabaque	658160	8402920

The Hydraulics Laboratory (LHUMSS) carried out 13 measurement campaigns by using an ADCP device to measure water depths, discharges and flow velocities. These campaigns run from 2015 until 2017, with observed discharges between 602 m^3/s up to 5176.80 m^3/s, being the second one the largest discharge ever measured by LHUMSS, corresponding to an event observed on 9/3/2017.

2.2.3. Sediment transport data. The sedimentometric data was collected from measurement campaigns carried out at the mentioned stations, the 13 campaigns reported by [6] also included suspended sediment transport measurements, the suspended sediment concentration observed at the stations ranges from 104 mg/L up to 4925 mg/L. This information was complemented by historic sediment transport data obtained from the PHICAB Project [8], the ORE-HYBAM database [2] and the review and analysis presented by [8].

2.2.4. Dimensions and geometry of the projected Dam. El Bala dam is a run-of-the-river dam, that allows the flow of water through the structure, generating power at the same time. The design and location are still preliminary (see Figure 3). However, this design used for the modelling purposes in order to estimate the effects of the dam. The height of the dam is 49 m and the length is 215 m. The crest of the dam is defined at 220 M.A.S.L. and the inundated area of the projected reservoir is depicted in Figure 3, reaching an extension of 65.44 km^2.

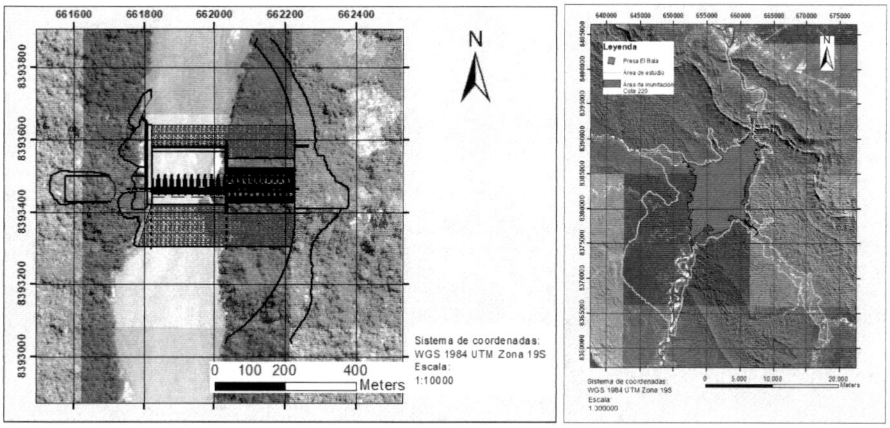

Figure 3. Projected preliminary design of El Bala run-of-the-river dam and the extension of the inundated area of the projected reservoir.

2.2.5. Mathematical modelling of hydrodynamics, sediment transport and morphology. Delft3D computational model was used for the modelling project. This model makes use of the depth-averaged flow equations derived from the Navier-Stokes equations [9,10], these equations are:

$$\frac{\partial \zeta}{\partial t}+\frac{\partial [hU]}{\partial x}+\frac{\partial [hV]}{\partial y}=S \tag{1}$$

$$\frac{\partial U}{\partial t}+U\frac{\partial U}{\partial x}+V\frac{\partial U}{\partial y}+\frac{\omega}{h}\frac{\partial U}{\partial \sigma}-fV=-\frac{1}{\rho_0}P_x+F_x+M_x+\frac{1}{h^2}\frac{\partial}{\partial \sigma}\left(\nu_V\frac{\partial u}{\partial \sigma}\right) \tag{2}$$

$$\frac{\partial V}{\partial t}+U\frac{\partial V}{\partial x}+V\frac{\partial V}{\partial y}+\frac{\omega}{h}\frac{\partial V}{\partial \sigma}-fU=-\frac{1}{\rho_0}P_y+F_y+M_y+\frac{1}{h^2}\frac{\partial}{\partial \sigma}\left(\nu_V\frac{\partial v}{\partial \sigma}\right) \tag{3}$$

where the horizontal pressure terms P_x and P_y are determined by the Boussinesq approximations, F_x and F_y are horizontal Reynolds stresses and are calculated using the eddy viscosity concept [10], M_x and M_y denote the contribution of external sources, S denotes the contributions of discharge of water, U and V denote the depth-averaged flow velocity components, ζ denotes the water level. These equations are solved by the ADI implicit scheme of finite differences [9].

Sediment transport is calculated by using the Van Rijn sediment transport equation for bed load and suspended load, these equations were proposed by [11,12] to be used with sand particles, which is the main particle present in the study area. While, the evolution of the river bed was accounted by the Exner equation for morphology, which calculates erosion and deposition due to flow currents and sediment transport [9], this equation is given by:

$$\left(1-\lambda_p\right)\frac{\partial z_b}{\partial t}=-\frac{\partial S_{b,x}}{\partial x}-\frac{\partial S_{b,y}}{\partial y}-E+D \tag{4}$$

where λ_p denotes the bed porosity, z_b denotes the bed level, $S_{b,x}$ and $S_{b,y}$ denotes the components of sediment transport, D corresponds to the deposition rate and E the erosion rate.

2.2.6. Setup of the model. The model was setup by using the data collected from the study area. The first step was the development of the numerical grid that covers all the area (see Figure 4), this grid is composed by 41957 grid cells with an average size of 20×40 m. The bathymetry obtained from the DEM was applied to the grid as depicted in Figure 4.

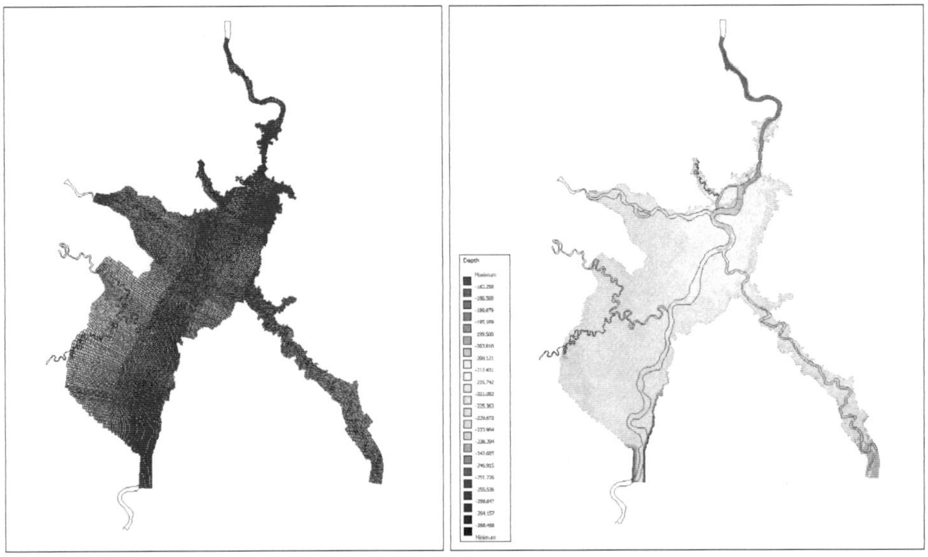

Figure 4. Numerical grid that covers all the model area and the bathymetry implemented in the model.

A "cold start" was prescribed as initial conditions for flow velocities and a uniform water depth of 5 m. These conditions were used as starting point for the simulations in order to reach a stable solution that was reached after 6 days and by using a time step of 1.5 s. which guarantees the numerical stability of the model according to the CFL stability condition. Additionally, constant discharges were prescribed as boundary conditions for the 3 tributaries that contribute to the projected reservoir:

Table 2. Boundary conditions prescribed at the inlets and outlet of the model area.

Boundary	Value	Description
Inlet	Q = 215.5 (m^3/s)	Quiquibey river
Inlet	Q = 4251.2 (m^3/s)	Beni river
Inlet	Q = 710.1 (m^3/s)	Tuichi river
Outlet	Water level = 192.23 M.A.S.L.	Rurrenabaque

Finally, bed roughness coefficients were estimated based on the soil use and vegetation cover maps obtained from the study area, where two main areas were identified: main channel of the river and floodplains Then, the prescribed Manning coefficients were 0.035 for main river channel and 0.12 for floodplains. Anyhow, these values were tested and tuned during the calibration process.

3. Results and Discussion

3.1. Hydrodynamic and sedimentologic calibration of the model

The hydrodynamic calibration had the purpose to reproduce an observed event in Beni river under natural conditions, i.e. without the implementation of the dam. The selected event corresponds to the measurement campaign executed on 09/03/2017, where the maximum flow discharge was observed. Two parameters were tuned during the calibration, i.e. Manning roughness coefficient of the main channel and the viscosity, leading to the execution of 20 runs, where the Manning coefficient was varied from 0.035 up to 0.049, and viscosity was varied from 2 m^2/s. up to 0.5 m^2/s.

The assessment of the model results for the calibration process were carried out based on the methodology proposed by [13,14], where a cost function is used as a global goodness of fit indicator that gathers other statistical indicators, i.e. RMSE, MAE, R^2, BIAS and Nash-Sutcliffe efficiency. This methodology makes use of weighting factors for every hydrodynamic variable, in this case 30% was applied to water levels and 70% was applied to flow velocities. Figure 5 depicts the variation of the cost function during the hydrodynamic calibration, showing an improvement up to Run 13 due to the variation of the Manning Coefficient, further improvements were not observed due to the variation of the viscosity. Figure 6 depicts the comparison of flow velocities at Bala and Suse stations.

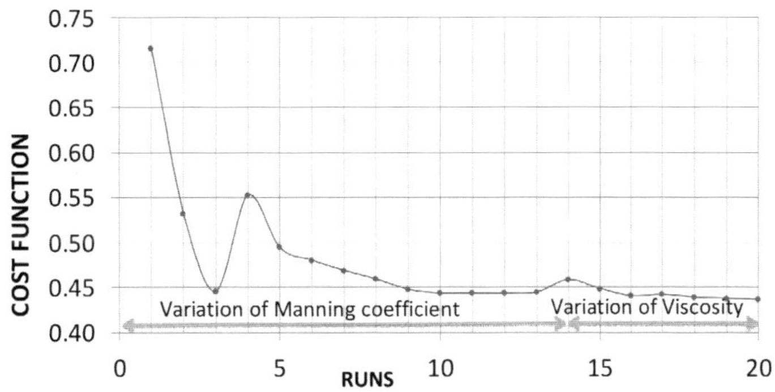

Figure 5. Variation of the cost Function during the hydrodynamic.

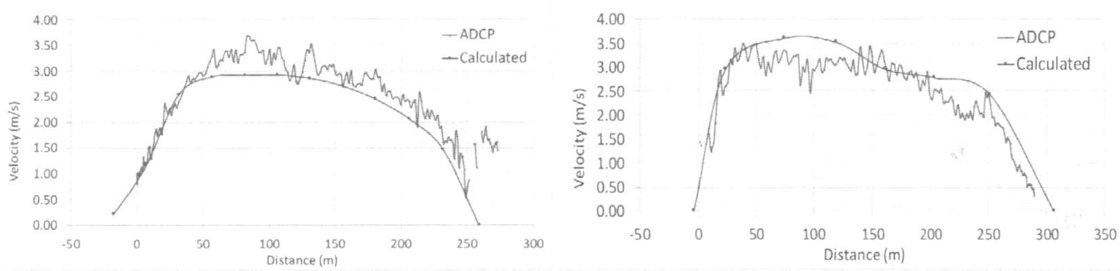

Figure 6. Comparison of measured vs. calculated flow velocities at Bala and Suse stations.

On the other hand, the calibration for sediment transport was aimed to reproduce the suspended sediment concentration observed at the measuring stations, where 25 runs were executed. The model was setup including the characteristic diameters of the bed material, [1] and [8] report that bed material of Beni river at the study area is characterized by presenting sand and coarse silt with d_{50} = 0.1 mm. The calibration parameters chosen for the sediment transport calibration are listed in Table 3.

Suspended sediment concentrations calculated at Bala and Suse stations were compared with the measurements, it was observed acceptable results for the mentioned stations: 4925 mg/L (observed) vs. 4957 mg/L (calculated) at El Bala, and 4629 mg/L (observed) vs. 4329 mg/L (calculated) at Suse.

Table 3. Calibration parameters for sediment transport.

Parameter	Range of values	Units
current-related reference concentration factor	1 – 11	-
current-related transport vector magnitude factor	1 – 14	-
horizontal eddy diffusivity	5 – 15	m^2/s
initial sediment layer thickness at bed	1 – 5	m
threshold sediment thickness	0.05 – 5	m

3.2. *Numerical modelling of hydrodynamics within the reservoir*

The next step in the study was the implementation of the run-of-the-river dam in order to simulate the hydrodynamics within the reservoir. A widening towards the right shore was applied at the section of the location of the dam, with the aim to reproduce the presence of the turbines of the projected hydropower plant. Additionally, a series of gates were placed at the main channel (see Figure 7). Two functionalities of Delft3D model were used for that purpose, i.e. "thin dams" and "discharges", to simulate the gates and turbines respectively.

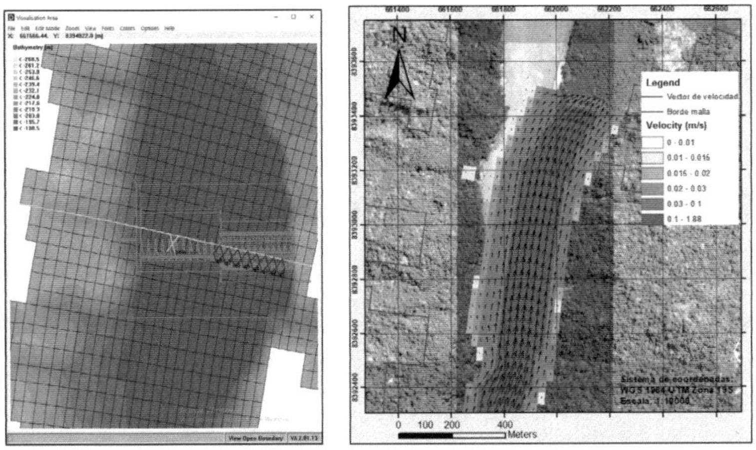

Figure 7. Implementation of the run-of-the-river dam and calculated flow patterns near the structure.

New boundary conditions were prescribed for the scenario, a discharge of 535.7 m^3/s was applied to Beni river, this discharge considers the further presence of Chepete Dam, which is located upstream of El Bala and both dams will operate in a cascade system, where Chepete will release the prescribed discharge. Furthermore, two additional discharges were prescribed for the two tributaries: 84.3 m^3/s for Quiquibey river and 278 m^3/s for Tuichi river. Additionally, since El Bala is a run-of-the-river dam, it is assumed that the discharge released by the dam should be the sum of the three main contributions of the upstream boundaries. Hence, a discharge of 898 m^3/s was prescribed at the outlet.

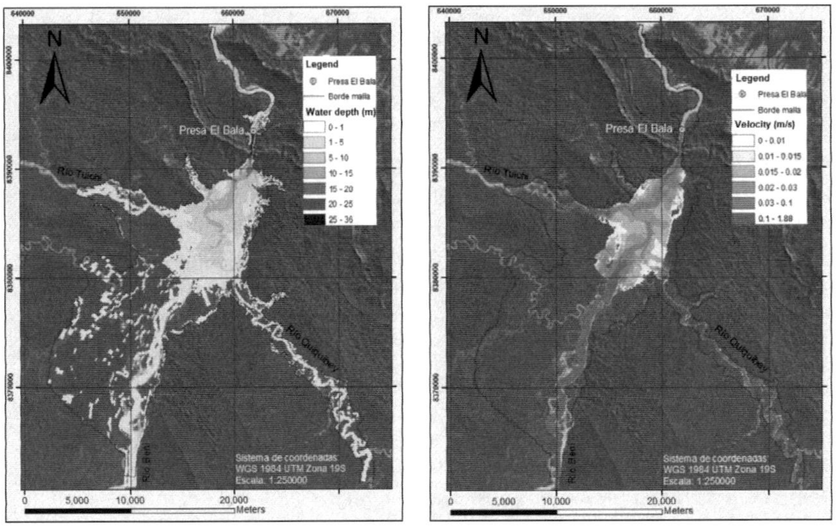

Figure 8. Water depths and flow velocities calculated for the scenario with the presence of the dam.

Figure 8 depicts the calculated water depth and flow velocities within the reservoir, with water depths up to 36 m upstream of the dam, and very low flow velocities due to the presence of the reservoir. Figure 9 depicts the calculated flow patterns at different locations within the reservoir, observing that the model was able to simulate the acceleration of the flow upstream of the dam and the formation of braided systems at the tributaries.

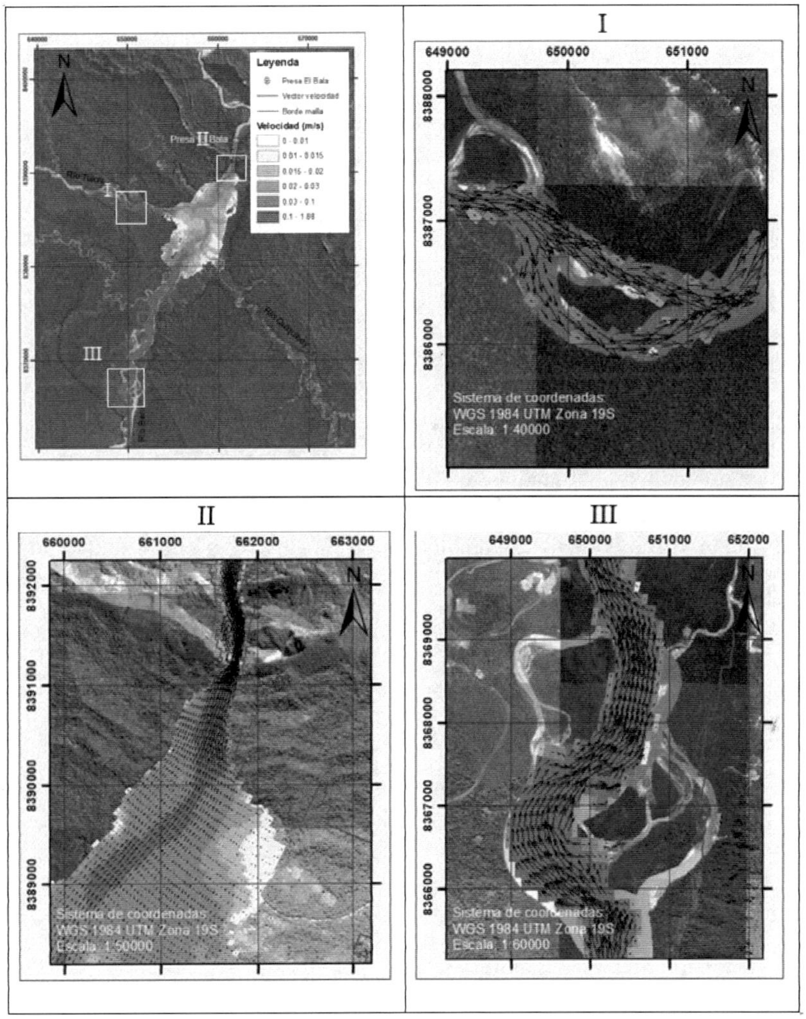

Figure 9. Water depths and flow velocities calculated for the scenario with the presence of the dam.

3.3. Numerical modelling of sedimentation and erosion scenarios

The final step in the modelling study was the simulation of erosion and deposition within the reservoir as a consequence of the implementation of the dam, due to the reduction of flow velocities, it is expected changes in the morphological activity of the river. This scenario considered the prescription of boundary conditions for sediment transport, in the case of Beni river, it was considered the presence of Chepete dam upstream, since this structure will retain the sediments that normally are transported by the river. It is not known the sediment concentration downstream of the Chepete dam after its implementation. Therefore, a "equilibrium condition" is prescribed to allow the river to transport sediment at its capacity.

Then, a 30-year morphological simulation was carried out to predict the erosion and deposition, the methodology proposed by [15] to model large-scale river morphology was applied for this purpose.

The simulation was set for 30 days and a "morphological factor" of 487 was applied to simulate the effect of 30-years of morphological activity and speed up the computational process.

Figure 10 depicts the cumulated erosion and deposition within the reservoir after 5 and 30 years. Erosion is observed at the tributaries, especially at Beni River, where it was calculated a cumulated erosion of about 5 meters after 30 years. This behaviour does not surprise, since the presence of Chepete dam upstream will affect the morphological activity of the river by retaining sediments and favouring the occurrence of erosion, helping Beni river to reach "again" its equilibrium condition.

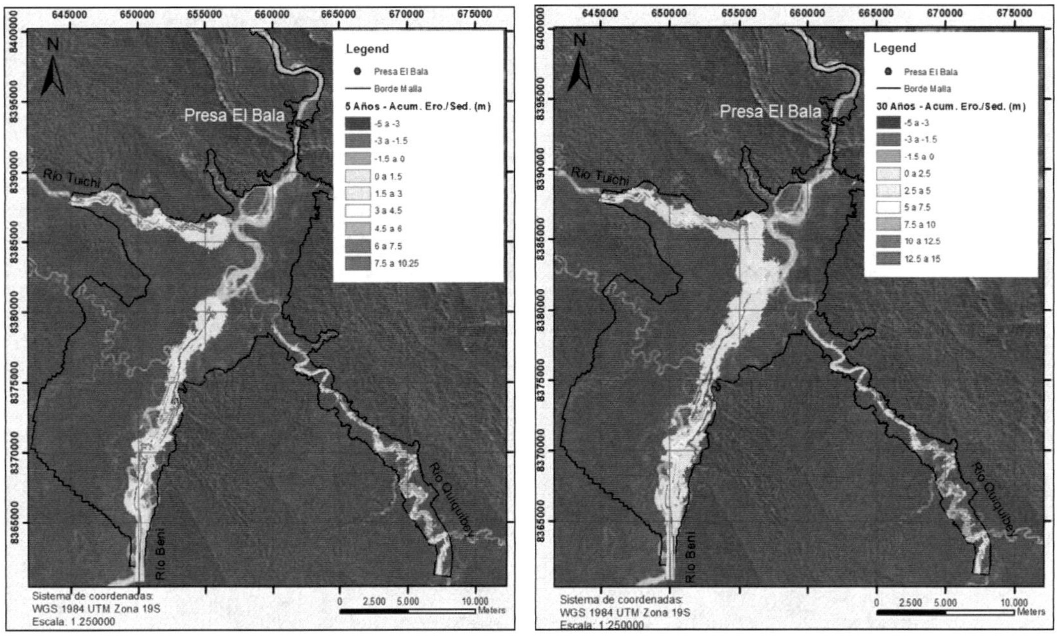

Figure 10. Cumulated erosion and deposition within de reservoir after 5 and 30 years.

A cumulated sedimentation is observed at the entrance of the reservoir, since very low velocities are observed, it is expected that sedimentation will take place at those areas, a very high sedimentation was calculated in the area of Tuichi river, about 15 m, it seems that this behaviour might be caused by a sort of numerical instability due to the presence of a "staircase" configuration of the main channel of Tuichi river, because a orthogonal structured grid was used. Finally, it was observed that no deposition was calculated in the areas close to the dam, because no finer sediment particles were prescribed in the model, the reason of this was the lack of information for the granulometry of the suspended sediments.

4. Conclusions

Delft3D computational model has been applied to simulate hydrodynamics and morphology in Beni river, based on the data collected from measurement campaigns and previous works, leading to an acceptable calibration of the model.

The computational model allowed to estimate the hydrodynamics and morphology within the reservoir area, where erosion and sedimentation were observed. Erosion was observed mainly in the tributary rivers, especially in Beni river, which might be a consequence of the implementation of Chepete dam upstream of El Bala. Hence, the equilibrium boundary condition for sediment transport seems to be a good option in this case. Additionally, it was observed that sedimentation takes place mostly at the mouths of the tributaries and not beyond, the reason for this behavior is the lack of information regarding the granulometric distribution of the suspended sediments, specifically for the finer fractions. It is suggested to collect more data for the finer fractions in order to estimate the morphological behavior upstream of El Bala dam.

The described application of a computational model to estimate the morphological behavior at Beni river, should be considered as a first step in the attempt to estimate the probable effects of implementing El Bala dam. Further work is necessary in order to obtain more information from the study area and improve the numerical model allowing the simulation of more scenarios that will help in the design.

References

[1] Guyot JL 1993 *Hydrogéochimie des fleuves de l'Amazonie Bolivienne* (Doctoral dissertation, Université de Bourdeaux, Institut Français de Recherche Scientifique pour le Développement en Coopération, France)
[2] Vauchel P et al 2017 A reassessment of the suspended sediment load in the Madeira River basin from the Andes of Peru and Bolivia to the Amazon River in Brazil based on 10 years of data from the HYBAM monitoring programme *Journal of Hydrology* **553** 35–48 http://dx.doi.org/10.1016/j.jhydrol.2017.07.018
[3] Campos J and Peña J 1992 Climatologic study of the Andean catchment of Beni river. In: Minutes of PHICAB symposium – Hydrologic and Climatologic Program for the Amazon Basin in Bolivia, *ORSTOM-SENAMHI-IHH-CONAPHI*, pp 33–40
[4] Gautier E, Brunstein D, Vauchel P, Jouanneau J, Roulet M, Garcia C, Guyot J and Castro M 2010 Channel and floodplain sediment dynamics in a reach of the tropical meandering Rio Beni (Bolivian Amazonia) *Earth Surface Processes and Landforms* **35** 1838–53 https://doi.org/10.1002/esp.2065
[5] Schwendel A, Nicholas A, Aalto R, Sambrook G and Buckley S 2015 Interaction between meander dynamics and floodplain heterogeneity in a large tropical sand-bed river: The río Beni, Bolivian Amazon *Earth Surface Processes and Landforms* **40**(15) 2026–40
[6] LHUMSS 2017 Measurement campaigns for discharge and sediment transport for the Technical Study of Pre-Investment – EDTP of the Hydropower Project El Bala *Technical Report* Hydraulics Laboratory, Universidad Mayor de San Simón, Bolivia
[7] Roche M 1988 The applications of PHICAB Project for the development of Bolivia. In: Minutes of PHICAB symposium – Hydrologic and Climatologic Program for the Amazon Basin in Bolivia, *ORSTOM-SENAMHI-IHH-CONAPHI*, pp 77–88
[8] Heredia M 2021 Update and analysis of the sedimentology of Beni River – Studies of river morphology in the Amazon basin in Bolivia *Technical Report LH/MHG/01* Hydraulics Laboratory, Universidad Mayor de San Simón, Bolivia
[9] Deltares 2022 Delft3D-Flow, Simulation of multi-dimensional hydrodynamics flows and transport phenomena including sediments *User Manual* Deltares, Delft
[10] Lesser GR, Roelvink D, van Kester JA and Stelling G 2004 Development and validation of a three-dimensional morphological model *Coastal Engineering* **51**(8–9) 883–915
[11] Rijn LC van 1984a Sediment transport, Part I: bed load transport *Journal of Hydraulic Engineering* **110**(10) 1431–56
[12] Rijn L C van 1984b Sediment transport, Part II: suspended load transport *Journal of Hydraulic Engineering* **110**(11) 1613–40
[13] Maximova T, Smolders S and Vanlede J 2015 Model calibration against different types of velocity data with a dimensionless cost function: application to the Scaldis model of the Scheldt estuary *Proc. 22nd Telemac and Mascaret Users Club,* STFC Daresbury Laboratory, United Kingdom
[14] Nash J and Sutcliffe J 1970 River flow forecasting through conceptual models part I - A discussion of principles *Journal of Hydrology* **10** 282–290
[15] Yossef MF, Jagers HRA, Van Vuren S and Sieben J 2008 Innovative techniques in modelling large-scale river morphology. In *River flow 2008, Proc. Int. Conf. on Fluvial Hydraulics*, Cesme, Izmir, Turkey, pp 1065–74.

Influence of dam break scenarios on flood wave characteristics. Case study – Paltinu Dam, Romania

Daniela-Elena Gogoașe Nistoran[1,2], Cristina Sorana Ionescu[1] and Ștefan-Mugur Simionescu[1,2]

[1] Department of Hydraulics, Hydraulic Machinery and Environmental Engineering, Faculty of Energy Engineering, University POLITEHNICA of Bucharest, 060042, Romania
[2] Author to whom any correspondence should be addressed

daniela.nistoran@upb.ro, s.simionescu@upb.ro

Abstract. Among the various constructive types of dams, the arch dams breaches are the largest and the quickest to occur, leading to the most catastrophic flood waves. Therefore, prediction of the shape, magnitude and arrival time of the breach waves is of utmost importance for flood risk management. The study analyses in GIS environment the influence of different dam breach scenarios on the evolution in time and space of the flood wave. As a study case it was chosen the 108m in height, Paltinu arch-dam (Romania), built on Doftana River and its about 19km river reach down to the confluence with Prahova River. Several breach widths and depths are considered in the analysed scenarios, coupled with selected inflow hydrographs and initial reservoir water elevation, to assess their influence on the dam break wave characteristics and downstream flooded areas. To model the flood wave routing following the dam brake, HEC-RAS software is used. Inundation boundaries, characteristic depths and velocities, as well as arrival time in selected control sections are compared and conclusions are drawn.

1. Introduction

Nearly 800,000 dams were built all over the world for various purposes, such as hydropower, flood protection, irrigation, water supply, fishing, leisure, etc. [1]. Over 58,000 of them are registered as large by the ICOLD (International Commission of Large Dams) in the World Register of Dams [2], having a height of at least 15 m, or a height between 10 and 15m and an impoundment volume of more than 3 million m^3.

According to the type of building material and their shape, dams are traditionally classified in: embankment and rock filled dams (made of earth and/or rock filled, which account for more than 78% of all the dams in the world), gravity dams (made of concrete, stone or other masonry), arch and buttress dams (made of concrete) [3-5].

Unfortunately, the history of dam construction coexists with the probability of their failure, causing catastrophic events. About 2,000 dam failures occurred since the 12th century [6], of which 10% took place during the 20th century, causing the loss of at least 8,000 lives and material damage worth millions of dollars [6], [7].

When a dam breaks, catastrophic quantities of water are suddenly released through the breach into the downstream area. The produced wave leads to flash flooding characterized by a huge discharge of short duration [8]. These dam break waves transporting large volumes of water, sweep up everything in the downstream area – buildings, bridges, roads and railways etc. [6].

In Romania, the increasing need of water for various purposes led to the development of over 2000 dams on the main Romanian rivers. Of these, about 241 are listed in the ICOLD group of large dams [2]. Most of the dams in Romania are built between 1950 and 1990 and have already had an operation period of 30 to 70 years. Also, since many of them are built at less than 100km from Vrancea epicentral major seismic area, they must be permanently monitored for displacements. The only important dam break accident in Romania was at Belci dam on Tazlău River (in 1991). Other minor one was in 2003 at Cornățel dam, from the Vedea River watershed [9].

It is very important to determine through hydraulic numerical modelling for a certain accident scenario the downstream flooded areas, flow depth, wave velocity, flood intensity, peak time of propagation in order to minimize human losses and material damage [6], [10]. From 1D to 3D approximations, numerical models have evolved in time, whereas the graphical interface for data management and visualisation of results have been incorporated under GIS environment. Therefore, present paper analyses the consequences of Paltinu arch dam failure on Doftana River (Romania), using different assumptions. The objective of the study is to analyse the influence of different dam break scenarios on the breach wave thus triggered and its progression in the downstream area. HEC-RAS software version 6.2 [11] is used for the numerical simulations under the 2D simplifying assumption [12], [13]. Data are processed and results are obtained under GIS environment, which makes easier to process the flood maps [14].

Main causes of dam breaks are: errors in design for flood events such as overtopping (34% of cases) [8], foundation rock instability and bank failures (30% of cases), piping/seepage (internal and underneath the dam, 20%), spillway structures failure/malfunction (10%), landslides [8], [15], structural failure, upstream dam failure, operation or maintenance errors, earthquake and sabotage. The failure modes may be through: overtopping, piping/seepage, cracking, sliding and overturning [16].

Breach progression time depends on dam material and failure mode. For embankment dams the rupture occurs slowly (order of hours), whereas concrete dams break quicker (order of minutes to tens of minutes). Of all types, the arch dams break in the shortest time, producing sharp hydrographs which can lead to catastrophic human losses and material damages [17]. The best- known case is that of Malpasset arch dam in France, which failed in 1959 due to geological issues in the left abutment [18]. The dam broke almost completely by overturning. 421 lives were lost and the damage was estimated at 68 million US dollars.

This is why scientists developed physical models to analyse the failure mechanisms and breach shape of different types of dams, including arch dams under different simulation scenarios [19]. For narrow canyon arch dams being subjected to earthquakes the physical model tests [20] developed diagonal cracks parallel to the side slope abutments and horizontal cracks, which subsequently led to trapezoidal - shape breaches.

Estimation of dam breach location, shape and dimensions and development time are of utmost importance for the flood risk assessment studies [21]. These parameters are mainly influenced by the type and construction material of the dam. The water discharge through/over the breach is computed with the same equations as the flow through an orifice or over a weir, whose head and opening area are increasing in time after an assumed mathematical relation.

In case of overtopping, the estimated dimensional discharge coefficient, C, of the breach (weir) directly affects the magnitude of the peak outflow, but its exact value cannot be known. USACE [21] suggests for arch dams being overflown values of $C = \frac{2}{3} C_d \sqrt{2g}$ in the range 1.1 ÷ 1.8 (where C_d is the non-dimensional weir discharge coefficient). As well, the average breach width of arch dams is $(0.8 \div 1)L$ (where L is the top length of the dam at the deck), its slope being equal or lower than the valley wall slope, whereas the failure time is equal or less than 10 min. Many researchers developed peak outflow equations from data of historic dam failures as a function of breach/weir head, water volume in the reservoir, height of dam, etc [21].

2. Site and data

Doftana is a 51 km in length River flowing in the Carpathians of Curvature region, down to the confluence with Prahova River, where the mean annual flow rate of Doftana is of 4.8 m^3/s. Doftana watershed has a 419 km^2 area, with a difference in elevation of 1543m [22] (Fig. 1). The only gauging station along Doftana River is Teşila, where the mean annual flow measured between 1950-1998 was 4.59 m^3/s [22]. Along the Doftana River and some of its 17 small tributaries, 13 micro and one hydropower plants were developed. 7 of them are located upstream and 6 - downstream of the Paltinu arch dam, which was built of concrete between 1966-1971 in a narrow valley.

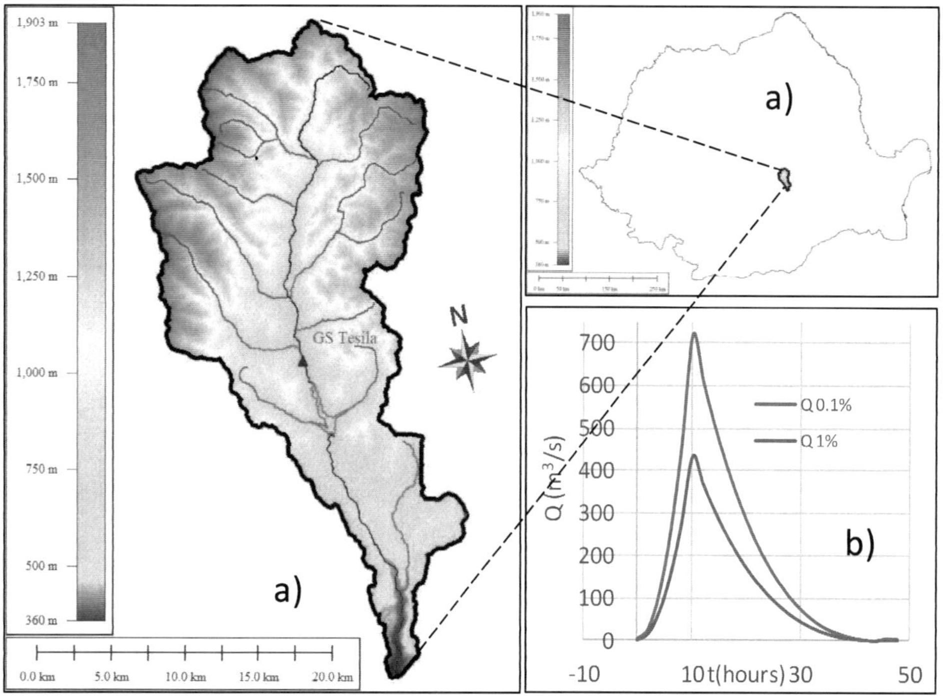

Figure 1: a) Location of Doftana watershed, Paltinu reservoir and Teşila Gauging Station; b) Inflow hydrographs for the1000-years (check) and 100-year (design) floods.

Paltinu arch dam is located at 18.4km upstream of the confluence with Prahova River. Its characteristic dimensions are a height of 108m, a length at deck of 460m and the widths at the deck and at the foundation of 6m and 22m, respectively. The reservoir length is 3.7km and its width 0.94km. In 2013 the reservoir had a volume of 49 million m^3, with a corresponding area of 182 ha at the Normal Operation Pool level (NOP), set at 649 m (m.a.s.l.). NOP is also the crest level of the spillway. The elevation-capacity curve is shown in Fig. 2.a.

The solution of an arch dam with double curvature with perimeter joint was chosen due to the geo-morphological conditions (faulted and brittle rock on the left wall slope, the bare rock unit consisting of sandstone with interlayered marl shales and clays, specific to a landslide-prone area). During the first reservoir filling, significant infiltrations (over 150l/s) at the foundation interface were observed together with small displacements of the structure. Therefore, in the next period additional consolidation work were carried out on the foundation and on the left abutment. Another issue of concern regarding Paltinu Dam stability is its proximity to the earthquake Vrancea epicentre (86km SW) with a 50-year return period of earthquakes of 8-degree intensity (MSK) and at 10-15km from the second in seismic activity epicentre area of Câmpulung – Făgăraş, with and intensity of 9 (MSK) and a return period of 100 years [23].

The project of the dam in the selected cross-section was carried out for a design flood having a maximum discharge of 435 m^3/s and corresponding to a 1% probability of exceedance and for a check flood having a maximum discharge of 718 m^3/s that corresponded to a 0.1% probability of exceendence, under natural regime flow. The main purposes of the reservoir are: water supply, hydropower, irrigations, flood control and recreation. The flood control storage of the reservoir is of 5 mil. m^3 for the 100-year flood and of 7.5 mil. m^3 for the 1000-year flood.

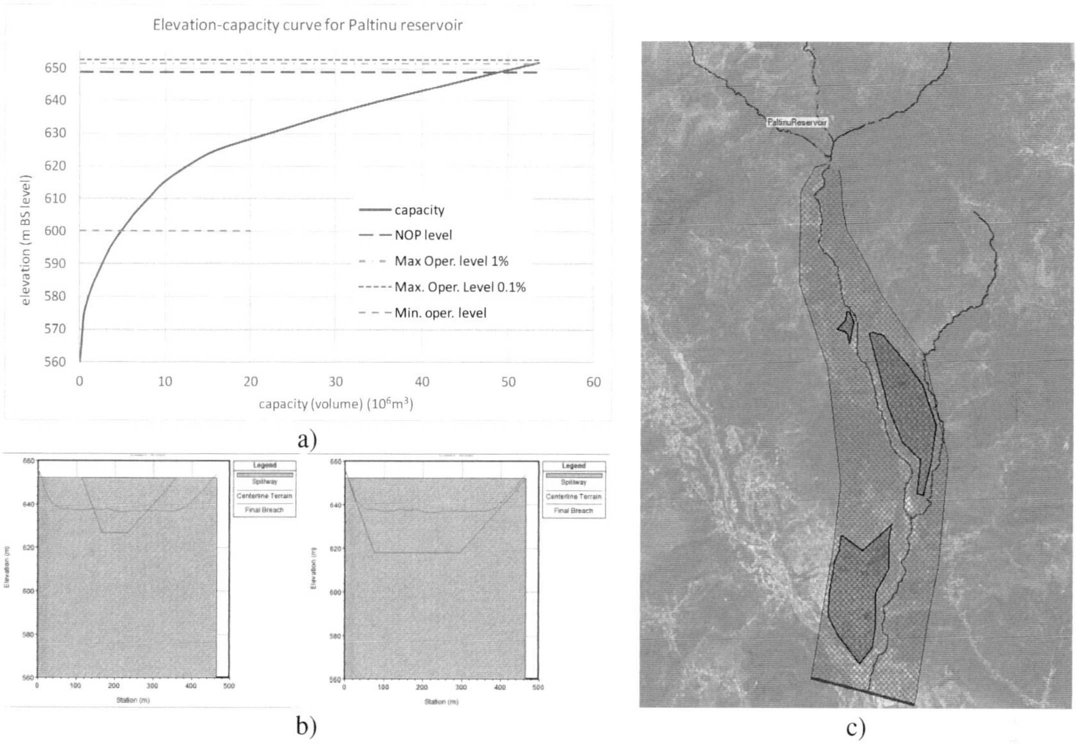

Figure 2: a) Elevation – capacity curve; b) Dam breach shape for scenarios S1 and S3; c) Numerical domain with reservoir storage area, dam, 2D computation mesh and different roughness for inhabited areas.

3. Methodology

3.1. Mathematical model

Due to the catastrophic consequences of a dam break, proper and timely estimation of the flood wave routing is of utmost importance in forestalling loss of life and material damages. Such events can hardly be accurately predicted, given their complex features with reference to how they evolve in time and space. Therefore, a mathematical model is needed to explore the way in which such flows progress downstream. To this end, one of the widely used models is the 2D shallow water equations (SWE), having as an essential hypothesis that the height scale of the flow is much smaller than its length scale. SWE are based on the fundamental conservation equations of mass and linear momentum and define the progression of water surface depth/ elevation as well as of the averaged velocities in x and y directions.

2D SWE can be solved using different numerical integration methods, the 2D approximation greatly reducing the numerical complexity, necessary data and computation time. The fully SWE equations (continuity and momentum) in which, for small scale analyses the Coriolis effect can be neglected are:

$$\begin{cases} \dfrac{\partial H}{\partial t}+\dfrac{\partial (hu)}{\partial x}+\dfrac{\partial (hv)}{\partial x}=q \\ \dfrac{\partial u}{\partial t}+u\dfrac{\partial u}{\partial x}+v\dfrac{\partial u}{\partial y}=-g\dfrac{\partial H}{\partial x}+v_t\left(\dfrac{\partial^2 u}{\partial x^2}+\dfrac{\partial^2 u}{\partial y^2}\right)-c_f u \\ \dfrac{\partial v}{\partial t}+u\dfrac{\partial v}{\partial x}+v\dfrac{\partial v}{\partial y}=-g\dfrac{\partial H}{\partial y}+v_t\left(\dfrac{\partial^2 v}{\partial x^2}+\dfrac{\partial^2 v}{\partial y^2}\right)-c_f v \end{cases} \quad (1)$$

In these equations, $t$is the time, $h(x,y,t)$ is the water depth, $z(x,y)$ is the bottom elevation, $H(x,y,t)=h(x,y\,,t)+z(x,y)$ is the water surface elevation, both with respect to the datum, u and v are the depth-averaged components of the velocity vector in the x and y Cartesian directions respectively, q is a source/sink flux term, c_f is the nonlinear bottom friction coefficient and v_t is the horizontal eddy viscosity coefficient.

3.2. Numerical model

Two types of data are needed for the study: spatial and hydrological. During the last decade, flood modelling is performed in GIS environment, so results can be integrated on maps or on digital terrain models. Therefore, a good accuracy of the top bathymetric data is extremely important.

For the spatial data, the free digital elevation model was downloaded from Copernicus Land Monitoring service [24], having a spatial resolution of 25m. Channel bathymetry was neglected, since the mean annual river discharge is very small. The reservoir has been modelled as a storage area with an elevation-capacity curve. The downstream 2D computation area has been delineated as the maximum possible flooded area along Doftana River, down to the confluence with Prahova River (Fig. 2b). The Manning roughness coefficient of the potentially flooded area has been approximated to an average value of 0.07 s/m$^{1/3}$ for the vegetated parts and of 0.11 s/m$^{1/3}$ for the inhabited areas, digitized on the map (Fig. 2c). The presence of other small dams, reservoirs or bridges, downstream of Paltinu was neglected. The reservoir tail was considered as the upstream boundary condition, and the confluence with Prahova River, the downstream one. Also, using its constructive features, the dam was introduced as a connection between the reservoir storage area and the downstream 2D computation zone.

The boundary conditions consist in: upstream - the inflow hydrographs at the tail of Paltinu reservoir (Fig. 1b) and a uniform flow condition along a cross section downstream (given by the terrain slope of about 1%), at the confluence of Doftana with Prahova River. The inflow hydrographs were considered to be those of the design and check floods, with the characteristics in Table 1. Other inflows from Doftana tributaries are neglected in the study, since the discharges are negligeable compared to those of the flood waves. The initial conditions are: the water elevation in the reservoir at the Normal Operation Pool Level and water elevation in the river valley below ground level.

Table 1: Characteristics of the Paltinu reservoir inflow hydrographs.

Case	$Q^{max}{}_{p\%}$ (m^3/s)	Probability of exceedance	T_{peak} time to peak (h)	T total duration (h)	compactness coefficient γ	Volume (mil. m^3)
1	718	0.1%	10	48	0.3	37.25
2	435	1%	10	48	0.3	22.55

The numerical method used by HEC-RAS to integrate the 2D SWE is finite volumes. For this study, the computation mesh for the potentially flooded area has mostly square 25x25m cells, with smaller polygons in the areas where break lines were inserted along the banks, bridges or small dams. A mesh with 78612 cells resulted, for which a small computation time of 1s was chosen to avoid instabilities specific to such rapid developing phenomena.

To route the inflow hydrographs through the reservoir, the level pool routing method is chosen, for which an elevation-capacity curve is used. The resulted dam-break hydrographs under different scenarios are routed along the downstream valley using the Shallow Water Equations, Eulerian-Lagrangian Method (SWE – ELM) module of HEC-RAS.

Numerical models are usually calibrated on experimental data. In the case of dam-breaks, such data can be obtained only in laboratory or from real-case accidents after their occurence. This is why software use benchmarking tests. Studies for scenarios of unhappened accidents cannot be calibrated. HEC-RAS verification, validation and benchmarking tests for 2D simulations [25], [26] include the case of Malpasset dam break and other flooding examples.

3.3. Scenarios

According to the previously mentioned physical modelling studies [18], [19], failure of Paltinu Dam was triggered by overtopping the deck of the dam (when the water reached the elevation of 652 m.a.s.l.). The breach progresses linearly to its maximum area in 6 minutes, and is considered to have a trapezoidal shape, and lateral slopes equal to the abutment ones: right slope 1:5, left slope 1:2 (Fig. 2b). Three breach scenarios of the dam were selected for simulations, based on recommendations of [21] and [27], having the characteristics presented in Table 2.

Table 2: Breach characteristics for the scenarios.

Scenario	Peak inflow at reservoir tail (m^3/s)	Initial level in reservoir	Breach width at base	Final breach elevation (m.a.s.l.)	Breach formation time (min)	Breach height
1	$Q_{0.1\%} = 718$	649	70	627	6	25
2	$Q_{0.1\%} = 718$	649	140	623	6	29
3	$Q_{0.1\%} = 718$	649	220	618	6	34
4	$Q_{0.1\%} = 435$	649	70	627	6	25
5	$Q_{0.1\%} = 435$	649	140	623	6	29
6	$Q_{0.1\%} = 435$	649	220	618	6	34

4. Results and discussion

The computation step was progressively diminished to avoid numerical instabilities. The results were outputted each 10 minutes. The main characteristics are presented in a table format (Table 3) as well as in graphical format (Fig. 3). Two cross-sections across the valley were considered as controls to investigate parameter evolution in space and time, one at 200 m from the dam and another at the confluence of Doftana with Prahova River.

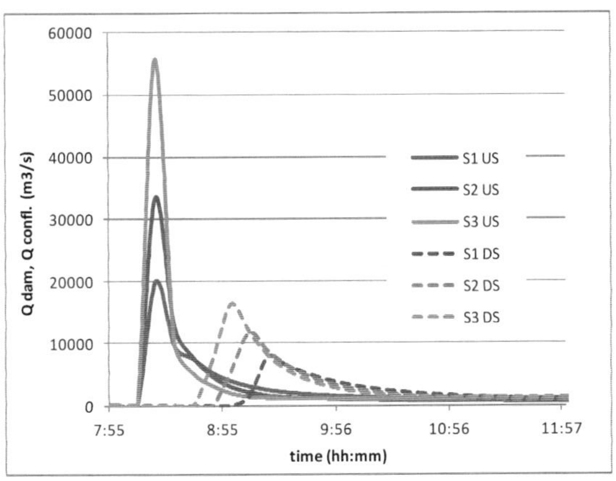

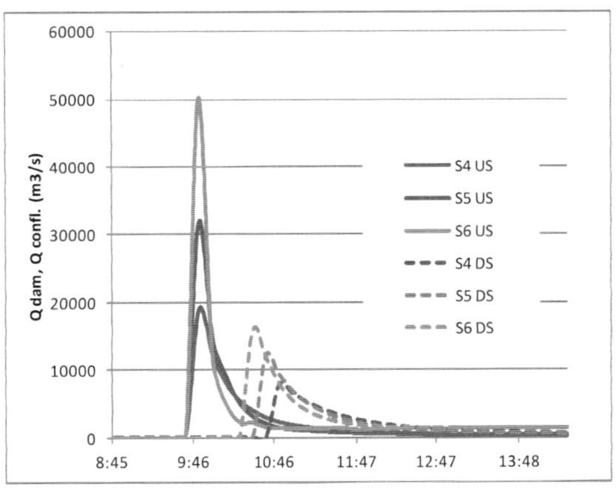

Figure 3: Flood attenuation: flow hydrographs at the dam and at the confluence with Prahova.

Table 3: Comparative results of the flood wave characteristics for the considered scenarios.

Characteristics / Scenario	Arrival time from dam break at the downstream confluence (min)	Peak Q and time at the dam / confluence, (m^3/s), (hh:mm)	Maximum width of the inundation boundary near confluence (m)	Peak flow attenuation $\frac{Q_{ds}^{max}}{Q_{us}^{max}}100$ (%)	Max. velocity *V*DS the dam (m/s)	Max. flood depth, *h* DS the dam (m)	Max. flood intensity V·h DS of the dam (m^2/s)
S1	60	19947 (8:20) / 7713 (9:20)	440	39	22.1	14.4	318
S2	50	33410 (8:20) / 11735 (9:10)	533	35	23.4	18.6	438
S3	40	55584(8:20) / / 16872 (9:00)	615	29	25.7	26.7	550
S4	60	18896 (9:50) / 7820 (10:50)	410	41	21	14	298
S5	50	31844 (9:50) / 12073 (10:40)	470	38	23	18	416
S6	40	50129(9:50) / 15781 (10:30)	520	31	25.6	26.2	540

Figures 4 and 5 show flood intensity for scenarios S1÷S6 and details of inundation boundaries for scenarios S3 and S4 under GIS environment.

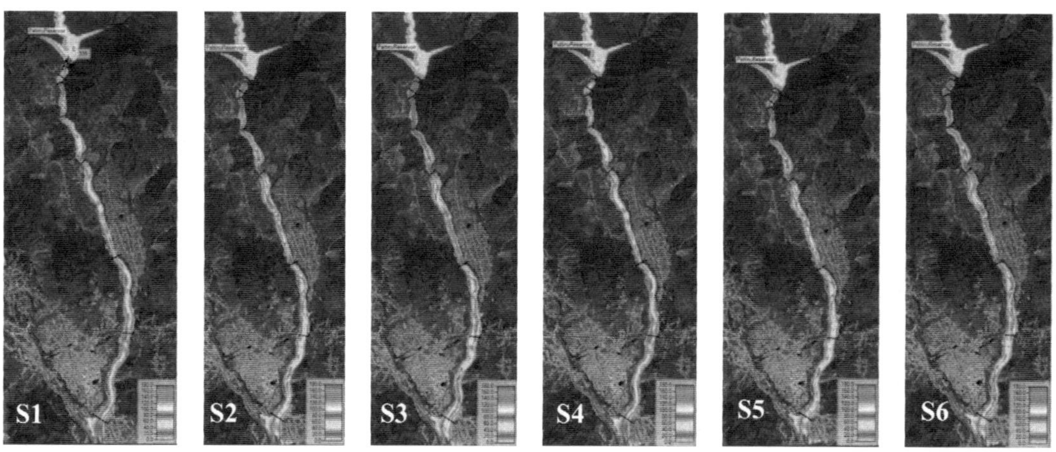

Figure 4: Maximum flood intensities ($V \cdot h$) for scenarios S1 – S6.

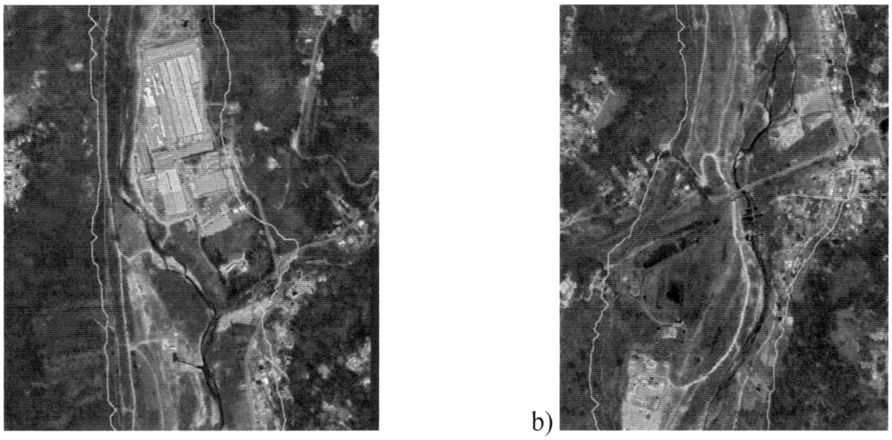

Figure 5: Details of inundation boundaries for scenarios S4 (red) and S3 (yellow) on the furniture factory Lemet from Câmpina (a) and Câmpina – Doftana bridge (b).

5. Conclusions

The present study represents an example of the influence of different dam breach scenarios and hydrologic events on the evolution in time and space of catastrophic flood waves released from a reservoir. As case study the Paltinu arch dam was selected, given its previous incidents and the short time arch dams break.

2D numerical simulations under GIS environment using HEC-RAS software along a 18.4 km reach of the Doftana River Valley, down to its confluence with Prahova River near Campina. Three breach scenarios were chosen for two extreme hydrologic events, the 100-year and 1000-year floods, considered as inputs at the reservoir tail. The following parameters of the computed dam break waves and inundation maps were analysed comparatively: maximum depth, velocity and flood intensity $(V \cdot h)$, peak discharge and arrival time in the control sections.

Values in Table 3 show that the peak inflows in the reservoir have little influence on the dam-break maximum discharges as long as the reservoir water level at the moment of failure and the breach characteristics are same. The dam breaks one hour and 30 minutes later for $Q_{1\%}$, than $Q_{0.1\%}$ case, for all three corresponding scenarios. This time gap is necessary for the reservoir to fill from NOP to deck level. As expected, the breach area has the greatest impact on the peak discharges at the dam. As a consequence, the arrival time of the dam-break wave at the downstream confluence with Prahova River, progressively reduces by 10 minutes, for each considered increase of the breach area.

The routed hydrographs downstream the dam at the confluence with Prahova River and presented in Fig. 3 show very similar shapes and close peak discharge values for the analogous breach scenarios in the two considered cases. This is due to the obviously resembling shape and magnitude of the dam-brake hydrographs entering the downstream 2D computational area. However, the peak flow attenuation (as defined in Table 3) is stronger for $Q_{0.1\%}$ case scenarios.

Values of flood intensity and velocity are so high, that qualify almost in all of the downstream flooded area for a total destruction ($V \cdot h > 7 m^2/s$ and $V > 1 m/s$). Exception might make the most lateral areas next to the inundation boundaries. Inhabitants of Seciuri and Lunca Mare villages will have to self-evacuate as quick as possible at the sound of sirens. Computed inundation boundaries can be overlapped onto maps or satellite images of inhabited areas (Fig. 5) to further develop vulnerability and risk maps, essential for flood risk management with the purpose to prevent loss of life and property damage. For more accurate results, Lidar topographic data and Corine land cover should be used [13, 28].

References

[1] International Rivers 2007 *Damming Statistics* https://archive.internationalrivers.org/damming-

statistics
[2] ICOLD – International Commision of Large Dams 2022 *World Register of Dams* https://www.icold-cigb.org/GB/world_register/database_presentation.asp
[3] Stematiu D and Ionescu Ş 1999 *Safety and Risk in Hydrotechnic Constructions* (in Romanian) (Bucharest, Romania: Editura Didactică şi Pedagogică)
[4] Deangeli C, Giani GP, Chiaia B and Fantilli AP 2009 Dam failures *WIT Transactions on State-of-the-Art in Science and Engineering* chapter 1 https://www.witpress.com/Secure/elibrary/papers/9781845641429/9781845641429001FU1.pdf?ref=patrick.net
[5] Tanchev L 2014 *Dams and Apurtnenant Hydraulic Structures* 2nd ed (CRC Taylor and Francis)
[6] Singh V P 1996 *Dam Breach Modeling Technology* (Dordrecht, The Netherlands: Kluwer Academic Publishers)
[7] Luino F, Tossatti G and Bonaria V 2014 Dam failures in the 20th century: Nearly 1000 avoidable victims in Italy alone *Journal of Environmental Science and Engineering*, **3**(1) 19–31 https://hdl.handle.net/11380/945091
[8] Adamo N, Al-Ansari N, Sissakian V, Laue J and Knutsson S 2020 Dam saftey and overtopping *Journal of Earth Sciences and Geotechnical Engineering*, **10**(6) 41–78
[9] Filotti A 2015 Human attitudes related to the dangers induced by extreme hydrologic events (in Romanian) *Noema* **XIV** http://noema.crifst.ro/ARHIVA/2015_3_01.pdf
[10] Nistoran Gogoaşe DE, Popovici DA, Savin B and Armaş I 2016 GIS for Dam-break flooding: Study area: Bicaz-Izvorul Muntelui (Romania) *Space and Time Visualization* ed Boştenaru DM and Crăciun C (Springer) pp 253–283
[11] U.S. Army Corps of Engineers 2022, *HEC-RAS Documentation* (Hydrologic Engineering Center) https://www.hec.usace.army.mil/confluence/rasdocs
[12] Kumar S, Jaswal A, Pandey A and Sharma N 2017 Literature review of dam break studies and inundation mapping using hydraulic models and GIS *International Research Journal of Engineering and Technology* **4**(5) 55–61
[13] Albu LM, Enea A, Iosub M and Breabăn IG 2020 Dam Breach Size Comparison for Flood Simulations. A HEC-RAS Based, GIS Approach for Dracs, ani Lake, Sitna River, Romania, *Water*, **12**, 1090; doi:10.3390/w12041090;
[14] Albano R, Mancusi L, Adamowski J, Cantisani A and Sole A 2019 A GIS Tool for Mapping Dam Break Flood Hazards in Italy, *International Journal of Geoinformation*, **8**, 250, 1-23;
[15] Adamo N, Al-Ansari N, Sissakian V, Laue J, Knutsson S 2020 Dam saftey: General considerations, *Journal of Earth Sciences and Geotechnical Engineering*, Vol.**10**, No.6, 1-20;
[16] Fiedler W 2016 Managing Dam Safety Risks Related to Hydraulic Structures, *Hydraulic Structures and Water System Management*, edts. B. Crookston & B. Tullis, 6th IAHR International Symposium on Hydraulic Structures, Portland, OR, 27-30 June, doi:10.15142/T3720628160853 (ISBN 978-1-884575-75-4);
[17] USBR, USACE 2019 Risk analysis for concret arch dams (Chapter E-4), *Best practices in Dam and Levee safety Risk Analysis*;
[18] Aureli F, Maranzoni, A and Petaccia G 2021 Review of Historical Dam-Break Events and Laboratory Tests on Real Topography for the Validation of Numerical Models, *Water*, **13**, https://doi.org/10.3390/w13141968;
[19] Morris MW, Hassan MAAM and Vaskinn KA 2007 Breach formation: field test and laboratory experiments, *Journal of Hydraulic Research*, **45**: S1, 9-17;
[20] USBR2002 Report, *Investigation of the Failure Modes of Concrete Dams -Physical Model Tests Dam Safety Office*, 2002;
[21] USACE 2014 HEC Engineering Center, *Using HEC-RAS for Dam break studies*, TD-39;
[22] Armas I 1999 *Doftana River Watershed* (in Romanian *Bazinul hidrografic Doftana*), Editura Enciclopedică;
[23] NWARW – National Waters Administration "Romanian Waters" 2012 *Project for special*

surveillance of Paltinu Dam (in Romanian: Proiect special de urmărire a barajului Paltinu)

[24] EU-DEM2016, version 1.1, European Environmental Agency, *EU Copernicus programme* https://land.copernicus.eu/imagery-in-situ/eu-dem/eu-dem-v1.1?tab=metadata;

[25] USACE 2018 Hydraulic Engineering Center, *HEC-RAS Verification and validation tests* ,https://www.hec.usace.army.mil/software/hec-ras/documentation/RD-52_HEC-RAS%20Verification%20and%20Validation.pdf;

[26] USACE 2018 Hydraulic Engineering Center, *HEC-RAS, Benchmarking of the HEC-RAS 2D hydraulic modeling capabilities*, https://www.hec.usace.army.mil/software/hec-ras/documentation/RD-52_HEC-RAS%20Verification%20and%20Validation.pdf;

[27] NRDWR Departament of Natural Resources Division of Water Resources 2010 *Guidelines for dam breach analysis*, February 10.

[28] Urzică A, Mihu-Pintilie A, Stoleriu CC, Câmpianu CI, Huțanu E, Pricop CI and Grozavu A 2021 Using 2D HEC-RAS modeling and embankment dam break scenario for assessing the flood control capacity of a multi-Reservoir System (NE Romania) *Water* **13** 57.

IOP Conf. Series: Earth and Environmental Science 1136 (2023) 012032 doi:10.1088/1755-1315/1136/1/012032

Methodology for determining hydromorphological indicators for rivers with hydropower uses

Stefan Ciprian Stanca[1], Alexandru Nicolae Dimache[1], Cornel Ilinca[1] and Cristian Gabriel Anghel[1]

[1]Technical University of Civil Engineering Bucharest, 124 Lacul Tei, Bucharest 020396, Romania

alexandru.dimache@utcb.ro

Abstract. The ecological condition of rivers is influenced by hydropower uses. The European Union encouraged the hydropower sector with installed capacities less than 10 MW, considering that their impact is small. Small hydropower plants influence the main hydromorphological indicators, namely the hydrological regime, the longitudinal connectivity of the river and the morphological conditions. This article proposes a methodology in which hydromorphological indicators take into account the dynamics of the phenomena and processes involved, due to the variation over time of the causes and effects. The methodology proposes a score based on linear interpolation, completing the threshold values, thus being able to differentiate the negative impact on rivers. The article proposes also a series of measures to reduce the negative impact of small hydropower plants.

1. Introduction

A main problem of the classifications and evaluation methods of the hydromorphology of rivers in Romania is represented by the differences from the categories, characteristics and attributes of the European standard EN 14614:2004 [1,2]. This issue is important because the Water Framework Directive requires European states to assess and classify the ecological status of rivers. Different national methodologies lead to significant differences in the classification of rivers into five classes of ecological status. Another problem is related to the subjectivism of giving the score corresponding to a class of ecological status of a river.

The methodology proposed in this article is in line with the European standard EN 14614: 2004. Thus, the categories, characteristics (indicators) and attributes related to river flows, respectively the geometry of the bankfull channel (table 1), which are the basis of the methodology, are well defined, representing a proposal that can be perfected in the future. Romania, but also other countries, had hydrotechnical works contested by environmentalists. Also, the hydromorphological assessment methodologies were an impediment in the unitary analysis of the impact of hydrotechnical works at national level. The impossibility of comparing heavily modified water bodies, at European level, is an evasion of European requirements, as EU countries have different methodologies that propose different classifications at the limit.

2. The proposed methodology for rivers in Romania

The European Commission's approach to the implementation of the Water Framework Directive (2000/60/EC) has a general approach and leads only to generic recommendations. This leads to delays at national level in the implementation of hydromorphological assessments, at required European standards.

The proposals of methodologies for determining the hydromorphological indicators in Romania have deficiencies justified by the impossibility of quantifying some attributes. In reality, using modern technologies and imposing the obligation of hydromorphology studies on water users (municipal water companies), for modified rivers, the deficiencies can be overcome. This makes it possible to assess short sectors of the river, making it possible for the relevant authorities to verify and concatenate the data obtained by water users. The article highlights the guidelines for the methodology for determining the hydromorphological indicators for rivers. The categories, characteristics and assessment attributes, which include a standard hydromorphological assessment of rivers according to the European standard EN 14614:2004 [1], adapted to the methodology proposed in this article are presented in the table 1.

Table 1. Categories, characteristics and attributes (based on EN 14614:2004 [1]).

Categories (Hydromorphological elements)	Characteristics (indicators)	Attributes
A. River Channel		
1.Geometry of river channels	1.1. Plan	1.1.1. Classification of river channels
	1.2. Longitudinal	1.2.1 Lengths and slopes on characteristic river sectors 1.2.2. Deviation from the state corresponding to the state of geomorphological equilibrium.
	1.3. Cross-section	1.3.1. Specific features 1.3.2. Hydraulic geometry 1.3.3. The evolution of the section
2. Substrate (river bed)	2.1. Granulometry	2.1. Riverbed 2.2. Evolution based on sediment balance on river sectors
3. River channel vegetation	3.1. Vegetation zoning	3.1.1. Depending on the hydrological regimes
4. Erosions/deposits in river channel	4.1. Erosive power	4.1.1. Effects of the erosion process on horizontal plane (erosions and deposits) 4.1.2. Effects of the erosion process on longitudinal profile (pits, sandbanks,)
5. Flow	5.1. Hydrological regime	5.1.1. Natural regime 5.1.2. Modified regime
6. Longitudinal continuity (river fragmentation)	6.1. Natural fragmentation 6.2. Artificial fragmentation	6.1.1. Waterfalls and cataracts 6.2.1. Cross-section hydrotechnical works
B. Shore and riparian zone		
7. Banks	7.1. Natural structure of the banks 7.2. Bank consolidations and protections	7.1.1. Granulometry and pedological profile 7.2.1. Types of works and affected areas
8. Bank vegetation	8.1. Vegetation zoning in cross section 8.2. Land use	8.1.1. Characteristic vegetation 8.2.1. Affecting the characteristic vegetation (density, agricultural crops, invasive vegetation, urban development)

C. Floodplain		
9. Floodplain land	9.1. Land use	9.1.1. Afforestation, agricultural land, urban development, etc.
	9.2. Moist terrain	9.2.1. Wetlands, meanders (active, inactive), irrigation canals, sand and gravel exploitation areas, etc.
10. Lateral connectivity (continuity of floodplain)	10.1. The degree of connectivity and lateral mobility given by longitudinal works	10.1.1. Dikes, etc 10.1.2. Permanent and non-permanent accumulations for flood mitigation
	10.2. The degree of connectivity and lateral mobility given by the flood attenuation	

Given the particularities of each watercourse, a detailed multicriteria analysis of all morphological and hydromorphological elements in the assessment, classification and determination of the degree of deviation from the natural conditions of the watercourse is required. The criteria for evaluating the hydromorphological elements that need to be analyzed are detailed below. The classification of river channel (1.1.1.) refers to the plane analysis of the river channel and its changes through natural (lateral erosion and sediment transport) and anthropogenic (hydrotechnical works, dikes etc.) processes. Also, the next three indices must be define: sinuosity index (Si) - the ratio between the modified and natural convolution coefficient for reference lengths depending on altitude [3], the braiding index (Bi) – number of meanders (active channels) separated from the river by deposits of alluvial materials and anabranching index (Ai) – number of meanders (active channels) separated from the river by vegetated islands [18]. The lengths and slopes on characteristic river sectors (1.2.1) are chosen depending on the altitude but also on the type of ecosystem. The criterion used may be the ratio between the global Taylor-Schwarz slope in the modified regime and that in the natural regime. The deviation from the state of geomorphological equilibrium (1.2.2.) can be performed based on the Ivanov index [16].

For characterizing the cross section (1.3.) we can use: the degree of incision [9], Rosgen entrenchment ratio [4,8], the cross-sectional shape coefficient defined by Lane (1935) [19], GINI coefficient [19,20], Knighton symmetry coefficients [19,21], the geometric method Wolman and Riley [5], and the hydraulic geometry defined by Leopold and Maddock [6]. The latter has a special importance because it establishes and analyzes the variation in time of the 3 main hydromorphological indicators (width W, depth D and velocity V) in the form of power-type ratios depending on the flow Q. These are expressed with the following relations [6,19]:

$$W = a \cdot Q^b \tag{1}$$

where W is the mean width and Q is the flow

$$D = c \cdot Q^d \tag{2}$$

where D is the mean depth and Q is the flow

$$V = e \cdot Q^f \tag{3}$$

where V is the mean velocity and Q is the flow, and the exponents (b, d, f) and parameters (a, c, e) must fulfil the following conditions:

$$a \cdot c \cdot e = 1 \tag{4}$$

$$b + d + f = 1 \tag{5}$$

Based on equations (1), (2), (3), the equation of continuity (equilibrium) is defined in which any change of one of the hydromorphological indicators leads to the recalibration of the rest (also known as "Lane's balance"), [19].

$$Q = W \cdot D \cdot V \tag{6}$$

It is generally recommended to carry out a Rosgen-type classification, which classifies rivers by taking into account all their characteristic elements and parameters (plane, longitudinal and cross-section), [12,13]. The rivers can be classified on four levels and nine typologies depending on the geomorphological criteria, the hydraulic geometry, the hydrological criteria and the direct on-site observations. Table 2 shows the classification in the nine typologies of rivers corresponding to Level I.

Table 2. Classification into nine typologies of rivers.

Criteria	Typologies								
	A_{a+}	A	B	C	D	DA	E	F	G
Entrenchment [R_s]	< 1.4	< 1.4	1.4 ÷ 2.2	> 2.2	-	< 4.0	< 2.2	< 1.4	> 1.4
Geometric ratio [B/H]	< 12	< 12	> 12	> 12	> 40	< 40	< 12	> 12	< 12
Convolution [C_s]	1.0 ÷ 1.1	1.0 ÷ 1.2	> 1.2	> 1.2	-	variable	> 1.5	> 1.2	> 1.2
Hydraulic gradient [i]	> 0.1	0.04 ÷ 0.099	0.02 ÷ 0.039	< 0.02	< 0.04	< 0.005	< 0.02	< 0.02	0.02 ÷ 0.039

where:

- Aa+ – typology characteristic of mountain streams, with high altitudes, very deep valleys, constrained, slightly meandering and with the presence of waterfalls;
- A – deep valleys, constrained streams, transport of sediment, erosion and deposition phenomena, slightly meandering;
- B – moderately constrained rivers, with moderate slopes, geomorphologically stable, slightly meandering;
- C – moderately constrained rivers, with low hydraulic gradients, the presence of inflection points, moderate B/H ratio;
- D – rivers with alluvial material deposits, bank erosion, intertwined channels, open valley;
- DA – anastomosed channels, with open valleys, abundant vegetation in the floodplain, relatively stable banks, low alluvial material deposits, strong washes;
- E – very meandering rivers with low hydraulic gradients, low B/H ratio, wide open valley;
- F – constrained meandering rivers, high B/W ratio, very unstable in plane with pronounced lateral erosion processes, with deep areas in the riverbed;
- G – constrained courses, with a morphology characterized by different altitudes of the riverbed, with unstable sections and erosions.

Level I consists in classifying watercourses according to geomorphological elements such as the entrenchment ratio, the ratio between the width of the bankfull channel and its average depth, convolution and slope. Level II consists of a subdivision of the typologies into six categories that vary and depend on the riverbed material: (1) bedrock, (2) boulders with a diameter greater than 250 mm, (3) stone with a diameter between 65- 250 mm, (4) gravel with a diameter between 4-65 mm, (5) sand, (6) mud and clays. If the distribution of the particle size is uneven, it is recommended to take into account the dominant diameter and not the average diameter D_{50}. Level III consists of an analysis and description of the river and its stability that also depends on the hydrological, biological, ecological, anthropogenic elements. Level IV is a stage of verifying the criteria of previous levels, through direct observations and measurements (sediment transport, erosion processes, degradation etc.), as well as

IOP Conf. Series: Earth and Environmental Science 1136 (2023) 012032 doi:10.1088/1755-1315/1136/1/012032

identifying the natural indicators of each phenomenon (presence of vegetation, ecological indicators, hydraulic geometry, etc.).

Substrate (river bed) category involves the analysis of the particle size composition of the riverbed, as well as the evolution based on the balance of sediments in river sectors, by determining the coefficient of mobility of sediments [7]. Since any change of the riverbed characteristics is based on changes in flow velocity, slope and hydraulic geometry, for this category is necessary to define and monitored in time the next five indicators: global Reynolds number (depending of mean velocity in section, hydraulic radius and kinematic viscosity of water), the local Reynolds number (depending of mean velocity in section, characteristic diameter of sediment particles and kinematic viscosity of water), the global Froude number (depending of mean velocity in section, hydraulic mean depth and gravitational acceleration), the local Froude number (depending of mean velocity in section, characteristic diameter of sediment particles and gravitational acceleration), bed shear stress (defined by the relationship between specific weight of water, hydraulic radius and water slope), [4,8,18].

The vegetation category of the river channel requires the analysis from the point of view of the zoning of the vegetation correlated with the hydrological regimes, a usual method being the Woodyer biological method [5]. The biological methods consist in identifying the limit of vegetation development, knowing that it cannot develop due to the continuous erosion of the superficial root-fixing material (substrate) during the transit of flows.

Erosions / deposits in the riverbed category consists in the analysis of the erosive power and the effects of the erosion process in plane (erosions and deposits), respectively the effects of the erosion process in longitudinal profile (sandbanks, pits). An attribute can be the volumetric ratio of sediments in the modified regime to the natural regime. Related to cross section analyses, all the hydraulic calculations based on empirical formulas, must be according with a sensitivity study of the roughness coefficient, using the values presented by Chow [22] or using other like Jarnett formula [19], Thorne and Zevenbergen formula [23], Bathurst equations [24], Ugarte and Madrid-Aris formula [25], Comiti et al. [26]. Flow category can be characterized very well for the natural regime by the multiannual average flow, the multiannual average monthly flows and the duration curve of the daily average flows. The modified hydrological regime requires the analysis of characteristic volumes of determined periods for the flows expressed as constant values and the curve of the normalized average daily flows (based on the flow differences). Analysis and variation of hydromorphological indicators with reference only to the multiannual average flow is not recommended because the value of the multiannual average flow usually corresponds to a probability of 30-40% of the duration curve of daily average flows, a value that loses its relevance in these types of analyses. Another problem is hydropeaking (washing of the riverbed), because of the pulsating regime of hydropower facilities with compensating basin or reservoir.

The connectivity of the surface water body with the ground water body in both natural and modified regime can be done using duration curves of hydrostatic levels in boreholes in reference sections or by using the average monthly hydrostatic levels, or using the duration curve of monthly average flow.

A serious problem, in the case of rivers with hydropower facilities, especially those with micro-hydropower plants, was the conditioning of the use of hydropower potential, of at least 70% and the granting of a higher technical score. This has led to an excessive increase in installed flows with the deterioration of the ecological status of water bodies.

In recent years, important steps have been taken in finding optimal solutions for improving the ecological status of rivers, among the most important being the appearance of G.D. no. 148 / 20.08.2020 for determining the ecological flow, flow with a variability over time which takes into account the hydrological regimes with the specific features of each watercourse. The ecological flow is characterized by three flow values for the period of large, medium and low flows, having values of approximately 35%, 25% and 15% from the average multiannual flow, resulting from a laborious calculation. Due to the disproportionate costs, most hydropower facilities cannot implement it [15].

Hydroelectric power plants with a capacity of more than 10 MW have been put into operation with very low salubrious/cleaning flows, i.e., 2-5% of the average multiannual flow. Micro-hydropower plants that have or have benefited from green certificates have salubrious/cleaning flows between 15-25% of the average multiannual flow. Thus, the impact for the flow category is very high for most hydropower plants. Hydroelectric power plants with storage reservoir have a strong effect because they lead to low flows and washing (purging) of the river banks, downstream of the hydroelectric power plant outlet. Small hydropower plants (SHPP) with compensating basins have a similar but smaller effect. Thus, the EU's decision to grant green energy aid schemes to new micro-hydropower plants with a capacity of less than 10 MW was right, as they have no compensating basins and have salubrious/cleaning flows close to the ecological flow values. For flows greater than 1.5-2 times the average multiannual flow, SHPPs discharge water, and for flows close to the salubrious/cleaning flow, in reality they operate only after exceeding that flow, to which a flow lower than the minimum machinable flow is added (approximately 5% of the installed flow) meaning 7-10% of the average multiannual flow; so the downstream flow of catchments has a variable dynamic over time compared to the flows downstream of water reservoirs.

Longitudinal continuity category requires an approach related to the existence of natural barriers (waterfalls etc). It is mandatory to map, describe and analyse the impact of all natural and artificial barriers on the watercourse. Hydraulic modelling calculations on ichthyofauna mobility works, where artificial barriers are provided with such constructions, are required to determine their efficiency for the purpose for which they were built, as well as to establish guidelines for their optimization/reconstruction. There are situations in which water and environmental authorities have imposed fish ladders on heavily fragmented areas naturally.

Shores category corresponding to the shore and riparian zone involve a comparison of the particle size curve and the pedological profile of the natural state with the modified state. In terms of shore reinforcements and their protections, a distinction must be made between those that fit into the landscape or even simulate natural conditions (fascines, wooden piles) and those that are made of concrete. A good score should be given to the protections that allow the growth of specific vegetation, to discourage the construction of bank defences that turn the river into artificial water channels.

For Shore vegetation category (8.1.1) corresponding to the shore and riparian zone, the Woodyer biological method can be used with a similar approach to category 3.

Floodplains category has two characteristics, land use and the type of wetlands exposed to flooding. An indicator can be based on the model FEM (Floodplain Evaluation Matrix).

3. Case study – Capra watercourse sector, Arges County, Romania

Capra River (cadastral code X.1.001) is located in Arges County, Romania, being part of the river basin of Arges River (cadastral code X.1).

Existing topographic measurements and especially the Digital Terrain Model were used to analyse the hydromorphological indicators. A two-dimensional flow modelling was made, using the Hec-Ras 6.2 program, open source. The hydraulic modelling method was chosen based on a preliminary analysis, using a series of selection criteria, namely:

The geometric characteristics of the riverbed, namely the width of the river, respectively the width of the bankfull channel, [10], land configuration data and plan route of the river. The quality of the hydraulic modelling results was dependent on the quality of the input data and the application of the methodology. To perform the hydraulic modelling, the following input data were used: digital terrain model in flt. format with 0.5 m grid, converted to hdf. format, surveys of hydrotechnical constructions on the river, land use map.

The following important work steps have been completed: For the 2D modelled area, cross sections were generated perpendicular to the flow direction of the river, in characteristic areas, so that they overlap from the point of view of the location with the transversal profiles from the topographic elevations, in order to check the DTM. Figure 1 shows the digital model of the land and some characteristic cross sections.

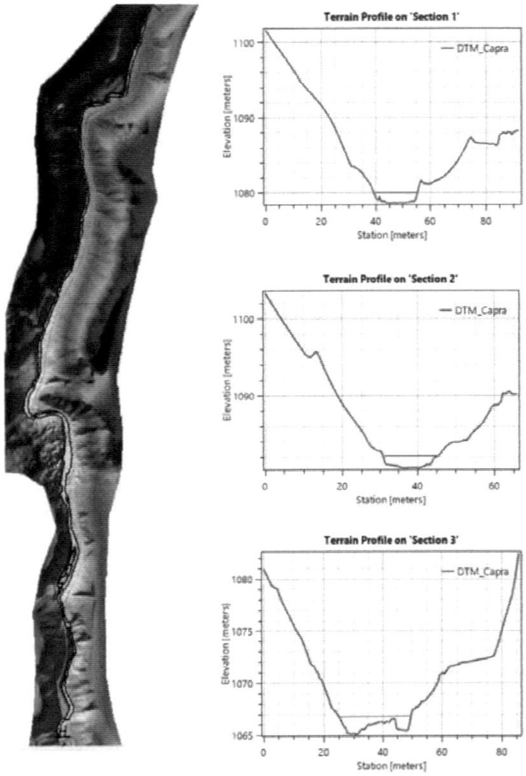

Figure 1. Digital terrain model (DTM) and characteristic cross sections.

The field analysis consisted in establishing the values of the roughness coefficients. The roughness coefficients taken into account in the hydraulic modelling were chosen based on the specialized literature, having values between 0.055 and 0.07, in correlation with Corine LandCover.

Regarding the delimitation of the analysed area and the realization of the calculation grid, it was done by dividing the area into cells whose dimensions were established according to the particularities of the area. The dimensions of the cells were set at 0.5 m, with a reduction in the area delimited by the bankfull channel. Based on this limit, all the files necessary for the two-dimensional modelling were made, namely: data on land elevation files, roads and access roads to objectives data, data on existing hydrotechnical constructions, data on roughness areas, etc.). 2D vectorization of the limits of the existing works was made based on the existing cartographic data (orthophoto plan etc.) or from the topographic situation plans.

The model was calibrated and the hydraulic parameters were adjusted (roughness, etc.). All the results of this calibration simulation were used as initial conditions for the hydraulic simulations for the probabilistic events considered. As upstream limit condition, the values of the flow's characteristic of these types of simulations were used, namely the average multiannual flow, the bankfull flow. The downstream condition is a normal flow-depth relationship on the last river sector. The calibration was done taking into account the results obtained from hydrometric campaigns by complying with the provisions of the international standard ISO748.

Based on the hydrometric measurement sheets, calculation summaries were prepared for each measurement, highlighting the sum of the mean square deviations of the errors, according to the international literature on the methodology for performing these types of measurements, the absolute errors being generally between 20 and 25% of the measured value.

Following these simulations, for the different hydrological regimes considered, the characteristic values of the hydromorphological indicators on the analysed river sector resulted.

4. Results and discussions

The main category that affects the ecological status of rivers by hydropower facilities, is Flow. In addition, for the facilities with storage reservoirs, the longitudinal connectivity, the substrate and the lateral connectivity are also important categories.

This article proposes a methodology for category 5 (Flow), more objective and relevant than that used in some studies in Romania [14]. In the methodology currently used, the average flow consumed and the average flow captured are not an indicator that fully reflects the ecological status of rivers. In the article is proposed to replace them with indicators derived from the average daily flow duration curve (CDZ-DCF).

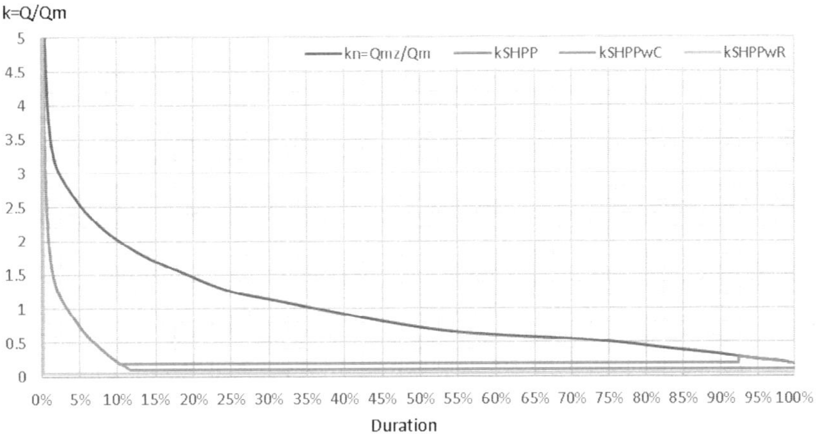

Figure 2. Duration curve of average daily flows for natural, SHPP without compensating basins, SHPP with compensating basins and HPP with a installed power greater than 10MW.

The volumes concerned are those from the period of deficient waters, the period of medium waters and the period of high waters estimated from the duration curve of the average daily flows. Thus, it is considered for Romania, regardless of the altitude level, that the period of high waters is generally 10% of the duration of an average hydrological year, the period of average waters is from 10 to 90% on the duration curve of flows, and the period of low waters is from 9 to 100% generally corresponding to flows lower than 2 - 30% of the average multiannual flow.

For example, on the Capra River, this means that the volumes in the natural regime on the river for an average hydrological year have a share of the average multiannual stock of about 28.5% during high water, 69% during medium water and only 2.5% during low water periods.

For a SHPP (small hydropower plant) without compensating basins it is estimated that downstream of the intake, volumes are 11% during high water, 16% during medium water and 2.3% during low water. For a SHPP with compensating basins and their typical salubrious/cleaning flow, it is estimated that downstream of the intake volumes are 11% during high water, 8% during medium and high water 1% during low waters. For an HPP (hydropower plant, with a installed power greater than 10MW) with storage lake with multi-year regularization and their typical salubrious flow (minimum flow downstream), it is estimated that the downstream catchment volumes are of 2.5% during high water, 4% during water average and 0.5% during water deficits.

It is noted that the use of indicators for these volumes defines much better the impact of the river by various types of hydrotechnical works, which would otherwise have obtained a very similar score.

The division into classes could be done by giving the score a weight of 15% for the period of high waters, 40% for the period of medium waters and 45% for the period of low waters, compared to the classification currently used which proposes 80% for the average flow consumed and 20% for maximum captured flow. The classification in classes is recommended for each of the three indicators, corresponding to the volumes of high, medium and low waters.

Table 3. Recommended classes.

Indicator	Class I	Class II	Class III	Class IV	Class V
High waters	>70%	70%-50%	50%-30%	30%-10%	<10%
Medium waters	>70%	70%-50%	50%-30%	30%-10%	<10%
Low waters	>70%	70%-50%	50%-30%	30%-10%	<10%

For an SHPP without compensating basin, it is estimated to be class III for high waters (11% / 28.5% = 38.6%), class IV for medium waters (16% / 69 % = 23.2%), class I for the period of low waters (2.3% / 2.5% = 92%). For an SHPP with compensating basin and their typical salubrious flow, it is estimated that it is class III for high waters (11% / 28.5% = 38.6%), class IV- a for medium waters (8% / 69% = 11.6%) and class III for the period of low waters (1% / 2.5% = 40%). For an HPP (hydropower plant, with a installed power greater than 10MW) with reservoir with multiannual regularization and their typical salubrious flow, it is estimated that it is class V for high waters (2.5% / 28.5% = 8.8%), class V for medium waters (4% / 69% = 5.8%), class IV for the period of low waters (0.5% / 2.5% = 20%).

In terms of longitudinal connectivity, natural fragmentation must be taken into account. An interesting case is a study submitted to the European Commission for infringing SHPPs, which makes a unfounded claim that fish ladders are not functional. In fact, the fish ladders provide longitudinal connectivity for existing species with one exception to Capra VI where there is fragmentation upstream and downstream at the tail of Vidraru Lake, which requires an ichthyofauna study to state pertinently whether a passage for Cottus Gobio (European bullhead) is needed. In fact, that study also states the need for passages that are not conducive to the area. In fact, the "National Administration Romanian Waters" has much bigger thresholds on that river and are not recorded. The existing methodology is non-unitary with non-intuitive scores and does not have rigorous criteria, just unfounded estimates.

It is noted that by amending the Flow Category (Criterion in the methodologies used at national level) it is necessary to amend the morphological conditions, which are currently estimated only against the multiannual average flow, not considering that in general the water width and depth must be insured during low waters, needed for ecological conditions.

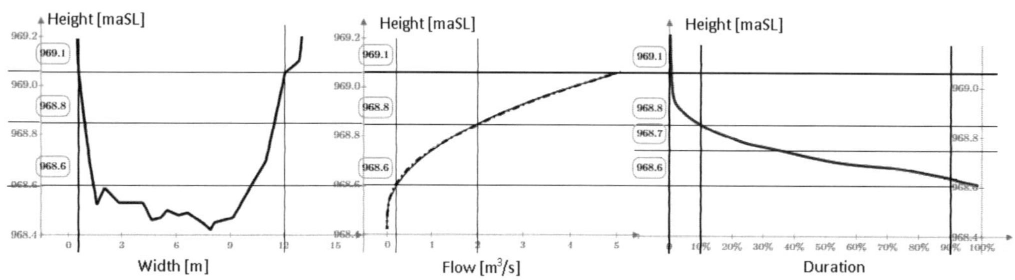

Figure 3. The relationship between water level and flow rate as a function of the duration curve of the average daily flow rates.

If the cross section presented in figure 3, it is observed that the width for high water and low water, respectively, does not differ significantly in the mountain area, and the depth of flow with a duration of 90% of duration flow curve is generally smaller but provides the necessary conditions for ichthyofauna. These are complied with by the newly built SHPPs but are completely ignored as a requirement by the authorities for large hydropower facilities.

5. Conclusions

The current Romanian methodology is in the process of improvement at national level and it is carried out by a small group of specialists from National Institute of Hydrology and Water Management. No consultation with the profile faculties was held, and public debates have generally been lacking lately due to pandemic restrictions.

The inclusion of more rigorous criteria in this methodology is necessary because improvements are required due to the implementation of measures to adapt existing hydrotechnical structures to the new environmental requirements. No significant progress has been made in Romania for rivers managed with old facilities, one reason being the impossibility of differentiating facilities by a score obtained with the current methodology. In order to propose new works and operating restrictions that are not cost-prohibitive but have positive environmental effects, a linear score within the ecological status classes is required.

References

[1] BS EN 14614:2004, BS 6068-5.36:2004 Water quality —Guidance standard for assessing the hydromorphological features of rivers

[2] Ilnicki P., Górecki K, Rzybowski M, Krzemińska A, Lewandowski P, Sojka M 2010 Ecological quality classes of river hydromorphology in Poland, *J. Water Land Dev.* **14**, 2010: 15–27

[3] Luna B. Leopold - A View of the River-Harvard University Press (1994), ISBN 0-674-93732-5

[4] United States Department of Agriculture, Natural Resources Conservation Service, Part 654 Stream Restoration Design National Engineering Handbook 2007 Rosgen Geomorphic Channel Design, directives.sc.egov.usda.gov

[5] Pawel Rowinski, Artur Radecki – Pawlik 2015 Rivers – *Physical, fluvial and environmental Processes, GeoPlanet: Earth and Planetary Sciences*

[6] Luna B. Leopold, M. Gordon Wolman, John P. Miller, Ellen Wohl 2020 *Fluvial processes in geomorphology*, Dover publications Inc., New York, ISBN 9780486845524

[7] Loghin V.,2009 Elemente de geomorfologie fluviatila, Valahia University Press, Targoviste, Romania

[8] Rosgen, D.L. 1994 A classification of natural rivers. Catena, 22, 169-199

[9] Stream Habitat Measurement Techniques, Day 2: Field Day, Copyright © 2018 Wildland Hydrology

[10] Federal government agencies - Stream Corridor Restoration_ Principles, Processes, Practices-Natl Technical Information (2006)

[11] Brice JC 1975 *Airphoto Interpretation of the Form and Behaviour of Alluvial Rivers* Final report (US Army Research Office) https://apps.dtic.mil/sti/citations/ADA008108

[12] Garde RJ 2006 *River Morphology* (New Age International (P) Ltd., Publishers)

[13] McCandless TL 2003 Bankfull discharge and channel characteristics of streams in the Allegheny Plateau and the Valley and Ridge hydrologic regions. Annapolis, MD: U.S. Fish and Wildlife Service CBFO-S03-01

[14] INHGA, EPC Consultanta de mediu SRL, Facultatea de Biologie - Universitatea Bucuresti 2019 *Raport cu concluziile rezultate din evaluările realizate în cadrul studiului și set de recomandări și propuneri de măsuri pentru reducerea eventualelor efecte negative asupra stării ecologice a corpurilor de apă, precum și asupra speciilor și habitatelor* (in Romanian)

[15] Gălie, A.C., Mătreață, M., Tănase, I. and Rădulescu, D. 2021 The Romanian Ecological Flow Method, RoEflow, Developed in Line with the EU Water Framework Directive. Concept and Case Studies. *Sustainability*, **13**, 7378. https:// doi.org/10.3390/su13137378.

[16] Zavoianu I 2006 *Hidrologie*, Editura Fundației România de Mâine

[17] Wang B et al 2020 Riverbed changes of the uppermost Atchafalaya River, USA - A case study of channel dynamics in large man-controlled alluvial river confluences, *Water*, **12**, 2139

[18] Rinaldi M, A.M. Gurnell, B.Belletti, M.I.Berga Cana et al. 2015 Restoring rivers for effective catchment management, REFORM, D6.2 Methods for HyMo Assesment, Part.1 Main Report

[19] Costel C.Cristescu 2015 Relaţii de ajustare a geometriei hidraulice a albiilor montane în bazine hidrografice mici, predominant forestiere. Raport de cercetare (in Romanian)
[20] Olson-Rutz K.M. and C. B. Marlow 1992 Analysis and interpretation of Stream Channel cross-sectional data, *North American Journal of Fisheries Management*, **12**:1, 55-61, DOI: 10.1577/1548-8675(1992)0122.3.CO;2.
[21] Knighton AD 1974 Variation in width-discharge relation and some implications for hydraulic geometry, *Geological Society of America Bulletin*; **85**, no. 7;1069-1076, doi: 10.1130/0016-7606(1974)85<1069:VIWRAS>2.0.CO;2
[22] Chow V.T. 1959 *Open-Channel Hydraulics* (McGraw-Hill Book Company)
[23] Biedenharn D.D., R.R.Copeland, C.R.Thorne, et al. 2000 *Effective Discharge Calculation: A Practical Guide* (Washington, DC: US Army Corps of Engineers) 20314-1000
[24] Bathurst J.C. 2002 At-a-site variation and minimum flow resistance for mountain rivers, *Journal of Hydrology* **269** 11-26
[25] Alfonso Ugarte S., M. Madrid-Aris 1994 Roughness Coefficient in Mountain rivers, *Hydraulic Engineering* vol **1** (New York: American Society of Civil Engineering)
[26] Comiti F., L.Mao, A.Wilcox, E.E.Wohl, M. A.Lenzi 2007 Field-derived relationships for flow velocity and resistance in high-gradient streams *Journal of Hydrology* **340**, 48–62.

IOP Conf. Series: Earth and Environmental Science 1136 (2023) 012033 doi:10.1088/1755-1315/1136/1/012033

Air entrainment and transport in a bottom outlet controlled by a cylindrical gate

Shicheng Li[1], James Yang[1,2] and Gustav Enmark[3]

[1] Civil and Architectural Engineering, Royal Institute of Technology, Stockholm, 10044, Sweden
[2] Vattenfall AB, R&D Hydraulic Laboratory, Älvkarleby, 81426, Sweden
[3] Fortum Sverige AB, Stockholm, 16968, Sweden

shicheng@kth.se

Abstract. This study deals with a bottom outlet that causes significant surges at both the inlet and exit. The bottom outlet comprises an intake, a vertical shaft and a tunnel underneath the rockfill dam and a submerged exit. In the intake tower, a cylinder gate controls hexagonal openings near the reservoir bottom. Operations during the dam construction show that discharges at a low reservoir water level result in upsurges of air-water mixture within the hollow gate cylinder and blowouts in the tailwater. To understand the flow behaviors, 3D CFD modeling is performed to examine air entrainment at the intake and transport down the waterway. Both low and high-water levels are simulated, and the air-water flow phenomenon is reproduced. In both cases, air pockets are generated, which undergo accumulation and breakup process. When moving downstream, they could cause severe surface fluctuations, even blowouts. Consequently, engineering solutions are required to address this issue. The aim of this study is to provide basis for risk assessment of outlet operations and potential rehabilitation measures.

1. Introduction

In Sweden, a significant number of flood release structures are low-level bottom outlets. They are however not used as often as surface spillways, which is mainly due to the concern for air entrainment and blowouts. In a study, the operation experiences of 38 outlets are reviewed, revealing the fact that about 80% of incidents are associated with air entrainment at the gates or intakes [1]. The entrained air is transported down the waterway, leading to undesirable flow fluctuations. To avoid safety risks, some outlets are already put out operation. However, in many facilities, the bottom outlet is the only way of emptying the reservoir when necessary. This means that engineering measures are necessary to guarantee their safe flood discharges. The bottom outlet examined in the study belongs to one of them.

2. Bottom outlet layout

The hydro-electric scheme in question was put into operation in 1973. The dam is a rockfill dam, with a crest length of 2200 m and a maximal structural height of 43 m. It has a sloping impervious core of moraine and supporting fills of gravel and rock. The core is founded on rock over a major part of its length. Legally, the full reservoir water level (FWL) is +653.50 m, and the lower limit is +630.00 m, implying that the reservoir regulation height is 23.5 m. According to the dam safety legislation, the dam belongs to a high-hazard class.

During the dam construction, the river is diverted through a tunnel in the right bank. The tunnel is then converted to a permanent bottom outlet, closed by a cylinder gate inside a concrete tower. At the far left of the dam, there is also an open spillway. The outlet is composed of an intake tower, a vertical shaft and a ~200 m tunnel running underneath the dam body and an exit in the tailwater. Figure 1 illustrates the layout of the bottom outlet.

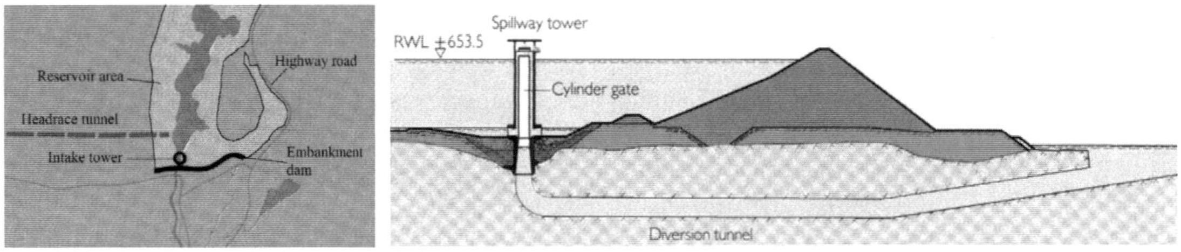

Figure 1. Bottom outlet layout - intake tower position in the reservoir and longitudinal profile with intake tower, cylindrical gate, vertical shaft, and unlined tunnel constructed by drill-and-blast method.

Figures 2 and 3 illustrates the layout of the intake. The intake tower is circular in planform, 6.00 m in inner diameter. The cylindrical gate is hydraulically maneuvered, 27.0 m high and 3.90 m in outer diameter. A circular annulus is therefore left between the tower and the gate shell; it measures 1.05 m in the radial direction and communicates freely with the reservoir.

Near the bottom of the reservoir in the tower, six intake openings are placed in a hexagonal form and controlled by the cylinder gate. The intake is grounded on bedrock; its sill elevation is Z_{in} = +627.00 m. The surrounding terrain is refilled with rock and gravel, at a 0.50 m lower elevation. The gate seals the intake at elevation Z_0 = +626.35 m. Each of the six openings is 2.40 high. The intake piers have a triangular shape, all with rounded nose corners. Three of them extend onto the bell-shaped profile; the other three are shorter and slightly wider and terminate with a rounded tail.

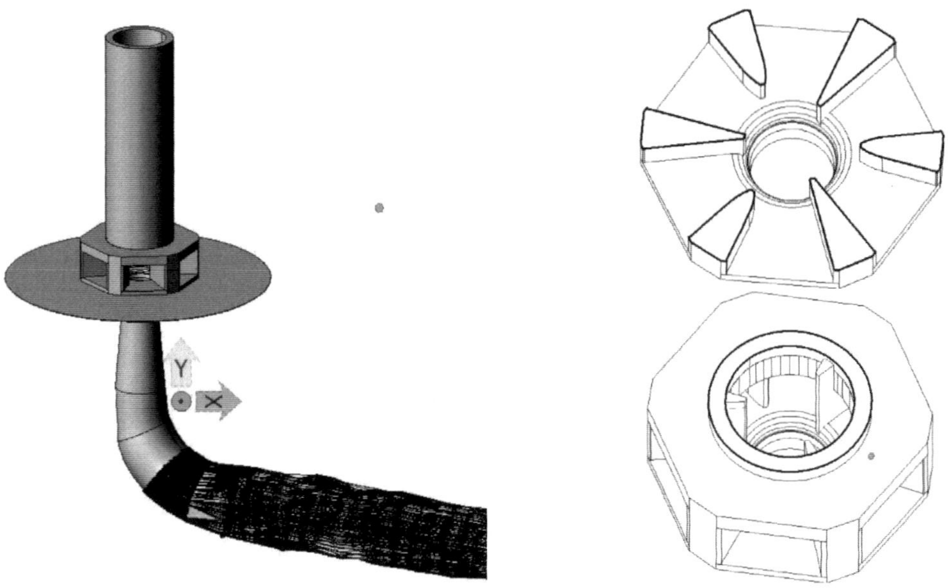

Figure 2. Intake layout with hexagonal intake openings.

The upper part of the circular shaft is grouted. Immediately below the gate, the shaft has the smallest diameter, 3.70 m. The concrete shaft then expands conically and terminates at el. +614.00 (~5.50 m in diameter). The lower part is excavated in rock and also expands downwards before bending for smooth transition to the tunnel. The transition is similar to the draft tube of a turbine unit.

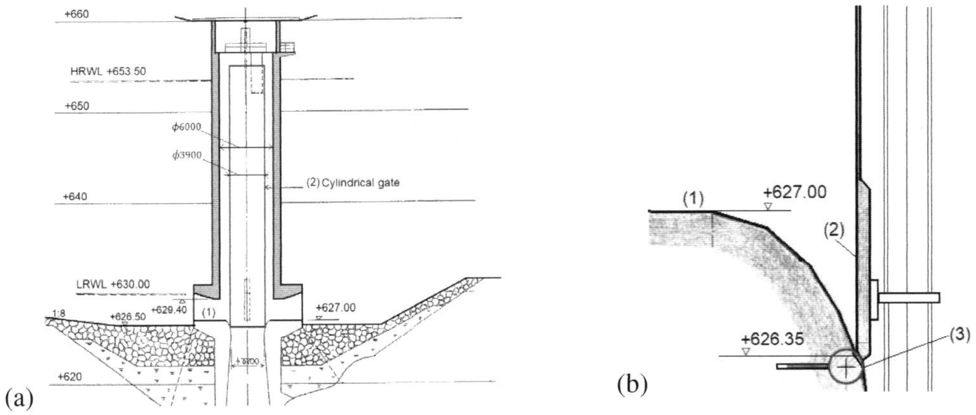

Figure 3. Intake layout: (a) cross-sectional profile; (b) cylinder gate at its closed position. (1) intake threshold (Z_{in} = +627.00 m); (2) cylindrical gate; (3) gate seal (Z_0 = +626.35 m).

The tunnel, excavated by drill-and-blast, is unlined. The upstream part is nearly horizontal and ~120 m long; its tunnel floor elevation drops 0.50 m, i.e., from +600.00 to +599.50 m over its entire length. The downstream part has a 1:7 adverse slope and is ~70 m long. With the transition part included, the total tunnel length amounts to ~200 m. As shown in Figure 4, the tunnel cross-section has a horseshoe shape, i.e., vertical walls with circular roof. It is 7.50 wide and 6.15 m high (area 40 m^2). As the tunnel is unlined, the actual profile at a given location is fairly "rough" and deviates from this ideal shape. The tunnel is usually filled with water.

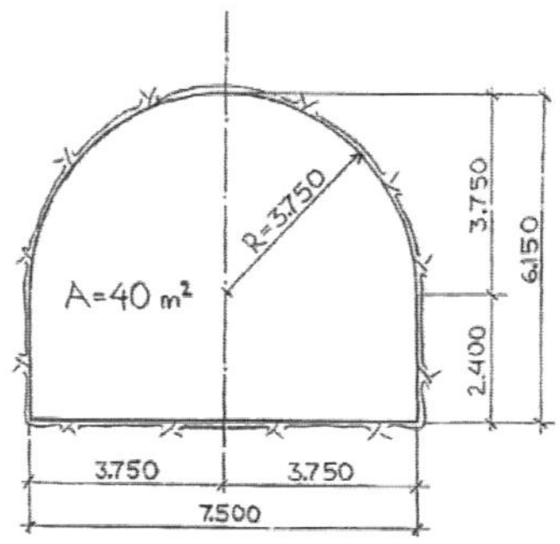

Figure 4. Horseshoe shape of the tunnel cross-section, 7.50 m wide and 6.15 m high.

The threshold of the open spillway is well above the river bed and cannot be used for reservoir emptying. In case of emergency, the reservoir level must be lowered, which is achieved only via the bottom outlet. Therefore, understanding the hydraulic behavior of the outlet and thereby avoid unexpected risks are essential for its safe operation. However, no documentation of hydraulic model tests is found. It is unclear if any model tests are at all performed previously for the outlet.

The records available are from the construction period, during which water is released at low reservoir levels. They show that air-water mixture surges up high in the tower and structural vibrations are recorded. Blowouts in the tailwater are also observed, indicating that air becomes entrained in the shaft and transported into the tunnel. Bearing this in mind, to open the outlet close to the FWL implies risks.

CFD modeling is therefore performed of flood discharge behaviors at both low and normal reservoir stages. For the former, qualitative comparison is made with the observation during the construction phase. For the latter, the intention is to predict the extent of the air entrainment and transport. The modeling is expected to provide basis for decision-making of the outlet operation and potential rehabilitation measures.

3. Prototype observations

The bottom outlet is used for flood diversion during the dam construction period. The construction work of the dam starts in 1970 and the scheme stands accomplished in the beginning of 1972. During the spring 1972, the reservoir starts its impoundment. First in 1973. The reservoir water stage reaches its FWL. During the construction, water is released twice through the bottom outlet, i.e., a spring flood early in 1971 and a planned discharge in connection with the impoundment in May 1972.

During the 1971 release, the intake tower is already built but the cylinder gate is not yet mounted. Near the tower, the water level elevation is +629.00 m; the flow rate is estimated at 80 m^3/s. The flow into the intake openings has free surface and nothing abnormal is observed, either within the tower or in the tailwater.

In April 1972, the bottom outlet is closed with the cylinder gate. One month later, when the reservoir level reaches +642.00 m, water is discharged through the outlet. Undesirable flow behaviors are observed. Already at several cm gate opening, air-water mixture surges up from inside the cylinder gate and the tower vibrates. When the gate lifts further to the 80–90 cm opening, very strong upsurges appear and hit the roof of the tower; water starts to flow through windows and doors. The pulsations seem to be irregular, sometimes at approximately 10 min interval. The concrete tower vibrates and oscillates somewhat sideward. In the tailwater, blowouts are observed. Figure 5 shows the picture of a blowout taken at the occasion. Because there is, at that time, no need to open the gate larger than 70 cm, the gate is mechanically blocked to the 70 cm upper limit. At this opening, the outlet discharge is estimated at ~30 and ~85 m^3/s at +630.00 and +642.00 m.

Figure 5. Blowout in the tailwater during water discharge April 1972, at the 80–90 cm gate opening and reservoir water level +642.00 m.

In July 1998, a test flood release is performed, at reservoir water level +651.30 m. The gate is lifted 10 cm from its closed position and kept open for 15–20 min. The estimated discharge is 9 m^3/s. It is noticed that air-water mixture pulsates at a 15 s frequency within the gate cylinder and blowouts also appear downstream. Concerning its operation, no further records are available. However, it is certain that the gate has never been opened to the upper mechanical limit (0.70 m opening) at reservoir levels close to the highest pool water level.

4. CFD Modeling

The major features involved in the bottom outlet include discharge rate, flow fluctuations and air entrainment and blowouts. As known, air entrainment in a physical model is not scaled as it does not follow the Froude law of gravity [2]. One resorts instead to CFD modeling. In FLUENT [3], the Volume of Fluid (VOF) model, in combination with the RNG k-ε turbulence model, is used to reproduce the flow behaviors. To solve two-phase flows of air and water, the model is a commonly used approach. For more description and its applications, one is referred to [4],[5].

The computational domain extends from the reservoir to the tailwater and is illustrated in Figure 6. The included reservoir area is circular, surrounding the entire tower and having a 25 m radius. This means that the upstream boundary in the reservoir is ~20 m from the facade of the tower. The geometrical layout being asymmetrical, all the gate openings are simulated. The prerequisite for reliable results is decent mesh quality. A structural grid of hexahedral shape is generated. For spatial discretization, the second-order upwind scheme applies with higher-order levels of accuracy achieved at cell faces. For each governing equation, it is possible to select a different discretization scheme.

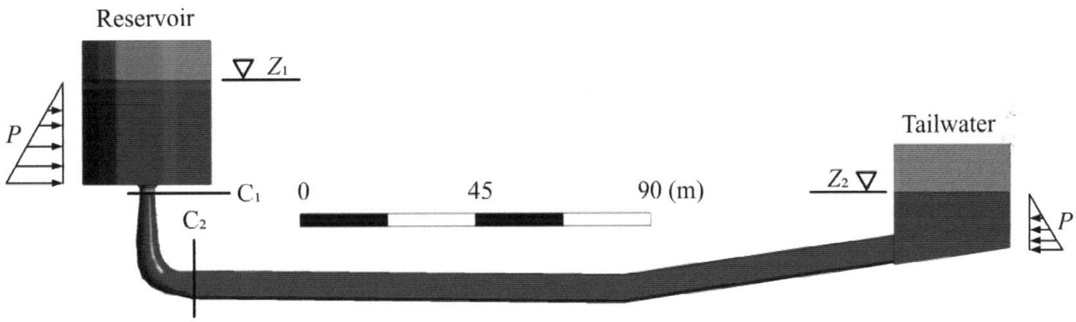

Figure 6. Schematics of the computational domain.

The numerical results must be grid independent, assessed usually by Grid Convergence Index (GCI). High mesh density is given at the intake openings, along the tunnel roof and along the free water surface. The evaluations indicate that the mesh with 0.84 million cells is sufficiently dense to model the aerator flow. The minimum cell size is 5 cm. Close to a solid boundary, a non-dimensional parameter usually measures the grid density's reasonability, $y^+ = u_\tau d/\mu_w$, where u_τ = shear velocity and d = distance of the first cell to the boundary. It should be in the range between 10 and 100. The resulting y^+ values fall within the range 40–60. The enhanced wall function is activated to cope with the viscous layer.

For both the reservoir and the tailwater, a water stage is given. At either boundary, the hydrostatical pressure defines the water depth (Figure 6). For the air part above the water, a velocity inlet is specified with zero velocity. The top surfaces in the shaft and the tailwater are given as pressure inlet at the atmospheric pressure, allowing free air flow in or out. The remaining boundaries are treated as walls, with no-slip condition irrespective of phase.

The flow simulation is transient. With a fully implicit scheme, the choice of time steps is not governed by any stability criterion. However, to solve a transient flow, the time step is usually set at least one order of magnitude smaller than the smallest time constant in the system. A common way to judge its choice is to count the number of iterations to a converged solution. 5–10 iterations per time step are ideal. With more iterations, the time step is too large and *vice versa*. The convergence is reached if the normalized residuals for each equation are lowered by at least three orders of magnitude. The check is made for each time step.

5. Results and discussions

Two case scenarios are set with different up- (Z_1) and downstream (Z_2) water levels (Figure 6). For Case 1, Z_1 = 642.0 m and Z_2 = 625.2 m; for Case 2, Z_1 = 653.0 m and Z_2 = 625.3 m (table 1). The gate opening is the same for both cases (70 cm). At the initial state (t = 0), the reservoir, shaft, and tunnel are filled with water (blue), and then the gate is lifted, which allows the water to be discharged downstream. The gate lifting process is not considered.

Table 1. Summary of the simulated scenarios.

Case no.	Upstream water level (m)	Downstream water level (m)	Gate opening (cm)
1	642.0	625.2	70
2	653.0	625.3	70

5.1. Air entrainment and conveyance

The typical flow pattern in the tunnel is shown in Figure 7 (Case 1). It includes a front view of the central plane and a top view of the tunnel, in which the blue and red represent water and air, respectively.

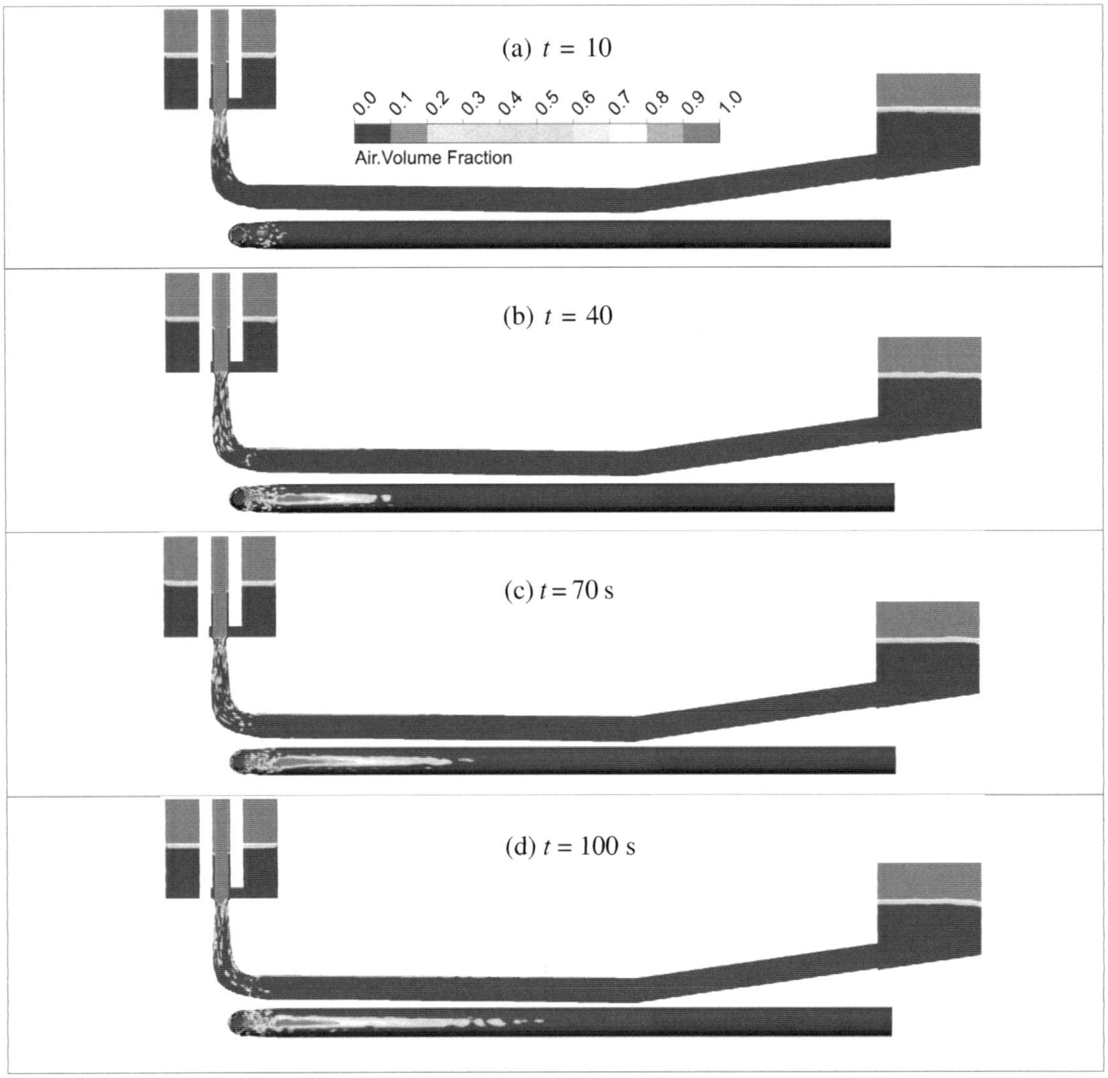

Figure 7. Volume fraction of air in Case 1. The front view is for central plane, and the top view is for the tunnel. The blue and red indicate water and air, respectively.

The airflow movement includes two stages: accumulation and breakup. When the gate is lifted, the water rapidly flows to the shaft and fills the space above the cross section C1 (Figure 6, the elevation of C1 is Z_2). In this process, a large amount of air is entrained into the water (red areas in Figure 7a). With the shaft becoming full, the air flow rate Q_a reduces, however the total volume of the air is still increasing. Therefore, at the joint of the shaft and tunnel, the air bubbles accumulate (top view in Figure 7b). As the flow constantly carries air to the shaft, the air contents in the tunnel grow and are transported downstream (Figure 7c). Due to velocity difference, the big air pocket is separated into small ones (Figure 7d).

5.2. Air flow discharge

To understand the air discharge process, Q_a at C1 and C2 is monitored, as shown in Figure 8 (Case 1).

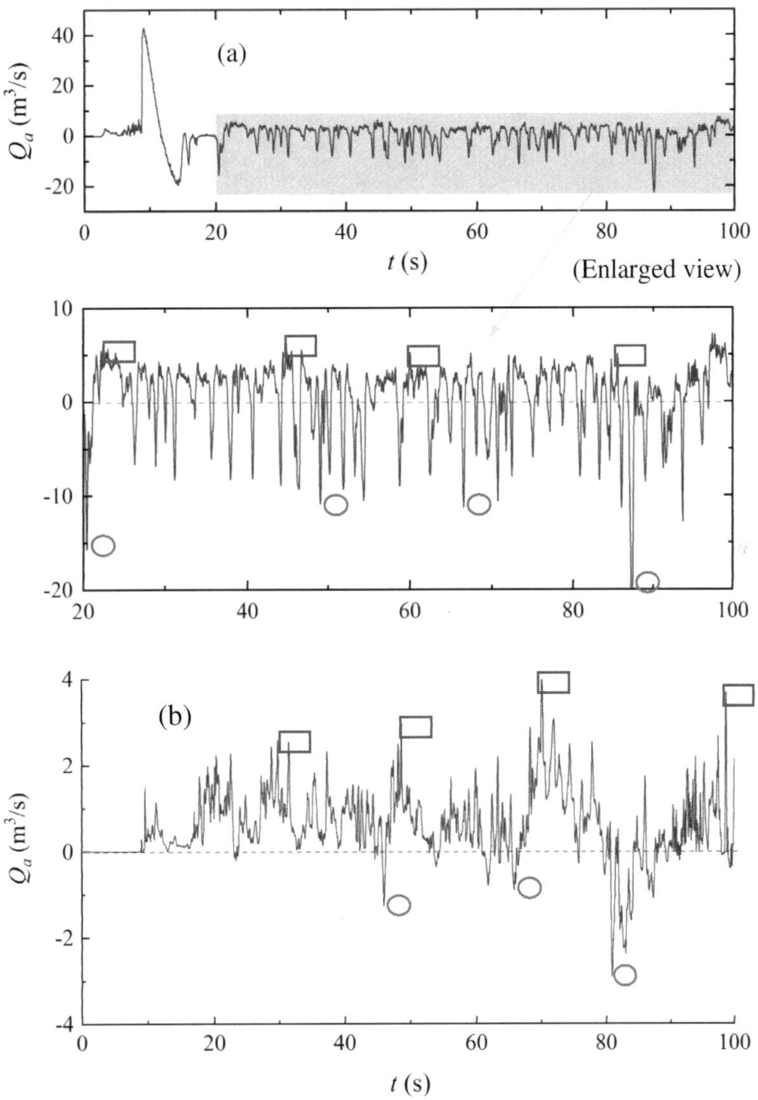

Figure 8. The variation of air flow rate Q_a for Case 1 over time t: (a) at C1 and (b) at C2.

A positive value indicates the air flow direction of up- to downstream, and a negative one represents the opposite. At C1, with the filling (t = 0–20 s) of the space above C1, Q_a shows an initial increase followed by a drop. After t = 20 s, the air flow reaches a normal state, where it shows periodical variations. From the enlarged plot (Figure 8a), multiple local extreme values are observed. For the positive ones (squared), the air flow moving downstream shows the largest rate. For the negative ones (circled), the entrained air is discharged upstream at the highest rate, which is due to the limited carrying capacity of the water. A large reverse flow might lead to remarkable surface fluctuations in the shaft. The time difference between the local extremums is 10–15 s. At C2, similar air flow variations are found. At this cross-section, the air is mostly transported downstream. At t = 80–90 s, a relatively large amount of air is moving upstream, which is presumable a result of the breakup of the air pockets. The largest reverse flow is $Q_a = 3$ m^3/s.

Figure 9 shows the variations of Q_a for Case 2 at cross-sections C1 and C2.

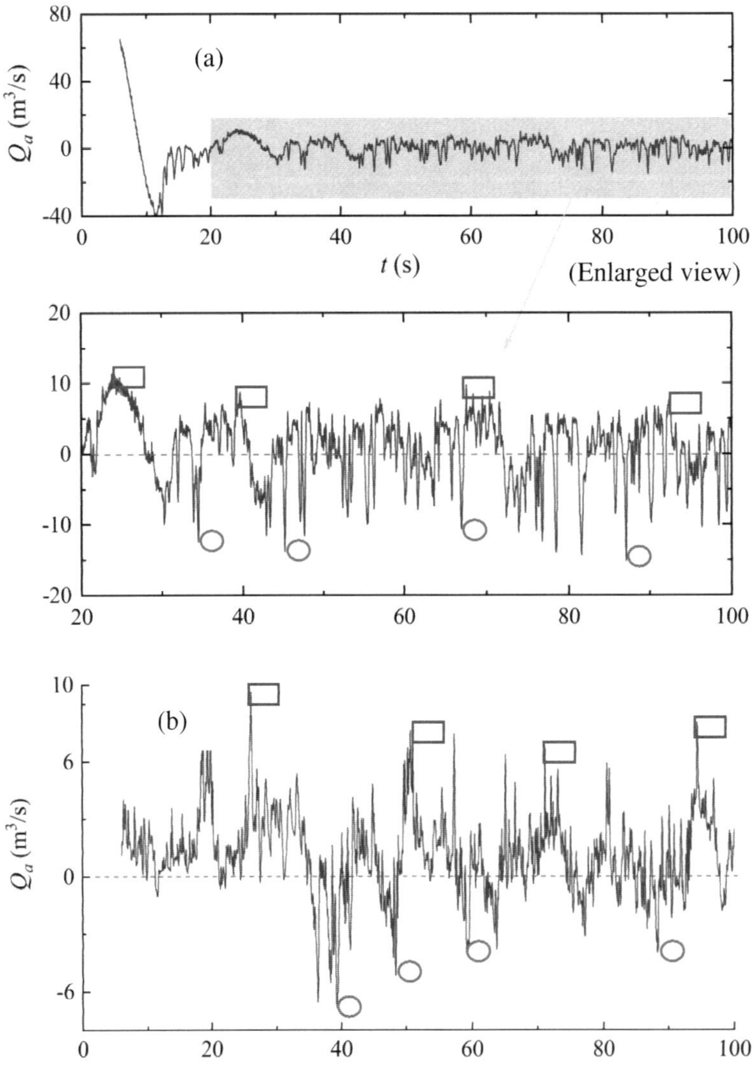

Figure 9. The variation of air flow rate Q_a for Case 2 over time t: (a) at C1 and (b) at C2.

At C1, it is noticed that the air flow changes rapidly its direction. Some air moves downstream and some travels upstream. The reverse flow causes surface variations in the shaft and the downstream moving one results in air accumulation in the tunnel. The largest reverse flow is Q_a = 15 m^3/s. At C2, after some time of accumulation (t = 0–35 s), the bubble clouds break up, and some of them move upstream, leading to negative Q_a values. The accumulation and breakup process continues, so that Q_a changes its direction constantly. The largest reverse flow at this cross section is Q_a = 7 m^3/s.

5.3 Comparison of air volume
To compare the total amount of air contents in both cases, Figure 10 plots the air volume V_a entrained through C2. In general, V_a in both cases shows an almost linear increase with t. The reduction in V_a corresponds to the reverse air flows. For instance, at t = 80–90 s, V_a drops slightly, which is a result of the negative Q_a values (Figure 8). Compared with Case 1, V_a in Case 2 is significantly larger. At t = 50 s, V_a in Case 2 is twice that of Case 1. At t = 100 s, V_a in Case 2 is 1.7 times that of Case 1. Based on the tendency of the curve, the difference of V_a in both cases is expected to grow if the simulation continues.

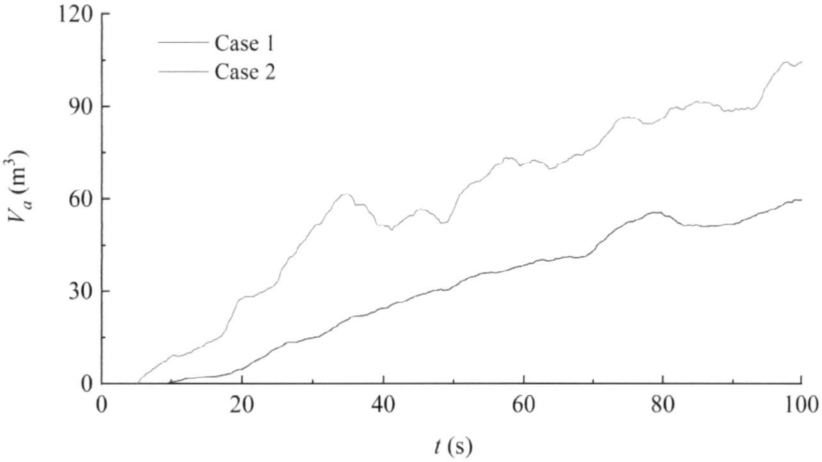

Figure 10. The volume of air V_a entrained through C2.

6. Conclusions
Many incidents in Swedish bottom outlets are caused by air entrainment at their intakes. Due to this, their use is in many cases governed by operation constrains. The outlet described in this study belongs to this category. The observations of air blowouts in the tailwater are briefed. With the purpose to understand its flow features and provide basis for countermeasures, numerical modeling is performed, in which the effects of reservoir stages and tunnel roughness are examined.

The air-water flow behaviors for two water levels are numerically reproduced. The water flow in both cases reaches a relatively stable state after t = 20 s. The water discharge for Case 1 and 2 is 80 and 95 m^3/s, which agree with the prototype values. The air entrained in the tunnel undergoes two processes: accumulation and breakup. When the gate is lifted, the air enters the shaft and accumulates at the joint of the shaft and the tunnel. When the air contents reach a certain level, the air pockets break up and move up- and downstream, leading to surface fluctuations in the shaft and exit, respectively. The air amount entrained into the tunnel shows an almost linear growth with time. The air volume in Case 2 is remarkably larger than that of Case 1. At t = 50 s and t = 100 s, the former is twice and 1.7 times the latter. To mitigate the blowout risks, countermeasures, e.g., a deaerator, are suggested to release the air contents in the tunnel.

Acknowledgments
The study is jointly funded by Fortum Sweden and Swedish Hydropower Centre (SVC). It falls within the frame of SVC research project "Quality and trust of numerical modeling of water-air flows for safe

spillway discharge". SVC was established by Swedish Energy Agency, Energiforsk, and Svenska Kraftnät together with Luleå University of Technology, Royal Institute of Technology, Chalmers University of Technology, and Uppsala University. Participating companies and industry associations include AFRY, Andritz Hydro, Boliden, Fortum Generation, Holmen Energi, Jämtkraft, Karlstads Energi, LKAB, Mälarenergi, Norconsult, Rainpower, Skellefteå Kraft, Sollefteåforsens, Statkraft Sverige, Sweco Energuide, Sweco Infrastructure, Tekniska verken i Linköping, Uniper, Vattenfall R&D, Vattenfall Vattenkraft, Voith Hydro, WSP Sverige, and Zinkgruvan.

References

[1] Dath J and Mathiesen M 2007 *Pre-studies of Hydraulic Design: Inventory and General Evaluations of Bottom Outlets in Swedish Dams* (Stockholm, Sweden: Elforsk AB)
[2] Shi QS 2007 High-velocity aerated flow *China Water & Power Press* (Beijing, China)
[3] ANSYS Inc. 2019 *ANSYS Fluent Theory Guide – Academic Research Release 19* (Canonsburg, PA, USA)
[4] Zhang J, Chen J, Xu WL, Wang Y and Li G 2011 Three-dimensional numerical simulation of aerated flows downstream sudden fall aerator expansion in a tunnel *Journal of Hydrodynamics* **23** 71–80
[5] Stamou AI, Mitsopoulos G, Rutschmann P and Bui MD 2018 Verification of a 3D CFD model for vertical slot fish-passes *Environmental Fluid Mechanics* **18**(6) 1435–1461.

Elements of hydraulic calculation of permanent movement at non-standard sewer collectors

Mihail Luca[1], Nicolae Marcoie[1], Alexandru-Lucian Luca[2] and Daniel Toma[1]

[1] Hydroamelioration and Environmental Protection Department, "Gheorghe Asachi" Technical University of Iasi, Iasi City, 700050, Romania
[2] Polias - Instal Company, Iasi City, 700585, Romania

mluca2015@yahoo.com

Abstract. An important issue in the design of the main visitable sewer collectors is the hydraulic calculation in permanent movement and non-permanent movement. Some main collectors that can be visited have non-standard flow sections, a situation that complicates the hydraulic calculation. Non-standard flow sections derive from standard sections (circular, ovoid, bell, rectangular and mixed) through natural and anthropogenic degradation processes. The initially designed section changes over time in terms of shape, dimensions and roughness, a situation in which determining the hydraulic parameters becomes very difficult. The research carried out on a series of visitable main collectors revealed that the flow section has geometrically transformed from curved shapes to mixed shapes (curves + right-hand segments).The hydraulic rehabilitation of degraded collectors requires the analysis of flow parameters in the case of geometrically modified sections. The non-standardized flow section can be considered as a "single bed", or as a "multiple bed". At the same time, the "single bed" type sections composed of both circular arcs and straight line segments can be hydraulically analyzed with a customized calculation program. The paper presents a model for calculating the geometric and hydraulic parameters of a sewer collector with a non-standard shape and in permanent uniform motion.

1. Introduction

Wastewater is collected and transported through open and closed canals. Closed channels are made of tubes of various shapes (circular, bell ovoid, etc.), in which the movement of water can be gravitational, pressure, vacuum or mixed. The most used type of wastewater movement in collectors in Romania is gravitational [1]. The movement of water in the collectors is generally non-permanent, given the variation in flow over time. On long sections of the collector and at a small variation of the flow, a non-uniform permanent movement of the varied gradual and varied rapid non-uniform type is accepted [1] [2].

The monitoring of the functional parameters of the sewage collectors is necessary for the management of the exploitation process. The exact knowledge of the hydraulic parameters (transported flow, average speed, and degree of filling) is priorities in the optimal management of sewage collectors. The main sewerage collectors (domestic and rainwater) located in urban areas has special operating characteristics. These collectors show a large flow variation, a situation that can put them under pressure and cause structural damage [3] [4]

Hydraulic calculation of the movement of water in a main sewer collector over a longer time interval can be admitted as permanent and non-uniform gradually varied, but also with sequences of varied speed. The calculation of the non-uniform movement requires the knowledge of the hydraulic parameters of the permanent uniform movement. Calculation of non-uniform movement in a channel requires knowledge of the movement regime on channel sections, as well as control sections with control elevations / depths. The control depths in the control sections, which have the function of boundary conditions, determine the evolution of the curves of the free surface of the water in the non-uniform motion [2].

The transportation of waste water with various degrees of loading with solid material is carried out in sections with a shape that allows the development of higher velocities towards the collector scraper. Technical constructive and hydraulic data for a limited number of transversal profiles are specified in the specialized literature for sewer collectors [1]. These tubes have prescribed dimensions and are classified into standard size classes.

In the specialized technical literature, analytical relations are presented for the calculation of the geometric and hydraulic parameters of the collector only for the full section and only for constant values of the roughness coefficient on the entire perimeter of the flow section [1], [2] [5]. For the calculation of the functional parameters of the sewer collectors, calculation programs were developed for a limited number of standardized sections [6] [7] [8]. Under certain conditions of execution or operation, sewer collectors with non-standard dimensions are also made in sewerage networks. They do not have a database for describing the geometric shape or specific calculation relationships for hydraulic parameters.

The present paper proposes an analysis model and a calculation program for the determination of hydraulic parameters on the main sewer collectors with non-standardized sections in uniform permanent movement.

2. Sewer collectors with non-standard sections

Sewage systems in Romania have been designed and executed for a long time with the use of various technologies and materials. Their design was made by using technical concepts specific to the stages of Romania's economic development. For gravitational transport of wastewater, cross-sections of the circular, ovoid and bell-shaped manifold are used (figure 1) [1]. In large-scale collectors, mixed geometric shapes consisting of curves and straight lines are also used (figure 2).

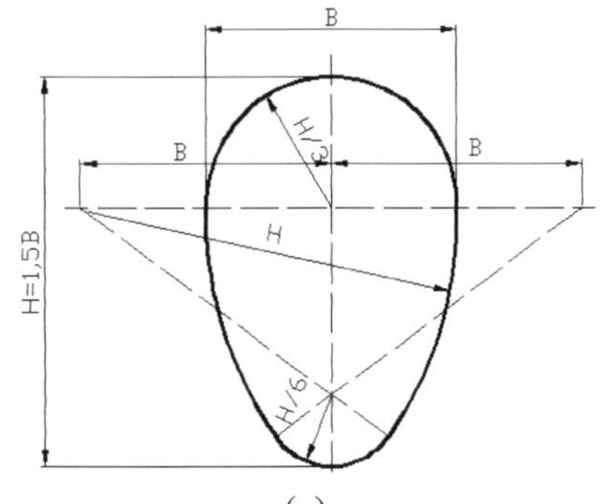

(a)

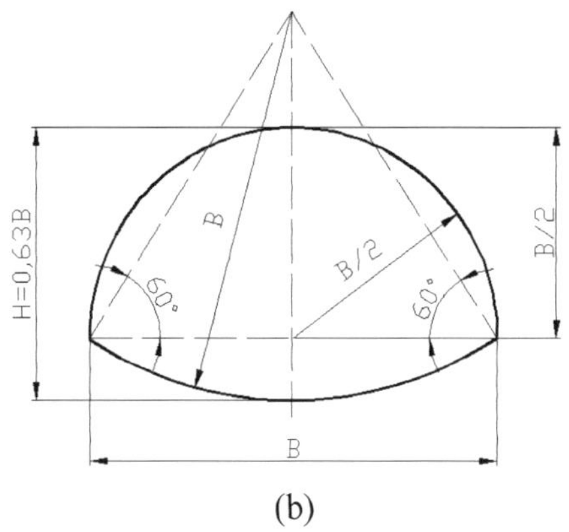

(b)

Figure 1. Profile sewer visitable collectors standardized: (a) ovoid type; (b) semi-circular bell type [1].

The materials and technologies for the execution of the sewerage networks were diverse. The most used materials were stone, basalt, brick, reinforced concrete and so on. Most large sewers were made of reinforced concrete [1]. They have the functions of main collector and exhaust collector in the sewage network. A sewer collector is considered visitable in Romania when the internal diameter (D) or the characteristic height (H) meets the condition $D/H > 1.50$ m [1].

The natural and anthropogenic factors in the site intervene on the collectors with a complex of actions, which influence their behaviour over time. Natural factors include geotechnical and hydrogeological characteristics, seismic action, land subsidence and so on. Anthropogenic factors acting on the collector show an evolution over time in terms of influence and impact. They influence the behaviour of the sewer collector through the design concept, the execution technology, the physico-chemical characteristics of the wastewater, the degree of application of the maintenance works and annual repairs and so on.

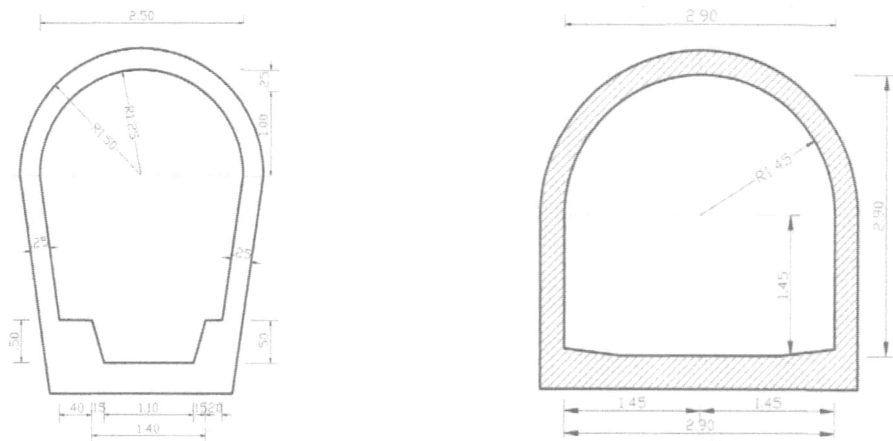

(a) (b)

Figure 2. Scheme of visitable sewer collectors with non-standard sections: (a) collector derived from an ovoid; (b) gallery type collector with mixed section (source author).

The most influential anthropogenic factors can be considered the degree of loading of wastewater, chemical parameters, speed of transport, water loss, and lack of maintenance and so on. Some of the natural factors in conjunction with the anthropic ones determine the appearance of some processes of continuous degradation of the collector structure and modify functional parameters in time (figure 3) [4], [9].

(a) (b)

Figure 3. Geometric modification of the clogging flow section at the visitable sewer collectors; (a) clogging in the concentration and cementing phase; (b) cemented clogging (source: author).

A series of situations that arose in the execution and exploitation process determined the formation of non-standardized flow sections. Important changes in the flow section are determined by the process of hydrodynamic erosion and clogging. The flow section is transformed geometrically, where the circle or ovoid curved shapes have changed into mixed shapes (curves + lines) (figure 4). The modified flow section and with a variable roughness on the perimeter, no longer allows an exact calculation of the functional parameters, a situation in which the operation process of the sewage collector becomes difficult.

Research has shown that the flow section of the collectors has been substantially altered from that designed [9]. Thus, a circular section can become a polycentric section through the erosion processes of the screed and the walls (figure 4.a). The same situation occurs in the ovoid type sections, where the sedimentation and cementing phenomena radically transformed the flow section, by transforming it into a polycentric section (figure 4.b).

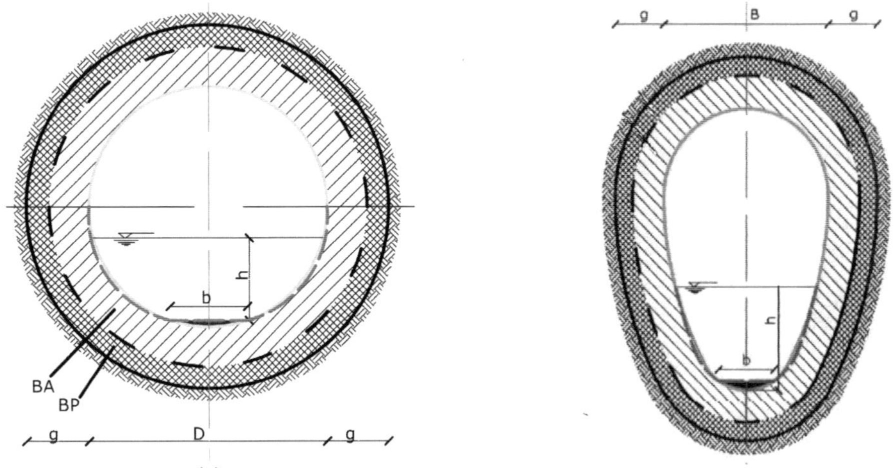

Figure 4. Visitable collector sections modified by erosion and clogging phenomena and technological rehabilitation processes; (b) - ovoid section (blue line-standard section; red line-section modification areas) (source [9] [10]).

The application of rehabilitation works to the collectors can geometrically transform the flow section by creating a transport or support area for the execution equipment. Monitoring of flow parameters (flow rate, minimum speed, and degree of filling) in non-standard sections can no longer be performed correctly. The geometric modification of the section determined the decrease of the speed and the transported flow by about 15… 28% [4], [9].

The phenomenon of hydrodynamic erosion has the effect of changing the roughness on the perimeter of the flow section. On the perimeter of the collectors and hydrotechnical galleries that transport water, there are areas with distinct roughness, on a wide range of values, a situation that required a complex analysis of this hydraulic calculation parameter [9].

3. The hydraulic-mathematical model

Obtaining correct data for waste water flow parameters in standardized / non-standardized visitable sewer collectors required the creation of hydraulic-mathematical models specific to each type of movement (permanent and non-permanent). To begin with, a model for permanent uniform motion was made, since with this form of motion any hydraulic calculation of closed-section channels begins. The hydraulic-mathematical model allows the determination of the hydraulic parameters of sewer collectors with non-standardized and standardized sections. The model can also analyze flow sections composed of straight lines and curves and consider the variation of roughness on the collector perimeter.

Next, models were developed to calculate the hydraulic parameters for non-uniform permanent movement. A non-uniform, gradually varied permanent movement in combination with a rapidly varied non-uniform movement can be admitted in a household sewage collector. This assumption is based on the reduced variation over a longer time interval of the flow carried by a main sewer collector. The non-uniform movement is composed of a sequence of curves of the free surface, which must comply with the flow conditions in sewer collectors (imposed minimum speed, imposed degree of filling). A non-permanent movement can be registered in a main storm water sewer in case of discharge of torrential rains.

The non-permanent movement in the collector can be simulated by using hydraulic models consisting of the following: a - the equations of movement; b - continuity equations; c - initial conditions; d – contour conditions. The initial conditions are represented by the parameters of the permanent movement by the forms non-uniform gradually varied, respectively non-uniform rapidly varied. To be able to approach any hydraulic simulation, the basic data of the sewer collector must be obtained, namely geometric and hydraulic parameters of the uniform permanent movement for any form of collector (standardized and non-standard).

The internal and flow section of a sewer collector is characterized by the following geometric parameters [1]: the shape of the section, the constructive height (D, H), the constructive width (B), the radii of curvature (r or R), the angles, the geodesic slope of the channel (I) and others. Hydraulic parameters considered in the calculation of the collector are the following [2]: water depth (h), flow (Q), flow rate (K), average speed (v), minimum speed (v_{min}) and maximum imposed (v_{max}), speed rate (W), degree of filling ($a = h/D$ or $a = h/H$), maximum degree of filling (a_{max}), wetted area of the flow section (A), wetted perimeter (P), hydraulic radius (R_h), roughness (k), roughness coefficient (n), hydraulic slope (J), the depth of the centre of gravity of the watered area (z_G) and others.

The main geometric and hydraulic characteristics of sewer collectors can be expressed by the functions [2]:

$$B = f_B(h),\ P = f_P(h),\ A = f_A(h),\ R_h = f_{Rh}(h),\ z_G = f_{zG}(h),\ W = f_W(h),\ K = f_K(h). \tag{1}$$

In order to generalize the hydraulic calculation of sewer collectors for the flow regimes existing in practice and for the type of collector section used (standardized and non-standardized), the M matrix was considered, in the form:

$$\mathbf{M} = [\mathbf{h}\ \ \mathbf{B}\ \ \mathbf{P}\ \ \mathbf{A}\ \ \mathbf{R_h}\ \ \mathbf{z_G}\ \ \mathbf{W}\ \ \mathbf{K}]. \tag{2}$$

The elements of the matrix **M** are N_C - dimensional column vectors, where N_C represents the number of distinct values assigned to the value pairs

$$\{h, B, P, A, R_h, z_G, W, K\}_i, (i = 1, 2, \ldots, N_C). \tag{3}$$

When simulating the flow of water along the length of a collector, it is more convenient to perform the substitution:

$$h = z - z_a, \tag{4}$$

where z_a is the apron elevation of the channel level; z - the elevation of the free water level. Thus, in the hydraulic characteristics (1) similar characteristics will be generated, but which present as an independent variable the elevation z, respectively

$$A = f_A(z),\ P = f_P(z),\ R_h = f_{Rh}(z),\ B = f_B(z),\ z_G = f_{zG}(z),\ W = f_W(z),\ K = f_K(z) \tag{5}$$

The main sewer collectors were considered on the calculation sections to be prismatic/cylindrical beds, with a constant slope. The connection of collector sections with different shapes and/or sizes, as well as the confluence with lower-order collectors, is achieved by using connection and intersection manholes. In this case, adequate contour conditions must be allowed in the manholes section.

IOP Conf. Series: Earth and Environmental Science 1136 (2023) 012034 doi:10.1088/1755-1315/1136/1/012034

To solve the mathematical models proposed for each type of water movement considered in the visitable sewer collector, a series of numerical calculation programs were developed in the MATLAB programming environment. So:

- to obtain the geometric and hydraulic parameters of non-standard visitable collectors in uniform permanent motion, the program Collector.Non-Standard_Uniform Motion_Unique Bed (**Col.NS_UM_UB**) was developed;
- to determine the hydraulic parameters of the non-uniform permanent movement in the visitable collectors, the program Collector_Permanent_Gradual_Variate_Movement (**Col_MPGV**) was developed;
- to obtain the contour conditions imposed by the manholes located on the route of the visitable collectors, manholes with large dimensions and various functions, calculation programs were developed for each structural and functional type of manhole;
- to determine the hydraulic parameters of the non-permanent gradual, varied movement in the visitable collectors, the Various_Gradual_Transient_Movement_Collector (**Col_VGTM**) program was developed.

Each calculation program was tested and calibrated on channels with open and closed section, after which they were used to obtain geometric and hydraulic parameters on non-standard flow sections [4], [9], [10], [11].

4. Results and Discussion

The calculation of functional parameters at the visitable sewer collectors with non-standard sections begins with the determination of geometric and hydraulic parameters using the program Col.NS_UM_UB. The program was tested and validated by running it on a classic collector section (standard circular bell with representative dimensions $B/H = 600/380$ mm and a uniform roughness on the perimeter with the value $n = 0.0135$). The data obtained were analyzed compared to those obtained by current calculation methods. The comparatively analyzed results showed a good agreement between the values calculated with the Col.NS_UM_UB program and those obtained with the classical model used in the design of standardized collectors.

The Calculation program requires field measurements of the current coordinates of the perimeter of the flow section, which has been modified by various actions. Measurements must be made on a collector section and then processed to obtain a characteristic calculation section [4], [9]. An important step in defining the primary calculation data is represented by the activity in the field, inside the visitable collector, for the analysis of the roughness on the perimeter of the flow section.

The flow section geometrically made of curved lines (circular arcs only) can be treated as "single bed". A section consisting only of straight line segments was treated as a "multiple bed". In the case of a flow section composed of both circular arcs and straight line segments, it was considered in the study to be a "single bed" (figure 4) [9].

In the analysis model it is assumed that the contour of the cross-section of the closed channel is completely determined by an odd number of points, N_M, conveniently established by a yOz coordinate system. The number of points is defined by the relationship

$$(y_C, z_C)_k, \quad k = 1, 2, \ldots, N_M \,. \tag{6}$$

The coordinates (6) are related to a Cartesian system of coordinate axes, yOz, defined as follows: the Oz axis, vertical included in the longitudinal plane of symmetry; the origin O at the point on the minimum height tube eraser, and the horizontal axis Oy, normal to the axis Oz, included in the transverse half-section of the tube.

To exemplify how to work with the calculation program Col.NS_UM_UB, the determination of the geometric and hydraulic parameters for a non-standard sewer collector is presented. The section of this collector consists of arcs of a circle and straight segments (figures 5).

The flow section describes a closed polycentric channel, where the geometric profile consists of curves (circle arcs) and straight line segments. At the bottom is a semicircle with radius $R_1 = D/2$ from centre O1(0, R_1). At the top is a rectangle with width B and height H_d, where the total height of the profile is $H = D/2 + H_d$. The numerical data adopted in the simulation were: $R = 0.35$ m, $B = 3.450$ m, $H = 2.600$ m.

The geometric profile is completely determined by 9 points (figure 5) of coordinates (y_C, z_C), according to equation (6), respectively:

$$(y_C, z_C) \in \left\{(0,0), \left(\frac{\sqrt{3}}{2}R_1, \frac{R_1}{2}\right), (R_1, R_1), \left(\frac{2R_1+B}{4}, R_1\right), \left(\frac{B}{2}, R_1\right), \left(\frac{B}{2}, R_1+\frac{H_d}{2}\right), \left(\frac{B}{2}, H\right), \left(\frac{B}{4}, H\right), (0, H)\right\} \quad (7)$$

At the sewer collectors that can be visited with sections composed of $N_a \geq 2$ arcs of a circle and/or line segments, the analytical expressions for each of the functions $A(h)$, $P(h)$, $R_h(h)$, $z_G(h)$, $W(h)$ and $K(h)$ must be rendered by N_a functions defined on portions of the perimeter, respectively:

$$A_i(h), P_i(h), R_{hi}(h), B_i(h), z_{Gi}(h), W_i(h), K_i(h), \text{ with } z_i - z_{C0} \leq h \leq (z_{i+1} - z_{C0}), i=1,\ldots,N_a \quad (8)$$

where z_{C0} is the height of the tube apron; z_i and z_{i+1} ordinates of the points that delimit the lower, respectively upper, circle arc or line segment i. $z_{C0} = 0$ is accepted in the calculation.

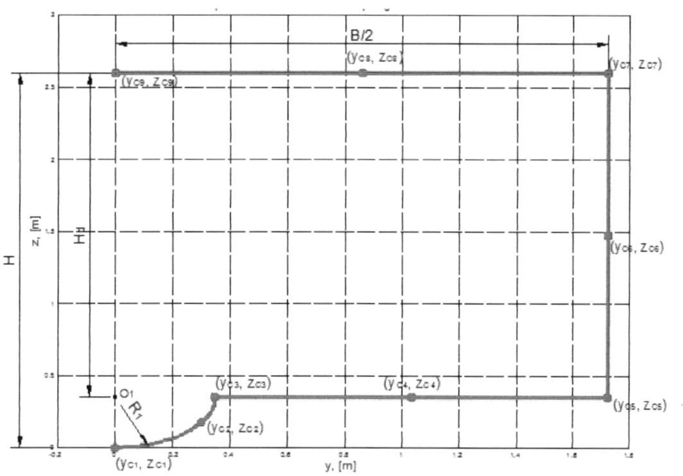

Figure 5. Calculation scheme for a non-standard sewer collector, "single bed" type consisting of arches and straight segments.

The determination of the functions (8) is carried out by going through the following stages:

A. Deduction of analytical expressions for circles and/or supporting lines of N_A circle arcs and/or line segments.

B. Determination of the N_A+1 points that border the lower and upper N_A circle arcs and/or straight line segments, (y_i, z_i), (i = 1, 2, ..., N_A+1).

C. Deduction of analytical expressions, by portions of section, for the hydraulic characteristics of sewer collectors.

The input data to the program, in a synthetic presentation, were: the calculation code for non-standardized polycentric section; calculation steps Dt_H = 0.02 m, Dt_B = 0.02 m, B = 3.450 m, H = 2.600 m, n = 0,017 – 0,019 (Manning) and $N_M = 9$.

The exit data from the program, in a synthetic presentation, were as follows:

1. Geometry of the contour: NA = 5, where the coordinates of the connection points y, z, on the contour of the collector are tabulated; the ordinate and abscissa of the centre of the support circle of the arcs, the abscissas and ordinates of the line segments.

2. The geometric and hydraulic-functional characteristics: N_C = 120. The N_C elements of each of the vectors **hC**, **BC**, **AC**, **RhC**, **zGC**, **WC** and **KC**, which constitute the columns of the matrix **M**,

are tabulated. Also, the values of the parameters for the full section h_f, B_f, A_f, R_{hp}, z_{Gf}, W_f and K_f are presented in the table. Due to the roughness coefficient n, which is uneven on the perimeter of the collector, the depths corresponding to the maximum values for the hydraulic radius R_h and the velocity modulus W are different.

3. Graphic representations for contour geometry and hydraulic and functional characteristics of the simulated collector: graphs for geometric characteristic curves, hydraulic characteristic curves and hydraulic-functional characteristic curves. So:

- geometric characteristic curves $B = f_B(h)$, $A = f_A(h)$, $z_G = f_{zG}(h)$ and hydraulic curves $P = f_P(h)$, $R_h = f_{Rh}(h)$ (figure 6);
- hydraulic-functional characteristics $W=f_W(h)$ si $K=f_K(h)$ (figure 7);
- the hydraulic and functional characteristics of the collector at partial filling, respectively $Q/Q_p=K/K_p=f_Q(h/H)$ şi $v/v_p=W/W_p=f_v(h/H)$ (figures 8).

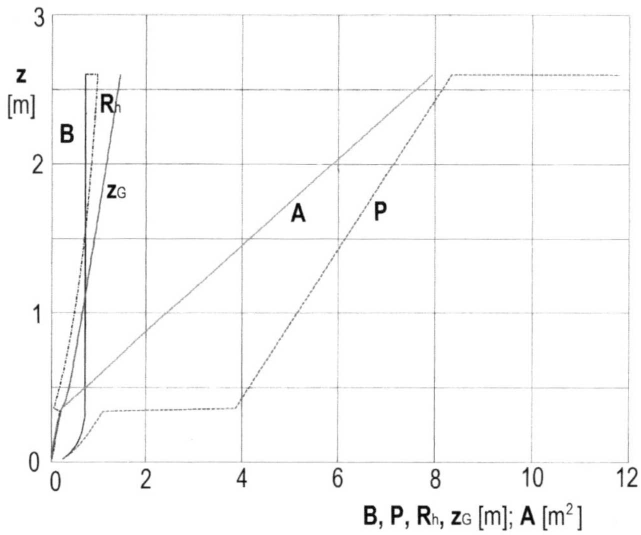

Figure 6. Characteristic curves $B = f_B(h)$, $A = f_A(h)$, $z_G = f_{zG}(h)$, $P = f_P(h)$, $R_h = f_{Rh}(h)$ for the non-standard collector with polycentric section.

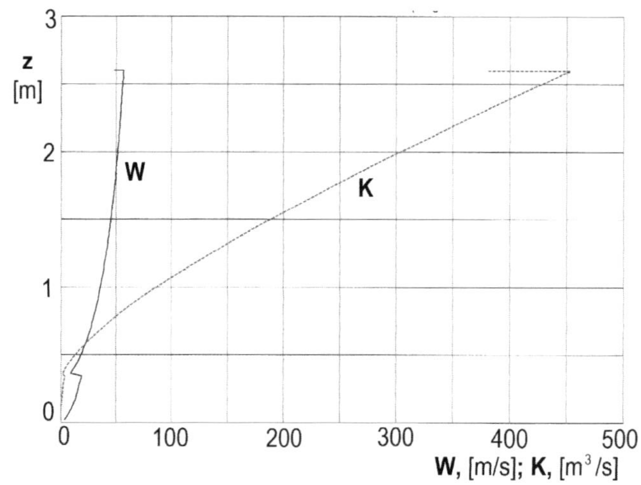

Figure 7. Hydraulic-functional characteristics $W = f_W(h)$ şi $K = f_K(h)$ at the collector non-standard with polycentric section.

The obtained results are represented by the analytical relations for each geometric and hydraulic parameter. The obtained equations were represented graphically for their generalization and rapid use in the calculation.

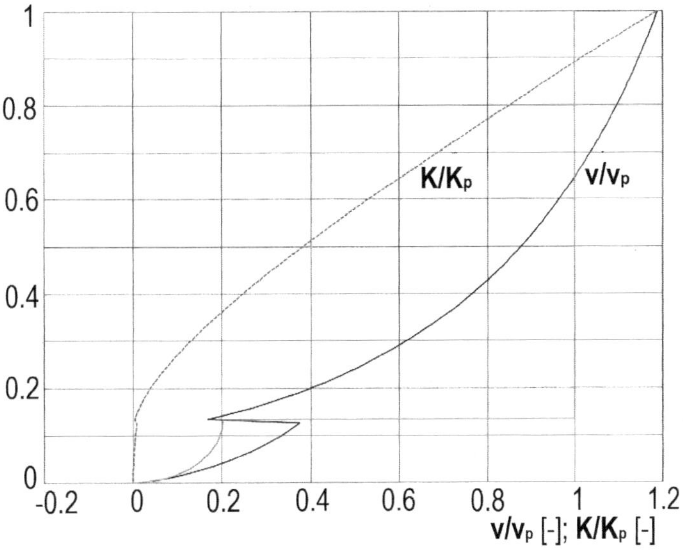

Figure 8. Partial filling hydraulic characteristics $Q/Q_p=K/K_p=f_Q(h/H)$ and $v/v_p=W/W_p=f_v(h/H)$ to the non-standard collector with polycentric section.

The results presented in the table are used as a database for the simulation of the gradually varied movement in the collector.

The program requires field measurements of the current coordinates of the perimeter of the flow section which has been modified by various actions. Measurements must be made on a collector section and then processed to obtain a characteristic calculation section [4]. An important step in defining the primary calculation data is represented by the activity in the field, inside the visitable collector, for the analysis of the roughness on the perimeter of the flow section. It is worth mentioning the change in roughness at the top of the sewer by degrading the material under the action of chemical corrosion caused by the transported wastewater.

The value of the equivalent roughness „k_e" and the roughness coefficient „n" after Manning will have to be analyzed and measured on characteristic perimeter lengths by research in the collector section. The data collected will be processed and entered into the input database. The rehabilitation works of the collector section have a great influence, which always changes its roughness. A condition for the application of this program is the continuous improvement of the collector database.

Based on the analysis of the data and the resulting conclusions, it can be stated that the mathematical model for the hydraulic-functional characteristics of sewer pipes with non-standard sections composed of two circular arcs can be used to solve the problems of sizing and checking for flows in uniform permanent motion. The data obtained with this program can be used to calculate the gradually varied and rapidly varied non-uniform motion formed in non-standard closed channels of this type.

5. Conclusions

Sewage collectors of the "visitable" type have a long service life, and under the action of natural and anthropogenic factors there are complex processes of degradation of the drainage section. This situation causes the flow section to change to a different shape than the designed one.

Visitable sewer collectors may have flow sections modified by hydrodynamic erosion phenomena, clogging processes, but also by technological repair and rehabilitation works. The degradation processes determine the shape and geometric dimensions of the flow section and implicitly the value of the hydraulic parameters.

The monitoring of the hydraulic parameters of the collector in the operation process requires their measurement and knowledge with high precision, a situation that cannot be achieved with collectors with non-standard shapes.

The elaborated calculation program (Col.NS_UM_UB) has the advantage that it allows to obtain the geometric and hydraulic parameters of the sewer collector with non-standardized section with a high precision and in a relatively short working time.

The elaborated calculation program (Col.NS_UM_UB) has the advantage that it allows to take into account the variation of the roughness on the watered perimeter on sectors with variable lengths.

The realized hydraulic-mathematical model can be generalized to obtain the geometric and hydraulic characteristics ($B = f_B(h)$, $P = f_P(h)$, $A = f_A(h)$, $R_h = f_{Rh}(h)$, $z_G = f_{zG}(h)$, $W = f_W(h)$ and $K = f_K(h)$) for polycentric closed sewer sections.

References

[1] Blitz E, Pislarasu I, Vladimirescu G 1970 Sewers *Hydrotechnical Engineer's Manual* vol 2 (Bucharest, Romania: Technical Publishing House) pp 984–992 (in Romanian).

[2] Luca M 1998 *Technical hydraulics - Permanent movement in canals* vol. 1 (Iasi, Romania: Tehnopress Publishing House) pp 107–118 (in Romanian).

[3] Brito RS, Almeida MC and Ribeiro A 2021 Enhancing hydraulic data reliability in sewers *Water Practice and Technology* **17**(1) 431–444 https://doi.org/10.2166/wpt.2021.093

[4] Luca M, Crețu FAC and Scripcariu CF 2017 Consideration on topographic parameters of rehabilitation sewer collectors *RevCAD Journal of Geodesy and Cadastre* **22** (Alba Iulia, Romania: „1 Decembrie" University of Alba Iulia, Aeternitas Publishing House) 237–242 http://revcad.uab.ro/upload/41_624_28_Luca_Cretu_Scripcariu_C.pdf

[5] Carlier M 1973 *Hydraulique générale et appliqueé,* (Paris Editure Eyrolles).

[6] Cazanescu S, Mărăcineanu F and Tadjer J 2010 Sewer network design using urbano Canalis version 7 release 2010 *Sci. Bull. „Politehnica" University Timișoara – Hydrotechnics Trans.* **55/69**(1) 81–86 https://www.ct.upt.ro/buletinhidro/Files/014-Severin_Caz.pdf

[7] Fuchs L, Beeneken T, Sponemann P and Scheffer C 1997 Model based real-time control of sewer system using fuzzy logic, *Water Sci. Technol.* **36**(8-9) 343–347 https://doi.org/10.1016/S0273-1223(97)00575-1

[8] Safari MJS, Aksoy H, Unal NE and Mohhamadi M 2016 Self-cleansing Design of Circular Urban Drainage Channels *Proc. 3rd Int. Conf. Water Across Time in Engineering Research* (23-25 June Constanța, Romania) pp 245–252

[9] Luca M, Tamaşanu F and Luca AL 2012 Flow modeling with free level in the polycentric channels *"Ovidius" University Annals of Constanța Series: Civil Engin*eering **14**, 155–162 https://revista-constructii.univ-ovidius.ro/doc/editii/2012.pdf

[10] Scripcariu CF. Contributions to the optimization of sewage network in order to increase operational safety [thesis]. „Gheorghe Asachi" Technical University of Iasi, Romania; 2018, (in Romanian).

[11] Luca M, Luca AIL, Scripcariu CF, Chirica S. 2019 Contributions to simulation non-permanent movement in sewerage collectors. IOP Conf. Series Materials Science and Engineering 586, 012037, IOP Publishing . https://doi: 10.1088/1757-899X/586/1/012037

Numerical simulations of flow over a side weir for diversion structures and water intakes

Daniela-Elena Gogoașe Nistoran[1,3], Ștefan-Mugur Simionescu[1,3], Gabriela Cîrciumaru[2] and Rareș-Andrei Chihaia[2]

[1] Department of Hydraulics, Hydraulic Machinery and Environmental Engineering, Faculty of Energy Engineering, University POLITEHNICA of Bucharest, 060042, Romania
[2] National Institute of Research and Development in Electrical Engineering ICPE-CA, Bucharest, 030138, Romania
[3] Author to whom any correspondence should be addressed

daniela.nistoran@upb.ro, s.simionescu@upb.ro

Abstract. Side weirs are hydraulic structures used to divert the excess flow from a main channel into a lateral one at diversion structures or intakes, in order to control the water surface elevation or reroute part of the discharge. Flow over a side weir is considered to be a typical case of spatially varied flow, with decreasing discharge along the main channel and varying water surface elevation along the lateral crest. The objective of the paper is to apply various numerical modelling approximations (1D, 2D and 3D) using HEC-RAS and ANSYS Fluent software to virtually reproduce the flow characteristics over the side weir along the diversion canal of the Valea Iasului hydropower plant (HPP), on Argeș River (Romania). Different geometries, mesh types and sizes and the appropriate initial and boundary conditions are used for two HPP operating scenarios (fully functional and completely shut down). The water surface elevation profile along the weir crest is computed, together with a qualitative comparison between the shape of the simulated flow nappe and in-situ visualizations. Quantitative results from the numerical simulation cases are given in terms of the computed upstream and downstream weir rating curves.

1. State of the art in experiments and numerical simulations

Side weirs, also known as lateral weirs, are hydraulic control structures used to divert a part of the flow from a main channel into a lateral channel to control the water surface elevation. These regulation structures used to divert the excess flow are mainly encountered at hydropower plant (HPP) intakes, dams or derivations, flood-control works, navigation and irrigation canals, urban drainage systems, treatment, sewerage and wastewater treatment facilities [1-3].

Numerous papers brought their contribution towards a better understanding of how the geometry of the main channel influences the hydraulics of flow over the side weir. This was accomplished through theoretical studies [4], [5], experimental investigations [6-13], numerical studies [14-18], or both experimental and numerical studies [19], [20]. The papers analyzed the flow through different cross-section shapes of the main channel, such as rectangular, triangular, trapezoidal, circular and U-shaped, fitted with various side weir types, as: sharp crested, broad crested, or round crested (ogee-shaped) profile with different opening shapes, such as: straight or oblique rectangular, trapezoidal, or triangular and even labyrinth or piano-key weir (used to increase the crest length). Most of the studies analyzed the weir discharge efficiency by finding the best equation for the discharge coefficient [18], [21-25], some analyzed the flow nappe shape as a function of the Froude number value in the main channel [26-28] or hydraulic parameters variation [20], whereas others looked for the best weir design [1]. The influence of the flow depth, velocity of approach and cross-section shape in the main channel, effect of the lateral outflow intensity and outflow angle were also investigated.

Flow over lateral weirs is acknowledged to be an example of spatially varied flow, with decreasing discharge along the main channel. However, usual theoretical assumptions assume simplified one-dimensional flow conditions in the main channel, neglecting the cross-section component of the velocity and the transverse variation of the free surface profile [20].

Depending on the flow regime (in terms of Froude number) in the main channel the water surface elevation in the weir opening increases along the crest for upstream subcritical flow ($Fr < 1$), decreases for upstream supercritical flow ($Fr > 1$) or may display a hydraulic jump for upstream critical flow ($Fr = 1$) [6].

Multiple numerical codes in different simplifying assumptions were used in the studies to reproduce the flow field over the side weirs. The HEC-RAS model [29] based on the 1D flow approximation was used by Un Ji et al [30] in 2013 to simulate the water surface level variation along the main channel due to the flow over side weir into a detention basin, during a flood wave, considering different discharge equations for the lateral flow. Simulation results were successfully validated with the help of experimental results, relative errors of the water levels being in the range of 2÷5%. For other 3D numerical studies, the ANSYS Fluent code [15], [16], and Flow 3D software [17], [31], [32] were mainly used to simulate the spatially-varied flow over the side weir, under different conditions.

The objective of the study is to simulate the flow over the side weir built on the approach channel of the Valea Iașului Hydropower Plant (HPP) on Argeș River (Romania), under different operation regimes and to analyse how some of the hydraulic parameters and flow characteristics vary in time and space. Simulations performed using the CFD ANSYS Fluent code are discussed comparatively with results from the 1D and 2D flow approximations with the HEC-RAS software.

2. Theoretical background

2.1. Flow hydraulics over the side weir

Figure 1 depicts a sketch of the simplified one-dimensional flow, in a trapezoidal channel over a lateral weir, whose crest length L is usually less than 3 times the channel top width, B. The approaching (upstream) flow, Q_1, divides into the lateral spilled (diverted) flow, Q_S, over the side weir and the onward (downstream) flow, Q_2 [3]. The water surface profile along the side weir depends on the flow regime in the main channel as follows:

- If the flow in the upstream mild-slope channel is subcritical ($Fr_1 < 1$) and the weir crest elevation, p, is higher than the critical depth, y_{c1}, corresponding to Q_1, ($y_0 > y_1 > p > y_{c1}$), then the longitudinal water surface along the side weir will increase in the downstream direction and the flow regime downstream the side weir will be subcritical as well. The flow is subcritical along the entire reach, and this is type I of side weir behaviour (Figure 1.a).
- If, however, the weir crest elevation is lower than the upstream critical depth, y_{c1}, then y_1 may be considered equal to y_{c1} and the upstream flow regime becomes critical ($Fr_1 = 1$). The longitudinal water surface along the side weir will decrease and the flow regime will be supercritical down to a cross-section where the flow will pass through a hydraulic jump into a subcritical regime. The position of this jump could advance into the side weir nappe area if the tailwater depth, y_t, increases. This is type II of side weir behaviour with mixed flow along the weir reach (Figure 1.b).
- On the other hand, if the upstream flow is supercritical ($Fr_1 > 1$, $y_1 < y_{c1}$), it also has to be uniform, and the weir crest is $p < y_{c1}$. The longitudinal water surface along the side weir will decrease and the flow regime will be supercritical all the way down. As a simplifying assumption, the cross-section specific energy, E, may be considered constant along the side weir reach. This is type III of side

weir behaviour with supercritical flow along the entire reach (Figure 1c).

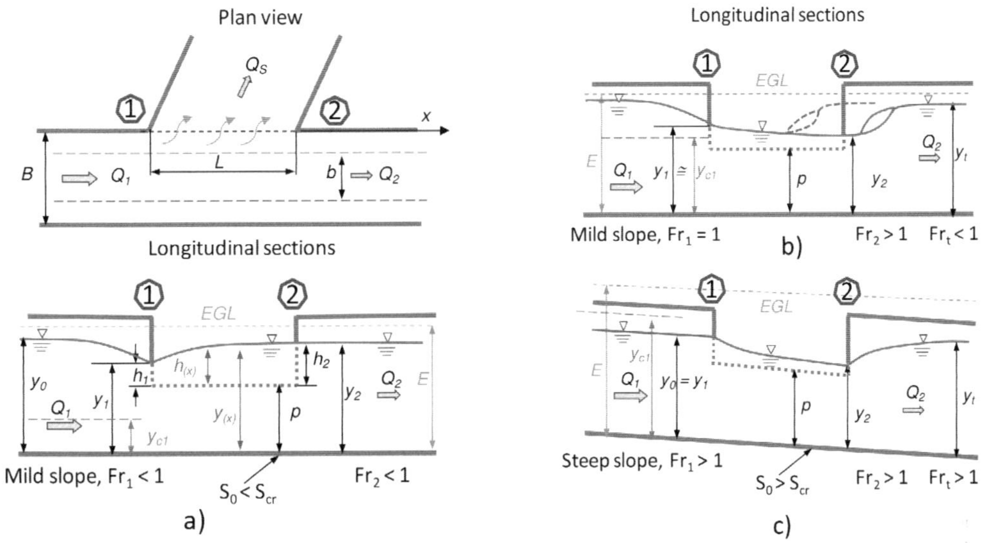

Figure 1. Types of longitudinal profile over the side weir as a function of upstream flow regime and downstream depth: a) type I (subcritical flow), b) type II (mixed flow), c) type III (supercritical flow).

2.2. Equations

To determine the discharge over the side weir, Q_S, in the 1D approximation, researchers looked for a similar equation to the one for the classical case of an inline weir, with the difference that the flow nappe has a variable head, h, along the flow direction, x. As a consequence, the lateral weir has two head-discharge characteristics (rating curves), one at the upstream end, $h_1 (Q_S)$ and another at the downstream end $h_2 (Q_S)$. Under simplifying assumptions, the specific flow rate over the side weir, $q(x)$, can be expressed as:

$$q(x) = \frac{Q_S}{L} = -\frac{dQ}{dx} = \frac{2}{3} C_d \sqrt{2g} [y(x) - p]^{\frac{3}{2}} = \frac{2}{3} C_d \sqrt{2g} h(x)^{\frac{3}{2}} \tag{1}$$

where Q is the discharge in the main channel, x is the distance from cross-section 1, upstream the side weir (Figure 1), dQ/dx is the specific discharge over 1 m length of the lateral opening, g is the acceleration of gravity, $y(x)$ is the depth, and $h(x) = (y(x) - p)$ is the variable head over the crest of the side weir with distance. C_d is the non-dimensional discharge coefficient of the side weir, depending on the upstream Froude number and different geometric ratios, as follows:

$$C_d = f\left(F_1 = \frac{V_1}{\sqrt{g y_1}}, \frac{L}{b}, \frac{L}{y_1}, \frac{p}{y_1}\right) \tag{2}$$

where V_1 is the stream velocity in the main channel in cross-section 1 upstream the side weir, y_1 is the upstream depth in the parent channel and p is the weir wall (Figure 1). The total diverted flow can be then calculated by integrating Equation (1) over the total length of the side weir, L:

$$Q_S = Q_1 - Q_2 = \frac{2}{3} C_d \sqrt{2g} \int_0^L (y(x) - p)^{\frac{3}{2}} \tag{3}$$

De Marchi [33] was the first scientist who solved the 1D dynamic equation of steady, gradually varied flow with non-uniform discharge for a horizontal, rectangular channel, assuming the specific energy, E, remains constant along the flow reach (dE/dx = 0), irrespective of the friction and found analytically an expression of C_d (De Marchi coefficient). Later on, other authors introduced simplified assumptions based on the shape of water surface profile along the side weir [22]. Many other authors [4], [11], [12], [16], [23] developed experimentally different equations for determining the discharge coefficient.

In the 2D approximation the Shallow Water Equations (SWE) are solved by the HEC-RAS software, in which, the Coriolis effect is neglected for small scale analyses.

The general case of a 3D two-phase flow can be solved through the Reynolds Averaged Navier Stokes equations with turbulence model. For the current study, the simulations are performed using the ANSYS Fluent code, solving the governing equations of the flow based on the finite volume method. The k - ω Shear Stress Transport (SST) turbulence model and the volume-of-fluid (VOF) method, which can simulate the flow of two or more immiscible fluids by solving a single set of momentum equations and tracking the volume fraction of each of the fluids throughout the flow domain. They allow both the calculation of turbulent kinetic energy and the length scale for turbulent structures.

2.3. Design

The design of side weirs is not as simple as for transverse weirs, the key parameters for the design of the side weir being: the crest elevation, z_c and the horizontal (recommended) crest length, L. Firstly, the water level at the upstream end of the weir is found from the gradually varied flow equation, when there is no lateral flow. This way the crest level is found. In the hypothesis of zero onward flow (closed penstocks of the HPP in the study case), the weir length is then calculated so that all the incoming discharge flows over the weir. Finally, one has to verify that the flow regime is subcritical at the upstream end of the weir, to avoid the risk of hydraulic jump forming. Researchers found that the angle of the lateral channel axis does not influence Q_S very much [1]. The direction of the velocity vector field over the side weir found from numerical simulations help engineers in designing the best angle.

3. Site and data

Argeş River is one of the main tributaries of Danube River, along which a series of 17 hydropower plants (HPPs) were built between 1960 and 1995. One component of these cascade HPPs is the run-of-river Valea Iaşului HPP, built on a diversion canal, in parallel with Arges River (Figure 2)

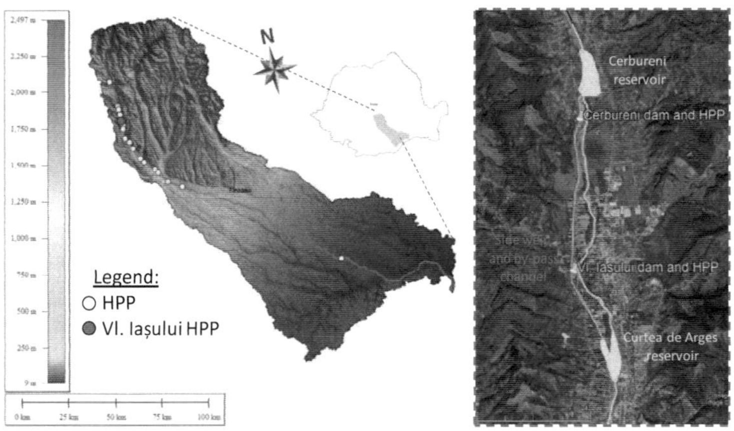

Figure 2. Location of Valea Iaşului HPP and side weir.

At the upstream end of the feeder canal is another HPP, Cerbureni, with the same installed discharge Q = 90 m^3/s and gross head, H_i = 20.5 m as Valea Iaşului HPP. Both HPPs are equipped with two Kaplan turbines, each of 45 m^3/s, leading to an installed power of 15 MW. The mean annual flow on this reach of Arges river is 14.5 m^3/s, at the Căpăţâneni gauging station, located upstream Valea Iaşului.

The derivation concrete-lined canal which connects the upstream Cerbureni HPP to the downstream Valea Iaşului HPP is 3.1 km long and has a trapezoidal-shape cross-section (Figure 3.a).

Figure 3. Valea Iaşului HPP feeder canal (a) and the side weir (b) (views from downstream).

The canal has an invert slope of 0.15% and side slopes of 1:2 H:V. It consists of two reaches: the 2.3 km in length escape canal of Cerbureni HPP, having a 20 m base width cross-section, and the 813 m in length feeder canal of Valea Iaşului HPP, having a 6 m base width cross-section, both connected by a 49 m long transition reach. The normal operating level (NOL) at the forebay of Valea Iaşului HPP is 444.5 m.a.s.l., which is the same as the crest level of the side weir.

When the downstream HPP has to shut down one or both its turbines, the discharge still coming from upstream will flow over the left-bank side weir (Figure 3.b). This discharged water flows into the Valea Danului River (a tributary of Argeş River whose channel underpasses the canal), which connects the HPP canal to the natural Argeş River channel (Figure 2). The total length of the side weir is of 75,18 m, and it is placed at a distance of 400 m from the forebay.

4. Methodology

4.1. Geometries and discretization mesh

HEC-RAS software uses for simulations under both 1D and 2D approximation the finite difference and finite volumes numerical methods, respectively. ANSYS-Fluent code uses for the 3D simulations the Volume of Fluid (VoF) method, to obtain the free surface at the interface.

For validation purposes and in order to set the correct conditions for the 3D simulations, a benchmark problem was performed: a comparison with a case from the literature [2], studying the flow in a rectangular free-surface channel with a side weir (Figure 4).

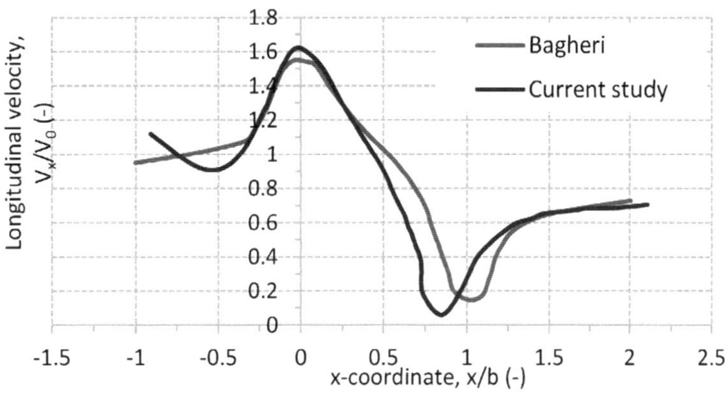

Figure 4. Comparison for validation: distribution of longitudinal velocity (V_x) along side weir.

Thus, a special geometry similar to the one in the specified work was built, so that the results can be compared. A good agreement was found in terms of outlet flowrate (less than 4% relative error), shape of the free surface curve over the weir, as well as the distribution of longitudinal velocity (V_x) along the side weir (Figure 4).

Table 1 presents the characteristics of the three flow domains and meshes used within the current research. To decrease the computational time of the numerical simulations, the 2D and 3D flow domains are shortened.

Table 1. Characteristics of the numerical domains and discretization meshes.

Type of numerical domain	Length of canal upstream the side weir (m)	Length of canal downstream the side weir (m)	Length of side canal downstream the weir (m)	Mesh size (cells)	Mesh type
1D	2755	350	0	38	50 m cross-section reaches
2D	1272	350	0	16300	rectangular, structured
3D	300	350	200	485000	hexahedral, unstructured

The 3D discretization mesh (Figure 5) has the following features:

- hexahedral, structured mesh cells along z, unstructured along x and y axis, with variable dimension ranging with a growth rate of 1.1, from 0.2 to 1 m;
- a size function was used in the vicinity of the weir, where the highest velocity gradients occur;
- 0.485 million cells with maximum shape skewness of 0.72.

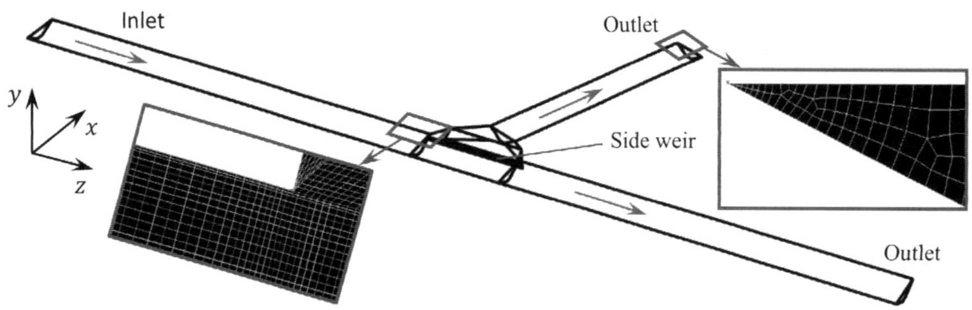

Figure 5. Numerical flow domain for the 3D case, with details of mesh.

4.2. *Initial and boundary conditions*

The boundary conditions for the numerical simulations are consistent with the worst-case scenario for the side weir, in which both turbines completely shut down from Q_i to 0, which means: constant flow hydrograph with $Q_i = 90$ m^3/s at the upstream end of canal, and a step flow hydrograph at the downstream canal end (at the forebay of Valea Iașului HPP).

The boundary conditions imposed for the simulations are presented in Table 2. The turbulence model used for the 3D simulations is k - ω SST. The qualitative and quantitative hydraulic parameters resulting from the simulations are compared for validation with the experiments and similar relevant results from the literature.

Table 2. Boundary conditions applied to the numerical flow domain.

Boundary type	Boundary conditions		
	1D	**2D**	**3D**
Inlet upstream	hydrograph $Q = ct = 90$ m^3/s	hydrograph $Q = ct = 90$ m^3/s	V = $ct.$ = 1.16 m/s constant level
Outlet downstream	hydrograph $Q_i = 90$; $Q_f = 0$ m^3/s	hydrograph $Q_i = 90$; $Q_f = 0$ m^3/s	$p_r = 0$ $\partial v_n/\partial n = 0$
Outlet lateral (weir)	free	free	$p_r = 0$ $\partial v_n/\partial n = 0$
Solid walls	v = v$wall$ = 0, Manning wall roughness 0.021 s/m$^{1/3}$	v = v$wall$ = 0, Manning wall roughness 0.021 s/m$^{1/3}$	v = v$wall$ = 0 stationary, smooth, impermeable walls
Interface	-	-	free-surface interface

5. Results and discussion

5.1. *1D case*

In Figure 6.a are shown the longitudinal profiles at the beginning of the simulation (HPP in operation and no water flows over the side weir) and after the HPP has completely shut down (and the flow is stabilised), which means all the upstream/incoming discharge flows over the side weir. Figure 6.b depicts the two rating curves at the upstream and downstream ends of the side weir.

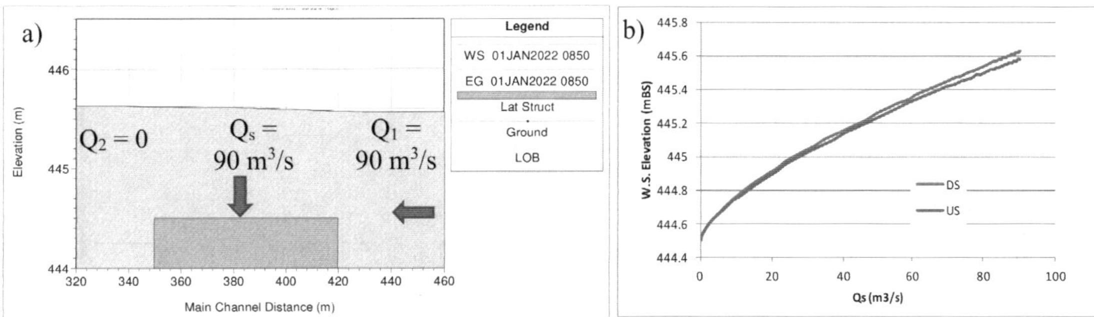

Figure 6. a) Long profile in the vicinity of the side weir after the shut down of the Valea Iaşului HPP; b) rating curves at both ends of the side weir.

5.2. *2D and 3D case*

Figure 7 shows the variation of the water surface level along the side weir, at the end of the simulation after the HPP has completely shut down and the flow is stabilised, which means all the upstream/incoming discharge flows over the side weir. One may see the water surface has a small increase along the flow (by about 5 cm).

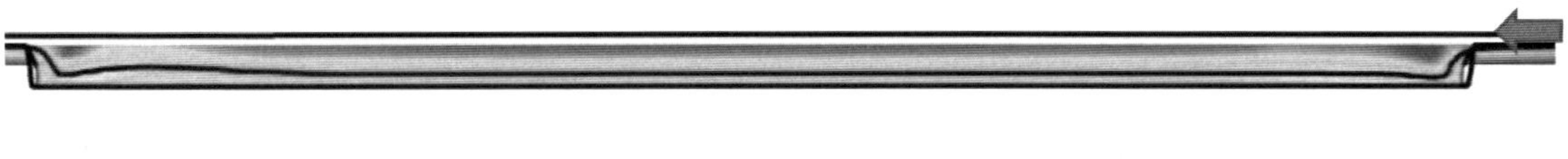

Figure 7. Variation of the water surface level along the side weir crest.

Figure 8 presents a comparative view of the water free surface in the area of the side weir, coloured by velocity magnitude, as it was obtained from the 2D and 3D numerical simulations, with the whole discharge flowing over the side weir. The streamlines and, respectively, velocity vectors, were plotted and make it easier to observe the direction change due to the side weir and the recirculation in the canal heading to the HPP. The free surface level in the area of the weir and over the weir crest, coloured by velocity magnitude, as computed from the 3D model, is shown in Figure 9. The aspect of the flow nappe and the flow behaviour downstream the lateral weir show a good agreement with the discharge, as observed in situ.

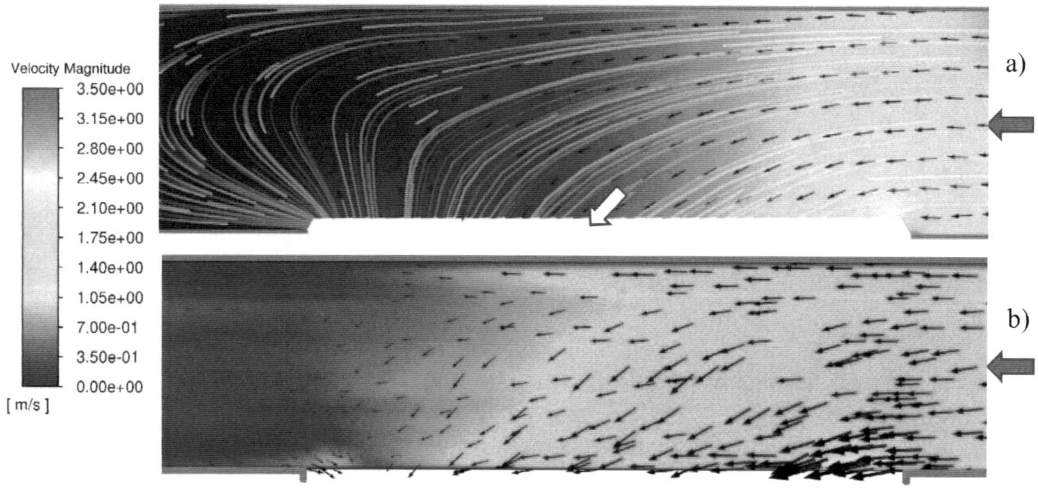

Figure 8. View from above of the flow in the main canal and over the side weir, overlapped with streamlines (2D model, a) and velocity vectors (3D model, b), after the downstream HPP has shut down.

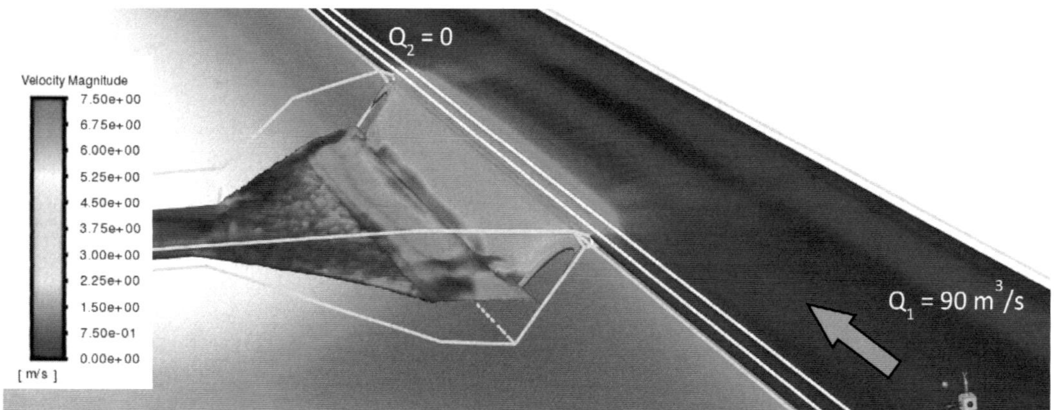

Figure 9. Free surface level over the weir crest, coloured by velocity magnitude (3D model).

6. Conclusions

One may observe from the computed rating curves in Figure 5.b that the nappe elevation is higher at the downstream end, which is in accordance with the type I behavior of the side weir under subcritical flow regime shown in Figure 1b. The water level increase of about 5 cm for Q_S = 90 m^3/s, towards the downstream end of the weir, may also be observed from the flow profile in Figure 1.a. Computed Froude numbers at the upstream and downstream end of the side weir are 0.13 and 0.06, respectively, which confirm the subcritical regime.

The computed discharge over the weir after the HPP is shut down resulted Q_S = 90 m^3/s in the 1D and 2D flow approximation. The computed flow nappe from the 3D CFD simulation has a similar shape with the real one.

The characteristics of the velocity vector field over the weir help engineers in designing the lateral channel axis direction. The mean value of this angle measured in Figure 7.b is of 65°, which is close to 68°, the one selected by the designer for the prototype.

Acknowledgments
This work was funded by a grant of the Romanian National Authority for Scientific Research, CNCS, UEFISCDI, through the project PN-III-P2-2.1-PED-2019-1444, "Eco-hybrid water intake, with behavioural barrier to reduce the impact on fish fauna and river morphology – Eco-WIBB".

References

[1] May RWP, Bromwich BC, Gasowski Y and Rickard CE 2003 *Hydraulic Design of Side Weirs* Thomas Telford, University Press, Cambridge
[2] Bagheri S and Heidrpour M 2012 Characteristics of flow over rectangular sharp-crested side weirs *Journal of Irrigation and Drainage Engineering* **138** 6
[3] Subramanya K 2009 *Flow in Open Channels (3rd ed.)* Tata McGraw-Hill Publishing Company Ltd. pp 398-407
[4] Subramanya K and Awasthy S C 1972 Spatially varied flows over side weirs *J. of Hydraulic Division* 98(1) 1-10
[5] Yilmaz M 2002 Lateral Weir flow model using a curve fitting analysis, *Journal of Hydraulic Engineering* **128**(7) 712-715
[6] Borghei SM, Jalili MR and Ghodsian M 1999 Discharge coefficient for sharp-crested side weir in subcritical flow *Journal of Hydraulic Engineering* **125** 10
[7] Emiroglu ME and Kaya N 2011 Discharge coefficient for trapezoidal labyrinth side weir in subcritical flow *Water Resource Management* **25** 1037-1058
[8] Al-Taee A 2011 Theoretical analysis of flow over the side weir using Runge Kutta method, *Annals of Faculty Engineering Hunedoara, International Journal of Engineering* **IX** Fascicule 2 pp 47-50 https://annals.fih.upt.ro/pdf-full/2011/ANNALS-2011-2-06.pdf
[9] Madjid D and Ebrazi BB 2014 An examination of rectangular side weir discharge coefficient equations under subcritical condition, *International Journal of Hydraulic Engineering* **3**(1) 24-34
[10] Lee OL, Yoon KS, Lee JS and Hong SH 2019 Estimates of discharge coefficient in levee breach under two different approach flow types *Sustainability* **11** 1-13 2374
[11] Bagheri S, Kabiri-Samani AR and Hidarpour M 2014 Discharge coefficient of rectangular sharp-crested side weirs, Part I: Traditional weir equation *Flow Measurement and Instrumentation* **35** 109-115
[12] Bagheri S, Kabiri-Samani AR and Hidarpour M 2014 Discharge coefficient of rectangular sharp-crested side weirs, Part II: Domínguez's method *Flow Measurement and Instrumentation* **35** 116-121
[13] Wang Y, Wang W, Hu X and Liu F 2018 Experimental and numerical research on trapezoidal sharp-crested side weirs *Flow Measurement and Instrumentation* **64** 83-89 https://doi.org/10.1016/j.flowmeasinst.2018.10.005
[14] Uyumaz A 1997 Side weir in U-shaped channels *Journal of Hydraulic Engineering* **111**(1) 639-646
[15] Azimi H and Shabanlou S 2017 Numerical simulation of free surface and flow field turbulence in a circular channel with the side weir in subcritical flow *Intl. J. of Nonlinear Sciences and Numerical Simulation* **18** (3-4) https://doi.org/10.1515/ijnsns-2017-frontmatter3-4
[16] Aydin MC 2016 New approach for estimation of rectangular side weirs discharge in subcritical flow *Journal of Irrigation and Drainage Engineering* 10.1061/(ASCE)IR.1943-4774.0001014

[17] Abassi S, Fatemi S, Ghaderi A and Di Francesco S 2021 The effect of geometric parameters of the antivortex on a triangular labyrinth side weir *Water* **13** 14 https://dx.doi.org/10.3390/w13010014
[18] Ghaderi A, Dasineh M, Abbasi S and Abraham J 2020 Investigation of trapezoidal sharp-crested side weir discharge coefficients under subcritical flow regimes using CFD *Applied Water Science* **10**(31) 1-12
[19] Uyumaz A and Muslu Y 1985 Flow over side weirs in circular channels, *Journal of Hydraulic Engineering* 111(1) 144-160
[20] Hager W H 1987 Lateral outflow over side weirs *Journal of Hydraulic Engineering* **113**(4) 491-504
[21] Ghodsian M 2003 Supercritical flow over a rectangular side weir *Canadian Journal of Civil Engineering* **30** 596-600
[22] Di Bacco M and Scorzini AR 2019 Are we correctly using discharge coefficients for side weirs? Insights from a numerical investigation *Water* **11** 2585 doi:10.3390/w11122585;
[23] Emiroglu ME, Agaccioglu H and Kaya N 2011 Discharging capacity of rectangular side weirs in straight open channels *Flow Measurement and Instrumentation* **22** 319-330
[24] Farzaneh N, Davood F and Ali NM 2015 Discharge coefficient for trapezoidal side weir *Alexandria Engineering Journal* **54** 595-605
[25] Parvaneh A, Parvaneh M, Rakhshandehroo GR, Ghazizadeh MRJ and Sadeghian H 2022 Discharge characteristics of a novel inclined-bed triangular side weir in subcritical flow *Journal of Irrigation and Drainage Engineering* **148** 2
[26] Castro-Orgaz O and Hager W 2012 Subcritical side-weir flow at high lateral discharge, *Journal of Hydraulic Engineering ASCE* **138**(9) 777-787
[27] Durga Rao KHV and Pillai CRS 2008 Study of flow over side weirs under supercritical conditions *Water Resource Management* **22** 131-143
[28] Oliveto G, Biggiero V and Fiorentino M 2001 Hydraulic features of supercritical flow along prismatic side weirs *Journal of Hydraulic Research* **39**(1) 73-82 DOI: 10.1080/00221680109499804
[29] Brunner G W 2010 *HEC-RAS, River Analysis System. Hydraulic Reference Manual* (U.S. Army Corps of Engineers, Hydrologic Engineer Center, Davis, CA)
[30] Ji U, Kim S, Yoon B and Kim S 2013 Analytical and experimental investigation of a side-weir detention basin in flood-level reduction in the Main Channel *Journal of Irrigation and Drainage Engineering* **139**(8) https://doi.org/10.1080/00221686.2020.1818304
[31] Azimi H and Shabanlou S 2018 Numerical study of bed slope change effect of circular channel with side weir in supercritical flow conditions *Applied Water Science* **8** 166 https://doi.org/10.1007/s13201-018-0816-5
[32] Maranzoni A 2021 3D CFD analysis of the performance of oblique and composite side weirs in converging channels *Journal of Hydraulic Research* **59**(4) 586-604
[33] De Marchi G 1934 Saggio di teoria del funzionamento degli stramazzi laterali *L'energia elettrica* **11**(11) 849-860
[34] Ferziger JH and Perić M 2002 *Computational Methods for Fluid Dynamics* Third Edition (Springer) DOI: 10.1007/978-3-642-56026-2
[35] Dănailă S and Berbente C 2003 *Metode Numerice în Dinamica Fluidelor* (in Romanian) (Bucharest: Editura Academiei Române)
[36] ANSYS Fluent [Computer Software] 2022 *User Guide* (ANSYS, Canonsburg, PA).

A POD-based reduced order model applied to 1D shallow water equations

Pablo Solán-Fustero[1], José Luis Gracia[2], Adrián Navas-Montilla[1] and Pilar García-Navarro[1]

[1] Fluid Dynamic Technology, I3A, University of Zaragoza, Zaragoza, 50009, Spain
[2] IUMA and Department of Applied Mathematics, University of Zaragoza, Spain

psolfus@unizar.es

Abstract. Many environmental problems involving free surface flow can be solved using the shallow water equations (SWE) often involving high computational costs due to the large spatial and temporal scales of the events. In recent times, reduced order models (ROM) techniques are increasingly used to improve the computational efficiency of simulation models. The Proper Orthogonal Decomposition (POD) method provides an orthogonal basis for representing a given set of data and constructing the ROM by means of the method of snapshots. In this work, a POD-based intrusive ROM strategy is applied to the 1D SWE. The main goal of this work is to build a simulation model able to reproduce realistic scenarios. We analyse the computational improvement and the accuracy of the ROM results with respect to those of the full-order model (FOM).

1. Introduction

The shallow water equations (SWE) are widely used as mathematical model to represent the time evolution of free surface flows in channels. The SWE can be solved computationally following several numerical methods [1-3], here called full order models (FOM) that may involve high computational costs in realistic cases, thus requiring the use of mathematical tools to speed up the computations.

Among the various tools currently developed, reduced order models (ROM) based on proper orthogonal decomposition (POD) [4] allow the resolution of partial differential equations more efficiently than FOM and with little loss of accuracy [5-7]. This requires the proper interval decomposition (PID) method to structure the snapshot method in different time windows [8,9]. The ROM strategy has a first stage, off-line part, in which the FOM is solved. The solutions obtained by the FOM are used to train the ROM and then solve it (on-line part). In this work, two FOM are used: one based on Lax-Friedrichs numerical flux due to its simplicity [2] and another following Roe's method due to its robustness and good performance [1]. They are followed by an intrusive ROM using the Galerkin method [10]. Due to the non-linear character of the equations, it is necessary to make use of time averages of the variables of interest following [11].

With this objective, we first compare the results obtained by two ROMs developed from the Lax-Friedrichs FOM, one standard (version 1) and the other with averages of the water velocity (version 2). Then the time averaging approach is applied to the Roe's method (version 3).

The second objective of this work is to determine the best set of ROM parameters in terms of accuracy and CPU time. This will be done by measuring the errors of the computed solutions compared to the exact solution [12] and the CPU times required to obtain them.

2. Mathematical model and numerical method

As mentioned in the introduction, the cases considered in this paper involve the 1D SWE in the framework of a frictionless channel of rectangular cross-section and constant unit width [1]

$$\frac{\partial}{\partial t}\begin{pmatrix} h \\ q \end{pmatrix} + \frac{\partial}{\partial x}\begin{pmatrix} q \\ q^2/h + gh^2/2 \end{pmatrix} = \begin{pmatrix} 0 \\ 0 \end{pmatrix}, \tag{1}$$

where h is the water depth and $q = hu$ is the water discharge per unit width, with u, the cross-sectional average water velocity; g is the gravitational acceleration.

Regarding the FOM numerical method, the computational domain is discretized by means of N_x volume cells of uniform length Δx and the positions of the centre and left and right interfaces of the i-th cell are x_i, $x_{i-1/2}$ and $x_{i+1/2}$, respectively, with $i = 1, \dots, N_x$. The time step $\Delta t = t^{n+1} - t^n$ is selected dynamically using the Courant-Friedrichs-Lewy (CFL) condition [13].

Two different FOMs are used to solve the SWE system (1) using Godunov's scheme

$$\frac{\boldsymbol{U}_i^{n+1} - \boldsymbol{U}_i^n}{\Delta t} + \frac{\boldsymbol{F}_{i+1/2}^{n,*} - \boldsymbol{F}_{i-1/2}^{n,*}}{\Delta x} = \begin{pmatrix} 0 \\ 0 \end{pmatrix}, \qquad i = 1, \dots, N_x, \tag{2}$$

$$\boldsymbol{U}_i^n = \begin{pmatrix} h_i^n \\ q_i^n \end{pmatrix}, \quad \boldsymbol{F}(\boldsymbol{U}_i^n) = \begin{pmatrix} q_i^n \\ \dfrac{(q_i^n)^2}{h_i^n} + \dfrac{1}{2} g (h_i^n)^2 \end{pmatrix}, \tag{3}$$

where $h_i^n \approx h(x_i, t^n)$, $u_i^n \approx u(x_i, t^n)$ and $q_i^n \approx q(x_i, t^n)$ are the cell average values of the water depth, velocity and discharge over the cell $(x_{i-1/2}, x_{i+1/2})$.

In the first case, the numerical fluxes are given by the Lax-Friedrichs method [2]

$$\boldsymbol{F}_{i+1/2}^{n,*} = \frac{1}{2}[\boldsymbol{F}(\boldsymbol{U}_{i+1}^n) + \boldsymbol{F}(\boldsymbol{U}_i^n)] - \frac{1}{2}\nu \frac{\Delta x}{\Delta t}(\boldsymbol{U}_{i+1}^n - \boldsymbol{U}_i^n), \tag{4}$$

with $CFL \leq \nu \leq 1$.

The second FOM is formulated using Roe's numerical flux

$$\boldsymbol{F}_{i\pm1/2}^{n,*} = \pm \sum_{j=1}^{2} \left(\tilde{\lambda}_j^{\mp} \tilde{\alpha}_j \tilde{\boldsymbol{e}}_j\right)_{i\pm1/2}^n, \qquad i = 1, \dots, N_x, \tag{5}$$

where $\left(\tilde{\lambda}_j^{\mp}\right)_{i\pm1/2}^n = \frac{1}{2}\left(\tilde{\lambda}_j \mp |\tilde{\lambda}_j|\right)_{i\pm1/2}^n$, $j = 1,2$, and where

$$\begin{cases} \left(\tilde{\lambda}_1\right)_{i+1/2}^n = \tilde{u}_{i+1/2}^n - \tilde{c}_{i+1/2}^n \\ \left(\tilde{\lambda}_2\right)_{i+1/2}^n = \tilde{u}_{i+1/2}^n + \tilde{c}_{i+1/2}^n \end{cases}, \quad \begin{cases} (\tilde{\boldsymbol{e}}_1)_{i+1/2}^n = \left(1, \left(\tilde{\lambda}_1\right)_{i+1/2}^n\right)^T \\ (\tilde{\boldsymbol{e}}_2)_{i+1/2}^n = \left(1, \left(\tilde{\lambda}_2\right)_{i+1/2}^n\right)^T \end{cases},$$

$$\begin{cases} (\tilde{\alpha}_1)_{i+1/2}^n = \dfrac{\delta h_{i+1/2}^n \left(\tilde{\lambda}_2\right)_{i+1/2}^n - \delta q_{i+1/2}^n}{\left(\tilde{\lambda}_2\right)_{i+1/2}^n - \left(\tilde{\lambda}_1\right)_{i+1/2}^n} \\ (\tilde{\alpha}_2)_{i+1/2}^n = \dfrac{\delta q_{i+1/2}^n - \delta h_{i+1/2}^n \left(\tilde{\lambda}_1\right)_{i+1/2}^n}{\left(\tilde{\lambda}_2\right)_{i+1/2}^n - \left(\tilde{\lambda}_1\right)_{i+1/2}^n} \end{cases}, \tag{6}$$

with $\delta h_{i+1/2}^n = h_{i+1}^n - h_i^n$ and $\delta q_{i+1/2}^n = q_{i+1}^n - q_i^n$, and the average velocities

$$\tilde{c}_{i+1/2}^n = \sqrt{g \frac{1}{2}(h_{i+1}^n + h_i^n)}, \quad \tilde{u}_{i+1/2}^n = \frac{q_{i+1}^n \sqrt{h_{i+1}^n} + q_i^n \sqrt{h_i^n}}{\sqrt{h_{i+1}^n h_i^n}\left(\sqrt{h_{i+1}^n} + \sqrt{h_i^n}\right)}. \tag{7}$$

3. Reduced-order model strategy

The POD-based ROM in this work is based on the snapshot method [14], which consists of the computation of a set of N_t time numerical solutions of SWE, (h_i^n, u_i^n), also called snapshots, as numerical approximations to h and u using the FOM. The snapshots are used to construct the snapshot matrices $M_h = (\boldsymbol{h}^1, \cdots, \boldsymbol{h}^{N_T}), M_u = (\boldsymbol{u}^1, \cdots, \boldsymbol{u}^{N_T})$ and $M_q = (\boldsymbol{q}^1, \cdots, \boldsymbol{q}^{N_T})$. The POD basis of functions is computed by applying the singular value decomposition of these matrices. The Galerkin method [10] and these bases are used to reconstruct the numerical solutions

$$h_i^n \approx \sum_{k=1}^{N_{POD}} \hat{h}_k^n \phi_{i,k}, \quad u_i^n \approx \sum_{k=1}^{N_{POD}} \hat{u}_k^n \varphi_{i,k}, \quad q_i^n \approx \sum_{k=1}^{N_{POD}} \hat{q}_k^n \Phi_{i,k}, \tag{8}$$

where $\phi_{i,k}$, $\varphi_{i,k}$ and $\Phi_{i,k}$ are the functions of the basis of each variable, being the number of POD modes $N_{POD} \ll N_x$.

The ROM of (1) is obtained by: i) introducing the Galerkin method (8) into the FOM (2) or (5); ii) multiplying each resulting equation by $\phi_{i,k}$ and $\Phi_{i,k}$, respectively; and iii) summing up over the cells. This procedure meets difficulties when the FOM scheme is nonlinear, as in the case of Roe's scheme. Zokagoa and Soulaïmani in [11] propose to use time averages of some variables where appropriate. Before applying this approach, the performance of two ROMs developed from the same Lax-Friedrichs FOM with and without time averages of$u$are compared.

With the aim of improving the resolution of non-linear problems, Zokagoa and Soulaïmani in [11] make use of the PID method, originally introduced by [8]. Following this method, the total simulation time T is partitioned into N_w non-overlapping time windows $[0, t_1] \cup [t_2, t_3] \cup \cdots \cup [t_{N_t-1}, t_{N_t} = T]$, so that as many snapshot matrices (and POD bases) are generated as there are time windows $M_{h_w} = (\boldsymbol{h}^{1_w}, \cdots, \boldsymbol{h}^{N_{T_w}}), M_{u_w} = (\boldsymbol{u}^{1_w}, \cdots, \boldsymbol{u}^{N_{T_w}})$ and $M_{q_w} = (\boldsymbol{q}^{1_w}, \cdots, \boldsymbol{q}^{N_{T_w}})$, with $w = 1, \ldots, N_w$. Hereafter the reference to time windows in the equations is omitted for the sake of clarity.

The vector formulations of three versions of the ROM of SWE are presented below. The first one (version 1) is based on the Lax-Friedrichs method without time averaging of u

$$\hat{\boldsymbol{h}}^{n+1} = \hat{\boldsymbol{h}}^n - \frac{\Delta t}{2\Delta x} A\hat{\boldsymbol{q}}^n + \frac{\nu}{2} B\hat{\boldsymbol{h}}^n, \quad \hat{\boldsymbol{q}}^{n+1} = \hat{\boldsymbol{q}}^n - \frac{\Delta t}{2\Delta x}(\hat{\boldsymbol{q}}^n)^T C\hat{\boldsymbol{u}}^n - \frac{g\Delta t}{4\Delta x} g\left(\hat{\boldsymbol{h}}^n\right)^T D\hat{\boldsymbol{h}}^n + \frac{\nu}{2} E\hat{\boldsymbol{q}}^n, \tag{9}$$

where the matrices, considering free boundary conditions (BCs), are

$$\begin{aligned}
A(k,p) &= (\Phi_{2,k} - \Phi_{1,k})\phi_{1,p} + \sum_{i=2}^{N_x-1} (\Phi_{i+1,k} - \Phi_{i-1,k})\phi_{i,p} + (\Phi_{N_x,k} - \Phi_{N_x-1,k})\phi_{N_x,p}, \\
B(k,p) &= (\phi_{2,k} - \phi_{1,k})\phi_{1,p} + \sum_{i=2}^{N_x-1} (\phi_{i+1,k} - 2\phi_{i,k} + \phi_{i-1,k})\phi_{i,p} + (\phi_{N_x,k} - \phi_{N_x-1,k})\phi_{N_x,p}, \\
C(q,k,p) &= (\Phi_{2,k}\varphi_{2,q} - \Phi_{1,k}\varphi_{1,q})\Phi_{1,p} + \sum_{i=2}^{N_x-1} (\Phi_{i+1,k}\varphi_{i+1,q} - \Phi_{i-1,k}\varphi_{i-1,q})\Phi_{i,p} + \\
&\quad (\Phi_{N_x,k}\varphi_{N_x,q} - \Phi_{N_x-1,k}\varphi_{N_x-1,q})\Phi_{N_x,p}, \\
D(q,k,p) &= (\phi_{2,k}\phi_{2,q} - \phi_{1,k}\phi_{1,q})\Phi_{1,p} + \sum_{i=2}^{N_x-1} (\phi_{i+1,k}\phi_{i+1,q} - \phi_{i-1,k}\phi_{i-1,q})\Phi_{i,p} + \\
&\quad (\phi_{N_x,k}\phi_{N_x,q} - \phi_{N_x-1,k}\phi_{N_x-1,q})\Phi_{N_x,p}, \\
E(k,p) &= (\Phi_{2,k} - \Phi_{1,k})\Phi_{1,p} + \sum_{i=2}^{N_x-1} (\Phi_{i+1,k} - 2\Phi_{i,k} + \Phi_{i-1,k})\Phi_{i,p} + (\Phi_{N_x,k} - \Phi_{N_x-1,k})\Phi_{N_x,p}.
\end{aligned} \tag{10}$$

The second one (version 2) is based on the same method, but considers time averages of the water velocity u, so that the Galerkin method is only applied to the water depth h and the water discharge q

$$\widehat{\boldsymbol{h}}^{n+1}=\widehat{\boldsymbol{h}}^{n}-\frac{\Delta t}{2\Delta x}A\widehat{\boldsymbol{q}}^{n}+\frac{\nu}{2}B\widehat{\boldsymbol{h}}^{n},\quad \widehat{\boldsymbol{q}}^{n+1}=\widehat{\boldsymbol{q}}^{n}-\frac{\Delta t}{2\Delta x}C\widehat{\boldsymbol{h}}^{n}-\frac{g\Delta t}{4\Delta x}g\left(\widehat{\boldsymbol{h}}^{n}\right)^{T}D\widehat{\boldsymbol{h}}^{n}+\frac{\nu}{2}E\widehat{\boldsymbol{q}}^{n}, \quad (11)$$

where the matrices are the same as in version 1, except for

$$\begin{aligned}C(q,k,p)=&\left(\phi_{2,k}(\bar{u}_2)^2-\phi_{1,k}(\bar{u}_1)^2\right)\Phi_{1,p}+\sum_{i=2}^{N_x-1}\left(\phi_{i+1,k}(\bar{u}_{i+1})^2-\phi_{i-1,k}(\bar{u}_{i-1})^2\right)\Phi_{i,p}\\&+\left(\phi_{N_x,k}\left(\bar{u}_{N_x}\right)^2\right.\\&\left.-\phi_{N_x-1,k}\left(\bar{u}_{N_x-1}\right)^2\right)\Phi_{N_x,p}.\end{aligned} \quad (12)$$

where $\bar{u}_{i+1}=\sum_{n=1}^{N_t}u_i^n/N_t$, with $i=1,\dots,N_x$.

The third version (version 3) arises from Roe's method and considers time averages of water velocity u and the water depth h in denominators or square roots of (7)

$$\widehat{\boldsymbol{h}}^{n+1}=\widehat{\boldsymbol{h}}^{n}+\frac{\Delta t}{4\Delta x}A\widehat{\boldsymbol{h}}^{n}+\frac{\Delta t}{4\Delta x}B\widehat{\boldsymbol{q}}^{n},\quad \widehat{\boldsymbol{q}}^{n+1}=\widehat{\boldsymbol{q}}^{n}-\frac{\Delta t}{4\Delta x}C\widehat{\boldsymbol{h}}^{n}+\frac{\Delta t}{4\Delta x}D\widehat{\boldsymbol{q}}^{n}, \quad (13)$$

where the matrices, considering free BC, are

$$\begin{aligned}A(k,p)=&\left(\phi_{2,k}-\phi_{1,k}\right)\bar{a}_{3/2}\phi_{1,p}+\sum_{i=2}^{N_x-1}\left[\left(\phi_{i+1,k}-\phi_{i,k}\right)\bar{a}_{i+1/2}-\left(\phi_{i,k}-\phi_{i-1,k}\right)\bar{a}_{i-1/2}\right]\phi_{i,p}\\&+\left(\phi_{N_x,k}-\phi_{N_x-1,k}\right)\bar{a}_{N_x-1/2}\phi_{N_x,p},\\B(k,p)=&\left(\Phi_{2,k}-\Phi_{1,k}\right)\bar{b}_{3/2}\phi_{1,p}+\sum_{i=2}^{N_x-1}\left[\left(\Phi_{i+1,k}-\Phi_{i,k}\right)\bar{b}_{i+1/2}+\left(\Phi_{i,k}-\Phi_{i-1,k}\right)\bar{b}_{i-1/2}\right]\phi_{i,p}\\&+\left(\Phi_{N_x,k}-\Phi_{N_x-1,k}\right)\bar{b}_{N_x-1/2}\phi_{N_x,p},\\C(k,p)=&\left(\phi_{2,k}-\phi_{1,k}\right)\bar{e}_{3/2}\Phi_{1,p}+\sum_{i=2}^{N_x-1}\left[\left(\phi_{i+1,k}-\phi_{i,k}\right)\bar{e}_{i+1/2}+\left(\phi_{i,k}-\phi_{i-1,k}\right)\bar{e}_{i-1/2}\right]\Phi_{i,p}\\&+\left(\phi_{N_x,k}-\phi_{N_x-1,k}\right)\bar{e}_{N_x-1/2}\Phi_{N_x,p},\\D(k,p)=&\left(\Phi_{2,k}-\Phi_{1,k}\right)\bar{d}_{3/2}\Phi_{1,p}+\sum_{i=2}^{N_x-1}\left[\left(\Phi_{i+1,k}-\Phi_{i,k}\right)\bar{d}_{i+1/2}+\left(\Phi_{i,k}-\Phi_{i-1,k}\right)\bar{d}_{i-1/2}\right]\Phi_{i,p}\\&+\left(\Phi_{N_x,k}-\Phi_{N_x-1,k}\right)\bar{d}_{N_x-1/2}\Phi_{N_x,p},\end{aligned} \quad (14)$$

with

$$\begin{aligned}&\bar{a}_{i+1/2}=\left(\frac{\left|\bar{\lambda}_1\right|\bar{\lambda}_2-\left|\bar{\lambda}_2\right|\bar{\lambda}_1}{\bar{c}}\right)_{i+1/2}, \qquad \bar{b}_{i\pm1/2}=\left(\frac{\bar{\lambda}_1^{\mp}-\bar{\lambda}_2^{\mp}}{\bar{c}}\right)_{i\pm1/2},\\&\bar{e}_{i\pm1/2}=\left(\frac{\bar{\lambda}_1\bar{\lambda}_2}{\bar{c}}\right)_{i\pm1/2}\left(\bar{\lambda}_1^{\mp}-\bar{\lambda}_2^{\mp}\right)_{i\pm1/2},\quad \bar{d}_{i\pm1/2}=\left(\frac{\bar{\lambda}_1\bar{\lambda}_1^{\mp}-\bar{\lambda}_2\bar{\lambda}_2^{\mp}}{\bar{c}}\right)_{i\pm1/2},\end{aligned} \quad (15)$$

where $\left(\bar{\lambda}_1\right)_{i+1/2}$, $\left(\bar{\lambda}_2\right)_{i+1/2}$, $\left(\bar{\lambda}_1^{\mp}\right)_{i+1/2}$, $\left(\bar{\lambda}_2^{\mp}\right)_{i+1/2}$ and $\bar{c}_{i+1/2}$ are computed from the time averages $\bar{u}_i$ and $\bar{c}_i$ following (7).

4. Numerical results

4.1. Case 1

Consider the 1D shallow water equations (1), where the spatial domain is $[0, L]$, where $L = 12$, and the final time is $T = 1.02$.

The initial conditions (ICs) are defined as a dam break time as

$$h(x,0) = \begin{cases} 2, & if\ x \leq 6 \\ 1, & if\ x > 6 \end{cases}, \quad u(x,0) = 0, \quad 0 \leq x \leq L, \qquad (16)$$

given in m and m/s, respectively; and free BCs are considered.

In this case, the two versions of the ROM, with and without time averages, developed from the Lax-Friedrichs method are compared in terms of accuracy and CPU time. The CFL number is 0.9 and $\nu = 0.9$.

The accuracy is measured by computing the error of h and q at the final time in the L_1 norm with respect to the exact solution [12]. They are denoted by $L_1(h)$ and $L_1(q)$, respectively. The CPU time required by each ROM is also measured to study their efficiency.

Following the PID strategy approach, the construction of the snapshot matrices has been divided into time windows with the same number of snapshots. Table 1 shows the number of time windows N_w arranged according to the number of snapshots per window N_s and per mesh refinement N_x. Each N_x corresponds to a value of N_t according to the CFL condition.

Table 1. Number of time windows $N_w = N_t/N_s$ for different values of N_t and N_s.

N_x	N_t	N_s			
		16	**8**	**4**	**2**
100	48	3	6	12	24
203	96	6	12	24	48
409	192	12	24	48	96
822	384	24	48	96	192

First, the influence of N_w and N_{POD} on the computation of the solution is studied. The goal is to find the optimal set of values of these ROM parameters.

Figures 1-4 show the errors in the L_1 norm of the solutions computed with version 1 (left) and version 2 (right) of the ROM in terms of h and q as a function of N_x.

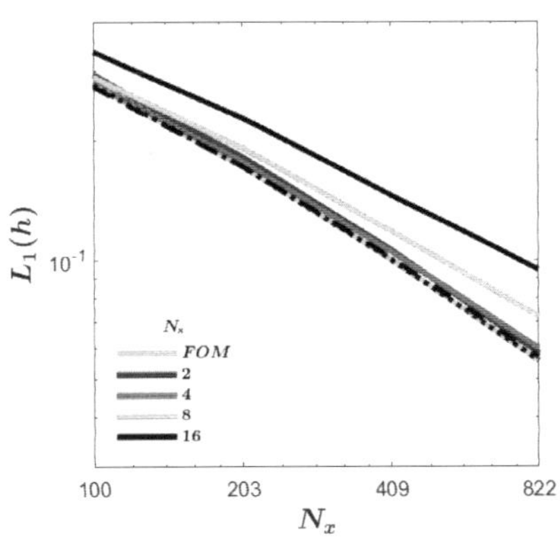

Figure 1. $L_1(h)$ of ROM v.1.

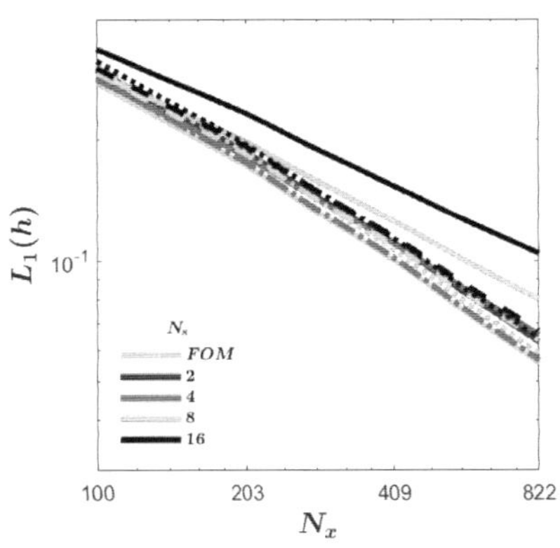

Figure 2. $L_1(h)$ of ROM v.2.

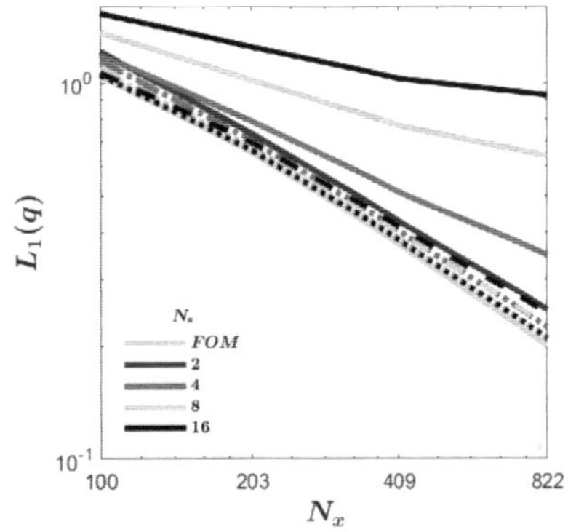

Figure 3. $L_1(q)$ of ROM v.1.

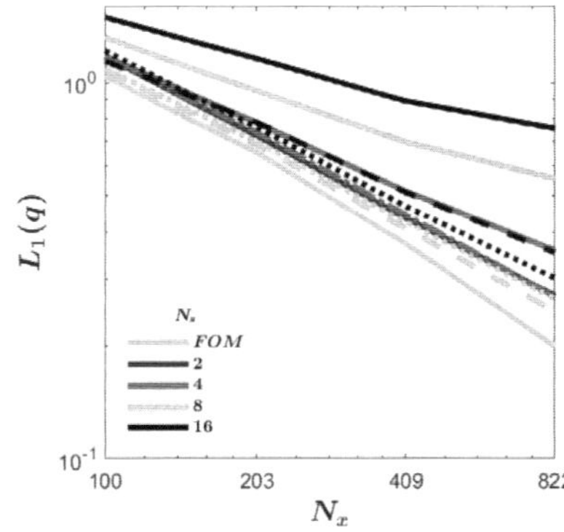

Figure 4. $L_1(q)$ of ROM v.2.

The yellow line represents the errors of the FOM solutions with respect to the exact solution. The rest of the colours represent the errors of the ROM solutions according to the number of snapshots in each time window N_S. The line type indicates the number of modes of the ROM: $N_{POD} = 2$ (solid line), 5 (dashed line) and 10 (dotted line).

As can be seen by comparing the figures of ROM version 1 with those of version 2, the ROM version 1 solutions are slightly more accurate for both variables h and q. In both cases, all ROM solutions are equally or less accurate than those of the FOM. On the other hand, the solutions obtained by solving only 2 POD modes (solid lines) are less accurate. The rest of the solutions with $N_{POD} = 5$ and 10 presented are slightly more accurate in ROM version 1.

Figures 5 and 6 show the CPU times required by versions 1 (left) and 2 (right) of the ROM. ROM version 1 results are clustered according to the number of modes solved N_{POD} regardless of the number of time windows, with $N_{POD} = 2$ (solid lines) being the best in terms of CPU time, achieving an order of magnitude improvement over FOM CPU times. However, the results of ROM version 2 do not show such clusters and are mixed without any pattern related to N_{POD} or N_S. In this case, the CPU time improvements reach two orders of magnitude. Nevertheless, the above study shows that the fastest results are the least accurate, so the CPU times against errors are plotted to obtain the optimal values of N_{POD} and N_W.

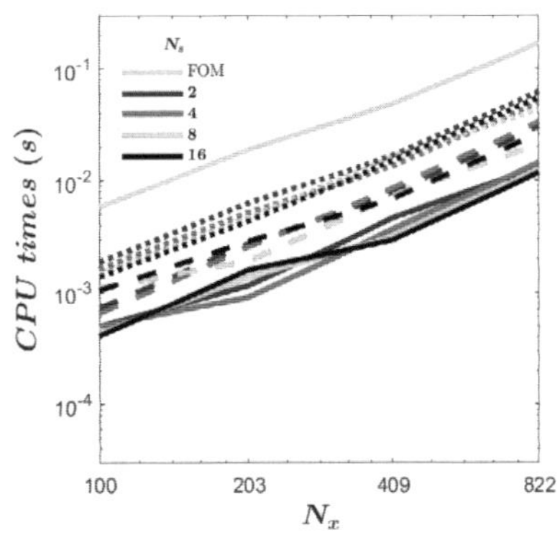

Figure 5. CPU times of ROM v.1.

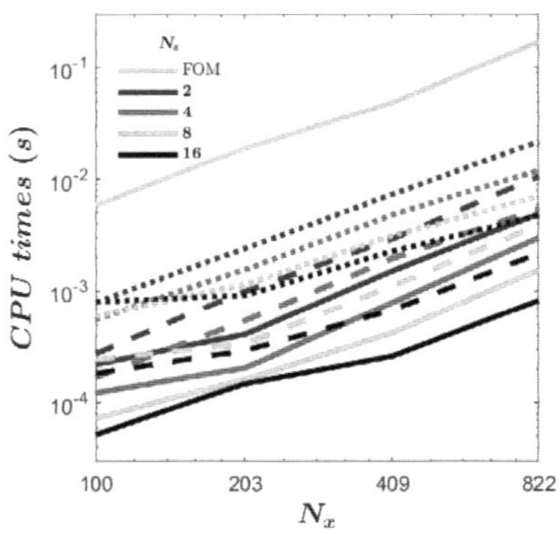

Figure 6. CPU times of ROM v.2.

Figures 7-10 show the errors in the norm against CPU times of versions 1 (left) and 2 (right) of the ROM. In general, the CPU times of ROM version 2 are better than those of version 1; otherwise, the errors obtained by both versions are very similar.

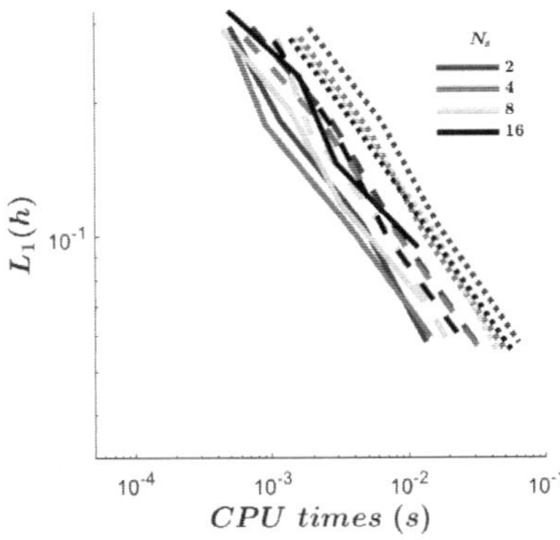

Figure 7. $L_1(h)$ vs CPU times of ROM v.1.

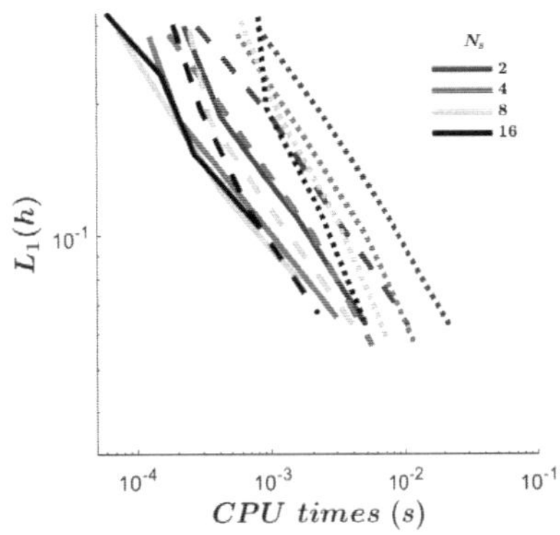

Figure 8. $L_1(h)$ vs CPU times of ROM v.2.

The best set of ROM parameter values is $N_s = 4$ and $N_{POD} = 5$, as it obtains very accurate final time solutions with the ROM, as can be seen in Figures 11 and 12, where the FOM and the ROM version 2 solutions at the final time T are represented together with the exact solution and the IC. Other ROM parameter values obtain results with inaccurate wave fronts.

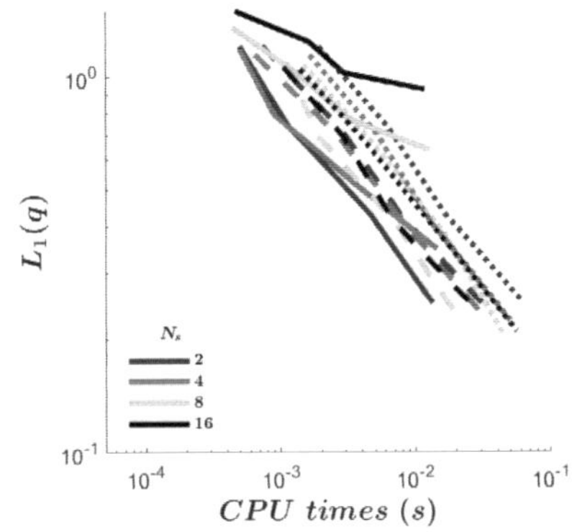

Figure 9. $L_1(q)$ vs CPU times of ROM v.1.

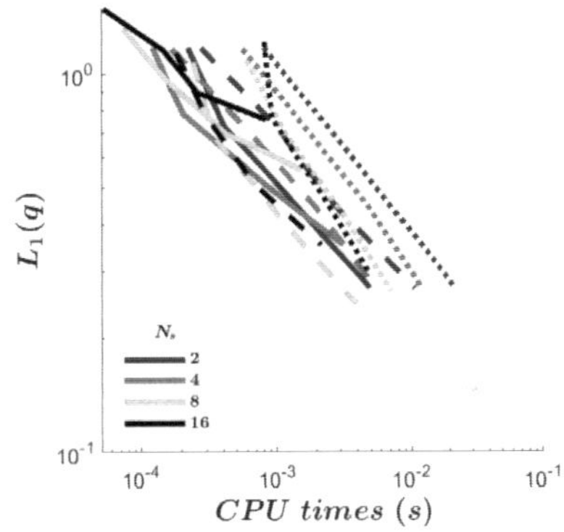

Figure 10. $L_1(q)$ vs CPU times of ROM v.2.

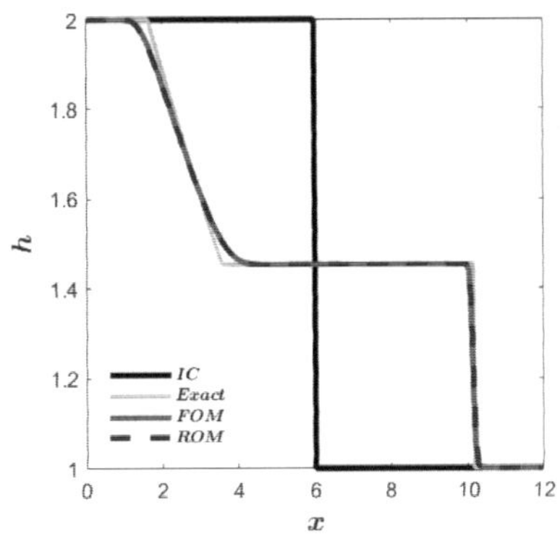

Figure 11. h computed with ROM v.2.

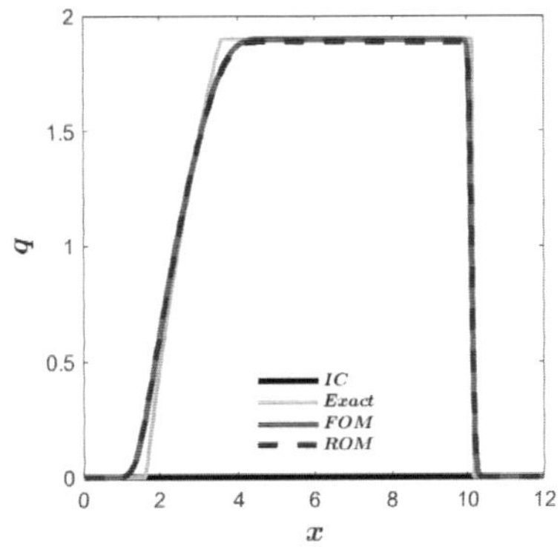

Figure 12. q computed with ROM v.2.

In conclusion, it can be said that ROM version 2 is more efficient with hardly any loss of accuracy despite the use of time averages of u.

IOP Conf. Series: Earth and Environmental Science 1136 (2023) 012036 doi:10.1088/1755-1315/1136/1/012036

4.2. Case 2

Once the validity of the time averaging approach has been checked, the same problem as in Case 1 is solved with the version 3 ROM developed from the Roe's method.

Figures 13 and 14 show the errors in the L_1 norm of h (left) and q (right) as a function of N_x. First of all, it should be noted that the solution computed by the FOM following the upwind method is more accurate than the FOM of the Lax-Friedrichs method. This makes some solutions obtained by version 3 of the ROM even better than those of the Lax-Friedrichs FOM.

In terms of water depth, the accuracy of solutions obtained with version 3 of the ROM becomes worse when only 2 POD modes are solved. It can also be observed that when two snapshots per time window are solved, worse results are obtained than in the case of $N_s > 2$. Considering the water discharge, the errors show more diversity. Altogether, $N_s = 4$ with $N_{POD} = 5$ and 10 are the most accurate.

Figure 13 shows the CPU times required by version 3 of the ROM, which reach two orders of magnitude better than the CPU times of the FOM. The CPU times required by the FOM based on the upwind method are practically the same as those of the FOM based on the Lax-Friedrichs. Some sets of ROM parameter values reach two orders of magnitude improvement over the FOM.

In terms of the measured error in the water depth, it could be considered that the best values of the ROM parameters would be $N_s = 8$ or 16 with $N_{POD} = 5$, as can be seen in Figure 16. However, these values present worse values of the measured error in the flow rate, as can be seen in Figure 17. Considering both figures, it is concluded that, as in case 1, the best set of ROM parameter values are $N_s = 4$ and $N_{POD} = 5$, as these give very high accuracy solutions, as can be seen in Figures 18 and 19.

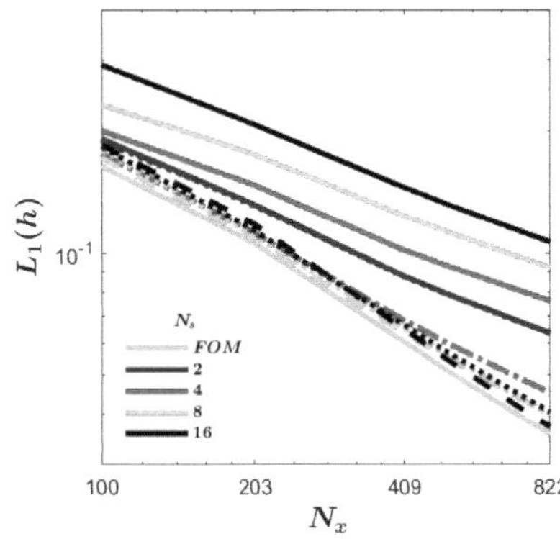

Figure 13. $L_1(h)$ of ROM v.3.

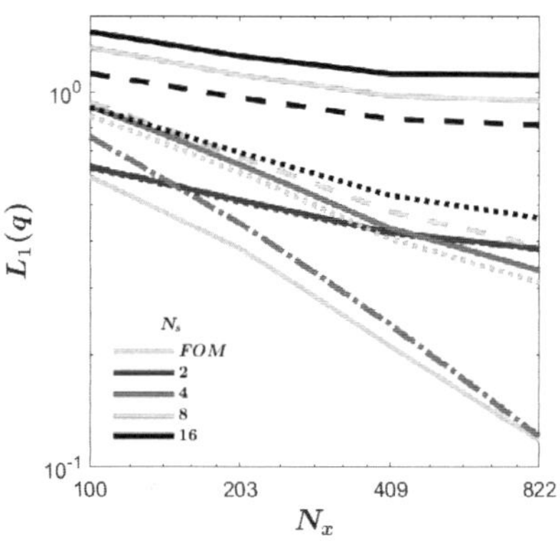

Figure 14. $L_1(q)$ of ROM v.3.

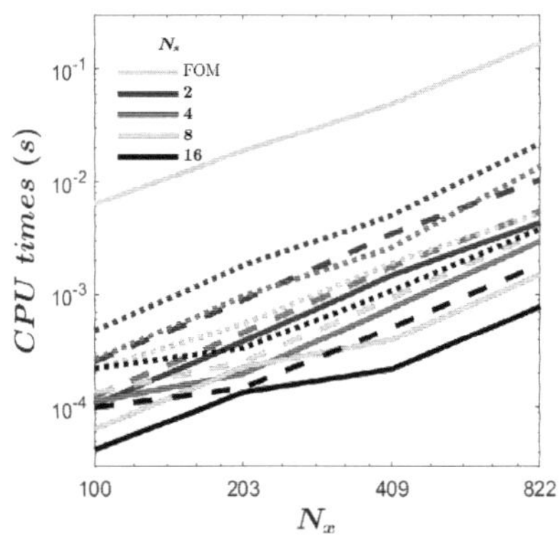

Figure 15. CPU times of ROM v.3.

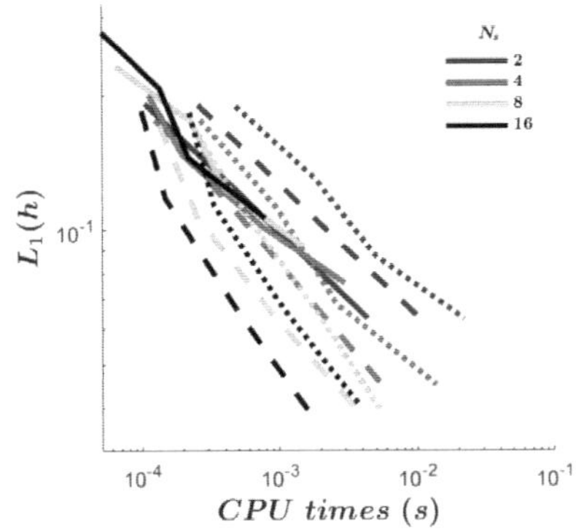

Figure 16. $L_1(h)$vs CPU times of ROM v.3

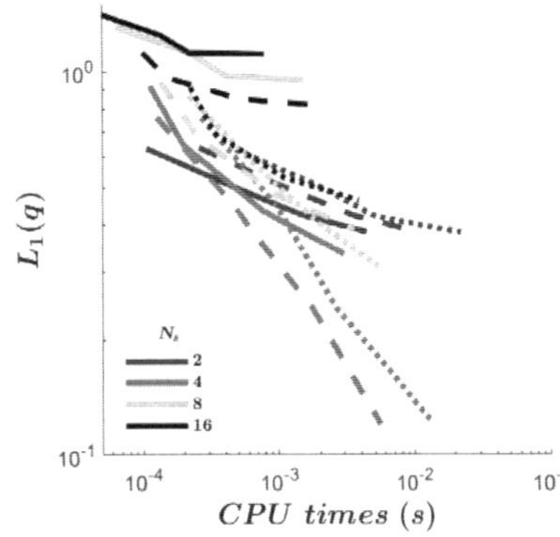

Figure 17. $L_1(q)$vs CPU times of ROM v.3.

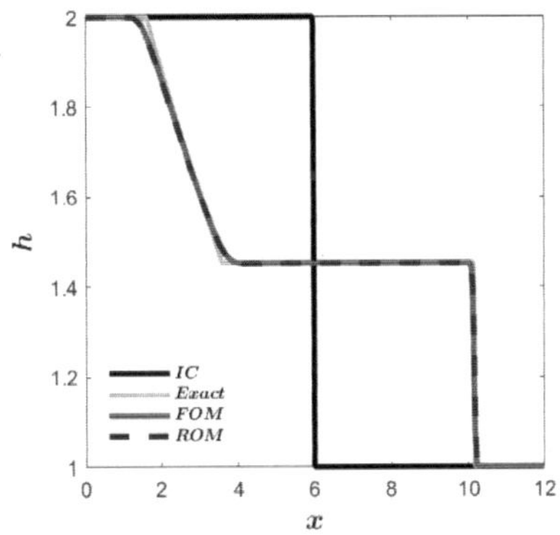

Figure 18. h of ROM v.3.

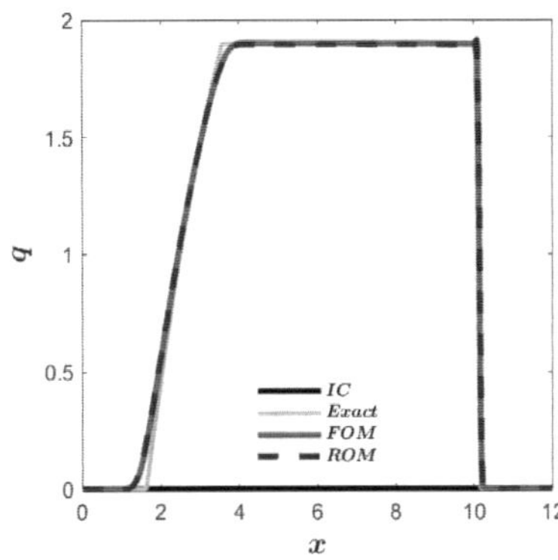

Figure 19. q of ROM v.3.

Table 2 shows the speed-ups of each of the ROM versions with respect to FOM. As can be seen, the versions with time averages 2 and 3 present higher speed-ups than version 1, which exceed the order of magnitude.

Table 2. Speed-ups of each version of the ROM.

N_x	100	203	409	822
ROM v.1	9.69	7.34	5.99	5.20
ROM v.2	37.18	34.77	25.86	31.85
ROM v.3	50.55	41.22	28.56	31.58

5. Conclusions

In this work a dam break problem has been solved with the 1D SWE using the ROM strategy. For this purpose, two numerical methods (FOM) of discretisation of the mathematical model have been considered, one based on the Lax-Friedrichs method and the other based on Roe's method. From these, three reduced order models have been studied.

With the two ROMs developed from the Lax-Friedrichs-based FOM, the validity of the time-averaging approach proposed in [9] has been tested. The results obtained show that the time-averaged ROM produces accurate solutions, as much as the standard ROM, and in a more efficient way. Furthermore, this time averaging approach has been applied to the development of the ROM from the FOM based on the Roe's method, obtaining satisfactory results.

The performance of the different ROMs has been studied against the variation of their own parameters, number of time windows (following the PID method) and number of POD modes solved, as well as the number of cells in the spatial domain. It can be concluded that the best values to obtain accurate and efficient results from the ROM are $N_s = 4$ and $N_{POD} = 5$, as they achieve good error rates and have low CPU times.

Acknowledgments
This work was funded by the Spanish Ministry of Science and Innovation under the research project PGC2018-094341-B-I00. This work has also been partially funded by Gobierno de Aragón through Fondo Social Europeo (T32-20R and E24-17R, Feder 2019-2021 "Construyendo Europa desde Aragón").

References

[1] García-Navarro P and Vázquez-Cendón ME 2000 On numerical treatment of the source terms in the shallow water equations *Comput. Fluids* **29**(8) 951–979 https://doi.org/10.1016/S0045-7930(99)00038-9
[2] Burguete J and García-Navarro P 2004 Improving simple explicit methods for unsteady open channel and river flow *Int. J. Numer. Methods in Fluids* **45**(2) 125–156 https://doi.org/10.1002/fld.619
[3] Murillo J and Navas-Montilla A 2016 A comprehensive explanation and exercise of the source terms in hyperbolic systems using Roe type solutions. Application to the 1D-2D shallow water equations *Adv. Water Resour.* **98** 70–96 https://doi.org/10.1016/j.advwatres. 2016.10.019
[4] Lumley JL 1967 The structure of inhomogeneous turbulent flows *Atmospheric Turbulence and Radio Wave Propagation* ed AM Yaglom and VI Tartarsky (Moscow: Publishing House "Nauka") pp 166–176
[5] Ahmed M and San O 2018 Stabilized principal interval decomposition method for model reduction of nonlinear convective systems with moving shocks *Comput. Appl. Math.* **37** 6870–6902 https://doi.org/10.1007/s40314-018-0718-z
[6] Ahmed SE, San O, Bistrian DA and Navon IM 2020 Sampling and resolution characteristics in reduced order models of shallow water equations: Intrusive vs nonintrusive *Int. J. Numer. Methods Fluids* **92**(8) 992–1036 https://doi.org/10.1002/fld.4815
[7] Akhtar I, Wang Z, Borggaard J and Iliescu T 2012 A new closure strategy for proper orthogonal decomposition reduced-order models *J. Comput. Nonlinear Dyn.* **7**(3) 034503 https://doi.org/10.1115/1.4005928
[8] IJzerman WL 2000 Signal Representation and Modeling of Spatial Structures in Fluids *PhD Thesis* University of Twente (Enschede, Netherlands) https://www.persistent-identifier.nl/urn:nbn:nl:ui:28-29660
[9] Borggaard J, Hay A and Pelletier D 2007 Interval-based reduced-order models for unsteady fluid flow *Int. J. Numer. Anal. Model.* **4**(3-4) 353–367 https://global-sci.org/intro/article_detail/ijnam/866.html
[10] Galerkin BG 1915 Rods and plates. Series occurring in various questions concerning the elastic equilibrium of rods and plates *Engineers Bulletin* **19** 897–908
[11] Zokagoa JM and Soulaïmani A 2018 A POD-based reduced-order model for uncertainty analyses in shallow water flows *Int. J. Comput. Fluid Dynamics* **32**(6-7) 278–292 https://doi.org/10.1080/10618562.2018.1513496
[12] Stoker JJ 2011 *Water Waves: The Mathematical Theory with Applications* (Wiley Classics Library)
[13] Courant R, Friedrichs K and Lewy H 1928 Uber die partiellen Differenzengleichungen der mathematischen Physik *Math. Ann.* **100**(1) 32–74 https://doi.org/10.1007/BF01448839
[14] Sirovich L 1987 Turbulence and the dynamics of coherent structures. I – Coherent structures. II – Symmetries and transformations. III – Dynamics and scaling *Q. Appl. Math.* **45** 561–571, 573–582, 583–590.

A model for computing thermally-driven shallow flows

Isabel Echeverribar[1,2], Sergio Martínez-Aranda[1], Javier Fernández-Pato[1,2], Reinaldo García[2], Pilar Brufau[1] and Pilar García-Navarro[1]

[1] Fluid Mechanics, I3A-Universidad de Zaragoza, Zaragoza, 50018, Spain
[2] Hydronia Europe, Madrid, 28046, Spain

echeverribar@unizar.es

Abstract. In many natural disasters such as overland oil spills or lava flows, physical fluid properties as density change when considering non-homogeneous spatial and time variable distributions of the temperature. This effect is even more remarkable when these flows show a non-Newtonian behaviour due to the sensitivity of their rheological properties as viscosity or yield stress to temperature. In these cases, temperature becomes a significant variable that drives the fluid behaviour, which must be solved using an energy equation coupled with the free surface flow system. Special attention is devoted to thermal source terms which must include all the heat fluid exchanges, and their modelling sometimes can govern the complete flow behaviour. Fluid density, viscosity and yield stress, also affected by temperature, must be recomputed every time step. Summarizing, this work presents a 2D free surface flow model considering density and temperature variations, which could even modify viscosity and yield stress, with heat transfer mechanisms. The model is applied to oil spill overland simulations and heating/cooling test cases are carried out to ensure the system energy balance. As conclusions, it can be said that the numerical results demonstrate the importance of the heat exchange effects and those of the density, viscosity and yield stress variations.

1. Introduction

When dealing with surface geophysical flows, such as oil spills, landslides, muddy slurries, lava flows, etc, the choice of the system of equations to solve depends on the type of phenomena to be reproduced. Generally, these kind of free surface non-Newtonian viscoplastic flows are widely addressed with the depth-averaged shallow-type models. Traditionally, these models assume a hydrostatic pressure distribution in the flow column and incompressible flow, dismissing density variations that might affect the flow dynamics. However, when the flow is governed by a non-uniform distribution of a driven property, such as temperature, the bulk properties of the flow column vary in time and space, and hence the incompressible flow assumption is no longer valid.

In these kinds of phenomena, the flow density, which depends on temperature, turns into an important variable which must be coupled with the rest of variables within the system of equations. On one hand, compressible shallow-flow models have been recently developed for mud/debris flows [1-4], and solid-laden currents [5,6], where the bulk density of the flow column is mainly controlled by the volumetric concentration of solid particles. As important differences were found between the results provided by incompressible and compressible models [4] when density is governed by particle concentration, temperature might have the same effects on fluids at high temperatures, such the oil of an oil spill. On the other hand, the temperature is often considered as a passively transported scalar with no influence on the fluid dynamics [7,8] or, when special attention is paid into the temperature, the system dynamics are coarsely modelled or estimated [9,10].

This work presents a model for incompressible flows developed for fluids with a high particle concentration but using temperature as the variable on which the density depends. The mathematical approach has unconditional hyperbolic character, which allows solving it as in [11]. This represents a decisive advantage in order to develop efficient numerical models able to deal with realistic large-scale long-term geophysical events.

2. Simulation model: equations and numerical scheme

2.1. System of equations

The mathematical model that describes the behaviour of the oil flow is compound by a mass conservation equation, two momentum equations in x and y directions and a final temperature transport equation that derives from energy conservation equation:

$$\begin{gathered}\frac{\partial(\rho h)}{\partial t}+\frac{\partial(\rho h u)}{\partial x}+\frac{\partial(\rho h v)}{\partial y}=0\\ \frac{\partial(\rho h u)}{\partial t}+\frac{\partial}{\partial x}\left(\rho h u^2+\frac{1}{2}\rho g h^2\right)+\frac{\partial}{\partial y}(\rho h u v)=-g\rho h\frac{\partial z_b}{\partial x}-\tau_{bx}\\ \frac{\partial(\rho h v)}{\partial t}+\frac{\partial}{\partial x}(\rho h u v)+\frac{\partial}{\partial y}\left(\rho h v^2+\frac{1}{2}\rho g h^2\right)=-g\rho h\frac{\partial z_b}{\partial y}-\tau_{by}\\ \frac{\partial(hT)}{\partial t}+\frac{\partial(huT)}{\partial x}+\frac{\partial(hvT)}{\partial y}=\boldsymbol{S}_T\end{gathered} \tag{1}$$

where ρ stands for the density, h represents water depth and u and v are the velocities in x and y directions, and T stands for fluid temperature. The temperature source term, S_T, considers the heat transfer with the environment as in Echeverribar et al. 2022. The closure equation relates density and temperature, and updates also viscosity and yield stress as

$$\begin{gathered}\rho(T)=\rho_0+K\,(T-T_0)\\ YS(T)=10^{\left(Ays\,T^2+Bys\,T+Cys\right)}\\ \mu(T)=A_\mu+\frac{B_\mu}{T}\end{gathered} \tag{2}$$

where parameters of the expressions come from experimental works to characterize oils. Finally, it is important to note that both, viscosity and yield stress, affect directly to friction stresses in momentum equations following a particular formulation. To compute stresses in equation (1) as a function of yield stress and viscosity of equation (2) a Simplified Bingham Model has been chosen in this case.

The source term is divided into two components as

$$S_T=\frac{\dot{Q}}{\rho h C_p}=\frac{\dot{Q}_{rad}+\dot{Q}_{conv}}{\rho h C_p} \tag{3}$$

where Q_{rad} stands for the solar radiation that falls upon the flow and Q_{conv} represents the convection heat losses, that are modelled as:

$$\dot{Q}_{conv}=h_c(T-T_{air}) \tag{4}$$

where h_c the convection coefficient in $Wm^{-2}K$ and Tair the air temperature, in K. The emitted solar radiation is neglected in this oil spill application, although it can be important when simulating other fluid such as lava.

2.2. Numerical scheme

To solve the system, the final model is written in terms of a normalized density, r, and a normalized temperature, T^{∇}, as:

$$\begin{gathered}\frac{\partial(rh)}{\partial t}+\frac{\partial(rhu)}{\partial x}+\frac{\partial(rhv)}{\partial y}=0\\ \frac{\partial(rhu)}{\partial t}+\frac{\partial}{\partial x}\left(rhu^2+\frac{1}{2}rgh^2\right)+\frac{\partial}{\partial y}(rhuv)=-grh\frac{\partial z_b}{\partial x}-\frac{\tau_{bx}}{\rho_0}\\ \frac{\partial(rhv)}{\partial t}+\frac{\partial}{\partial x}(rhuv)+\frac{\partial}{\partial y}\left(\rho hv^2+\frac{1}{2}rgh^2\right)=-grh\frac{\partial z_b}{\partial y}-\frac{\tau_{by}}{\rho_0}\\ \frac{\partial(hT^{\nabla})}{\partial t}+\frac{\partial(huT^{\nabla})}{\partial x}+\frac{\partial(hvT^{\nabla})}{\partial y}=0\end{gathered} \tag{5}$$

while the final temperature is updated decoupled with the source term with:

$$\frac{\partial T}{\partial t}=\frac{\dot{Q}}{\rho hC_p}=\boldsymbol{S}_T \tag{6}$$

Therefore, the updating scheme for the conserved variables follows:

$$U_i^{n+1}=U_i^n-\frac{\Delta t}{A_i}\sum_{k=1}^{NE}\mathbf{R}_k^{-1}\boldsymbol{\mathcal{F}}_k^{\downarrow -}l_k \tag{7}$$

where $\boldsymbol{\mathcal{F}}_k^{\downarrow -}$ is a defined augmented flux vector through the kth wall, which includes the convective fluxes and the momentum source term contributions as:

$$\boldsymbol{\mathcal{F}}_k^{\downarrow -}=\mathbf{F}(\widehat{\mathbf{U}}_i^n)+\sum_{m-}\left[(\tilde{\lambda}_m\tilde{\alpha}_m-\tilde{\beta}_m-\tilde{\sigma}_m)\tilde{\mathbf{e}}_m\right]_k^n \tag{8}$$

where $\left(\tilde{\lambda}_m\right)_k^n$ are the eigenvalues of the Jacobian matrix of the local Riemann problem, which stand for the wave celerities of the system; $(\tilde{\mathbf{e}}_m)_k^n$ are the eigen vectors associated to that values; $(\tilde{\alpha}_m)_k^n$ are the wave strengths; and $\left(\tilde{\beta}_m\right)_k^n$ and $(\tilde{\sigma}_m)_k^n$ denote the source strengths including bed slope and basal resistance, respectively. Since an explicit numerical scheme is being used, the time step size is restricted by the Courant-Friedrichs-Levy (CFL) number, which lead to a stability condition as

$$\Delta t=\mathrm{CFL}\,\underset{k}{\mathrm{MIN}}\left(\frac{\mathrm{MIN}(A_i,A_j)}{l_k\,\underset{m}{\mathrm{MAX}}\left(|\tilde{\lambda}_m|\right)_k^n}\right) \tag{9}$$

This CFL number must be bounded between 0 and 1 for 2D triangular meshes.

3. Circular dam-break test case

3.1. Test case description

A 2D circular dam break is simulated on a 10 km square flat domain. The initial cylindrical fluid column has a diameter of 1000 m and a height of 52 m. The properties relating temperature and density following equation (2) are a reference temperature of 20ºC (T_0), a density at that temperature of 860 kg/m^3 (ρ_0), and a proportionality constant, *K*, of -1. Two different initial temperatures are considered to stablish two different cases: 35 and 50 ºC. The relationship between viscosity and yield stress with temperature is modelled by pairs of values on a table instead of using expression (2). The values can be found in Table 1.

The case is simulated on an unstructured triangular mesh of 105182 elements. The simulation runs for 360 s with a CFL of 0.95. No thermal source terms have been considered in this case.

IOP Conf. Series: Earth and Environmental Science 1136 (2023) 012037 doi:10.1088/1755-1315/1136/1/012037

Table 1. Variations of yield stress and viscosity with temperature.

Temperature [°C]	Yield stress [Pa]	Temperature [°C]	Viscosity [Pa s]
35	2000	35	0.362
40	200	55	0.138
45	20	60	0.110
50	0.02	65	0.086

3.2. Numerical results

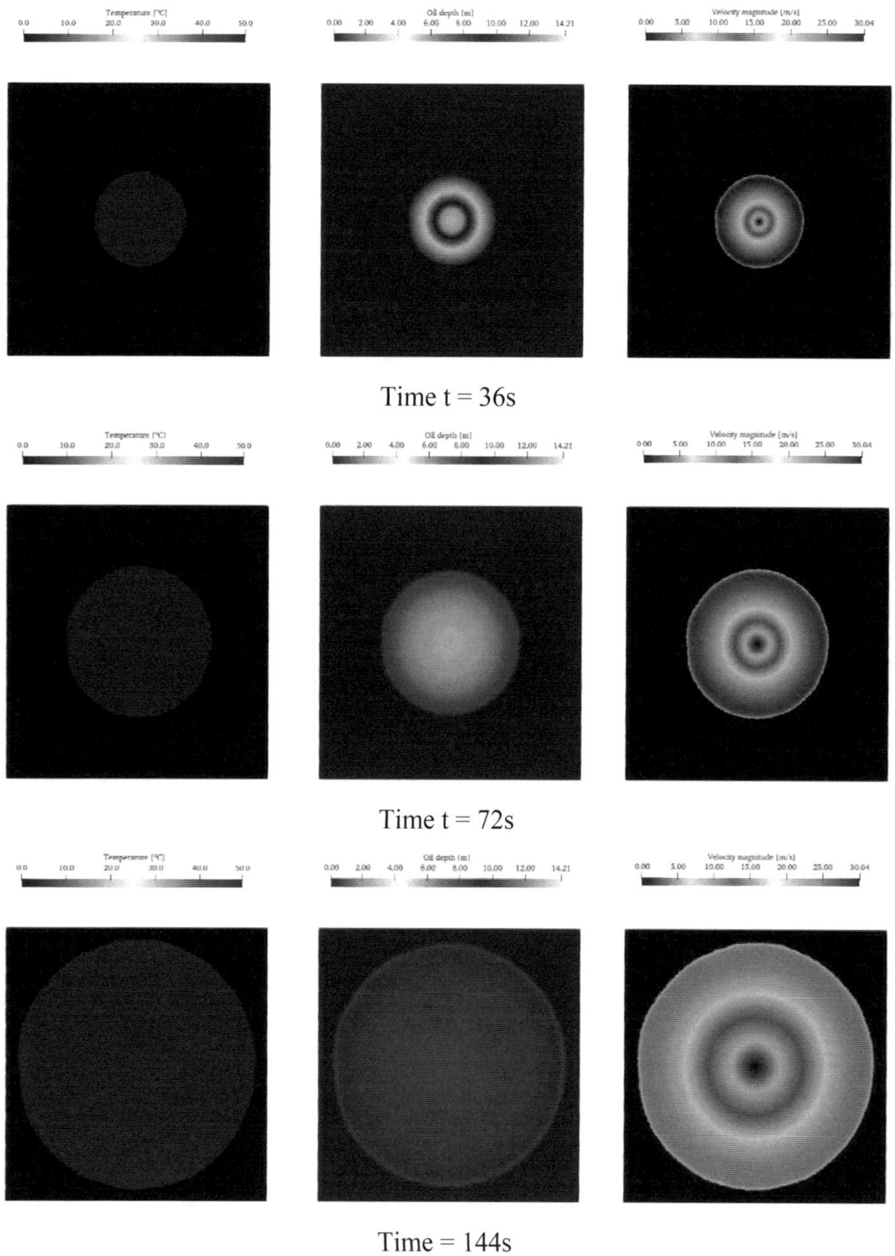

Figure 1: Spatial distribution of temperature (left), oil depth (centre) and velocity magnitude (right) for an oil at 50°C.

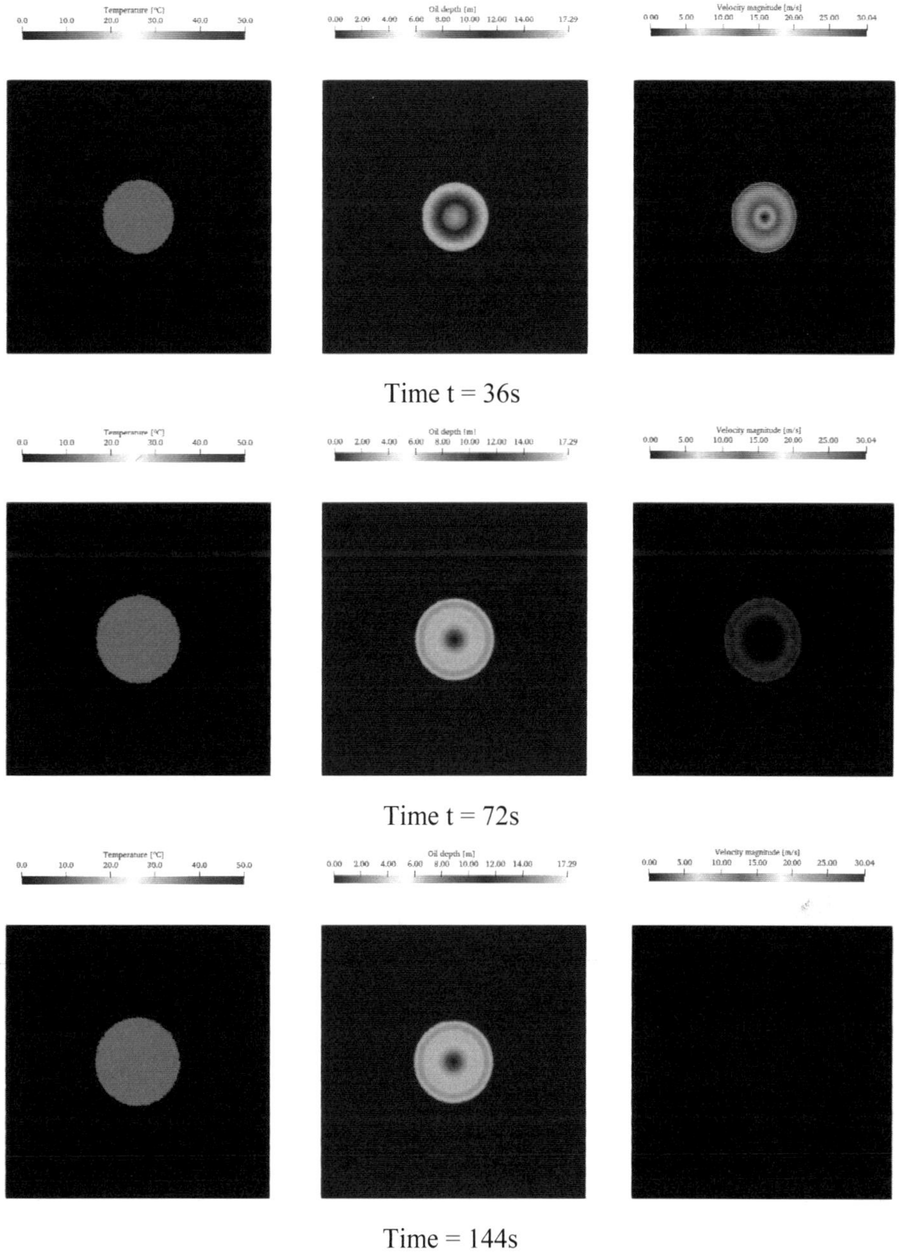

Figure 2: Spatial distribution of temperature (left), oil depth (centre) and velocity magnitude (right) for an oil at 35°C.

The initial condition is the same in both cases except for the temperature, which causes variations in the rheological properties of the fluid and its density. Figure 1 shows the results of dam break for an initial temperature of 50 °C at different times. The left column of pictures shows the fluid temperature, constant at 50°C due to the absence of heat source terms. In the centre is represented the fluid depth, which remains thicker in the centre while in the periphery begins to reduce. The right column shows the magnitude of the velocity, which is lower in the centre and increases radially, decreasing just at the periphery due to the advance of the dry front. It is observed that at 72 and 144 s the stain advances maintaining a radial symmetry and the same behaviour. At 144 s, in the fluid depth, an over-elevation of the surface begins to be observed in the periphery together with a decrease in the velocity caused by the yield stress.

Figure 2 shows the same type of results as in Figure 1 for an initial temperature of 35 ºC. Being lower than in the previous case, yield stress, density and viscosity are much higher, following equation (2) and Table 1. Thus, it is observed that the velocities are much lower and the expansion of the oil slick is much smaller. A comparison between cases in terms of fluid depth can be seen on Figure 3, where the longitudinal profile of oil depth is represented for both cases at different time levels. In both figures, it is seen how for an initial temperature of 35ºC the yield stress reaches such high values that the fluid is completely stopped at 144 s without a large expansion.

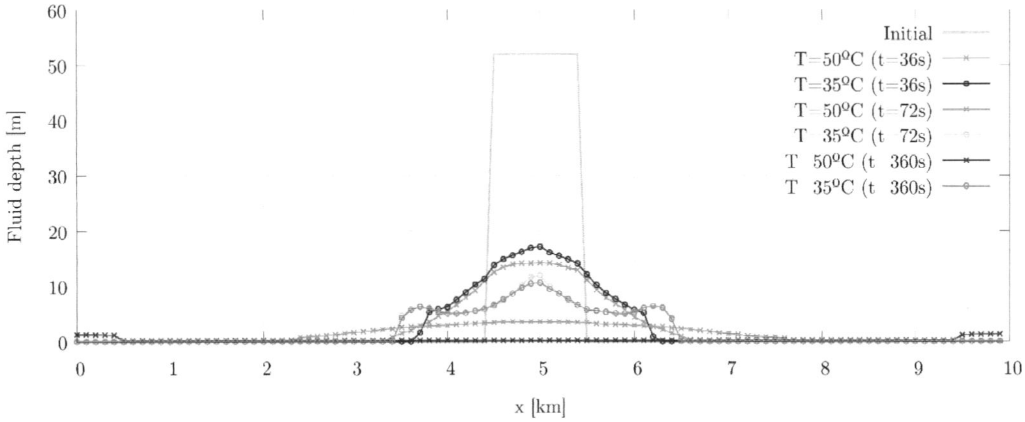

Figure 3: Longitudinal profile of oil depth in the circular dam-break for different times and different temperatures: 35ºC (-o-) and 50ºC (-×-).

4. Algeciras overland oil spill test case

4.1. Test case description

A digital terrain model of the Algeciras port, where exist an oil station, has been used to simulate an eventual overland oil spill. The model has been prepared to deal with wet/dry fronts to simulate overland oil spills. The necessity of considering the spatial distribution of temperature, density and heat transfer is clearly demonstrated on the results. Two different cases are simulated: environment A, with no wind (low convection coefficient), normal air temperature and solar radiation; and environment B, with high wind speed, low air temperature and no solar radiation.

Table 2. Environmental characteristics of the simulation cases.

	Environment A	**Environment B**
Initial temperature	85ºC	85ºC
Air temperature	40ºC	10ºC
Radiation	1000 W/m^2	0 W/m^2

4.2. Numerical results

The spill is simulated for 20 h and the flooding slick is analysed then as well as its temperature distribution. The triangular mesh contains 35581 elements and the simulations are carried out with a CFL of 0.9.

Figure 4 shows the different temperature distribution provoked by two oil sources at the same temperature spilled on two different environments. In this case, the differences on the flooded area are provoked by differences on yield stress and viscosity –both influenced by temperature, which changes depending on the environment-. Colder environments provoking important heat transferences lead to high yield stress, cold oil and reduced flooded areas.

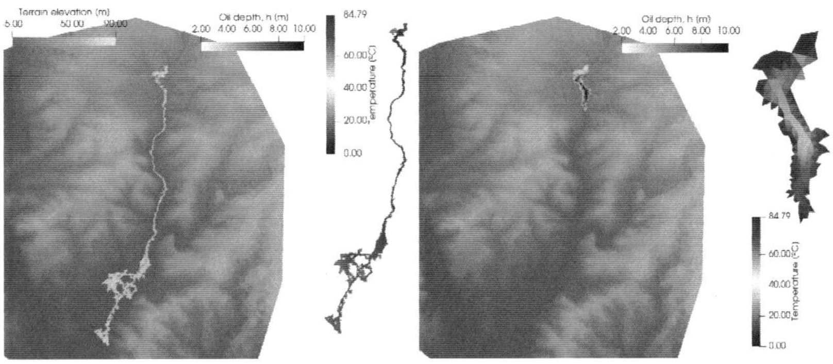

Figure 4: Flooded area and temperature at $t = 20$ h of an overland oil spill for the same oil into environment A (left) and B (right).

5. Conclusions

When dealing with high temperature non-Newtonian flows, the treatment of density and temperature variation considerations entails a difference on the results. A model of temperature-dependent flows has been developed and applied to oil spill simulations. It has been proved how a combination of both heat transfer and density/temperature changes affect the rheological properties of the flow, which affect the friction law when using models such as the Bingham model. The model has been applied first to a theoretical dam break to show a different evolution depending on the fluid temperature, although remaining constant. Additionally, a real test case of an oil spill has been simulated in different surrounding conditions to test the influence of heat transfer. It has been shown that initial inertial forces provoke a fast movement that later slows down due to a shear stress that depends on temperature, which varies depending on the heat transfer. The necessity of these models to simulate thermally-driven flows has been proved.

The sensitivity to temperature of the model allows to simulate different rheological flows such as crude oil or lava flows. However, further research must be done to develop more sophisticated heat transfer models or take evaporation into account.

Acknowledgments

Authors wish to acknowledge Hydronia Europe for the collaboration. This work was partially funded by the Spanish Ministry of Science and Innovation under the research project PGC2018-094341-B-I00. This work has also been partially funded by Gobierno de Aragón through Fondo Social Europeo (T32-20R, Feder 2014–2020 "Construyendo Europa desde Aragón").

References

[1] Kowalski J and McElwaine 2013 Shallow two-component gravity-driven flows with vertical variation *Journal of Fluid Mechanics,* **714** pp 434-462 https://doi.org/10.1017/jfm.2012.489

[2] Xia C, Li J, Cao Z, Liu Q and Hu K 2018 A quasi single-phase model for debris flows and its comparison with a two-phase model *Journal of Mountain Science,* **15**(5) pp 1071-1089 https://doi.org/10.1007/s11629-018-4886-5

[3] Martínez-Aranda S, Murillo J and García-Navarro P 2020 A robust two-dimensional model for highly sediment-laden unsteady flows of variable density over movable beds *Journal of Hydroinformatics* **22**(5) pp 1138-1160 https://doi.org/10.1016/j.enggeo.2021.106462

[4] Martínez-Aranda S, Murillo J and García-Navarro P 2021 A GPU-accelerated Efficient Simulation Tool (EST) for 2D variable-density mud/debris flows over non-uniform erodible beds *Engineering Geology* **296** pp 106462 https://doi.org/10.1016/j.enggeo.2021.106462

[5] Murillo J, Latorre B and García-Navarro P 2012 A Riemann solver for unsteady computation of 2D shallow flows with variable density *Journal of Computational Physics* **231** pp 4775-4807 https://doi.org/10.1016/j.jcp.2012.03.016
[6] Cao Z, Xia C, Pender G and Liu Q 2017 Shallow water hydrosediment-morphodynamic equations for fluvial processes *Jounral of Hydraulic Engineering* **143**(5) pp 02517001 https://doi.org/10.1061/(ASCE)HY.1943-7900.0001281
[7] Kim K S and Chapra S C 1997 Temperature model for highly transient shallow streams *Journal of Hydraulic Engineering* **123**(1) pp 30-40
[8] Cea L, Bermúdez M, Puertas J, Bladé E, Corestein G, Escolano E, Conde A, Bockelmann-Evans B and Ahmadian R 2016 IberWQ: new simulation tool for 2D water quality modelling in rivers and shallow estuaries *Journal of Hydroinformatics* **18**(5) pp 816-830 https://doi.org/10.2166/hydro.2016.235
[9] Ramírez-Camacho J G, Carbone F, Pastor E, Bubbico R and Casal J 2017 Assesing the consequences of pipeline accidents to support land-use planning *Safety Scince* **97** pp 34-42 https://doi.org/10.1016/j.ssci.2016.01.021
[10] Chen C, Li C, Reniers G and Yang F 2021 Safety and security of oil and gas pipeline transportation: A systematic analysis of research trends ans future needs using wos *Journal of Cleaner Production* **279** pp 123583 https://doi.org/10.1016/j.jclepro.2020.123583
[11] Echeverribar I, Morales-Hernández M, Brufau P and García-Navarro P 2019 2D numerical simulation of unsteady flows for large scale floods prediction in real time *Advances in water resources* **134** p 103444 https://doi.org/10.1016/j.advwatres.2019.103444

IOP Conf. Series: Earth and Environmental Science 1136 (2023) 012038 doi:10.1088/1755-1315/1136/1/012038

Experimental and numerical investigations of the jetting regime

Nicoleta-Octavia Tanase[1], **Ioana Rasuceanu**[1] **and Corneliu Balan**[1]

[1] REOROM Laboratory, Dept of Hydraulics, Hydraulic Machinery and Environmental Engineering, University POLITEHNICA of Bucharest, Bucharest, 060042, Romania

octavia.tanase@upb.ro

Abstract. In this paper we highlight the first stages of liquid injection through a capillary nozzle prior to the jet formation regime. We emphasize the asymmetry of water-air interface at the beginning of the injection of liquid and the axial symmetry which follows when the jet is fully formed. Numerical simulations are performed in order to estimate the relaxation of the velocity distribution of the liquid as it leaves the capillary nozzle. The plug flow condition for low-viscosity liquid jets is investigated in relation with the parabolic velocity distribution in the nozzle. The numerical data is also used to characterize the flow during the drop formation and pinch-off, both driven by the action of surface tension.

1. Introduction

A liquid column injected through a capillary nozzle displays a tendency to break up into drops. The transition between the dripping and jetting regime occurs when a critical Weber number ($We = \rho V_0^2 D/\sigma$) is exceeded [1], where ρ and σ are the density and surface tension of the injected liquid, D is the diameter of the nozzle and V_0 is the mean injection velocity. These regimes and instabilities have been studied intensively since 1880 [2] and their investigations are continued in the present [3-5] The jetting regime is characterized by detachment of drops far downstream from the capillary nozzle [6]. Liquid jets have been a subject of interest as they display many applications in various industrial processes such as ink-jet printing [7] and cooling or heating systems [8].

In this study, the stages of the jetting regime are investigated experimentally and numerically by a computational fluid dynamics model implemented in ANSYS Fluent. Numerical simulations have been an important tool and have supplied extensive information about various studied phenomena [9,10], including liquid jets and their breakup [11,12]. By use of the CFD model, we determine the velocity distributions during drop formation and pinch-off at various sections throughout the liquid jet. A comparison between the experimental and numerical results is presented and differences due to asymmetry during the first stages of jet formation are observed.

2. Methodology

For the experimental part we inject deionized water through a capillary nozzle of 1.8 mm in diameter positioned downward vertically in a gaseous environment (ambient air at 25°C). Water is being injected at a flow rate of 100 ml/min, with a Harvard syringe pump, ensuring the formation of a liquid jet. A high-speed camera, working at 5000 fps, was used to record the first stages of the jet formation process.

The surface tension of water was measured with a pendant drop technique and was found to be 72 mN/m. All experiments and measurements are performed at 20°C. The flow is characterized by two non-dimensional numbers: the Reynolds number (Re = 1170) and the Weber number (We = 10).

The geometry used for the numerical simulations is two-dimensional and axial symmetric. The inner diameter of the capillary nozzle was set at 1.8 mm. A rectangular domain of 40 mm (radial) × 120 mm (axial) was used with a structured mesh. The mesh for the whole domain (water and air) has 1 931 720 cells (quad-element, type map).

The numerical solution of the free liquid jet is computed in Fluid Flow analysis using Fluent solver of Workbench 2021 R1 with a laminar solver and the Volume of Fluid (VOF) method. The VOF multiphase model can be used for two or more fluids, aiming to determine the surface between the two fluids by solving a single set of equations of motion for all fluid phases. The flow separation water-air line is traced using the *Geo-Reconstruct* procedure and explicit scheme The reference computed interface line is drawn through the cells which contain 50% of each phase.

The boundary conditions (Fig. 1) for the numerical simulations are: (i) entrance water: the velocity is kept constant $V_0 = 0.65$ m/s, (ii) the entrance of the air: constant atmospheric pressure, $p = p_0$; (iii) stationary walls and zero shear stress.

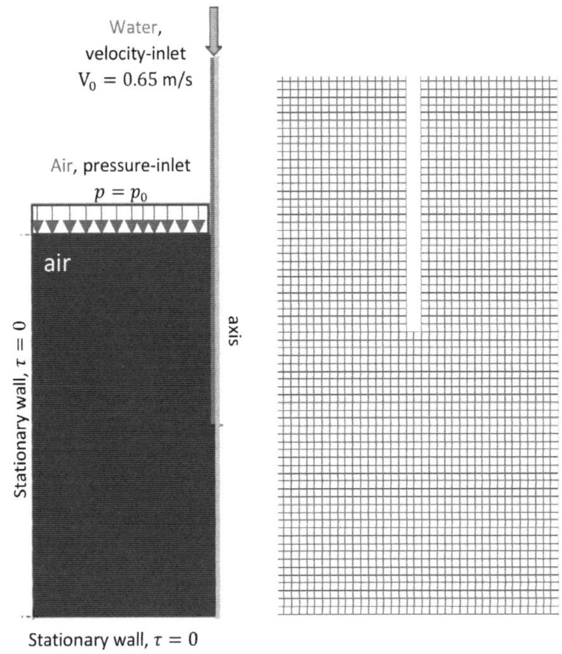

Figure 1: Numerical computational domain, initial phases configurations, the boundary conditions and mesh detail.

The solvers are run under unsteady conditions with influence of surface tension $\sigma = 0.072$ N/m and a precision of 10^{-6}, with the time step 10^{-5} and maxim iteration for time step 200.

3. Findings

We observe asymmetry in the first stages of liquid jet injection. Differences between the experimental findings and the numerical simulation appear and are presented in figure 2. The detachments of the first three drops occur at later moments in time, almost double than the ones obtained in the simulation and at shorter distances from the capillary nozzle.

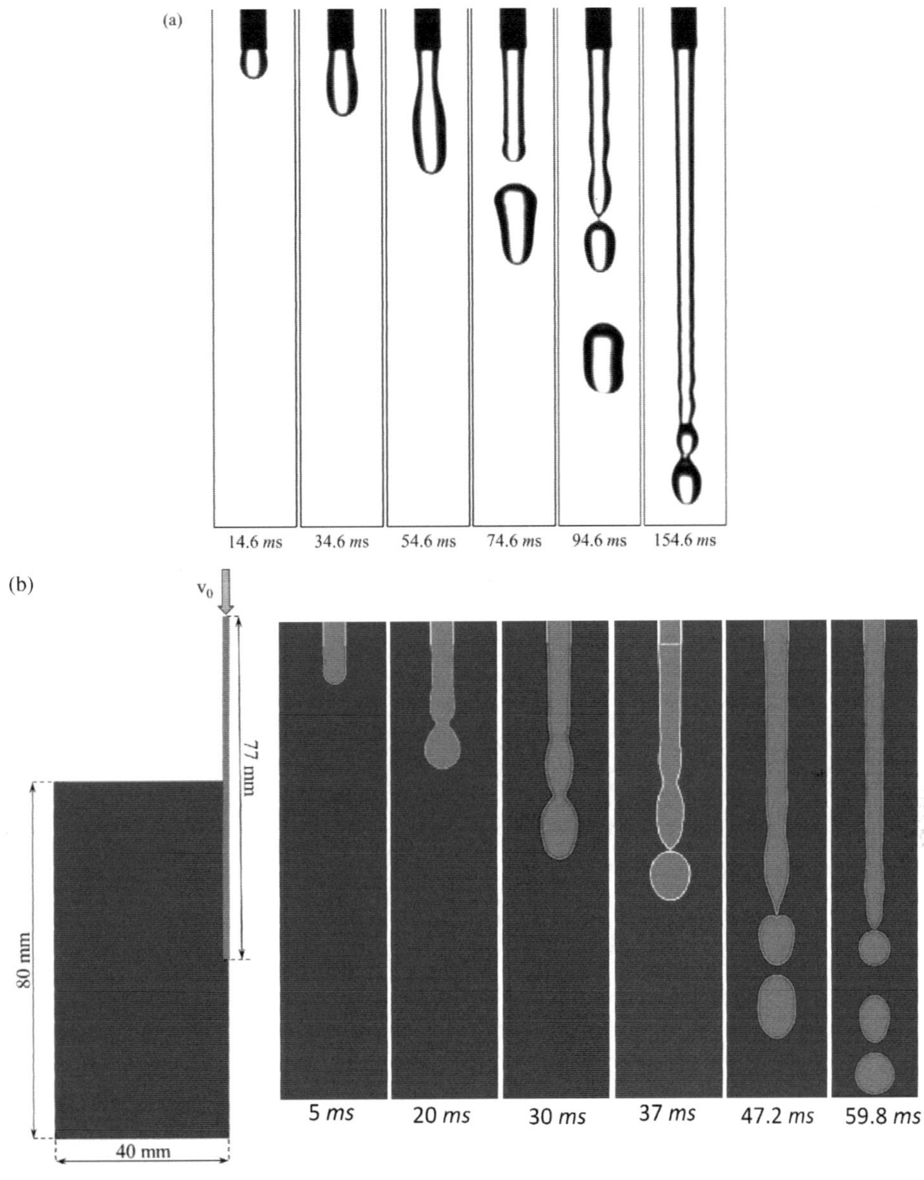

Figure 2: (a) First stages of liquid jet formation; (b) Numerical working model and numerical simulation of the shape of a water jet.

The shapes of the jets and drops can also be distinguished as the first drop detaches. The jet appears to have a smoother surface and the drop has a larger volume and an elongated shape than predicted. The diameter of formed drop is approximately 4.56 mm and numerically it was found to be 3.57 mm. The second drop detaches later than simulated, after a time of 24.2 ms, more than double the 10.2 ms obtained numerically. As the interface develops into an almost symmetric form, the configurations of the jet and drop during the second pinch-off resemble the ones predicted. The separation of the third drop occurs after an almost equal time to the one determined by simulations (12.4 ms experimentally and 12.6 ms numerically). However, the same differences regarding the time and place of the detachment are found in both cases. The diameters of second and third detached droplets are measured and compared with the ones obtained in the simulation. The results are found to be in agreement, being approximately 3 mm both experimentally and numerically. We observe similar configurations of the jets and the formation of capillary waves on account of the detachment of the third drop, which can be predicted via numerical simulations (figure 3).

The cause of the observed discrepancy between simulations and experiments can be attributed to the extra inertial forces, that can be introduced via the concept of added mass. These supplementary inertial forces are needed to accurately describe the first moments of liquid motion, in this case, the first moments of liquid injection. Also, small deviations in the nozzle's geometry (such as a small deviation from a circular cross-section) can also contribute to the observed asymmetry. There is also the problem of the nozzle-surface roughness residing at the contact between the three phases. This will affect the contact angle and with it the jet's development. All the above have not been implemented into the numerical procedure making the predictions less accurate only for the first moments of liquid injection.

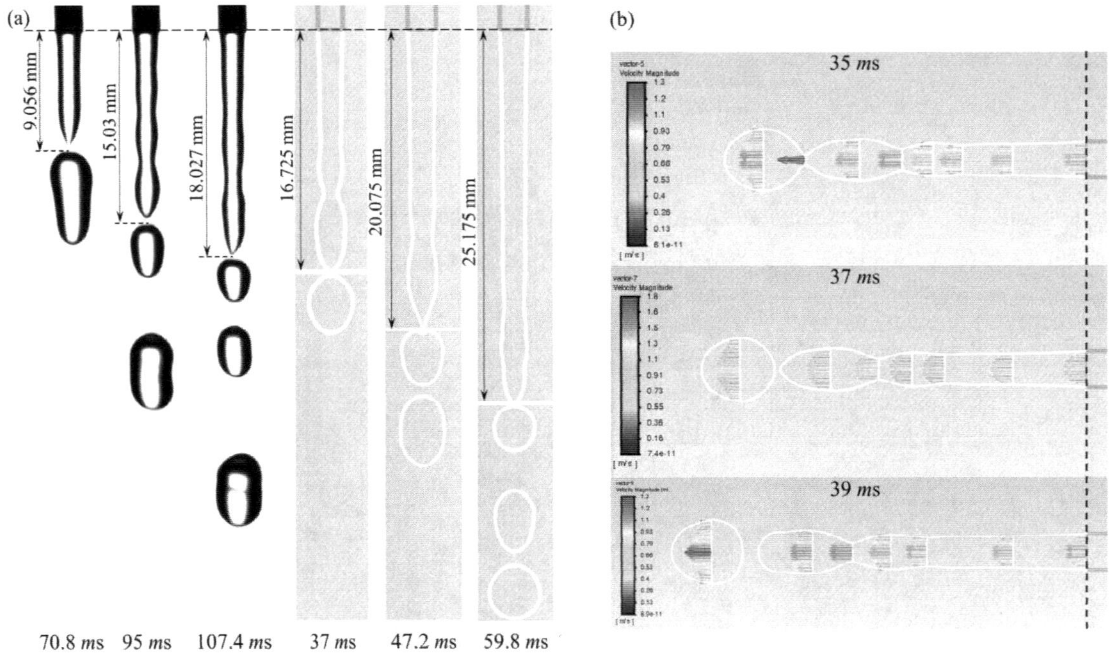

Figure 3: (a) Experimental and numerical data for the detachments of the first three drops; (b) Numerical data of the velocity vectors before, during and after pinch-off.

The mean injection velocity for the flow rate of 100 ml/min and the capillary nozzle of 1.8 mm diameter was found to be 0.65 m/s. Velocity vectors and magnitudes are determined numerically 1 millisecond before the detachment of the first drop. We present velocity distributions at 8 different sections of the jet in figure 4, starting with the exit of the capillary nozzle where the parabolic velocity distribution is observed. As the sections advance throughout the jet, the relaxation of the distribution develops. A maximum velocity of 1.3 m/s is reached at the point of detachment, which is approximately double the mean injection velocity. After pinch-off the relaxation of the velocity distribution is observed at the equatorial section of the drop. In Fig. 5 are represented the velocity magnitude distributions in the jet at different time steps.

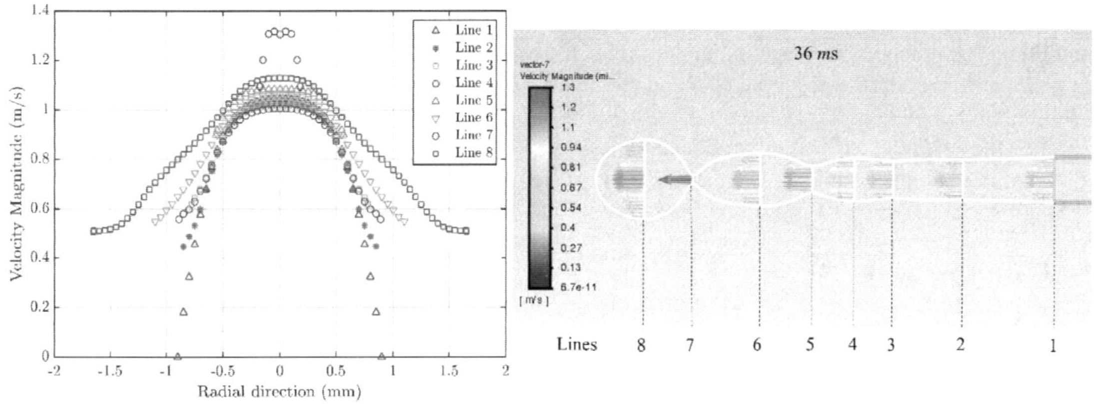

Figure 4: Numerical data for the velocity magnitude at t = 36 ms for various sections.

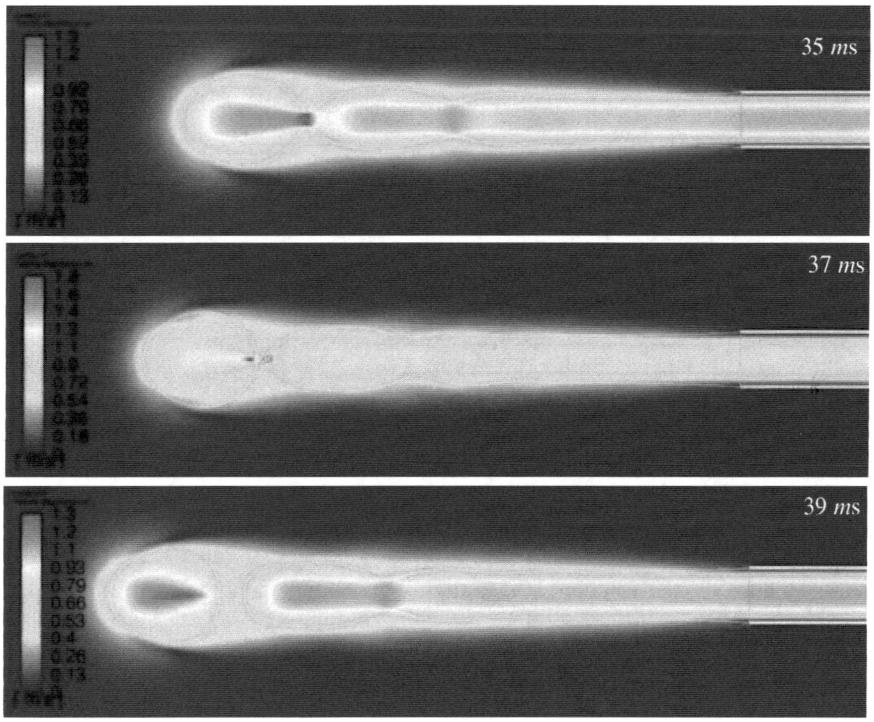

Figure 5: The evolution of the velocity magnitude contours at different time steps.

4. Closing

The paper tackles the problem of liquid jet injection and transition regime seen when the liquid is first pumped out of the capillary nozzle. This rich regime remains largely unexplored both experimentally and theoretically. In this study we emphasize the first stages of the jetting regime and droplet generation experimentally and numerically. Due to an asymmetric water-air interface we observe differences between the results and the numerical simulations which use an axial-symmetric geometry. As the interface settles, the configuration of the liquid jet resembles the one obtained in the simulations. The formation of capillary waves at the detachment of the third drop is observed.

Velocity distributions before the first detachment are determined throughout different sections of the liquid jet. The parabolic distribution at the exit of the nozzle relaxes along the axis of the formed column. At pinch-off, maximum velocity is reached and it is approximately double the average injection velocity. Inside the formed droplet, relaxation of the distribution is observed.

The study shows the limitations of an axial-symmetric numerical approach in characterizing the first stages of liquid injection. Even in seemingly simple experiments such as this one, the low viscosity and the high surface tension of the liquid bring asymmetry to the interface's shape and breakup.

Acknowledgments
The experimental investigations were possible due to European Regional Development Fund through Competitiveness Operational Program 2014-2020, Priority axis 1, Project No. {P-36-611}, MySMIS code 107066, Innovative Technologies for Materials Quality Assurance in Health, Energy and Environmental - Center for Innovative Manufacturing Solutions of Smart Biomaterials and Biomedical Surfaces – INOVABIOMED. The authors acknowledge the support of CHIST-ERA – 19 – XAI – 009 MUCCA project, by the founding of EC and The Romania Executive Agency for Higher Education, Research, Development and Innovation Funding - UEFISCDI, grant COFUND-CHIST-ERA MUCCA no. 206/2019. The authors would like to thank Claudiu Patrascu for fruitful discussions.

References
[1] Clanet C and Lasheras JC 1999 Transition from dripping to jetting *Journal of Fluid Mechanics* **383** 307–326 https://doi.org/10.1017/S0022112098004066
[2] Rayleigh L 1879 On the stability, or instability, of certain fluid motions *Proceedings of the London Mathematical Society* **1**(1) 57–72
[3] Patrascu C and Balan C 2021 Decay of stationary capillary waves on impinging liquid jets *Physics of Fluids* **33**(2) 022109 https://doi.org/10.1063/5.0038906
[4] Broboana D, Bratu AM, Magos I, Patrascu C and Balan C 2022 Kinematics of the viscous filament during the droplet breakup in air *Scientific Reports* **12**(1) 1–9 https://doi.org/10.1038/s41598-022-05839-y
[5] Patrascu C and Balan C 2020 The effect of curvature elasticity on Rayleigh–Plateau instability *Eur. J. Mech. B-Fluid* **80** 167–173 https://doi.org/10.1016/j.euromechflu.2019.11.005
[6] Cordero ML, Gallaire F and Baroud CN 2011 Quantitative analysis of the dripping and jetting regimes in co-flowing capillary jets *Physics of Fluids* **23**(9) 094111 https://doi.org/10.1063/1.3634044
[7] Martin GD, Hoath SD and Hutchings IM 2008 Inkjet printing-the physics of manipulating liquid jets and drops *J. Phys.: Conf. Ser.* **105**(1) 012001 doi 10.1088/1742-6596/105/1/012001
[8] Oliphant K, Webb BW and McQuay MQ 1998 An experimental comparison of liquid jet array and spray impingement cooling in the non-boiling regime *Experimental Thermal and Fluid Science* **18**(1) 1–10 https://doi.org/10.1016/S0894-1777(98)10013-4
[9] Simionescu ŞM, Tănase NO, Broboană D and Bălan C 2017 Impinging air jets on flat surfaces at low Reynolds numbers *Energy Procedia* **112** 194–203 https://doi.org/10.1016/j.egypro.2017.03.1083
[10] Tănase NO, Enache A, Broboană D and Bălan C 2021 Experimental and numerical studies of the free surface flow over the patterned weir *2021 12th Int. Symp. Advanced Topics in Electrical Engineering (ATEE)* pp 1–4 doi: 10.1109/ATEE52255.2021.9425294
[11] Shinjo J and Umemura A 2010 Simulation of liquid jet primary breakup: Dynamics of ligament and droplet formation *International Journal of Multiphase Flow* **36**(7) 513–532 https://doi.org/10.1016/j.ijmultiphaseflow.2010.03.008
[12] Farvardin E and Dolatabadi A 2013 Numerical simulation of the breakup of elliptical liquid jet in still air *Journal of Fluids Engineering* **135**(7) 071302 https://doi.org/10.1115/1.4024081

IOP Conf. Series: Earth and Environmental Science 1136 (2023) 012039 doi:10.1088/1755-1315/1136/1/012039

Comparison of model-based techniques for pipe burst location in water distribution networks

Bruno Ferreira[1], **André Antunes**[2], **Nelson Carriço**[1], **Dídia Covas**[3]

[1] INCITE, Barreiro School of Technology, Polytechnic Institute of Setúbal, Barreiro, 2839-001, Portugal
[2] Sustain.RD, Setúbal School of Technology, Polytechnic Institute of Setúbal, Setúbal, 2914-508, Portugal
[3] CERIS, Instituto Superior Técnico, Universidade de Lisboa, Lisboa, 1049-001, Portugal

bruno.s.ferreira@estbarreiro.ips.pt

Abstract. The current paper compares the performance of three model-based techniques for the automatic location of pipe burst events in water distribution networks. The first technique is based on nodal pressure sensitivities, the second solves an inverse analysis problem and the third technique uses hydraulic simulation to train a classifier. A real case study is used and a set of artificial measurements is generated for a number of pipe burst scenarios, with fixed burst location and variable pressure and flowrate noise levels and burst sizes. The performance of each technique is assessed based on the determined burst distance to the real burst location. Obtained results are discussed and the most relevant conclusions are drawn.

1. Introduction

The control of water losses is a major concern for water utilities, as these imply an excessive depletion of water resources, energy consumption, and chemicals in water treatment plants, whilst also posing a health risk to the population due to risks of bacteria and pollutant contamination [1]. The global volume of real water losses in Portugal in 2019 is estimated at 168 million cubic meters, which was enough to supply the entire country for nearly 4 months. An important source of water loss in water distribution networks (WDN) is pipe burst events [2]; in addition to the previously mentioned consequences, these events may also cause service disruption and third-party damage. So, the early detection, location, and repair of pipe bursts is critical to reduce the lost water volume, repair costs and lower the risk of contaminant intrusion. Thus, the earlier the repair is carried out, the safer the distribution system is.

Some pipe burst events cause service disruption and may even be visible at street level; these burst events are promptly reported by users to the water utility, which, in turn, repairs them and restores the water supply. The problem lies in the burst events which are not visible at the street level; these events can go unreported for months. In these situations, water utilities should adopt a set of systematic strategies to locate these events as soon as possible.

Traditional water loss control techniques include minimum night flow monitoring, acoustic signal analysis, and water balance calculation in network sectors [2,3]. Although efficient in leakage assessment (i.e., quantifying the amount of lost water), these techniques might not be suited for an early response regarding the burst location; in these situations, more efficient and lower–cost burst location methods are required. Such methods have been actively developed by the scientific community and can be roughly divided into three groups, depending on the type of data and analysis being used [4]: transient-based approaches [5–8], model-based approaches [9–11], and data-driven approaches [12–15].

The application of transient-based techniques in real and complex networks may be unfeasible since it requires an accurate transient model and high-frequency data obtained through the installation of a great number of sensors with high accuracy and frequency [3]. Data-driven approaches are found to be effective in burst detection, but often fail to successfully locate those burst events in the WDN [16]. Furthermore, they require a vast amount of historical data (i.e., pressure and flow measurements) regarding the WDN hydraulic behavior, both in the base operation scenario and during pipe burst events (as well as some information regarding the repair intervention itself). Due to the lack of human resources, technical, and financial capacity, it is usually not possible for water utilities to systematically generate the required information through daily routines [17], which directly compromises the implementation and efficiency of such techniques.

Innovations in real-time hydraulic modeling allow using techniques based on steady-state or extended period hydraulic simulations to locate burst events in real-time. Distinct approaches have been developed using real-time hydraulic simulation models to locate pipe burst events in WDN. These approaches can be based on solving the inverse analysis problem [9,11,18,19], calculating the network sensitivities [10,20], focusing on error-domain model falsification [21,22], or using a classifier [23,24]. Although model-based approaches can correctly locate leaks, these require the hydraulic model to be accurately calibrated so that flowrate and pressure-head estimations are realistic reflections of the operating state of the WDN. Furthermore, their results are also highly sensitive to the uncertainty of model parameters, such as demands, valve states, and pipe characteristics [4]. Thus, model-based approaches have not reached the maturity required for mainstream adoption by water utilities [9].

The current paper presents the comparison of three distinct model-based techniques to the automatic location of pipe burst events in WDN. The three techniques are applied to a real WDN in Lisbon metropolitan area using a calibrated hydraulic model. Obtained results are discussed and the most relevant conclusions are drawn.

2. Methods

2.1. General approach

The current section describes the three techniques that are compared herein for the location of pipe burst events in WDN. It is assumed that the pipe burst event was already detected, for instance, by using data-driven techniques, such as early presented, though it was not located. Consider that, after the pipe burst detection, pressure data are collected from a given number of N_p pressure sensors installed in the WDN:

$$\boldsymbol{P}^{real}(\boldsymbol{k})=\left[P^{real}_{sensor\,1}(k)\ \ldots\ P^{real}_{sensor\,N_p}(k)\right] \tag{1}$$

where $\boldsymbol{P}^{real}(\boldsymbol{k})$ is the set of pressure measurements for N_p pressure sensors and with respect to time k; $P^{real}_{sensor\,1}(k)$ corresponds to the real pressure measurement obtained by the 1st pressure sensor with respect to time k; $P^{real}_{sensor\,N_p}(k)$ corresponds to the real pressure measurement obtained by the N_p-th pressure sensor with respect to time k.

Similarly, flowrate data are obtained by a given number of N_q flowrate sensors installed in the inlets of the WDN as follows:

$$\boldsymbol{Q}^{real}(\boldsymbol{k})=\left[Q^{real}_{sensor\,1}(k)\ \ldots\ Q^{real}_{sensor\,N_q}(k)\right] \tag{2}$$

It is assumed that neither pressure nor flowrate data contain anomalous values, except the artificially introduced noise.

A set of artificial measurements is generated for a number of pipe burst scenarios by using the emitter coefficient C, with varying noise levels and burst sizes. The burst flowrate Q of each scenario, assuming a pipe burst event of fixed size, can be computed as follows:

$$Q = C\,P^{\alpha} \tag{3}$$

The performance of each technique is assessed based on the Euclidean distance to the real burst location (expressed in meters). The following sections briefly present the three techniques used for the automatic location of pipe burst events in WDN.

2.2. Technique based on nodal pressure sensitivities

The first technique is based on the works of Casillas et al. [10]. The pipe burst location is found by assessing the effect of a pipe burst event (with fixed size) in every possible network location and taking into account the set of pressure sensors installed in the WDN.

The first step requires the computation of the pressure sensitivity matrix of nodal pressure variations to a pipe burst event (of fixed size) in each WDN location (nodes) with respect to time k. Pipe burst events are simulated using equation (3). The pressure sensitivity matrix can be computed using the finite differences method [25], resulting in the matrix $S(k)$ with dimensions number of nodes × number of sensors. Each element $S_{i,j}(k)$ represents the variation of the pressure in node j with respect to the variation in the pipe burst size in node i. For a total of N_n nodes and N_p pressure sensors, the matrix $S(k)$ is obtained:

$$S(k) = \begin{bmatrix} \partial P_{sensor\ 1}(k)/\partial C_{node\ 1} & \cdots & \partial P_{sensor\ N_p}(k)/\partial C_{node\ 1} \\ \cdots & \cdots & \cdots \\ \partial P_{sensor\ 1}(k)/\partial C_{node\ N_n} & \cdots & \partial P_{sensor\ N_p}(k)/\partial C_{node\ N_n} \end{bmatrix} \tag{4}$$

The second step requires the calculation of the residual vector with respect to time k, $\boldsymbol{R}(\boldsymbol{k})$. Each residual element can be computed as the difference between real pressure data and its simulated value obtained by hydraulic simulation assuming a no burst scenario.

The third step evaluates the cosine distance between the residual vector $\boldsymbol{R}(\boldsymbol{k})$ and each column of the leak sensitivity matrix $S(k)$. The network dynamics can be captured by computing the average angle for a number of T time-steps and for all N_n possible locations. The pipe bust location is finally determined as the node which presents the smallest angle.

2.3. Technique based on the inverse analysis

The second technique is inspired on the works of [9,11,18]. The pipe burst location is found by solving an inverse analysis problem, that is, by finding the values of the unknown parameters in the hydraulic model (burst size and location) that minimize the discrepancies between simulated and field measurements. In other words, the object is to determine the location (node) and respective value of the emitter coefficient C that produces $\boldsymbol{P}^{real}(\boldsymbol{k})$ and $\boldsymbol{Q}^{real}(\boldsymbol{k})$.

Consider vector $\boldsymbol{M}$, containing the indexes of all possible nodes where the pipe burst event could be located (for instance, all network nodes). A single-objective optimization problem is formulated to solve the inverse analysis problem. Two decision variables are considered, namely, the pipe burst location (defined as a discrete variable containing the different possible locations in $\boldsymbol{M}$) and the burst size (i.e., the value of the emitter coefficient, defined as a continuous variable). Two types of constraints are considered, namely, a set of implicit type constraints considering hydraulics systems (which are guaranteed by the hydraulic solver) and an additional constrain which bound the algorithm solution search space. The optimization problem can be formulated as follows:

$$\begin{aligned} &Search\ for: X=(L_i,C_i);\ \ L_i\in \boldsymbol{M} \\ &Minimize: F(X) \qquad Subject\ to: P_i>0 \end{aligned} \tag{5}$$

where L_i and C_i are the node index and emitter coefficient value for a pipe burst event in location i, $\boldsymbol{M}$ is the vector containing the indexes of all possible nodes where the pipe burst event could be located, and P_i is the pressure (in a burst scenario) in location i.

The objective function $F(X)$ is defined as the weighted sum of squared differences between real and numerically simulated values (assuming a pipe burst in location L_i and size C_i) for both pressure and flowrate as follows:

$$Minimize: F(X)=\sum_{k=1}^{T}\left[\sum_{s=1}^{N_p}\frac{1}{N_p}\left(\frac{P^{real}_{sensor\ s}(k)-P^{sim}_{sensor\ s}(k)}{P_{pnt}}\right)^2+\sum_{q=1}^{N_q}\frac{1}{N_q}\left(\frac{Q^{real}_{sensor\ q}(k)-Q^{sim}_{sensor\ q}(k)}{Q_{pnt}}\right)^2\right] \tag{6}$$

where $P^{real}_{sensor\ s}(k)$ and $P^{sim}_{sensor\ s}(k)$ are, respectively, the field and simulated (with burst in location L_i and size C_i) pressure values in sensor s with respect to time k. Similarly, $Q^{real}_{sensor\ q}(k)$ and $Q^{sim}_{sensor\ q}(k)$ are the field and simulated flowrate values in sensor q with respect to time k. P_{pnt} and Q_{pnt} are the sensors' measurement error and are used so that pressure and flowrate values are converted into dimensionless values which can be summed.

2.4. *Technique based on Classifier*

The third technique is based on the works of Zhang et al. [23]. The idea is to use the WDS hydraulic model to obtain nodal pressure and pipe flowrate measurements associated with a variety of leakage scenarios (of distinct locations), which are then used to train a pattern recognition classifier.

Firstly, the network is divided into zones. This is essential in large networks to reduce the number of possible burst locations (and therefore labels) for the classifier. In this study, nodes are grouped by their topological information, namely, the shortest distance of pipes (in meters) between each node to the N_p pressure sensors. To this end, the following matrix D is considered:

$$D=\begin{bmatrix} d_{1,1} & \cdots & d_{N_p,1} \\ \cdots & \cdots & \cdots \\ d_{1,N_n} & \cdots & d_{N_p,N_n} \end{bmatrix} \tag{7}$$

where $d_{1,1}$ is the shortest distance of pipes (in meters) between the first node and the first sensor, while d_{N_p,N_n} is the shortest distance between the last node and the sensor N_p. Note that each row represents a possible burst location and is characterized by N_p variables (the shortest distance to each pressure sensor).

The objective is to cluster the N_n possible burst locations (rows) in matrix D into a given number of k clusters (or k leakage areas). Furthermore, a representative location should be selected from each cluster (with a total of k representative nodes, one for each leakage area). These representative locations will be used to simulate pipe burst events and train the classifier. The Affinity Propagation clustering method [26] is used in this study. This non-hierarchical method based on graph theory works by identifying the so-called "exemplars" among data points and by forming clusters of data points around these exemplars. It operates by simultaneously considering all data points as potential exemplars and by exchanging messages between data points until a good set of exemplars and clusters emerges [26]. The formal mathematical formulation of the updating functions can be found in [26]. As a result of the clustering process, a set of k clusters (or leakage areas) is obtained, as well as a set of k exemplars nodes (which will be used as representative nodes of each leakage area).

The second step consists of the generation of training data. To this end, pipe burst scenarios (for instance 25% of the total inlet flowrate) are simulated in each representative node by using equation (3); each scenario leads to a set of pressure and flowrate simulated values (at sensor locations) and is related to a specific label (containing the identification of the burst location). So, k datasets of pressure and flowrate (and a similar number of classification labels) are obtained. Note that at this point, each label contains a single set of measurements (or a single observation), which may affect the performance of a pattern recognition classifier. The following procedure is used to increase the number of observations for each label and to account for the uncertainty in model parameters and metering errors: 1) the mean perceptual error between simulated (using the hydraulic model) and measured values, *MPE*, should be estimated (e.g., 5%); 2) r copies of each label's dataset are generated by introducing random uniform noise in each measurement of $\pm MPE$; 3) the measurements of each sensor (across all datasets) are normalized into [0,1].

The third step focus on the classifier definition, parameterization, and training. Artificial Neural Networks (ANN) are used in this study, composed of a flattening layer, an active dense layer of size 128, and a *softmax* activation function for final classification. The $r \times k$ datasets (and respective labels) are used during the training phase, which is carried out for a given number of epochs (e.g., 5000).

Finally, the trained classifier can be used to identify the likely leakage zone (by identifying the representative node) based on measured flowrate and pressures during a pipe burst event. The received measurements of each sensor should be normalized into [0,1] before classification (using the limits already obtained during the data generation step).

3. Application: Results and discussion

3.1. *Case study description*

A real case study located in the metropolitan area of Lisbon, Portugal, is used herein for the comparison of burst location methods. The hydraulic simulation model is developed in EPANET and presents one storage tank, 1,261 nodes, 390 valves, and 967 pipes, with a total network length of 35 km. The WDN presents an average nodal pressure-head of 37.5 m and an average inlet flowrate of 31 L/s. For this study, it is assumed that eight pressure sensors are installed in the network (whose location was optimized using the multi-objective methodology presented in [27,28]), recording values on an hourly basis. The inlet flowrate of the storage tank is measured with the same time step.

3.2. *Effect of burst size and noise level*

Different pipe burst scenarios of a single burst are considered. The pipe burst and pressure sensor locations are not changed, varying only the size of the burst and the noise level of flowrate and pressure measurements. A set of artificial pressure and flowrate measurements is generated for each pipe burst scenario based on a hydraulic simulation of the WDN. Hydraulic simulations are carried out in EPANET [29] by using the Water Network Tool for Resilience (WNTR) Python package [30]. A total duration of 24 hours ($T = 24$) is considered in the analysis as well as a time step of one hour; thus, a total of 24 values is obtained for each of the nine sensors. The pipe burst location is chosen to occur in a looped part of the WDN, and in such a way that pressure sensors are located both downstream and upstream of the event. Figure 1(a) contains the spatial representation of the WDN, with the eight pressure sensors depicted in green triangles, the flowrate sensor in a blue square, and the pipe burst in a red triangle.

Three distinct pipe burst sizes are considered, corresponding to 5%, 20%, and 33% of the total average inlet flowrate (equivalent to flow rates values of 1.5 L/s, 6 L/s, and 10 L/s). For each of these burst sizes, three distinct noise levels associated with pressure and flowrate measurements are considered. This noise is generated using a uniform distribution of ±1% and ±5% of the real value to take into account the uncertainty in certain model parameters (such as nodal demands) as well as sensor reading errors. The ideal scenario (i.e., absence of noise) is also considered. A total of 9 pipe burst scenarios are considered for the location presented in figure 1(a).

Figure 1(b) depicts the artificially generated pressure and flowrate measurements for a pipe burst event of medium size (6 L/s) and a low noise level (1%) in dashed lines, and measurements for a no-burst scenario depicted in continuous lines. A generalized reduction of pressure is observed in the different sensors, which is associated with an increase in friction losses due to the increase in the flowrate.

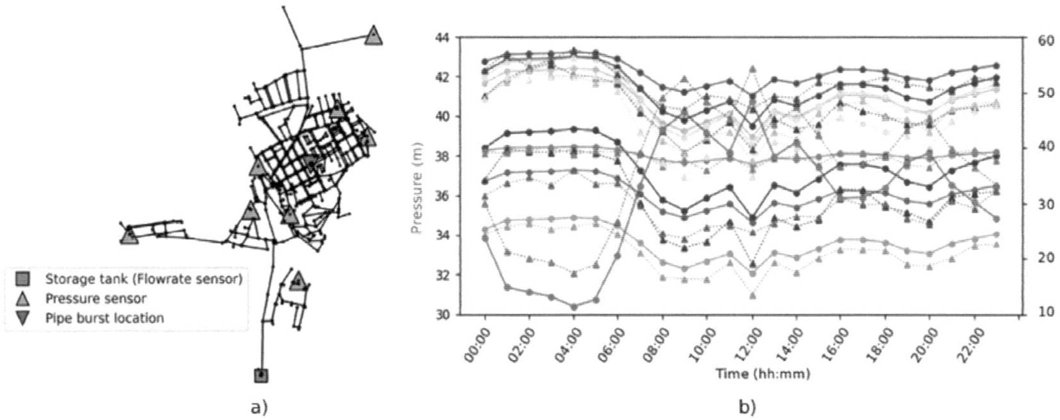

Figure 1. (a) Spatial representation of the WDN and; (b) artificially generated pressure data (in blue and green) and flowrate (in red), with the no-burst scenario and the pipe burst scenario with added noise depicted, respectively, in continuous and in dashed lines.

Each pipe burst scenario is generated 10 times to account for the heuristic nature of the genetic algorithms as well as the randomness of the introduced noise, leading to a total of 90 runs for each technique. The average geographical distance is calculated for each burst scenario and each technique.

Table 1 presents the average distance to the real pipe burst location obtained by using the technique based on nodal pressure sensitivities. All 1,261 network nodes are considered as potential burst locations and the pressure sensitivity matrices $S(k)$ are calculated using a ∂C value equivalent to an average pipe burst of 3 L/s. Results show that this technique is no capable of finding the exact burst location, even in the ideal scenarios (absence of noise). This is due to the fact that the sensitivity matrices $S(k)$ are calculated with a distinct emitter coefficient value (equivalent to 3 L/s) from the one used to characterize the pipe burst events in the different scenarios (and which are equivalent to 1.5 L/s, 6 L/s, and 10 L/s). The burst size has a direct influence on the technique's performance: larger burst events (10 L/s and 6 L/s) are naturally easier to locate, as noted by the smaller distance to the real burst location (and for the same noise level). The influence of the noise level is also relevant in scenarios of small bursts (1.5 L/s) or high noise level (5%) with either small or medium pipe burst sizes (1.5 L/s and 6 L/s).

Table 1. Results for the first technique: geographical distance (in meters) between the identified burst and the real burst.

Burst size	Distance		
	0% noise	1% noise	5% noise
33% (10 L/s)	97.4 m	105.2 m	99.9 m
20% (6 L/s)	124.4 m	123.6 m	208.2 m
5% (1.5 L/s)	124.0 m	261.8 m	699.1 m

Table 2 presents the distance to the real pipe burst location obtained by using the technique based on the inverse analysis. The optimization problem is solved using Genetic Algorithms [31] with the Pymoo Python package [32]. Every node associated with valves, as well as the storage tank, is excluded from the analysis, thus reducing the 1,261 possible locations to just 481. This second technique, and in opposition to the first technique, can find the precise pipe burst location. Nonetheless, this is significantly influenced by the burst size and noise level. In the absence of noise, both the medium and large-size bursts (6 L/s and 10 L/s) are correctly located. The small size burst (1.5 L/s), however, is not correctly located (with the 85.3 m being nonetheless lower than the 124 m found by the first technique). This small size burst leads to an average pressure drop of just 0.2 m and only impacts pressure measurements, on average, by around 0.50%; these small pressure changes at lower flows can result in a non-uniqueness problem, that is, when multiple combinations of decision variable values (burst location and size) produce equally fit solutions but inaccurate burst locations [9]. It is also possible to notice an increase in distance, as a result of increasing noise levels: low noise levels of 1% refer to maximum variations of ±0.375 m, which account for up to 47% of pressure variation due to a medium-size burst (with an average pressure drop of 0.8 m). This effect is enhanced in higher noise levels or smaller burst sizes, for instance, when considering an average pressure drop of 0.2 m (small size burst) and a maximum noise variation of ±1.875 m (high noise level). In this scenario, pressure variations generated by the burst are smaller than the introduced noise, thus hiding any burst signature in pressure measurements.

Table 2. Results for the second technique: geographical distance (in meters) between the identified burst and the real burst.

Burst size	Distance		
	0% noise	1% noise	5% noise
33% (10 L/s)	1.7 m	52.2 m	82.9 m
20% (6 L/s)	2.2 m	54.0 m	244.5 m
5% (1.5 L/s)	85.3 m	218.2 m	590.5 m

Table 3 presents the distance to the real pipe burst location obtained by using the technique based on a classifier. The WDN is divided into 23 leakage areas as a result of the network division process through clustering, leading to k=23 representative nodes being selected for possible pipe burst locations (as well as a similar number of classification labels). A total of 9 classification models are created, one for each burst size and noise level.

Table 3. Results for the third technique: geographical distance (in meters) between the identified burst and the real burst.

Burst size	Distance		
	0% noise	1% noise	5% noise
33% (10 L/s)	133.9 m	133.9 m	205.9 m
20% (6 L/s)	133.9 m	229.7 m	330.9 m
5% (1.5 L/s)	133.9 m	263.1 m	468.3 m

The ANN are developed with TensorFlow [33] Python package. This technique, in opposition to the ones previously analyzed, considers a small number of possible burst locations spread around the WDN; as a result, 133.9 m is the smaller distance that is possibly identifiable by this method (with the 23 representative nodes and the real burst location). So, it fails to find the precise location of the pipe burst event even in situations of absence of noise and large size bursts. On the other hand, as it directly takes into account the expected level of random noise, it provides the best results (amongst the three techniques) for situations of small size leak (1.5L/s) and high noise level (5%), with 468.3m. Furthermore, this technique is the one that requires the smaller number of hydraulic simulations, with just 23 simulations in total (against, for instance, the 1261 hydraulic simulations performed by the first technique).

Summarizing, the performance of model-based techniques for pipe burst location is affected by the models' calibration accuracy and by the ability of these models to correctly represent the WDN behaviour under a pipe burst scenario. As noted by Sophocleous et al. [9], current WDN models are calibrated to simulate observed pressures within ±1 m, which makes it hard to detect anomalies when pressure variations are smaller than the calibration error (such as small leaks with minor head losses). So, to directly improve the performance of model-based techniques, a stricter calibration criterion should be adopted by water utilities (for instance, ±0.2 m).

The technique based on inverse analysis obtained the best overall results and should be the subject of further studies. Uncertainty in model estimations should be directly considered in the analysis to reduce the burst distance found by the technique in situations of high noise level (e.g., 5%). The technique based on the nodal pressure sensitivities is deeply affected by the emitter coefficient value used to compute the sensitivity matrices; this may jeopardize the performance of this technique to find precise locations even in situations of absence of noise. The technique based on a classifier is promising, as it has the native ability to consider the expected mean perceptual error between simulated (using the hydraulic model) and measured values. Nonetheless, dividing the WDN into leakage areas offers a trade-off between location precision and problem complexity: a smaller number of leakage areas translates into a smaller number of classification labels (and, hence, a less computationally expensive and complex problem). This is obtained at the cost of less precise locations (as a smaller number of possible burst locations is considered). On the other hand, a greater number of leakage areas would result in a greater precision at the cost of solving a more complex classification problem (as a greater number of classification labels are considered). This method will be improved in further studies by considering distinct ANN architectures, distinct clustering methods, and a distinct number of leakage areas.

4. Conclusions

The current paper presented a performance comparison of three distinct model-based techniques for pipe burst location in water distribution networks. The first technique works by comparing the changes in pressure measurements (using the angle distance) caused by a single leak with respect to the nodal pressure sensitivities of a single leak in every possible leak location. The second technique is based on solving an inverse analysis problem (where the unknown variables are the burst size and location). The third technique uses hydraulic simulations to train a classifier.

A set of artificial measurements is generated for a number of pipe burst scenarios of fixed location with varying noise and size levels. The performance of each technique is assessed based on the distance to the real pipe burst location. The technique based on the inverse analysis presented the best overall results and has the ability to find the precise pipe burst location. Nonetheless, this ability is significantly influenced by the burst size and noise level. As demonstrated by this paper, a stricter calibration criterion (illustrated by a smaller noise level) should be adopted by water utilities (for instance, ±0.2 m) in order to improve the performance of automatic model-based techniques for pipe burst location.

Acknowledgments

The authors would like to thank the Fundação para a Ciência e Tecnologia (FCT) for funding this research through the WISDom project (grant number DSAIPA/DS/0089/2018), the Ph.D. Research Studentship of Bruno Ferreira (grant number SFRH/BD/149392/2019), and the research unit CERIS (grant number UIDB/04625/2020).

References

[1] Fontanazza CM, Notaro V, Puleo V, Nicolosi P and Freni G 2015 Contaminant intrusion through leaks in water distribution system: Experimental analysis *Procedia Eng* **119** 426–33 https://doi.org/10.1016/j.proeng.2015.08.904

[2] Puust R, Kapelan Z, Savic DA, Koppel T, Savić D and Koppel T 2010 A review of methods for leakage management in pipe networks *Urban Water J.* **7**(1) 25–45 https://doi.org/10.1080/15730621003610878

[3] Li R, Huang H, Xin K and Tao T 2015 A review of methods for burst/leakage detection and location in water distribution systems *Water Sci. Technol. Water Supply* **15** 429–41 https://doi.org/10.2166/ws.2014.131

[4] Hu Z, Chen B, Chen W, Tan D and Shen D 2021 Review of model-based and data-driven approaches for leak detection and location in water distribution systems *Water Supply* **21**(7) 3282–3306 https://doi.org/10.2166/ws.2021.101

[5] Soares AK, Covas DIC and Reis LFR 2011 Leak detection by inverse transient analysis in an experimental PVC pipe system *J. Hydroinformatics* **13** 153–66 https://doi.org/10.2166/hydro.2010.012

[6] Covas D and Ramos H 2010 Case studies of leak detection and location in water pipe systems by inverse transient analysis *J. Water Resour. Plan. Manag.* **136** 248–57 https://doi.org/10.1061/(ASCE)0733-9496(2010)136:2(248)

[7] Liggett JA and Chen L 1994 Inverse transient analysis in pipe networks *J. Hydraul. Eng.* **120** 934–55 https://doi.org/10.1061/(ASCE)0733-9429(1994)120:8(934)

[8] Covas D, Ramos H and de Almeida AB 2005 Standing wave difference method for leak detection in pipeline systems *J. Hydraul. Eng.* **131** 1106–16 https://doi.org/10.1061/(asce)0733-9429(2005)131:12(1106)

[9] Sophocleous S, Savić D and Kapelan Z 2019 Leak localization in a real water distribution network based on search-space reduction *J. Water Resour. Plan. Manag.* **145** 04019024 https://doi.org/10.1061/(ASCE)WR.1943-5452.0001079

[10] Casillas Ponce M V., Garza Castañón LE and Cayuela VP 2014 Model-based leak detection and location in water distribution networks considering an extended-horizon analysis of pressure sensitivities *J. Hydroinformatics* **16** 649–70 https://doi.org/10.2166/hydro.2013.019

[11] Wu ZY, Sage P and Turtle D 2010 Pressure-dependent leak detection model and its application to a district water system *J. Water Resour. Plan. Manag.* **136** 116–28 https://doi.org/10.1061/(ASCE)0733-9496(2010)136:1(116)

[12] Xu W, Zhou X, Xin K, Boxall J, Yan H and Tao T 2020 Disturbance extraction for burst detection in water distribution networks using pressure measurements *Water Resour. Res.* **56** 1–17 https://doi.org/10.1029/2019WR025526

[13] Romano M, Kapelan Z and Savić DA 2011 Burst detection and location in water distribution systems *World Environ. Water Resour. Congr. 2011*, Reston, VA: American Society of Civil Engineers pp 1–10 https://doi.org/10.1061/41173(414)1

[14] Loureiro D, Amado C, Martins A, Vitorino D, Mamade A and Coelho ST 2016 Water distribution systems flow monitoring and anomalous event detection: A practical approach *Urban Water J.* **13** 242–52 https://doi.org/10.1080/1573062X.2014.988733

[15] Wu Y, Liu S, Smith K and Wang X 2018 Using correlation between data from multiple monitoring sensors to detect bursts in water distribution systems *J. Water Resour. Plan. Manag.* **144**(2) https://doi.org/10.1061/(ASCE)WR.1943-5452.0000870

[16] Wu Y and Liu S 2017 A review of data-driven approaches for burst detection in water distribution systems *Urban Water J.* **14** 972–83 https://doi.org/10.1080/1573062X.2017.1279191

[17] Carriço N, Ferreira B, Barreira R, Antunes A, Grueau C, Mendes A et al 2020 Data integration for infrastructure asset management in small to medium-sized water utilities *Water Sci. Technol.* **82** 2737–44 https://doi.org/10.2166/wst.2020.377

[18] Pudar RS and Liggett JA 1992 Leaks in pipe networks *J. Hydraul. Eng.* **118** 1031–46 https://doi.org/10.1061/(ASCE)0733-9429(1992)118:7(1031)
[19] Blocher C, Pecci F and Stoianov I 2020 Localizing leakage hotspots in water distribution networks via the regularization of an inverse problem *J. Hydraul. Eng.* **146** 04020025 https://doi.org/10.1061/(ASCE)HY.1943-7900.0001721
[20] Geng Z, Hu X, Han Y and Zhong Y 2019 A novel leakage-detection method based on sensitivity matrix of pipe flow: Case study of water distribution systems *J. Water Resour. Plan. Manag.* **145** 04018094 https://doi.org/10.1061/(ASCE)WR.1943-5452.0001025
[21] Goulet J-A, Coutu S and Smith IFC 2013 Model falsification diagnosis and sensor placement for leak detection in pressurized pipe networks *Adv. Eng. Informatics* **27** 261–9 https://doi.org/10.1016/j.aei.2013.01.001
[22] Moser G, Paal SG and Smith IFC 2018 Leak detection of water supply networks using error-domain model falsification *J. Comput. Civ. Eng.* **32** 04017077 https://doi.org/10.1061/(ASCE)CP.1943-5487.0000729
[23] Zhang Q, Wu ZY, Zhao M, Qi J, Huang Y and Zhao H 2016 Leakage zone identification in large-scale water distribution systems using multiclass support vector machines. *J. Water Resour. Plan. Manag.* **142** 04016042 https://doi.org/10.1061/(ASCE)WR.1943-5452.0000661
[24] Quiñones-Grueiro M, Ares Milián M, Sánchez Rivero M, Silva Neto AJ and Llanes-Santiago O 2021 Robust leak localization in water distribution networks using computational intelligence *Neurocomputing* **438** 195–208 https://doi.org/10.1016/j.neucom.2020.04.159
[25] Lansey KE, El-Shorbagy W, Ahmed I, Araujo J and Haan CT 2001 Calibration assessment and data collection for water distribution networks *J. Hydraul. Eng.* **127** 270–9 https://doi.org/10.1061/(ASCE)0733-9429(2001)127:4(270)
[26] Frey BJ and Dueck D 2007 Clustering by passing messages between data points *Science* **315**(5814) 972–6 https://doi.org/10.1126/science.1136800
[27] Ferreira B, Carriço N and Covas D 2021 Optimal number of pressure sensors for real-time monitoring of distribution networks by using the hypervolume indicator *Water* **13** 2235 https://doi.org/10.3390/w13162235
[28] Ferreira B, Antunes A, Carriço N and Covas D 2022 Multi-objective optimization of pressure sensor location for burst detection and network calibration *Comput. Chem. Eng.* **162** 107826 https://doi.org/10.1016/j.compchemeng.2022.107826
[29] Rossman L 2000 *EPANET 2 User's Manual* (Cincinnati, OH, USA: U.S. Environmental Protection Agency)
[30] Klise KA, Bynum M, Moriarty D and Murray R 2017 A software framework for assessing the resilience of drinking water systems to disasters with an example earthquake case study *Environ. Model. Softw.* **95** 420–31 https://doi.org/10.1016/j.envsoft.2017.06.022
[31] Goldberg D 1989 *Genetic Algorithms in Search, Optimization, and Machine Learning* (New York, USA: Addison Wesley)
[32] Blank J and Deb K 2020 Pymoo: Multi-objective optimization in Python *IEEE Access* **8** 89497–509 https://doi.org/10.1109/ACCESS.2020.2990567
[33] Abadi M, Agarwal A, Barham P, Brevdo E, Chen Z, Citro C et al 2016 *TensorFlow: Large-Scale Machine Learning on Heterogeneous Distributed Systems* https://doi.org/10.48550/arXiv.1603.04467

Understanding district metered area level leakage using explainable machine learning

Matthew Hayslep[1], **Edward Keedwell**[1], **Raziyeh Farmani**[1], **Joshua Pocock**[2]

[1] College of Engineering, Mathematics and Physical Sciences, University of Exeter, Exeter, EX4 4PY, United Kingdom
[2] DWS Water Resources & Water Efficiency, South West Water, Exeter, EX2 7HR, United Kingdom

mh989@exeter.ac.uk

Abstract. Understanding the various interrelated effects that result in leakage is vital to the effort to reduce it. This paper aims to understand, at the district metered area (DMA) level, the relationship between leakage and static characteristics of a DMA, i.e. without considering pressure or flow. The characteristics used include the number of pipes and connections, total DMA volume and network density, as well as pipe diameter, length, age, and material statistics. Leakage, especially background and unreported leakage, can be difficult to accurately quantify. Here, the Average Weekly Minimum Night Flow (AWM) over the last 5 years is used as a proxy for leakage. While this may include some legitimate demand, it is generally assumed that minimum night flow, strongly correlates with leakage. A data-driven case study on over 800 real DMAs from UK networks is conducted. Two regression models, a decision tree model and an elastic net linear regression model, are created to predict the AWM of unseen DMAs. Reasonable accuracy was achieved, considering pressure is not an included feature, and the models are investigated for the most prominent features related to leakage.

1. Introduction

It has been consistently noted throughout the literature that leakage is a pervasive and important challenge for water companies which has an economic, environmental and sustainability impact [1]. To combat this and reduce leakage, water distribution networks in the UK are commonly divided into District Metered Areas (DMAs) [2, 3]. These DMAs have flow sensors at the inlets and outlets, providing a measure of how much water is used in that DMA. There are multiple ways of measuring and analyzing leakage [4]. This paper uses Average Weekly Minimum Night Flow (AWM) as a proxy for leakage which focuses this study on background and unreported leakage (see Section 2). In addition, Minimum Night Flows (MNFs) are part of the reporting required of water companies in the UK by the regulator OFWAT [4, 5], meaning that the data necessary to undertake this study is widely available.

Other research has estimated non-revenue water ratios and leakage rates [6-9]. Many of these works use principal component analysis and artificial neural networks [6-9]. These methods, while more sophisticated, are less explainable than the methods used in this paper. Explainability is important from the perspective of water companies to support decision making about leakage mitigation methodologies to implement and areas on which to focus resources. Explainable models allow us to understand the relationship between input features and outputs (e.g. predicted leakage), furthering our understanding of the problem itself and what factors impact leakage the most. For water companies, this means additional factors can be considered, or better understood, when deciding on actions, company policies, and management.

2. Methods

This paper investigates two models, a decision tree and an elastic net. Using explainable models is important because it allows the exploration of the *reasons* why a certain prediction is made. This means that the models are useful and informative outside of their direct application.

In this paper, leakage is defined as Minimum Night Flow (MNF), measured in m^3h^{-1}, and is the flow during the hour of least flow between midnight and 4am. MNF is recorded, ideally, every day for each DMA. The Weekly MNF is the lowest MNF recorded during a Monday-Sunday period. The use of Weekly MNF smoothes out what can be an extremely noisy signal and will also mask bursts or other major leakage events that do not last for at least a full week, which focuses this work on background and unreported leakage. Weekly MNF data covers a 5-year period from the beginning of 2017 to the beginning of 2022 for 866 real DMAs in the UK. The availability of Weekly MNFs for any single DMA varies from a few examples to almost total coverage of the 5-year period. To further reduce this problem from a time series forecast problem into a regression problem we use the Average Weekly MNF (AWM) over the last 5 years. Each model takes as input the static characteristics of a DMA and outputs the predicted AWM for that DMA. A prediction of, for example, 10 m^3h^{-1} would indicate that the average, over the last 5 years of operation, Weekly MNF for that DMA would be 10 m^3h^{-1}.

The aim of this paper is to predict the AWM of unseen DMAs based on their static characteristics using explainable, data-driven methods. Emphasis is put on predicting AWM, which is derived from MNF, *without* a hydraulic model, pressure, or flow information. By predicting AWM in this way, we can further our understanding of the causes of leakage. Accurate predictions of the expected AWM may also inform new leak detection methodologies, which often rely on some understanding of 'normal' operation.

Predictions are based on the number of customer connections in a DMA and various pipe features. The models learn from a subset of the real DMAs and are tested on the remaining subset of completely unseen DMAs. There is considerable variety amongst each of the 866 real DMAs in terms of total length (less than 1km to over 10km), pipe materials (100% metal to nearly 100% plastic), number of customer connections (100s to 1000s), etc. In total, more than 80 features were created. Most of these features were calculated using the pipe data available for each DMA. Each pipe provides the following information: installation date (used to calculate age), length, diameter, and material. Various materials were grouped together into 'metal' (various types of iron and steel), 'plastic' (PE, PVC, etc.) and 'other' (mainly asbestos cement). From this, various features were created such as 'total length', which is the sum of all pipe lengths in a DMA, 'proportion of metal pipes', which is the ratio of the number of metal pipes to all pipes, 'proportion by volume of pipes older than 20', which is the ratio of the volume of all pipes older than 20 years to the volume of all pipes, and so on. The volume of a pipe is calculated, based on the assumption that a pipe is a cylinder, using its length and diameter. Additional naming conventions are that large/small refers to diameter while long/short refers to length. Therefore, the smallest pipe diameter refers to the *diameter* of the pipe with the smallest diameter, whereas shortest pipe age refers to the *age* of the pipe with the shortest length. In addition, there are several 'weighted age' metrics which are calculated as follows:

$$f \text{ weighted age} = \frac{\sum_{i=1} f(p_i) \cdot a(p_i)}{\sum_{i=1} f(p_i)} \quad (1)$$

where p_i is the ith pipe in a DMA, f is the parameter of interest, for example diameter, length, or volume, and a is the age. Additional features of note include the number of customer connections and network density (customer connections per meter of pipe). A full table of the features investigated can be found in the appendix (table A1).

Each DMA is randomly placed into the training set or test set. The training set, which consists of 60% of the DMAs, is used to train the models. The test set, consisting of the remaining 40%, is used to report the performance of the models, these DMAs are not seen during training. The same test-train split was used consistently across all models, i.e., the same DMAs were always part of the training set and vice-versa for the test set. This is important because our purpose is to find the best model, not the most representative training set.

Many machine learning models make assumptions about the normalization of data, making it important to preprocess the DMA data. Therefore, features which are not proportions (since these are already normalized) are normalized. This is done by removing the median and dividing by the interquartile range, i.e., the range between the 1st and 3rd quartile. The median is used to scale instead of the mean because the data is not normally distributed. Therefore, using the median is more robust to outliers. To ensure that all features lie in the range 0 to 1, the features which are not proportions are then rescaled to fall within that range. The precise values for these transformations are found using the training set, rather than the entire dataset. This is important to avoid 'leaking' information about the test set back into the training set. This method of normalization is similar to the 'Z-score' in [6, 9] but replaces the mean and standard deviation with the median and interquartile range.

Cross-validated grid search is used to finetune the hyperparameters of the decision tree and elastic net. Cross-validation takes a random subset of the training data away, without replacement, to serve as an intermediate test set, known as the validation set. This process is repeated, 8 times in this paper, until every training sample has been in the validation set exactly once. Each validation set is used to test different model hyperparameters, with the average over the repeated random subsets being used to determine the best hyperparameter values.

Each model is assessed primarily using the R^2 metric. Explained variance and mean absolute error (MAE) have also been reported for completeness. R^2 values range from $-\infty$ to 1, with 1 being the best possible score. A model which simply returns the average value of all samples would get an R^2 of 0.

2.1. Decision Tree

Decision trees [10] are simple models where predictions are made based on a series of if-then-else decisions, similar to a flow chart. Each decision, or node, is based on a threshold on a single feature. At the end of a series of decisions, a leaf node determines what value is outputted. The depth of a tree, i.e., the maximum number of decisions before reaching a leaf node, determines its complexity. At each node, the decision (which feature and what threshold) which minimizes the mean squared error (MSE) in each child node is chosen. This is repeated until the tree is complete. In this paper, the leaf nodes on the decision trees output a constant value. The hyperparameters of the decision tree that were searched using cross-validated grid search were: maximum tree depth, minimum samples required in a leaf node, and minimum samples required to split a node.

2.2. Elastic Net

Elastic Nets [11] are linear regression models that use both l_1 and l_2 regularization. Regularization adds additional components to the loss function, which is the function that is minimized during training to determine the coefficients of the model. While l_1 regularization adds a penalty based on the absolute value of the coefficients, l_2 regularization adds a penalty based on the squared value of the coefficients. Elastic net tries to combine the benefits of lasso linear regression, which prefer to have fewer features with non-zero coefficients by using l_1 regularization, and ridge linear regression, which penalizes the magnitude of coefficients by using l_2 regularization. Elastic net is particularly useful when there are groups of correlated features [11], which is certainly the case in this instance. When given correlated features, other linear regression models would pick one of the correlated features to be the only important one, whereas an elastic net spreads the importance amongst the correlated features. The hyperparameters of the elastic net that were searched using cross-validated grid search were: alpha, which controls ratio between the l_1 and l_2 regularization terms and lambda, which scales both the l_1 and l_2 regularization terms.

3. Results and Discussion

This section discusses the results of the two methods that were used. Ten different runs, including cross-validation for hyperparameter tuning, with different random seeds were conducted, as these models are not deterministic. The following sections detail the results of the best model for each method from these runs. Table 1 shows the overall results for each of the best models. From this table, we can see that the elastic net model significantly outperforms the decision tree. This is to be expected because of the simplicity of the decision tree and due to its fundamental limitations when used as a regressor, i.e., discrete single value outputs. Section 3.1 details the decision tree results while Section 3.2 details the elastic net.

Other models, such as a neural network, support vector machine and random forest, which are not detailed here, were tested. Overall, the elastic net model gave the best results, but most models achieved an explained variance and test R^2 of at least 0.6. It is possible that the performance of the elastic net is near the best achievable with the current set of features. Better performance might be achievable with more or different explanatory factors.

Table 1. Overall results for predicting AWM.

Method	Explained Variance	Test R^2	Train R^2	Overall R^2	Test MAE	Train MAE	Overall MAE
Decision Tree	0.572	0.571	0.632	0.606	3.031	2.730	2.851
Elastic Net	0.680	0.680	0.698	0.691	2.741	2.465	2.575

3.1. Decision Tree

Decision trees are simple, but explainable, models with the limitation of only being able to output discrete values, which is not ideal in a regression context. Therefore, it is common to use many different decision trees in an ensemble, i.e., a random forest. However, in this case the performance of the lone decision tree is good enough to justify examining it. Figure 1 shows the full decision tree with the thresholds transformed back into their original space. The sample sizes and MSEs are based on the training set. The decision tree model splits the data using 'total length' three times, 'number of customer connections' and 'shortest pipe age' twice, and 'diameter weighted age' and 'total volume of metal pipe' once.

The root node splits roughly two thirds of the training set away based on total length. The right-hand subtree then splits again on a much larger total length value with DMAs with more than 33km of pipe being assigned the highest AWM. If a DMA does not have more than 33km of pipe, then the number of customer connections is considered. DMAs with fewer customer connections are assigned a lower AWM than those with more. Overall, the right-hand subtree can be summarized as 'larger DMAs have a higher AWM' using different metrics.

Meanwhile, the left-hand subtree splits on shortest pipe age. The shortest pipes in DMAs in this particular dataset are often less than 1m long and might be better understood as representing junctions. Therefore, one way of interpreting the significance of this metric is that the age of the shortest junction in DMAs is an important factor. If the DMA has an older shortest pipe, the assigned AWM is dependent on size metrics, where larger DMAs are assigned a higher AWM. If the DMA has a younger shortest pipe, then the diameter weighted age and shortest pipe age are considered. Overall, the left-hand subtree can be summarized as 'smaller, younger infrastructure leaks less than older, larger infrastructure', i.e. the importance of infrastructure condition [4] and size.

Splitting on the number of customer connections and total length could be interpreted as the importance of network density, in terms of connections per meter of pipe. However, network density was a metric included as a feature which the decision tree did not consider important. This might indicate that the number of customer connections and total length are important in and of themselves rather than as a ratio of each other.

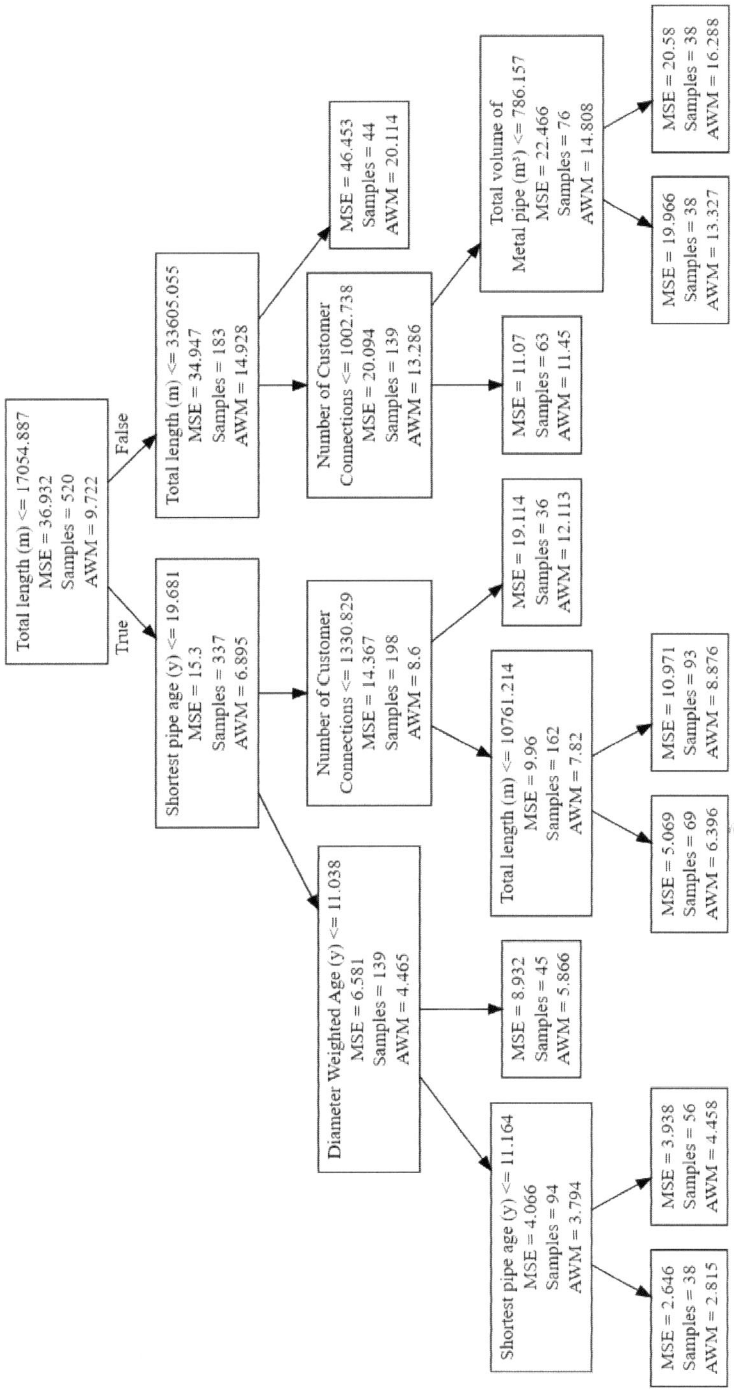

Figure 1. Decision tree model. The MSE, number of samples and AWM shown are based on the training set. The decision thresholds are shown transformed back into their original space.

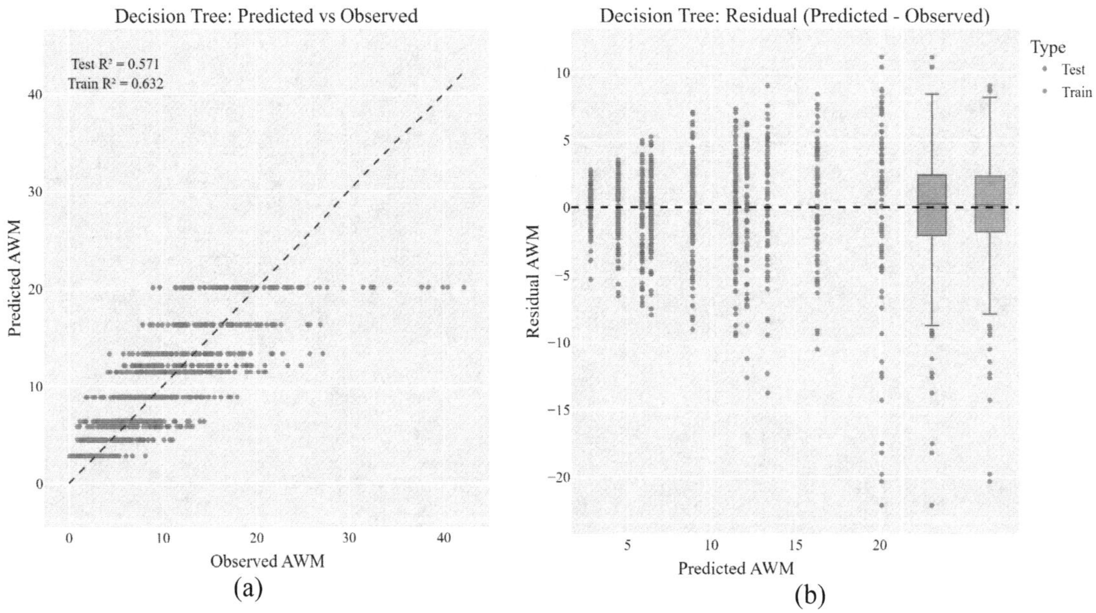

Figure 2. Decision tree comparison between (a) predicted and observed AWM, and (b) predicted and residual AWM (prediction - observed). The dotted line in (a) and (b) shows the zero-error line, i.e., perfect predictions. Figure 2b also shows the distribution of samples along the residual axis as a box plot.

Figure 2 shows the predictions of the decision tree across both the test and training set. The discrete outcomes of the decision tree can be seen clearly in both figure 2a and 2b. While there is some overlap between a DMA's predicted AWM and observed AWM, the box plot in figure 2b shows that most predictions lie within ~2.5 m^3h^{-1} of their observed value. The absolute median percentage error is 25.7% and 30.0% for the training and test set respectively. In addition, the decision tree is not significantly less accurate on the test set compared to the training set, which is confirmed by the R^2 of each set, and indicates that the model did not overfit. However, the model does have difficulty predicting the AWM of extreme outlier DMAs, although this is not completely unexpected behavior.

Overall, the performance of the decision tree model is good, especially considering its simplicity and limitations. In addition, the decision tree as shown in figure 1 allows straightforward assessment of more DMAs without using of the model directly.

3.2. Elastic Net

The Elastic Net model results are detailed in this section. Unlike the decision tree, the elastic net model has a continuous output because, fundamentally, it is a linear model. Therefore, we find that this model outperforms the decision tree model. This comes at the cost of additional complexity, though it is still a linear model. Figure 3 shows the predictions of the elastic net model across both the test and training set. Figure 3b again shows that most predictions lie within ~2.5 m^3h^{-1} of their observed value. The absolute median percentage error is 23.8% and 25.0% for the training and test set respectively. Furthermore, the accuracy of the model is similar on the test set and the training set, which is confirmed by the R^2 of each set, indicating the model has not overfitted. The elastic net model, as with the decision tree, has more difficulty predicting the AWM of extreme outlier DMAs. Again, this is not surprising as they are outliers.

As a linear regression model, the elastic net model derives a set of coefficients which allow us to understand why the model makes certain predictions. Table 1 lists the coefficients of the elastic net linear model. These coefficients are for the transformed data. They allow us to understand which features the elastic net has determined to be the most important for predicting the AWM of a DMA. In general, coefficients of less than 1 are insignificant, but have been reported for completeness.

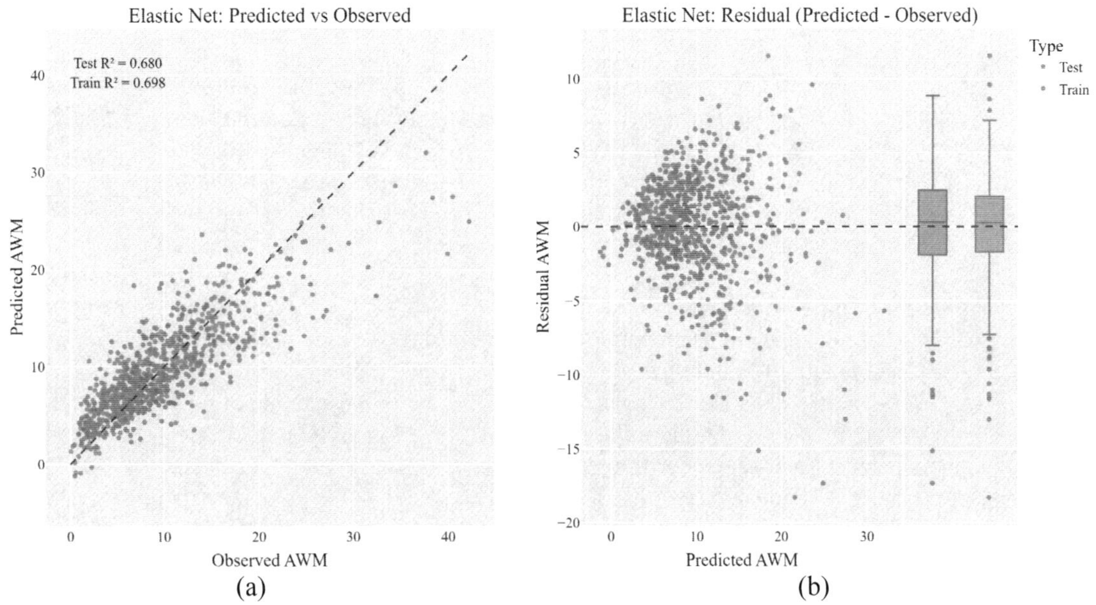

Figure 3. Elastic net comparison between (a) predicted and observed AWM, and (b) predicted and residual AWM (prediction - observed). The dotted line in (a) and (b) shows the zero-error line, i.e., perfect predictions. Figure 3b also shows the distribution of samples along the residual axis as a box plot.

From table 2 we can see that the number of customer connections is by far the most important factor and correlates with an increase in AWM. This is consistent with research on the leakage at service connections and water mains, where service connections, due to the high density of joints and fittings, are responsible for more leakage than their total length might indicate [4]. This is also consistent with the practice of reporting leakage in relation to the number of customer connections in an area [4]. This may also reflect the legitimate night usage of customers. The next most important coefficients are 'total length of pipes larger than 100mm' and 'total length', both correlating with an increase in AWM. Both of these features are closely correlated with each other because the pipes smaller than 100mm are usually very short. Again, this is consistent with the literature, in that the more pipe there is, the more leakage there is.

Table 2 also shows the factors which correlate with a decrease in AWM. Of these, the most important is 'smallest pipe diameter'. This suggests that, if the diameter of the smallest pipe in a DMA is larger, then the AWM is smaller. Bernoulli's principle states that a decrease in pipe diameter will decrease pressure and increase velocity. This increased velocity may result in more wear and tear, leading to more leakage, in smaller pipes compared to larger pipes. This increased velocity may be at its height during the MNF, as this is usually the period with the highest pressure, unless pressure management policies are in place. Furthermore, there is an interplay between the positive coefficients: 'largest pipe diameter' and 'total length of pipes larger than 100mm', which indicate that larger pipes cause more leakage, and the negative coefficient 'smallest pipe diameter' which indicates that larger pipes cause less leakage. One interpretation is that this describes the relationship *between* pipes of different sizes in a DMA. Specifically, those significant changes in pipe diameter within a DMA, such as stepping down from a 200mm pipe to a 20mm pipe, and the resulting pressure changes, are a significant underlying cause of these coefficients.

Table 2. Coefficients for the elastic net model in descending magnitude. Left shows positive coefficients, right negative. Prop. is the abbreviation of proportion.

Feature	Coeff.	Feature	Coeff.
Number of Customer Connections	9.53	Smallest pipe diameter (mm)	-3.14
Total length of pipes larger than 100mm (m)	3.54		
Total length (m)	3.31	Prop. of Metal pipes	-1.30
Oldest pipe length (m)	2.79	Longest pipe length (m)	-1.22
Prop. by volume of pipes older than 20	2.67	Total length of Plastic pipes older than 20 (m)	-1.20
Total volume of Metal pipe (m^3)	2.51		
Smallest pipe length (m)	2.17	Prop. by length of Plastic pipes older than 20	-0.87
Prop. by volume of Metal pipe	2.09		
Shortest pipe age (y)	2.08	Total length of pipes older than 20 (m)	-0.84
Prop. by length of pipes longer than 400m	2.01		
Largest pipe diameter (mm)	1.85	Prop. by volume of Other pipe	-0.79
Total length of Metal pipe (m)	1.84		
Number of pipes	1.67	Total volume of pipes longer than 400m (m^3)	-0.71
Prop. by length of Other pipe	1.56		
Average Metal pipe diameter (mm)	1.39	Total volume of Plastic pipe (m^3)	-0.59
Largest pipe age (y)	1.05		
Total length of Plastic pipe (m)	1.03	Longest pipe diameter (mm)	-0.55
Prop. by length of Metal pipe	0.88		
Total volume of pipes larger than 100mm (m^3)	0.88	Prop. by volume of pipes longer than 400m	-0.55
Total length of Metal pipes older than 20 (m)	0.77		
Prop. of Plastic pipes	0.76	Prop. by volume of pipes larger than 100mm	-0.39
Diameter Weighted Age (y)	0.66		
Total length of Other pipe (m)	0.65	Number of Metal pipes	-0.34
Prop. by length of pipes larger than 100mm	0.58		
Prop. of Other pipes	0.56	Number of Other pipes	-0.33
Network Density (mc^{-1})	0.37		
Average Metal pipe length (m)	0.33	Prop. of pipes longer than 400m	-0.26
Prop. by length of Metal pipes older than 20	0.33		
Length Weighted Age (y)	0.30	Prop. by length of Plastic pipe	-0.22
Number of Plastic pipes	0.27		
Oldest pipe diameter (mm)	0.12	Number of pipes longer than 400m	-0.08
Average pipe age (y)	0.07		
Prop. of pipes larger than 100mm	0.07	Total volume of Other pipe (m^3)	-0.05
Average Other pipe diameter (mm)	0.03		
Total length of pipes longer than 400m (m)	0.03	Number of pipes older than 20	-0.03
Shortest pipe length (m)	0.02		

Table 2 also shows the impact of age on AWM with 'oldest pipe length', 'proportion by volume of pipes older than 20' and 'shortest pipe age' all correlating with an increase in AWM. In addition, the impact of different materials is also present, though more difficult to decipher. For example, the proportion of metal pipes (by number) in a DMA correlates with a decrease in AWM while the proportion *by volume* of metal pipe correlates with an increase in AWM. This might mean that for smaller/shorter pipes metal is better than other materials, but for larger/longer pipes other materials are better than metal. This may be because of the mechanical joints of metal pipes which allow more minor leaks than, for example, a welded plastic pipe [4]. In addition, there is 'total length of plastic pipes older than 20' and 'proportion by length of plastic pipes older than 20', both of which correlate with a decrease in AWM. This could either mean that as plastic pipes age they do not degrade as badly as other pipe materials, or it could mean that the type of plastic used more than 20 years ago is better than the type of plastic used since. All the factors that affected the decision tree are present as coefficients for the elastic net model, though not necessarily as prominently as their importance in the decision tree might suggest.

Many of these features are related to others. For example, 'smallest pipe diameter' is related to 'total length of pipes larger than 100mm', which is also related to the total length of all pipes. If the smallest pipe diameter is 100mm then 'total length of pipes larger than 100mm' is equal to the total length of all pipes. Because of this, the interactions between the different coefficients are more complex in some cases than a cursory glance might suggest, and it becomes important to interpret which features are related to each other. Therefore, it is much harder to answer a question such as 'how would the AWM change if we replaced the oldest and largest metal pipe with a plastic one' without calculating the new statistic for the theoretical DMA and running it through the model. This is one of the main downsides of this linear model compared to the decision tree: while the elastic net is more accurate, it is also more difficult to interpret.

4. Conclusion

This paper has presented two models for predicting Average Weekly MNF of DMAs in a real-world data-driven case study. By doing so we hoped to further our understanding of the causes of long-term leakage. The results have illustrated the main factors that contribute to these predictions for each model and attempts have been made to interpret why these factors are important. The models and results are based on a data-driven approach which relies on little to no domain knowledge. The results presented support and demonstrate many of the ideas on what influences leakage presented in the literature. Specifically, the importance of DMA size, the number of customer connections and service pipe sizes, the different material properties for different pipe sizes and the interplay between different materials and their age all have an effect on Average Weekly MNF.

The simplicity and explainability of the methods used should allow these results to be repeated and improved upon. Further work may include the addition of more features. In particular, pressure may be the most obviously missing metric. Adding pressure or other features may help to improve the accuracy of these models even further. In addition, increasing the size of the dataset with more DMAs may also improve the accuracy of the predictions. Finally, further work may also incorporate these predictions as part of a leak detection methodology.

Acknowledgments

The first author would like to acknowledge the financial support from South West Water in collaboration with the University of Exeter Centre for Resilience, Environment, Water and Waste (CREWW). Special thanks are given to James Mercer for his assistance in accessing the data used in this paper. This work was undertaken as part of the first author's pursuit of a PhD.

Appendix

Table A1. Table of all features, in no specific order, created for each DMA.

Features		
Number of pipes	Diameter Weighted Age (y)	Length Weighted Plastic Age (y)
Total length (m)	Average pipe diameter (mm)	Prop. of pipes larger than 100mm
Oldest pipe age (y)	Largest pipe diameter (mm)	Average Other pipe diameter (mm)
Average pipe age (y)	Longest pipe diameter (mm)	Number of pipes longer than 400m
Longest pipe age (y)	Average Other pipe age (y)	Average Metal pipe diameter (mm)
Largest pipe age (y)	Average Metal pipe age (y)	Total length of Plastic pipe (m)
Prop. of Metal pipes	Smallest pipe diameter (mm)	Number of pipes larger than 100mm
Prop. of Other pipes	Shortest pipe diameter (mm)	Total volume of Plastic pipe (m^3)
Number of Metal pipes	Youngest pipe diameter (mm)	Average Plastic pipe diameter (mm)
Youngest pipe age (y)	Prop. of pipes older than 20	Prop. by length of pipes older than 20

Shortest pipe age (y)	Average Plastic pipe age (y)	Prop. by volume of pipes older than 20
Smallest pipe age (y)	Prop. by length of Other pipe	Total length of pipes older than 20 (m)
Number of Other pipes	Prop. by volume of Other pipe	Total volume of pipes older than 20 (m^3)
Network Density (mp^{-1})	Average Other pipe length (m)	Prop. by length of pipes longer than 400m
Oldest pipe length (m)	Length Weighted Other Age (y)	Prop. by volume of pipes longer than 400m
Prop. of Plastic pipes	Number of pipes older than 20	Prop. by volume of pipes larger than 100mm
Number of Demand Points	Prop. by length of Metal pipe	Total length of pipes longer than 400m (m)
Volume Weighted Age (y)	Prop. by volume of Metal pipe	Prop. by length of pipes larger than 100mm
Average pipe length (m)	Length Weighted Metal Age (y)	Total volume of pipes longer than 400m (m^3)
Longest pipe length (m)	Average Metal pipe length (m)	Total length of pipes larger than 100mm (m)
Largest pipe length (m)	Total length of Other pipe (m)	Total volume of pipes larger than 100mm (m^3)
Length Weighted Age (y)	Total length of Metal pipe (m)	Prop. by length of Other pipes older than 20
Number of Plastic pipes	Total volume of Other pipe (m^3)	Prop. by length of Metal pipes older than 20
Total volume of DMA (m^3)	Prop. of pipes longer than 400m	Total length of Other pipes older than 20 (m)
Shortest pipe length (m)	Total volume of Metal pipe (m^3)	Total length of Metal pipes older than 20 (m)
Smallest pipe length (m)	Prop. by length of Plastic pipe	Prop. by length of Plastic pipes older than 20
Youngest pipe length (m)	Prop. by volume of Plastic pipe	Total length of Plastic pipes older than 20 (m)
Oldest pipe diameter (mm)	Average Plastic pipe length (m)	

References

[1] Puust R, Kapelan Z, Savic D A and Koppel T 2010 A review of methods for leakage management in pipe networks *Urban Water Journal* **7**(1) 25–45 https://doi.org/10.1080/15730621003610878

[2] Romano M 2021 Review of techniques for optimal placement of pressure and flow sensors for leak/burst detection and localisation in water distribution systems *ICT for Smart Water Systems: Measurements and Data Science* **102** ed A Scozzari et al (Cham: Springer International Publishing) chapter 2 pp 27-63 https://doi.org/10.1007/698_2019_405

[3] Savić D and Ferrari G 2014 Design and performance of District Metering Areas in water distribution systems *Procedia Engineering* **89** 1136–43 https://doi.org/10.1016/j.proeng.2014.11.236

[4] Farley M and Trow S 2005 *Losses in Water Distribution Networks: A Practitioners' Guide to Assessment, Monitoring and Control* vol 4 (London, UK: IWA Publishing) https://doi.org/10.2166/9781780402642

[5] Ofwat 2018 *Targeted review of common performance commitments final report* (Ofwat, UK) pp 281 https://www.ofwat.gov.uk/publication/targeted-review-common-performance-commitments-final-report/

[6] Jang D and Choi G 2017 Estimation of non-revenue water ratio for sustainable management using artificial neural network and Z-score in Incheon, Republic of Korea *Sustainability* **9**(11) 1933 https://doi.org/10.3390/su9111933

[7] Jang D and Choi G 2018 Estimation of non-revenue water ratio using MRA and ANN in water distribution networks *Water* **10**(1) 2 https://doi.org/10.3390/w10010002

[8] Jang D, Park H and Choi G 2018 Estimation of leakage ratio using principal component analysis and artificial neural network in water distribution systems *Sustainability* **10**(3) 750 https://doi.org/10.3390/su10030750

[9] Kizilöz B 2021 Prediction model for the leakage rate in a water distribution system *Water Supply* **21**(8) 4481–92 https://doi.org/10.2166/ws.2021.194

[10] Breiman L, Friedman JH, Olshen RA and Stone CJ 1984 *Classification and Regression Trees* (New York: Routledge) p 368 https://doi.org/10.1201/9781315139470

[11] Zou H and Hastie T 2005 Regularization and variable selection via the elastic net *Journal of the Royal Statistical Society: Series B (Statistical Methodology)* **67**(2) 301–20 https://doi.org/10.1111/j.1467-9868.2005.00503.x

IOP Conf. Series: Earth and Environmental Science 1136 (2023) 012041 doi:10.1088/1755-1315/1136/1/012041

A multi-utility and dynamic approach for the upgrade of an aged water distribution network

Amin Minaei[1,2], **Enrico Creaco**[3] **and Robert Sitzenfrei**[4,5]

[1] Austrian Academy of Sciences (ÖAW), dr Ignaz Seipel-Platz 2, 1010 Viena
[2] Unit of Environmental Engineering, Department of Infrastructure Engineering, University of Innsbruck, Technikerstrasse 13, 6020 Innsbruck, Austria
[3] Dipartimento di Ingegneria Civile e Architettura, Via Ferrata 3, 27100 Pavia, Italy
[4] Unit of Environmental Engineering, Department of Infrastructure Engineering, University of Innsbruck, Technikerstrasse 13, 6020 Innsbruck, Austria
[5] Author to whom any correspondence should be addressed

robert.sitzenfrei@uibk.ac.at

Abstract. Water Distribution Networks (WDNs) along with other utilities are among the most important assets in every industrialized society. The rehabilitation and upgrade of WDNs refer to a very complex and multi-criteria problem which needs a comprehensive decision support system providing optimal renewal plans for the asset managers. There is a multitude of real-world problems faced with such activities. In this regard, this study aims to propose a novel method to cope with three practical challenges in WDNs rehabilitation activities including 1- budget limitation, 2- hydraulic deficiency and 3- correlation of WDNs with other infrastructures bringing the risk of cascading failures to a multiplex network. This results in a multi-objective optimization problem with three objectives which are cost, hydraulic and decoupled reliabilities. The problem is solved dynamically with the contribution of a nature inspired optimization algorithm, the Non-Dominated Sorting Genetic Algorithm (NSGA-II). The method is applied to a deteriorated water pipe network, and the results are compared with the ones obtained only by the consideration of two objectives, costs and hydraulic reliability showing how much the objectives are conflicted with each other.

1. Introduction

Conflict of water infrastructures with other buried networks, phasing design and budget constraints are among the most important practical considerations in solving water distribution problems through optimization algorithms [1].

Many studies have been conducted to show the importance of spatial correlation and interdependency of different infrastructures in terms of asset management perspective. Sitzenfrei, Mair [2] examined the risk of cascading failure for a water and drainage network and showed that dependency of infrastructures increases cascade vulnerability of networks that have highly correlation with each other. Continuously, through graph analysis of WDN, cascade failures either self or with other critical infrastructures have been the target of many recent studies [3-6].

Phasing design and construction of WDNs as well as budget constraints play a major role in bringing applicable upgrade programs to the authorities. Herein, many studies have been carried out to give solutions for solving WDNs problems concerning the dynamic behaviour of such systems. For this, some authors have taken such transitions into account to present new methods of design, based on phase-by-phase design and construction [7-10]. Owing to the multi-phase design approach, designs become significantly costly, which may not suit the budget availability of upgrade projects. In this regard, Minaei, Haghighi [11] through comparing different approaches related to static and dynamic design and/or upgrading of WDNs, proposed a kind of dynamic approach bringing about a variety of efficient programs to overcome funding constraints. Looking at the literature, there is a research gap about how an aged WDN can be upgraded dynamically and economically such that not only the hydraulic performance of the network be upgraded but also network becomes decoupled with adjacent systems decreasing the risk of cascade failures under failure events.

2. Material and Methods

To implement upgrading programs in which upgrade actions make safe spatial decoupled between important elements, first an accurate graph analysis of the neighbor networks is done. The next steps are making dependency models and formulating the optimization algorithm for generating dynamic upgrade programs.

2.1. Graph analysis of networks

To identify the topological properties of a multiplex network, a general undirected network graph can be constructed by a set nodes/vertex (n) and edges (e) connecting the vertices. For example, for WDNs, edges and nodes are links (pipes, pumps, valves) and junctions (demand nodes, reservoirs, tanks etc.), respectively. This is true for all graph-based networks and thereby each graph can be represented by the symmetric, $n\times n$, adjacency matrix **A** (equation 1) in which each member a_{ij} shows the connectivity between pairwise nodes i and j through an edge, where a_{ij} ($i \neq j$) is either 1, if there is a link between nodes i and j, or 0 otherwise, and $a_{ii} = 0$.

$$\mathbf{A} = \begin{bmatrix} a_{11} & a_{12} & \cdots & a_{1j} & \cdots & a_{1n} \\ a_{21} & a_{22} & \cdots & a_{2j} & \cdots & a_{2n} \\ \vdots & \vdots & \vdots & \vdots & \vdots & \vdots \\ a_{i1} & a_{i2} & \cdots & a_{ij} & \cdots & a_{in} \\ \vdots & \vdots & \vdots & \vdots & \vdots & \vdots \\ a_{n1} & a_{n2} & \cdots & a_{nj} & \cdots & a_{nn} \end{bmatrix} \tag{1}$$

2.2. Making the dependency model

Each network can be illustrated by an adjacency matrix **A** and therefore it is possible to form such a matrix for a multiplex network that involves K number of graphically interconnected infrastructures. This is a binary matrix in which the members are either 1 or 0 and its total number of nodes (n_{tot}) and edges (e_{tot}) are equal to $\sum_{m=1}^{K} n_m$ and $\sum_{m=1}^{K} e_m$ respectively, where n_m and e_m are the number of nodes and edges of network m. Hence, the adjacency matrix of multiplex network is defined as $\mathbf{M} = [a_{ijm}]$ ($i = 1 \ldots n_{\mathrm{tot}}$, $j = 1 \ldots n_{\mathrm{tot}}$ and $m = 1 \ldots K$), in which $M_{n_{\mathrm{tot}} \times n_{\mathrm{tot}} \times 1}$ indicates the layout matrix $\mathbf{A}^{\mathrm{WDN}}$ of WDN and is shown as follows:

$$\mathbf{A}^{\mathrm{WDN}} = \begin{bmatrix} a_{111} & a_{121} & \cdots & a_{1j1} & \cdots & a_{1n_{\mathrm{tot}}1} \\ a_{211} & a_{221} & \cdots & a_{2j1} & \cdots & a_{2n_{\mathrm{tot}}1} \\ \vdots & \vdots & \vdots & \vdots & \vdots & \vdots \\ a_{i11} & a_{i21} & \cdots & a_{ij1} & \cdots & a_{in_{\mathrm{tot}}1} \\ \vdots & \vdots & \vdots & \vdots & \vdots & \vdots \\ a_{n_{\mathrm{tot}}11} & a_{n_{\mathrm{tot}}21} & \cdots & a_{n_{\mathrm{tot}}j1} & \cdots & a_{n_{\mathrm{tot}}n_{\mathrm{tot}}1} \end{bmatrix} \tag{2}$$

where, a_{ij1} ($i \neq j$) is either 1, if there is a link between nodes i and j, or 0 otherwise (hereinafter a_{ij1} is showed by a_{ij}). In other words, each member in $\mathbf{A}^{\mathbf{WDN}}$ refers to a specified location (a_{ij}) over the layout of multiplex network that can be either empty or occupied by a link of WDN. This is not a symmetric matrix meaning that if a_{ij} gets 1, a_{ji} is 0 as the matrix representing an undirected graph. This requires the detailed datasets about the location coordinates of elements in different infrastructures. In this regard, a geographic information system is used to capture spatial data of elements to determine co-locating of elements. The dependency of a_{ij} is defined through its co-locating with the components of other infrastructures. For each location, the co-locating degree ($\mathrm{D}_{\mathrm{co}-a_{ij}}$) is introduced by counting the number of co-located elements with a_{ij} and it is evaluated by equation (3):

$$\mathrm{D}_{\mathrm{co}-a_{ij}} = \sum_{m=2}^{K} a_{ijm} \tag{3}$$

With respect to equation (3), if $\mathrm{D}_{\mathrm{co}-a_{ij}} = 0$, a_{ij} is not co-located with other infrastructures' edges, and if $\mathrm{D}_{\mathrm{co}-a_{ij}} \geq 1$, it is co-located at least with one element. For example, a_{ij} may refer to a specified street in which a pipe correlates with a conduit, gas and power line, as a consequence, $\mathrm{D}_{\mathrm{co}-a_{ij}} = 4$ (street, conduit, gas and power). High degree for each location means that there are many correlated components and needs an attention for asset management activities.

The average correlation for every pipe site is defined by equation 4:

$$\mathrm{Avg}\left(\mathrm{D}_{\mathrm{co}-a_{ij}}\right) = \frac{\sum \mathrm{D}_{\mathrm{co}-a_{ij}}}{\max(\mathrm{D}_{\mathrm{co}-a_{ij}})} \tag{4}$$

where the maximum number of correlations for every pipe site of WDN is the maximum possible co-located neighbour elements. For example, if a WDN has only correlation with sewer and road infrastructures, the maximum number of correlations for every WDN pipe equals two; in this regard, if a pipe has correlation only with a road element, the average correlation of that pipe is 0.5. As the result, the correlation index and decoupled reliability of WDN are calculated by equations (5) and (6):

$$\mathrm{Cor}_{\mathrm{WDN}} = \mathrm{mean}\left(\sum \mathrm{mean}\left(\mathrm{Avg}\left(\mathrm{D}_{\mathrm{co}-a_{ij}}\right)\right)\right) \tag{5}$$

$$\mathrm{Rel}_{\mathrm{dec}} = 1 - \mathrm{Cor}_{\mathrm{WDN}} \tag{6}$$

2.3. *Phased and multi utility approaches for dynamic design of renewal plans*

To solve an upgrade problem dynamically, construction and design period is divided into multiple phases; this is unlike the traditional approach of design, which only considers one construction phase and demand scenario estimated on the year of planning horizon (figure 1).

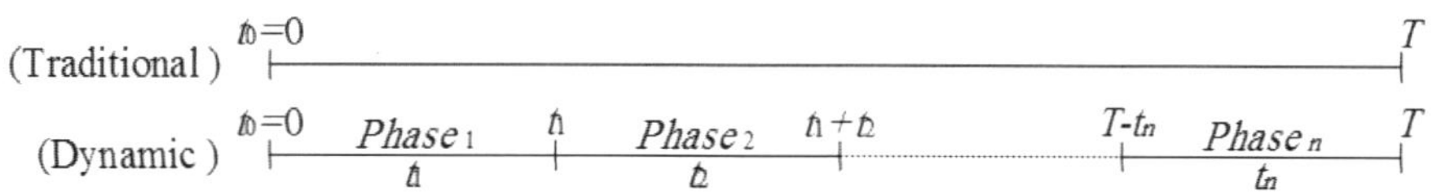

Figure 1. Phasing design and construction period in both traditional and dynamic approaches.

In this regard, [11] introduced different approaches for generating upgrade programs through optimization engines. It was proved that using dynamic design and construction approach in a single-phase optimization run performs better than other approaches in terms of flexibility of the designs and optimization efficiency. Hence, this approach is used in this paper.

As regards the multi-utility approaches, this study defines two categorizes which are asynchronized and synchronized are where in the first approach the upgrade actions are only applied to WDN elements and in the second one the adjacent networks could be upgraded together with WDN. This study uses the first approach for multi utility upgrade designs and constructions.

2.4. *Formulating the optimization algorithm of the problem*

In every phase, the optimization problem is formulated as follows:

$$\text{Minimize}\left(\text{Cost}_k\left(\Gamma_k(r,d)\right), -\text{Rel}_{\text{hyr},k}, -\text{Rel}_{\text{dec},k}\right) \quad (7)$$
Subject to,

$$h_i \geq H_{\text{min-s}},\ i = 1:n_{n,k} \quad (8)$$

$$r_{ij} \in \{0,1,2,3\},\ i = j = 1:n_{\text{WDN},k} \quad (9)$$

$$d_{ij} \in \{1, \text{D}_\text{c}\},\ i = j = 1:n_{\text{WDN},k} \quad (10)$$

where, Cost_k is the rehabilitation cost of upgrade program Γ, indicating the budget of upgrading network at the beginning of phase k, and evaluated by equation (11); $\text{Rel}_{\text{hyr},k}$ and $\text{Rel}_{\text{dec},k}$ are the hydraulic and decoupled reliability indices of the WDN estimated by equations (12) and (6) for the phase k; h_i is the actual head of node i, $H_{\text{min-s}}$ is the minimum pressure head below which the water demands cannot be supplied. The inequality constraint of (8) emphasizes on that any upgrade program must be feasible with respect to $H_{\text{min-s}}$ at the end of the construction phase k; $n_{n,k}$ is the number of nodes in phase k related to the layout of WDN. The decision variables r_{ij} and d_{ij} are respectively the upgrade indicator and new pipe diameter size for the location a_{ij} explained in table 1.

$$\text{Cost}_k\left(\Gamma_k(r,d)\right) = \sum_{i=1}^{n_{\text{WDN},k}} \sum_{j=1}^{n_{\text{WDN},k}} a_{ij} c_{ij} L_{ij} \quad (11)$$

In equation (11), a_{ij} refers to a specified location in the multiplex network, which is either 1, if there is a link, or 0 otherwise; c_{ij} is the unit cost of the commercial diameter size assigned to the pipe of location a_{ij} with length L_{ij}.

Table 1. Decisions for upgrade of an aged water distribution network.

Decision ID	Action Indicator	Explanation
1	$r_{ij} = 0$	The old pipe is removed from location of a_{ij}
2	$r_{ij} = 1$	The old pipe is remained to continue supplying consumers
3	$r_{ij} = 2$	The old pipe is replaced with the new one which has d_{ij} diameter
4	$r_{ij} = 3$	The old pipe gets parallel pipe with d_{ij} diameter

As shown in table 1, there are four upgrade decisions for every old pipe. The first decision contributes to make the network robust against interconnectivity with adjacent networks. There is a pre-check for applying first decision to every old pipe; if a pipe is in the shortest path (where the weight is length over diameter) from a source to a demand node of the network's graph, it is in the set of critical sites and only can be acted by other decisions (decisions 2, 3 and 4) as it has the main role to supply the associated node.

The second decision contributes to save upgrade cost (assuming the neglection of operating cost). The third and fourth decisions are for improving the hydraulic aspect of the network. In this regard, while laying parallel pipes is in favour of hydraulic reliability of network, this makes the upgrade program costly and the network weak against cascading failures. Hence, there are serious trade-offs between optimization objectives and decision variables.

$\mathrm{Rel}_{\mathrm{hyr},k}$ is a measure to evaluate the network hydraulic reliability, which is a hybrid index introduced by [11]. This is calculated in each phase k using the following relationship:

$$\mathrm{Rel}_{\mathrm{hyr},k} = \begin{cases} \mathrm{Rel1}_k & \mathrm{Rel1}_k < 1 \\ 1 + \mathrm{Rel2}_k & \mathrm{Rel1}_k = 1 \end{cases} \tag{12}$$

where $\mathrm{Rel1}_k$ is a measure to evaluate the network reliability in weak conditions in which the minimum service pressure head cannot be met in all demand nodes in all construction phases of k and after. For $\mathrm{Rel1}_k$ a surrogate index (between 0 and 1) according to equation (13) is used to count the number of failures in meeting the minimum service pressure head in all $n_{\mathrm{WDN},k}(n-k+1)$ node-periods. Also, $\mathrm{Rel2}_k$ is a measure to evaluate the network reliability in robust conditions in which the network has surplus power. For $\mathrm{Rel2}_k$ a version of resilience index (between 0 and 1) is used as presented in equation (14). This index was originally introduced by Todini [12] and then adapted by Prasad et al. [13] for the degree of uniformity of pipe diameters entering demand nodes. This index is popular for evaluating the reliability of pipe networks which are quite feasible and operated in robust conditions so that all demand nodes in all construction phases have surplus pressures. In other words, $\mathrm{Rel2}_k$ is evaluated if and only if $\mathrm{Rel1}_k = 1$.

To evaluate the network reliability first $\mathrm{Rel1}_k$ is estimated from equation (13). If $\mathrm{Rel1}_k < 1$, the network is in weak conditions and $\mathrm{Rel}_{\mathrm{hyr},k} = \mathrm{Rel1}_k$. If $\mathrm{Rel1}_k = 1$, the network is in robust conditions and its hydraulic performance can be reinforced by increasing the network resilience so that $\mathrm{Rel}_{\mathrm{hyr},k} = 1+\mathrm{Rel2}_k$. Consequently, the value of $\mathrm{Rel}_{\mathrm{hyr},k}$ places between 0 and 2.

$$\mathrm{Rel1}_k = 1 - \frac{\sum_{i=k}^{n}\sum_{j=1}^{n_n} \max\left(0, -\mathrm{sign}\left(H_{ij} - H_{\mathrm{min-s}}\right)\right)}{n_n\left(n-k+1\right)} \tag{13}$$

$$\mathrm{Rel2}_k = \frac{1}{\left(n-k+1\right)}\sum_{i=k}^{n}\left[\frac{\sum_{j=1}^{n_n} C_{ij} \times q_{ij}\left(H_{ij} - H_{\mathrm{min-s}}\right)}{\left(\sum_{l=1}^{nr_i} Qr_{il} \times Hr_{il} + \sum_{m=1}^{pn_i} \frac{P_{im}}{\gamma}\right) - \sum_{j=1}^{n_n} q_{ij} H_{\mathrm{min-s}}}\right] \tag{14}$$

where q_{ij} and H_{ij} are respectively the design demand and actual head at node j at the end of phase i; nr_i is the number of reservoirs in phase i; Qr_{il} and Hr_{il} are respectively the discharge from and head at reservoir l in phase i; pn_i is the number of pumps in phase i and P_{im} is the power of pump m at the end of phase i. Also, C_{ij} is a weighting coefficient associated with the uniformity of the loop diameter of pipes connected to node j in phase i as follows:

$$C_{ij} = \frac{\sum_{r=1}^{np_{ij}} D_r}{np_{ij} \times \max\left\{D_r\right\}} \tag{15}$$

where np_{ij} and $\max\{D_r\}$ are respectively the number of pipes and the maximum pipe diameter size connected to node j in phase i. According to the above equation, $C_{ij} = 1$ if only one pipe is connected to node j or all pipes connected to that node have the same diameter and, $C_{ij} < 1$ if pipes connected to node j have different diameters, Prasad et al [13].

Using the hybrid reliability index (equation (12)), the upgrade program becomes more flexible and manageable so that the system can be gradually upgraded from a weak state ($\mathrm{Rel}_{\mathrm{hyr},k} < 1$) to a normal state ($\mathrm{Rel}_{\mathrm{hyr},k} \approx 1$) and then to a robust state ($\mathrm{Rel}_{\mathrm{hyr},k} > 1$), depending on the money invested for the project and the reliability expected.

In each construction phase k, the improved version of a multi-objective optimization algorithm [14], Non-Dominated Sorting Genetic Algorithm (NSGA-II) [15] is used to upgrade the network in phase k, while the hydraulic simulation is done by EPANET 2.2. Finally, the optimum Pareto front containing a wide range of upgrading programs Γ_k is achieved for phase k. From the Pareto fronts the desirable solution is selected and constructed in phase one. At the beginning of phase two, the network is updated based on the constructed plan in phase one and all the mentioned process for phase one is repeated for phase two to generate the upgrade programs. This is a three dimensional (3D) Pareto fronts and is compared with the two-dimensional (2D) Pareto fronts in [11].

3. Application

3.1. Case Study

To investigate the proposed approach, an aged water distribution network, Baghmalek network (figure 2), located in the southwest of Iran is considered for upgrading. The network has 90 pipe sites and 72 consumption nodes and is fed by a reservoir (node 1) located at the highest elevation of the region. The network is more than 30 years old and its current hydraulic performance turns out to be in an urgent need for rehabilitation and upgrading. It is assumed that the network only has correlation with sewer and road networks. The multiplex system is conceptualized in figure 2, representing the correlations and the network configuration in which the Hazen-Williams coefficients of pipes as well as nodal demands are estimated for year zero. Other specifications of the network are in [11] and available upon request.

The network is first analyzed for the existing conditions in year zero. Currently the hydraulic performance of the WDN is weak and only 36% of the consumption nodes meet the service pressure of 18 m required for the network according to the national regulations ($\mathrm{Rel}_{\mathrm{hyr}}$ = 36%). Also, the correlation of WDN with the adjacent infrastructures is about 63% ($\mathrm{Cor}_{\mathrm{WDN}}$ = 0.63) and therefore the decoupled reliability is about 37% ($\mathrm{Rel}_{\mathrm{dec}}$ = 0.37). Hence, the hydraulic and decoupled reliabilities of network should be improved through upgrade actions. There are some main assumptions for upgrading the current network as follows:

- A 25-year design period is considered for upgrading the network (T = 25);
- During the design period, the network has no extension in plan. The network layout is fixed with time;
- A list of polyethylene pipes containing 14 diameter sizes as shown in table 2 is used for upgrading the network;
- For all new pipes, the Hazen–Williams coefficient in year zero is 130. Also, the Hazen–Williams coefficients are supposed to change linearly with time with the reduction rate of -0.6% yearly;
- The installation of a parallel pipe is more difficult and expensive than replacing a new pipe. To take this issue into account the unit cost of parallel pipes is increased by 20%;
- The network consumption is supposed to change with time linearly. The annual rate of consumption increasing is estimated 0.0332 l/s/year;
- The design period is divided by 5 construction phases, 5-year periods;
- The correlation of a water pipe with a street (road element) is assumed if the water pipe is located under or in the 3-meter distance from the margin of the street;
- The correlation of a water pipe with a sewer network element is assumed if they are located in 3-meter radius distance from the axis line of each other (this assumption comes from the possible cascade failure at the earthquake circumstances);
- In this study, only the effect of WDN element failures on the adjacent networks' performances are considered and vice versa is not the target. This means that while the pipe breakage could

cause failure to a street performance, the failure in street does not have impact on the WDN performance.

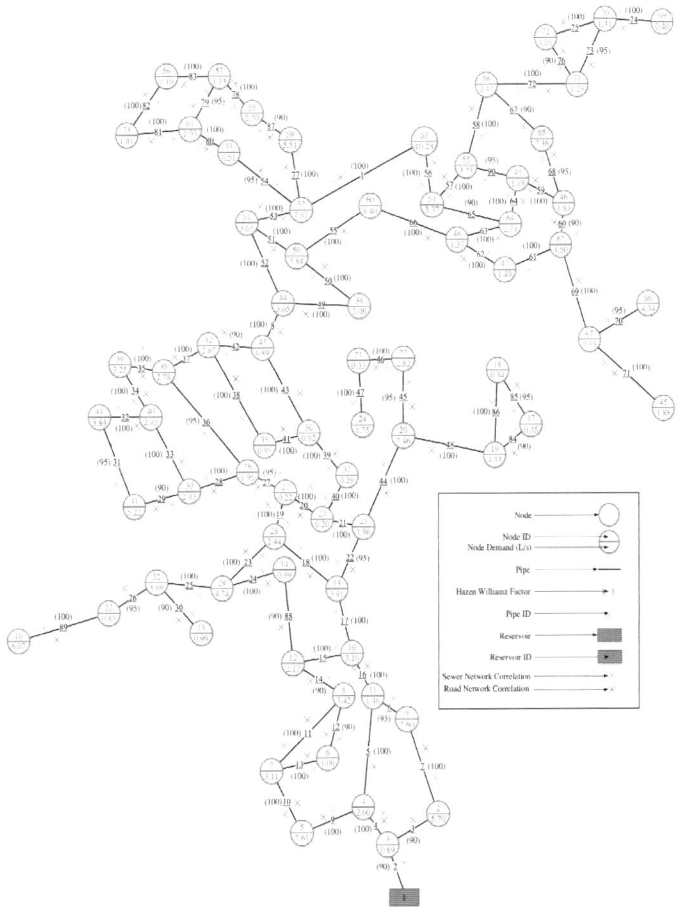

Figure 2. Baghmalek water distribution network and correlation with adjacent networks' elements.

Table 2. Commercial polyethylene pipes with their unit construction cost in year zero; D_C: Commercial diameter, D: Internal diameter.

D_C (mm)	D (mm)	Unit cost (Rials/m)	D_C (mm)	D (mm)	Unit cost (Rials/m)
63	53.60	163473.90	250	213.20	1147234.00
75	63.80	192204.30	315	268.20	1752000.00
90	76.60	232484.70	400	341.20	2734489.00
110	93.80	300636.10	450	383.89	3455312.00
125	106.60	374078.20	630	537.50	6466257.00
160	136.40	543685.60	710	605.77	8102474.00
200	170.60	769448.30	800	682.58	10187126.00

3.2. Multi Objective Optimization Results

Figure 3 shows the Pareto fronts obtained from NSGA-II optimization algorithm whose critical parameters are tuned through some initial trial-and-error runs and sensitivity analysis. For the sake of parsimony, the Pareto fronts of two phases upgrading are shown. In figure 3(a), a 2D Pareto front referring to the solutions of two objectives optimization problem is shown [11], while in figure 3(b), a 3D Pareto fronts containing the solutions of current study problem is shown. In [11], there are budget constraints and the client cannot allocate more than approximately 5,500 and 7,000 Million Rials for upgrading Baghmalek WDN in phases one and two, respectively. As observed in 2D Pareto fronts, after construction of upgrade plans for phases 1 and 2, the hydraulic reliability reaches 0.93 and 1.57, meaning that the network hydraulic performance has been improved gradually from weak to robust conditions.

To achieve the mentioned reliability values through the multi-utility upgrade approach, the network upgrade plans become costlier (figure 3(b)) emphasizing the need for doing synchronized multi utility asset upgrading. This could be due to the strategy of removing pipe as this activity makes the WDN network hydraulically weak and this should be compensated by larger diameter pipe replacement and laying parallel pipes.

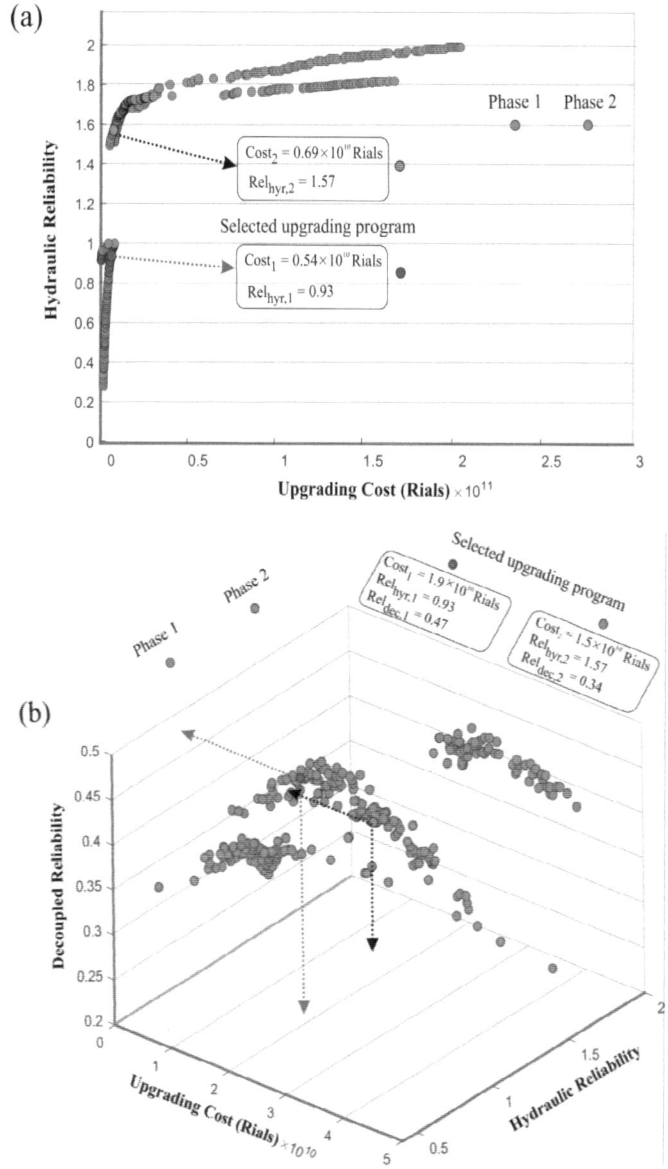

Figure 3. 2D and 3D Pareto fronts for upgrading Baghmalek WDN.

Figure 4 (a) and (b) shows the upgrade plan of selected solutions. Even though the upgrade design of network has become more expensive through the asynchronized multi-utility approach, the decoupled reliability improves to 47% in phase 1. In attempt to make the resilient network in phase 2 with 1.57 hydraulic reliability (robust condition), the decoupled reliability decreased to 34% due to the many parallel pipes laid in this phase; having said that, removing pipe strategy still contributes to keep roughly the initial interdependency network states.

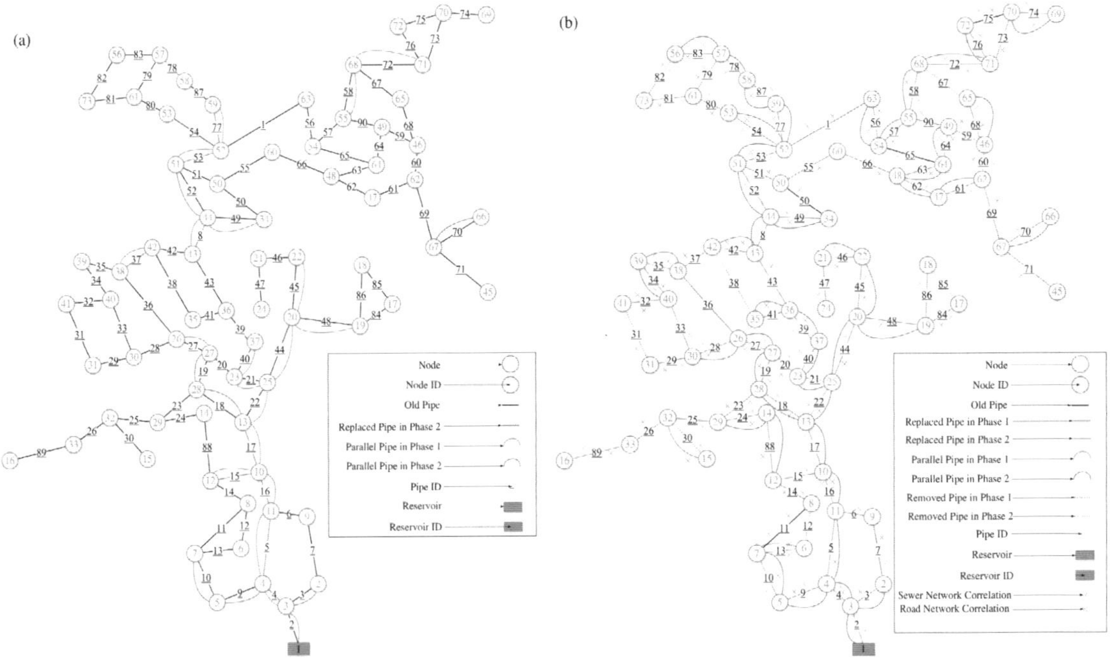

Figure 4. Constructed upgrade plans selected from 2D and 3D Pareto fronts.

4. Summary and Conclusions

This study paid attention to two important practical considerations for upgrading aged water distribution networks, 1- the dynamic behavior of such networks and transition conditions (gradual budget allocation and increase in the consumers' water demand), 2- the interdependency of WDNs with other adjacent networks such as sewer and road infrastructures increasing the risk of cascade failures under hazard-based circumstances. For this, a multi objective optimization problem with three objectives (cost, hydraulic and decoupled reliabilities) was solved by the help of an improved version of NSGA-II optimization engine. The upgrade results obtained by the multi-utility approach were compared by the ones obtained only by consideration of two objectives (cost and hydraulic reliability). The summary of the results are as follows:

- Making a decoupled WDN through a multi-utility approach is necessary in terms of asset management perspectives. There could be some upgrade actions in this regard. Current study proposed removing WDN elements from highly correlated sites;
- Considering only the upgrade of one utility in the multi-utility asset upgrade approach could make the plans costly. Hence, such plans should be synchronized with other assets upgrade plans;
- All the upgrade programs should be designed and constructed dynamically due to the change in the conditions including the funding constraints and desired reliabilities by the decision makers.

Acknowledgements
This study was partly funded by the Austrian Science Fund (FWF): P 31104-N29 and largely funded by Austrian Academy of Sciences (ÖAW) fund: DOC Fellowship.

References

[1] Walski T 2014 How does water distribution design really work? *World Environmental and Water Resources Congress 2014* ed WC Huber (ASCE) pp 375–82 https://doi.org/10.1061/9780784413548.040
[2] Sitzenfrei R, Mair M, Möderl M and Rauch W 2011 Cascade vulnerability for risk analysis of water infrastructure *Water Sci. Technol.* **64**(9) 1885–91 doi: 10.2166/wst.2011.813
[3] Diao K, Butler D and Ulanicki B 2021 Fractality in water distribution networks: application to criticality analysis and optimal rehabilitation *Urban Water J.* **18**(10) 885–95 https://doi.org/10.1080/1573062X.2021.1948076
[4] Diao K, Jung D, Farmani R, Fu G, Butler D and Lansey K 2021 Modular interdependency analysis for water distribution systems *Water Res.* **201** 117320 https://doi.org/10.1016/j.watres.2021.117320
[5] Bross L, Bäumer J, Voggenreiter I, Wienand I and Fekete A 2021 Public health without water? Emergency water supply and minimum supply standards of hospitals in high-income countries using the example of Germany and Austria *Water Policy* **23**(2) 205–21 https://doi.org/10.2166/wp.2021.012
[6] Shuang Q, Huang CR and Wang J 2020 Optimization of water distribution network design for resisting cascading failures *IEEE Access* **8** 128856–65 doi: 10.1109/ACCESS.2020.3008911
[7] Creaco E, Franchini M and Walski TM 2014 Accounting for phasing of construction within the design of water distribution networks. *J. Water Res. Pl.* **140**(5) 598–606 https://doi.org/10.1061/(ASCE)WR.1943-5452.0000358
[8] Creaco E, Franchini M and Walski T 2015 Taking account of uncertainty in demand growth when phasing the construction of a water distribution network *J. Water Res. Pl.* **141**(2) 04014049 https://doi.org/10.1061/(ASCE)WR.1943-5452.0000441
[9] Kang D and Lansey K 2014 Multi-period planning of water supply infrastructure based on scenario analysis *J. Water Res. Pl.* **140**(1) 40–54 https://doi.org/10.1061/(ASCE)WR.1943-5452.0000310
[10] Zischg J, Rauch W and Sitzenfrei R 2018 Morphogenesis of urban water distribution networks: A spatiotemporal planning approach for cost-efficient and reliable supply *Entropy* **20**(9) 708 https://doi.org/10.3390/e20090708
[11] Minaei A, Haghighi A and Ghafouri HR. Computer-aided decision-making model for multiphase upgrading of aged water distribution mains *J. Water Res. Pl.* **145**(5) 04019008 https://doi.org/10.1061/(ASCE)WR.1943-5452.0001070
[12] Todini E 2000 Looped water distribution networks design using a resilience index based heuristic approach *Urban Water* **2**(2) 115–22 https://doi.org/10.1016/S1462-0758(00)00049-2
[13] Prasad TD, Hong S-H and Park N 2003 Reliability based design of water distribution networks using multi-objective genetic algorithms. *KSCE J. Civ. Eng.* **7**(3) 351–61 https://doi.org/10.1007/BF02831784
[14] Minaei A, Sabzkouhi AM, Haghighi A and Creaco E 2020 Developments in multi-objective dynamic optimization algorithm for design of water distribution mains *Water Resour. Manage.* **34** 2699–2716 https://doi.org/10.1007/s11269-020-02559-8
[15] Deb K, Pratap A, Agarwal S and Meyarivan T 2002 A fast and elitist multiobjective genetic algorithm: NSGA-II *IEEE T. Evolut. Comput.* **6**(2) 182–97 doi: 10.1109/4235.996017.

A flexible, leak crew focused localization model using a maximum coverage search area algorithm

Brett Snider[1], Gareth Lewis[1], Albert Chen[1], Lydia Vamvakeridou[1,2] and Dragan Savić[1,2]

[1] Centre for Water Systems, University of Exeter, Exeter, EX4 4QF, UK
[2] KWR Research Institute, Nieuwegein, Netherlands

b.snider@exeter.ac

Abstract. Buried watermains are deteriorating and pipe failure is increasing in many cities. In response, advanced leak location models have been developed to help identify where a leak is occurring – which allows utilities to react quickly to pipe bursts and reduce the impact of the leak. This paper develops a new leak location model that is designed to identify optimal search areas for leak crews using a random forest classification model and the maximum coverage location problem algorithm. The model, when compared with other machine learning and clustering localization predictions, reduces the search space by over 35%, allowing utilities to confirm leak location and mitigate its impact more efficiently. The new model is also highly customizable, able to adjust the number of search areas and search size quickly and easily to meet leak crews' requirements.

1. Introduction

Beneath all cities lay a network of underground watermains that deliver drinking water to their residents and businesses. Many of these buried watermains have experienced significant deterioration, resulting in a major increase in pipe breaks in recent years [1]. The impacts from these pipe breaks can be severe, resulting in a loss of service, possible contamination, and impact to nearby infrastructure. Therefore, it is a priority for water utilities to locate and repair these leaks as soon as possible to help mitigate their impact. However, locating a leak is usually a very difficult task since the length of buried watermains that must be checked can be extensively long (for example London, England's water distribution system contains over 20,000 kms of buried watermains [2].

In recent years a variety of leak location models have been developed that can help predict where a leak is likely occurring within the network. These tools aim to improve leak crews' response time, by narrowing their focus, allowing them to locate the leak sooner and mitigate the leak's impact. However, from an operations point of view, these models still leave a lot of room for improvement. First, many of the location models developed attempt to pinpoint an exact location of the leak. However, leak crews are still required to search in-person to confirm the location of the leak. If the leak is not at the predicted location, which is often the case, crews are left wondering where to search next. Second, none of the existing models are customized to fit the utilities leak crews' preferences, such as search area size and/or number of search areas, which may be highly dependable on staff availability for instance.

The model presented in this paper addresses these concerns by designing a leak location model that identifies optimal search areas for leak crews to confirm leak location. The predicted search area size

IOP Conf. Series: Earth and Environmental Science 1136 (2023) 012042 doi:10.1088/1755-1315/1136/1/012042

and number of search areas can be customized to fit the utilities demands at that time of leak detection. The model developed applies the Maximum Coverage Location Problem algorithm (MCLP) [3] to a node probability prediction from a Random Forest Machine Learning (RF) classifier model. This new leak location modelling approach is compared with other machine learning and clustering approaches to leak localization to evaluate and ultimately demonstrate its effectiveness.

The remainder of this paper analyses the effectiveness of the proposed leak location model by first providing a quick overview of the literature regarding leak location models. A detailed methodology outlines the model's development. A case study is then used to highlight the impacts of the model's customizable parameters and provide a comparison between other leak location models.

2. Leak Location Models

A digital transition is occurring within the water industry. Many water utilities have begun to realize the numerous benefits associated with digitization of their assets and water systems. Remote sensors are becoming more popular and leak management models more common. In recent years a large increase in leak location models has been developed. The leak location models that have been developed use a variety of methods to locate leaks within a water distribution system and can be broadly classified into three main approaches: transient based, data driven, or model driven methods [4].

Transient based methods use high frequency pressure sensors within the water distribution system to detect transient pressure waves resulting from a leak occurrence. By analysing this pressure waves and identifying the signal variation between sensors, transient based models attempt to identify the location of the leak within the WDS. A variety of transient based models have been developed and described in detail by various literature reviews [5], [6]. However, transient based models require high frequency sensors, which can be very costly and may not be applicable for real-time application or for use in large complex water networks [5], [6].

Data driven leak location models do not require any specific knowledge regarding the water distribution system. Instead, signal processing and statistical analyses are used to detect and locate leaks [7]. Most data driven leak models only focus on leak detection, however, it is possible to develop a leak location model using data driven techniques[4], [8]. Overall, data driven methods require significant amount of historic sensor readings to develop an accurate prediction model and may be inaccurate when water network experiences seasonal or festive variation that is common within water networks [7]. Due to these drawbacks, the focus of this paper is on the development of model driven leak location methods.

The most popular approach to leak location prediction is via model driven methods. These methods rely on hydraulic models of the water distribution system (WDS), comparing simulation reads to recorded sensor measurements to help identify where a leak is occurring within the network. Model driven methods are able to be calibrated without requiring significant historic sensor readings and have been shown to have a high degree of accuracy when hydraulic models are well calibrated [4]. The various model driven methods used to detect leak location include sensitivity matrix-based approaches, mixed model-based approaches, optimization-calibration approaches, and error-domain falsification-based approaches [4]. Instead of focusing on the specific modelling-based method, this paper concentrates on the output these models generate, as this has a direct impact on how the model is utilized by the utility/leak crews.

2.1. Leak Location Prediction Output

Typically, the leak location models predict a specific geographical point, either through hydraulic model node classification, or x-y coordinates using a regression analysis [9]. Using these methods, leak crews are sent to the predicted location and if the leak is not there, they are left without any guidance on where to search next.

A few leak location models have been designed to predict a zone within the network that is likely to contain the leak. The leak zone prediction models have been developed in two distinct ways;

through classification of predefined leak zones / district metered areas [9], [10], or through iterative cluster / graph theory partitioning [11], [12]. By splitting up the network into predefined leak zones, these location models predict which zone the leak is occurring in. Utility leak crews are then sent to search that leak zone and confirm the leak location. These leak zone approaches provide boundaries to help guide the leak crews. However, they do not place limits on leak zone size, resulting in some leak zones being very large and would take a long time to search.

Lacking within the literature is a leak location model specifically developed for leak search crews. A location model designed for leak crews should identify search areas, based on the crews preferred search radius and number of search locations that contain the highest probability of leak. By developing this type of model, leak crews would be able to locate and respond to the leak quicker, minimizing the overall impact of the leak on the customers, utility, and nearby infrastructure. This paper sets out to fill this literature gap by applying the MCLP algorithm to a nodal leak location prediction from a RF classification model. This approach allows the specific search area radius and number of search areas to be defined by the utility, while outputting the largest probability of including the leak within these parameters.

3. Methodology

The following section describes the MCLP leak location prediction model developed as well as the development of four other leading leak location prediction used evaluate the effectiveness of MCLP modelling approach.

The overall development of the leak location model involves three main steps: developing a training dataset, training the RF classification model, and applying the MCLP algorithm to predict leak locations. The overall development of the model is highlighted in Figure 1 with inputs and outputs depicted as well.

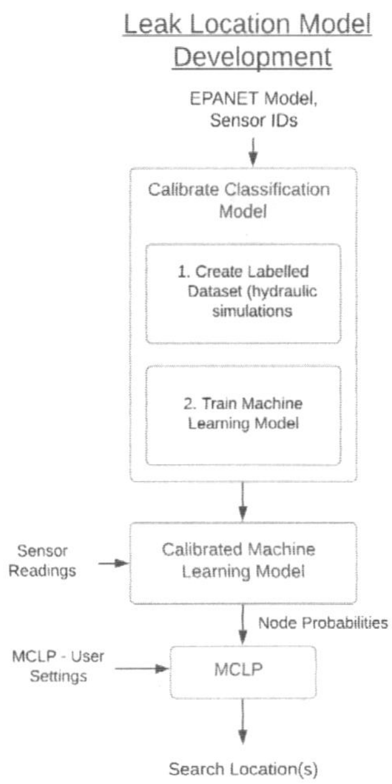

Figure 1: MCLP leak location model development.

3.1. Creating a labelled dataset

The first step in developing the MCLP location model is to create a labelled dataset to train the RF classification model. The dataset is generated by performing several leak simulations for every node

IOP Conf. Series: Earth and Environmental Science 1136 (2023) 012042 doi:10.1088/1755-1315/1136/1/012042

within the hydraulic model, storing only the readings associated with sensor values (i.e., flow and/or pressure) and the location, or node, of the simulated leak. The simulated sensor values are converted into residuals (or z-score) based on a no-leak scenario, as highlighted in Equation 1 below:

$$z_{i,t} = \left(x_{i,t} - xavg_{i,t}\right) / \sigma_{i,t} \quad (1)$$

where z represents the calculated residual, x the sensor measurement (flow or pressure), i the specific sensor, t the reported time step; $xavg_{i,t}$ and $\sigma_{i,t}$ are the average and standard deviation for sensor readings calculated from the non-leak scenario for that particular sensor and time period. For each leak simulation a six-hour window of residuals are stored, starting six hours before detection, and ending when the burst is detected (which is randomly chosen to occur between 1-4 hours after burst occurrence).

3.2. Training a random forest classification model

A random forest machine learning classifier is trained using the labelled dataset. Here each leak node represents a class, and the input variables are the residuals for each simulated sensor reading during the six-hour detection window. A random forest machine learning algorithm is chosen due to its high performance reported in other leak location models [9], [13] and its ease of hyper-parameter tuning and robustness against data outliers [14].

The hyper-parameters are tuned using a grid-search approach for tree-depth and number of trees. The tuning parameters are optimized to maximize classification accuracy using a hold-out validation dataset. The RF machine learning classifier is then used to predict the probability score for each class (or in this case node within the hydraulic model). The predicted probability for each node is used as an input to the Maximum Coverage Solution Problem in order to identify the optimal search area(s).

3.3. Maximum coverage location problem

The Maximum Coverage Location Problem (MCLP) is a type of geographical location set covering solution that was developed in 1974 by Church & Revelle [3] but has garnered significant attention in recent years as large GIS datasets have been developed and used to solve unique problems [15].

To identify the optimal search areas for leak detection crews, the MCLP is applied to the output from the RF classification model. Since the RF classification model predicts the probability of the leak occurring at each node within the network, the MCLP can be used to group these results geographically, using the node's geographic coordinates, and locate the search areas that maximizes the predicted likelihood of containing the leak event.

The MCLP algorithm adapted to leak location is as follows:

1. $Maximize\ z = \sum_{i \in I} a_i y_i$,
2. $Subject\ to: \ y_i \le \sum_{j \in N_i} x_j \quad i \in I$,
3. $\sum_{j \in J} x_j = p$,
4. $0 \le y_i \le 1, \ i \in I$,
5. $x_j \in \{0,1\}, \ j \in J$,

where:

i, I the index and set of EPANET nodes (i.e. possible anomaly locations);
j, J the index and set of search area centroids;
a_i the predicted probability at node i;
$d_{i,j}$ the shortest distance from node i to search area centroid j;
S the search area radius;
N_i $\{j \mid d_{i,j} \le S\}$ = the nodes j that are within a distance of S to node i;
p the number of search areas to be determined;

x_j a binary variable that equals one when the search centroid is located at the j-th node;

y_i a binary variable which equals one if node i is within one or more search areas.

To solve the MCLP a list of possible search area centroid locations is required. However, search area centroid locations are typically not defined by a utility / leak detection crew since crews are often flexible enough to go to any location throughout the network. Therefore, a grid of possible search areas that cover the entire network is created using the hydraulic model's nodes with maximum and minimum latitude and longitude coordinates to create a grid boundary. A grid of possible search area centroids within this boundary is then generated using a set distance between grid points.

In addition to the grid of search areas centroids, preferred search area radius and number of search areas must be provided to solve the MCLP. Various grid spacing, search area radiuses, and number of search areas are compared and discussed in the results section.

4. Case study

An EPANET hydraulic model representing a UK water distribution system is used to evaluate the impact grid size, search radius and number of search areas have on the MCLP approach as well as the model's overall effectiveness.

A schematic of the case study's EPANET water network is presented in Figure 2 below.

Figure 2: Example EPANET schematic.

The EPANET network is a gravity fed network, supplied by a single reservoir. The hydraulic model is set-up to run with 15-minute extended period analysis and is assumed to have a weekly cycle (i.e., steady-state is reached with a weekly pattern of flows, pressures and demands). The overall attributes of the model are highlighted in **Table 1** below:

Table 1: Greater Torrington hydraulic model attributes.

Network attribute	Value
# of Nodes	1005
# of Pipe Segments	783
Total length of Pipes (km)	25.25
Avg. Pipe Diameter (mm)	118
Avg. Daily Demand (LPS)	13.43
Avg. Pressure (m)	48.0
Geographic Area (km^2)	12

Hypothetical pressure and flow sensors are modelled within the EPANET at thirteen locations throughout the network. Sensor readings are obtained by performing a hydraulic analysis using the EPANET hydraulic model, with the readings corresponding to the link or node id of the sensor for every 15-minute time increment.

4.1. Leak simulations
To build a comprehensive training dataset, six leaks are simulated at every node with random start times. The leaks are simulated using EPANET's emitter equation:

$$Q = C\,P^{\gamma} \tag{2}$$

where Q is the leakage flow, C is the emitter coefficient, P is the nodal pressure and γ is the emitter exponent. The emitter coefficient is randomly selected between 0.5 to 3 and the emitter exponent set to 0.9.

To reflect realistic scenarios, noise is added to user demand and sensor readings. Demand noise is added to each node by multiplying a noise factor to the original demand at that time step. The noise factor is selected from a normal distribution with an assumed standard deviation of 0.25. Noise is also added to the simulated sensor readings to reflect measurement inaccuracy that is typical with all sensors. The sensor noise is added to the sensor reading and is assumed to follow a normal distribution with standard deviation of 0.5.

The sensor readings simulated during the leak scenario are converted into residuals using the residual equation (Equation 1). A six-hour window of residuals is stored for each simulation, starting six hours before leak detection awareness. For testing purposes, the leak detection time is randomly chosen to occur between 1 to 4 hours after burst occurrence.

5. Results

5.1. Random forest machine learning classification model
To tune the RF hyper-parameters, a grid search methodology was employed, and results were tested using a validation dataset which included leak scenarios for 100 randomly selected leak nodes. The tuning parameters assessed include:

- Tree depth: [100, 200, 300, 400, 500, 600];
- Number of trees: [25, 50, 100, 150, 200, 250, 300].

The results from this analysis suggest the optimal tuning parameters were a tree depth of 250 with 300 decision trees developed. The overall accuracy of the classification model tested on an independent test dataset was 7.5%. These results suggest the model can identify important variations in sensor data to detect the leak node location, since the model's accuracy is roughly 75 times greater than chance. However, this is still a relatively low accuracy from a leak detection perspective, highlighting the importance of identifying leak search areas (that are more likely to contain the leak), instead of specific location predictions.

5.2. MCLP sensitivity analysis
A sensitivity analysis is performed for each of the MCLP parameters, which include grid size, search radius and number of search areas. All sensitivity analyses were performed using an independent test dataset set of 100 randomly generated leak scenarios.

Table 2 reflects the impact of grid spacing on predicting leaks located by MCLP approach assuming two search areas, each with a search radius of 50m. Table 2 highlights the trade-off between computation time and accuracy of locating leaks within the MCLP search area by adjusting the grid spacing of search area centroids. As the grid spacing becomes smaller, more search areas must be evaluated by the MCLP algorithm. This improves the likelihood of identifying the optimal search location and its likelihood of containing the leak but results in greater computation time. It is up to the requirements of the utility to determine the grid spacing that works best for them (weighing the trade-off between computational requirements and leak location accuracy).

Table 2: Impact of grid spacing on predicted leaks located by MCLP.

Grid spacing (m)	Computation time (sec)	Predicted probability covered	Leaks located
10	278	40.5%	64%
15	131	39.6%	64%
25	51	38.8%	63%
40	23	36.1%	60%
60	11	34.4%	60%
80	8	31.2%	52%

The second parameter analysed is the search radius. This analysis is performed using the 100-leak test set for two search areas, with 25 m grid spacing between centroids and various search radii.

As shown in Figure 3, as the search radius increases the likelihood of the search area containing the leak increases as well. However, a large search radius would require leak crews to search a larger area before locating the leak. Therefore, choosing a search radius will depend on a utility's own preference for accuracy and total search area (i.e., a trade-off between resolution and accuracy).

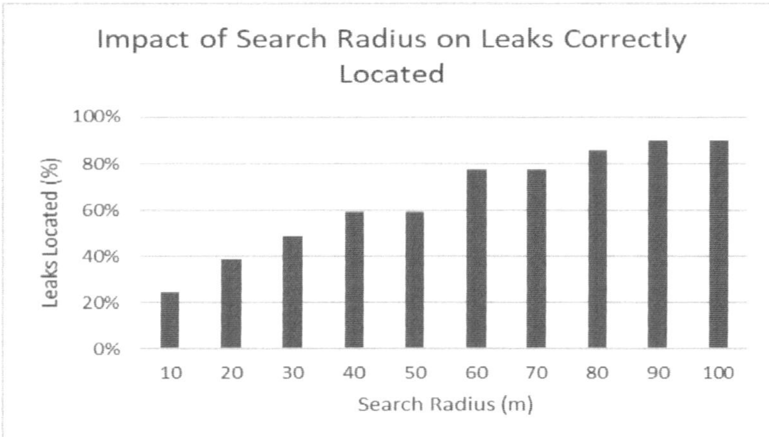

Figure 3: Search radius versus leaks located within search areas.

Lastly, the MCLP approach was analysed using a various number of search areas. A utility with several search crews may prefer several search areas to be identified to increase the speed of which the leak location is confirmed. The following table highlights how the number of search areas impact the accuracy of identifying the leak location using the 100-leak test set, with 25m grid search and 50m search radius. As the number of search areas increase the accuracy of detecting the leak increases as well. However, again a trade-off between total search area and accuracy must be made by the utility.

Table 3: Impact of search areas on anomaly detection.

Number of search areas	Predicted probability covered	Leaks detected	Total search area (m^2)
1	25%	45%	5,027
2	40%	59%	10,053
3	51%	65%	15,080
4	59%	73%	20,106

Overall, the sensitivity analysis performed highlights one of the major strengths of the MCLP approach – that is its customizability. The MCLP leak location model can be adjusted to fit the utilities specific computation and/or labour resources.

5.3. Model comparison

The Random Forest MCLP model developed is compared with other machine learning and clustering localization models. Specifically, the Random Forest MCLP approach detailed in this paper is assessed using one search area (MCLP-1 SA) and two search area (MCLP – 2 SA), expanding the search radius until 50% of the simulated leaks within the test set are accurately identified within the search area(s). The required search areas are then compared with three other leading location models:

1. A classification random forest model (Classification - 1 SA) - where the node with the highest predicted probability being chosen as the centroid for the search area. A single circular search area is expanded out from this centroid until 50% of the leaks from the test dataset are within the search areas.
2. A regression based random forest model (Regression - 1 SA) - where the model predicts the latitude and longitude of the search area. A single circular search area is expanded outwards from the predicted latitude and longitude until 50% of the leaks from the test dataset are within the search areas.
3. An agglomerative hierarchical clustering model (Clustering - 2 SA) – where a random forest model classification model is trained to predict probabilities of the leak occurring at each node for each simulation. Using this output, an agglomerative hierarchical clustering algorithm is performed using a max linkage parameter set to the desired search diameter [16]. The search diameter is expanded until two clusters include 50% of the leaks from the test dataset.

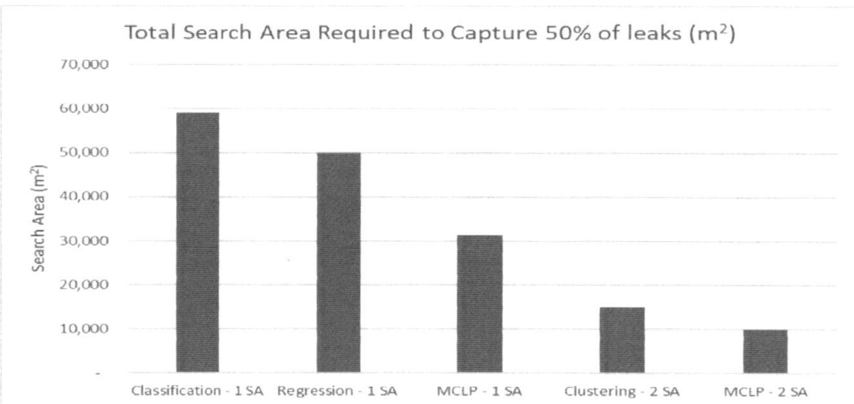

Figure 4: Comparing search areas for various location methods.

Figure 4 identifies the geographical search area required by each model to accurately locate 50% of the simulated leaks within the test dataset. The results indicate that the Random Forest MCLP approach requires the smallest single search area (35% less than leading regression-based model) and the smallest two search area (34% less than the clustering model). Overall, these results indicate the Leak location model developed in this report is effective at reducing the search space required to locate the leak and outperforms the other leak location models assessed.

6. Conclusion

Watermains throughout the world are deteriorating and pipe breaks appear to be increasing [1]. Leak location models that are able to predict the location of leaks can be a very useful tool to help utilities respond to pipe bursts quickly and mitigate their impact. For these tools to be effective they must be designed with the end-user in mind, which in this case would be leak crews that are tasked with confirming the leak location and begin leak mitigation response.

This paper develops a new leak location model, specifically designed to reduce the leak search area required by leak crews to locate and mitigate the leak. Specifically, a MCLP approach is applied to the output from a RF machine learning classification to predict ideal search area(s) for leak crews to locate a leak.

The new model is compared with other machine learning and clustering localization models and the results indicate the MCLP approach significantly reduces the search space required for leak crews to locate the leak. By reducing the search space, the leak crews will be able to locate the leak quicker and mitigate the impacts to the customer and the utility. The MCLP approach also has the important benefit of being highly customizable, being able to adjust the size of the search area, and/or the number of search areas to meet leak crews' preferences.

Overall, the MCLP model is a promising approach to leak location. Future research will continue to advance this applicability of this model. Specifically, developing a heuristic approach to identifying the ideal search area centroid for the MCLP approach could improve the accuracy and computation requirement for the leak localization model. Also, a more in-depth analysis into the model's sensitivity to hydraulic models' inaccuracies and sensor reading inaccuracies would help identify level of calibration and sensor accuracy is needed before the model is implemented within a water distribution system. The current approach described in this paper assumed a set level of noise/inaccuracy with sensors and hydraulic model. Adjusting this inaccuracy level until the model fails to predict the location would allow a utility to understand the accuracy needed before implementing the leak localization model [17]. Also worthwhile to investigate would be the impact of background losses, or small ongoing leaks, on the localization model's accuracy, since all water distribution systems have some level of background water loss. Analyzing the impact these background losses can have on the leak localization accuracy will allow water utilities to determine whether this approach is viable for their particular system.

Acknowledgment
The work presented in this paper was funded by the ongoing European Commission H2020 Fiware4Water (GA 821036) and aqua3S (GA 832876) projects.

References

[1] Folkman S 2018 Water main break rates in the USA and Canada: A comprehensive study *Mech. Aerosp. Eng. Fac. Publ.* **174** https://digitalcommons.usu.edu/mae_facpub/174
[2] Thames Water 2021 *Your Water Services* https://www.mendeley.com/reference-manager/library/all-references
[3] Church R and Revelle C 1974 The maximal covering location problem *Pap. Reg. Sci.* **32**(1) 101–118
[4] Hu Z, Tan D, Chen B, Chen W and Shen D 2021 Review of model-based and data-driven approaches for leak detection and location in water distribution systems *Water Supply* **21**(7) 3282–3306 doi: 10.2166/WS.2021.101
[5] Adedeji KB, Y Hamam, BT Abe and AM Abu-Mahfouz 2017 Towards achieving a reliable leakage detection and localization algorithm for application in water piping networks: An overview *IEEE Access* **5** 20272–85 doi: 10.1109/ACCESS.2017.2752802
[6] Colombo AF, P Lee and BW Karney 2009 A selective literature review of transient-based leak detection methods *J. Hydro-Environment Res.* **2** 212–227 doi: 10.1016/j.jher.2009.02.003
[7] Chan TK, CS Chin and X Zhong 2019 Review of current technologies and proposed intelligent methodologies for water distributed network leakage detection *IEEE Access* **6** 78846–67 doi: 10.1109/ACCESS.2018.2885444
[8] Murvay PS and I Silea 2012 A survey on gas leak detection and localization techniques *J. Loss Prev. Process Ind.* **25**(6) 966–973 doi: 10.1016/J.JLP.2012.05.010

[9] Ares-Milián MJ, M Quiñones-Grueiro, CC Corona and O Llanes-Santiago 2021 Clustering-based partitioning of water distribution networks for leak zone location *Lect. Notes Comput. Sci.* (including *Subser. Lect. Notes Artif. Intell. Lect. Notes Bioinformatics*) **12702** LNCS 340–350 doi: 10.1007/978-3-030-93420-0_32

[10] Zhang Q et al 2016 Leakage zone identification in large-scale water distribution systems using multiclass support vector machines *J. Water Resour. Plan. Manag.* **142**(11) 04016042 doi: 10.1061/(ASCE)WR.1943-5452.0000661

[11] Chen J, X Feng and S Xiao 2020 An iterative method for leakage zone identification in water distribution networks based on machine learning *Structural Health Monitoring* **20**(4) 1938–56 https://doi.org/10.1177/1475921720950470

[12] Shekofteh M, M Jalili Ghazizadeh and J Yazdi 2020 A methodology for leak detection in water distribution networks using graph theory and artificial neural network *Urban Water J.* **17**(6) 525–533 https://doi.org/10.1080/1573062X.2020.1797832

[13] Lučin I, B Lučin, Z Čarija and A Sikirica 2021 Data-driven leak localization in urban water distribution networks using big data for random forest classifier *Mathematics* **9**(6) doi: 10.3390/MATH9060672

[14] Breiman L 2001 Random Forests *Machine Learning* **45**(1) 5–32 https://doi.org/10.1023/A:1010933404324

[15] Elhabyan R, W Shi and M St-Hilaire 2019 Coverage protocols for wireless sensor networks: Review and future directions *J. Commun. Networks* **21**(1) 45–60 doi: 10.1109/JCN.2019.000005

[16] scikit-learn 2021 "2.3. Clustering – scikit-learn 1.0.2 documentation" https://scikit-learn.org/stable/modules/clustering.html#hierarchical-clustering (accessed Apr. 03, 2022)

[17] Sophocleous S, Savić D and Kapelan Z 2019 Leak localization in a real water distribution network based on search-space reduction *J. Water Resour. Plan. Manag.* **145**(7) 04019024 doi: 10.1061/(ASCE)WR.1943-5452.0001079.

IOP Conf. Series: Earth and Environmental Science 1136 (2023) 012043 doi:10.1088/1755-1315/1136/1/012043

Topological analysis of water distribution networks for optimal leak localization

Débora Alves[1,3,5], Joaquim Blesa[1,2,4,5], Eric Duviella[3] and Lala Rajaoarisoa[3]

[1] Supervision, Safety and Automatic Control Research Center (CS2AC) of the Universitat Politècnica de Catalunya, 08222 Terrassa, Barcelona, Spain
[2] Institut de Robòtica i Informàtica Industrial (CSIC-UPC), 08028 Barcelona, Spain
[3] IMT Nord Europe, Université de Lille, CERI Digital Systems, F-59000 Lille, France
[4] Serra Húnter Fellow, Automatic Control Department (ESAII), Universitat Politècnica de Catalunya (UPC), 08019 Barcelona, Spain
[5] Author to whom any correspondence should be addressed

adeboracris@gmail.com, joaquim.blesa@upc.edu

Abstract. This paper introduces two methodologies to provide an optimum sensor deployment layout, one based on a model-based approach and the other entirely data-driven. The first method is formulated as an integer optimization problem, an optimization criterion consisting of minimizing the average topological distance. The second method is a new methodology to provide an optimum sensor placement regarding how many sensors to install without using hydraulic information but just exploiting the knowledge of the topology of the Water Distribution Networks. The method uses the Girvan-Newman clustering algorithm to ensure complete coverage of the network and the study of the installation of pressure sensors in the central nodes of each group, selected according to different metrics of topological centrality. The approach is illustrated in the Modena network.

1. Introduction

Water Distribution Networks (WDN) are large-scale and complex systems composed with thousands of pipes and characterized by strong nonlinearities due to hydraulic behavior. In WDN, one of the main problems is leaks that may appear in the system due to different factors (weak joints, water hammers, utility construction or excavation, seasonal temperature changes, etc.). Therefore, a significant effort has been made to reduce the impact of leakage on the network, such as the leak localization study that indicates which area in the WDN there may be a leak.

In literature, leak localization methods were released by applying model-based approaches and demand-driven (DD) hydraulic simulators Common approaches are based on the analysis of pressure residues [1], or on AI techniques using hydraulic models. Even if the performance of hydraulic model-based approaches is excellent, the main difficulty concerns the calibration of accurate models and data availability for all possible complex scenarios.

Thinking about these challenges, recent studies have proposed analyzing the problem using data-driven methods [3,4] that combine the use of standard operation data and topological information. The method in [5] studies the effect of the extra flow when a leak occurs in the pressure sensors presented in the network. It aims at developing a relative incidence of a leak using network topology correlated with the flow and pressure measurement.

An element that has the potential to significantly improve the localization of leaks is the sensor pressure configuration. In the last years, several strategies that tackle the optimal sensor placement problem in WDNs for the localization of leaks have been proposed. As a branch and bound searches [6], Genetic Algorithms [7], feature selection techniques [8], and game theory approaches [9].

This work presents two new methodologies for sensor placement in the WDN, one using hydraulic simulation formulated as an integer optimization problem solved with a Genetic Algorithm (GA). And the other uses only the topological network information to improve the leak localization methods that use residual analysis [1,3,5]. The second approach aspires to make easier the sensor placement problem by suppressing the calibration of the hydraulic water models and reducing the computational burden. The methodology is based on the complex network theory applying the graph approach with the hydraulic information to represent the WDN.

In this document, the leak localization methodology is presented in Section 2. Section 3 presents algorithms dedicated to the sensor placement. The application and the performances of the proposed methods are highlighted with a real water distribution network in Section 4. A conclusion and some perspectives are provided in Section 5.

2. Leak localization

Using pressure measurements, the proposed methodology aims to detect and isolate leaks in WDN. Normally, leak detection methods are done using inlet flow analysis and are triggered only when a leak is detected. Considering that inlet pressure and flow sensors and other pressure sensors in inner nodes are installed in the WDN, the localization of leak can be performed by employing the analysis of pressure residuals generated by the comparison of inner pressure measurements and leak-free pressure estimations as

$$r_i = \hat{p}_i(c) - p_i(c) \qquad i = 1,\dots,s \tag{1}$$

where r_i, $\hat{p}_i(c)$ and $p_i(c)$ are the residual, leak-free pressure estimation, and pressure measurement at inner node i; c is the operating condition defined by inlet measurements, with$s$ the number of inner sensors installed in the WDN. Leak-free pressure estimations $\hat{p}_i(c)$ can be computed by physical models or through of historical data. If a physical model of the WDN is available, a leak sensitivity matrix Ω

$$\Omega = \begin{pmatrix} \frac{\partial r_1}{\partial f_1} & \cdots & \frac{\partial r_1}{\partial f_n} \\ \cdot & \cdots & \cdot \\ \frac{\partial r_s}{\partial f_1} & \cdots & \frac{\partial r_s}{\partial f_n} \end{pmatrix} \tag{2}$$

where $\frac{\partial r_i}{\partial f_j} = \hat{p}_i(c) - \hat{p}_i^j(c) \quad i = 1,\dots,s \quad j = 1,\dots,n$ with $\hat{p}_i^j(c)$ is the pressure in node i considering a leak in node j denoted as f_j. Then, leak localization can be formulated as the maximum correlation between the observed residuals and the different leak hypothesis

$$\underset{j\in\{1,\dots,n\}}{\arg\max} \frac{\omega_{\bullet j}\cdot \boldsymbol{r}}{\|\omega_{\bullet j}\|\|\boldsymbol{r}\|} \tag{3}$$

where $\omega_{\bullet j}$ is the jth column of a sensitivity matrix (2) and $\boldsymbol{r}$ is the residual, whose components are computed in equation (1). Alternatively, if it is not available any hydraulic model of the WDN the leak localization method can be formulated as the maximum residual component

$$\underset{i \in \{1,\ldots,s\}}{\arg\max} \; \{r_i\} \tag{4}$$

The main disadvantage in the use of leak localization in equation (4) compared with leak localization in equation (3) is that the result is a cluster related to one of the s inner pressure sensors, and not a node. But as it is very simple, and it does not require any physical model, it is a good reference point to develop new data-driven leak localization methods. As in [1,5], where topological information was used to formulate equation (4) at the node level. In this work equation (4) is used to be the leak localization method.

3. Sensor placement

An additional objective is to develop an approach for placing a fixed number of sensors, s, in a WDN to obtain a sensor configuration leading to a maximization of the leak localization performance. In order to cope with the combinatory complexity of the sensor placement problem, following the ideas of [10], a two-step suboptimal search algorithm is proposed:

STEP 1: Divide the nodes of the WDN into s clusters $C = \{C_1, \ldots, C_s\}$, i.e., a cluster for every sensor to be installed in the WDN.

STEP 2: Choose a node among all nodes of a cluster as the optimal place to install a sensor.

For STEP 1, the WDN can be represented as a directed graph $G=(V, E)$, with V as the set of vertices that represents the n connections between the components of the network (junctions, reservoirs, and tanks), and E, the edges, which represent the m links (pipes, valves, and pumps) in the network. Edges E are associated with a cost value that are determined with the friction loss in pipes based on the Hazen-Williams formula, that consists in guaranteeing a model closer to the real behavior of the WDN by dividing the pipe length by the pipe diameter. The Girvan-Newman (GN) clustering method [11] is proposed for STEP 1. GN clustering is an algorithm that focuses on edges mostly between communities. From the original graph, clusters are created by considering the edge betweenness based on the computation of shortest paths that run through them.

For STEP 2, two methods were developed: the first being the model-based approach which uses hydraulic models formulated as an integer optimization problem. This approach is only possible to use if the hydraulic model has high credibility. Furthermore, the second is a data-driven approach. It aims at locating sensors at the topologically most essential nodes of each cluster by ensuring a spatially uniform distribution.

3.1. Model-based approach

STEP 1 gives the information of the $C = \{C_1, \ldots, C_s\}$ clustering with the numbers of nodes that make up each one

$$\begin{aligned} X_1 &= \{1, \ldots, C_1^n\} \\ X_2 &= \{C_1^n + 1, \ldots, C_1^n + C_2^n\} \\ &\vdots \\ X_s &= \left\{\sum_{i=1}^{s-1} C_i^n + 1, \ldots, \sum_{i=1}^{s-1} C_i^n + C_s^n\right\} \end{aligned} \tag{5}$$

where $X_1, \ldots, X_s$ are the groups of nodes that contain the cluster C, they are organized in crescent order with C^n represent the last node of each clustering and C_s^n value is equivalent to the total n node numbers of WDN. The goal in STEP 2 is to choose one node in each X to be the place to install a sensor. A performance index, Average Topological Distance (ATD) is used to display the information on distance between nodes, and more specifically the distance between the node predicted as leaking and the actual node with the leak. Hence, the optimal sensor placement using the equation (3) can be performed by minimizing the ATD.

The ATD is calculated with a matrix that contains the minimum topological distance (in nodes), $D \in \mathbb{R}^{n \times n}$. The performance of equation (3) is assessed thanks to the confusion matrix Γ (Table 1). In this matrix, the rows correspond to the leak scenario and the columns to which the leak is located by proposed method.

Table 1. Confusion matrix Γ.

	$\hat{l}_1$	$\cdots$	$\hat{l}_i$	$\cdots$	$\hat{l}_n$
l_1	$\Gamma_{1,1}$	$\cdots$	$\Gamma_{1,i}$	$\cdots$	$\Gamma_{1,n}$
$\vdots$	$\vdots$	$\vdots$	$\vdots$	$\vdots$	$\vdots$
l_i	$\Gamma_{i,1}$	$\cdots$	$\Gamma_{i,i}$	$\cdots$	$\Gamma_{i,n}$
$\vdots$	$\vdots$	$\vdots$	$\vdots$	$\vdots$	$\vdots$
l_n	$\Gamma_{n,1}$	$\cdots$	$\Gamma_{n,i}$	$\cdots$	$\Gamma_{n,n}$

In this way, the ATD can be calculated as:

$$f(x) = \frac{\sum_{i=1}^{n}\sum_{j=1}^{n}\Gamma_{i,j}D_{i,j}}{\sum_{i=1}^{n}\sum_{j=1}^{n}\Gamma_{i,j}} \tag{6}$$

where $x = \{x_1, \ldots, x_s\}$ is the set of s sensors, with the constraint of $x_1 \in X_1, \ldots, x_s \in X_s$. In addition, the optimal value is the minimal distance. Based on the performance index $f(x)$, an integer optimization problem is formulated to deal with the sensor placement problem, as

$$\begin{aligned} \min_x : \quad & f(x) \\ s.t: \quad & x_i^l \le x_i < x_i^u, \quad i = 1, \cdots, s \\ & x \in \mathbb{Z}^n \end{aligned} \tag{7}$$

with x_i^l, x_i^u, the lower and upper bounds based on the equation (5). It should be noticed that the solution of the previous optimization algorithm provides the best sensor location when the operating conditions are similar to the one used to evaluate residuals in equation (1). The optimality of the sensor location is no longer guaranteed if the operating conditions are different. In fact, it could vary according to them.

3.2. Data-driven approach

As explained in the previous section, STEP 2 aims to select nodes in each $X_1, \ldots, X_s$ defined in equation (5). The data availability being often limited or not suitable, the sensor localization must be performed without any hydraulic simulations. In addition, it reduces the computational burden.

Thus, a criterion is defined to approach the sensor placement problem even when hydraulic information on the network is unavailable or incomplete. The assumption on which the topology most central nodes of each cluster are considered suitable sensors locations is made. For this purpose, three indicators of the importance of the nodes were used to select the positioning of the sensors:

- Closeness centrality: the inverse sum of the distance from a node to all other nodes in the graph is determined; More central a node is, the closer it is to all other nodes.

$$c_c(i) = \frac{n-1}{\sum_{j=1}^{n} d(i,j)} \tag{8}$$

with $d(i,j)$ the distance between vertices i and j; n the number of nodes in the graph/clustering;

- Betweenness centrality: it consists in computing how often each graph node appears on a shortest path between two nodes in the graph.

$$c_{st}(i) = \sum_{s \neq t \neq i} \frac{\sigma_{st}(i)}{\sigma_{st}} \tag{9}$$

with σ_{st} the total number of shortest paths from node s to node t. $\sigma_{st}(i)$ is the number of those paths that pass through i;

- Eigenvector centrality: the eigenvector corresponding to the largest eigenvalue of the graph adjacency matrix is employed.

$$A_s x = \lambda_{\max} x \tag{10}$$

with A_s the adjacency matrix of the subgraph G and $\lambda_{\max}$ the largest eigenvalue. According to the coordinates $e_c(v)$ of the eigenvector $x = e_c$ the centrality for a node based on the centrality of its neighbors is computed in association with the largest eigenvalue of $\lambda_{\max}$ matrix A_s.

3.3. *Performance indicators*

The proposed leak localization performance indicators to assess the sensor placement optimality are based on:

- Average topological distance (ATD) is the distance between the centroid predicted as leaking with the true node leak. Thus, smaller the ATD index is, better is the leak localization. The ATD requires the matrix with the minimum topological distance in nodes (see equation (6)). This index is a suitable parameter for analyzing the improvement of the leak localization method.
- F1-score is the weighted average of precision and recall. The precision is the analysis of all positive predictions and their numbering. The recall is the study of real positive cases and their numbering. The F1-score is a good indicator of imbalanced data [11].
- Cohen's Kappa represents the degree of accuracy and reliability. It corresponds to the difference between the observed overall accuracy of the model and the overall accuracy obtained by chance. It is a more practical measure to use on problems with an imbalance in the classes. The kappa has a range from −1 to +1. Values between 0.6 and 0.8 are considered good [12].

The presented indexes are used to analyze the performance of sensor placement, and to improve it by using the average topological distance of the coverage area where the sensor can identify a leak. Moreover, the ATD index is an excellent index to measure the improvement of leak localization methods. Furthermore, the F1-score and the Cohen's Kappa were suggested to analyze the GM clustering method and the node importance centroids according to the residues, being the higher these indexes are, the better the GM performance.

4. Case study

The real water distribution network of the Italian city Modena is selected to illustrate the proposed methods and to show their performance. It is a well-known WDN that is composed with 268 junctions (nodes) connected through 317 pipes. This large-scale network is also equipped with four reservoirs.

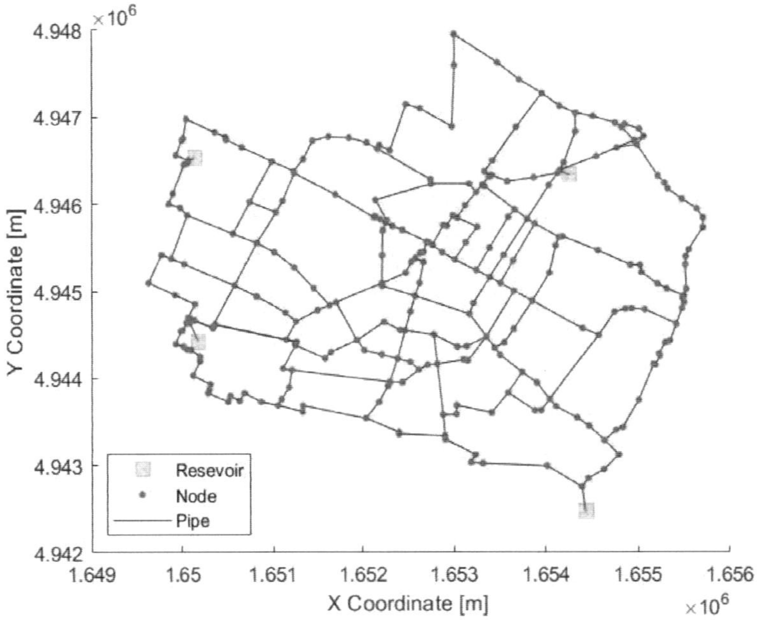

Figure 1. Simplified Modena topological WDN.

The Modena WDN has been modelled using the EPANET hydraulic simulator (figure 1). A leak scenario of 72h has been generated to analyze the performances of the proposed methods. The following simulation conditions were used:

- samples were collected every 10 minutes and filtered to 24 hours values. It aims at reducing the uncertainty in the data.
- nominal demand is considered with an uncertainty of 10% as a normal distribution.
- the leak size was randomly selected, with 3 to 6 l/s representing 1 to 2.5% of the network consumption.

The minimization of the optimization presented in section 3.1 is carried out using Genetic algorithms (GA) based on principles of natural genetics and natural selection [13,14]. The near-optimal placement of sensors in WDN for leak localization can be obtained by using GA by associating a chromosome to a possible presence or absence of a sensor at a given node.

Table 2 shows the results obtained in three different scenarios: with 3, 5, and 10 possibilities of sensor placement. The number of nodes chosen to have a sensor is exhibited in all scenarios for the GA solution and each node importance method. As the optimization of equation (7) is based on the optimization of the ATD index, the results presented in Table 2 show that the solution obtained by the GA is the best in any case. Even if the proposed data-driven methodology does not lead to the optimal solution, the obtained values are close to those of GA. Note that even getting the best optimal value of the ATD, the solution obtained by the GA does not guarantee the best result for the F1 score and the kappa. The Eigenvector centrality is the results that present the worst results, principally in the scenario with three sensors, having the Cohen's Kappa with the worst value, inferior a 0.6.

On the other hand, the Betweenness and the Closeness centrality had a similar result. However, analyzing the ATD index, the Betweenness metric improves the scenario with 3 and 5 sensors. In the scenario with ten sensors, the Closeness centrality has the better performance. Whereas in a general case, the Betweenness centrality is the best choice in the data-driven methodologies.

Figure 2 shows the result of the second scenario with 5 sensors. Each cluster C is highlighted with a different shade. The position of the sensors obtained with the proposed methods is shown with a circle with different colours.

Table 2. Average evaluation metrics.

Criterion	Nodes with sensors	ATD	F1 score (%)	Kappa
		3 sensors scenario		
GA	**88 109 207**	**6.82**	**82.43**	**0.74**
Closeness	91 147 207	7.50	82.01	0.71
Betweenness	7 63 109	7.37	79.63	0.70
Eigenvector	4 83 119	8.05	67.48	0.48
		5 sensors scenario		
GA	**5 41 80 110 164**	**5.80**	**47.01**	**0.71**
Closeness	49 91 135 147 207	6.50	82.01	0.75
Betweenness	7 63 49 91 135	6.54	81.91	0.75
Eigenvector	4 83 92 119 134	6.70	75.64	0.65
		10 sensors scenario		
GA	**10 31 47 78 129 171 129 183 225 258**	**4.45**	**28.97**	**0.62**
Closeness	11 31 49 83 91 105 137 129 180 258	4.70	73.36	0.68
Betweenness	1 11 31 35 49 83 91 129 135 180	4.68	69.60	0.64
Eigenvector	1 4 31 83 92 102 119 129 134 180	4.71	83.36	0.79

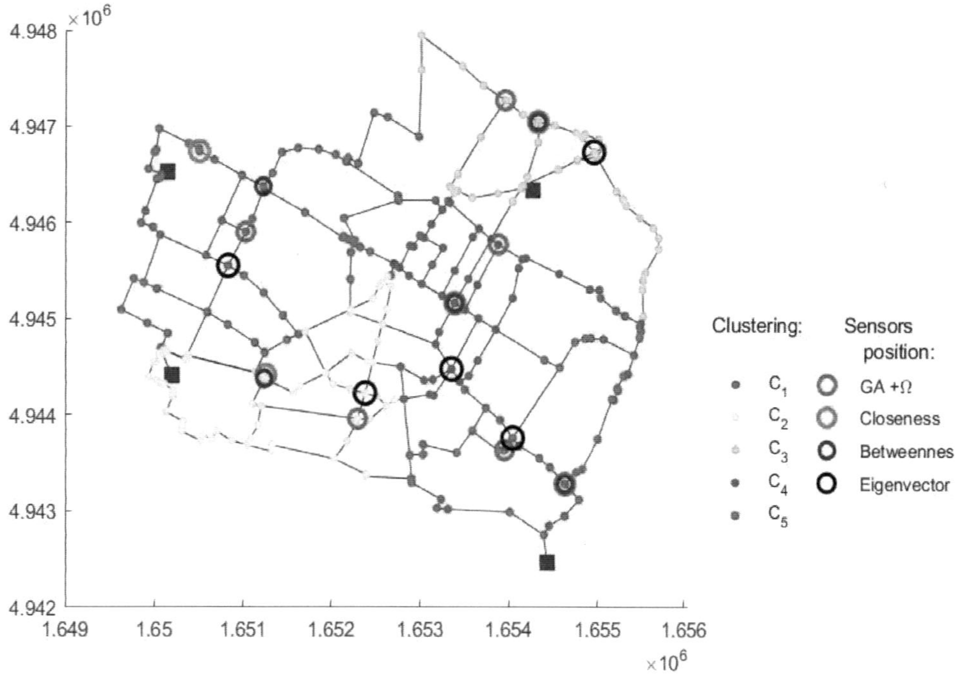

Figure 2. WDN of Modena and the four sensor layouts according to the three topological centrality metrics and GA solution.

Clustering generated with GN provides a division focused on edges between communities that do not guarantee a homogeneous distribution of nodes. As the nodes are more communicated, the effect on the residue in a leak between these nodes will affect more the sensor installed in that region.

5. Conclusions

The quality of sensor placement in WDSs impacts leak identification efficiency, and sensor-placement optimization remains one of the top issues in related research. In this work, a new full sensor placement method utilizing just the topological information of the WDN based on the high connection density of the graph representation of the system in association with the node importance has been presented in this study. It employs only the topological characteristics of a WDN, combining the identification of clusters of nodes and topological centrality metrics for the design. It does not require any hydraulic simulation. It is well suited for real cases where only partial information about the network is available.

In addition, a new sensor placement approach is proposed to minimize the average topology distance of leak isolability. It is formulated as an integer optimization problem. However, the method uses a hydraulic model to simulate all node leaks based on the system's demand pattern and uncertainties.

The proposed approaches have been described by using a simplified version of the Modena Water Distribution Network. They demonstrate that the methodology using only the system topology information obtained a good result, ideal for cases where partial details on the system are available.

Future work will investigate how different objective functions can improve the selection of placements for sensors. The effect of other stressing conditions (i.e., sensor failures) in the network that can change the result of sensor placement will be investigated too.

Acknowledgments

This work has been partially funded by L-BEST project (Ref. PID2020- 115905RB-C21) from the Spanish Ministry of Science.

References

[1] Soldevila A, Fernandez-Canti RM, Blesa J, Tornil-Sin S, Puig V. 2016 Leak localization in water distribution networks using model-based Bayesian reasoning, *European Control Conference (ECC)* 2016 Jun 29 (pp. 1758-1763). IEEE.

[2] Javadiha M, Blesa J, Soldevila A, Puig V. 2019Leak localization in water distribution networks using deep learning, 6th *International Conference on Control, Decision and Information Technologies (CoDIT)* 2019 Apr 23 (pp. 1426-1431). IEEE.

[3] Soldevila A, Boracchi G, Roveri M, Tornil-Sin S, Puig V. 2022 Leak detection and localization in water distribution networks by combining expert knowledge e and data-driven models, *Neural Computing and Applications.*; **34**(6):4759-79.

[4] Sun C, Parellada B, Puig V, Cembrano G. 2019 Leak localization in water distribution networks using pressure and data-driven classifier approach. *Water.* **12**(1):54.

[5] Alves D, Blesa J, Duviella E, Rajaoarisoa L. 2021 Robust data-driven leak localization in water distribution networks using pressure measurements and topological information. *Sensors.* **21**(22):7551.

[6] Sarrate R, Blesa J, Nejjari F, Quevedo J. 2014 Sensor placement for leak detection and location in water distribution networks. *Water Science and Technology: Water Supply.* **14**(5):795-803.

[7] Steffelbauer DB, Fuchs-Hanusch D. 2016 Efficient sensor placement for leak localization considering uncertainties. *Water resources management.* **30**(14):5517-33.

[8] Soldevila A, Blesa J, Tornil-Sin S, Fernandez-Canti RM, Puig V. 2018 Sensor placement for classifier-based leak localization in water distribution networks using hybrid feature selection. *Computers & Chemical Engineering.* **108**:152-62.

[9] Arbesser-Rastburg G, Fuchs-Hanusch D. 2019 Serious Sensor Placement—Optimal Sensor Placement as a Serious Game. *Water.* **12**(1):68.

[10] Blesa J, Nejjari F, Sarrate R. 2015 Robust sensor placement for leak location: analysis and design. *Journal of Hydroinformatics.* **18**(1):136-48.

IOP Conf. Series: Earth and Environmental Science 1136 (2023) 012043 doi:10.1088/1755-1315/1136/1/012043

[11] Landis JR, Koch GG. 1977 An application of hierarchical kappa-type statistics in the assessment of majority agreement among multiple observers. *Biometrics*. **1**:363-74.
[12] Romano M, Woodward K, Kapelan Z. 2017 Statistical process control based system for approximate location of pipe bursts and leaks in water distribution systems. *Procedia Engineering*. **186**:236-43.
[13] Koza JR. 1995 Survey of genetic algorithms and genetic programming. *Wescon conference record* (pp. 589-594). Western Periodicals Company.
[14] Girvan M, Newman ME. 2002 Community structure in social and biological networks. *Proceedings of the national academy of sciences*. **99**(12):7821-6.

IOP Conf. Series: Earth and Environmental Science 1136 (2023) 012044 doi:10.1088/1755-1315/1136/1/012044

Using WDN-tailored complex network centrality metrics to assess isolation valve system reliability

Daniele Biagio Laucelli[1], Francesco G Ciliberti[2], Luigi Berardi[2], Giuseppe Mauro[3] and Orazio Giustolisi[1]

[1] Department of Civil, Environmental, Land, Building Engineering and Chemistry; Technical University of Bari, 70126 Bari, Italy
[2] Department of Engineering and Geology; University "G. D'Annunzio" of Chieti Pescara, 65127 Pescara, Italy
[3] Department of Civil Engineering, Energy, Environment and Materials; Mediterranean University of Reggio Calabria, 89100 Reggio Calabria, Italy

danielebiagio.laucelli@poliba.it

Abstract. Mechanical reliability Water Distribution Networks (WDNs) depends on the Isolation Valve System (IVS). In fact, in the event of accidental disruptions or planned maintenance works, segments of the WDNs need to be temporarily disconnected by closing isolation valves to allow pipes repair. This maneuver it modifies the WDN topology and relevant hydraulic behavior. In this paper a recent approach for IVS reliability based on the Complex Network Theory (CNT) has been applied. It couples the construction of the graph of segments and isolation valves generated by the IVS with the risk of segment interruptions, using several WDN-tailored centrality metrics. The study is demonstrated on a real WDN using a Digital Water Service developed ad hoc for enabling technicians and water utilities in targeting the most critical valves in case of failure and thereby defining a lineup for valves inspections and maintenance.

1. Introduction

The reliability of Water Distribution Networks (WDNs) represents a classic topic for researchers [1] as well as a long-standing challenge for the water utilities from all over the world. As reported in [2], it generally addresses to the degree to which the network can provide and adequate supply to consumers, under both normal and abnormal operating conditions. Reliability can be summarized into two classes [3]: the mechanical reliability, which measures the ability of the system to ensure an adequate level of service even under unplanned events interrupting one or more pipes, and the hydraulic reliability, which reflects how WDNs hydraulic behavior can face consumer demand variations or deterioration of components. Previous studies have tried to quantitively assess the reliability of WDNs. The number of papers on this topic is quite large some only some example works are cited here. [4] and [5] applied the graph theory and statistical metrics; [6] reported several techniques for analyzing WDNs reliability, including minimum cut sets and event or fault tree analysis. [7] defined the resilience index for measuring the WDN reliability, which has been shown to be correlated to hydraulic and mechanical reliability [8], later adapted by [9] for accounting the network resilience, and generalized by [10] for network with multiple sources. [11] extended the Todini's index to the pressure-driven analysis, while [12] reported other measures involving reliability aspects in WDN design.

Previous works toke the assumption of the *N valve rule* [13], which means that isolation valves are located at both ends of each pipe in order to isolate it. Nonetheless, such configuration is usually not adopted since it requires high cost for maintenance and increases the probability of valve malfunctioning. Therefore, only a limited amount of isolation valves is installed in the network and when planned or unplanned operations ask for closing valves a portion of the network comprising two or more pipes, namely "segments", cannot be served with water. Moreover, isolation valve system (IVS) represents itself an element of failure: each valve works few times during its service life, so the IVS functionality is uncertain and hence can degrade the expected mechanical reliability.

[14] evaluated the WDNs reliability considering the hydraulic modeling of the supplied demand to customers during unplanned conditions due to valves shutdown, and the failure probability of each isolated segment. It has been found out that the optimal IVS, with respect to the network connectivity, minimizes the risk of disconnection of the segments. Recently, [15] reported a IVS analysis using the metrics of the Complex Network Theory (CNT) for ranking the importance of the segments and the IVS. It is based on drawing a bipartite graph composed by two sets of vertices, representing the segments bounded by the IVS and the IVS itself, respectively. A modified CNT betweenness centrality [16], which reflects the intrinsic relevance of the vertices, has been applied for the analysis of the IVS bipartite graph to rank the importance of the segments and the IVS.

Based on the recently digital transformation paradigm for WDNs [17], this work proposes a Digital Water Service (DWS), named *DigitalWaterIVS_Vulnerability*, as a tool for enabling users, even not expert in CNT, to get information about IVS vulnerability of WDNs. It embodies the methodology defined by [15] and extends the topological analysis of the networks through the evaluation and representation of three main WDN-tailored centrality metrics [16], mostly adopted for studying real networks [18]. The service, developed as a QGIS plugin and completely interoperable with the WDNetXL platform [19], has been applied to a real case study for supporting the management strategies for IVS planning.

2. Methods

The strategy adopted by [15] has conceived into two steps, which holds:

- the definition of the graph of segments and isolation valves generated by the IVS;
- the analysis of the graph of segments and isolation valves using the relevance-embedded centrality metrics, as defined by [16].

2.1. IVS Bipartite graph of segments and isolation valves

Each isolation valve can be conceived as a conceptual cut in the main graph of the WDN, close to a node, so the IVS divides the network into segments. The segments can hold: one or more pipes and nodes; a single node, such as a reservoir or a node separated by isolation valves at every pipe joining the node; or a single pipe, as in the case of the *N valve rule* [13].

The IVS defines the bipartite graph where the vertices are: the isolation valves and the segments, represented as "segment vertices" located in the barycentric position of each segment. The edges of that graph connect each segment vertex with the isolation valves bounding that segment.

2.2. Analysis of the graph of segments and isolation valves using WDN-Tailored Centrality metrics

The intrinsic relevance of the centrality metrics [16] is related to the function $f(R_s, R_t)$, which allows weighting the shortest paths between couples of vertices and it comprises several formulations, as reported below:

$$f(R_s,R_t)=\{(R_s+R_t)/2; R_sR_t; \max(R_s,R_t); R_s; R_{path}\} \quad (1)$$

The classic metrics and the corresponding ones tailored by [16], for a network N composed by V nodes and E links, are defined as follows:

- Degree centrality. The degree centrality [20] is defined as the number of links incident upon a node and describes the local connectivity of the network. It can be expressed as:

$$d(s)=\sum_t A_{st} \tag{2}$$

where A_{st} is the adjacency matrix of the network. The standard formulation of Equation (2) has been modified by [18] with Equation (1) for embedding the importance of the connection of the vertex s with its neighbours:

$$d(s)=\sum_t A_{st} f(R_s,R_t) \tag{3}$$

- Harmonic centrality. It was introduced by [21] and represents a variant of the closeness centrality for extending the original formulation to the unconnected graphs. The harmonic version of the closeness $C(s)$ centrality of a generic vertex s is:

$$C(s)=\sum_t \frac{1}{dist_{st}} \tag{4}$$

where $dist_{st}$ is the shortest distance between vertices s and t. The intrinsic relevance version of that metric, as reported in [18], is:

$$C(s)=\sum_t \frac{f(R_s,R_t)}{dist_{st}} \tag{5}$$

- Betweenness Centrality. The standard version of the betweenness centrality assigns the number of shortest paths traversing a vertex v for all couples of vertices s and t of the network. Similarly, the edge betweenness, $EB(e)$, is the sum of the fractions for all the couples of vertices s and t of the network traversing and edge e. The formulation of both metrics is reported below:

$$B(v)=\sum_{s\neq v\neq t\in N} \frac{\sigma_{st}(v)}{\sigma_{st}} \tag{6}$$

$$EB(e)=\sum_{s\neq t\in N} \frac{\sigma_{st}(e)}{\sigma_{st}} \tag{7}$$

The tailored versions proposed by [16] are:

$$EB(e)=\sum_{s\neq t\in N} f(R_s,R_t)\frac{\sigma_{st}(e)}{\sigma_{st}} \tag{8}$$

$$B(v)=\sum_{s\neq v\neq t\in N} f(R_s,R_t)\frac{\sigma_{st}(v)}{\sigma_{st}} \tag{9}$$

3. The *DigitalWaterIVS_Vulnerability* service

Digital Water Services (DWSs) are designed to bring the opportunities of digital transformation in WDN sector close to technicians and decision makers [17]. The main idea behind this paradigm is to provide end users with tools that tackle specific tasks among the common activities in WDNs planning and management, which integrate the representation of the system in GIS platform usually adopted by technicians with advanced analysis procedures, including hydraulic modelling, artificial intelligence, optimization and CNT tailored for WDN. In this framework, the Digital Twin of a WDN is intended as a virtual object integrating the WDN topological configuration, asset features of pipes as well as the hydraulic behaviour based accurate hydraulic modelling.

The main innovation transfer paradigm used herein exploits the data structure and the modelling capabilities in the WDNetXL platform to generate the WDN Digital Twin, as ESRI® shapefiles, to be gathered, analyzed and visualized in the QGIS software.

The *DigitalWaterIVS_Vulnerability* service has been developed as QGIS plugin and it is designed for performing the IVS vulnerability analysis by applying the strategy reported above.

The user can load the Digital Twin of the WDN through the Graphical User Interface (GUI) in Figure 1 and visualize on the GIS map canvas all the features and devices of the network (Figure 2).

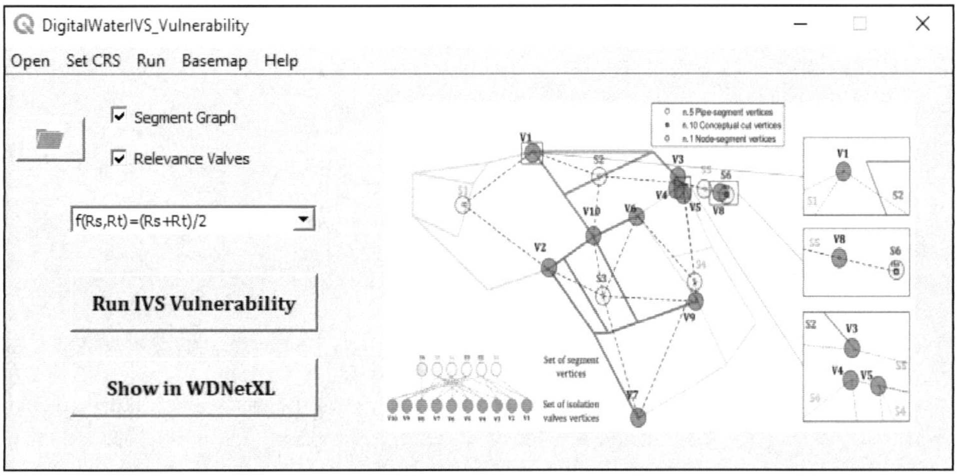

Figure 1. *DigitalWaterIVS_Vulnerability* Graphical User Interface (GUI).

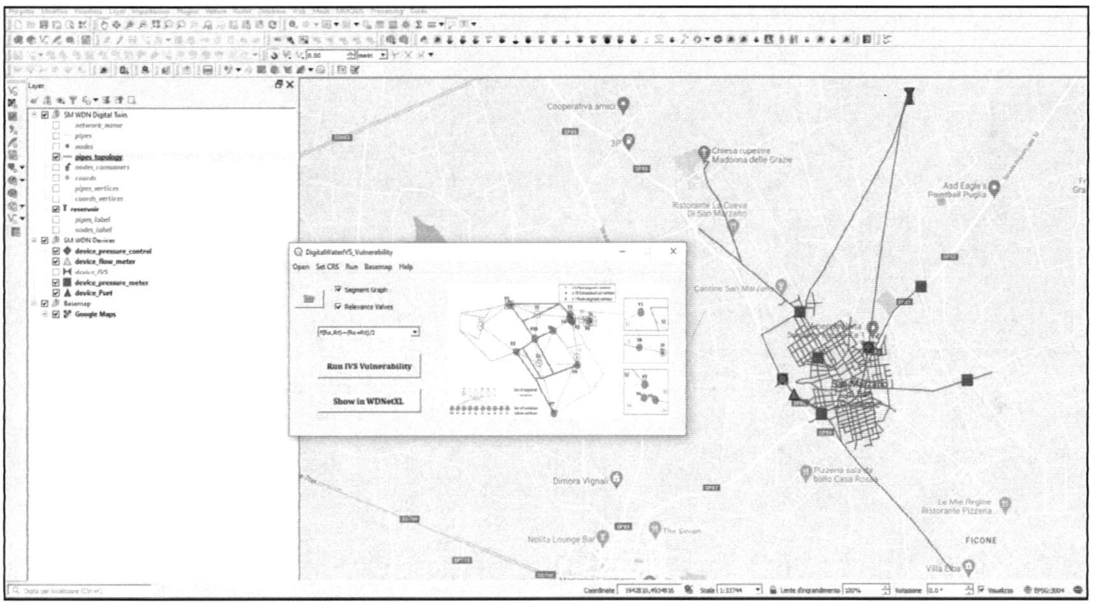

Figure 2. Representation of the digital twin of WDN on the GIS map canvas.

After that, users can choose from among several different types of calculations of the function of the relevance (Figure 3), as reported in Equation (1), for the three metrics in the Equations (3) (5) (8) (9), or using the classic approach, as expressed in the Equations (2) (4) (6) (7), and then perform the analysis and visualize the results. An explicit representation of results performed by the DWS is reported in the next section.

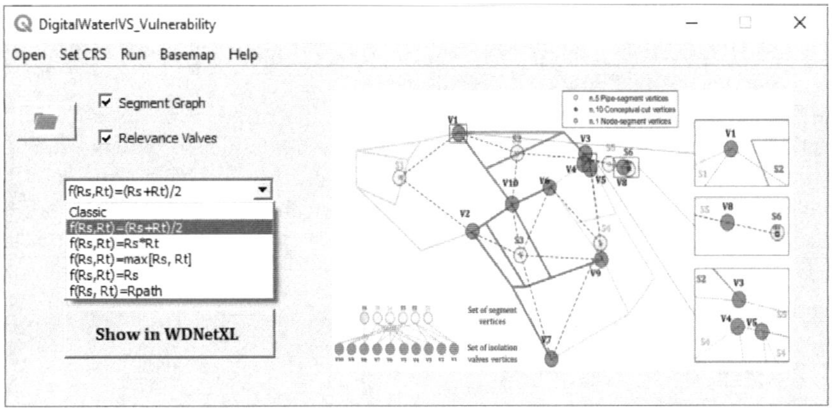

Figure 3. Selection of the relevance formulation type.

4. Case study

The *DigitalWaterIVS_Vulnerability* service has been tested on a real WDN, whose model includes 898 pipes, 729 nodes, 630 isolation valves and one reservoir.

The WDN Digital Twin, built and calibrated in WDNetXL platform, has been loaded into the QGIS software through the *DigitalWaterIVS_Vulnerability* GUI.

In this work, the analysis of the intrinsic-embedded metrics has been performed using the formulation $f(R_s, R_t) = (R_s + R_t)/2$ of the relevance. Indeed, it was demonstrated to realistically render the role of the intrinsic relevance in assessing WDN reliability [18]. Figure 4a shows the layout of the layer of segments, which amounts to 898 segments, and Figure 4b shows the layer of the IVS bipartite graph of segments and solation valves, drawn as a black dotted line. It is worth to recall that, as explained in [18], the intrinsic relevance for WDNs is assigned equal to the water demand for each vertex corresponding to a demand node and the sum of supplied water for the source node.

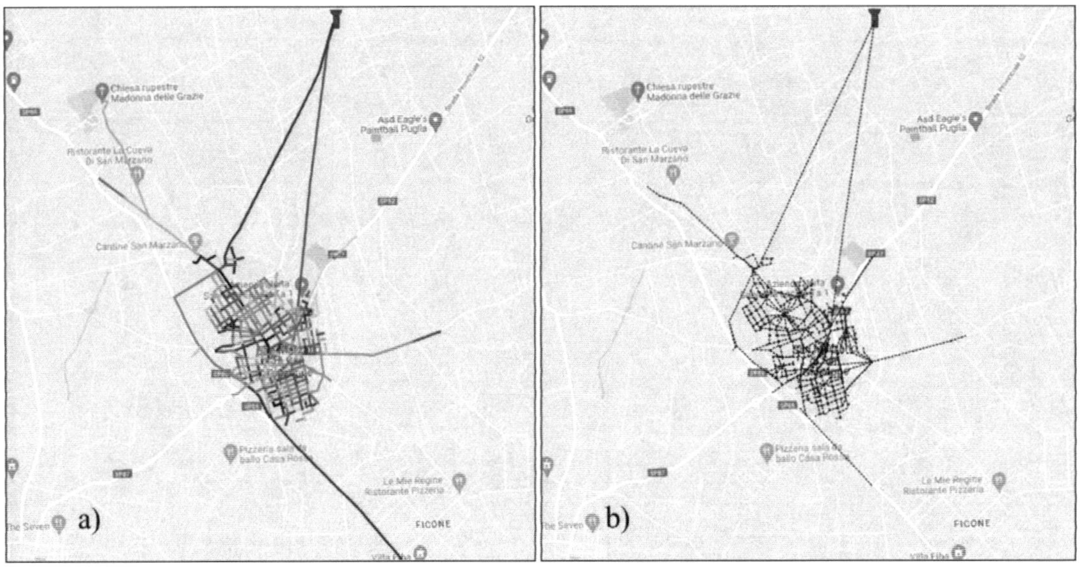

Figure 4. Representation of segments layer (on the left) and IVS bipartite graph (on the right).

Figure 5 represents the WDN-tailored betweenness with respect to valves. It shows that higher values, between 90% and 100%, of such metric are assigned to valves located in the main supplying paths of the network. This means that, in case of malfunctioning of such valves it would require the isolation of most (or all) of the segments. Other valves, located in the looped part of the distribution network, are assigned quite lower values of the metric meaning that, in case o valve failures, there are other alternative paths to feed other part of the WDN. Figures 6 and 7 show the representation of ranking of valves importance according to the tailored degree and harmonic metrics, respectively. In the case of the WDN-tailored degree (Figure 6), it is worth noting that higher values are assigned only to the valves that are closer to the source of network, while in the case of the WDN-harmonic centrality (Figure 7), higher values are referred to some bounding valves of the network. These results are mostly conditioned by the topological configuration of the IVS and show that, in this case, the two latter metrics are not able to well represent the IVS failure criticality compared to the WDN-tailored betweenness centrality.

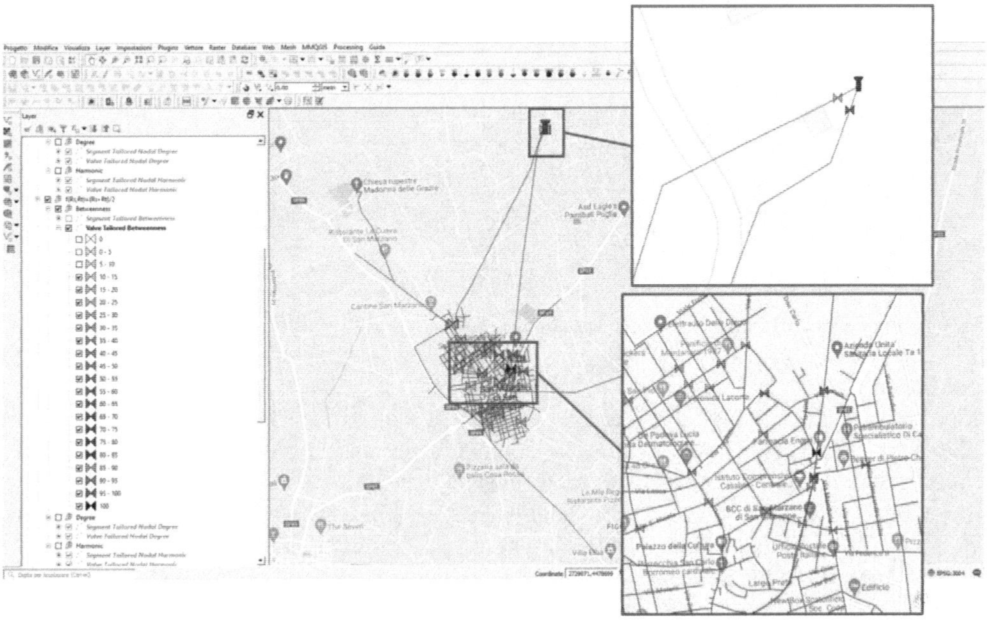

Figure 5. WDN-tailored betweenness of valves visualization on GIS map canvas.

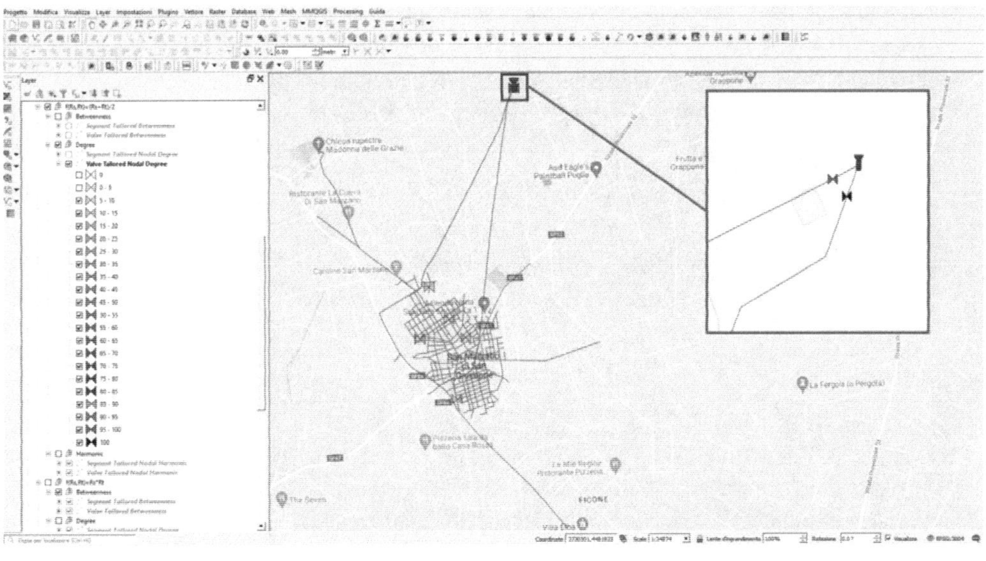

Figure 6. Visualization of WDN-tailored valves degree layer on GIS map canvas.

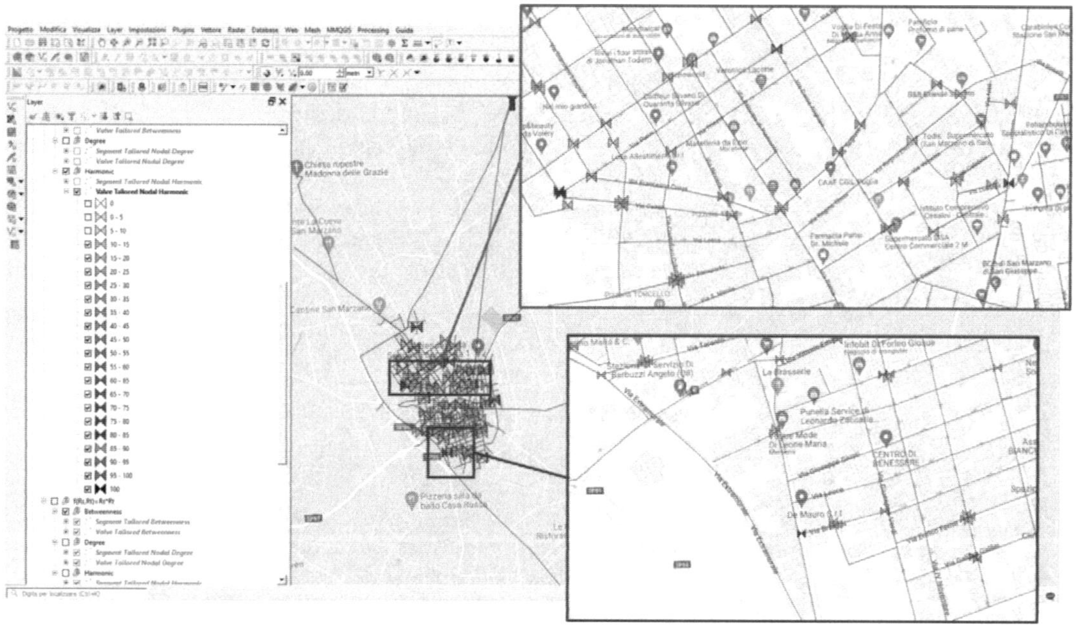

Figure 7. Visualization of WDN-tailored valves harmonic layer on GIS map canvas.

5. Conclusions

This work presents a novel Digital Water Service, named *DigitalWaterIVS_Vulnerability*, for the IVS vulnerability assessment of WDNs, within the context of the digital transformation paradigm [16]. It allows implementing the strategy for WDN mechanical reliability analysis with respect to the IVS [15]. The strategy exploits metrics from CNT which has been properly tailored for the WDNs.

The digital water service, is developed as a plugin for QGIS software, allows loading of the digital twin of WDN, performing and comparing several CNT analyses considering the classic formulation of the centrality metrics or the a version accounting for the intrinsic relevance of vertices. Following the framework behind the digital transformation paradigm, the DWS is conceived to expose results of direct relevance for technicians and decision makers.

The distribution of IVS relevance provides a methodology for ranking the importance of the isolation valves in the network, thus supporting the identification of the most impacting failure scenarios and enabling the WDN management planning for IVS inspections and retraining.

The ease of use in a well known working environment enables technicians, which are not required to be expert in CNT in water systems, to get effective results and to compare alternative analyses (i.e. using different metrics) according to their expertise on WDN management.

Acknowledgments

This paper was partly financed by European Union - PON Ricerca e Innovazione 2014-2020 E FSC, call D.D. 13 July 2017 n. 1735, Project "Energidrica - Efficienza energetica nella gestione delle reti idriche"(code ARS01_00625).

References

[1] Gheisi A, Forsyth M and Naser G 2016 Water distribution systems reliability: A review of research literature *Journal of Water Resources Planning and Management* **142**(11) 04016047
[2] Xu C and Goulter IC 1999 Reliability-based optimal design of water distribution networks. *Journal of Water Resources Planning and Management* **125**(6) 352–362
[3] Atkinson S, Farmani R, Memon FA and Butler D 2014 Reliability indicators for water

distribution system design: Comparison *Journal of Water Resources Planning and Management* **140**(2) 160–168

[4] Jacobs P and Goulter IC 1988 Evaluation of methods for decomposition of water distribution networks for reliability analysis *Civil Engineering System* **5**(2) 58–64

[5] Yang SL, Hsu NS, Louie PW and Yeh WW 1996 Water distribution network reliability: Connectivity analysis *Journal of Infrastructure Systems* **2**(2) 54–64

[6] Tung YK 1985 Evaluation of water distribution network reliability *Hydraulics and Hydrology in the Small Computer Age* **1** 359–364

[7] Todini E 2000 Looped water distribution networks design using a resilience index based heuristic approach *Urban water* **2**(2) 115–122

[8] Farmani R, Walters GA and Savic DA 2005 Trade-off between total cost and reliability for Anytown water distribution network *Journal of Water Resources Planning and Management* **131**(3) 161–171

[9] Prasad TD and Park NS 2004 Multiobjective genetic algorithms for design of water distribution networks *Journal of Water Resources Planning and Management* **130**(1) 73–82

[10] Jayaram N and Srinivasan K 2008 Performance-based optimal design and rehabilitation of water distribution networks using life cycle costing *Water Resources Research* **44**(1)

[11] Creaco E, Franchini M and Todini E 2016 Generalized resilience and failure indices for use with pressure-driven modeling and leakage *Journal of Water Resources Planning and Management* **142**(8) 04016019

[12] Liu H, Savić DA, Kapelan Z, Creaco E and Yuan Y 2017 Reliability surrogate measures for water distribution system design: comparative analysis *Journal of Water Resources Planning and Management* **143**(2) 04016072

[13] Walski TM, Weiler JS and Culver T 2008 Using criticality analysis to identify impact of valve location *Water Distribution Systems Analysis Symposium 2006* pp 1–9

[14] Giustolisi O 2020 Water distribution network reliability assessment and isolation valve system *Journal of Water Resources Planning and Management* **146**(1) 04019064

[15] Giustolisi O, Ciliberti F G, Berardi L and Laucelli DB 2022 A novel approach to analyze the isolation valve system based on the Complex Network Theory *Water Resources Research* e2021WR031304

[16] Giustolisi O, Ridolfi L and Simone A 2020 Embedding the intrinsic relevance of vertices in network analysis: the case of centrality metrics *Scientific Reports* **10**(1) 1–11

[17] Ciliberti FG, Berardi L, Laucelli DB and Giustolisi O 2021 Digital Transformation paradigm for asset management in water distribution networks *2021 10th Int. Conf. Energy and Environment (CIEM)* Bucharest pp 1–5 https://doi.org/10.1109/CIEM52821.2021.9614864

[18] Borgatti SP 2005 Centrality and network flow *Social Networks* **27**(1) 55–71

[19] WDNetXL-WDNetGIS 2022 http://www.idea-rt.com/digital-water-latest-innovation

[20] Freeman LC 1977 A set of measures of centrality based on betweenness *Sociometry* pp 35–41

[21] Marchiori M and Latora V 2000 Harmony in the small-world *Physica A: Statistical Mechanics and its Applications* **285**(3-4) 539–546.

IOP Conf. Series: Earth and Environmental Science 1136 (2023) 012045 doi:10.1088/1755-1315/1136/1/012045

WDN-tailored Complex Network Centrality Metrics for water distribution networks domain analysis

Francesco G Ciliberti[1], Luigi Berardi[1], Daniele B Laucelli[2], Giuseppe Mauro[3] and Orazio Giustolisi[2]

[1] Department of Engineering and Geology; University "G. D'Annunzio" of Chieti Pescara, 65127 Pescara, Italy
[2] Department of Civil, Environmental, Land, Building Engineering and Chemistry; Technical University of Bari, 70126 Bari, Italy
[3] Department of Civil Engineering, Energy, Environment and Materials; Mediterranean University of Reggio Calabria, 89100 Reggio Calabria, Italy

francesco.ciliberti@unich.it

Abstract. Water Distribution Networks (WDNs) are spatially organized infrastructures, whose hydraulic behavior greatly depends on their connectivity properties, i.e., the topological domain. The Complex Network Theory (CNT) provides a wide range of useful functions and metrics that have been tailored for identifying WDNs behavior, even before using hydraulic models. This work exploits the recent studies on the WDN domain, which couple the CNT metrics of centrality tailored for WDN with the intrinsic relevance of the spatial elements of the networks. The study accounts for various combinations of metrics and intrinsic relevance functions aimed at supporting WDN analysis for operational and management tasks. A Digital Water Service (DWS) is used to enable technicians and water companies to replicate the same domain analysis in the ongoing context of digital transformation in WDN sector.

1. Introduction

Water Distribution Networks (WDNs) represent essential infrastructures for urban life and development. Water utilities plan several technical and operative activities for achieving asset management and leakage reduction along the networks, in order enhance the reliability of the system and ensure adequate quality of service. The technical performances of their management activities depend on the ability to understand and control the hydraulic system behavior. Overall, WDNs are complex infrastructures, involving interconnected pipes, tanks and devices, which are irregular in space and evolving in time. Therefore, WDN hydraulic behavior is strongly constrained by the topological domain in terms of their structure and shape [1].

For extremely large and complex WDNs, planning management activities becomes even more difficult to achieve, especially considering the issues related to hydraulic modeling and calibration. Using tools returning the most relevant properties of the topological domain might provide a valuable support on various tasks in WDN analysis and management.

Complex Network Theory (CNT) inspired powerful procedures and metrics for understanding the characteristics of WDNs, as explained by [2] [3] in the case of assessment of WDNs vulnerability and structure, for optimizing WDNs solution [4] or for constraining the district design [5].

In fact, CNT allows to describe and capture the emergent behavior of networked systems [6], such as social networks [7], computing networks [8], biological networks [9] and spatial infrastructures [10]. The aim is to analyze the structure of the systems by assuming the relationship between the single elements of the systems: each element is represented by a node, while the link connects two elements related to each other. The centrality metrics, first introduced by [11], represents the measure for better describing the structure of the network by ranking the importance of the nodes [12], such as degree, closeness [11], eigenvector [13], Katz centrality [14], or PageRank [15].

For spatial networks [10], centrality metrics point out that the concept of the shortest path, i.e., the optimal path between two nodes with fewer links, is relevant compared to the information on flows (e.g., water flow in hydraulic systems). Furthermore, WDNs represent a peculiar type of spatial networks, limited by urban constraints as buildings and streets which affects the shape and the nodal degree distribution [16]. This fact, in addition to the need of embedding the hydraulic status information into the metrics, have motivated the development of some tailored metrics, such as [16] for assessing the local and global centrality metrics accounting for the nodal relevance, while [17] suggested to modify the edge betweenness for analyzing the spatial domain of the networks.

Recently, [18] proposed an enhanced version of three most suited centrality metrics for studying spatial networks [10], i.e., degree, betweenness and harmonic, by embedding the intrinsic relevance of edges and vertices. In this way, [18] showed that those metrics can be weighted, considering several formulations of the relevance function, constrained by the type of network, to better capture the interplay between the network topological structure and the intrinsic relevance of nodes.

Based on the digital transformation paradigm explained in [19], several services aimed at supporting technicians and water utilities in solving several WDN life-cycle management issues have been developed.

In this work, a novel Digital Water Service (DWS), named *DigitalWaterDomain_Analyzer*, is presented herein for analysing spatial domain of WDNs based on several metrics taken from CNT. It allows the topological analysis of nodes and links of WDNs by computing three main WDN-tailored centrality metrics for analysing and WDNs behaviour: (i) betweenness, either nodal or edge; (ii) nodal; and (iii) harmonic degree. The service, developed as QGIS plugin and completely interoperable with WDNetXL platform [20], has been applied to a case study for evaluating the hydraulic behavior of the network, identifying the main supplying paths along the network and the most relevant pipes.

2. Centrality Metrics

The DWS reported in this work consider the possibility to perform several analyses of network topological domain, which accounts both the classic definition of the metrics and the proposed tailored metrics by [18].

The formulations of the classic metrics and the corresponding ones adopted by [18], considering a network N composed by V nodes and E links, are shown below:

- Degree centrality. The degree centrality [11] is defined as the number of links incident upon a node and describes the local connectivity of the network. Considering a network N composed of V vertices and E edges, it can be expressed as:

$$d(s) = \sum_t A_{st} \tag{1}$$

where A_{st} is the adjacency matrix of the network.

- Harmonic centrality. Introduced by [21], it represents a variant of the closeness centrality for extending the original formulation for unconnected graphs. The harmonic version of the closeness $C(s)$ centrality of a generic vertex s is:

$$C(s) = \sum_t \frac{1}{dist_{st}} \tag{2}$$

where $dist_{st}$ is the shortest distance between vertices s and t.

- Betweenness Centrality. The standard version of the betweenness centrality assigns the number of shortest paths traversing a vertex v for all couples of vertices s and t of the network. Similarly, the edge betweenness, $EB(e)$, is the sum of the fractions for all the couples of vertices s and t of the network traversing and edge e. The formulation of both metrics is reported below:

$$B(v) = \sum_{s \neq v \neq t \in N} \frac{\sigma_{st}(v)}{\sigma_{st}} \tag{3}$$

$$EB(e) = \sum_{s \neq t \in N} \frac{\sigma_{st}(e)}{\sigma_{st}} \tag{4}$$

The tailored version proposed by [18] holds:

$$d(s) = \sum_{t} A_{st} f(R_s, R_t) \tag{5}$$

$$C(s) = \sum_{t} \frac{f(R_s, R_t)}{dist_{st}} \tag{6}$$

$$EB(e) = \sum_{s \neq t \in N} f(R_s, R_t) \frac{\sigma_{st}(e)}{\sigma_{st}} \tag{7}$$

$$B(v) = \sum_{s \neq v \neq t \in N} f(R_s, R_t) \frac{\sigma_{st}(v)}{\sigma_{st}} \tag{8}$$

In expressions (5)-(8), the function $f(R_s,R_t)$ is related to the intrinsic relevance, which allows weighting the shortest paths between couples of vertices [18] and it holds to several formulation, as:

$$f(R_s, R_t) = \begin{cases} (R_s + R_t)/2 \\ R_s \cdot R_t \\ \max(R_s, R_t) \\ R_s \\ R_{path} \end{cases} \tag{9}$$

3. The Digital Water Domain Analyzer service

The *DigitalWaterDomain_Analyzer* service has been developed consistently with the framework of digital transformation introduced by [19], to provide innovative processes as products for technicians and researchers. In more details, this service has been developed as QGIS plugin and it is designed for performing the network analysis using the reported metrics as described above.

It allows analyzing the DT of WDN by applying the tailored metrics cited above, irrespectively on hydraulic modelling of the system. The main purpose is to enable the understanding of the underlying network behavior which can be useful for various tasks in WDN analysis and management. For instance, it can support the identification of the main pipeline affecting WDN pressure status which deserve detailed calibration, or the identification of pipelines that are likely to have major impact on water supply service, e.g., for designing possible changes in diameters.

The nodes relevance represents an indicator of the main supplying nodes as well as the strategic consumers of the network (e.g., fire stations, hospitals), which cannot be identified by the classic node centrality metrics, as well as the most vulnerable junctions of the network, as they represent the vertices where a higher number of shortest paths pass through.

IOP Conf. Series: Earth and Environmental Science 1136 (2023) 012045 doi:10.1088/1755-1315/1136/1/012045

Users can load the virtual layer of the WDN, as generated by the WDNetXL platform, through the Graphical User Interface (GUI) in Figure 1 and visualize on the GIS map canvas the features and devices represented by the Digital Twin of the WDN, as shown in Figure 2. The user can choose several different types of calculations of the relevance function (Figure 3), as reported in Eq. (9), for the three metrics explained above, or using the classic approach, as expressed in equations (1) (2) (3) (4), and then performs the analysis and visualizes the results as separate layers.

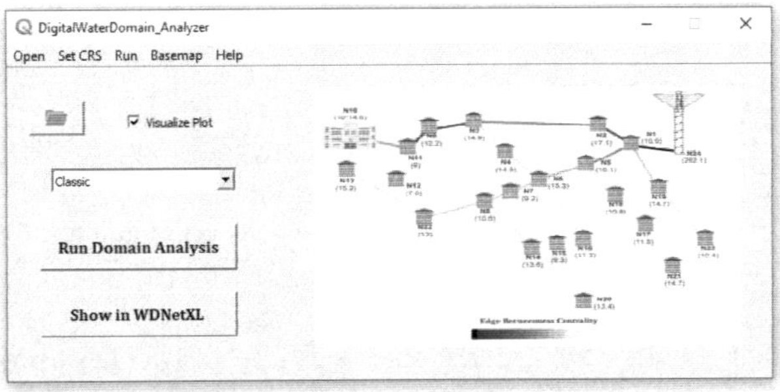

Figure 1. Graphical User Interface (GUI) of the *DigitalWaterDomain_Analyzer* service.

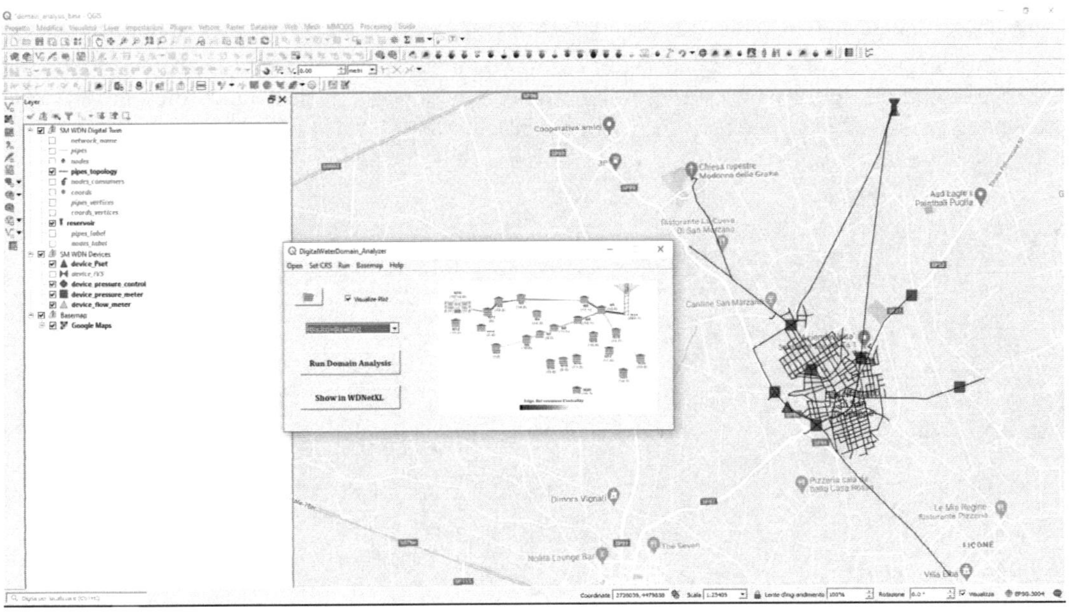

Figure 2. Visualization of WDN Digital Twin on QGIS map canvas.

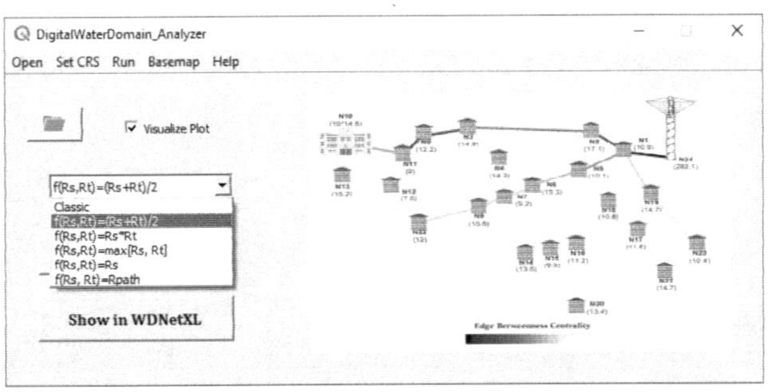

Figure 3. Combo box for the selection of the relevance formulation type.

4. Case Study

The *DigitalWaterDomain_Analyzer* service has been tested on a real WDN, whose model include 898 pipes, 729 nodes and one reservoir. The WDN model has been exported as Digital Twin by the WDNetXL platform and loaded into *DigitalWaterDomain_Analyzer* service. The analysis has been performed considering either the classic centrality metrics or the intrinsic relevance methodology for each type of formulation of $f(R_s, R_t)$, as in equation (9).

In this work, it has been considered the intrinsic-embedded metrics with the formulation $f(R_s, R_t) = (R_s+R_t)/2$ of the relevance only, because it was demonstrated to realistically render the role of the intrinsic relevance [18].

Figure 4, Figure 5, Figure 6 and Figure 7 report nodes and pipes ranking according to various metrics. In addition, as explained in [18], the intrinsic relevance for WDNs is assigned equal to the water demand for each vertex corresponding to a demand node and the sum of supplied water for the source node. Figure 4 reports the ranking of the intrinsic-embedded edge betweenness along the network, which shows higher values of metrics on main supplying paths.

Figure 5, which represents the distribution of nodal betweenness along the network, as described in equation (8), confirms that the intrinsic-embedded relevance betweenness provides consistent results with Figure 4.

Figures 6 and 7 show that the nodal degree and nodal harmonic formulations, as in equations (5) and (6), even accounting for intrinsic-embedded relevance are not able to provide useful information on WDN functioning. In more details, in nodal degree case, this happens because the WDN is quite looped and all nodes have very similar degrees and demands. In case of nodal harmonic formulation, this happens because the distance, in terms of shortest paths, form each node to the others is quite close among various nodes; therefore, the highest centrality metric value is assigned to the outlet node feeding the other network.

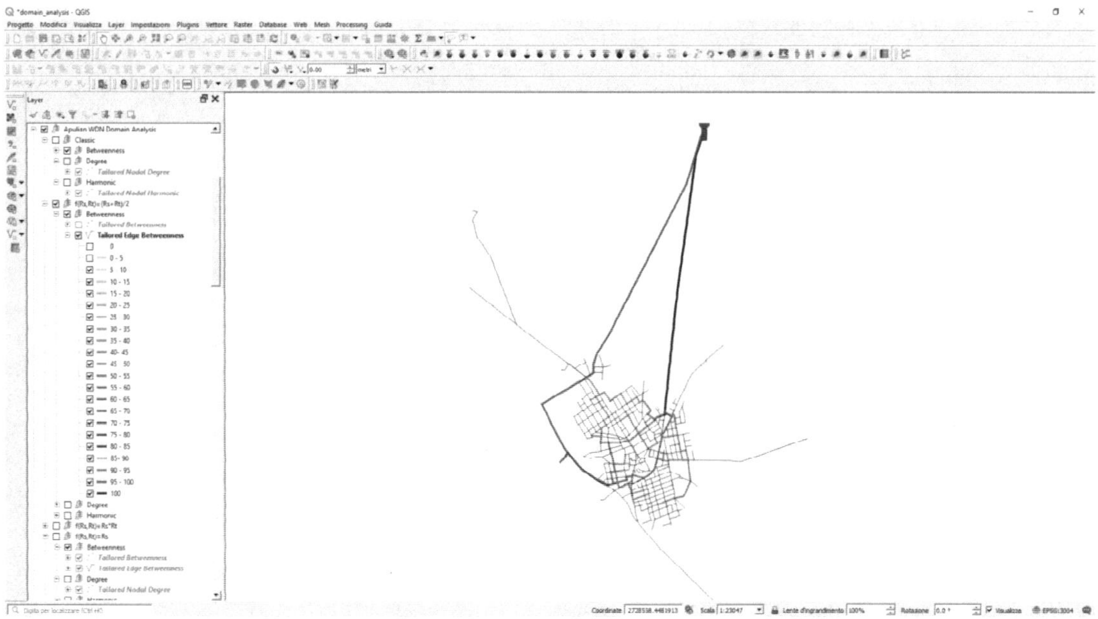

Figure 4. Visualization of the intrinsic-embedded edge betweenness for the case study.

Figure 8 reports the comparison between the results obtained using the edge betweenness calculated with the classic approach and the corresponding formulation with the intrinsic-embedded relevance.

The WDN-tailored edge betweenness (Figure 8b) correctly identifies the two main paths delivering water to the network, i.e., pipes linked to the reservoir, assigning the highest values of about 60% and 100%. Besides, using the intrinsic-embedded relevance formulation allows identifying a pipeline, represented as a red bold line on the bottom of the network, from where the analysed WDN feeds another network. This means that the water supplied to the other network (i.e., represented as a lumped demand in a single outlet node) increases the intrinsic relevance of that path.

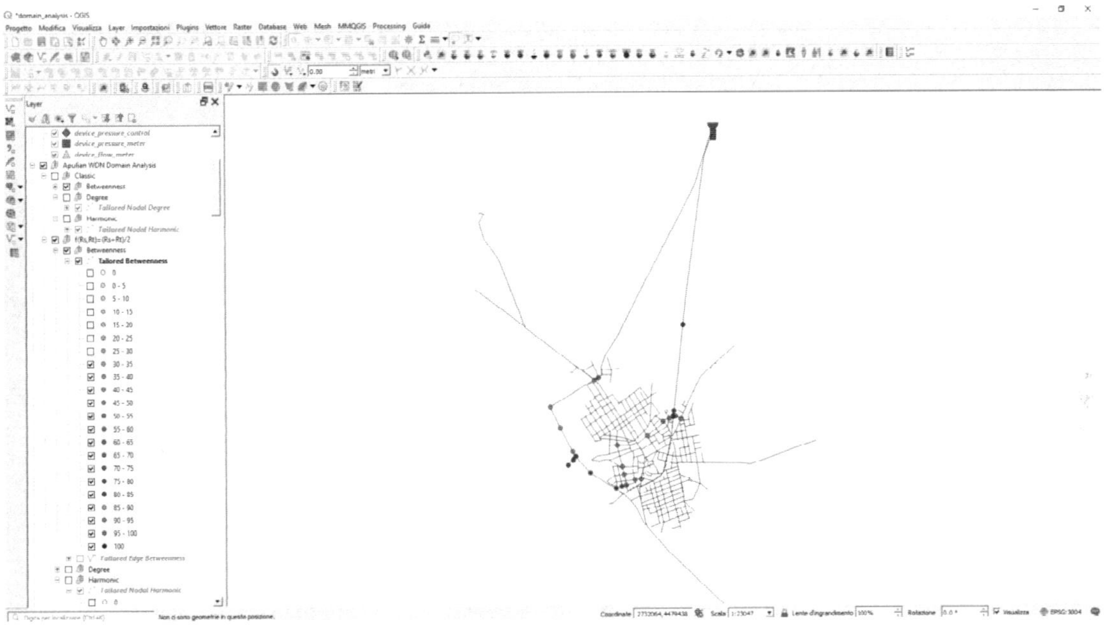

Figure 5. Visualization of the intrinsic-embedded relevance betweenness of nodes for the case study.

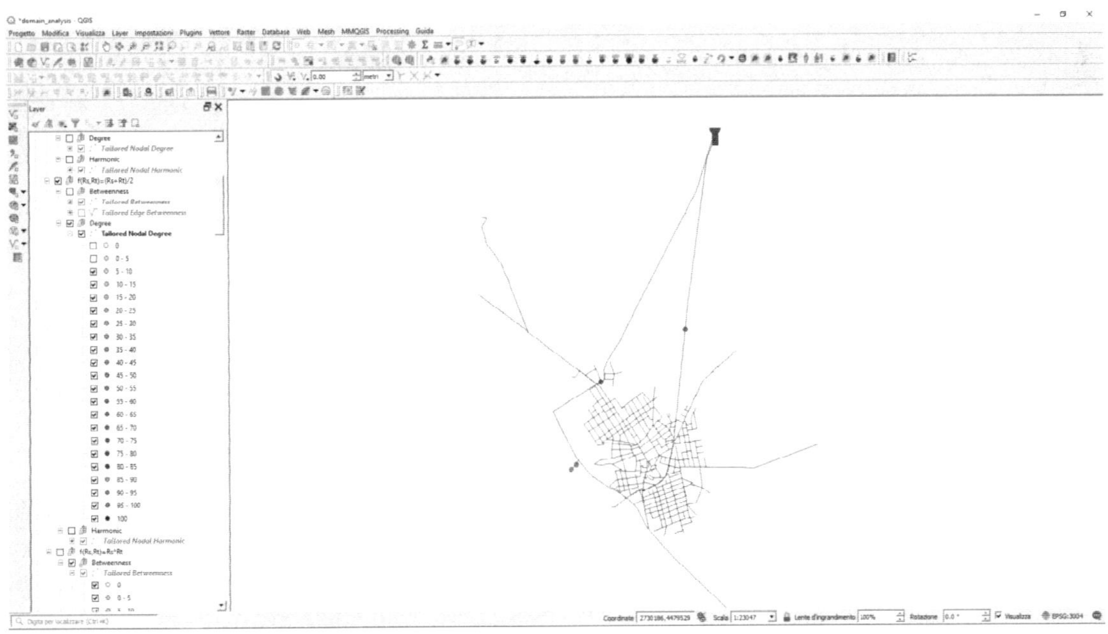

Figure 6. Visualization of the intrinsic-embedded relevance degree of nodes for the case study.

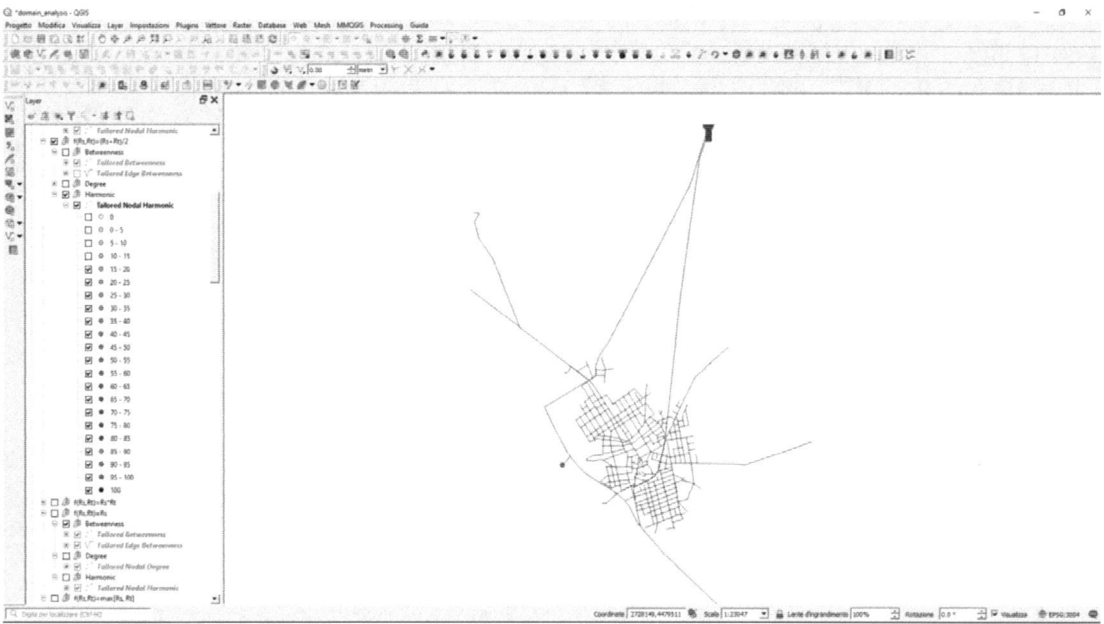

Figure 7. Visualization of the intrinsic-embedded relevance harmonic degree of nodes for the case study.

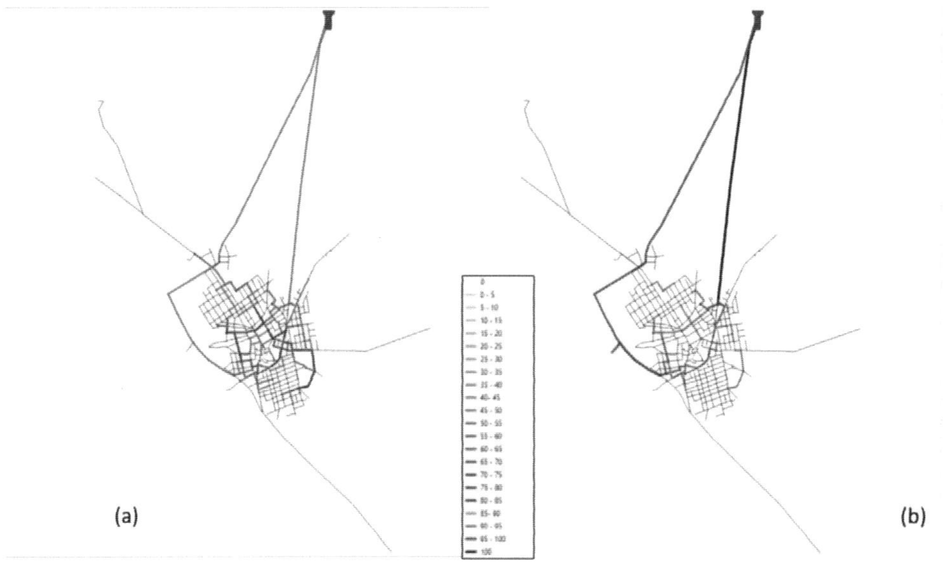

Figure 8. Edge betweenness using classic formulation (a) and the embedded-intrinsic relevance formulation (b).

Conversely, the classic edge betweenness (Figure 8a) cannot capture the real behaviour of the WDN, i.e, such type of network is highly looped and hence shows high redundancy, this turns to have a higher probability of the same shortest paths for pipes with different hydraulic conditions. Furthermore, the classic formulation, conversely to the intrinsic-embedded formulation of equation (8), neglects the distribution of demands.

5. Conclusions

In this work a novel DWS for the domain analysis of WDNs, *DigitalWaterDomain_Analyzer*, is presented. It has been exploited by the digital transformation paradigm [19] and encompasses the functionalities and methodologies taken from CNT for research transfer to technicians and water utilities.

The service, developed as a plugin for QGIS software, allows loading of the digital layer of WDN, performing several CNT analyses considering the classic formulation of the centrality metrics or the tailored with intrinsic relevance, and then visualizing the results in the GIS map canvas.

The results confirm the usefulness of the service for the WDN domain analysis: the evaluation of the centrality metrics can represent a strategic tool for the analysis and visualization of the main components of the WDN, emphasizing the role of the centrality metrics for a preliminary assessment of the WDN hydraulic behavior. Although these metrics cannot replace the hydraulic simulation, it can provide useful indications to drive model validation, model calibration, maintenance works planning as well as detecting the most relevant pipes for rehabilitation planning.

Acknowledgments

This paper was partly financed by European Union - PON Ricerca e Innovazione 2014-2020 E FSC, call D.D. 13 July 2017 n. 1735, Project "Energidrica - Efficienza energetica nella gestione delle reti idriche"(code ARS01_00625). Data for figures were provided by IDEA-RT (www.idea-rt.com).

References

[1] Strogatz SH 2011 Exploring complex networks *Nature* **410**(6825) 268–276
[2] Yazdani A and Jeffrey P 2010 Robustness and vulnerability analysis of water distribution networks using graph theoretic and complex network principles *Water Distribution Systems Analysis 2010* 933–945 https://doi.org/10.1061/41203(425)85
[3] Yazdani A and Jeffrey P 2011 Complex network analysis of water distribution systems. *Chaos: An Interdisciplinary Journal of Nonlinear Science* **21**(1) 016111
[4] Sitzenfrei R, Wang Q, Kapelan Z and Savić D 2020. Using complex network analysis for optimization of water distribution networks *Water Resources Research* **56**(8) e2020WR027929
[5] Laucelli DB, Simone A, Berardi L and Giustolisi O 2017 Optimal design of district metering areas for the reduction of leakages *Journal of Water Resources Planning and Management* **143**(6) 04017017
[6] Simone A, Ridolfi L, Berardi L, Laucelli D and Giustolisi O 2018 Complex network theory for water distribution networks analysis *EPiC Series in Engineering* **3** 1971–78
[7] Carrington PJ, Scott J and Wasserman S 2005 *Models and Methods in Social Network Analysis* (Cambridge University Press)
[8] Pons P and Latapy M 2005 Computing communities in large networks using random walks *Int. Symposium on Computer and Information Sciences* (Springer, Berlin, Heidelberg) 284–293
[9] Girvan M and Newman ME 2002 Community structure in social and biological networks *Proceedings of the National Academy of Sciences* **99**(12) 7821–26
[10] Barthélemy M 2011 Spatial networks *Physics Reports* **499**(1-3) 1–101
[11] Freeman LC 1977 A set of measures of centrality based on betweenness *Sociometry* 35–41
[12] Newman M 2018 *Networks* (Oxford University Press)
[13] Bonacich P 1987 Power and centrality: A family of measures *American Journal of Sociology* **92**(5) 1170–82
[14] Katz L 1953 A new status index derived from sociometric analysis *Psychometrika* **18**(1) 39–43
[15] Page L, Brin S, Motwani R and Winograd T 1998 The PageRank citation ranking: Bringing order to the web *Proc. 7th International World Wide Web Conference* 161–172
[16] Giustolisi O, Simone A and Ridolfi L 2017 Network structure classification and features of

water distribution systems *Water Resources Research* **53**(4) 3407–23
[17] Simone A, Ciliberti FG, Laucelli DB, Berardi L and Giustolisi O 2020 Edge betweenness for water distribution networks domain analysis *Journal of Hydroinformatics* **22**(1) 121–131
[18] Giustolisi O, Ridolfi L and Simone A 2020 Embedding the intrinsic relevance of vertices in network analysis: the case of centrality metrics *Scientific Reports* **10**(1) 1–11
[19] Ciliberti FG, Berardi L, Laucelli DB and Giustolisi O 2021 Digital Transformation paradigm for asset management in water distribution networks *2021 10th Int. Conf. Energy and Environment (CIEM)* Bucharest pp 1–5 https://doi.org/10.1109/CIEM52821.2021.9614864
[20] WDNetXL-WDNetGIS 2022 http://www.idea-rt.com/digital-water-latest-innovation
[21] Marchiori M and Latora V 2000 Harmony in the small-world *Physica A: Statistical Mechanics and its Applications* **285**(3-4) 539–546.

Determining the spatio-temporal relationship between water quality monitors in drinking water distribution systems

Killian Gleeson[1], Stewart Husband[1], John Gaffney[2] and Joby Boxall[1]

[1] The Department of Civil and Structural Engineering, The University of Sheffield, Sheffield, S10 2TN, United Kingdom
[2] Siemens UK, Manchester, M20 2UR, United Kingdom

kgleeson1@sheffield.ac.uk

Abstract. A novel method to both assess the strength of connectivity and determine hydraulic transit times between water quality monitors from time series data is reported. It was developed using a network of over 50 mobile multi-parameter sensors deployed for 18 months across a UK drinking water distribution system, and then validated using a network of 18 sensors from a different UK utility. Correlation coefficients are calculated at different time shifts for each possible sensor pair, with strength of connectivity represented by the highest correlation coefficient, and with the temporal lag of this highest correlation also designates the transit time. The results demonstrate the potential to derive valuable spatio-temporal information, with potential to increase understanding of system performance and connectivity. This information can be used to assist with further analytics such as tracking water quality events and improving hydraulic and disinfection residual decay modelling.

1. Introduction

Drinking water quality is typically monitored by periodic discrete sampling that fulfils regulatory purposes, but this can only provide limited information in understanding daily system performance or water quality deterioration processes that are known to occur between treatment and tap, such as discolouration [1] and residual disinfection decay. Technological advances now make it possible to deploy high-frequency (between 1 and 15 samples/minute) water quality monitors along drinking water distribution networks (DWDS), to monitor variables such as turbidity and residual free chlorine. The UK is currently pioneering the deployment of such sensor networks, largely driven by performance targets that include reducing discolouration customer contacts. The resulting data has focussed on identification and recording of events with little analysis to extract greater network understanding and hence inform network management.

There are several challenges and barriers that must be overcome to enable these datasets to be transformed into actionable information, many of which stem from the difficulties in obtaining good quality data. Water quality sensors measuring parameters like free chlorine and turbidity are sensitive scientific instruments that can produce erroneous data when deployed remotely within DWDS. This has led to poor quality data, limiting what can be done with analysis. Determining optimal sensor deployment strategy [2] and how best to analyse the subsequent datasets are also major challenges. To date, water quality analytics has mainly focused on event detection [3–5] and little work has been done to understand how the spatio-temporal combination of water quality sensor data can be used to enhance DWDS water quality data analysis.

The topic of how simultaneously recorded time series are related to each other spatio-temporally has been studied in areas such as seismology [6], astronomy [7], ultrasound imagine [8], and psychology [9]. Cross-correlation is the most commonly used method for determining the strength of relationship and time lag between two time series signals [10]. This involves shifting one time series relative to another and calculating a correlation coefficient at each step, with the step giving the highest correlation taken as the time lag. Pearson's correlation coefficient (PCC) is the most used coefficient as it measures the linear relationship between two variables. Many variants on cross correlation, such as detrended cross correlation analysis have been developed to deal with non-stationarity and the presence of unwanted periodicity [11].

The aim of this work is to demonstrate the suitability of applying cross-correlation analysis to DWDS water quality time series data, to determine strength of relationship and transit times between sensors. To achieve this, a method is developed to calculate the cross-correlation coefficient between water quality time series and is tested on multiple real-world UK DWDS datasets. As cross-correlation analysis is particularly susceptible to the presence of outliers and erroneous data such as flatline periods, suitable data quality assessment is needed for the correlations to be meaningful.

2. Method

A process involving sensor data quality checks [12] and subsequent cross-correlation analysis was written in Python, primarily using open-source library Pandas [13]. The flow chart in Figure 1 sets out the functionality of the code. This method was initially written to analyse a network of over 50 turbidity and chlorine multi-parameter sensors deployed for 18 months across a UK network. Data quality checks were developed for turbidity and chlorine data to overcome issues experienced by scientific monitors and remote communications from field deployed instruments. Cross-correlation is then used to determine the strength of relationship and transit time between two water quality sensors. These sensors must have sufficient good-quality data in common for the correlations to be meaningful, set as 50% of total window length for this work. This 50% commonality limit was selected to ensure that correlations were meaningful while also allowing for the long periods of missing or low-quality data experienced. PCC's are calculated at different time shifts for each possible sensor pair, with strength of connectivity represented by the highest correlation coefficient, and with the temporal shift of this highest correlation also designating the transit time. The transit time is only meaningful if the maximum PCC is sufficiently high. For this work, a threshold of 0.7 was used as any values above this are widely accepted to indicate a strong correlation [14].

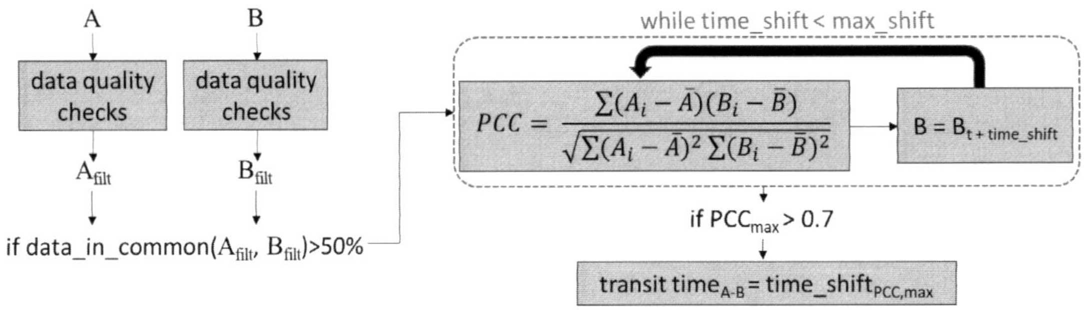

Figure 1. Process flowchart for calculating strength of connectivity and transit time between time series A and time series B.

2.1. Data quality assessment
A rules-based data quality assessment process was developed for turbidity and chlorine to detect the presence of specific anomalies and trends [12]. These are listed in Table 1, and result in the data being categorised into 'removed', 'flagged' or 'remaining' classes. Time stamp errors refer to datapoints that have an unintended sampling interval, compared to the previous datapoints, which can be problematic for analysis and may indicate malfunctioning instrumentation. For mobile sensors as used in this study, missing data is likely due to battery or communication issues, automated checks applying these rules can provide notification to minimise ongoing errors. Single point outliers refer to values that are unrepresentative relative to data before and after. These can occur in turbidity sensor data due to the presence of air bubbles or single highly reflective particles and as potentially unrepresentative are removed before further analysis. Flatlining data occurs when sensors return the same value repeatedly and would not be expected for sensitive instruments in a dynamic environment. Extended periods above a threshold can indicate a sensor error or external interference but could also be a real event so are flagged but not removed. For chlorine sensors, extended periods below a threshold can equally indicate a sensor error and so are flagged. Excessive noise was an issue identified as specific to the turbidity sensors during this trial when the sensor fluctuated between two distinct data points (considered unlikely for turbidity; as a result of data assessment analysis this was later identified as a power-cycle related fault). Drift can occur in turbidity sensors due to optical lens fouling from material accumulation or deterioration of membranes, usually manifesting itself in a slow gradual baseline increase over several weeks.

Table 1. Rules-based data quality assessment.

Turbidity	Chlorine	Removed/Flagged
Time stamp errors	Time stamp errors	Removed
Missing Data	Missing Data	Removed
Single Point Outliers		Removed
Flatlining Data	Flatlining Data	Removed
Extended periods above a threshold	Extended periods above a threshold	Flagged
Excessive noise	Extended periods below a threshold	Flagged
Drift	Drift	Flagged

An example of these rules applied to a turbidity and chlorine sensor is shown in Figure 2. As the legend indicates, any periods greater than 1 hour where the turbidity data was above 1 NTU were highlighted. Likewise, any periods greater than 1 hour where the chlorine data was below 0.2 or above 1 mg/l were highlighted. Any periods where the value did not change for at least 6 hours were highlighted as flatlining data. In this example, this sensor produced very little useful data despite monitoring for over a year, highlighting the multiple issues that can occur. From these results later investigation identified problems with the monitor fitment and sample line flows.

2.2. Cross-correlation analysis
Figure 3 demonstrates how cross-correlation can be used to determine the transit time between two sensors. The top plot shows two synthetic time series, over the course of four days. Time Series B is a copy of Time Series A, with a lag of 4 hours and an offset applied. The bottom plot displays the cross-correlation curve, the peak of which is the time shift which results in the strongest correlation. The maximum correlation coefficient was found to occur for a time shift of 4 hours (indicated by the dotted red vertical line). This method was tried out on multiple real water quality parameters, to determine which are most suited, and the calculated transit times were inspected visually on the time series data. In some cases, high quantities of missing data meant that there was insufficient data in common between two sensors for the cross-correlation to be meaningful. To prevent this, a 50% total

window length commonality was added as a process step following data quality assessment as shown in Figure 1. If the two time series are over a year in length, cross-correlation analysis was done over shorter monthly periods and reporting an average result for the entire length. This is done to avoid the correlations being dominated by seasonal trends shared by many unrelated locations, with shorter time frames more likely to produce correlations that are meaningful in terms of hydraulic connectivity.

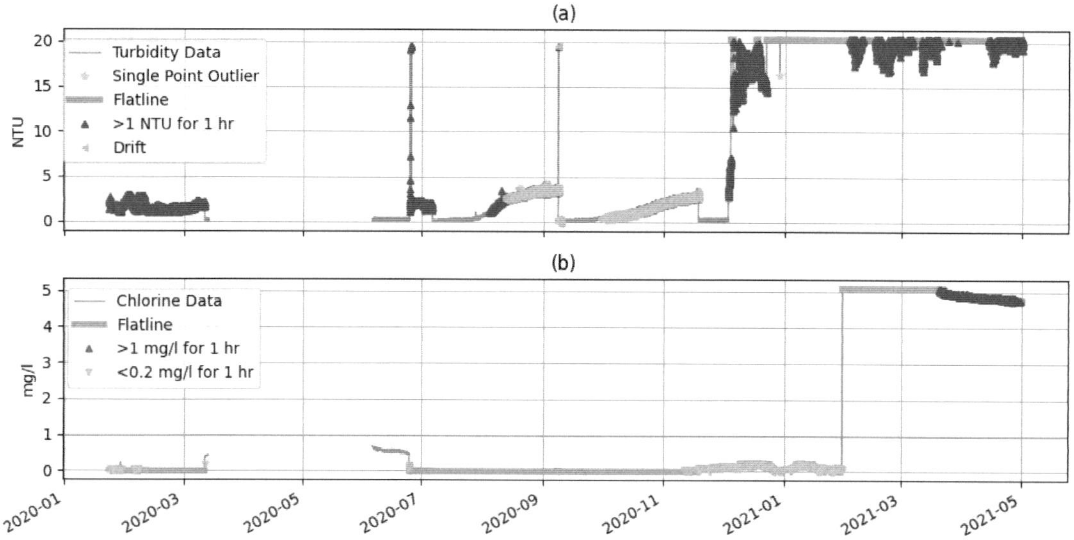

Figure 2. Data-quality rules applied to turbidity (a) and chlorine (b) time series.

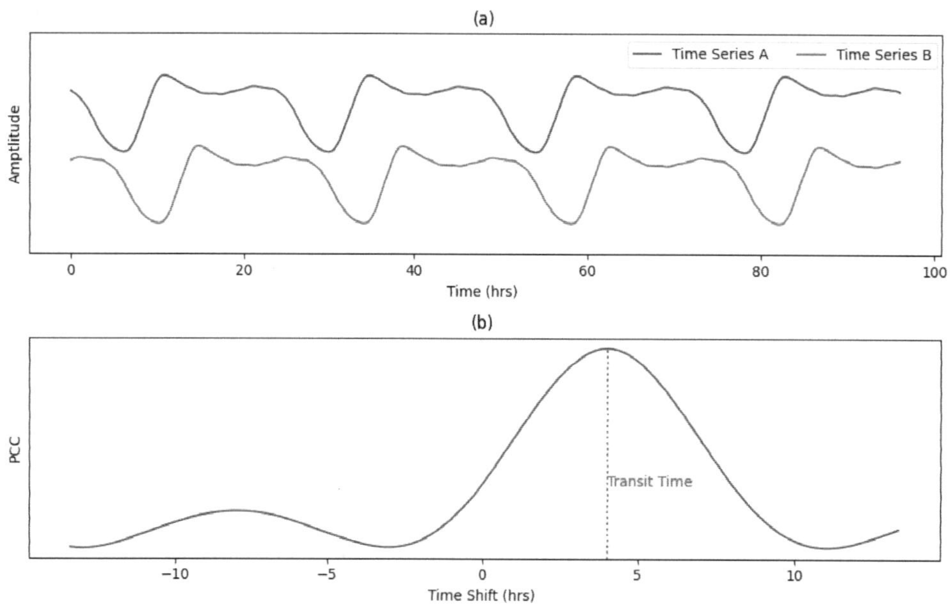

Figure 3. Two example time series (a) and their corresponding cross-correlation curve (b).

IOP Conf. Series: Earth and Environmental Science 1136 (2023) 012046 doi:10.1088/1755-1315/1136/1/012046

3. Results

Cross-correlation analysis using PCC was done for over 60 sensors, across 2 DWDS from different UK water utilities. Selected results are presented in figures 4 and 5.

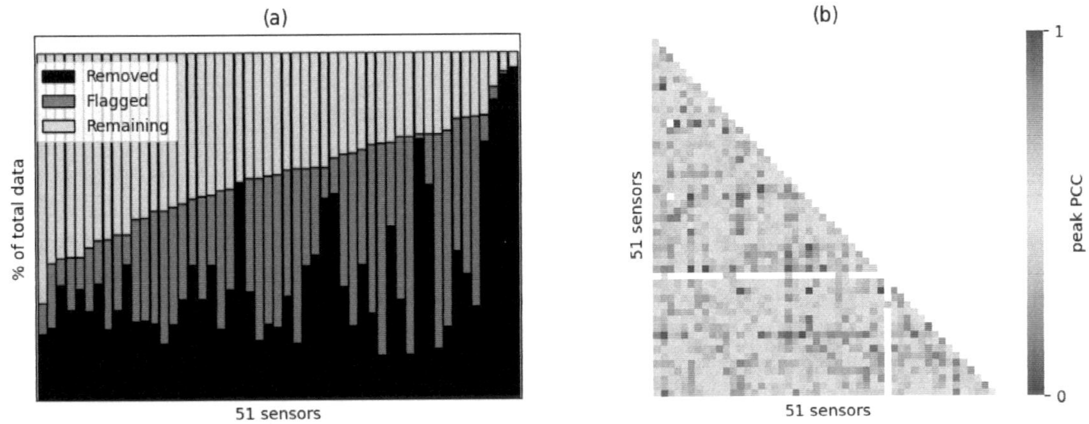

Figure 4. (a) Bar chart showing data quality rules applied to 51 sensors; (b) Heatmap showing peak PCC for same set of 51 sensors.

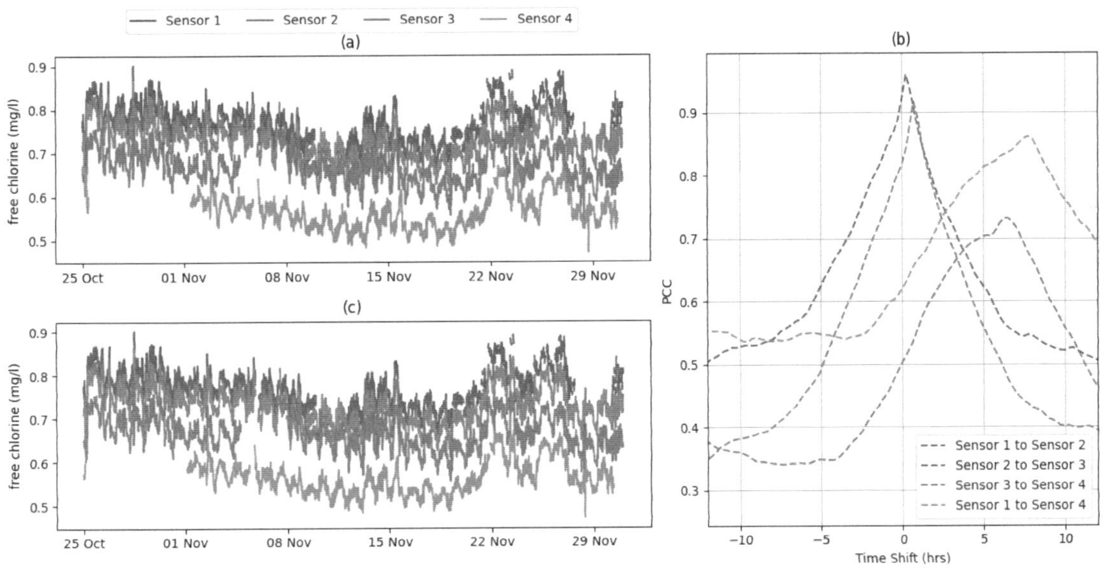

Figure 5. (a) Four interconnected chlorine time series; (b) Cross-correlation curves between the four chlorine sensors; (c) Sensors 2, 3 and 4 synced relative to sensor 1 using the calculating transit times.

Figure 4(a) shows data quality from 51 sensors deployed over 18 months, with 4(b) the corresponding heatmap showing the highest calculated PCC coefficient between each possible sensor pair, with red indicating sensors that were highly correlated and blue weakly correlated. The deployment of instrumentation was not bespoke for this research, but usefully involved deploying multiple sensors across a network resulting in a dataset of variable data quality. After performing the data quality assessment stage for the entire set of time series data, cross-correlation analysis was done for each month of data, with median PCC values used to obtain the results presented here. Chlorine was the parameter used for this analysis and Figure 4b indicates which sensors are hydraulically connected to each other and is a good starting point before more detailed cross-correlation analysis is done to determine transit times in specific network sections. The empty cells indicate sensor pairs that did not have at least 50% of data in common, after erroneous data was removed, with one sensor having very little data. As the flow chart in Figure 1 explains, transit times are only taken to indicate connectivity and useful information for PCC above 0.7, but the heatmap allows for all PCC's to be visualised. Despite the data quality issues with this dataset, cross-correlation analysis was still able to identify sections of strong connectivity and transit time information.

Figure 5 presents results from four sensors that were found to be closely interlinked through the initial cross-correlation analysis presented in figure 4. The upper left plot shows the four chlorine time-series profiles for just over a month, with the bottom left plot showing the effects of time-syncing the time-series relative to Sensor 1. The right-hand side plot shows the cross-correlation curves for each section, with the transit times calculated to be 15 minutes (Sensor 1 to Sensor 2), 45 minutes (Sensor 2 to Sensor 3) and 6 hours, 45 minutes (Sensor 3 to Sensor 4). As these sensors were sampled every 15 minutes, the transit times cannot be calculated to a higher degree of precision. Though there was some missing data in these sensors, there was enough data commonality (i.e. > 50%) to accurately calculate the cross-correlation.

Figure 6 presents data from a second different set of instrument deployment elsewhere in the UK. The utility had 18 monitors deployed and following this analysis it was found many had data quality issues and some incorrectly sited such that they were not hydraulically connected as planned.

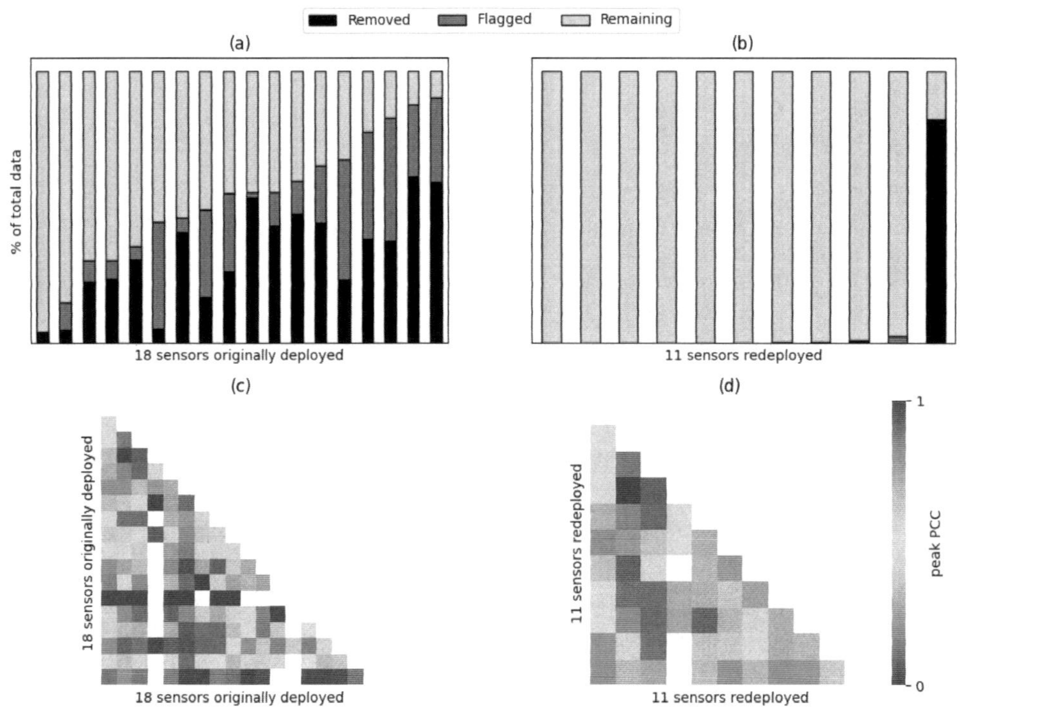

Figure 6. (a) Bar plot showing data quality of 18 sensors originally deployed over course of 12 months; (b) Bar plot showing data quality from 2 months of 11 sensors being redeployed following analysis; (c) Heatmap showing median monthly peak PCC values between 18 sensors originally deployed over 12 months; (d) Heatmap showing peak PCC values in 2 months between 11 sensors redeployed after analysis.

Based on initial analysis and the results in figure 6(a), it was decided to revise the project with a more focused monitor deployment, consisting of 11 monitors in a believed-to-be connected network section. As figure 6 shows, using data from a 12-month window pre and a 2-month period post the focussed sensor deployment, data quality was greatly improved, as fewer sensors now needed to be maintained. For the initial 12-months of data, monthly cross-correlations were done, with median PCC values reported for each pair. Of the connectivity illustrated by the second cross-correlation heatmap (figure 6(d)), there is an overall improvement, and 5 sensors can be observed as highly connected. The remaining 6 sensors were less connected, with 2 redeployed in locations considered of specific water company interest but with this analysis indicating that these sites display distinctly different behaviour. This demonstrates a benefit of cross-correlation analysis to determine if monitors deployed at correct location and if system connectivity is as expected.

4. Discussion

Cross-correlation is shown as an effective method for determining strength of connectivity and hydraulic transit times between water quality sensors. It relies on there being a detectable pattern in each time series. If the time series profile is too flat, the correlation will be difficult to compute due to noise. The method is sensitive to outliers so careful pre-processing and data preparation is required. Data quality checks are an essential prior step to cross-correlation, with outliers and erroneous data such as periods of flatlining removed so they do not negatively impact the correlations. This is particularly essential for water quality monitors which are prone to measurement errors when deployed within DWDS. If a single time series has a strong autocorrelation (i.e., correlates with itself on some time lag, often a diurnal pattern) then the transit time can also not be trusted.

The lessons learned from the initial 51-sensor deployment were taken and applied to the 18-sensor deployment in a different UK utility. The results shown in figures 5 and 6 show how an initial cross-correlation can be applied to large datasets, with the results allowing for a more detailed investigation into connected network sections. Figure 4 demonstrates how a combination of effective data quality assessment and cross-correlation can be used to enhance not only the data quality achieved but the amount of inter-connected high-quality data. This enables a move from single sensor analysis to multi-sensor analysis that can diagnose entire network sections.

Window size is an important consideration, and this work has shown that month long window sizes were effective. This protects against seasonal trends dominating each correlation, while also making the process more robust against periods of erroneous data that have not been removed by the data quality assessment stage. Even short erroneous periods could interfere with a correlation between long time series but would only appear in one of the smaller windows. Shorter windows also allow for changes in transit times over time to be looked at. Though not done in this work, overlapping windows like those used in detrended cross correlation analysis could be employed to increase the temporal resolution of cross-correlation outputs even further.

The methods developed in this work are designed to be agnostic to sensor manufacturer but depends heavily on what parameter is considered. Chlorine is a well-suited parameter for cross-correlation as it retains a similar time series profile, even several hours downstream. Turbidity was not as well-suited due to their being strong local variations even in locations very closely linked. Fortunately, chlorine is a key parameter for DWDS water quality monitoring with companies wanting to optimise amount dosed (cost and concerns with disinfection by-product formation) whilst aiming to retain a residual. The absolute value of chlorine is dictated by the quality and frequency of calibration, but cross-correlation is affected by patterns not absolute values. This gives it a unique data quality requirement that focuses on finding and removal erroneous sub-sequences rather. This work shows that detrending may not be necessary for cross-correlation to be effective on chlorine time series data, but the use of overlapping windows could help understand how transit times are changing, while also reducing the influence of erroneous data. The availability of flow data could be used to explain transit time variances seen for different time periods. For example, increased transit times between sensors could be explained by a drop in average flow rates.

The connectivity and transit time results can help bridge the current gap between data and actionable information. Examples include:

- Identification of network connectivity, determined through strength of correlation. Useful as utilities can be surprised by which locations are not actually connected and have entirely different water quality profiles.
- Improved sensor data quality assessment, which is vital for the successful deployment of such sensitive traditionally lab-based instruments in remote locations. By comparing connected sensors, erroneous data can be identified and separated from genuine events with greater accuracy.
- Derived transit time information could be used to improve the calibration accuracy of DWDS hydraulic models, typically calibrated using pressure data. This could be particularly useful when adding water quality functionality to hydraulic models, as higher standards of calibration would be required [15].
- Use in characterisation of discolouration events. For example, an event could be described as local to a specific sensor, or global and seen by multiple sensors. Knowing the connectivity and transit times is necessary to be sure about such conclusions. Global events that travel through the network can be assessed with knowledge of hydraulic transit times, which could help in characterising an event.
- Calculation of disinfection residual decay rates for different network sections. As this is difficult to accurately model [16], decay rates for specific sections can be determined, potentially highlighting regions with excessive chlorine decay and improving disinfection decay modelling.

5. Conclusions

This work demonstrates how cross-correlation between water quality sensors can be used to identify DWDS spatio-temporal connectivity and transit times. Data quality assessment is an essential first step to ensure the cross-correlation results are meaningful. The results show that chlorine is a well-suited parameter for such cross-correlation analysis. The derived connectivity and transit time information can be used to enhance data quality assessment and water quality event tracking; and can improve hydraulic and disinfection residual decay modelling.

Acknowledgments

This research has been supported by an Engineering and Physical Sciences Research Council (EPSRC) studentship as part of the Centre for Doctoral Training in Water Infrastructure and Resilience (EP/S023666/1) with support from industrial sponsor Siemens UK, Anna Taliana and Paul Gaskin of Welsh Water, Jez Downs of Southern Water, and Derek Leslie of ATi.

References

[1] Husband PS and Boxall JB 2011 Asset deterioration and discolouration in water distribution systems. *Water Res.* **45**(1) 113–24

[2] Klise KA, Nicholson BL and Laird CD 2017 *Sensor Placement Optimization using Chama* [Internet] (Albuquerque, NM and Livermore, CA; United States) available from: http://www.osti.gov/servlets/purl/1405271/

[3] Hart D, McKenna SA, Klise K, Cruz V and Wilson M 2007 CANARY: A water quality event detection algorithm development tool *Proc. World Environmental and Water Resources Congress 2007* [Internet] (Reston, VA: American Society of Civil Engineers) pp 1–9 available from: http://ascelibrary.org/doi/10.1061/40927%28243%29517

[4] Klise KA and McKenna SA 2008 Multivariate applications for detecting anomalous water

quality *Proc. Water Distribution Systems Analysis Symposium 2006* [Internet] (Reston, VA: American Society of Civil Engineers) pp 1–11 available from: http://ascelibrary.org/doi/abs/10.1061/40941%28247%29130

[5] Mounce S, Machell J and Boxall J 2012 Water quality event detection and customer complaint clustering analysis in distribution systems *Water Sci. Technol. Water Supply* **12**(5) 580–7

[6] Vandecar JC and Crosson RS 1990 Determination of teleseismic relative phase arrival times using multi-channel cross-correlation and least squares *Bull.-Seismol. Soc. Am.* **80**(1) 150–69

[7] Peterson BM, Wanders I, Horne K, Collier S, Alexander T, Kaspi S et al 1998 On uncertainties in cross-correlation lags and the reality of wavelength-dependent continuum lags in active galactic nuclei *Publ. Astron. Soc. Pacific* **110**(748) 660–70 http://iopscience.iop.org/article/10.1086/316177

[8] Bonnefous O 1986 Time domain formulation of pulse-Doppler ultrasound and blood velocity estimation by cross correlation *Ultrason. Imaging* **8**(2) 73–85 https://linkinghub.elsevier.com/retrieve/pii/0161734686900015

[9] Boker SM, Rotondo JL, Xu M and King K 2002 Windowed cross-correlation and peak picking for the analysis of variability in the association between behavioral time series *Psychol. Methods* **7**(3) 338–55 http://doi.apa.org/getdoi.cfm?doi=10.1037/1082-989X.7.3.338

[10] Benesty J, Chen J and Huang Y 2004 Time-delay estimation via linear interpolation and cross correlation *IEEE Trans. Speech Audio Process* **12**(5) 509–19 http://ieeexplore.ieee.org/document/1323087/

[11] Horvatic D, Stanley HE and Podobnik B 2011 Detrended cross-correlation analysis for non-stationary time series with periodic trends *EPL (Europhysics Lett.)* **94**(1) 18007 https://iopscience.iop.org/article/10.1209/0295-5075/94/18007

[12] Gleeson K, Boxall J, Husband S and Gaffney J 2021 Automated data quality assurance for water quality sensors in drinking water distribution systems *Proc. Water Quality Technology Conf.* (Tacoma, WA: AWWA)

[13] McKinney W 2010 Data structures for statistical computing in Python *Proc. Python in Science Conf.* (Austin, Texas: SciPy)

[14] Schober P, Boer C and Schwarte LA 2018 Correlation coefficients *Anesth. Analg.* **126**(5) 1763–1768 http://journals.lww.com/00000539-201805000-00050

[15] Boxall JB, Saul AJ and Skipworth PJ 2004 Modeling for hydraulic capacity *J. Am. Water Works Assoc.* **96**(4) 161–9 https://onlinelibrary.wiley.com/doi/10.1002/j.1551-8833.2004.tb10607.x

[16] Speight V and Boxall J 2015 Current perspectives on disinfectant modelling *Procedia Eng.* **119** 434–41 https://linkinghub.elsevier.com/retrieve/pii/S187770581502576X

IOP Conf. Series: Earth and Environmental Science 1136 (2023) 012047 doi:10.1088/1755-1315/1136/1/012047

Predicting iron exceedance risk in drinking water distribution systems using machine learning

Ehsan Kazemi[1], Grigorios Kyritsakas[1], Stewart Husband[1], Katrina Flavell[2], Vanessa Speight[1] and Joby Boxall[1]

[1] Department of Civil and Structural Engineering, The University of Sheffield, Mappin Street, Sheffield S1 3JD, UK
[2] Yorkshire Water Services Limited, Western House, Halifax Road, Bradford BD6 2SZ, UK

e.kazemi@sheffield.ac.uk

Abstract. A Machine Learning approach has been developed to predict iron threshold exceedances in sub-regions of a drinking water distribution network from data collected the previous year. Models were trained using parameters informed by Self-Organising Map analysis based on ten years of water quality sampling data, pipe data and discolouration customer contacts from a UK network supplying over 2.3 million households. Twenty combinations of input parameters (network conditions) and three learning algorithms (Random Forests, Support Vector Machines and RUSBoost Trees) were tested. The best performing model was found to be Random Forests with input parameters of iron, turbidity, 3-day Heterotrophic Plate Counts, and high priority dead ends per District Metered Area. Different exceedance levels were tested and prediction accuracies of above 70% were achieved for UK regulatory concentration of 200 µg/L. Predicted probabilities per network sub-region were used to provide relative risk ranking to inform proactive management and investment decisions.

1. Introduction

It is recognised that increased iron concentrations in potable water are associated with customer contacts for discoloured water. Past research indicates that high iron concentrations in drinking water distribution systems (DWDS) are related either to insufficient water treatment or to corrosion phenomena in distribution pipes [1]. While there has been progress understanding the role of iron concentrations and discolouration, most of these phenomena are addressed by water utilities in a reactive way. This is mainly due to the cost issues – such as that water utilities have to focus activities where the reward (or more likely penalty avoided) is greatest – as well as the complex interactions and processes occurring within DWDS and the vast scale and individual complexity of networks together with a lack of predictive tools. It is therefore important to research and develop methods that when applied could transform this reactive approach into a proactive one by providing information that could inform decision making regarding strategical interventions and investment within DWDS.

Water utilities collect a vast number of discrete samples from their customer taps to measure various water quality parameters and regulatory compliance. While this is a huge effort and creates large datasets it is still sparse (spatially and temporally) data compared to the size and complexity of DWDS. Prior work on discolouration that applied machine learning (ML) methodologies to this kind of data indicated that these methods have the potential for both informing the causes of discolouration and estimating the risk of iron exceedance in the DWDS [2, 3].

In this paper, a classification ML methodology for the prediction of iron exceedance in DWDS sub-regions of a water utility in the UK is presented. The methodology uses as inputs the annual averages of various water quality parameters in previous years at the district meter area (DMA) scale for the prediction of iron exceedance risk in DMAs for the following year. In addition, the methodology provides information regarding the probability of iron exceedance per DMA and thus facilitates creating a DMA risk ranking that can direct and prioritise interventions.

2. Data and approach

The datasets used in this study were obtained from a large drinking water distribution network in the UK, owned and operated by Yorkshire Water, supplying over 2.3 million households collected over a 10 year (pre-COVID) period. The dataset includes 134,803 regulatory water quality samples (including iron, manganese, turbidity, aluminium, chlorine and many others, but not always the same parameters for every sample), 62,695 water quality customer contacts, disinfection type per Water Supply Zone (WSZ) and static asset pipe data (material, diameter, length, and age).

Firstly, a qualitative analysis using an unsupervised ML technique called Self-Organising Map (SOM) was used to analyse the raw data and thus identify the major relationships and key parameters influencing iron exceedance. Then, a risk model was developed based on supervised ML algorithms to quantify the risk of iron exceedance in the network. This was performed by mapping input parameters (identified by the SOMs as the major factors influencing iron exceedance) to the output parameter (the occurrence of iron exceedance) and then predicting probability of exceedance.

3. Identification of key parameters

SOM is a type of data clustering technique, suitable for sparse and incomplete data, especially when the relationships within data are complex and highly non-linear [4]. It uses unsupervised learning algorithms to train the model, i.e., all parameters are fed into the model as input, and through producing a two-dimensional representation of the high dimensional data set (due to multiple variables), the linkages between them are qualitatively and visually investigated.

Table 1. Variables used in the SOM analysis example and as predictors in the risk model.

Variable (unit)	Data source	Comments
Iron (μg/l)	Water quality data	Regulatory samples
Manganese (μg/l)	Water quality data	Regulatory samples
Total chlorine (mg/l)	Water quality data	Regulatory samples
Turbidity (NTU)	Water quality data	Regulatory samples
3-day heterotrophic plate count (HPC) (no/ml)	Water quality data	Regulatory samples
Number of customer contacts per DMA per year	Customer contact data	-
DMA-clustered customer contacts per DMA per year	Calculated	Each cluster includes a minimum of 5 customers contacts per DMA over a minimum of a 2-day period. This parameter is the sum of these clusters per DMA per year
High priority dead-ends per DMA	Standalone data	Total number of dead-end pipes that, according to water utility's estimations, are at higher risk of discolouration per DMA

To uncover the important relationships in the present dataset and identify the key parameters, various combinations of the parameters introduced in Section 2 were tested with SOM. It was found that, for example, the level of iron within samples is not linked to properties of the pipe (material, diameter, age) closest to the location of the property where the water quality sample was collected and that the level of iron within samples is linked to iron, manganese, turbidity, 3-day heterotrophic plate count (HPC), low chlorine, residual type (chlorine and chloramine), number of customer contacts per District Metered Area (DMA) per year, the DMA clustered customer contacts per DMA per year, and high priority dead ends per DMA. These parameters were thus suggested to be used in the risk model to predict iron exceedance. Some of the parameters used in the SOM analysis are presented in Table 1 and an example output is shown in figure 1.

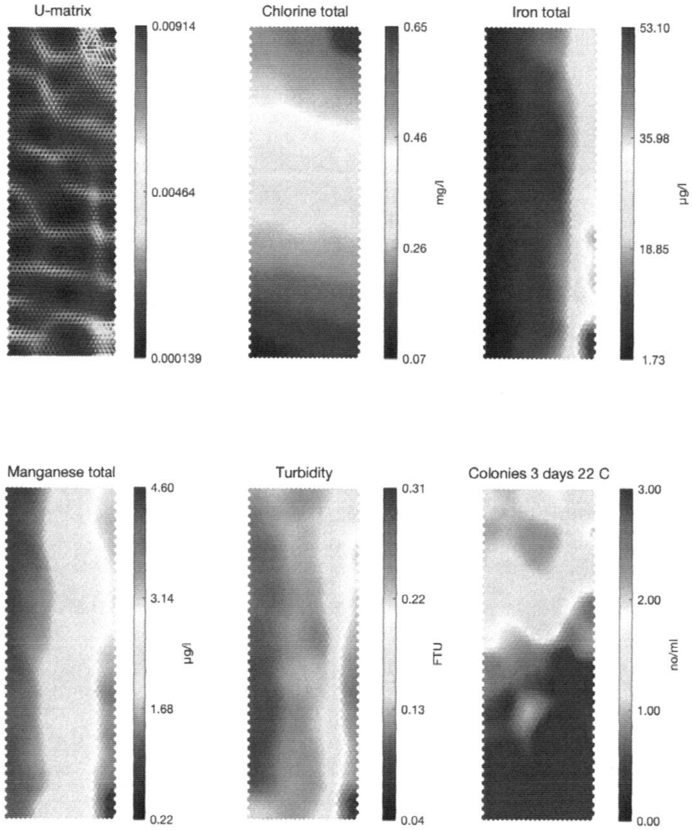

Figure 1. An example SOM developed in this study. Lattices from top left represent the U-matrix, chlorine, iron, manganese, turbidity, and 3-day HPC, respectively.

Figure 1 shows one example of the many SOMs performed in this study. On the lattices, each hexagonal cell (neuron) represents a group of clustered observations; the spatial location of a cell in a topographic map corresponds to a particular domain or feature drawn from the input data; colours show the value of the variables (red: high, blue: low) and each cell in the same position on different maps corresponds to the same cluster of observations/data. In addition to the maps of parameters, an additional lattice called unified distance matrix, or 'U-matrix' is provided (see Figure 1, top left) which shows the strength of the clusters. Blue areas on the U-matrix correspond to the clusters and red lines correspond to where the clusters are separated. This example shows that iron is strongly linked with manganese and turbidity and that high iron is also correlated with high 3-day HPC numbers and low chlorine concentrations in the customers' taps.

4. Risk model

Three supervised ML models were developed from classification Random Forests (CRF) [5], Support Vector Machines (CSVM)[6], and Boosted Trees (CBT) learning algorithm based on Random Undersampling Boosting (RUSBoost) [7]. These models were selected as they provide 'white box' approaches, such that the role of different parameters in the predictions can be understood and appreciated by end users aiding acceptance and uptake. The analysis was performed at a DMA scale, with a one-year time lag between input and output. The parameters identified by the SOMs as important were averaged per DMA per year and used as input parameters; and occurrence of iron exceedance above a certain threshold in DMAs in a year ahead was set as the output parameter. The output parameter consists of two classes of 'exceedance (E)' and 'non-exceedance (N)', referring to whether there was at least one event of iron exceedance above the threshold in the DMA, or not.

The data of years 2009 to 2018 was employed to train the models, and the data of 2019 was used as a validation set to test the performance of the model in predicting unseen data. For assessing the performance of the models and the accuracy of the predictions, True Positive Rate (*TPR*), True Negative Rate (*TNR*), Accuracy (*ACC*), and Matthews Correlation Coefficient (*MCC*) were measured as defined in the following equations.

$$TPR = \frac{TP}{TP + FN} \tag{1}$$

$$TNR = \frac{TN}{TN + FP} \tag{2}$$

$$ACC = \frac{TP + TN}{TP + TN + FP + FN} \tag{3}$$

$$MCC = \frac{TP \times TN - FP \times FN}{\sqrt{(TP + FP)(TP + FN)(TN + FP)(TN + FN)}} \tag{4}$$

where *TP*, *TN*, *FP* and *FN* are True Positive, True Negative, False Positive and False Negative, respectively. Positive and Negative denote exceedance and non-exceedance events, respectively. For example, True Negative means actual observational data is a non-exceedance (Negative), and the model predicted it correctly (True). *TPR* represents the probability that the model will correctly predict positive class values (exceedances); *TNR* represents the probability that the model will correctly predict negative class values (non-exceedances); *ACC* is the proportion of samples, positive or negative, predicted correctly; and *MCC* is a measure of summarising performance even when there is a skew in class sizes.

In the following, the ML models are compared (using a threshold of 150 μg/L to define the output parameter, iron exceedance). Then, the model which performs best for the current dataset is used to investigate various effects such as bias in the data (imbalance between exceedance and non-exceedance events), various combinations of input parameters, and exceedance level threshold. Finally, the best configuration is employed to estimate risk of iron exceedance. All the calculations were performed in MATLAB 2019b.

4.1. Comparison of ML models

CRF is an ensemble learning method for classification which combines a collection of decision trees. It improves the prediction by introducing a split on a random subset of features, i.e., it randomly selects observations and variables to build multiple decision trees and then averages all the trees. This makes it robust in dealing with non-linear data with outliers. CSVM uses classification algorithms to find a hyperplane in a multi-dimensional space (due to multiple variables) that distinctly classifies the data points into two groups. It has the advantages of high speed and good performance with datasets where a limited number of observations is available. CBT-RUSBoost is a type of decision trees suitable for imbalanced datasets. It uses boosting algorithms combined with a random under-sampling method to reduce the size of majority class and thus improve the performance of the model.

The three ML models are compared in terms of training (with the data of 2009 to 2018) and prediction of unseen data (data of 2019), and their performance is summarized in Table 2. CSVM shows very poor performance. It calculated most of the events as non-exceedance (high *TNR* and low *TPR* in training, and *TPR* of zero in validation) due to the highly unbalanced data. Therefore, it is not considered as a suitable method for the present application. RUSBoost and CRF algorithms, although showing better performance than CSVM, still performed poorly, especially in terms of *TPR* and *MCC*. This is also due to the high imbalance in the size of classes of the output parameter (rare exceedance events). The total number of DMA yearly data used in this analysis is 17,507. With an exceedance threshold of 150 μg/L, only 250 of these belong to class 'E'. This indicates that the bias in the data, if defined as M_N/M_E, where M_N = number of non-exceedances, and M_E = number of exceedances, is equal to 69, a situation which cannot be handled by ML algorithms, even with the RUSBoost which is specifically designed for imbalanced datasets.

Table 2. Performance metrics of training of the ML models (with data of 2009 to 2018) and testing it with the validation set (data of 2019). 'NaN' indicates 'Not a Number'.

	Training set			Validation set		
ML model	*TPR*	*TNR*	*MCC*	*TPR*	*TNR*	*MCC*
CSVM	0.127	1.0	0.354	0	1	NaN
RUSBoost	0.697	0.739	0.118	0.440	0.819	0.075
CRF	0.667	0.769	0.123	0.360	0.864	0.073

To overcome the issue, two methods were explored: i) use of synthetic data to improve the balance between exceedance and non-exceedance classes by creating examples of the former; and ii) a random removal of the non-exceedance class events. Due to concerns that the linear extrapolation in the synthetic data generation may be distorting the true underlying patterns in the data it was not considered appropriate; therefore, the random removal of the non-exceedance events was employed for the present analysis.

4.2. Bias in data

The bias was reduced by random down sampling of the majority class (non-exceedances), i.e., reducing M_N. Six tests were performed with M_N/M_E ranging from 69 (original data) to 1 (complete removal of skew).

Tables 3 and 4 present the *TPR* and *MCC* calculated by the CRF and RUSBoost models for training and validation sets, respectively.

Table 3. *TPR* of training and validation of the CRF and RUSBoost models.

		Training set		Validation set	
M_N/M_E	N_t	RUSBoost	CRF	RUSBoost	CRF
69	17507	0.697	0.667	0.440	0.360
10	2783	0.771	0.709	0.423	0.385
6	1771	0.822	0.797	0.412	0.471
4	1265	0.863	0.783	0.630	0.519
2	759	0.667	0.714	0.421	0.421
1	506	0.395	0.649	0.475	0.480

As the imbalance in the data decreases, the size of data (N_t) also decreases, but performance of both models improves. *MCC* constantly increases from M_N/M_E = 69 to 1, especially with CRF, for both

IOP Conf. Series: Earth and Environmental Science 1136 (2023) 012047 doi:10.1088/1755-1315/1136/1/012047

training and validation sets, but *TPR* tends to decrease for $M_N/M_E < 4$. The highest value of *TPR* is at $M_N/M_E = 4$, but *MCC* is poor at this value, particularly for the validation set (i.e., potential for overfitting), while $M_N/M_E = 1$ provides more reliable predictions. In summary, CRF with $M_N/M_E = 1$ performs best and is selected as the best model for the present application.

Table 4. *MCC* of training and validation of the CRF and RUSBoost models.

		Training set		Validation set	
M_N/M_E	N_t	RUSBoost	CRF	RUSBoost	CRF
69	17507	0.118	0.123	0.075	0.073
10	2783	0.359	0.315	0.166	0.137
6	1771	0.408	0.354	0.168	0.132
4	1265	0.453	0.389	0.255	0.191
2	759	0.522	0.455	0.261	0.192
1	506	0.422	0.483	0.359	0.367

4.3. Input parameter combinations

While the number of possible input parameters was reduced by the SOM, there are still a great number of combinations of these parameters that can be used for the risk model. The parameters identified by SOMs were manganese, turbidity, 3-day HPC, low chlorine, residual type, number of customer contacts, DMA clustered customer contacts, and high priority dead ends per DMA. The best model from the last section (CRF with $M_N/M_E = 1$) was employed to test twenty combinations out of these parameters in order to find the best one which gives the highest accuracy of predictions. For brevity, the details of the tests are not presented. It was found that using four of these parameters as input, iron, turbidity, 3-day HPC and number of high priority dead ends, provided the highest accuracy. The performance of the model using these parameters is presented in Table 5.

Table 5. CRF model performance with best combination of input parameters (iron, turbidity, 3-day HPC, and high priority dead ends per DMA).

Metric	Training set	Validation set
TPR	0.612	0.609
TNR	0.874	0.857
ACC	0.739	0.727
MCC	0.502	0.478

4.4. Exceedance level

Using the best input parameter combination, iron exceedance threshold was explored to investigate its effect on the performance of the model. Four levels were tested: 200, 150, 100 and 50 μg/L. The details of the tests and the performance metrics calculated for training and validation sets are presented in Table 6.

For higher iron exceedance thresholds, the size of data is smaller due to rarer exceedance events; however, model fitting to the data is more accurate. For example, when the threshold is set to 200 μg/L, *MCC* of training is 0.704, while for a threshold of 50 μg/L, it is 0.326. For the validation set, the accuracy is highest for the 150 μg/L threshold. However, the misclassification error is also higher. Figure 2 shows the CRF misclassification error calculated for these tests. For exceedance level of 200 μg/L, the error is lower, denoting that more samples contributed to model training, thus the result is more reliable. This analysis shows that the model, including parameters combination, is well suited to identifying the higher levels of iron risk.

Table 6. CRF model performance for different levels of iron exceedance

		Training set			Validation set		
Exceedance level (μg/L)	N_t	*TPR*	*TNR*	*MCC*	*TPR*	*TNR*	*MCC*
200	254	0.866	0.838	0.704	0.500	0.833	0.354
150	485	0.612	0.874	0.502	0.609	0.857	0.478
100	984	0.571	0.797	0.379	0.436	0.906	0.357
50	3202	0.680	0.646	0.326	0.496	0.748	0.252

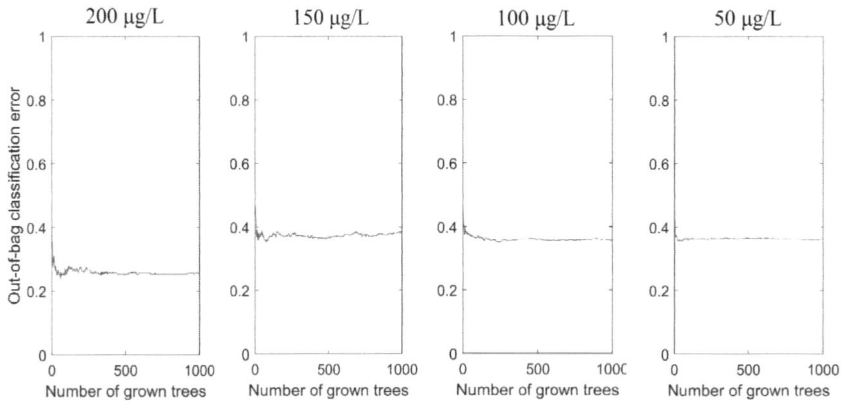

Figure 2. CRF out-of-bag classification error for exceedance levels of 200, 150, 100 and 50 μg/L, from left to right, respectively.

4.5. Performance of the preferred model

Summarising the results of the tests, the CRF model with DMA yearly averaged iron, turbidity, 3-day HPC and number of high priority dead ends as input, and with a balanced data (i.e., $M_N/M_E = 1$) gives the highest accuracy of the prediction of iron exceedance in a year ahead. For an exceedance level of 200 μg/L (the regulatory threshold), the *TPR* and *TNR* of predictions are 0.866, 0.838 for the training set, and 0.5 and 0.833 for the validation set. Therefore, this setup is selected as the preferred configuration for the current dataset and used for the calculation of probability of iron exceedance.

Figure 3 presents the confusion matrices of the model training and validation with the best setup. These show how many of the 'E' and 'N' classes were predicted correctly, or incorrectly. For example, for the validation set, half of the exceedances and most of the non-exceedances (10 out of 12) were predicted correctly.

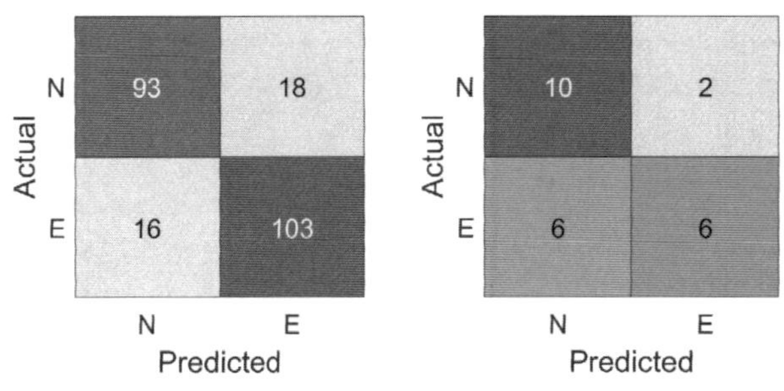

Figure 3. Confusion matrices of model training (left) and validation (right) for the preferred setup.

Figure 4 shows the predictor importance estimates calculated by the CRF model for the best setup. It indicates that the two most important input parameters in the analysis were high priority dead ends and iron, respectively. It means these two had the highest correlation with the output parameter, which is a year ahead iron exceedance, while turbidity and 3-day HPC had smaller contributions.

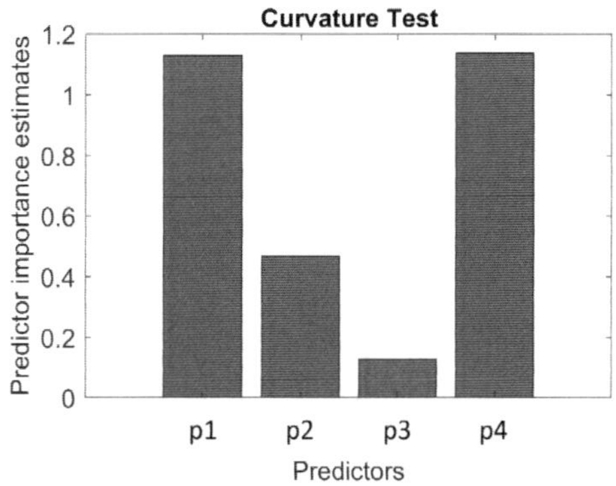

Figure 4. CRF predictor importance estimates for best setup. Input parameters, p1 to p4, are DMA yearly average of iron, turbidity, three-day HPC, and high priority dead ends per DMA, respectively.

4.6. Risk of iron exceedance

CRF grows a collection of individual decision trees. Each tree produces a class prediction ("votes" for a class). Then the forest chooses the class having the most votes over all the individual trees. There are two classes in our application, 'E' and 'N', denoting exceedance and non-exceedance events. Thus, if the fraction of trees that vote for an event to be in the class 'E' is above 0.5, that event is considered as an exceedance event, and if it is below 0.5, the event is a non-exceedance. Probability of a class is then obtained by counting the fraction of trees in the forest that voted for that certain class. The probability calculated for the class 'E' is thus inferred as the probability of iron exceedance, which represents the likelihood an event falls in the class exceedance.

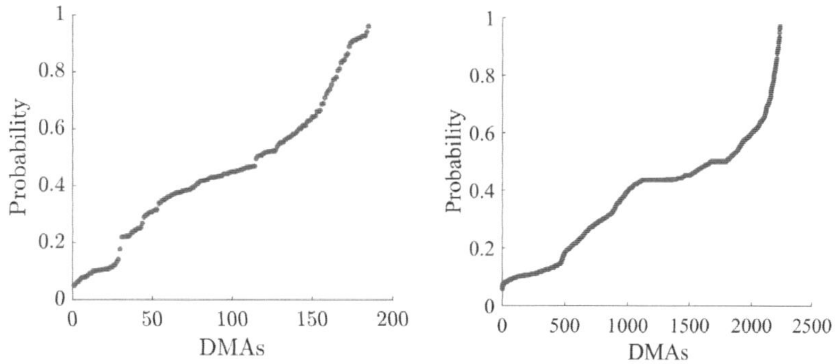

Figure 5. Prediction of relative iron exceedance probability above 200 μg/L of DMAs for the training set (left) and all the DMAs in 2020 (right), sorted by probability values.

The CRF model with the best setup was used to calculate probability of iron exceedance above 200 μg/L for the DMAs in the training set and the result is presented in Figure 5-left, where the DMAs are sorted by the calculated values of probability. The model was then employed to predict exceedance probability for all the DMAs in 2020. For this purpose, data from 2019 was fed into the model and the probability of iron exceedance in 2020 was estimated. The result of this test is presented in Figure 5-right, where DMAs are sorted ascendingly by probability. The graph exhibits a 'kick' upwards at higher probabilities. The vast majority (90% and over) of DMAs have probabilities of exceedance < 0.5, while < 10% have values above 0.7~0.8. This suggests that targeting interventions and resources to these DMAs should provide the greatest returns for managing iron exceedance risk.

It is noted that the calculated probability is a relative measure of iron exceedance risk in a DMA compared to all the other DMAs in the analysis, rather than being taken as an absolute value of risk. As a result, it is suitable for the purpose of risk ranking and informing proactive management.

5. Conclusions

Machine learning was used to analyse a large dataset of water quality samples, discolouration customer contacts and static asset data from a drinking water distribution network to understand the causes of high iron concentrations at a sub-region (DMA) level, and to thus develop a risk model giving predictions of relative probability of iron exceedance above a certain threshold.

A preferred setup was identified in terms of accuracy and reliability for risk modelling across the network. The setup included a CRF model trained with yearly DMA averaged iron, turbidity, 3-day HPC, and high priority dead ends per DMA as input parameters, with output a year ahead prediction of relative probability of iron exceedance above 200 μg/L.

The trained model was used to estimate and rank likelihood of iron exceedance for all DMAs in year 2020. This ranking showed that less than 10% of DMAs pose a risk compared to the entire system, suggesting that targeting these for interventions should provide the greatest return for managing iron exceedance risk.

The developed model can inform proactive management and is easily applicable to investigate other discoloration parameters such as turbidity, and transferable to other datasets or regions.

Acknowledgments

For the purpose of open access, the author has applied a creative commons attribution (CC BY) license to any author accepted manuscript versions arising.

References

[1] Vreeburg J and Boxall J 2007 Discolouration in potable water distribution systems: A review *Water Research* **41**(3) 519–529 https://doi.org/10.1016/j.watres.2006.09.028

[2] Speight V, Mounce S and Boxall JB 2019 Identification of the causes of drinking water discolouration from machine learning analysis of historical datasets *Environ. Sci. Water Res. Technol.* **5** 747–55 https://doi.org/10.1039/C8EW00733K

[3] Mounce SR, Ellis K, Edwards JM, Speight VL, Jakomis N and Boxall JB 2017 Ensemble decision tree models using RUSBoost for estimating risk of iron failure in drinking water distribution systems *Water Resour. Manag.* **31**(5) 1575–89 https://doi.org/10.1007/s11269-017-1595-8

[4] Kohonen T 1990 The Self-Organizing Map *Proc. IEEE* **78**(9) 1464–80 https://doi.org/10.1109/5.58325

[5] Breiman L 2001 Random forests *Machine Learning* **45** 5–32 https://doi.org/10.1023/A:1010933404324

[6] Hastie T, Tibshirani R and Friedman J 2008 *The Elements of Statistical Learning* 2nd edition (New York: Springer)

[7] Seiffert C, Khoshgoftaar T, Hulse J and Napolitano A 2008 RUSBoost: Improving classification performance when training data is skewed *Proc. 19th Int. Conf. Pattern Recognition* 1–4.

A data-driven model for the prediction of chlorine losses in water distribution trunk mains

Grigorios Kyritsakas[1], **Vanessa Speight**[1] **and Joby Boxall**[1]

[1] Department of Civil and Structural Engineering, The University of Sheffield, Mappin Street, Sheffield S1 3JD, UK

gkyritsakas1@sheffield.ac.uk

Abstract. A data-driven model that uses 4 different machine learning (ML) algorithms (Feed forward artificial neural networks (ANN), Nonlinear autoregressive exogeneous (NARX) ANN, support vector machine and Random Forest) was designed for the prediction of chlorine loss events in water distribution trunk mains. The model, firstly, identifies past chlorine loss events and their associate flow or temperature events. Then, the detected past flow events and their associate past chlorine loss events are used to train the ML algorithms. The model was tested in 3 trunk mains of the same drinking water distribution system with similar diameter but with different characteristics, using each time a different combination of parameters (flow (input) - past chlorine losses (output) or flow, temperature, and chlorine (input) - past chlorine losses (output)) and machine learning algorithms. Results indicate that the model could predict a future chlorine loss event with a period between 2 to 10 hours depending on the parameter and ML algorithm used and the trunk mains' hydraulic characteristics.

1. Introduction

Chlorine (Cl) is the most common disinfectant used to minimise the bacteriological activity in the drinking water exiting the treatment processes and prevent bacterial regrowth in the water distribution systems (WDS). Chlorine concentration in the water exiting the works is particularly high; however, due to chlorine reactions in the bulk water and in the pipe walls, it decreases during the water travel to the customer taps through the WDS [1]. Therefore, it is required by the water utilities to monitor the chlorine residual and, if possible, to accurately predict the chlorine behaviour in the WDS.

Traditional approaches in predicting chlorine decay in the WDS are made using a combination of processed based models and hydraulic models. Generally, these models produce accurate results; however they require a large number of accurate data and a good understanding of the physical chemical and hydraulic characteristics of the WDS for their implementation. In addition, these models require a determination and a continuous recalibration of a number of parameters (e.g. chlorine decay constant) in order to accurately simulate the chlorine behaviour. This process could be complicated and sometimes it requires a vast amount of computational time for its completion.

Data-driven models that apply machine learning (ML) algorithms for understanding the trends over the collected data and predict future behaviour could be an alternative solution for the prediction of chlorine decay in the WDS. In the past, data-driven models have been applied in various projects in the water sector for analysing the quality of the source water [2], improving the water treatment processes [3] and predicting turbidity events in trunk mains [4, 5]. As regards the chlorine decay prediction, Gibbs et al. presented a promising data-driven model using artificial neural networks (ANNs) to predict chlorine concentrations in customer taps [6].

In this paper, the model that Kazemi et al. [4] developed for the prediction of turbidity events in the water distribution trunk mains is redeveloped and adapted to predict chlorine loss events. More specifically, the model, presented here, is a regression model that predicts chlorine loss events, in terms of periods with low chlorine concentrations, at the end of the water distribution trunk mains up to several hours ahead. The model initially detects past chlorine loss events in the dataset and then it uses these events in combination with past water quality data related to these events (temperature, flow and chlorine concentration) as inputs in 4 ML algorithms (Feed Forward ANN (FF-ANN), non-linear autoregressive exogeneous ANN (NARX-ANN), random forest (RF) and support vector machines (SVM)) for the prediction of future chlorine loss events. The model was tested in 3 real world trunk mains of the same WDS and the prediction performance of the various ML algorithms was compared.

2. Site details and available datasets

The datasets used in this work were created for a case study on the impact of hydraulic interventions for managing the discolouration risk in the trunk mains [7]. The study area (figure 1) is a WDS located in the north of the UK consisting of 3 different trunk mains (TM-1, TM-2, TM-3) fed with water from the same water treatment plant. Chlorine concentrations were monitored with a frequency of 15 minutes at the plant outlet and at the outlet of each one of the trunk mains. Water temperature was also measured at the end of each trunk main, and flow was measured at the start of each trunk main with the same frequency. The main pipe characteristics for all the trunk mains are similar (table 1), however, during the study period, their hydraulic characteristics differed for the purposes of the case study [7].

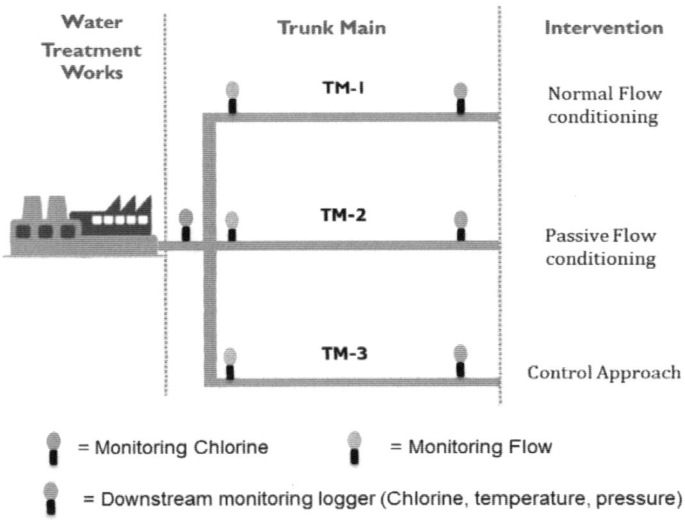

Figure 1. Schematic of the DWDS trunk mains [7].

Table 1. Water distribution trunk mains characteristics.

Trunk main	Mean internal diameter [mm]	Pipe material	Velocity [m/s]	Length from WTW outlet to downstream logger [km]
TM-1	304.8	Unlined CI (25% lined)	0.6	6.4
TM-2	406.8	Unlined CI	0.3	5.6
TM-3	304.8	Unlined CI	0.4	5.9

For our work, we assumed that there is no leakage in the trunk mains, and therefore for each trunk main and for each time step there were no changes in the monitoring flow during the travel of the water through them. Overall, three different datasets, one for each water distribution trunk main, are collected. Each dataset consists of 15 minutes time chlorine, flow and temperature data for a period of 7 months. During this period, there are some months with no available chlorine or temperature data that are excluded from the analysis.

3. Methodology

3.1. Data preparation

3.1.1. Removing data anomalies and outliers. In the collected datasets there were several outliers, either spikes or zero and negative values that could influence the methodology's results. These were identified and removed using a gradient algorithm created in MATLAB (version 2019b). More specifically, the algorithm followed these steps:

- It identified and replaced the zero and the negative chlorine values with missing values;
- It calculated the gradient between a data point and its previous one and if this was found greater than 0.4 mg/l it was replaced with a missing value;
- If the consecutive missing values were more than 12 (3 hours without measurement, the points were ignored, otherwise these were filled using the spline interpolation.

3.1.2. Smoothing the data. The last data preparation step was to smooth the data for removing the noise that could influence the training of the data-driven methodology. For the smoothing of the dataset the cubic spline function was used in MATLAB (version 2019b). The smoothing length was set equal to $2\Delta t$ (30 minutes) due to the increased noise in the Chlorine dataset. Figure 2 shows the original and the smoothed Cl, flow, and temperature data in a small part of the TM-3 dataset.

3.2. Chlorine loss event detection

A model was designed to capture and extract the chlorine loss events from the dataset. These events are identified as the periods of the dataset where unusually low chlorine concentrations are measured. The model extracted the chlorine loss events by following these steps:

- The event peaks were identified as the local minimum Chlorine values;
- For each peak the event starting and finishing points were detected using the gradient of the timeseries before and after the event. The model traced the data up to certain hours before the event to find a gradient that surpassed the gradient threshold of 0.02mg/l per 15 min (0.02mg/l per timestep). If no gradient higher than the threshold was found the peak was ignored. If a gradient higher than the threshold was found, the chlorine event was detected. The detected event consisted of equal datapoints prior the peak and after the peak depending on the defined forecasting period. For example, if the forecasting period was set as 4 hours ahead, the event consisted of datapoints 2 hours before and 2 hours after the peak;
- A "base value" was set as equal to the pre-event chlorine concentration in the dataset and baseline that connected all the base values was generated;
- The event magnitude was calculated as the difference between the peak and the base value of each event. The final events were then selected as those that exceeded the magnitude threshold of -0.15mg/l.

3.3. Flow and temperature event detections

Once the final chlorine events were detected a similar model was used to detect their associate flow or temperature events. These events were detected by searching the dataset over a period of certain hours before the detected chlorine event. This detection model identified the associate flow or temperature events by following 5 steps. The first four were same steps as the ones that the chlorine loss event detection model was using. The magnitude threshold in the fourth step was set equal to 10l/s ($0.01m^3/s$) when the flow dataset was used and to 0.3°C when the temperature dataset was used. The final step in this model was to remove all the Chlorine loss events that no flow event or no temperature event was found.

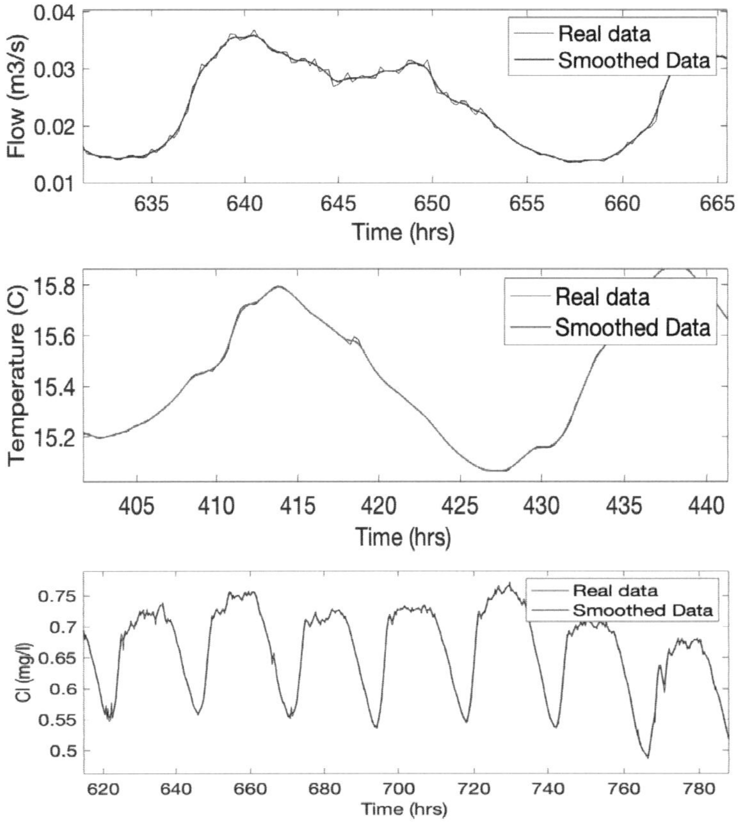

Figure 2. Original and smoothed flow (top), temperature (middle) & chlorine (bottom) data in TM-3.

The model was applied in all three trunk mains and on average there were 30 chlorine loss events associated with high flow events per trunk main in comparison with 12 chlorine loss events associated with temperature events. Therefore, the events used in the predictive model were chlorine loss events as outputs and their associate flow events as inputs. Figure 3 shows two detected chlorine events with their associated high flow events for TM-1.

3.4. Predictive model
A multistep data-driven prediction model was constructed in MATLAB (version 2019b) for the prediction of future chlorine loss events. In the predictive model, 4 ML algorithms (RF, SVM, FF-ANN and NARX-ANN) were selected for the chlorine loss forecasting and a comparison between their performances was made.

3.4.1. ML algorithms. The selected ML algorithms tested here were as follows:

Random Forest (RF) [8]: RF is an ensemble decision tree algorithm that uses the outputs of large number of equal weak decision trees for making predictions. Each weak tree makes its splitting decision using randomly a small part of training dataset and part of the available input parameters. The final RF prediction is equal to the mean of the predictions made by the week trees. In this predictive model the number of weak trees was set equal to 1000 and the minimum number observations per tree were set to 5.

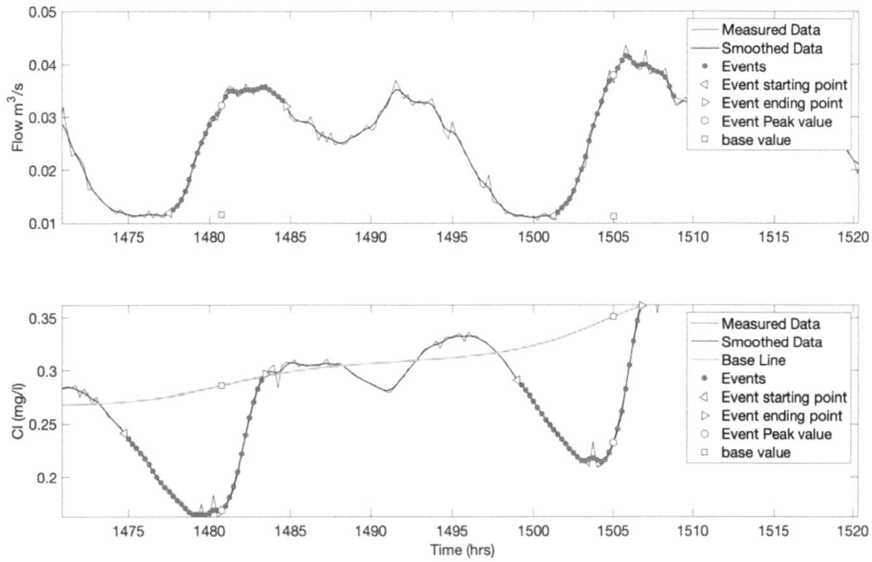

Figure 3. Detected chlorine and their associated flow events in TM-1.

Support vector machine (SVM) [9]: SVMs is trained to find the maximum margins between the boundaries that separate the input dataset and thus reducing the generalisation error. In this predictive model, the 5-Fold cross-validation was used for the data calibration and the Gaussian kernel for its training.

FF-ANN [9]: ANNs were constructed to simulate the human brain function. The FF-ANN is an ANN where the information is moving only forward from the input layer to the output layer through the hidden layers. The FF-ANN used here had one hidden layer with 14 neurons and it was using the regularization backpropagation function for training.

NARX-ANN [10]: NARX-ANN is an ANN that uses the non-linear version of the autoregressive exogenous model. There are two versions of the algorithm architecture, the open and closed loop. The former uses the past and presents inputs and the past targets for predicting the future targets and the latter uses past and present inputs and past targets and predicted past targets by the model for the prediction of future targets. The NARX-ANN selected here was using a hidden layer with 10 neuros and the input and feedback delays were set to 3 empirically.

3.4.2. Predictive model input variables. The model initially collected the past chlorine loss events and their associated flow events. Then, either the flow measurements or the flow with the temperature and the chlorine measurements during the flow event period were used as inputs in the predictive model and the chlorine loss events were used as targets (outputs).

The model is trained using a few past events combinations (10-20 events) to predict a future chlorine loss event given new unseen input data (flow or flow, chlorine, and temperature) during the current flow event. To remove the seasonality effect in the dataset, instead of the absolute chlorine values, the predictive model uses the magnitude of the event i.e. the difference between each data point of the event and the its base value. A simplified schematic of the predictive model that used 10 past chlorine events and all the inputs in their associated flow events is presented in figure 4.

3.4.3. Performance metrics. Three performance metrics were selected for quantifying the performance of the predictive model and justify which of the ML models was better, the mean absolute error

(MAE), the root mean squared error (RSME) and the Nash-Sutcliffe Model Efficiency Coefficient (NSE). MAE is used as an indication of the overall agreement between predicted and targeted values, RMSE is used as a metric for the identification of high errors in the prediction and NSE is used for the comparison of the predictive skills of the model with the mean of the observed data.

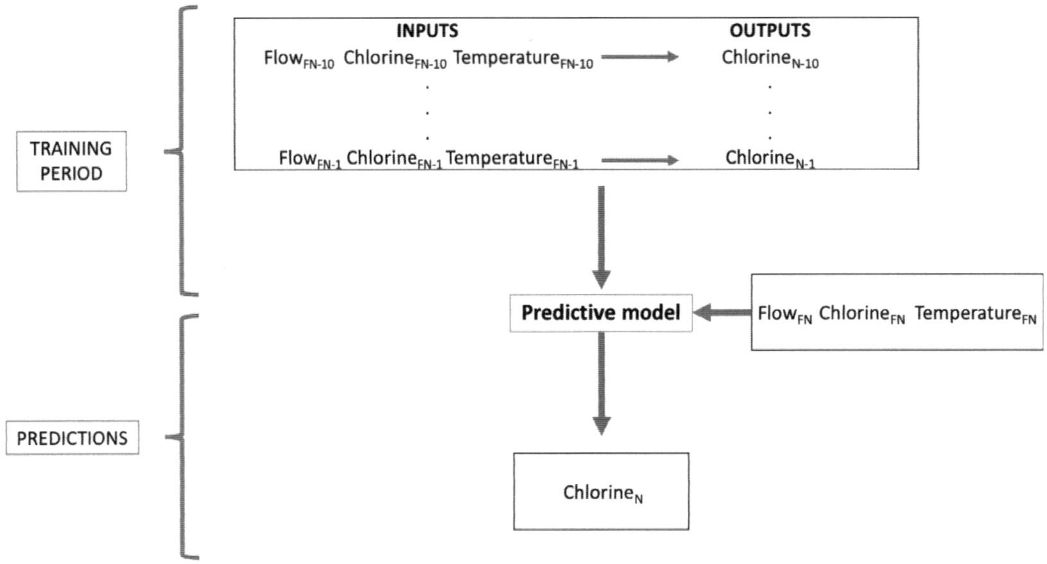

Figure 4. Schematic of the predictive model for the prediction of the Nth Chlorine event using for training the data from 10 past chlorine events (N-10…N-1) as outputs and all the inputs (chlorine, temperature, flow) from their associated flow events (FN-10…FN-1).

4. Results and discussion

The model was tested using 8 different combinations of ML algorithms and input variables (NARX-flow, NARX- flow/temperature/chlorine, FF-flow, FF - flow/temperature/chlorine, SVM-flow, SVM-flow/temperature/chlorine, RF-flow, RF-flow/temperature/chlorine) in each trunk main and for different forecasting periods (4-10 hours). In each trunk main different combinations of inputs and outputs were used for training and testing. Overall, the model managed to have accurate predictions with a forecasting horizon of up to 8 hours in TM-1 and up to 10 hours in TM-2 and TM-3. In table 2, the average performance results from all the model's tests in all 3 trunk mains are presented.

The model performance metrics as presented in table 2, indicate that the RF algorithm is the best out of all the tested ML algorithms and the NARX - ANN is the worst one. More specifically, NARX outputs produced only negative NSE results when tested in all three trunk mains while the RF, using flow, chlorine, and temperature as input variables, produced the highest NSE average and lowest MAE and RMSE average in TM-2 and TM-1. In addition, in one of the test events in TM-3, RF produced the highest NSE (0.92) of all simulations. This result disagrees with the finding that Kazemi et al found using the NARX algorithm with timeseries data in their work [4]. However, in their case their aim was to predict turbidity timeseries events instead of chlorine events.

Figure 5 shows a 9-hour event in TM-2 with 10 hours ahead forecasting period and figure 6 shows a 12-hour event in TM-3 with 10 hours ahead forecasting period. Both figures 5 and 6 shows that RF, FF and SVM follow the measured data changing trend in comparison with NARX which, especially in TM-3, produced completely different to the measured data outputs.

Regarding the other models, SVM with only the flow parameter used as input was the second-best predictive model in TM-1 and TM-2 with an NSE of 0.3 and 0.425 respectively. However, in TM-3 the SVM with all input parameters outperformed the SVM with flow input.

Table 2. Performance of the model using different ML algorithm-input combinations.

Trunk Main	ML Method	Prediction horizon	Parameters used	MAE	RMSE	NSE	Highest NSE
TM-1	RF	8	cl flow temperature	0.0205	0.02645	0.543	0.579
			flow	0.0275	0.03475	0.280	0.386
	SVM		cl flow temperature	0.02625	0.03375	0.274	0.54
			flow	0.02425	0.03125	0.300	0.5
	FF		cl flow temperature	0.02895	0.03075	0.296	0.65
			flow	0.03025	0.04025	-0.002	0.473
	NARX		cl flow temperature	0.2545	0.35175	-2.878	-1.62
			flow	0.07825	0.12355	-1.225	-1.09
TM-2	RF	10	cl flow temperature	0.0245	0.031	0.448	0.796
			flow	0.028	0.0365	0.32	0.327
	SVM		cl flow temperature	0.04	0.046	-0.1475	0.378
			flow	0.0245	0.033	0.425	0.591
	FF		cl flow temperature	0.0425	0.0465	0.403	0.74
			flow	0.0275	0.0355	0.3535	0.449
	NARX		cl flow temperature	0.145	0.196	-1.85	-1.8
			cl flow	0.1285	0.1925	-0.55	0.33
TM-3	RF	10	cl flow temperature	0.028	0.036	0.530	0.920
			flow	0.0385	0.048	0.308	0.569
	SVM		cl flow temperature	0.0273	0.035	0.587	0.644
			flow	0.033	0.046	0.395	0.440
	FF		cl flow temperature	0.119	0.138	-0.410	0.780
			flow	0.0395	0.0485	0.313	0.320
	NARX		cl flow temperature	0.164	0.1895	-1.865	-1.800
			flow	0.3985	0.4615	-1.350	-1.300

FF also produced good results, but the algorithm had completely different predictions in each TM depending on the inputs used. In TM-1 the FF algorithm when all the inputs were used, outperformed the one that used the flow as input in contrast with the other 2 TMs where the opposite occurred. Because all 3 trunk mains had different hydraulic characteristics during the study period, this finding indicates that both SVM and FF algorithms are influenced by the noise in the data due to the different hydraulic characteristics of the 3 trunk mains. In contrast, the predictive model that used the RF algorithm with either the flow or all the parameters as inputs, produced good outputs in all three trunk mains which indicates its ability to absorb the noise in the data. Moreover, in all the trunk mains the RF with all the outputs outperformed the RF where only the flow was used as input parameter which potentially indicates that the more the relevant parameters are given to RF the better the results would be. Overall, this investigation demonstrated the potential of this model in predicting chlorine events in this dataset with a forecasting horizon of up to 10 hours ahead. However, further work is required in other water distribution networks with larger available datasets that cover a period of at least 2 years. This investigation will aim to examine how the model behaves in periods with different hydraulic conditions and its adaptation to seasonality changes.

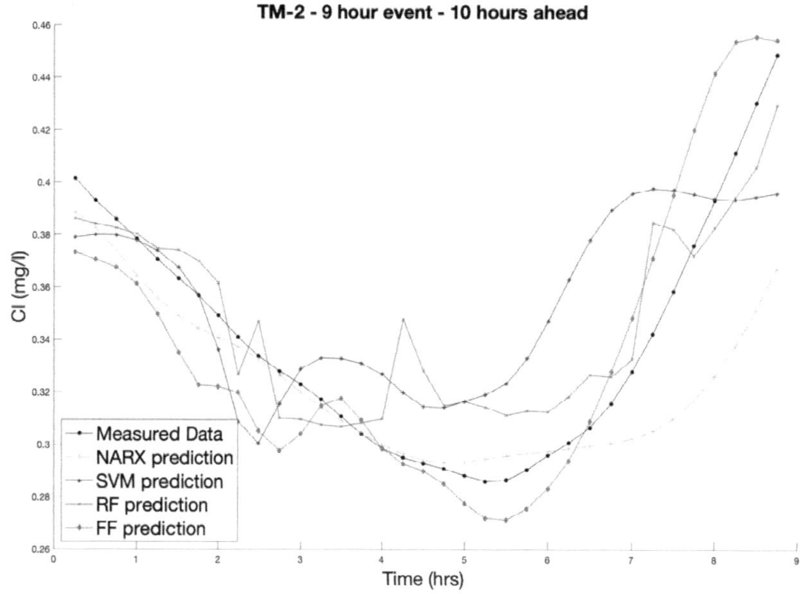

Figure 5. Chlorine predicted vs Measured data of a chlorine loss event in TM-2.

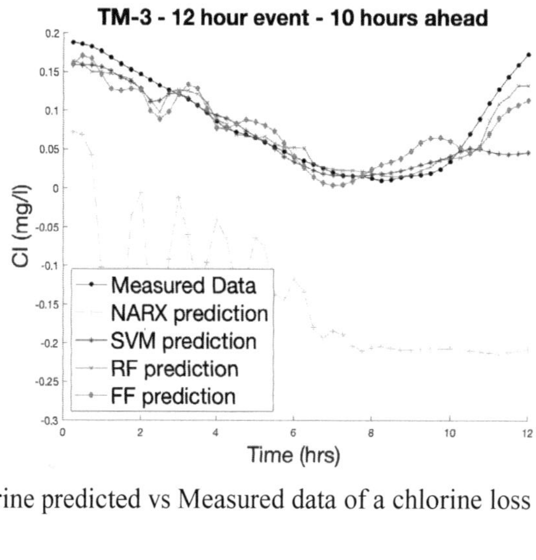

Figure 6. Chlorine predicted vs Measured data of a chlorine loss event in TM-3.

5. Conclusions

A data-driven methodology for the forecasting of a future chlorine loss event at the end of a water distribution trunk main is presented. This methodology, firstly, identifies past chlorine loss events and their associated flow events. Then, it imports either the flow, or the flow, the temperature, and the chlorine measurements during the flow event as inputs and the chlorine events as outputs in a predictive model that uses one of the following ML techniques: NARX, FF, SVM and RF. The methodology was tested in a WDS that consists of 3 trunk mains (TM-1, TM-2, TM-3). The aim was to investigate the predictive ability of the model when different input parameters (flow or flow, temperature chlorine) - ML algorithms (NARX, SVM, FF, RF) combinations were used.

For the quantification of the model's performance, 3 main performance metrics were used, MAE, RMSE and NSE. The main conclusions are as follows:

- The predictive model managed to predict accurately a future chlorine loss event with a period of 8 hours ahead in TM-1, 10 hours ahead in TM-2 and 10 hours ahead in TM-3;
- The performance metrics (MAE, RMSE, NSE) indicated that best overall algorithm was the one that was using the RF algorithm with all the parameters as inputs;
- NARX was the worst algorithm in all three datasets for this predictive model;
- FF and SVM models had acceptable performances but both models appeared to be influenced by the different noises in the 3 datasets caused by the different hydraulic conditions in each trunk main;
- This predictive model has the potential of becoming an accurate supporting tool for supporting water utilities decision making regarding proactive intervention. However, as the available data for this work were taken for a small period (overall less than 5 months of data were available if we include the spikes and the missing data), further research is required using larger datasets - with at least 2 year of data - to investigate the ability of the model to adapt in the seasonality changes and different hydraulic conditions.

Acknowledgments

The authors would like to thank the Engineering and Physical Sciences Research Council (EPSRC), the UKRI and Scottish Water for providing the financial support through STREAM IDC and PODDS for the data provision and permission to use it in this work. For the purpose of open access, the authors have applied a creative commons attribution (CC BY) license to any author accepted manuscript versions arising.

References

[1] Parsons SA and Jefferson B 2006 *Introduction to Potable Water Treatment Processes* (Blackwell Publishing Ltd) pp 1–179 doi:10.1002/9781444305470

[2] Liu P 2019 Analysis and prediction of water quality using LSTM deep neural networks in IoT environment *Sustainability* **11**(7) 2058 https://doi.org/10.3390/su11072058

[3] Li L, Rong S, Wang R and Yu S 2021 Recent advances in artificial intelligence and machine learning for nonlinear relationship analysis and process control in drinking water treatment: A review *Chem. Eng. J.* **405** 126673 https://doi.org/10.1016/j.cej.2020.126673

[4] Kazemi E, Mounce S, Husband S and Boxall J 2018 Predicting turbidity in water distribution trunk mains using nonlinear autoregressive exogenous artificial neural networks *Proc. 13th Int. Conf. Hydroinformatics* (1-4 July, Palermo, Italy)

[5] Meyers G, Kapelan Z and Keedwell E 2017 Short-term forecasting of turbidity in trunk main networks *Water Res.* **124** 67–76 doi: 10.1016/j.watres.2017.07.035

[6] Gibbs M et al 2006 Investigation into the relationship between chlorine decay and water distribution parameters using data driven methods *Math. Comput. Model.* **44**(5-6) 485–498 https://doi.org/10.1016/j.mcm.2006.01.007

[7] Sunny I, Husband S, Moore G, Drake N, Mckenzie K and Boxall J 2017 Discolouration risk managememnt and chlorine decay *Proc. 15th Int. Conf. CCWI* (5-7 Sept., Sheffield, UK)

[8] Breiman L 2001 Random forests *Machine Learning* **45** 5–32 https://doi.org/10.1023/A:1010933404324

[9] Hastie T, Tibshirani R and Friedman J 2008 *The Elements of Statistical Learning* 2nd edition (New York: Springer)

[10] Lin T, Horne BG, Tino P and Giles CL 1996 Learning long-term dependencies in NARX recurrent neural networks *IEEE T. Neural Networ.* **7**(6) 1329-51 doi: 10.1109/72.548162

IOP Conf. Series: Earth and Environmental Science 1136 (2023) 012049 doi:10.1088/1755-1315/1136/1/012049

Analysis of diffusion and dispersion processes in water distribution networks through the use of the Péclet number threshold

Stefania Piazza[1], Mariacrocetta Sambito[1] and Gabriele Freni[1]

[1] School of Engineering and Architecture, University of Enna "Kore", Enna, 94100, Italy

stefania.piazza@unikore.it

Abstract. To model the water quality within the distribution networks, simplified models are used (such as EPANET), which use an advective-reactive approach and neglect relevant phenomena, such as diffusion and dispersion, for low values of the Reynolds number. Although valid in the presence of a purely turbulent flow regime, as they do not produce substantial modelling errors, these simplifications are not applicable equally to all distribution networks. Therefore, the present study defines a threshold of the Péclet number, which relates the effectiveness of mass transport by advection with the effectiveness of mass transport by dispersion or diffusion, in order to establish a priori the behaviour of the water distribution network and find the most suitable model to use. The numerical analysis was applied to a real case study (network of Oreto-Stazione) using EPANET advective model and EPANET-DDR (Dynamic-Dispersion-Reaction) model, developed by the authors.

1. Introduction

Within the water distribution systems, the speeds are significantly variable during the day and this causes laminar, transition and turbulent flow regimes to occur simultaneously which determine different advective and diffusive-dispersive contaminant transport mechanisms.

Currently, to model the water quality within water networks, EPANET [1] model is used which solves the contaminant transport equation by introducing simplifications, based on solving the mass balance equation of the plug flow substance and on a simplified reaction kinetics. This is advantageous from a computational point of view as it does not produce particular significant errors in purely turbulent flow regimes. Nevertheless, diffusive-dispersive processes assume great importance when low velocities and laminar/transition flow regime take part and affect the positioning of the monitoring system, as highlighted in Piazza et al. (2020) [2].

To evaluate whether the network has a prevalent advective or diffusive-dispersive behaviour, the Péclet number can be used, which relates the effectiveness of mass transport by advection with the effectiveness of mass transport by dispersion or diffusion.

Several studies [3] state that, if the Péclet number is small, diffusive mass transport predominates over advective mass transport; otherwise, if the Péclet number is high, the advection is strongly dominant and diffusive processes can be considered negligible.

In the literature, numerous definitions of the Péclet number [4] are available that do not allow to identify a single threshold in order to identify a sharp transition between the two contaminant transport mechanisms.

This study aims to provide a rapid decision-making tool that can establish, *a priori,* the behaviour of a distribution network, defining a threshold for the Péclet number beyond which the simple advective model is not efficient for water quality modelling and it is necessary to use a more complete diffusive-dispersive approach. The real distribution network Oreto-Stazione (Palermo) was used as a reference to demonstrate the proposed approach.

2. Materials and Methods

The Oreto-Stazione network is located in the third district of the city of Palermo (Italy). It was completely rebuilt in 2002 even if the old cast iron supply pipes remain in service. Therefore, all the geometric characteristics of the network are precisely known, as well as the number and distribution of user connections, the volumes of water supplied and measured, the pressure and flow values in some important nodes.

The network is fed by a source and a tank positioned at different heights, with a storage capacity of 40000 m^3. Overall, it has a length of about 40 km and is made up of 264 junctions, having variable elevations in a range of 3 ÷ 47 m above sea level, and 396 polyethylene pipes with diameters between 110 and 225 mm. The network was designed to supply around 400 L/capita/day, even if the real average consumption is around 260 L/capita/day. In 2019, AMAP (Municipalized Aqueduct Company of Palermo) carried out an experimental campaign on this network lasting 24 hours, in which 2 mg/L of chlorine were inserted into the upstream tank (red box), monitoring some specific nodes highlighted in red (Figure 1 nodes 3, 16, 49, 54, 58, 162).

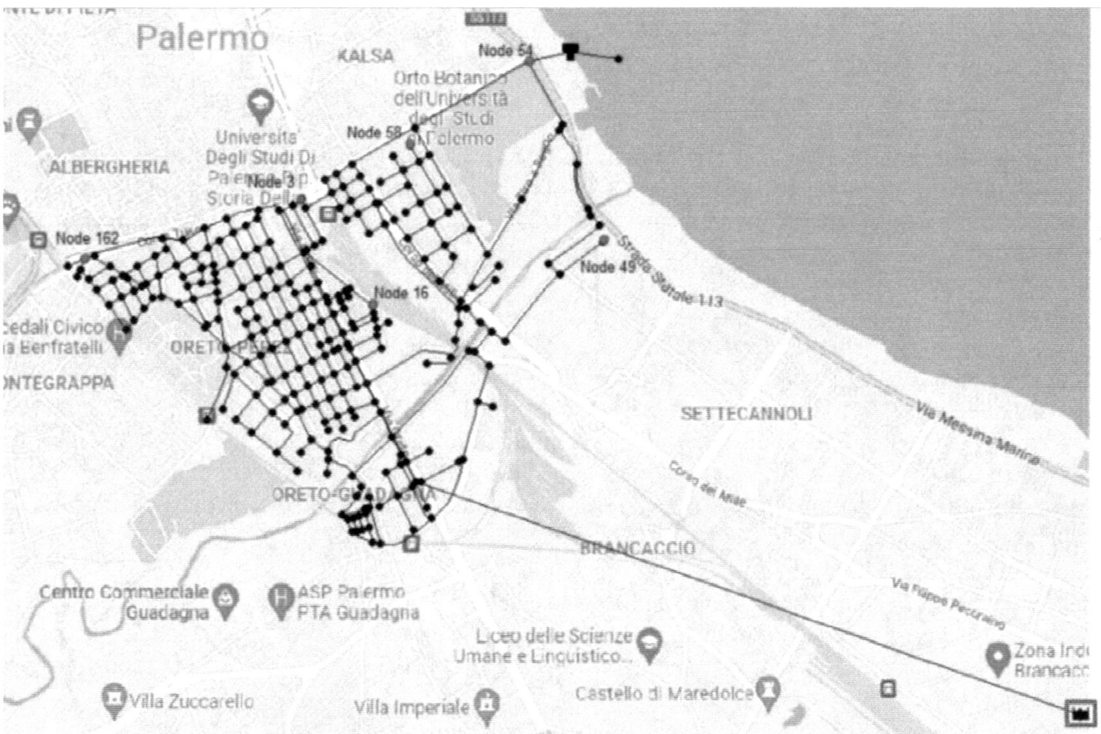

Figure 1. Layout of the Oreto-Stazione network.

The determination of the Péclet number thresholds was carried out using two models: the advective EPANET model [1] and the new dynamic model proposed by the authors EPANET-DDR (Dynamic-

Dispersion-Reaction) which solves the equations under quasi-steady flow conditions, solving the hydraulic problem under steady flow conditions with the EPANET-MATLAB-Toolkit [5] and the advection-diffusion-dispersion-reaction equation under dynamic flow conditions in the two-dimensional case with the classical random walk method [6], implementing the diffusion and dispersion equations proposed by Romero-Gomez and Choi (2011) [7]. This model allows to determine the position of the contaminant particles in the two-dimensional case as a function of the different flow regimes that occur within the network, as shown in equations (1) and (2), where u_x and u_y are the velocities along the two x and y axes respectively, dt is the duration of the contamination event, d is the pipe diameter, E_f and E_b are the forward and backward diffusion coefficients, respectively, as defined by Romero-Gomez and Choi (2011), and k is the reaction coefficient. In equation (1) the dispersion coefficients assume a backward or forward value as a function of the negative or positive flow direction. The contaminant concentration was determined by increasing the concentration at the previous time by an amount that corresponds to the concentration per unit of particles $(C \cdot n)$ passing through the control $(L/\Delta x \cdot \pi\ d\char`^2/4)$, where L is the length of the pipe, Δx is the section number of the pipe and $\pi\, d^2/4$ is the cross-sectional area of the pipe.

$$x = x + \frac{3}{2}u_x\left(1 - \left(\frac{y}{\frac{d}{2}}\right)^2\right)dt + \sqrt{2 \cdot E_{f\ or\ b} \cdot dt} \pm k \cdot dt \tag{1}$$

$$y = y + u_y dt + \sqrt{(E_f + E_b) \cdot dt} \pm k \cdot dt \tag{2}$$

The Péclet number was determined as the product of the Schmidt number $Sc = \upsilon/Diff$ and the Reynolds number $Re = (V \cdot D)/\upsilon$, show in equation (3), which are dependent on the kinematic viscosity of water $\upsilon = 10\char`^(-6)$ m^2/s, on the chloride diffusion coefficient $Diff = 14.44$ m^2/s, on the velocity of the current V m/s and from the diameter of the pipe D m.

$$Pe = Sc \cdot Re \tag{3}$$

Both models have been applied to the Oreto-Stazione network considering three different scenarios in which the pipe lengths vary in a range of 2.8 ÷ 2000 m, the diameters vary in a range of 0.025 ÷ 0.50 m and the flow rate is variable in the range of 0.00002 ÷ 0.0045 m^3/s. The network was contaminated by a constant concentration pattern of chlorine equal to 2 mg/L allocated to the upstream tank.

3. Results

Figure 2 shows the Péclet number thresholds determined for the Oreto-Stazione network using both models (EPANET and EPANET-DDR), varying respectively the length of the pipes, the diameter and the flow rate, respectively in range between 1 ÷ 2000 m, 0.01 ÷ 0.50 m and 0.00002 ÷ 0.0044 m^3/s. For both models applied, the thresholds are perfectly coincident. It is noted that using the chlorine diffusivity coefficient equal to 14.44 m^2/s, the threshold values of the Péclet number are lower than unity.

Comparing the results obtained in the present study with those determined previously [8], in which the Péclet number thresholds for the laboratory network of the University of Enna "Kore" were determined using a conservative tracer (sodium chloride having diffusivity $1.2 \cdot 10^{-9}$ m^2/s), it is observed that the thresholds trend determined in this study reflects that determined in a previous study. Furthermore, it is observed that by changing the type of tracer the threshold values change considerably. In fact, in [8] the threshold values are in the order of $10^5 \div 10^8$, in contrast to what was determined in the present study.

In Figure 2a, it is observed that keeping the flow rate and pipeline diameter values constant and varying its length in a range between 1 ÷ 2000 m, the threshold value of the Péclet number presents a constant trend having a value equal to 0.00016. Figure 2b shows the trend that links the Péclet number as a function of the diameter. It is observed that in correspondence with the minimum value of the diameter, the maximum threshold value of the Péclet number equal to $Pe = 0.00176$ is obtained, while it is obtained the minimum threshold value of the Péclet number equal to $Pe = 0.00004$ corresponding to the maximum value of the considered diameter. Finally, Figure 2c shows a linearly increasing trend of the Péclet number as a function of the flow rate having respectively minimum and maximum values equal to 0.000008 and 0.001724 in correspondence with the minimum and maximum flow rates.

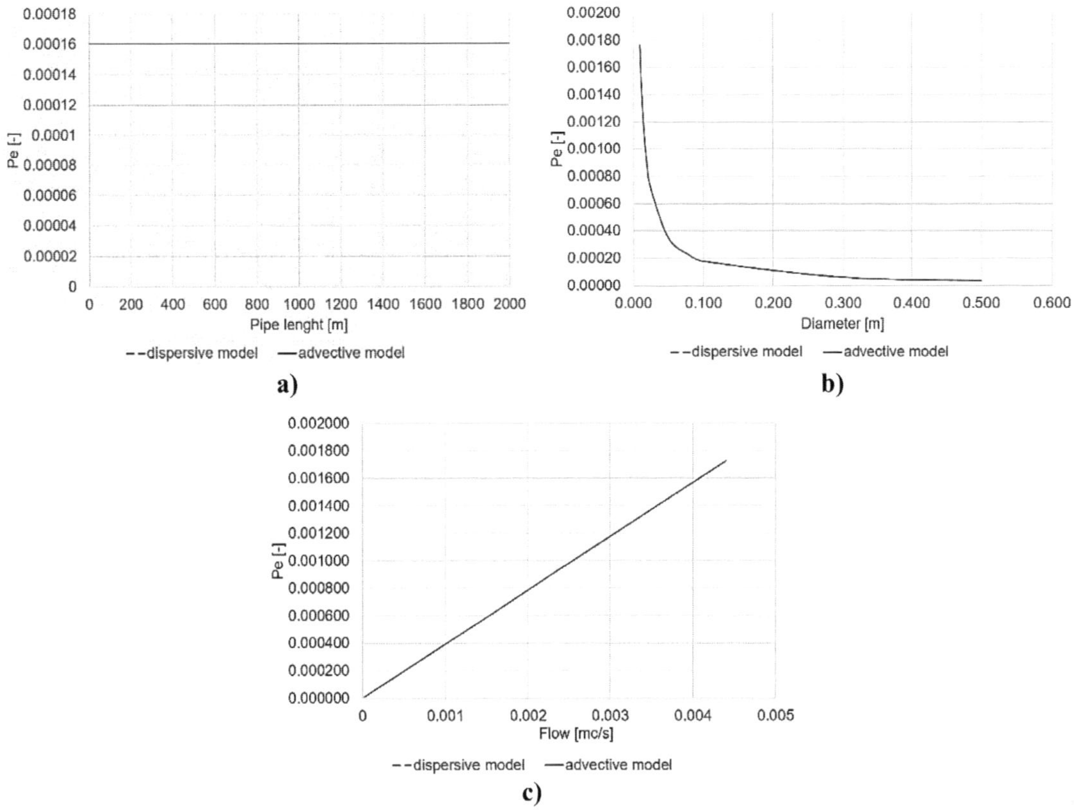

Figure 2. Thresholds of the Péclet Number by varying the pipes length a), the diameter b) and the flow rate c) respectively.

The Table 1 shows the geometric characteristics of the network sections upstream of the monitored nodes and the relative values of the Péclet number. Note that in the sections where the flow regime is purely turbulent (nodes 3, 16, 49, 54), the values of the Péclet number obtained exceed the three determined thresholds.

A preliminary analysis shows that the network has 271 pipelines in which the flow regime is purely turbulent, 45 in which the flow regime is transitional and 80 in which the laminar flow regime occurs.

Table 1. Geometric and hydraulic characteristics of the pipes upstream of nodes 3, 16, 49, 54, 58, 162 of the Oreto-Stazione network.

	Node 3	Node 16	Node 49	Node 54	Node 58	Node 162
Pipe length [m]	43.44	111.56	212.95	282.25	72.79	45
Flow [m^3/s]	0.01002	0.00052	0.00051	0.00043	0.00014	0.00022
Diameter [m]	0.225	0.110	0.110	0.110	0.110	0.110
Reynolds [-]	56250	6600	5500	14300	1100	2200
Peclet [-]	0.003894	0.000457	0.000381	0.000990	0.000076	0.000152
Flow Regime	Turbulent	Turbulent	Turbulent	Turbulent	Laminar	Laminar

By modelling the chlorine behaviour within the network using the two models (advective-reactive and advective-diffusive-dispersive-reactive) in which the respective coefficients have been appropriately calibrated (bulk coefficient = -0.10 sec^{-1} and wall coefficient = -0.40 m/sec for the first model, and bulk coefficient = -0.10 sec^{-1}, wall coefficient = -0.40 m/sec, backward coefficient = 0.01 m^2/s and forward coefficient = 0.022 m^2/s for the second) and comparing the latter with the real data, it is observed that lower values of the RMSE (Root Mean Square Error) were obtained through the use of the diffusive-dispersive model, obtaining percentages ranging between 4.02% and 10.45% (Table 2). RMSE values of 0 mean that the model returns results that match the real data.

Table 2. Chlorine modelling efficiencies using advective-reactive and diffusive-dispersive-reactive model through statistical parameters RMSE.

	Node 3	Node 16	Node 49	Node 54	Node 58	Node 162
RMSE % Advective model	10.45	13.51	12.16	8.94	21.09	18.14
RMSE % DDR model	10.45	7.06	4.62	4.02	4.77	5.52

This is also confirmed by Figure 3 and Figure 4, which show that the advective-reactive model is able to provide a better representation of the measured data.

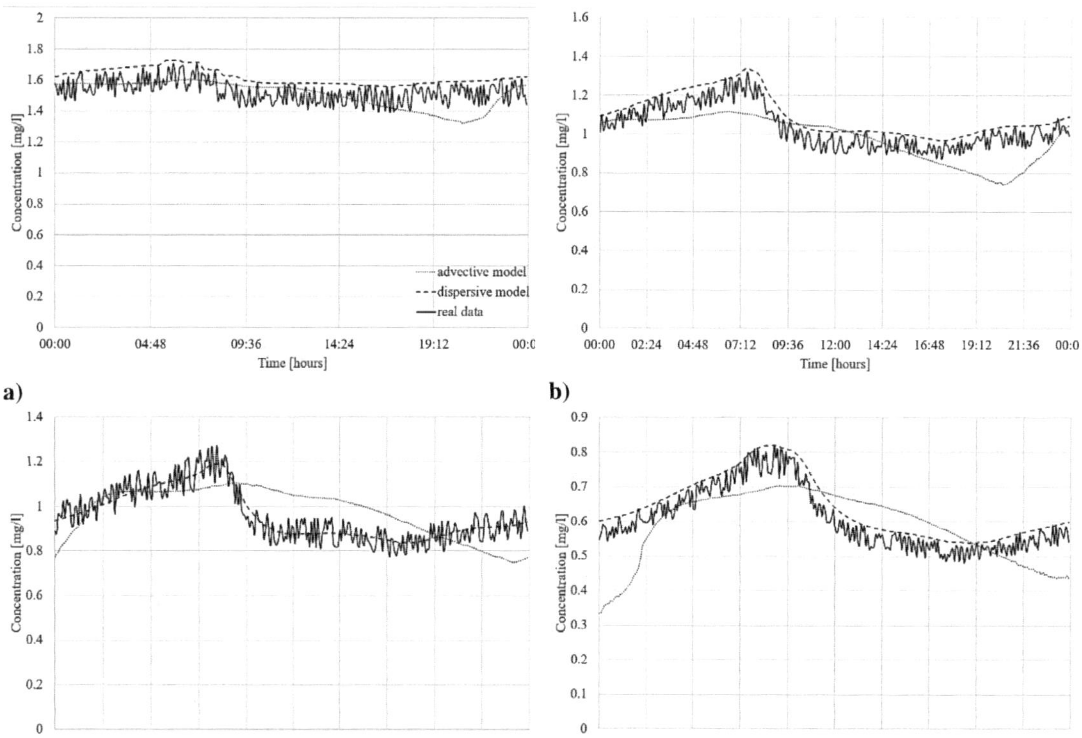

Figure 3. Comparison between real measurements and numerical simulations using the advective-reactive model (bulk coefficient = -0.10 sec^{-1} and wall coefficient = -0.40 m/sec) and the diffusive-dispersive model (bulk coefficient = -0.10 sec^{-1}, wall coefficient = -0.40 m/sec, backward coefficient = 0.01 m^2/s and forward coefficient = 0.022 m^2/s) for nodes: 3 (a), 16 (b), 49 (c) and 54 (d) of the Oreto-Stazione network.

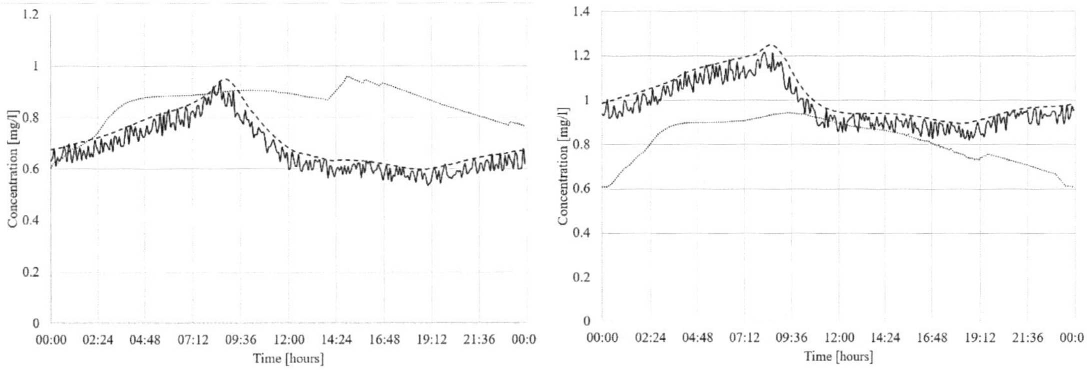

Figure 4. Comparison between real measurements and numerical simulations using the advective-reactive model (bulk coefficient = -0.10 sec^{-1} and wall coefficient = -0.40 m/sec) and the diffusive-dispersive model (bulk coefficient = -0.10 sec^{-1}, wall coefficient = -0.40 m/sec, backward coefficient = 0.01 m^2/s and forward coefficient = 0.022 m^2/s) for nodes: 58 (a) and 162 (b) of Oreto-Stazione network.

4. Conclusion

The present study allowed to determine three thresholds of the Péclet number for the Oreto-Stazione network actually existing in Palermo.

The use of the three selected thresholds has different impact on simulations depending on the diameter, length of the pipeline and the flow rate inside each of them. It can be observed that these modelling behaviours are in agreement with those determined in a previous study, developed on the laboratory network of the University of Enna "Kore", in which a conservative tracer was used.

Furthermore, depending on the contaminant considered, the threshold values of the Péclet number change considerably. In fact, in the previous study in which a conservative tracer was used with diffusivity equal to $1.2 \cdot 10^{-9}$ m^2/s, the threshold values of the Péclet number are higher than unity and are indicatively in the order of $10^5 \div 10^8$.

Considering chlorine as a tracer, having diffusivity equal to 14.44 m^2/s, it is observed that the threshold values are lower than unity.

EPANET-DDR model and original EPANET model were used to determine the thresholds, perfectly reproducing the same results.

The analysis on the network allowed to determine its behaviour according to the transport mechanisms involved. It is observed that it is predominantly advective, since a purely turbulent flow regime occurs in 271 pipeline sections of the network, while laminar flow regime occurs only in 80 pipelines. This was confirmed by the comparison between the values of the Péclet number, calculated for the single sections of pipeline monitored upstream of some selected nodes, and the threshold values, since the values relating to the purely turbulent flow regime all fall above of the thresholds.

By evaluating the RMSE parameter for the monitored nodes, it is observed that despite the predominant advective behaviour of the network, the diffusive-dispersive processes must be taken into consideration in order to best model the quality of the water.

References

[1] Rossman R 2000 *EPANET 2 Users Manual* EPA/600/R-00/057 (Cincinnati, OH: U.S. Environmental Protection Agency)

[2] Piazza S, Blokker EJM, Freni G, Puleo V and Sambito M 2020 Impact of diffusion and dispersion of contaminants in water distribution networks modelling and monitoring *Water Supply* **20**(1) 46–58 https://doi.org/10.2166/ws.2019.131

[3] Rapp BE 2017 Conservation of momentum: The Navier-Stokes equation *Microfluidics:*

Modelling, Mechanics and Mathematics Chapter 11 pp 273–289 (Elsevier)

[4] Huysmans M and Dassargues A 2005 Review of the use of Péclet numbers to determine the relative importance of advection and diffusion in low permeability environments *Hydrogeology Journal* **13** 895–904 https://doi.org/10.1007/s10040-004-0387-4

[5] Eliades DG, Kyriakou M, Vrachimis S and Polycarpou MM 2016 EPANET-MATLAB Toolkit: An open-source software for interfacing EPANET with MATLAB *Proc. 14th Int. Conf. Computing and Control for the Water Industry (CCWI)* Amsterdam pp 1–8

[6] Delay F, Ackerer P and Danquigny C 2005 Simulating solute transport in porous or fractured formations using random walk particle tracking: A review *Vadose Zone Journal* **4**(2) 360–379 https://doi.org/10.2136/vzj2004.0125

[7] Romeo-Gomez P and Choi CY 2011 Axial dispersion coefficients in laminar flows of water-distribution systems *Journal of Hydraulic Engineering* **137**(11) 1500–1508 https://doi.org/10.1061/(ASCE)HY.1943-7900.0000432

[8] Piazza S, Sambito M and Freni G 2022 Definition of a numerical prior indicator to identify the relevance of diffusion and dispersion processes in water distribution networks *Proc. 39th IAHR World Congress* Granada, Spain pp 3711–3717.

Rehabilitation of the water supply network of Timişoara City from Timiş County in Romania

Alina Stanciu[1], **Vasile Lucian Pavel**[1], **Daniel Toma**[1] **and Cristina Mihaela Vîrlan Toma**[1]

[1] Faculty of Hydrotechnical Engineering, Geodesy and Environmental Engineering; "Gh. Asachi" Technical University of Yassy, Iaşi, 700050, Romania

alinaagafitei@yahoo.com

Abstract. In such an urban centre like Timisoara (the largest one from the West of Romania), the problem of water treatment for drinking, respectively wastewater collection and treatment, have always been issues of strategic importance to the community. The policies of the European Union to support the financing of the member states in the field of environmental engineering and protection represented the basic pillar in the development of the water supply network also in this case. In light of the increasing and pressing need to efficiently manage scarce water resources, there has been renewed interest by water distribution network owners to develop and implement water management strategies and tools that would assist in the integrated and automated management of those networks. The modernizations made at the treatment plants can be summarized in a few basic ideas: the automatic control of processes, lower costs, and higher safety. The most commonly used rehabilitation techniques for reducing both types of water losses are pressure management and pipeline rehabilitation. In Timişoara City, this modernization, and the rehabilitation of the water supply network led to a significant reduction of the environmental pollution (water, soil, air), and to a better protection of human health. Investments in modern technologies for water treatment, which allow an automatic monitoring of the process, translate into more safety in exploitation, and, implicitly, a better-quality control. The main objectives pursued in this study aim at: improving the quality of drinking water supplied to the population, ensuring continuous supply of drinking water, compliance with the regulations in force, both Romanian and European, and ensuring the access for the majority of the population at this system. There were followed the rehabilitation systems for street drinking water supply networks, also those for street sewerage networks, with modern constructive solutions.

1. Introduction

Timişoara was and is the largest city in the western part of the country. The social and economic life of the city has experienced, over time, an important dynamic. In such an urban centre, the problem of wastewater treatment and collection have always been issues of strategic importance to the community.

We can talk about modern water supply and sewerage systems only since the beginning of the 20th century, the first wastewater treatment plant in Romania being in operation in Timişoara, on October 26, 1912, and the first water plant of the city, on June 1, 1914. The water supply of this city in a centralized system was made in 1914, with the installation in operation of the treatment plant, called Plant no. 1, and a drinking water distribution network of 87.4 km. Plant no. 1 was located in the south-eastern extremity of the city. The treatment plant had a spray aeration stage, a pre-filtration stage, and a filtration stage. They were thus eliminated iron and manganese salts in water.

The main improvements of the water supply and sewerage system from Timişoara and from the County, made since 1990s, have been possible because Aquatim (Timişoara's Water Company) focussed, at the right time, on attracting external sources of funding. Metering, modernization of water plants and the new station of the city, completed in 2011, are example of such investments, which benefit, today, all the people of Timişoara, and which were made with European funds and money borrowed by Aquatim Society from the international banks.

The modernizations made at the treatment plants can be summarized in a few basic ideas: automatic control of processes, lower costs, and higher safety. In Timişoara City, the water pressure in the distribution network is achieved and maintained with the help of existing pumping stations at the treatment plants. No tanks were provided in the distribution network, consumption variations being solved by starting or stopping some pumps in pumping stations or, in the past, by partially closing the valves of repression. This operation mode led at large variations in the system pressure, generating high energy consumption.

2. Experimental

Timişoara is the residence city of Timiş County, from Banat, Romania (figure 1). It is located in western Romania, close to the borders with Hungary and Serbia, on the banks of the Bega River. In 2011, Timişoara had 319,279 inhabitants, and was the third largest city of our country. Timişoara is supplied with water from two sources, about 25-30% comes from drilling, and 70-75% from the source of surface water, and the flow captured from both sources reach over 4,260,000 l/h.

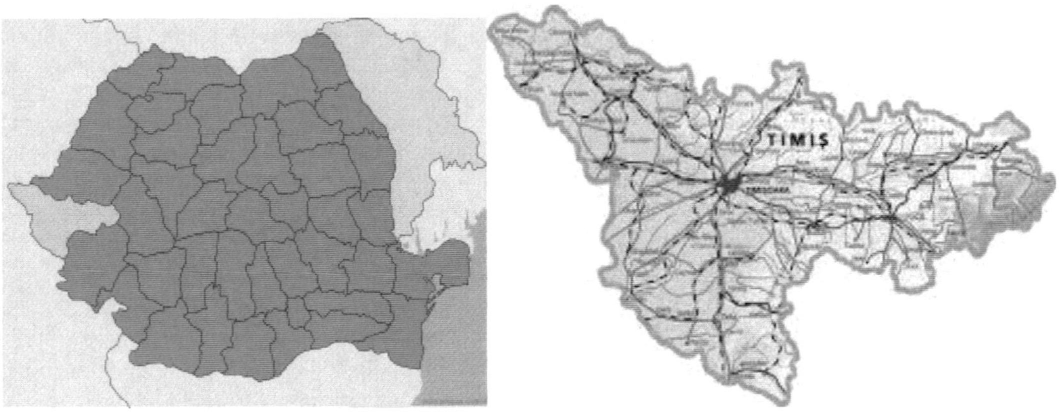

Figure 1. Location of Timişoara from Timiş County, Romania.

The surface water is captured from the treated Bega River, and distributed by the Bega Water Treatment Plant, and groundwater is captured by sea drilling depth, located in the East, South-east and West of the Timişoara City, treatment and distribution being provided by Urseni and Ronaţ Water Treatment Plant. Bega is, therefore, the main source of water for the city. However, the percentage of groundwater in the water destined for treatment increased, from 18% in 1999 to over 28% in 2011.

In fact, the increase was more pronounced until 2004, after which the ratio remained relatively constant until present. The Aquatim Company did not aim to develop water capture underground, despite its important properties (taste, smell) better for drinking water. Here's why the water operator choose to supply Timişoara with water mainly from Bega. When establishing the water supply of drinking water strategy for a city, it is taken into account, first of all, the safety of water supply.

In Timişoara, groundwater resources cannot provide the wate necessary, not even on a limited period of time, being low. They must be protected. Uncontrolled overexploitation of an aquifer may have immediate adverse effects.

Figure 2. The Water Supply Treatment Plant of Timişoara City, Romania.

In the previous years, there have been severe decreases of hydrostatic levels in the Timişoara area, because of the prolonged drought.

Through the existing catchment front, in Timişoara, a catchment flow of 450-500 l/s can be ensured, covering the mentioned percentage, of about 25%, under the conditions of maintaining the aquifer.

The need to develop hydraulic models as accurately as possible results from the fact that, by creating and calibrating the hydraulic model of a system, a variety of system problems can be identified, solutions can be mitigated to them by addressing the problem of rehabilitation, expansion or design of new systems.

The program used for networks dimensioning, and for creating a calibrated hydraulic model is an independent one. It is used separately for water supply systems, also for sewage networks. We also can use some hydraulic modelling programs that include both components. In the second case, hydraulic calculus and running of calibrated models is done individually. The statistical and numerical processing of our data was performed taking into account the already existing numerical data, and their interpretation, but also by their numerical conversion for all sets and types of field data. Different types of media, computing consoles and editing programs were used to process the data sets, also CAD type programs for processing data obtained from topographical surveys, and 3D processing of the topographical surveys data.

Data extracted from GIS database has been converted to numeric tabular data for water supply system objects. In order to achieve the hydraulic models for our research system, it was completed the GIS database with additional measurements made in the field. The results can be tracked both at the scale of time, and in real time by simulating the functioning of the system at these time scales.

Moreover, the development of existing or of new catchment fronts involves investments with huge initial costs, unjustified in relation with other investment priorities, and their future exploitation will be always limited by resource control.

On the other hand, if deep water is characterized by a relatively constant quality over time, water from Bega has the advantage of ensuring a constant quantity, the Timiş - Bega hydrotechnical node being a safety factor, and in terms of ensuring the necessary water. In addition, if necessary, water from Surduc Lake (figure 3) can be transferred to the Bega River.

Figure 3. Surduc Lake, from Timiş County, Romania.

Another important aspect is that the use of two sources for water supply of a City like Timişoara offers more security and flexibility, by the fact that they can be used alternatively, in case of need.

In terms of investments for water quality, the Aquatim Company has chosen to invests in modern technologies for water treatment systems, which allow the automatic monitoring of the process (for example, more safety in exploitation and, implicitly, better quality control) [1], [2].

The existing water treatment facilities were developed after 1945 and new facilities were established as well. The supply and collection systems were extended, as the industry developed. New technologies and equipment replaced the old ones.

Since 1993, process parameters (pH, turbidity, chlorine concentration, etc.) are automatically adjusted on the technological flow. In the water quality laboratory, more than 20 quality parameters are tested daily, before pumping into the city network.

Water samples are also checked weekly in the water distribution network, taken from 32 control points from the city, established by mutual agreement with the Public Health Institution [3].

There is also an online monitoring system network that allows measurement of the chlorine content in a few key points.

Analyses have performed for drinking water (pH, turbidity, *Clostridium perfrigens*, residual chlorine, etc.), and for wastewater (CBO_5, CCO-Cr, pH, ammonium, nitrates, etc.).

The monitored parameters are measured by instruments installed on the pipes, the information is then transmitted via GSM network to a central server, recorded in a database and processed in various reports.

Reports are accessed and followed, at any time, on the central dispatcher, with a computer monitor. The latest monitoring system implemented allows more control over the network operation, which means even more efficiency in interventions. Automatic monitoring of the chlorine concentration in water (from the distribution network) brings extra security consumers.

Systematic and continuous rehabilitation of the water network represents an essential component of the water Company's investment programs. With the city's development, the gray cast iron pipes could no longer ensures the necessary flow for new consumers. In these pipes, with a diameter of 50 –

150 mm, other problems have arisen, such as shearing and breaking the pipe material as a result of temperature changes.

The damage thus generated is significant, given the weight, of about 30% of them, in the City's network. Inside these pipes, deposits are produced, which modify the roughness coefficient, influencing negative hydraulic flow parameters, the flow and pressure.

Successive stops and starts, as well as any change of the flow regime, leads to the dislocation of the deposits, and the staining of the water. After each one stop it is necessary to wash the pipes, to remove water with suspensions.

Presence of suspensions it also has a negative impact on consumer perception, raising doubts about water quality.

Modern materials from which pipes are made aimed at eliminating the disadvantages, with the main purpose of extending their life period, reduction of assembly work, disposal execution errors, due to the human factor, and maintaining the quality for the water transported through these pipes [4], [6].

For sewerage systems modernization in the County, Aquatim Company has a lot of work to do, complete with about 200 km of new sewer pipes, works financed from European funds. The evolution of water and sewerage networks length is shown in figure 4.

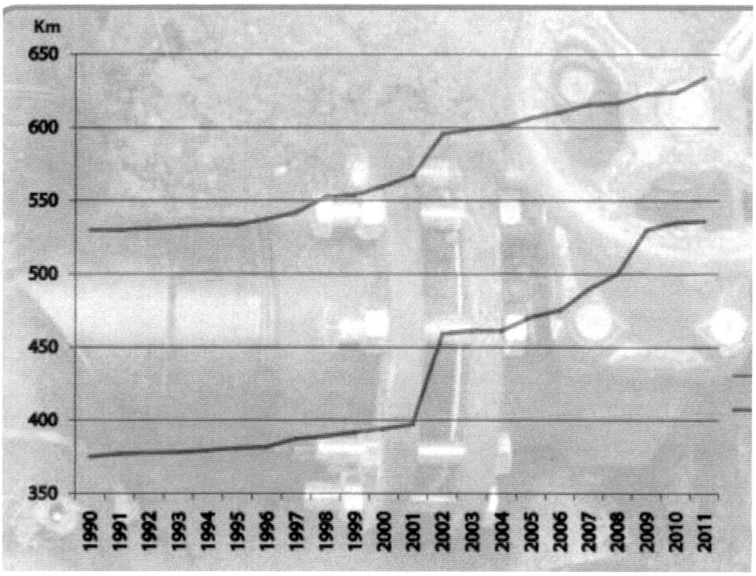

Figure 4. The evolution in time of water (--) and sewerage networks length (--).

Losses detection is done planned, with equipment specialized and competent staff. Very helpful for streamlining the operation and network maintenance is the implementation project of the Geographic Information System (GIS), used for storing and processing the geographical data.

This involves collecting field data, with topographic equipment specialized and subsequent saving in digital format. With the implementation of GIS system, the resources required for asset maintenance will be significantly reduced.

Consequently, Aquatim Company became a regional company, operating in more than 40 localities throughout the county. 119 million Euros, 300 km of water supply and sewerage pipes, three new water treatment plants, seven new waste water treatment plants and 390,000 customers who will benefit from improved services – this is the summary of the investments planned by Aquatim for the region [1], [5], [7].

The coverage rate of the water supply and sewerage services in the localities included in the project proposed for Timiş County's water network will raise up to 99% and the quality of the services will be in accordance with the European Union requirements.

Figure 5. Drills made in Timiş County, Romania (at Jimbolia and Ciacova).

3. Conclusions

An efficient rehabilitation plan should usually improve all performance metrics, but different researchers have addressed specific performance metrics.

The efficient and effective management of existing water distribution systems faces challenges related to aging of infrastructure, population growth, extended urbanization, climate change impacts and environmental pollution.

It is obvious the need to implement the planning process of investments in water-sewer infrastructure, necessary in Timiş County, investments related to the programming period (2014 – 2020), from European non-reimbursable funds, taking into account the territorial coordination of the interventions.

After our researches regarding the performance of the existing water distribution system (including the hydraulic modelling), it is necessary to follow a sustainable development, an unique investment planning framework in this County, with the possibility of using different structural tools or other available sources of funding, territorial approach of planning based on the concept of agglomeration, emphasizing performance and monitoring of results, and a more good funds management.

References

[1] Agafiţei A and Gabor V 2018 *Environmental Engineering – Course* (in Romanian: *Ingineria Mediului – Curs*) (Iaşi, Romania: Editura PIM) ISBN 978-606-13-4228-0, 312 p.

[2] Agafiţei M, Agafiţei A and Marcoie N 2000 Software solutions for dimensioning of water supply networks *Bulletin of the Polytechnic Institute of Iasi, Section Hydrotechnics* **XLVI**(L) fascicula 1-4 pp 65–72

[3] Geml J 2018 *Alt-Temesvar im letzten Halbjahrhundert 1870-1920* (Timişoara, Romania: Editura Helicon)

[4] Ilieşiu N 2003 *Historical Monograph of Timişoara* (in Romanian: *Timişoara Monografie Istorică*) 2nd Edition (Timişoara, Romania: Editura Planetarium)

[5] Toma D, Marcoie N and Boboc V 2016 Monitoring and controlling the operational and energy-related parameters from the Boldeşti re-pumping station within the „Boldeşti – Cotnari" water supply system *Bulletin of the Polytechnic Institute of Iasi, Section Hydrotechnics* **62**(64) 1-2 pp 21–35

[6] Zănescu Al 1974 *Pages from the History of Timişoara Water Supply* (in Romanian: *Pagini din istoria alimentării cu apă a oraşului Timişoara*) (Timişoara, Romania: Editura Tibiscus)

[7] Vlaicu I and Haţegan I 2012 *Timişoara Water Sypply – History, Present and Perpectives* (in Romanian: *Alimentarea cu apă a Timişoarei – Istorie, prezent şi perspective*) (Timişoara, Romania: Editura Brumar) https://www.aquatim.ro/media/1041/alimentarea-cu-apa-a-timisoarei.pdf

IOP Conf. Series: Earth and Environmental Science 1136 (2023) 012051 doi:10.1088/1755-1315/1136/1/012051

Performance of water valves required by water supply network digitalization process

Carmen Anca Safta[1], Valentin Nicolae Cococi[1], Constantin Călinoiu[1] and Alexandru Marin[1]

[1] Faculty of Energy Engineering, University POLITEHNICA of Bucharest, Bucharest, 060042, Romania

carmen.safta@upb.ro

Abstract. Digitalization process has the capacity to increase productivity and efficiency, to improve the business and reducing costs. Working with big data, using sensors, meters, digital controls and analytic tools, or artificial intelligence in smart water management, the water industry has known a profound transformation with the aim to improve the water services and consumption in the context of our day's water threats. The water supply networks are exposed permanently to more leaks when water pressures are high which mean higher energy costs and more water losses. For this reason, water supply systems need to monitor in real time the main hydraulic parameters and to control and drive all the hydraulic equipment, to predict consumption and to anticipate the problems. The new water technology is using valves which regulate the water inlet to a system and receiving real-time wireless pressure values. Staring from the quality of services in water distribution systems, the present paper presents a pneumatic actuator used to control and drive water valves. The pneumatic actuator is controlled by a proportional valve. The numerical simulation results highlight the pneumatic actuator performance regarding the dynamic behaviour.

1. Introduction

With the fourth Industrial Revolution (4IR) new concepts, like Artificial Intelligence, Internet of Things, Big Data, Virtual Reality, Augment Reality, have been developed. Digital transformation and automation can erase the limits between physical objects and turn them into interconnected and interdepended elements of complex systems [1].

Digital transformation that marks the economic activity of many sectors has begun to influence the water industry, too. There are benefits of digitalization in water industry as the improvement of connectivity through a high level of using different kinds of sensors to a better monitoring, controlling and management of devices, processes, and operations [2-7]. But the benefits came with high risks, too, as could be seen in aircraft and automotive industry [2].

For this reason, in the development of water digitalization process the human operators must in the process loop and must have a very good background in automation training, in order to be capable to answer quickly with manual feedback in the system, if is necessary [2].

Because the digitalization can help not only to collect a lot of data parameters in the technology process of drinking water and wastewater, but also to analyze, optimize and produce a better decision-making using SCADA (Supervisory Control and Data Acquisition) System, the cybersecurity of any digital process is in fact the backbone of digitalization [8]. Using smart sensors and other Internet of

Things technologies that can use historical data, new applications are developed for the monitoring of different operations in water industry or in optimization of the resources use and management of the maintenance equipment. [1,9,10].

In water supply, systems valves are used in distribution water or wastewater network. Valves operate to start or stop the flow, to control the flowrate or the pressure in the system, to control flow direction or to adjust the flow rate. The resilience of a water distribution network or pipeline network depends on the valve's status [10].

The purpose of this paper is to determine the dynamic performance of a valve used in flow control, in a water supply system when pressure disturbances occur in the system. The valve is operated by a pneumatic actuator which in turn is controlled by a proportional pneumatic distributor. After a brief presentation of the requirements of water supply systems in the concept of digital transformation, the second section presents the flow characteristics specific to the valves and their mode of operation in a water supply system. The third section presents the pneumatic servomechanism operated with proportional distributor and the numerical simulation scheme used. The fourth section presents the results of the numerical simulations for different reference flows, at a given disturbance. The conclusions and the bibliography conclude the paper.

2. Flow characteristics of the valves

Valves are mechanical devices that control the flow (in pressure, flow rate and direction) in any kind of installation [9,10]. Valves are everywhere! With a simple design the main part of a valve is the obturator knows as disk, plug or ball, depends on the geometrical form. The obturator is moved in the fluid current by a stem that has a linear or rotary movement been powered by an actuator. In an automatic drive system are used electric, hydraulic, or pneumatic actuators.

It is known that in a water network, the pressure must be kept at the low values required by the consumers. A higher pressure leads to higher costs because old pipes of the distribution system are affected. In Figure 1 is represented the variation of the operating point in a water pumping system when the pressure is kept constant, and when the flow rate is constant, too.

If in the water distribution network, the flow rate increase with ΔQ, the pump characteristic must be changed by increasing the speed of rotation, so that the pressure remains constant. The flow rate in the pipe is increasing to respond to the consumer needs.

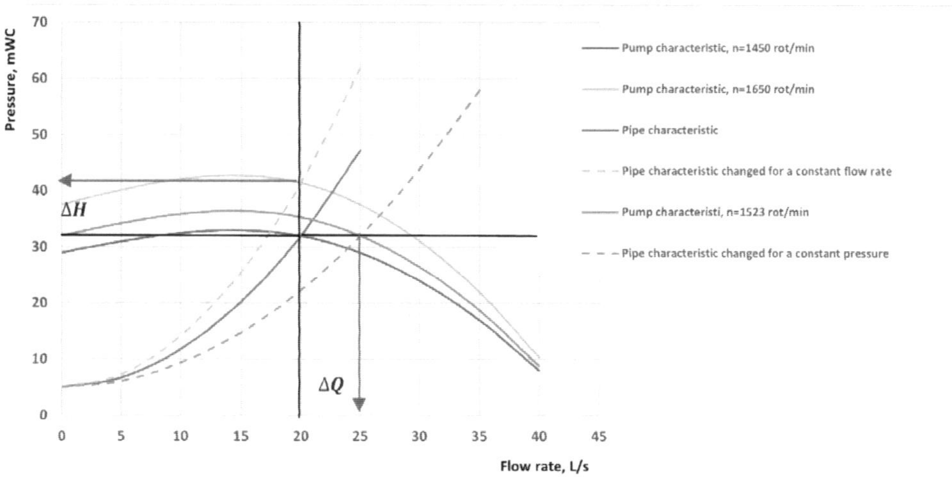

Figure 1. Operating points in a pumping system.

When the water supply system is operating by keeping the flow rate constant to a reference value x_{ref} the water valve is driven by using the system control represented in Figure 2. The controller C will transmit to the system an electrical or digital signal $x_{control}$ to act the actuator system of the valve and it will change the obturator position in the valve. The flow rate will be changed until the flow rate in the system measured with the flow meter FRT will be the same as the reference required by the system.

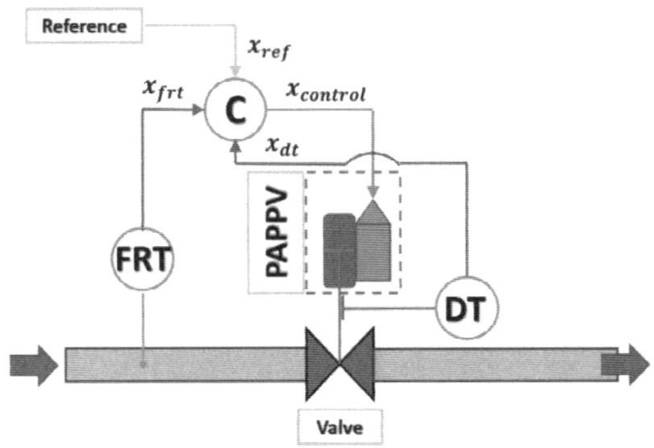

Figure 2. Sketch of an automatic flow control system: FRT – flow meter, C – controller, DT – displacement transducer, PAPPV – pneumatic actuator with proportional pneumatic valve.

A valve is a hydraulic resistance from the hydraulic point of view. Three factors are important when sizing a valve: the flow characteristic, C_v coefficient and R rangeability. Valve flow characteristic is the relationship between the valve opening position and the C_v value when the valve opening position varies from 0% to 100%. It is called usually "inherent flow characteristic" [11].

The C_v is the flow coefficient and represents the maximum flow rate that can pass through the valve. The flow equation used for liquids and respecting the measurement units is [11,12]:

$$C_v = 1.17Q\sqrt{\frac{G}{\Delta p}} \tag{1}$$

where Q is the flow rate (m^3/h), G specific gravity (-), Δp pressure drop on the valve (kg/cm^2).

The inherent flow characteristic is a theoretical characteristic, determined in laboratory test bench, when a constant Δp across the valve can be obtained. In a real hydraulic system, Δp cannot be held constant and the inherent flow characteristics are replaced with the installed flow characteristics. There are three types of inherent flow characteristics: quick opening, linear and equal percentage characteristic. In Figure 3 are plotted the linear and equal percentage flow characteristics.

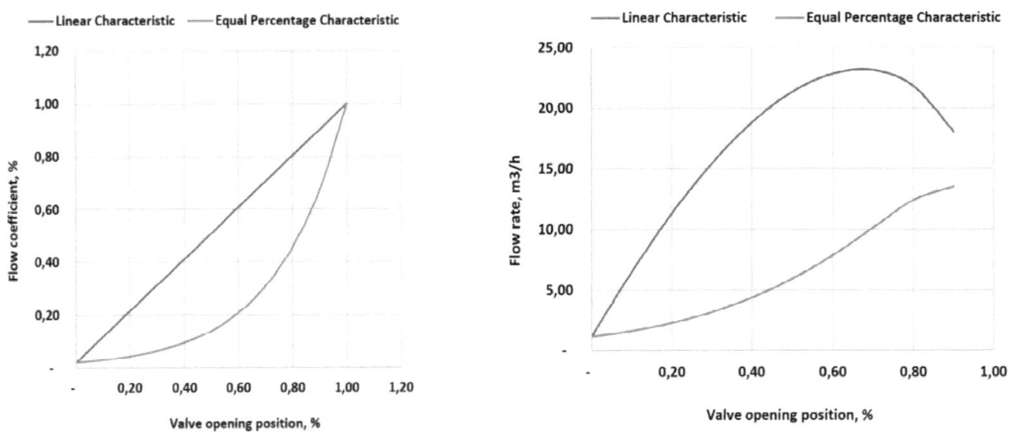

a) b)

Figure 3. Flow characteristics: a) inherent flow characteristics, b) installed flow characteristics.

For a linear characteristic, the valve opening position and the flow rate are proportional and $C_v = Ks$ with K a constant and s the valve opening position. For an equal percentage characteristic is a proportion between the ratio of the change in flow rate to the change in valve opening position (dC_v/ds) and the flow rate before the change C_v expressed by [11]:

$$\frac{dC_v}{ds} = K \cdot C_v \qquad \text{and} \qquad C_v = e^{K(s-1)}. \tag{2}$$

So, the flow characteristics can be expressed as a relationship between the flow rate and the valve opening position, Figure 3.b.

The rangeability is defined as the ration of the maximum flow rate C_{vmax} to the minimum flow rate C_{vmin} that can be controlled by the valve. This is the third factor used in valve sizing [12]. In Figure 4 are represented the influence of C_{vmax} in the case of a linear flow and equal-percentage characteristics.

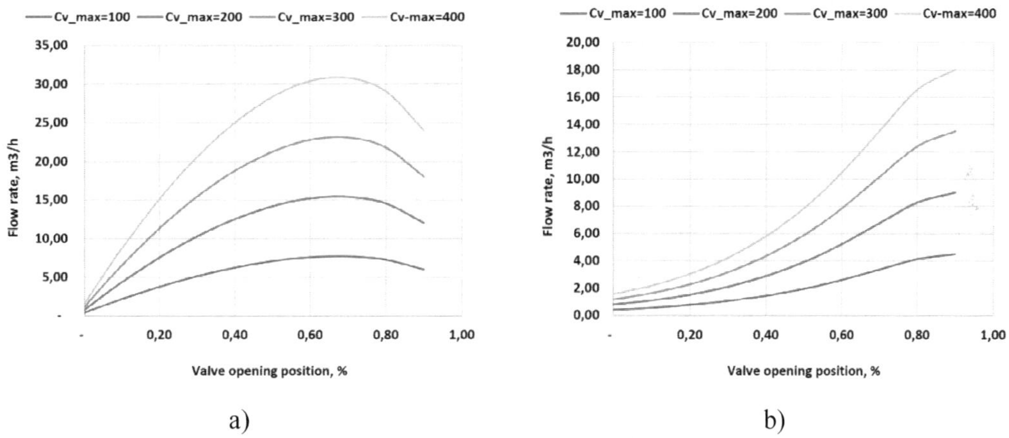

Figure 4. The influence of C_{vmax} in flow characteristics: a) linear, b) equal-percentage.

From the flow characteristics it is observed that for a linear characteristic the flow increases rapidly with the opening of the valve, while for equal-percentage curves the increase of the flow is almost linear. Such a feature is very advantageous for flow control. Also, on the linear characteristic, can be seen that with the increase of C_{vmax} the values of flow at different openings are increasing. At large openings the flow curve flattens out and the controllability deteriorates. Therefore, in the control processes, the operation of the valve at large openings will be avoided if the C_{vmax} has high values.

3. Pneumatic actuator characteristics

Control valves are actuated using pneumatic actuators. The most popular types in use are spring and diaphragm pneumatic actuators. Pneumatic piston actuator is double-acting and provides maximum force in both directions and fast movement speed [13].

Control valves are installed in many applications and their performances of speed response, stability and accuracy are improved by using electronic devices: controllers (DCS-distributed control system, PLC-programmable logic controller), sensors and transducers, limit switches. So, digital valve controllers are developed [13]. They are control valves with a position control servosystem. More than controlling the position of the valve obturator, the microprocessor inside the digital valve can analyze and store the diagnostic tests of the valve that are communicate with the control system via a digital signal [13].

The development of pneumatic drive equipment currently allows the use of more efficient drive systems that can be used in water valves control. In papers [14,15] the authors proposed a pneumatic positioning servomechanism that is a pneumatic cylinder driven by a proportional distributor operating in a closed loop. Numerical and experimental results show that the use of the proportional pneumatic distributor offers a good dynamic behavior for different reference signals applied to the system, even in the presence of a disturbing force. Based on the results of previous research [14,15], in this paper we analyze, by numerical simulation, the behavior of the pneumatic servomechanism with proportional distributor used as a valve actuation system for a water distribution systems. It is studied the behavior of the pneumatic actuator for a step signal type reference applied in the presence of a disturbing force that act through the water valve against the pneumatic actuator. The disturbing forces in water distribution systems are pressure forces and occur when operating conditions change.

The pressure forces in the water valve are acting as normal forces on valve obturator. Considering a friction coefficient with 0.1 value, the resistance forces taken by the piston of the pneumatic actuator are calculated, Figure 5. In the case of a pipe with inner diameter of 3.5 inches (≈ 89 mm) the static pressure force is 2500 N and the resistance force 250 N.

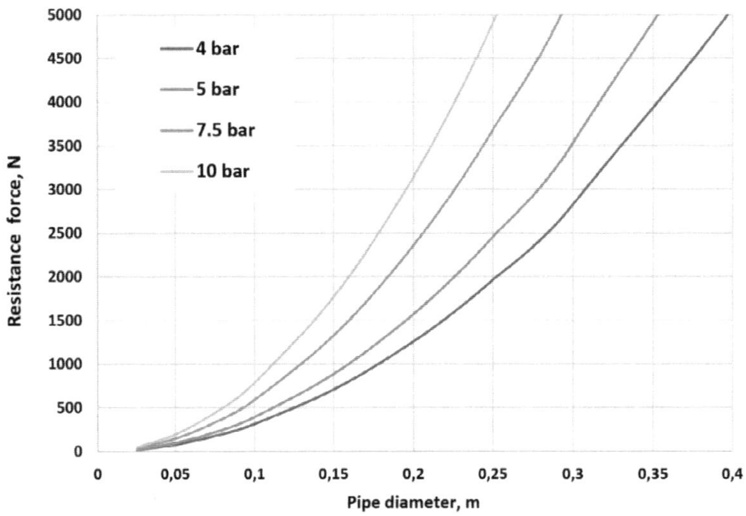

Figure 5. Resistance force on the actuator piston for different pipe diameter size.

We assume that it is used a globe valve to control the flow rate in the water distribution network. The valve is acted with a pneumatic actuator that is a double-acting pneumatic cylinder powered and controlled by means of a proportional pneumatic distributor with four ports and three positions (4/3). Because the pneumatic actuator is operating in closed loop it is used a proportional controller having the coefficient gain K_p =1. The pneumatic cylinder is supplied with 7 barA compressed air and the numerical simulation scheme is represented in Figure 6.

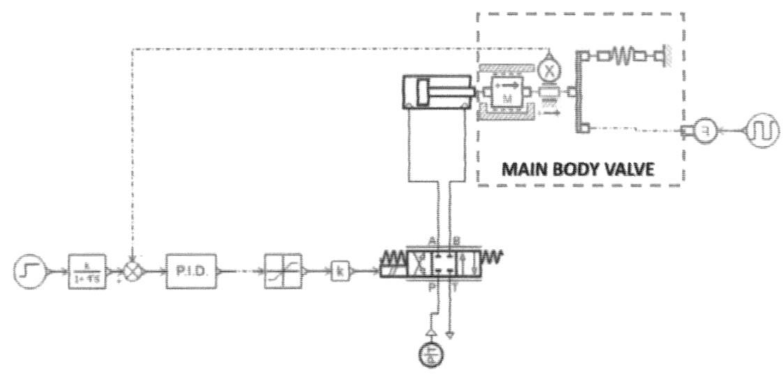

Figure 6. Numerical simulation scheme (in AMESim environment software) of the pneumatic servomechanism used to act a water control valve.

In Table 1 are given the parameters of the pneumatic equipment used in numerical simulations.

Table 1. Numerical model parameters.

Elements of the pneumatic water valve drive	Parameter	Value
Double-acting pneumatic cylinder	Active piston area	490.87 mm^2
	Length stroke	200 mm
	Viscous friction coefficient	150 Ns/m
	Thermal exchange coefficient	50 $J/m^2/K/s$
	External temperature	293.15 K
Proportional pneumatic distributor (4/3)	Area of control orifices	28.27 mm^2
	Flow control coefficient	0,72
	Valve natural frequency	70 Hz
	Valve rate current	10 mA
	Inertial mass	2 kg
	Elastic constant	1000 N/m
	Gain of the displacement sensor	50 V/m

The input reference was considered a step with an amplitude of 2 V, 5 V and 8 V and a delay of 0.5 seconds, Figure 7.

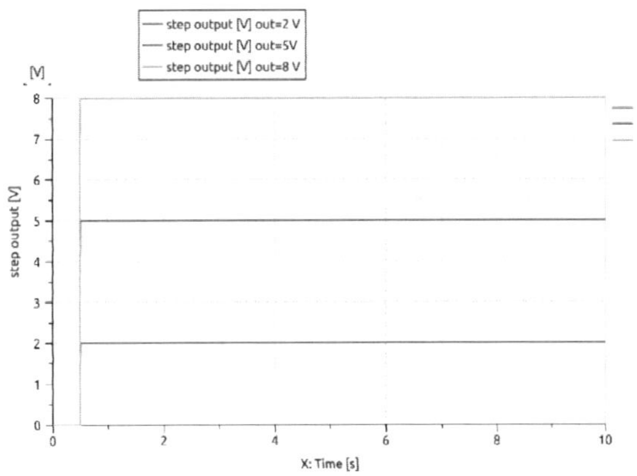

Figure 7. The reference step input signal.

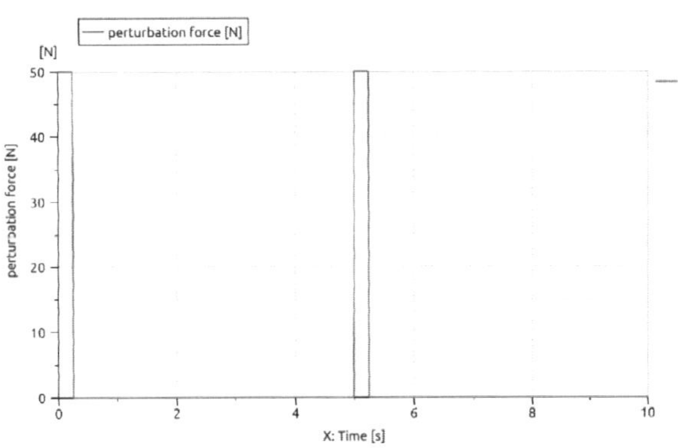

Figure 8. Perturbation force.

IOP Conf. Series: Earth and Environmental Science 1136 (2023) 012051 doi:10.1088/1755-1315/1136/1/012051

The perturbation force is plotted in Figure 8 and is a variable perturbation force in time, having an amplitude of 50 N which appears for 0.24 s at each 5 s.

4. Numerical simulations

The aim of this study is to determine the dynamic performance of a drive system with proportional pneumatic servomechanism for operating a valve used in water supply to control the flow in the distribution network.

By using the numerical simulation model, parameters, input signals and perturbation force are obtained different responses from the pneumatic actuator. The mentioned input signals were modified by using a numerical block with a delay of first order having a constant $T = 0.1$ as in the numerical simulation scheme Figure 6. In such manner it was obtained a PT1 input type that is a softer input, Figure 9. Physically, PT1 is obtained as a digital signal or electronic by using a capacitor. Regarding the perturbation force, it was considered an overlap of the disturbing force with an elastic force. This type of disturbance simulated the appearance of pressure waves in the piping system, Figure 10.

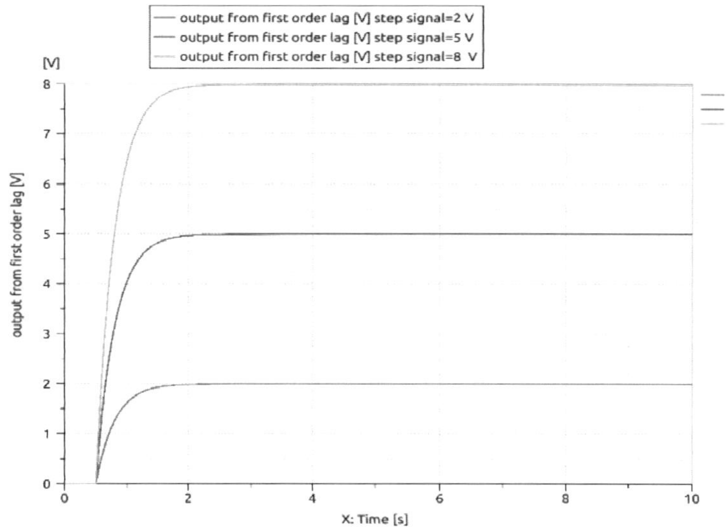

Figure 9. The modified input signal step in a PT1 input.

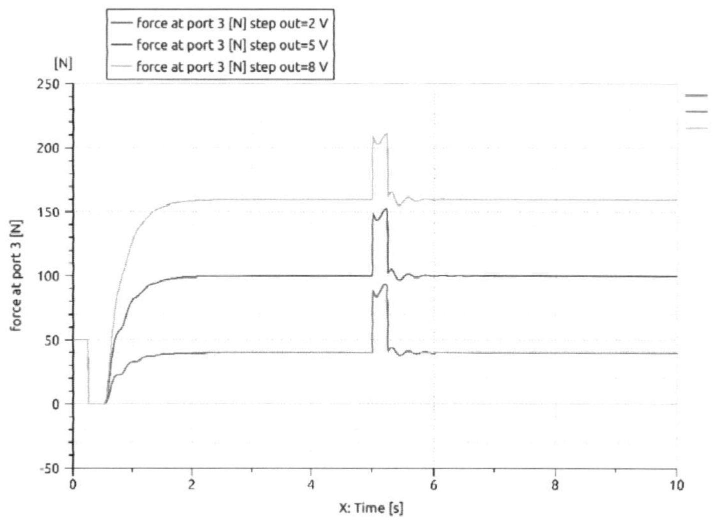

Figure 10. The modified perturbation forces.

In Figure 11 is plotted the piston displacement of the pneumatic cylinder, considering that the piston starts moving from the middle of the total stroke of the cylinder. For different amplitudes of the input signals the displacements are proportional with the input amplitude and the response answer is asymptotic stable in time. The system is overdamped because the overshoot is zero. Even in this case it is observed that over the delay of the input system is added the delay of the pneumatic servomechanism response [16].

In Table 2 are mentioned the parameters obtained for the dynamic behavior of the pneumatic actuator and proportional pneumatic distributor used as positioning servomechanism system.

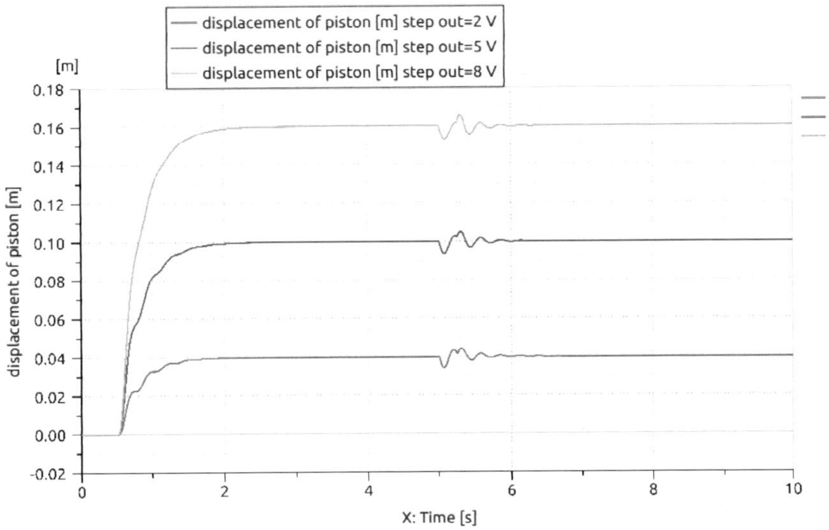

Figure 11. Piston displacement on step input signals.

Table 2. Dynamic response of the system.

Parameters	The reference (input signal)		
	2 V	5 V	8 V
Stationary piston's displacement (m)	0.04	0.1	0.16
Delay answer (s)	0.52	0.52	0.52
Time response on the input signal(s)	2.93	2.64	2.3
Time response on force perturbation (s)	1.4	1.5	1.5
Maximum amplitude of the piston displacement to force perturbation (m)	Min 0.153 Max 0.166	0.093 0.105	0.034 0.044

The response of the system to perturbation force is seen on all the diagrams represented and the resilience of the system in taking over this disturbance is observed.

Comparing the disturbing signal of perturbation force, Figure 10, with the system response shows that the response is prompt, without delay in taking over the disturbance, and the amplitude of the disturbance reflected in the piston movement has small values, even at a low intensity input signal.

In the speed variation it is noticed that the perturbation force is taken away in one second, no matter the amplitude of input signal, and the system movement is continued with the same speed, Figure 12. In the case of the air pressure variation in the cylinder chambers, Figure 13, is observed that the pressure in the active chamber has a value with 0.5 bar bigger after the perturbation than before, probable because of the friction force in the cylinder.

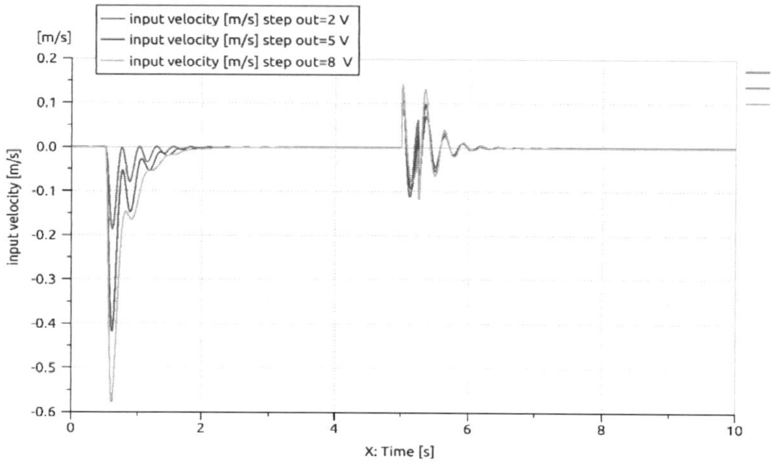

Figure 12. Piston velocity.

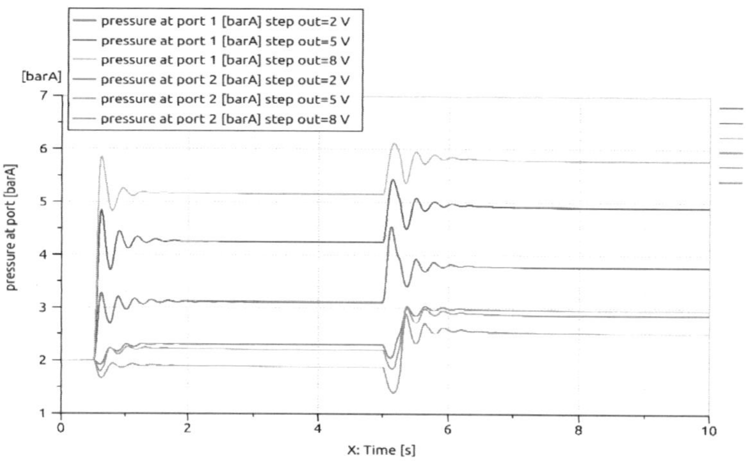

Figure 13. Pressure variation in the cylinder chambers.

5. Conclusions

Digital transformation in the field of water engineering needs not only smart flow meters, sensors, and devices necessary for remote communication of data, but also to develop new drive systems that support an electronic or digital interface in the control of classical water valves.

In this paper it was proposed a closed loop pneumatic system for operating a valve that must keep constant the flow rate at a reference value in the water supply system. The proposed drive system is a pneumatic actuator with a double-acting cylinder driven by a proportional distributor. The dynamics of the positioning pneumatic servomechanism having an input signal reference with different amplitudes and 0.5 s delayed from the initial moment was studied. The behavior of the servomechanism has also been studied if an external impulse force interferes at some point for a short time in the operation of the system.

The results of the numerical simulations show a good dynamic behavior of the pneumatic servomechanism, with a delay response for the step signal and an overdamping. The servomechanism positions the valve stem obturator to the prescribed value according to the input signal and responds to the external force disturbance by annihilating it. Thus, when the disturbance occurs, the system takes over the effect of the disturbance, remains stable and maintains its position according to the input signal.

From the analysis of the pressure variation, an increase of the pressure in the active chamber of the cylinder was observed by 0.5 bar compared to the existing pressure before the application of the disturbing force. This result requires an analytical study of the mathematical model associated with the studied simulation model to see how the frictional forces influence the behavior of the system.

The results obtained in this study encourage us to recommend the use of pneumatic positioning systems with cylinder and proportional distributor to operate the water valves, regardless of the type of flow characteristic. The type of actuation presented in this paper also has the advantage of a digital control which facilitates their integration in any digitization process.

References

[1] Stankovic M, Hasanbeigi A and Neftenov N 2020 *Use of 4IR Technologies in Water and Sanitation in Latin America and the Caribbean* (Technical report) Inter-American Development Bank https://www.fondosdeagua.org/content/dam/tnc/nature/en/documents/latin-america/Use-of-4IR-Technologies-in-Water-and-Sanitation-in-Latin-America-and-the-Caribbean.pdf

[2] Savić D 2022 Digital water developments and lessons learned from automation in the car and aircraft industries *Engineering* **9** 35–41 doi.org/10.1016/j.eng.2021.05.013

[3] *** 2021 *Digitalization in Water Industry* https://www.azom.com/article.aspx?ArticleID=20323

[4] Maderic J and Patel A 2021 Demystification digital transformation in rafineries *Hydraulic & Pneumatics* September pp 14

[5] Cassidy J, Barbosa B, Damião M, Ramalho P, Ganhão A, Santos A and Feliciano J 2021 Taking water efficiency to the next level: digital tools to reduce non-revenue water *J. Hydroinform.* **23**(3) 453–465, 10.2166/hydro.2020.072

[6] Kiliç R 2021 The strategic development for water loss prevention *Applied Water Science* **11** paper 28, https://doi.org/10.1007/s13201-020-01322-z

[7] Di Nardo A, Boccelli D L, Herrera M, Creaco E,Andrea Cominola A, Sitzenfrei R and Taormina R 2021 Smart Urban Water Networks: Solutions, Trends and Challenges *Water* **13** 501 https://doi.org/10.3390/w13040501

[8] Public Utilities Board Singapore 2016 Managing the water distribution network with a Smart Water Grid, *Smart Water* **1** paper 4 DOI 10.1186/s40713-016-0004-4

[9] Pérez-Padillo J, Morillo J G, Ramirez-Faz J, Roldán M T and Montesinos P 2020 Design and implementation of a pressure monitoring system based on IoT for water supply networks *Sensors* **20** 4247 doi:10.3390/s20154247

[10] Blokker M, Di Cristo C, Gentile A, Gargano R, van Laarhoven K, Leopardi A, Quintiliani C and Vertommen I 2021 Optimal valve operation for restoring functionality of WDN during critical events *Environ Sci Proc 2020* doi:10.3390/environsciproc2020002032

[11] Marinoiu V, Poșchină I, and Stoica M 1980 *Control Valves* (in Romanian) (Bucharest: Editura Tehnică)

[12] Yokoyama K 2017 *Control Valve Characteristics (1707)* PreFEED Corporation https://www.prefeed.com/common/pdf/1707E.pdf

[13] *** 2019 *Control Valves Handbook* 5th edition (Marshalltown, USA: Emerson Ficher) https://www.emerson.com/documents/automation/control-valve-handbook-en-3661206.pdf

[14] Cococi V N, Călinoiu C, and Safta C A 2021 Pneumatic actuator controlled by proportional valve. Experimental results *E3S Web of Conferences* **286** 04010 https://doi.org/10.1051/e3sconf/202128604010 TE-RE-RD 2021

[15] Cococi V N, Safta C A, and Călinoiu C 2020 Numerical simulation approach of a pneumatic actuator to force perturbation *EPE Proc. 11th Int. Conf. and Exp. on Electrical and Power Engineering* 9305558 pp. 589–593

[16] Cococi V N, Safta C A and Călinoiu C 2021 Parameter tuning process for a closed-loop pneumatic actuator *IOP Conf. Ser.: Earth Environ. Sci.* **664** 012030, doi:10.1088/1755-1315/664/1/012030

IOP Conf. Series: Earth and Environmental Science 1136 (2023) 012052 doi:10.1088/1755-1315/1136/1/012052

Review of leakage detection in water distribution networks

Ahmed Negm[1], Xiandong Ma[1] and George Aggidis[1,2]

[1] Lancaster University Energy Group, Engineering Department, Lancaster University, Lancaster, LA1 4YW, United Kingdom

[2] Author to whom any correspondence should be addressed

G.Aggidis@lancaster.ac.uk

Abstract. The conservation of water is, justifiably, a huge concern to water companies, regulatory bodies, environmentalist and more but, unfortunately, water loss through leakage can remain undetected for long periods of time without effective leakage detection. Leakage detection in water distribution networks has taken many forms through investigating varying properties of leakage. Understanding the characteristic leakage types and properties introduces the different emerging technologies. Even though some methodologies have gained popularity in the past decade, the need to establish a complete, economical leakage detection solution that effectively identifies background leakage as well as burst events persists. The benefits and limitations of the aforementioned technologies has often confused water utilities on adapting the most suitable method. Therefore, there is an arising need to classify and benchmark leakage detection practices. This paper reviews technologies in leakage detection contrasting hardware & software, intrusive & non-intrusive, steady state & transient, single & hybrid methods. A particular focus will be placed on scoping the projected direction of leakage detection and localisation. As anticipated, the various techniques refined over the last two decades introduce different capabilities, conditions, and constraints[1]. Assessing and comparing those methods will provide a deeper novel understanding of the research area thus paving the way for novel solutions.

1. Introduction

The search for a robust leakage detection and localisation method has been the interest of the water industry for the past two decades making it a well-developed research area. This is mainly due to the economic loss that water utilities incur through leakage and the resulting non-revenue water. On average, water networks leak 20% to 30% of the water distributed through them totalling around £7 billion of revenue loss through direct and indirect damage [2]. Additionally, the abusive arm of leakage extends to the environment through increased greenhouse gas emissions from pumping the water across the network. Contaminations from leakage often causes the water quality to decrease beyond acceptable levels thereby risking the health of the public. The occurrence of leakages in WDNs relies heavily on pipe infrastructure such as age-induced corrosion, inadequate fittings, and other sorts of pipe deterioration failures [3]. Operational factors that cause disturbances in the flow within the pipes increase the probability of leakage and burst events. These hydraulic shortcomings can arise from cyclic pressure loading or transient surges [4].

Despite the availability of various hardware and software-based detection equipment, the varying limitations and technical challenges exposes a gap in this field.

2. Leakage Overview

Unreported leakage can be broadly identified into two types: Burst and Background leakage [5]. Burst leakages are often detected through their clear properties such as acoustic emissions (AE) and significant pressure reduction [6],[7]. In contrast, background leaks are often small water loss through fittings, creeping joints or small cracks which do not have inherent detectable qualities. As a result, background leakage often run for longer causing adverse losses to the network [5]. Across the literature, the terms burst events and burst leakage are interchangeable whilst background leakage is referred to as leakage [7]. The detection of leakage can be summarised to three phases that dictate important objectives: Identify, Localise and Pinpoint (ILP) [2,8]. The identification phase is concerned with successfully differentiating leak signals from other network signals, such as fire hydrants, to determine the presence of a leak in the network with little or no false alarms [2]. The second phase of the ILP approach is the localisation stage. This focuses on finding the general section of the network such as a DMA [9]. Pinpointing attempts to site the exact location of the leak down to a radius of 20cm. Formerly, the pinpoint phase were two separate phases (locating and pinpointing, ILLP) where locating signifies estimating the leak location to a 30cm radius. The 10 cm difference makes merging the two phases logical [2].

The detection of leaks requires a meaningful understanding of its hydraulic anatomy and detectable properties. Habitually, a leak induces a sudden pressure decline at its location that spreads through the pipes in a set of waves which could be detected through negative pressure wave (NPW) strategies [10]. This pressure anomaly is difficult to detect for background leakage events and could be an indication of unaccounted demand (e.g., fire hydrants) nevertheless pressure fluctuations have been the basis of several leakage detection techniques. The inversely proportional relationship between pressure and flow rate dictates that this decrease in upstream pressure will trigger a decrease in downstream flow rates. Pressure and flow changes are the most exploited characteristics of burst events [11]. Another measurable leak quality is the resulting acoustic emissions released by the loss of water. These vibrations display several wave properties such as reflection, refraction, absorption, and diffraction that can be exploited to identify and locate burst events [12]. These waves can be collected through a variety of sensors such as dynamic transducers, accelerometers, or microphones [13]. Temperature anomalies in the vicinity of a leak arise making it another identifier that could aid in its detection [7].

Considerable research has been taken to investigate detection strategies in water networks making it a diverse field. Therefore, it is necessary to divide these approaches into appropriate categories. A novel classification tree was formulated to help navigate new readers to the research area (figure 1). The easiest discrimination of leakage detection methodologies in literature can be between hardware-based and software-based. Hardware leakage detection highlights the different sensing methods to identify and locate leakage in a network that exploit the characteristics (acoustic, pressure, flow, temperature, etc). These can be further refined into intrusive, robotic, in-pipe systems or non-intrusive, out-of-pipe systems. However, software-based detection is more concerned with the computational and data analysis of network parameters to extract leakage information. It exceeds hardware methods in its ability to assess leakage for steady state and transient flows.

3. Hardware detection methods

In this section we uncover the varying hardware technology developed for leakage recognition and localisation in literature. This is defined as any form of device that can be used to sense or collect data that can identify leakage anomalies described in Section 2. Some of these hardware detection methods are often paired with data processing software method to provide a more accurate leakage detection strategy.

3.1. In-pipe inspection devices

Intrusive devices are an underdeveloped subsection of hardware detection methods that revolve around inspection devices that enter the pipe networks to explore leaks. Robotic inspection devices vary greatly depending on their system characteristics which include their driving method, sensing technology and level of autonomy [14].

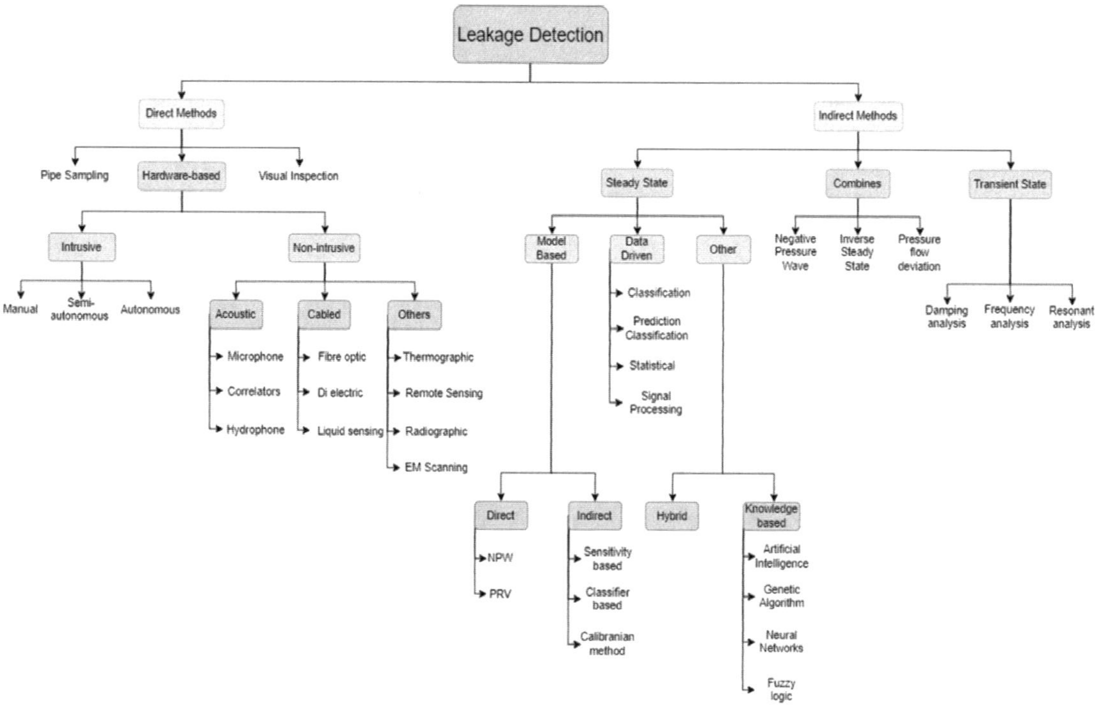

Figure 1. Leakage detection classification tree.

3.2. Driving methods

Generally, moving mechanisms can be defined as passive or active where passive approaches rely on the flow of water to inspect the pipe and active approaches are equipped with their one or more actuators to achieve the desired motion. Passively propelled inspection robots are often named PIGs (pipeline inspection gauge) [15]. PIGs are highly effective, safe and economical devices due to their simple nature and navigation system [14,15]. They have proven useful in clearing deposits gathering on the interior of the pipes in addition to their main role to assess the pipe status and detect leakages. PIG inspection systems utilise one or more sensors such as ultrasonic and eddy current sensors [16]. The navigation is usually achieved through odometers, visual sensors and inertial measurement units [15]. PIGs' motion can be problematic at higher flows where it is hard to halt their motion and when passing through corners in pipe infrastructure however intelligent PIGs might present better speed control capabilities. Commercially available PIGs include the 'smart PIG' by NORSEN GROUP; 'Smartball' by Pure Technology [17] and 'Remoted PIG' by Jiutai Technology [18,19].

Active driving methods are categorised into wheel, track, inchworm, walking and snake mechanisms. Wheel propulsion is generally coupled with a spring mechanism to press against the pipe walls to smoothly adapt to the in-pipe topology. Wheel-based prototypes are available in the following literature [20,21]. Wheeled robots can also be screw-driven which usually entails a stationary and rotational section. For soft or cracked ground motion, wheeled robots are often outperformed by track-driven systems. However, this drive is rarely used due to its higher complexity and energy requirements [22]. An example of legged inspection robots can be found in Bradbeer et al.'s work [23]. Worm-like movements are produced by the cooperation of a clamper and extensor modules to push the robot through the pipe. This motion is often used in foreign environments where caution is a priority. Further work on inchworm movements has been conducted by [24,25], [26], [27]. Like the inchworm, snake robots have proven more adaptable in abnormal environments. They are comprised of several connected modules capable of planar movement [28]. Reptile movement for inspection robots is not widely due to the complexity of increasing degrees of freedom with every adjoining module

3.3. Level of autonomy

Intrusive hardware inspection techniques belong to one of three classes of autonomy: No autonomy, semi-autonomous, or fully autonomous. Most robotic inspection devices lie in the first category of non-autonomous hardware however the introduction of autonomy provides freedom from user interference [14,29].

Fully operated robots are usually controlled through a tether cable by trained users or through a wireless link. The operator examines the inside of the pipe in real time using the incoming sensor data as the robot moves along the network [14]. The tether cable is preferred because it enables a smoother recovery as shown in the study [30]. This is a popular method because, as of now, there are no cost effective solutions that can navigate the varying scenarios inside water networks [30].

Semi-autonomous inspection is achieved through implementing automatic control modules that can remove some of the user's duties such as navigation or pipe condition assessment. This shifts some of the users' responsibilities and introduces higher accuracy. Prototypes that belong in this category include the PIRAT [31] and Karo [32] robots.

Fully automated robots are those voids of any user interaction. They are able to navigate, assess and communicate pipe condition through their sensor payload in real-time without being lost in the system. The challenges faced by autonomous inspection devices are numerous but the most prominent is energy and communication in long-time and long-range applications [14]. Kirchner and Hertzberg developed the Kurt robot that uses a map of the pipe network to collect video graphic, ultrasound and gradient data through automatic operation [33].

In this review, we have already specified some smart PIGs and robotic detection prototypes that highlight the range of in-pipe inspection including Makro [21,34], Karo [32], PIRAT [31], Kurt [33] and Smartball [17]. Finding a complete intrusive inspection robot is difficult due to the design challenges they must overcome. Despite that, pipe inspection gauges can be an appealing solution to some network with successful case studies shown [35].

3.4. Non-intrusive methods

Leakage detection systems can also be labelled as dynamic or static methods. Whilst intrusive methods are referred to as dynamic due to their motion throughout the network to investigate inner pipe conditions, non-intrusive methods depend on mounted sensors that collect data used to infer leakage making it a static method [2]. Static methods carry the advantage of identifying a leak immediately whereas dynamic detection is often deployed after a leak is expected/identified to pinpoint the leak area [36,37]. The research scene has been marginally focused on the static strategies of leakage detection for the last two decades due to their more tangible benefits and their ability of real-time management [2]. The most prominent technologies exploit acoustic or pressure properties accounting to more than 50% of the literature [2]. Other technologies rely on flow sensors, ground penetrating radars (GPR), tracer gas detection, infrared thermography which will be mentioned in this section.

3.4.1. Acoustic technique. Acoustic based leakage detection and localisation can be traced back to the early 1990s in water and oil networks [38]. The localisation of leak events through acoustic methods can be classified into time-of-flight-based or attenuation-based. Attenuation based relies on the decrease of signal amplitude as the acoustic signals travel across the pipeline while time based monitors the increase of signal transit time [39]. Acoustic emissions result from turbulent pressure fluctuations at the leak, vapor bubbles forming at high velocities and imploding as shock waves on pipe walls. The frequency of those acoustic emissions (AE) varies depending on the source where turbulent flows produce low frequency signals and cavitation bursts cause high them in plastic pipes requires a denser distribution of the sensors [40]. These instruments could be geophones (electrical or mechanical), hydrophones, listening sticks, accelerometers, or correlators.

3.4.2. Fibre optics. These systems measure temperature anomalies inherent in leaks throughout the pipe. In comparison to oil and gas pipelines, the leak induced temperature change in water pipelines is smaller and harder to detect [41]. Daily and seasonal temperature fluctuations increase the difficulty of fibre optics leakage detection [41]. Optical fibres could alternatively monitor the strain in the pipe wall due to leaks [42,43]. Developments to increase the use of fibre optics for pipeline monitoring of leak-induced temperature or strain include the use of Raman Distributed Temperature Sensor (RDTS), Fibre Bragg Gratting (FBG) [41] and Brilloun Optical Time Domain Reflectometry (BOTDR) [6]. The use of fibre optics exceeds other methods in its immunity to electrical noise, corrosion resistance and stability [7] however their high initial and operating costs make them less desirable to water utilities.

3.4.3. Infrared thermography. Like fibre optics, infrared thermography exploits the thermal effects of leakage in pipelines to identify and locate the event. IR cameras have been applied to asses pipe conditions [44,45] as well as leakage detection in water networks [46,47]. Despite it being a scarce research topic, thermography could provide a cost-effective, efficient, non-destructive way to monitor large areas of water networks. Capturing the thermal anomalies translated to the surface above the leakage is affected by several factors [48]. IR should be measured during times where the ambient temperatures is closer to equilibrium, hence increasing thermal visibility. Pre-sunrise and post-sunset hours have been suggested as the most suitable periods for thermography [48,49]. Soil moisture tends to hinder the investigation which can be problematic in rainy countries such as the UK despite the moist soil's superior heat transfer ability [49].

3.4.4. Ground Penetrating. Radars Ground penetrating radars have gained some interest in the leakage research community [50,51]. Their ability to utilise the electromagnetic irregularity of water leakage in infrastructure to identify and locate the failure. This imaging method excels in its applicability to both metal and plastic pipes regardless of the material and size [2]. GPRs are easy to use and transport making it possible to survey large areas with less manpower [52]. Despite that, the disadvantages of deploying GPRs outweigh the possible benefits. Its inability to discriminate between leak-induced irregularities and soil inhomogeneity increases the false alarm rate [38]. This strategy is limited to pipes buried less than 5m deep and highly influenced by soil types. The reliability of this method can be improved through the aid of decision support systems [53] and evolutionary search algorithms to obtain accurate leakage detection.

3.4.5. Tracer gas. Gas injection utilises inert, non-toxic, insoluble, traceable gases such as halogens, ammonia, and helium to pinpoint leakage sites. As these gases seep through faulty infrastructure, operators detect their location by surveying the suspected area [54].This requires a proficient knowledge of the network's flows to limit the gas flow to the suspected area by blocking other routes to exit the system. Tracer gas is able to detect both background leakage and burst events with low false alarm rate [7,46]. The fast and accurate response of this technique is crippled by its expense especially in large, low-pressure networks that require higher volumes of gas and the implementation cost of in-built sensors for monitoring and possible filtering stages makes this method unrealistic [7,55].

3.4.6. Magnetic induction. Magnetic induction is an accurate detection technique that establishes a communication link between two sets of sensors. One of the sets captures the flow, pressure and acoustic properties of the suspected leak from within the pipe whilst the other assesses the external factors such as humidity, temperature and soil properties outside the pipeline [56]. Through a current-modulated signal, the coils of the magnetic transmitter induces the current to the receiver [57]. This communication link enables real time control of leakage detection hence increasing the response rate of utilities in harsh underground conditions. However, this strategy incurs high implementation costs making it undesirable.

4. Software detection Methods

Background leakage poses a large threat to water networks as they often go undetected by the conventional hardware methods accumulating more losses over time. Therefore, the application of software methods is essential to counter this prevailing issue. The long-time savings that accompany software techniques are usually offset with the initial costs of installing sensors throughout the network making it a less popular option to utilities *[58]*. Water distribution systems often operate their leakage detection techniques under the premise of steady-state flow *[59]*. This method compares the behaviour of the actual network in comparison to the expected performance to detect anomalies that are often caused by leakage or blockage. An abundance of real data (historic and live) from the network suggests a data-driven approach to leakage detection however, in networks where data is scarce, a model-based approach takes precedence given that the hydraulic model is available.

4.1. Model based

As the name suggests, model-based detection relies heavily on the model's resemblance to the actual network, the data analysed and the arithmetic techniques employed [1]. Therefore, building a realistic accurate model is crucial to the success of this technique, often having a separate calibration stage to compare the model and the network. Zaman et al. [1] display a useful general framework for the model-based leakage detection method. a used Developing a reliable replica on hydraulic simulation machines (e.g., EPANET, LOOP) should include input information of a leak-free system through different streams of information such as Supervisory Control and Acquisition (SCADA), Geographic Information Systems (GIS) and more. Some model platforms have an inherent leakage detection module such as WaterGEMS that exercises a genetic algorithm (GA) to signify potential leak nodes.

As soon as the model is complete, it is necessary to validify the model through several techniques to ensure that the model tracks the real-life example. In some cases, model pre-processing is performed before calibration to decrease the potential candidates [59]. The calibration techniques used are steady-state and extended period simulation (EPS) and contrast them to the field data.

Once the model has been calibrated successfully, several leakage detection strategies can be applied to the model to predict and inform of possible leak locations and their corresponding sizes.

Several detection strategies have been developed for the model-based approached. They exploit the simulated parameters and field data to locate possible leak areas However, these leakage detection models often suffer with the unaccounted ageing properties of the pipe causing the pipe diameters to decrease [6]. A simple method for investigating leakage is applying the conservation of mass calculation. Balancing the mass in and out of nodes can uncover unaccounted for loss hinting at a possible leak. Whilst this approach, works well for steady-state, they are prone to disturbances and pipeline dynamics resulting in false alarms [60]. A different method called pressure residual vector (PRV) exploits the leak-induced pressure changes in the real system and compares it to the leak-free model from their subsequent locations on the network [61]. When the disparity between the modelled and actual pressure exceeds a pre-determined threshold, set through uncertainty analysis and statistical considerations, the area is flagged as a potential leak location and investigated [62,63].

Indirect methods for model leakage detection can be classified into three types as shown in figure 1. Calibration-based methods rely on optimising the model calibration stage by infusing it with leakage information. This information is obtained by modelling leakage as a pressure demand. Genetic algorithms (GA) have proved as a useful evolutionary search algorithm (EAs) to investigate possible leak location through calibration [64]. EA has been widely used in the optimisation of water distribution system design for both single objective and multi-objective as highlighted by the comprehensive literature review [65]. Sensitivity-based analysis is another method that exploits network models [61,66] through investigating the pressure sensitivity of nodes in the model under leak and non-leak conditions. Combining the sensitivity matrix with the corresponding pressure residual vector can more accurately indicate potential leakage. This is represented in the study [67] with the aid of the angle based method. To develop that further, [68] introduces a classifier-based method to detect leakage. Exploiting statistical classifiers, greatly improves fault localisation with comparison to the angle method demonstrated in [67] especially regarding demand uncertainties. Classifiers are often used as a data-driven approach however, their used in model-based detection has proved rewarding.

5. Data-driven

Using abundant data, leak detection can navigate complex, heterogenous, large water distribution networks by bypassing the complications of hydraulic modelling. This makes it a more reliable and accurate technique due to its reliance on real data at the cost of increased sensitivity to faulty sensors. These methods rely on their ability to reveal aberrant signals/patterns in the monitoring data received that could suggest the existence of a leak.

5.1. Data pre-processing

Data-based detection often engineers one or a combination of flow, pressure, and demand readings. Consumer demand being the least probable data source can be rationalised by their uncertainty in the localisation stage [68] and its relative insensitivity to smaller leak flow rates [69]. The data used might differ in source, sample source (1-15 minutes) and length of time series which are crucial aspects to consider [67]. These sensor readings are often raw and require pre-processing before they can be implemented to any leakage detection algorithm. Data pre-processing often involves sorting, filtering, and transforming the incoming data making it a tedious task. Other issues such as uncertainty and variability must be considered when employing real data, yet this could be avoided in the instances data is extracted from models. Pre-processing is essential to filter erroneous data, filling gaps in the time-series and arranging the results for assessment making a critical step in data-driven leakage detection [1].

5.2. Detection methods

In our classification tree (figure 1), data-driven techniques were divided into four types depending on their technical procedure. A different way to organise these techniques is to match their data source and data types. The technical procedures highlighted in this review include statistical, classification, prediction, signal processing. These methods are also used for transient leakage detection.

5.2.1. Pressure/flow monitoring. The simplest data-driven techniques utilise pressure monitoring such as negative pressure wave (NPW) and pressure point analysis (PPA). The NPW method detects the propagating pressure fluctuation at both sides of the leak through the use of transducers [70]. Localising the leak is established by contrasting the time difference between the reading on both sensors through cross-correlation. Applying the NPW approach practically is challenging particularly for long-range pipes [6]. Another limitation of NPW is its high false alarm rate resulting from its sensitivity to transient flows in networks. In order to increase the method's reliability, study [71] proposes several improvements regarding false alarm reduction. NPW hybrid leakage detection techniques are encourage to justify the alarms such as represented in [57]. Pairing pressure transducers to compare their leak results is an alternative method to reduce false alarms [71]. The last recommendation uses the aid of pattern recognition to distinguish leakage-induced pressure fluctuations from valve-induced variations [71]. Other ways to improve NPW, include implementing an adaptive threshold and improving data quality which could be achieved through filtering background noise and advanced data processing. Pressure point analysis (PPA) developed by EFA technologies ltd. is commercially available technique that statistically analyses the mean pressure measurements along a pipe [55]. Similar to the other pressure-based leakage detection strategies, PPA issues an alarm when the mean pressure value drops beyond an established threshold. This method is straightforward and economic but lack credibility under transient conditions and cannot localise the leaks [6].

5.2.2. Statistical. Statistical analysis techniques for leakage detection are methods that have no classification or prediction stage and depend completely on statistical theory [69]. Statistical Process Control (SPC) is a central method in this category often using control charts to monitor measurement variations. They are also used for data pre-processing [69]. The differences between univariate and multivariate SPC methods can be found in Jung et al. work [72] The univariate methods elicited are Western Electric Company (WEC) rules, Cumulative Sum (CUMSUM) and Exponentially Weighted Moving Average (EWMA). WEC rules can only consider the past eight readings whilst EWMA has largest memory [72]. The multivariate methods described where Hotelling T2 control chart with elliptical control and the multivariate versions of the CUMSUM and EWMA methods. Other statistical strategies include Principle and Independent Component Analysis (PCA, ICA) that are used to reduce the state space of the data without decreasing its value. ICA can be considered an extension to PCA that considers higher order statistics [73]. Newer methods of statistical procedures rely on data clustering [74], support vector machine (SVM), artificial neural networks (ANN) [75] and newer versions of the multivariate methods mentioned earlier.

5.2.3. Classification. Classification based strategies build models that can effectively distinguish (classify) normal and outlier data. The simplest form of a classification technique calculates the absolute mean difference between expected and recorded hydraulic measurements [1]. More commonly used models are trained using sets of labelled normal and abnormal hydraulic data to successfully detect bursts. An example of this is the comparative study conducted by Mounce and Machell on burst detection through flow reading analysis using static and time-delay artificial neural networks (ANN) [76]. The different architectures displayed a different relationship with the inputs causing improved detection due to its more dynamic nature. It is clear that the performance of the classification model relies heavily on the abundance of normal and outlier real data to train the model and the quality of the inputs used. Using a leak function and a self-organising map (SOM) ANN, the classification model can output a value from 0-1 to identify the probability of a leak at a node [77]. This method is more adept in distinguishing leak data without supervision or labelled training data [77]. The need of adequately labelled and balanced training data for both normal and outlier conditions is the main disadvantage of this technique therefore unsupervised learning is a logical step for further research. In addition to that, poorly trained classification models often lead to high false positive rates (FPR) which is another concern for classification-based leakage detection.

5.2.4. Prediction-classification. Unlike classification techniques, prediction-based method introduces a preliminary stage of outlier data prediction hence enabling the classification model to be built effectively with normal hydraulic data alone. An additional stage of data selection is required to achieve this which often utilises some of the statistical methods mentioned earlier [78,79]. A linear Kalman Filter (LKF) could be trained using normal historical data to provide a statistical description of the current system [80]. This is an efficient method that can extract the prediction using live data alone. Expert systems such as Fuzzy Interference Systems (FIS) and Bayesian Interference Systems (BIS) provide reliable detection results however they can be developed further by using historical data, evolutionary algorithm (EA) and expectation maximisation (EM) to optimise their parameters. Mounce et Al. employs a combination of an artificial neural network called mixed density network (MDN) in the prediction stage [78] followed by a FIS in the classification stage to improve burst detection in the form imitate human cognition [81]. Following a prediction stage, support vector regression (SVR) was used to classify deviations in the input data for leakage detection [82]. Changes in historical data used for prediction-classification propagates the data uncertainty decreases the accuracy of leakage detection and requires a data selection stage.

5.2.5. Signal Processing. Digital signal processing (DSP) is a commonly used technique to improve leak detection and localisation using pressure or acoustic signals due to sharper transitions than traditional techniques such as NPW. This benefit is often offset by the bandwidth restrictions that these methods introduce through pipe resonance [36].

The use of time and frequency response analysis of acoustic emissions for leakage feature extractions has been fruitful leading researchers to explore hybrids that can exploit their benefits. This initiated the need for time-frequency analysis to obtain valuable information from both domains. Several types of Fourier transforms have been used to capture leakage characteristics, but short-term Fourier transform (STFT) has been the primary interest of researchers. STFT introduces a time variable to the spectrum by slicing the signal using a time window function. The short time segments (called frames) are then input to a discrete Fourier transform (DFT) to produce a time-frequency analysis. This method has been justified in multiple occasions and compared to fast Fourier transform (FFT) in [83] where it outperforms FFT in the uncertainty analysis. Li et al.'s proposed methodology of wavelet denoising and STFT combination has shown higher accuracy than other signal processing methods such as wavelet decomposition, gaussian mode, recurrence plot, Wigner-Ville distribution (WVD), Wigner-Hough Transform (WHT), and empirical mode decomposition (EMD) [84]. Fast Fourier transform has also been validated in the study [85] through a fault detection and isolation (FDI) system of underground plastic pipes.

The use of STFT and other time-frequency analysis methods has been overshadowed by the application of wavelet transforms (WT) for leakage detection and localisation. STFT has the disadvantage of a strict resolution limit due to its limited window size which does not exist when using WT. Therefore, using this method better fits the multi-resolution nature of leakage and burst events [86]. WT can be used in multiple areas of signal processing such as denoising [84], decomposition [84] recognition, classification, and feature extraction. The effectiveness of WT relies heavily on the selection of the mother wavelet used is shifted and scaled across the signal to create daughter wavelets. This was emphasised in the work of Ahadi and Bakhtiar and their comparison of Haar and db8 mother wavelets [87]. Studies [86,87] further prove the benefits of WT over STFT. Wavelet transforms, however, are limited by the length of the mother wavelet and its non- adaptive nature [12]. Examples of mother wavelets used also include Meyer, Morlet, Daubechies and Mallet functions [1]. WT has proven to reduce noise and better locate sharp transitions in the leak signals [1]. In study [88], wavelet features pf pressure signals were compared and outperformed by two Multi-Layer Perceptron Neural Networks (MLPNN) for feature extraction and leakage classification that are then fused by the Dempster-Shafer (D-S). The neural networks have outperformed the wavelet feature methodology in correct classification rate (CCR%) where D-S classifier fusion method resulted in 95.11%, wavelet at 86.94% and statistical features trailing behind at 64.56% [88].

6. Conclusion and future work

The leakage detection field spare and multi-directional filled with researchers attempting to equip water companies with a solid method to tackle the heterogenous nature of this issue. Reviewing the field is challenging but this paper should offer guidance to the multiple research areas and possible novelties that can be uncovered in each. Every section offers a further breakdown of the research area and a light comparison between technologies within to collate the most research findings and methodologies.

The robotic platforms reviewed include many platforms that vary in driving methods, sensing capabilities and autonomy. The active driving method sin intrusive inspection devices include wheeled, screw-driven, track-driven, worm, snake and legged showing that they all vary in applicability depending on the nature of the pipe and its environment. The autonomy level of the Smart PIGs and robots introduce an interesting trade-off between recoverability and less manpower. However, there are no economical devices capable of autonomously adjusting to all the scenarios present in all water networks and effectively communicating with the users with no intervention.

Non-intrusive hardware detection consists of a range of sensors that detect leak-induced anomalies to identify and locate events. The most common are acoustic sensors which include microphones, geophones, hydrophones, accelerometers, leak noise loggers and correlators. Other sensors mentioned are magnetic induction, infrared thermography, fibre optics, ground penetrating radar and tracer gas. These are often insufficient alone and will benefit further if paired with signal processing methods.

Software leakage detection draws from two general methodologies. Model-based leakage detection requires the analyst to construct an accurate model of the water distribution network using hydraulic analysis software and compare expected pressures/flows to the actual measurements to find outliers. In comparison, data-driven techniques include varying methods of collecting, pre-processing, and analysing data to directly find leak-induced outliers. Data-driven methods are classified into several categories based on their nature including statistical, classical, prediction and signal processing. Using models or data-driven methods rely solely on the availability of sensor data and their accuracies but can often be used together to provide a better hydraulic analysis of the system. These methods rely heavily on computational efforts to detect leakage and can benefit majorly from the rise of data engineering and artificial intelligence breakthroughs.

It is common for researchers and industry to use a hybrid between the methods reviewed to draw on their advantages and this should be explored further to use our current knowledge to bridge the faults of these methodologies. It is also crucial to explore new venues for model leakage prediction which can help identify background leakage. This can benefit from the emergence of neural networks as function approximators especially graph neural network due to their similar data types. Furthermore, neural networks models should be tested for transfer learning applications hence reducing the training time required for neural networks to model water distribution networks.

Acknowledgments
This work is funded by Lancaster University through European Regional Development Fund, Centre of Global Eco-Innovation, and industrial partners DNS Ltd. The authors declare that they have no known competing financial interests or personal relationships that could have appeared to influence the work reported in this paper.

References
[1] Zaman D, Tiwari MK, Gupta AK and Sen D 2020 A review of leakage detection strategies for pressurised pipeline in steady-state *Eng. Fail. Anal.* **109** 104264
[2] El-Zahab S and Zayed T 2019 Leak detection in water distribution networks: an introductory overview *Smart Water* **4**
[3] Barton NA, Farewell TS, Hallett SH and Acland TF 2019 Improving pipe failure predictions: Factors effecting pipe failure in drinking water networks *Water Res.* **164**
[4] Rezaei H, Ryan B and Stoianov I 2015 Pipe failure analysis and impact of dynamic hydraulic conditions in water supply networks *Procedia Eng.* **119** 253–62
[5] Adedeji KB, Hamam Y, Abe B and Abu-Mahfouz AM 2017 Leakage detection algorithm integrating water distribution networks hydraulic model *SimHydro 2017* pp 1–9
[6] Adedeji KB, Hamam Y, Abe BT and Abu-Mahfouz AM 2017 Towards achieving a reliable leakage detection and localization algorithm for application in water piping networks: An overview *IEEE Access* **5** 20272–85
[7] Chan TK, Chin CS and Zhong X 2018 Review of current technologies and proposed intelligent methodologies for water distributed network leakage detection *IEEE Access* **6** 78846–67
[8] Hamilton S 2009 ALC in low pressure areas - It can be done *5th IWA Water Loss Reduction* Cape Town, South Africa
[9] El-Abbasy MS, Mosleh F, Senouci A et al 2016 Locating leaks in water mains using noise loggers *J. Infrastruct. Syst.* **22** 04016012
[10] Abdulshaheed A, Mustapha F and Ghavamian A 2017 A pressure-based method for monitoring leaks in a pipe distribution system: A review *Renew. Sustain. Energy Rev.* **69** 902–11
[11] Abdulla MB and Herzallah R 2015 Probabilistic multiple model neural network based leak detection system: Experimental study *J. Loss Prev. Process Ind.* **36** 30–8
[12] Adnan NF, Ghazali MF, Amin MM and Hamat AMA 2015 Leak detection in gas pipeline by acoustic and signal processing - A review *IOP Conf. Ser. Mater. Sci. Eng.* **100** 012013
[13] Khulief YA, Khalifa A, Mansour RB and Habib MA 2012 Acoustic detection of leaks in water pipelines using measurements inside pipe *J. Pipeline Syst. Eng. Pract.* **3** 47–54
[14] Tur JMM and Garthwaite W 2010 Robotic devices for water main in-pipe inspection: A survey *J. F. Robot.* **27** 491–508

[15] Guan L, Gao Y, Liu H et al 2019 A review on small-diameter pipeline inspection gauge localization techniques: Problems, methods and challenges *2019 Int. Conf. Communications, Signal Processing and their Applications (ICCSPA)* pp 1–6
[16] Bickerstaff R, Vaughn M, Stoker G et al 2002 Review of sensor technologies for in-line inspection of natural gas pipelines *Sandia Natl. Lab. Albuquerque*
[17] SmartBall 2022 *Leak and Gas Pocket Detection* https://puretechltd.com/technology/smartball-leak-detection/
[18] Ismail IN, Anuar AS and Sahari KSM 2012 Developments of in-pipe inspection robot: A review *2012 IEEE Conf. Sustainable Utilization and Development in ENgineering and Technology (STUDENT)* pp 310–5
[19] Roslin NS, Anuar A, Jalal MFA and Sahari KSM 2012 A review: Hybrid locomotion of in-pipe inspection robot *Procedia Eng.* **41** 1456–62
[20] Roh SG and Choi HR 2005 Differential-drive in-pipe robot for moving inside urban gas pipelines *IEEE Trans. Robot.* **21** 1–17
[21] Kolesnik M and Streich H 2002 *Visual orientation and motion control of MAKRO – Adaptation to the sewer environment* https://kolesnik.leute.server.de/papers/pdf/sab2002.pdf
[22] Roman HT, Pellegrino BA and Sigrist WR 1993 Pipe crawling inspection robots: An overview *IEEE Trans. Energy Convers.* **8** 576–83
[23] Bradbeer R, Harrold S, Nickols F and Yeung LF 1997 Underwater robot for pipe inspection *Proc. Annu. Conf. Mechatronics Mach. Vis. Pract. MViP* 152–6
[24] Bertetto AM and Ruggiu M 2001 In-pipe inch-worm pneumatic flexible robot *IEEE/ASME Int. Conf. Adv. Intell. Mechatronics, AIM* **2** 1226–31
[25] Lim J, Park H, Moon S and Kim B 2007 Pneumatic robot based on inchworm motion for small diameter pipe inspection *2007 IEEE Int. Conf. Robot. Biomimetics, ROBIO* 330–5
[26] Choi C, Jung S and Kim S 2004 Feeder pipe inspection robot with an inch-worm mechanism using pneumatic actuators *Int. J. Control. Autom. Syst.* **4** 87–95
[27] Menciassi A, Park JH, Lee S et al 2002 Robotic solutions and mechanisms for a semi-autonomous endoscope *IEEE Int. Conf. Intell. Robot. Syst.* **2** 1379–84
[28] Liljebck P, Pettersen KY, Stavdahl O and Gravdahl JT 2012 A review on modelling, implementation, and control of snake robots *Rob. Auton. Syst.* **60** 29–40
[29] Liu Z and Kleiner Y 2013 State of the art review of inspection technologies for condition assessment of water pipes *Measurement* **46** 1–15
[30] Moraleda J, Ollero A and Orte M 1999 A robotic system for internal inspection of water pipelines *IEEE Robot. Autom. Mag.* **6** 30–41
[31] Kirkham R, Kearney PD, Rogers KJ and Mashford J 2000 PIRAT – A system for quantitative sewer pipe assessment *The International Journal of Robotics Research* **19**(11) 1033-1053 doi:10.1177/02783640022067959
[32] Kuntze HB and Haffner H 1998 Experiences with the development of a robot for smart multisensoric pipe inspection *Proc. IEEE Int. Conf. Robot. Autom.* **2** 1773–8
[33] Kirchner F and Hertzberg J 1997 A prototype study of an autonomous robot platform for sewerage system maintenance *Auton. Robot. 1997 44* **4** 319–31
[34] Rome E, Hertzberg J, Kirchner F et al 1999 Towards autonomous sewer robots: the MAKRO project *Urban Water* **1** 57–70
[35] Pure Technologies 2022 *Lyon inspects water main for leaks with SmartBall tool* (Xylem)
[36] Cataldo A, Persico R, Leucci G et al 2014 Time domain reflectometry, ground penetrating radar and electrical resistivity tomography: A comparative analysis of alternative approaches for leak detection in underground pipes *NDT E Int.* **62** 14–28
[37] Lee PJ, Vítkovský JP, Lambert MF et al 2005 Frequency domain analysis for detecting pipeline leaks *J. Hydraul. Eng.* **131** 596–604
[38] Gupta A and Kulat KD 2018 A selective literature review on leak management techniques for water distribution system *Water Resour. Manag.* **32** 3247–69

[39] Lee MR and Lee JH 2000 Acoustic emission technique for pipeline leak detection *Key Eng. Mater.* **183–187** 887–92
[40] De Silva D, Mashford J and Burn S 2011 *Computer Aided Leak Location and Sizing in Pipe Networks* http://www.urbanwateralliance.org.au/publications/UWSRA-tr17.pdf
[41] Jacobsz SW and Jahnke SI 2019 Leak detection on water pipelines in unsaturated ground by discrete fibre optic sensing https://doi.org/10.1177/1475921719881979 **19** 1219–36
[42] Inaudi D and Glisic B 2008 Long-range pipeline monitoring by distributed fiber optic sensing *Proc. Bienn. Int. Pipeline Conf. IPC* **3**(B) 763–72
[43] Davila M, Davila Delgado JM, Brilakis I and Middleton C 2016 Distributed monitoring of buried pipelines with Brillouin fiber optic sensors *Proc. Int. Conf. Smart Infrastructure and Construction* 33–38
[44] Joung OJ and Kim YH 2006 Application of an IR thermographic device for the detection of a simulated defect in a pipe *Sensors* **6** 1199–208
[45] Gross W, Hierl T, Scheuerpflug H et al 1999 Quality control of heat pipelines and sleeve joints by infrared measurements *Thermosense XXI* **3700** 63–9
[46] Hunaidi O, Chu W, Wang A and Guan W 2000 Detecting leaks in plastic pipes *J. Am. Water Works Assoc.* **92** 82–94
[47] Khawandi S, Daya B and Chauvet P Automated monitoring system for fall detection in the Elderly *Int. J. Image Process.* 476
[48] Bach PM and Kodikara JK 2017 Reliability of infrared thermography in detecting leaks in buried water reticulation pipes *IEEE J. Sel. Top. Appl. Earth Obs. Remote Sens.* **10** 4210–24
[49] Huang Y, Fipps G, Maas SJ and Fletcher RS 2010 Airborne remote sensing for detection of irrigation canal leakage *Irrig. Drain.* **59** 524–34
[50] Demirci S, Yigit E, Eskidemir IH and Ozdemir C 2012 Ground penetrating radar imaging of water leaks from buried pipes based on back-projection method *NDT E Int.* **47** 35–42
[51] Wai-Lok Lai W, Dérobert X and Annan P 2018 A review of ground penetrating radar application in civil engineering: A 30-year journey from locating and testing to imaging and diagnosis *NDT E Int.* **96** 58–78
[52] Hamilton S and Charalambous B 2013 Leak detection: Technology and implementation *Water Intell. Online* **12**
[53] Kiss G, Konez K and Melinte C 2007 WaterPipe project: An innovative high resolution ground penetration imaging radar for detecting water pipes and for detecting leaks and a decision support system for the rehabilitation management of the water pipeline *IWA Water Loss Conference* (Bucharest, Romania) pp 622–31
[54] KVS Tracer Gas Detection
[55] Geiger G 2006 State-of-the-Art in leak detection and localisation *Pipeline Technology* (Hannover, Germany) pp 1–25
[56] Boaz L, Kaijage S and Sinde R 2014 An overview of pipeline leak detection and location systems *Proc. 2nd Pan African Int. Conf. Sci. Comput. Telecommun. PACT 2014* 133–7
[57] Sun Z, Wang P, Vuran MC et al 2011 MISE-PIPE: Magnetic induction-based wireless sensor networks for underground pipeline monitoring *Ad Hoc Networks* **9** 218–27
[58] Farley B, Mounce SR and Boxall JB 2010 Field testing of an optimal sensor placement methodology for event detection in an urban water distribution network *Urban Water J.* **7** 345–356
[59] Perez R, Sanz G, Puig V et al 2014 Leak localization in water networks: A model-based methodology using pressure sensors applied to a real network in Barcelona [Applications of Control] *IEEE Control Syst.* **34**(4) 24–36
[60] Wan J, Yu Y, Wu Y et al 2012 Hierarchical leak detection and localization method in natural gas pipeline monitoring sensor networks *Sensors* **12** 189–214
[61] Pérez R, Puig V, Pascual J et al 2011 Methodology for leakage isolation using pressure sensitivity analysis in water distribution networks *Control Eng. Pract.* **19** 1157–67

[62] Sousa J, Muranho J, Sá Marques A and Gomes R 2014 WaterNetGen helps C-Town *Procedia Eng.* **89** 103–10
[63] Ishido Y and Takahashi S 2014 A new indicator for real-time leak detection in water distribution networks: Design and simulation validation *Procedia Eng.* **89** 411–7
[64] Sophocleous S, Savić DA, Kapelan Z and Giustolisi O 2017 A two-stage calibration for detection of leakage hotspots in a real water distribution network *Procedia Eng.* **186** 168–76
[65] Mala-Jetmarova H, Sultanova N and Savic D 2018 Lost in optimisation of water distribution systems? A literature review of system design *Water (Switzerland)* **10**
[66] Geng Z, Hu X, Han Y and Zhong Y 2018 A novel leakage-detection method based on sensitivity matrix of pipe flow: Case study of water distribution systems *J. Water Resour. Plan. Manag.* **145**(2)
[67] Casillas MV, Garza-Castanon LE and Puig V 2013 Extended-horizon analysis of pressure sensitivities for leak detection in water distribution networks: Application to the Barcelona network *2013 Eur. Control Conf. ECC 2013* 404–9
[68] Ferrandez-Gamot L, Busson P, Blesa J et al 2015 Leak localization in water distribution networks using pressure residuals and classifiers *IFAC-PapersOnLine* **48** 220–5
[69] Wu Y and Liu S 2017 A review of data-driven approaches for burst detection in water distribution systems *Urban Water J.* **14** 972–83
[70] Silva RA, Buiatti CM, Cruz SL and Pereira JAFR 1996 Pressure wave behaviour and leak detection in pipelines *Comput. Chem. Eng.* **20** S491–6
[71] Tian CH, Yan JC, Huang J et al 2012 Negative pressure wave based pipeline Leak Detection: Challenges and algorithms *Proc. IEEE Int. Conf. Serv. Oper. Logist. Informatics, SOLI 2012* 372–6
[72] Jung D, Kang D, Liu J and Lansey K 2015 Improving the rapidity of responses to pipe burst in water distribution systems: a comparison of statistical process control methods *J. Hydroinformatics* **17** 307–28
[73] Westra S, Brown C, Lall U and Sharma A 2007 Modeling multivariable hydrological series: Principal component analysis or independent component analysis? *Water Resour. Res.* **43** 6429
[74] Wu Y, Liu S, Wu X et al 2016 Burst detection in district metering areas using a data driven clustering algorithm *Water Res.* **100** 28–37
[75] Zhou X, Tang Z, Xu W et al 2019 Deep learning identifies accurate burst locations in water distribution networks *Water Res.* **166** 115058
[76] Mounce SR and Machell J 2007 Burst detection using hydraulic data from water distribution systems with artificial neural networks **3** 21 http://dx.doi.org/10.1080/15730620600578538
[77] Aksela K, Aksela M and Vahala R 2009 Leakage detection in a real distribution network using a SOM *Urban Water J.* **6** 279–89
[78] Mounce SR, Khan A, Wood AS et al 2003 Sensor-fusion of hydraulic data for burst detection and location in a treated water distribution system *Inf. Fusion* **4** 217–29
[79] Mounce SR, Boxall JB and Machell J 2009 Development and verification of an online artificial intelligence system for detection of bursts and other abnormal flows *J. Water Resour. Plan. Manag.* **136** 309–18
[80] Ye G and Fenner RA 2010 Kalman filtering of hydraulic measurements for burst detection in water distribution systems *J. Pipeline Syst. Eng. Pract.* **2** 14–22
[81] Mounce S, Boxall JB and Machell J 2007 An Artificial Neural Network/Fuzzy Logic system for DMA flow meter data analysis providing burst identification and size estimation *Proc. Water Management Challenges in Global Change* 313–320
[82] Mounce SR, Mounce RB and Boxall JB 2011 Novelty detection for time series data analysis in water distribution systems using support vector machines *J. Hydroinformatics* **13** 672–86
[83] Lay-Ekuakille A, Vendramin G, Trotta A and Vanderbemden P 2009 STFT-Based spectral analysis of urban waterworks leakage detection *Proc. XIX IMEKO World Congress* 2172–76.

[84] Li H, Li H, Pei H and Li Z 2019 Leakage detection of HVAC pipeline network based on pressure signal diagnosis *Build. Simul.* **12** 617–628 https://doi.org/10.1007/s12273-019-0546-0
[85] Kadri A, Yaacoub E and Mushtaha M 2014 Empirical evaluation of acoustical signals for leakage detection in underground plastic pipes *Proc. Mediterr. Electrotech. Conf. - MELECON* 54–8
[86] Wu R, Liao Z, Zhao L and Kong X 2008 Wavelets application on acoustic emission signal detection in pipeline *Can. Conf. Electr. Comput. Eng.* 1211–4
[87] Ahadi M and Bakhtiar MS 2010 Leak detection in water-filled plastic pipes through the application of tuned wavelet transforms to Acoustic Emission signals *Appl. Acoust.* **71** 634–9
[88] Zadkarami M, Shahbazian M and Salahshoor K 2017 Pipeline leak diagnosis based on wavelet and statistical features using Dempster–Shafer classifier fusion technique *Process Saf. Environ. Prot.* **105** 156–63.

Parametrization of a wastewater hydraulic model under incomplete data constraint

Yassine Belghaddar[1,2,3,4], **Carole Delenne**[1,4], **Nanée Chahinian**[1], **Abderrahmane Seriai**[3] **and Ahlame Begdouri**[2]

[1] HSM, University of Montpellier, CNRS, IRD, Montpellier, 34090, France
[2] LSIA, Sidi Mohamed Ben Abdellah University, Fez, Morocco
[3] Berger-Levrault, Pérols, 34477, France
[4] Inria Lemon, CRISAM – Inria Sophia Antipolis – Méditerranée, 34095, France

yassine.belghaddar@berger-levrault.com

Abstract. Hydraulic simulation represents a powerful tool for studying wastewater networks. In order to achieve this target, hydraulic software require a set of parameters such as pipe slopes, roughness, diameters, etc. However, these pieces of information are rarely known for each and every pipe. Moreover, underground networks are frequently expanded, repaired and improved and these changes are not always reported in databases. The task of completing the required data represents the most time-consuming part of model implementation. In this context, we present algorithms that complete missing data required by hydraulic software. We automated this data insertion and transformation in SWMM© format to make it quicker and easier for the user. This automated solution was compared with manually estimated inputs. The simulation results show a coherent hydraulic behaviour.

1. Introduction

Urban stormwater and wastewater networks play an environmental role in many ways. The former, draining rainwater, constitute a protection against floods whereas the latter are intended to route wastewater to treatment systems, when they exist, or directly into the receiving bodies. In the framework of flood and pollution risk management, the hydrodynamic modelling of such networks is essential. However, getting accurate and updated information on the underground wastewater networks is a cumbersome task, especially in cities undergoing urban expansion. Data related to wastewater networks can be divided into two categories: spatial data representing the geo-referenced position of the objects such as pipes or manholes, and attribute data associated to each object such as the diameter of a pipe.

The most common challenge when managing environmental systems is missing data [1-3]. For wastewater networks, incomplete databases have various consequences such as delays in repair and collateral damage on the networks thus reducing the quality of the services provided to the citizens and increasing expenses.

To help managers acquire precise information about the position of the buried objects, diverse solutions have been published. In [4], the authors propose a Bayesian mapping model to integrate information collected using Ground Penetrating Radar, surveyed manholes and available statutory records. A methodology has been developed in a previous project for an automatic mapping procedure in data scarce settings [5,6] using manhole covers detection in aerial images [7].

Even if the networks' maps obtained through these procedures are in good agreement with the actual networks in terms of topology, they cannot be used directly by hydraulic modelling software. Indeed, no or very little information is available on the main attributes of the network: pipe shapes and dimensions, roughness, slopes, etc. To address the challenge of missing attributes related to wastewater networks, in [8], the authors use Missing Value Imputation techniques to estimate missing values of pipe diameters and the number of valves. Although the results were encouraging for some methods, this study was conducted on a small percentage of missing values (between 2% and 12%), which is not always the case in real life situations.

The present work aims at automatically completing the database associated to the mapped network, using the little data available and rules based on classical guidelines for the construction of such networks. This paper is organized as follows: Section 2 present the methodology and the materials used in this study. The results are described and discussed in Section 3. Section 4 concludes this paper.

2. Materials and methods

We assume that the map of a wastewater network is available in the form of a polyline shapefile which may be obtained either from the local stakeholder or from the previously developed mapping process. In this work, we use the wastewater network of Prades-le-Lez, a small town located in Southern France, produced in [5]. We already know that no or little data is available in the attribute table associated to the underground network. Moreover, other information may be available from national or open access geographical databases. In France, they correspond to commercial products developed by the French Geographical Institute IGN: Bd-Topo® for roads and buildings and RGE-alti 1m® for elevation.

The minimum information required to run hydraulic software are: inlet positions, depth and associated input discharges, pipe geometries, slopes and roughness. Inlet positions are assumed to be known from the map. In a first approximation, pipe roughness can be assumed uniform given the most probable material used in the city. Pipe geometry can also be assumed circular with little impact on the results leaving one parameter to be estimated: their diameters. Finally, if the absolute depth of inlet is not mandatory, the corresponding pipes slope is a parameter of great importance, as classical hydraulic models cannot compute gravity fed flow on counter-slopes, unless pumping stations are installed.

We thus present the methodology used to automatically assign diameter and slope values to each pipe of the network and to assess input discharges.

2.1. Diameter estimation

Using minimal and maximal value bounds chosen by the user, the algorithm allocates the pipe diameters according to Strahler's order [9], thus ensuring the general increase of the diameters from upstream to downstream. When a minimum of 20% of the attributes are available, one may use a semi-supervised learning method to predict the missing values as described in [10].

2.2. Slope estimation

The developed algorithm first allocates to pipes nodes the ground elevation value minus the minimum burying depth of pipes, i.e. 0.8m in France. Starting from the outfall, it associates an upstream and downstream point, and thus a direction to each pipe, assuming gravity fed flow. However, even if ground elevation has proven to be well related to manhole depths, the precision of this simple estimation is not sufficient and requires modifications to ensure the conveyance of flow in the correct direction.

The algorithm proposed here can be summed up as follows:

- Separate the network into "branches" where all the points can have one upstream and one downstream point at most;
- For each branch starting from the general outfall:
  - Check the coherence of the elevations and correct if needed to ensure a general decrease from upstream to downstream;

- If several corrections are to be carried out on a same point, choose the one with minimum elevation.

Figure 1 describes the main steps of the slope estimation.

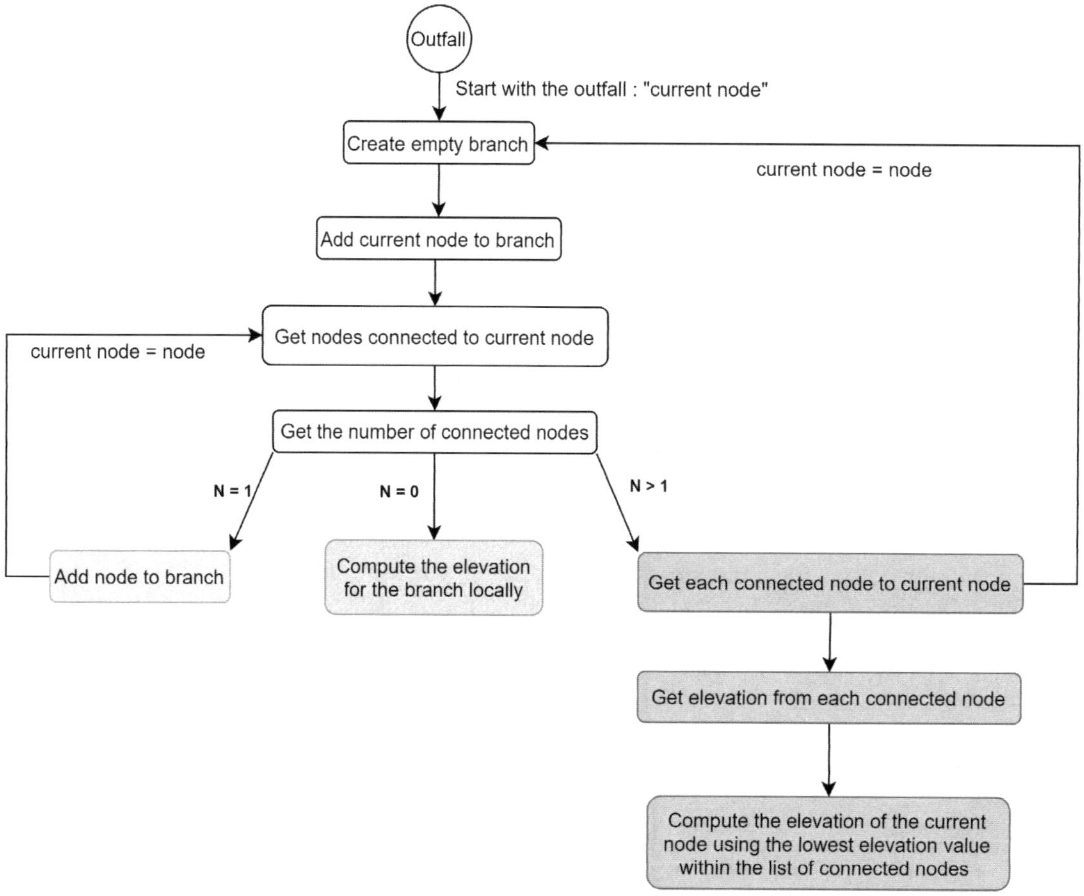

Figure 1. Slope estimation steps.

2.3. Input discharges

At each node, input discharge is estimated from the following pieces of information: the average number of inhabitants in the surrounding houses and their mean daily sewage discharge assessed from drinking water consumption. We assume that each construction is more likely to be connected to the closest pipe in terms of Euclidean distance. Figure 2 presents an illustration of the process based on the IGN building database (BD-Topo©). Each Polygon is linked to the closest node and an input discharge is estimated for each node.

SWMM© [11] is a Windows-based desktop program, used for planning, analysis, and design related to stormwater runoff, combined and sanitary sewers, and other drainage systems. We use SWMM© to run the hydraulic simulation. The proposed process should result in the automatic creation of all the attributes required by this software. This will ensure considerable gain in preparation time for the user.

Tests were carried on the wastewater network of Prades-le-Lez. The database contains approximately 800 pipes and almost no associated characteristics, namely no elevation.

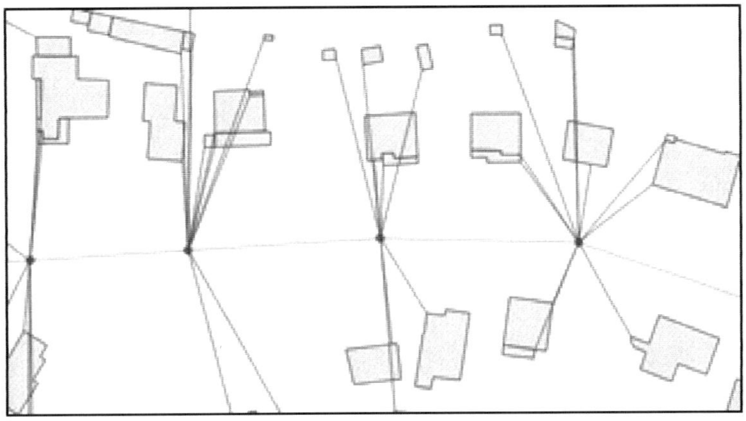

Figure 2. Linking each building to the closest network node.

3. Results and discussion

The procedure succeeded in completing the database and running hydraulic simulations without any errors. As no validation dataset is presently available for a better hydraulic validation, the attributes' values automatically estimated by the above-mentioned algorithms are compared with those manually estimated by a hydraulic engineer. For the elevation estimation the correlation is almost equal to 1 (figure 3). Moreover, the simulation run on the entire dataset shows coherent hydraulic behaviour (figure 4).

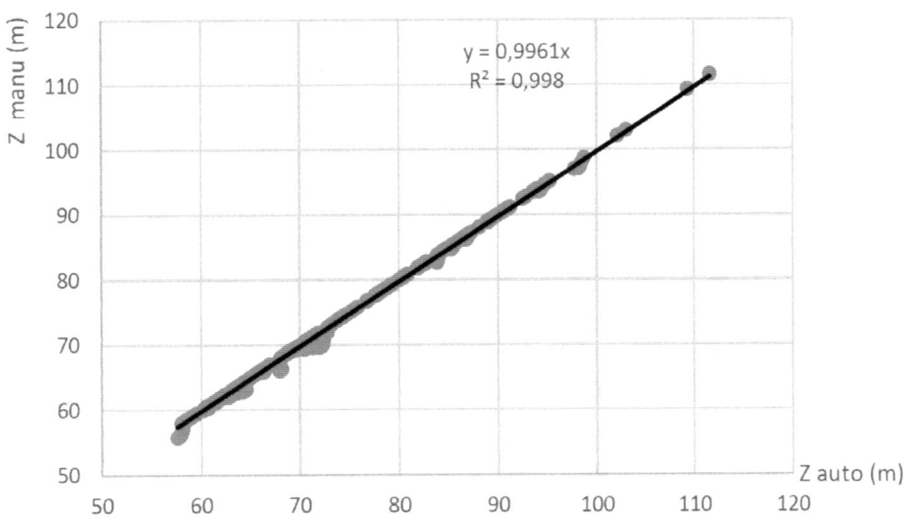

Figure 3. Automatic elevations (x-axis) and manual elevations (y-axis).

Figure 5 shows an example of the hydraulic simulation using SWMM©, where the various colours indicate water flow in each pipe and manhole.

It should be reminded that the network used in this study is the result of the map created in [5], where mapping errors may occur and that our propositions to compute the missing values are carried out under the worst conditions i.e. where no attributes values are provided in advance. By running the hydraulic simulation, we were able to detect some mapping errors during this process, precisely on the slope estimation step.

We noticed that assigning a slope value based on the elevation and the structure of the network may produce abnormal values such as a too big difference of altitude between two nodes of the same pipe. Such error impacts directly the entire network, thus the simulation directly. We put forward a tracing algorithm to identify the pipes where the errors are initiated. After comparing the reconstructed map with the map provided by the operator, we noticed that all the pipes where the errors are identified are not part of the ground truth network and are due to detection or mapping errors. Figure 6 shows an example of such a situation, where the difference between the node identified by "657" and "738" is of almost 15 meters. The pipes from the reconstructed network (in blue) having the endpoints ("657","738") and ("738","688") do not exist in the actual network (in red).

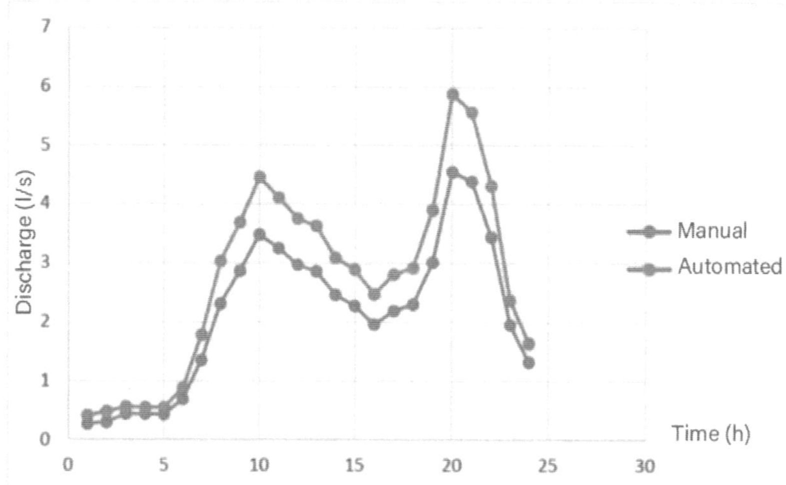

Figure 4. Comparison between outputs hydrographs from automatic and manual inputs.

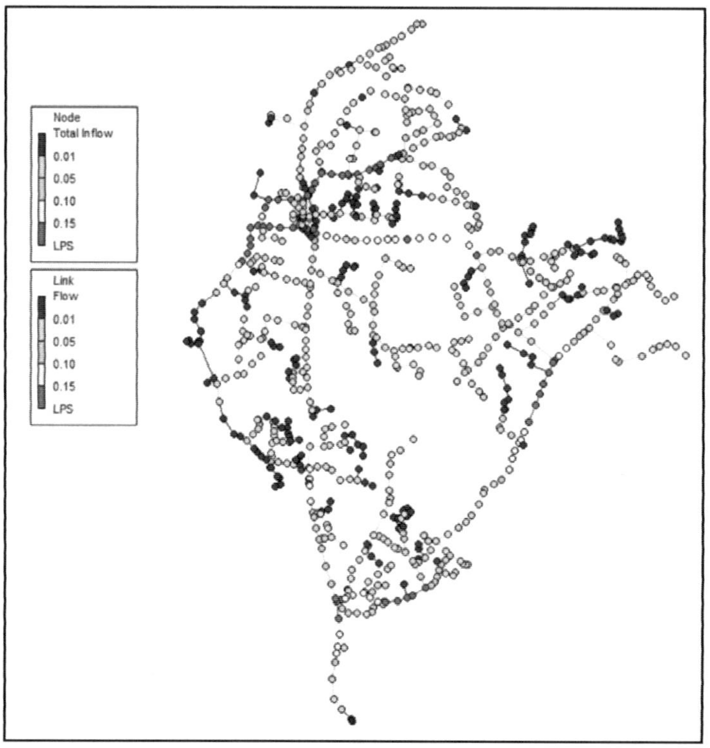

Figure 5. Simulation example using the SWMM© software with the proposed estimation process.

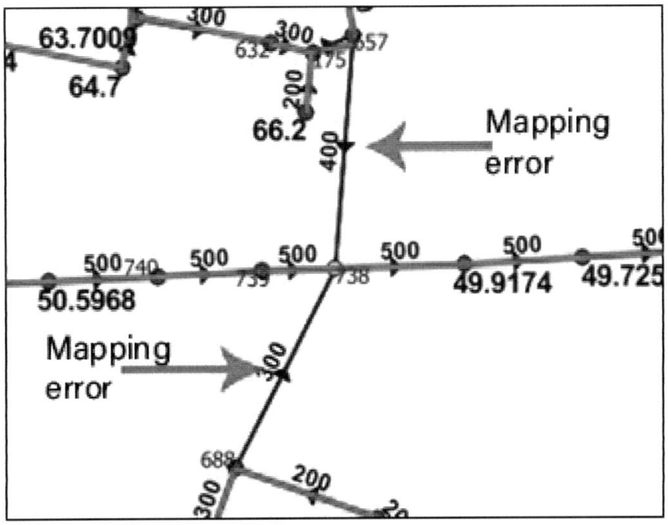

Figure 6. Identification of mapping errors.

4. Conclusions and perspectives

In this study we proposed a process to estimate the missing data needed to run hydraulic simulations using a different approach for each required parameter. We based our algorithms on information that are often available: the ground elevation, the average number of inhabitants per house and the mean daily discharge. In addition to being quicker and easier to use, the results accomplished by the automated solution indicate a coherent hydraulic behaviour when comparing to manually estimated inputs.

The estimated data are not accurate and do not systematically reflect the reality on the ground. Consequently, the simulation results are inevitably uncertain and imprecise, and they can lead to wrong conclusion despite running without errors. In addition, estimation approaches are used when official recorded data are unavailable, and they are not intended to substitute for good quality field data. Thus, currently we are working on improving our process with the aim to account for the accuracy of the data that the user may define in the input map. Indeed, when the values of several pipe characteristics are known with certainty, they should be maintained by the algorithm and propagated to the surrounding pipes. For example, if a pipe's diameter is known, the algorithm should be able to modify other diameters to ensure the general increase of pipe conveyance from upstream to downstream. Furthermore, since information about the networks can be collected from different sources, using various techniques such as Radars and images, a data fusion framework, which includes all the available data, could help the managers surpass the problem of missing data and thus improve the accuracy of the simulations.

Acknowledgments

This work was carried out within the framework of the CIFRE-France/Morocco Program and the project Cart'Eaux funded by the European Regional Development Fund (ERDF) and the Languedoc Roussillon Region.

We thank Mustapha Derras for the general supervision of the research group in Berger-Levrault and administrative support.

References

[1] Kofinas DT, Spyropoulou A and Laspidou CS 2018 A methodology for synthetic household water consumption data generation *Environ. Model. Softw.* **100** 48–66 https://doi.org/10.1016/j.envsoft.2017.11.021

[2] Schneider T 2001 Analysis of incomplete climate data: Estimation of mean values and covariance matrices and imputation of missing values *J. Clim.* **14** 853–871 https://doi.org/10.1175/1520-0442(2001)014<0853:AOICDE>2.0.CO;2

[3] Junninen H, Niska H, Tuppurainen K, Ruuskanen J and Kolehmainen M 2004 Methods for imputation of missing values in air quality data sets *Atmos. Environ.* **38**(18) 2895–2907 https://doi.org/10.1016/j.atmosenv.2004.02.026

[4] Bilal M, Khan W, Muggleton J, Rustighi E, Jenks H, Pennock SR, Atkins PR and Cohn A 2018 Inferring the most probable maps of underground utilities using Bayesian mapping model *J. Appl. Geophys.* **150** 52–66 https://doi.org/10.1016/j.jappgeo.2018.01.006

[5] Chahinian N, Delenne C, Commandré B, Derras M, Deruelle L and Bailly J-S 2019 Automatic mapping of urban wastewater networks based on manhole cover locations, *Comput. Environ. Urban* **78** 101370 https://doi.org/10.1016/j.compenvurbsys.2019.101370

[6] Commandre B, Chahinian N, Bailly J-S, Chaumont M, Subsol G, Rodriguez F, Derras M, Deruelle L and Delenne C 2017 Automatic reconstruction of urban wastewater and stormwater networks based on uncertain manhole cover locations *Proc. ICUD 14th IWA/IAHR Int. Conf. Urban Drainage* (10-15 Sept., Prague, Czech Republic) pp 2345–52 https://hal.archives-ouvertes.fr/hal-01584776

[7] Pasquet J, Desert T, Bartoli O, Chaumont M, Delenne C, Subsol G, Derras M and Chahinian N 2016 Detection of manhole covers in high-resolution aerial images of urban areas by combining two methods *IEEE J. Sel. Top. Appl.* **9**(5) 1802–1807 doi: 10.1109/JSTARS.2015.2504401

[8] Kabir G, Tesfamariam S, Hemsing J and Sadiq R 2020 Handling incomplete and missing data in water network database using imputation methods *Sustainable and Resilient Infrastructure* **5**(6) 365–377 https://doi.org/10.1080/23789689.2019.1600960

[9] Strahler AN 1957 Quantitative analysis of watershed geomorphology *Eos T. Am. Geophys. Un.* **38** 913–920 https://doi.org/10.1029/TR038i006p00913

[10] Belghaddar Y, Chahinian N, Seriai A, Begdouri A, Abdou R and Delenne C 2021 Graph convolutional networks: Application to database completion of wastewater networks *Water* **13**(12) 1681 https://doi.org/10.3390/w13121681

[11] U.S. Environmental Protection Agency *Storm Water Management Model – EPA SWMM* https://www.epa.gov/water-research/storm-water-management-model-swmm

Assessment of heavy metals removal in a wastewater using aquatic ferns

Amir Hajiali[1] **and Lăcrămioara Diana Robescu**[1]

[1] Department of Hydraulics, Hydraulic Machinery and Environmental Engineering, University POLITEHNICA of Bucharest, Bucharest, RO-060042, Romania

a.hajiali.env@gmail.com

Abstract. Aquatic ferns, which due to their rapid growth on the surface of the water in aquatic areas in many countries as an invasive species threaten the aquatic life and ecosystem of the regions, have the ability to remove heavy metals in wastewater. In this study, their use in removal of existing heavy metals in a domestic wastewater was experimented and the results proved their remarkable efficiency in treating such a wastewater. This research was conducted in two steps; laboratory and semi-industrial. Applying a jar testing apparatus was the first attempt in laboratory scale and optimization of the following four different factors by a routine procedure was the further goal: pH-value, reaction time, absorbent, pollutant concentration. Afterwards, it was scaled up by a real wastewater and removal efficiency was evaluated under optimum conditions designed by the basic experiments. In the second stage, the reactor was designed and constructed on a semi-industrial scale and the removal rate of contaminated metals was achieved. Focusing on the energy efficiency after the heavy metal removal in this method in comparison with another method (active sludge) can be considered to be suggested for treating domestic wastewaters which include considerable contents of heavy metals.

1. Introduction

Nowadays, the contamination of water resources has led to serious problems because of the indiscriminate disposal of heavy metals. The adsorption technique has successfully removed toxic heavy metals from the aqueous solution over several decades [1],[2]. The use of biological purification methods is due to advantages such as low costs, availability, low amount of sludge, high yields, heavy metals recovery and the possibility of reuse of biological mass [3], [4].

In this research active sludge and ferns were provided using biological method in the laboratory and in a semi-industrial process to remove heavy metals from wastewater. The effect of pollutant concentration was studied at constant temperature [5]. Removal of zinc and iron is therefore effectively performed using the adsorbents and low concentrations were evaluated. Increasing initial concentration of the heavy metal could enhance production of metal ions in the solution so competition for absorption increased between ions [6], [7]. Furthermore, all active sites on the absorbent surface activated and removal efficiency increased.

2. Materials and Method

The evaluation and process regarding the heavy metals removal by aquatic ferns is as followings:

- Evaluation of the removal of Zn and Fe metal by ferns adsorbent from synthetic wastewater in pilot where 600 grams of adsorbent was prepared and 100 liters of synthetic wastewater containing about 52 ppm iron and 2 ppm zinc at pH 5 for 120 minutes in a mixer tank with 30 rpm and 50, 100 and 150 rpm cycles for 20 minutes (time needed for separation absorber phase) was dissolved and mixed, then the solutions were smooth and the amount of metal ions was analyzed by a spectrophotometer.

- Evaluation of the removal of Zn and Fe metal by ferns adsorbent from the actual effluent in the pilot where 600 grams of adsorbent and 100 liters of actual waste containing about 56 ppm iron and 1.5 ppm of zinc with a pH of about 6.8 for 120 minutes in a mixer tank with 30, 50, 100 and 150 rpm cycles for 20 minutes (the time required to remove the adsorbent phase from the solution) was carried out and mixed, then the solutions were smooth and the metal ions were analyzed by a spectrophotometer.
- Evaluation of the removal of zinc and iron metal by dry sludge adsorbent from synthetic wastewater in pilot where 500 grams of absorbent was prepared and 100 liters of synthetic wastewater containing about 52 ppm iron and 2 ppm of zinc at pH 5 for 90 minutes in a mixing reservoir with 30, 50, 100 and 150 rpm cycles for 20 minutes (time to separate absorber phase) were dissolved and mixed, then the solutions were smooth and the amount of metal ions was analyzed by a spectrophotometer.
- Evaluation of the removal of zinc and iron metal by a dry sludge adsorbent from the actual effluent in the pilot plant where 500 grams of adsorbent was prepared and 100 liters of actual waste containing about 56 ppm iron and 1.5 ppm of zinc with a pH of about 6.8 for 90 minutes in a mixer tank with 30, 50, 100 and 150 rpm cycles for 20 minutes (the time required to remove the adsorbent phase from the solution) was carried out and mixed, then the solutions were smooth and the metal ions were analyzed by a spectrophotometer.

2.1. Used equipment and materials

The equipment and chemicals which were used in the laboratory can be seen in table 1:

Table 1. The equipment and chemicals used in the laboratory.

Machine Name	Chemicals
Digital Scale (Accuracy of 0.001 gr)	Concentrated Chlorophyll
Spectrophotometer	($ZnSO_4$. $7H_2O$)
Aqualytic Jarring Test	Sulfuric Acid (H_2 SO_4)
pH Measurer (limit of 4-9)	Hydrochloric Acid (HCl)
Stirrer (0-1500 rpm)	Sodium Hydroxide(NaOH)
Filter Paper	Iron / Zinc Detector

Pilot components according to the design and studies carried out and the cost of conducting was also accomplished, in table (2) the technical specifications of the equipment and a brief description of their performance are given.

The pilot setup steps are as followings:

- Construction of pilot components (cutting, bending, welding)
- Installing the shaft on the electro-motor
- Installing the inverter frequency
- Installing the valves in the tank
- Construction and installation of mechanical stirrers
- Installing the electromechanical switches

Table 2. Technical specifications of pilot components.

Components	Technical Specifications	Functions	Construction Steps & Explanations
Tank	Tank's Material: Polyethylene Height: 0.6 m Diameter:0.5 m Volume :100 l	Sample collection site (wastewater) and mixing to perform heavy metal removal operations Reasons for choosing Teflon tank: - Low weight -Suitable Price -No reaction to solution -No corrosion	The following steps were taken to prepare the reservoir: 1. Turning (Tank interface to valves) 2. Drilling 3. Adhesive 4. Connecting the valves
Funnel	Opening Diameter:0.5 m Height: 0.3 m	Sample collection site (wastewater) and mixing to perform heavy metal removal operations	The end of the tank is equipped with outlet valves for sampling and drainage is funnel-shaped.
Valves	Material: UPVC Valve Type: Ball Diameter : 0.5 in Quantity: 2	-One valve to drain the sludge -One valve for sampling	Due to the lightness and easy installation, this kind of valve was used.
Supports	Height: 1.68 m Length :0.65 m Width: 0.53 m	In order to place the mechanical reservoir and stirrer in it, and the connection of the electromotor and other components.	The following steps were taken to build it: Measuring, Cutting, Rolling (Bending), Welding, Locating, Perforation.
Electromotor	700 rpm 3 Phase Power: 0.1 kw Current: 0.8 Amp	To convert electrical energy to mechanical energy in order to turn the stirrer.	Three-phase alternating current motors (AC) were used in the wastewater plant. Steps to install the motor: - Perforation - Installing the shaft on the electromotor - Installing the inverter frequency.
Engine Speed Changer	Inverter Frequency iG5A	Smartly detects the load on the engine and it fits the engine at the same time accordingly.	The inverter frequency was used. It can be used to change the speed of the three-phase motors.
Electrome-chanical Switch	Input Flow ON/OFF Key 35 Amp /500 w	Setting up the electromotor and protecting it during the work.	It was mounted on the supports to control the flow of work.
Mechanical Stirrer	Diameter :0.14 m Width: 0.3 meter Quantity of Fins:3	Stirring the effluent.	Mechanical stirrer was designed in motion so that its height can be adjusted according to the sample volume and concentration.

Shaft	Material: Steel Bar Length: 1m Height: 0.75 m	Connection Stirring rod with electromotor	Shaft was designed in motion so that its height can be adjusted according to the sample volume and concentration.

3. Results and Discussion

3.1. The result of changes in the initial concentration of iron

In general, studies have shown that in the absence of other parameters, increasing the initial iron concentration leads to a reduction in the removal efficiency [8].

According to the above results, with the improvement of other removal related parameters including pH, contact time and appropriate amount of adsorbents, a good and acceptable removal efficiency was obtained and no significant reduction in the concentration of iron removal was observed. The highest percentage of removal was found at 50 ppm concentration for iron removal (99.8% and 96.8%) in ferns and dry sludge respectively (figure 1).

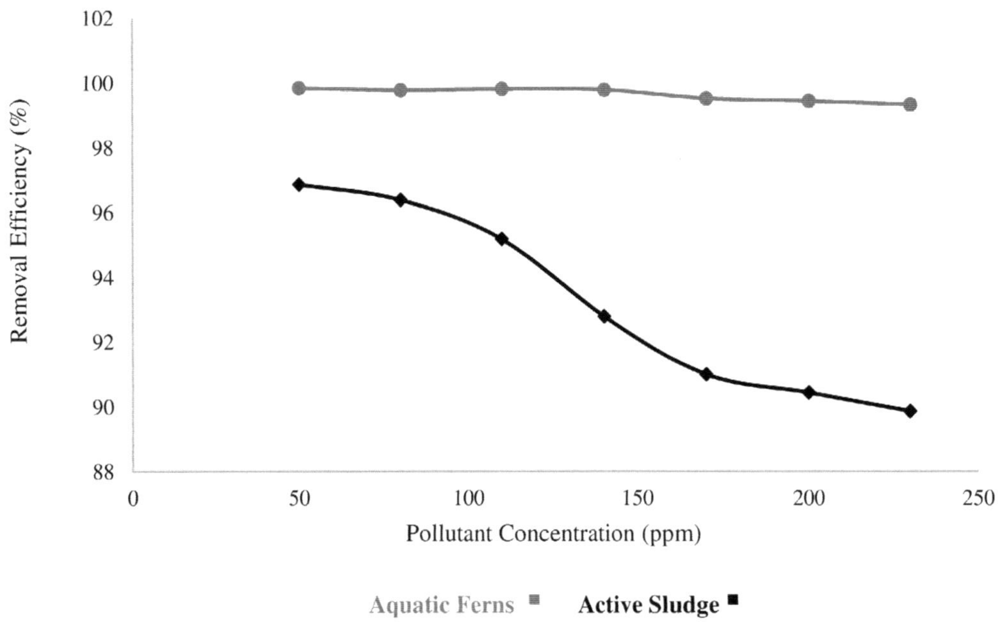

Figure 1. Effect of pollutant concentration (Fe) at optimum condition by aquatic Ferns (concentration: 6 g/l) and active sludge (concentration: 5 g/l).

3.2. The result of changes in the initial concentration of zinc

In the studies, it is observed that zinc concentrations are often found to occur even in a plating wastewater between 10 and 20 ppm and in general in low concentrations. In this research, low concentrations were evaluated due to the use of wastewater from the wastewater treatment plant in the next stages.

By increasing the initial concentration of metal, the number of metal ions in the solution increases and the competition for access to the binding sites increases at the contact surface of the adsorbent, and the entire level of biological absorption of the cell wall functional groups is activated by contact with the metal and the absorption rate of the metal. And of course, the metal absorption increases according to the initial concentration. With respect to both adsorbents, changes in concentrations in this range are not significantly altered in the higher concentrations of contaminants [9].

The highest removal at the concentration of 5 ppm in the presence of ferns and dry sludge is shown to be equivalent (99.6%) and (93.2%) respectively (figure 2).

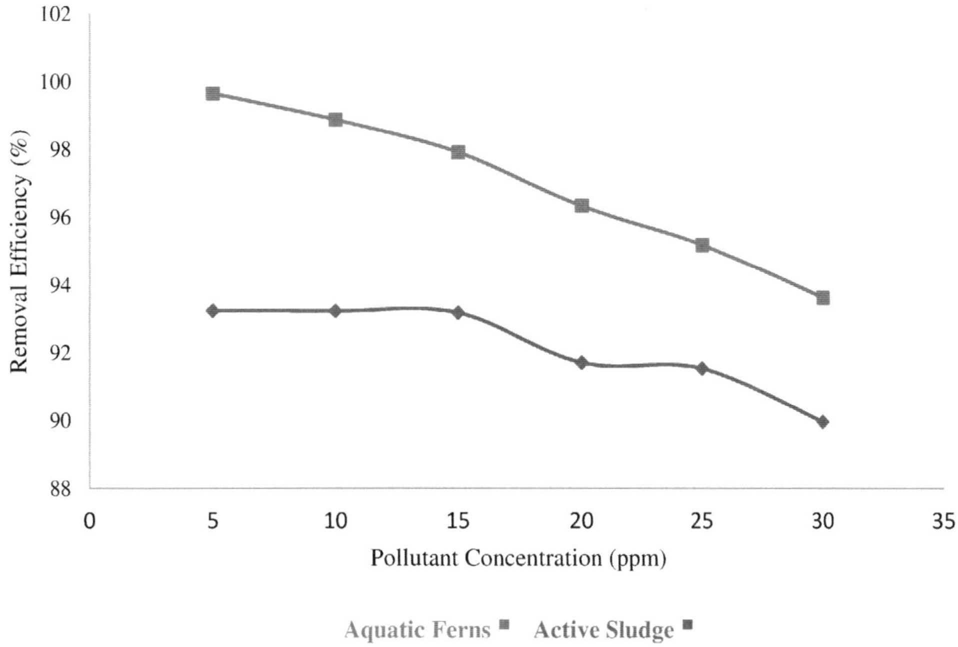

Figure 2. Effect of pollutant concentration (Zn) at optimum condition by aquatic Ferns (concentration: 6 g/l) and active sludge (concentration: 5 g/l).

3.3. The economic estimate and efficiency

Since this project is done on an industrial scale, it is essential to take advantage of the costs. At each filtration, approximately 90% of the wastewater is refined and 10% will be wasted.

Among the two biological adsorbents mentioned above, ferns have a higher removal capacity and at the first level, the ability to remove more pollutants in the actual wastewater was by more than 99%. Recovery of the adsorbent was carried out to prevent its loss. The optimum removal time for ferns is 120 minutes. If this is done only during the treatment hours of the treatment plant, an average of 8 treatments per day are performed (ferns absorbent is used twice a day) and 90% of the volume of wastewater will be purified.

It should be noted that for a wastewater treatment plant, the wastewater treatment used to cost $ 13.8 per 1000 gallon while it costs $ 2.15per 1000 gallon now. A dry sludge absorber can also be exploited with the same calculations. However, it is preferable to use this adsorbent once to achieve higher efficiencies. According to figure (3) after the seventh month, the price of the treated water will be almost 5 times cheaper than the current cost per 1000 gallon of the wastewater, and reaching the end point at the end of a year, the price will be even less ($ 2.15 per 1000 gallon) that is approximately 6.5 time less which shows a great efficiency as a matter of cost and of course energy.

4. Conclusion

The results of this study showed that the used adsorbent could be considered as a potential adsorbent in order to remove the two existing heavy metals in the studied wastewater. In the case of both absorbents there weren't any significant changes in the aforementioned evaluated ranges and it was observed that removal efficiency decreased by rising pollutant concentration and showed the best results for zinc with aquatic ferns and active sludge by 99.6 % and 93.2 % removal efficiency, respectively.

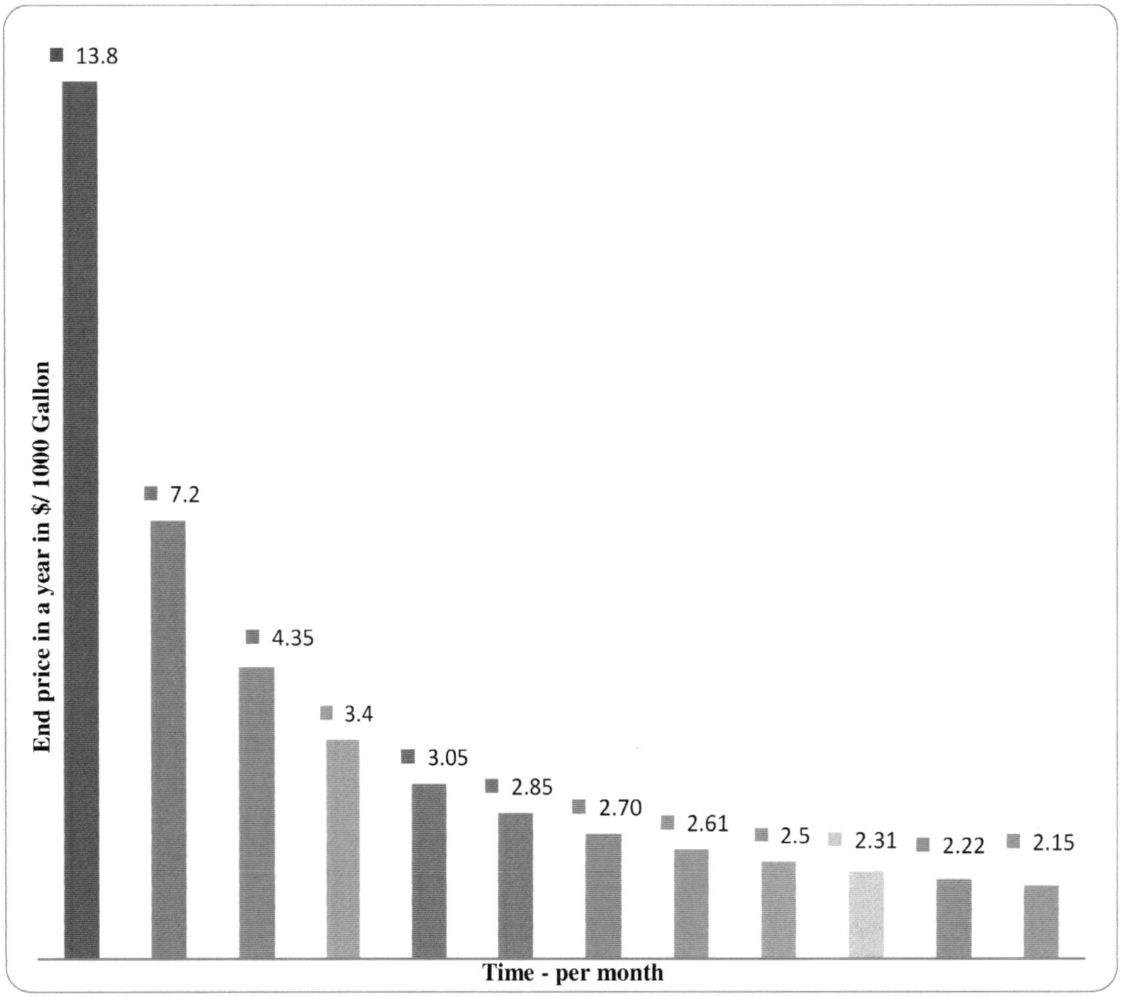

Figure 3. The cost in USD per 1000 gallon of a wastewater treatment within one year in the wastewater treatment plant.

As the literature displayed increasing pollutant concentration had an adverse effect on removal efficiency. As it was recorded for aquatic ferns and active sludge, 99.8 % and 96.8 % removal efficiency obtained respectively for 56 ppm Fe (III) ions concentration. The economic estimation of the performed project also showed a great energy efficiency as the cost is not comparable after a year with the previous expenses for treating such a wastewater.

References

[1] Fatemeh Gorzin and Bahri Rasht Abadi 2018 Adsorption of Cr (VI) from aqueous solution by adsorbent prepared from paper mill sludge: Kinetics and thermodynamics studies, *Adsorption Science & Technology* **36**(1–2) 149–169

[2] Ijaz Ali, Danish Khan, Tariq Mehmood and Ayub Khan 2021 Removal of Uranium U(IV) from aqueous solution using acid treated spent tea leaves *Current Research in Green and*

Sustainable Chemistry **4** 100197

[3] Amir Hajiali, Pirumyan G, Sharif Vaghefi HR and Shahmiri M 2012 Treatment of pulp factories wastewater containing Chorophenolic compounds with ozonation *Asian Journal of Chemistry* **24**(12) India

[4] Amir Hajiali 2017 Total and suspended solids removal evaluation in effluent from anaerobic bioreactor and effluent from cyclic ozonation biotreatment in dyeing industry wastewater treatment *IEEE Proc. Int. Conf. on Energy and Environment (CIEM)* pp 115–119

[5] Zhao M and Duncan JR 1998 Removal and recovery of nickel from solution and electroplating rinse effluent using Azolla filiculoides *Process Biochem.* **33**(3) 249–255

[6] Zhao M, Duncan JR and van Hille RP 1999 Removal and recovery of Zinc from solution and electroplating effluent using Azolla filiculoides *Wat. Res.* **33**(6) 1516–1522

[7] Bennicelli R, Stepniewska Z, Banach A, Szajnocha K and Ostrowski J 2004 The ability of Azolla caroliniana to remove heavy metals (Hg(II), Cr(III), Cr(VI)) from municipal waste water *Chemosphere* **55** 141–146

[8] Shokoohi R, Saghi MH, Ghafari HR and Hadi M 2009 Biosorption of iron from aqueous solution by dried biomass of activated sludge *J. Environ. Health. Sci. Eng.* **6**(2) 107–114 Iran

[9] Moradi H and Lohrasbi S 2011 *Water Plant Aquarium* 1st edition (Iran: Moalef Publication).

IOP Conf. Series: Earth and Environmental Science 1136 (2023) 012055 doi:10.1088/1755-1315/1136/1/012055

AI-based validation of wastewater treatment plant sensor data using an open data exchange architecture

Siddharth Seshan[1], Dirk Vries[1], Maarten van Duren[2], Alex van der Helm[2] and Johann Poinapen[1]

[1] KWR Water Research Institute, Groningenhaven 7, 3433 PE, Nieuwegein, the Netherlands
[2] Waternet, PO Box 94370, 1090 GJ, Amsterdam, the Netherlands

siddharth.seshan@kwrwater.nl

Abstract. Typically, within the context of treatment plant-wide data, the quality of data can be impacted by sensor faults, sensor calibration issues, fouling of and obstruction to the sensors and connectivity problems between sensors, actuators and the data management system, therefore hampering advanced data driven monitoring and control of (critical) water operations. Here, a smart data validation scheme is proposed that validates sensor data from a wastewater treatment plant and is tightly integrated with the open-source data exchange system called FIWARE, an EU supported framework. The data validation application and FIWARE setup are integrated, tested and deployed at the water utility, Waternet. The validation scheme is based on an anomaly detector using (statistical) threshold techniques and a data reconciliation part that aggregates deep learning based autoencoder model predictions whenever an anomaly is detected. The autoencoder models proved to have a high accuracy and good reconciliation performance considering the variability of the signal. Furthermore, (near) real-time validated and raw data signals are relayed towards a dashboard. Finally, the validated data can be used as a screening for data ingested by another AI-based model that enables monitoring and smart control of the wastewater treatment plant in order to minimise greenhouse gas emissions and energy consumption while meeting effluent water quality standards.

1. Introduction

Wastewater treatment infrastructures get increasingly complex due to the need to meet stringent effluent quality standards such as the European Directive 91/271 on urban wastewater, but also due to resource recovery technologies and the increasing use of sensor data for advanced process monitoring and control of key variables. This leads to a stronger dependency on data. Therefore, ensuring high quality data has become even more relevant. For example, a water utility can be faced with (many) anomalous sensor data, i.e., missing values, outliers, irregular trend breaks in time series, bias due to connection errors, sensor fall-outs or wrong sensor calibrations, drifting due to sensor wear, fouling or obstruction. In addition, the multitude of sensors and their different characteristics and the availability of external data sources not only provide opportunities for intelligent data driven operation of wastewater treatment plants (WWTPs), but also a challenge in keeping the operability among different devices up-to-date. Analogously to the interconnection of smart devices as is the case in the Internet of Things (IoT), as IoT scales up, interoperability needs to be safeguarded in order to maintain operational performance and up-to-date information [1].

In addition, WWTPs typically rely on multiple purification and sometimes resource recovery technologies which in turn are characterised by multigranular, interrelated nonlinear biological, physical and chemical processes. Furthermore, these systems are very dependent on environmental and operational conditions, and often have cyclic behaviour due to recycle streams. It is recognised that operation of WWTPs has evolved from sensor signal processing and using statistics and process identification in the seventies towards knowledge-based systems where sensor data is increasingly fed into data mining techniques and predictive analytics in order to capture complexity with the aim to support, possibly multi-criteria optimised, plant-wide control and decision support [2]. However, the quality of data is frequently hampered by the intrinsically challenging measurement conditions of the aquatic environment, which makes that on-line sensors can be affected by many faults [3].

Hence, the reliance on validated and clean data is increasingly becoming apparent and, moreover, open data management platforms are becoming attractive from (i) a software lifecycle assessment point of view and (ii) the prospect of managing heterogeneous data sources more efficiently.

Here, we present an automated data validation and reconciliation (DVR) framework which is embedded in an open data management system called FIWARE [4]. The DVR is one of the key steps in achieving intelligent, data-driven plant-wide control of a large WWTP called Amsterdam West, situated nearby Amsterdam, the Netherlands and operated by Waternet, the public water cycle utility of Amsterdam and surrounding areas. The Amsterdam West WWTP has a capacity of 1 million population equivalent and serves the city of Amsterdam. Currently, the control loops of Amsterdam West WWTP are for a large part locally distributed and dedicated to a single treatment process unit. To illustrate the context: the objective of the smart control application (not discussed in this work, see [5], [6] for more details) is to minimise energy consumption and N_2O emissions while meeting effluent quality criteria. N_2O has a large impact on greenhouse gas emissions and consequently, climate footprint.

In the following, we demonstrate the extension and coupling of FIWARE to the existing legacy data management system of Waternet in order to improve the utilisation of (near) real-time plant data for DVR of key sensor signals. For the Amsterdam West WWTP case, the sensor data of two key process parameters, nitrate (NO_3^-) and ammonium (NH_4^+), in the aerobic tank of the research lane's bioreactor unit were considered to demonstrate the prospect of using AI-based DVR and FIWARE as a data exchange platform.

2. Data management system

2.1. *Distributed control system, information management and FIWARE4Water setup*

Amsterdam West WWTP is real-time controlled by a distributed control system (DCS). The DCS stores and handles measured sensor data, e.g., online water quality and flow measurements, actuator data from e.g., valves and pumps, control-related data such as setpoints and local control settings. It also logs events, alarms and operator changes. Operators and water process engineers access the DCS via a graphical user interface.

Data is sent through the (legacy) Process Information Management System (PIMS) which acts as a historian and logs all process data, alarms and events. PIMS also hosts laboratory measurements, e.g., water quality and data signals manually validated by plant operators. All data and information are collected (near) real-time and can be accessed from office automation applications. Selected sensor data, i.e., NO_3^- and NH_4^+ concentrations, is forwarded to the FIWARE4Water setup which consists of FIWARE components, a dashboard and the DVR application, see Figure 1.

In turn, the FIWARE4Water setup consists of Docker™ containers running on a Windows VM in a Microsoft Azure™ cloud environment, each having dedicated predefined functionalities. The Docker containers contain the following components and are linked as follows:

- the IoT Agent JSON, ingests data via the DVR routines manager (see section 4.1. for more details) from the legacy system into the FIWARE setup and is coupled with the Orion-LD context broker;

- the Orion-LD Context Broker routes data between the IoT agent, the DVR application and notifies the Cygnus NGSI-LD component when new values are available
- the Cygnus NGSI-LD component writes all data that passes through Orion-LD to a PostgreSQL database;
- a PostgreSQL database stores data which has been ingested in FIWARE, as well as data generated by the DVR application;
- a Grafana dashboard queries the data from the PostgreSQL database which is visualised in real-time;
- a MongoDB database is used to store data models, FIWARE devices, FIWARE subscriptions, and other FIWARE configuration settings;
- a DVR docker containing the routines manager which handles FIWARE specific management tasks, e.g., creation and checks of data entities, subscriptions, etc. Furthermore, the routines manager serves as an application programming interface (API): it calls the relevant key components such as the anomaly detection methods, the trained autoencoder models to make a prediction and the reconciliation algorithm. As a final step, it outputs anomaly flags, predictions and reconciled data which is in turn passed through the IoT agent to the PostgreSQL database.

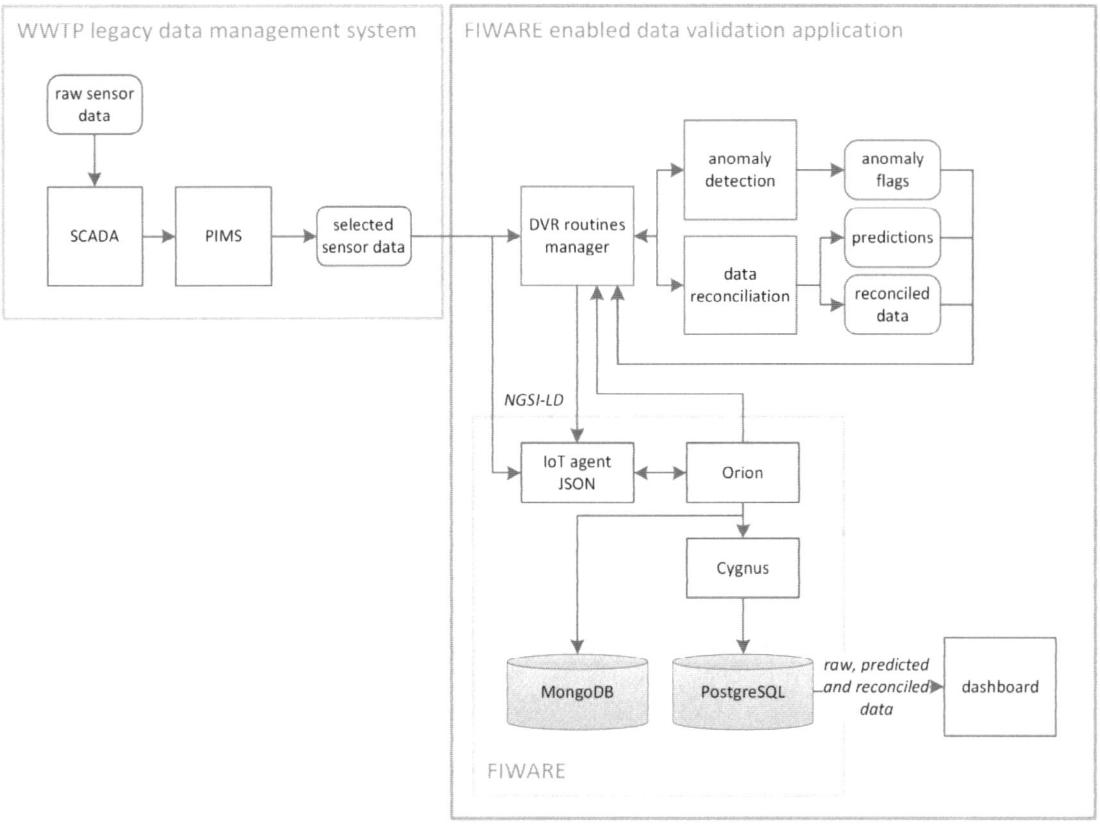

Figure 1: Schematic overview of the integration of FIWARE with the legacy control system.

Finally, raw, predictions from the models, and reconciled data from the DVR are visualised by the analytics and interactive visualisation web application Grafana by querying the PostgreSQL database.

2.2. FIWARE4Water data models

FIWARE relies on the NGSI-LD based data standard [7]. Smart data models for wastewater treatment have been specifically developed for this case and used in the current FIWARE setup, but have generic properties. Specifically, the following two NGSI-LD data models have been implemented for the DVR:

- a *WasteWaterTank*, an entity type used to describe the physical tanks in the bioreactor unit within the research lane. The properties of the entity are important process parameters, e.g., NO_3^- and NH_4^+, that are measured using online sensors.
- *WasteWaterSimulationResults*, a NGSI-LD entity that has been developed to describe the results provided from (data-driven) models. Note that this entity type represents digital-based solutions instead of wastewater treatment assets.

More details can be found in the smart data model repository available on GitHub [8]. In practice, the raw data as received from the legacy system is initially translated to NGSI-LD and then relayed to the IoT Agent JSON and the Orion-LD Context Broker. The data values specific to NGSI-LD entities are then accessed by the DVR routines manager. Similarly, once the DVR routines manager receives the output of the various DVR analytics, the data is once again transformed into the NGSI-LD data models, which are then sent back to the IoT Agent JSON.

3. Data Validation and Reconciliation

3.1. Procedure

The data validation consists of two components, first being the anomaly detection step where the raw data values are assessed as to being faulty or correct, and the second component includes the reconciliation of the anomalous values using predictions from trained, deep neural network models. The handling of data from and to the FIWARE4Water setup is taken care of by the FIWARE enabled DVR Routine Manager (FW-DVR). While streaming data is near real-time being ingested into FIWARE, the FW-DVR Routines Manager regulates when to trigger a given data validation routine, based on the availability of new data in FIWARE.

DVR follows a sequence of steps (Figure 2):

- The selected raw sensor data (NO_3^- and NH_4^+) is inputted initially into the anomaly detector to specifically flag faulty single values. The anomaly detector consists of two autoencoder models for each sensor signal, one representing short-term dynamics (5-minute) and one for long-term dynamics (30-minute). Furthermore, (historical) time series are inputted into the autoencoder models to conduct predictions that will be used in the reconciliation process. The autoencoder models are further explained in section 3.2.
- The single value anomaly detection is followed by flatline detection and flagging.
- The anomaly flags aid the data reconciliation process. Reconciliation is preceded by aggregation of predictions from the autoencoder models using exponential smoothening.
- Finally, three data outputs are provided by the data validation application, i.e., the raw sensor data, the autoencoder model predictions and reconciled data. The reconciled data constitutes raw sensor data during anomaly-free periods as well as aggregated, predicted values by autoencoder models during anomaly events.

3.2. Autoencoder models

Feedforward neural network models known as autoencoders were used for the purpose of data reconciliation. Autoencoders consists of two modules, an encoder and a decoder, where the encoder is trained for a given input of sensor data, to learn the underlying features and is represented in a reduced dimension. The decoder then reconstructs the input data, which is also the target variable [9], [10]. Conventionally, autoencoders are trained using only a single layer for the encoder and decoder each, however utilising deep autoencoders have known to yield better performance in the cases of non-linear and complex systems. Within wastewater treatment, autoencoder models are used for the purpose of reconstructing the input data which allows for such models to be used as soft sensors to predict the sludge volume index [11] and key effluent parameters such as biological oxygen demand [12]. Additionally, autoencoders have been used to denoise data from WWTPs, prior to its ingestion to data-driven based control strategies [13]. For the purpose of data validation, stacked denoising autoencoders have been used to detect faulty data and perform reconciliation for influent wastewater quality parameters [14] and additionally, deep autoencoders been used to reconcile missing or faulty values in wastewater treatment operations data [15]. In both cases, the prediction error between the reconstruction from the autoencoder model and the data evaluated was used for the fault detection.

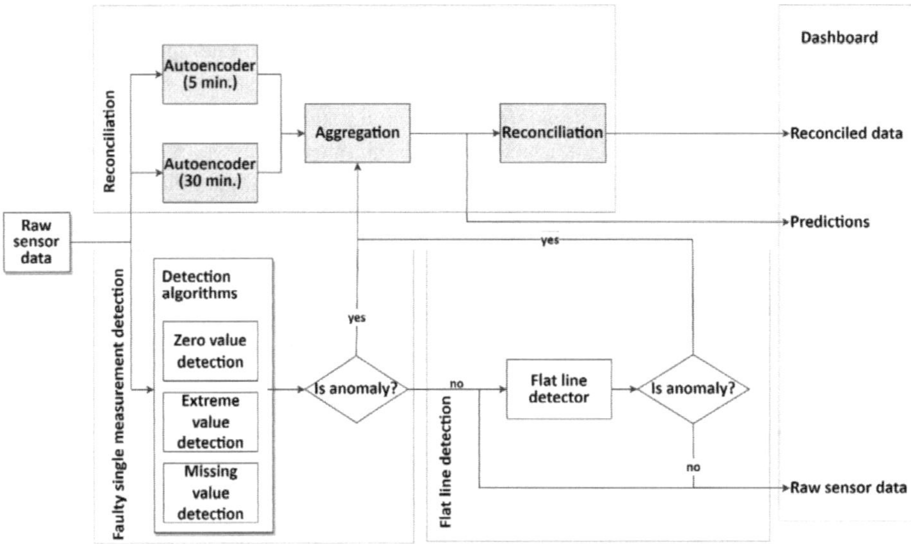

Figure 2: AI-based FIWARE enabled DVR Routines Manager.

In this study, over 70 different deep autoencoder models with varying architecture and hyperparameter combinations were investigated prior to model selection. The most relevant model structures (48 models) and the tuned hyperparameters are provided in [16]. The Python library TensorFlow was applied for development and training of the models. Initially, the model training was conducted to identify the type of layers that will be required within the model architecture, to be able to reconstruct the sensor behaviour with some degree of accuracy, where complexity was added with every iteration if deemed necessary. An architecture with only dense layers could not capture the non-linear effects present in the data. Hence, the model architecture was extended with layers that contain memorising units, i.e. more specifically, Long Short-Term Memory (LSTM) units. The use of LSTM layers resulted in an instant increase in prediction accuracy, which seems to confirm the autoregressive nature of the process. To mitigate the computational time involved when training LSTM models, the following was taken into account: (i) LSTM layers followed by dense layers in order to increase the depth of a model but keep the number of trainable parameters as small as possible, (ii) length of the input sequences, and (iii) use of dropout regularisation [17].

The resulting model structure settings are shown in Table 1 and training settings in Table 2. More details are provided in [16].

4. Real-time FIWARE-based data validation

4.1. Implementation of anomaly detection and reconciliation

For the purpose of recovering and reconciling anomalous NH_4^+ and NO_3^- signals, two granularities of 5-minute and 30-minute resolutions were chosen to perform the developed analytics. As a result, the

FW-DVR routines manager triggers the data validation methods every five or thirty minutes. Initially, the single value anomaly detection step is triggered, where the last available five or thirty data points are resampled and then assessed for being a faulty data point, thereby providing a flag for the current data point. Subsequentially, the flatline detection is triggered. For this use case, a minimum length of 3 consecutive data points has been used as a threshold (as available in the metadata for the sensor signal) to determine whether the points are part of a flat line. To perform this method, the manager therefore queries raw data points from the last 1.5 hours, that are resampled to the required resolutions separately, and inputted into the flat line detection. The method will provide a flag as an output which assesses the current data point to be part of a flat line, based on the values of the previous data points. The flags combined are used to judge whether the raw values should be reconciled using the predictions from the autoencoder models.

Table 1: Amount of historical input used for training of the autoencoder models.

Autoencoder Model	Resolution (minutes)	Historical Input (hours)	Historical Input (no. of timesteps)
5 minutes	5	3	36
30 minutes	30	24	48

Table 2: Model training settings.

Hyperparameter	Value	Comment
# of epochs	35	Decision made based on results obtained from a learning curve.
Optimiser	Adam	-
Learning rate	0.00001	-
Loss Function	Mean Squared Error	-
Activation Function	ReLU	Same activation function used for all layers
Batch size	112 (5 min. autoencoder), 14 (30 min. autoencoder)	Based on a sensitivity analysis, it was concluded that a batch size representing 2 weeks of data yielded the best results.
Dropout rate (p)	NO_3^- – 0.06 NH_4^+ – 0.03	-

Faulty raw data values are reconciled by the predictions of autoencoder models. To perform a prediction with the autoencoder models, a certain period of historical data is necessary as input, in order to make a one-step ahead prediction. For the case of the 5-minute autoencoder, the last 3 hours of data are needed (36 data points) and for the 30-minute autoencoder, the last 24 hours of data (48 data points). The FW-DVR Routines Manager queries the historical data which are needed for the model predictions. Model predictions are then pushed back into FIWARE via the IoT Agent, which are then subsequently stored in the database under the relevant table. The autoencoder models make use of recursive predictions in case an anomaly is detected to prevent recursive propagation of anomalous data. The procedure is as follows: raw data values flagged as non-faulty and model predictions which would replace faulty data points are stored in intermediate processed data streams and pushed into the IoT Agent. Each (5-minute and 30-minute) data stream is linked to its own intermediate processed data stream, and the data stream is used as an input to the autoencoder models. In the final step, predictions from both the autoencoder models are used in the reconciliation of the data stream per sensor signal, via exponential smoothing and resampling to a targeted resolution of 15 minutes. Finally, the calculated reconciled data points are pushed to the IoT Agent and stored in the PostgreSQL database. The various outputs from the data validation application, namely the predictions from the autoencoder models and the reconciled signals are (near) real-time visualised in the Grafana dashboard.

4.2. Performance of the anomaly detection method

The performance of the autoencoder models is discussed in detail in [16]. The accuracy of the methods, as well as the autoencoder prediction performance are summarised below.

The determination scores of the autoencoder models (Table 3) show that a very high prediction accuracy was achieved, indicating the sensor data signals can accurately be reconstructed. Such a performance subsequently provides adequate confidence to use the predictions as provided by the autoencoder models for the purpose of reconciling anomalous values and to therefore realise validated NO_3^- and NH_4^+ data signals.

The predictions of the 30-minute NH_4^+ autoencoder for the test set are shown in Figure 3 for illustration purposes.

Table 3: R^2 Score of autoencoder models for NO_3^- and NH_4^+ signals.

	NO_3^- autoencoder model		**NH_4^+** autoencoder model	
	Train	*Test*	*Train*	*Test*
5 Minutes	0.97	0.95	0.90	0.90
30 Minutes	0.99	0.99	0.98	0.98

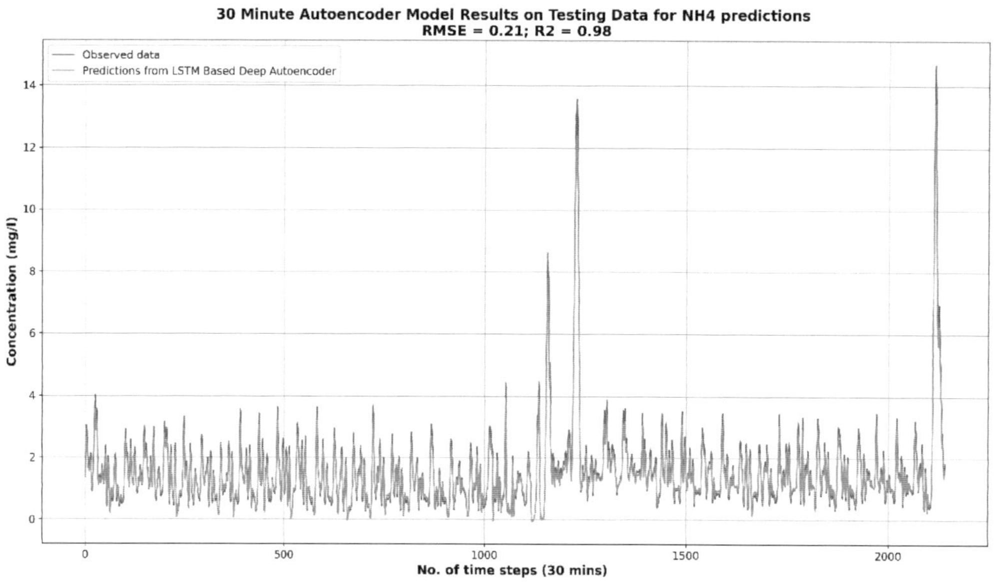

Figure 3: 30-minute autoencoder model predictions for the NH_4^+ signal (orange) and measured data (blue).

4.3. Visualisation

The DVR application has been running successfully near real-time. Yet, further tests are needed with synthetically generated anomalies and by running the DVR for an extended period of time to further test the reliability of the method. Early results indicate that reconciliation of long duration (more than 2 hours) anomalous events fail because of error propagation. Possible improvements are (i) extending the DVR with another autoencoder specifically trained on slow dynamics by using a larger than 30-minute time resolution, (ii) and/or including other sensor data signals. Future directions in development are to extend the DVR with a technique for classification of anomalies, and to deploy the DVR as a screening layer for soft sensors of key wastewater process variables [5] and the control agent for minimising the climate footprint while meeting regulatory effluent standards [6].

Measurement data of NO_3^- and NH_4^+, as well as predictions and reconciled values, are shown in **Figure 4** where the DVR is running in production in real-time at Waternet.

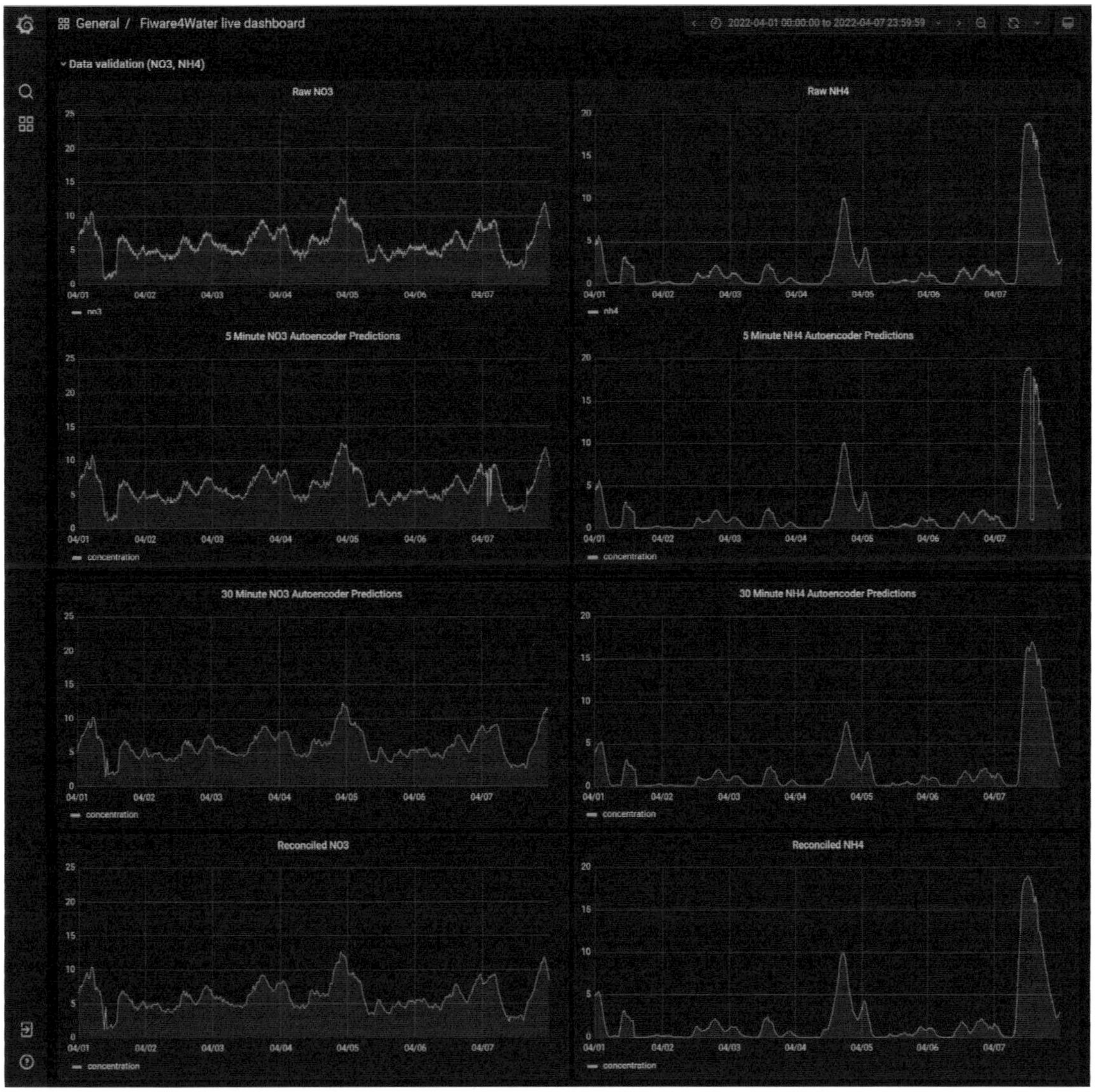

Figure 4: Grafana dashboard of the DVR of $NO3^-$ (left) and $NH4^+$ (right). Upper panels: raw signal, second row: 5-minute autoencoder predictions, third row: 30-minute autoencoder predictions, bottom row: reconciled data.

5. Concluding remarks

In this work, crucial sensor data for monitoring the performance of a WWTP, i.e., ammonium and nitrate, are checked for errors and anomalous data values are reconciled by model predictions using an advanced data validation and reconciliation (DVR) method which makes use of recurrent neural network (RNN) models and exponential smoothing. The DVR application is deployed in a cloud environment where sensor data is retrieved from the (legacy) WWTP's process information management system and relayed towards a setup using the open data exchange platform FIWARE. The FIWARE setup is extended to allow wastewater treatment specific data and data is successfully communicated, stored and processed near real-time to a DVR application.

The following can be concluded:

- The RNN autoencoder models are able to detect extreme or unexpected sensor values, as well as flatlines with differing duration;

- Reconciliation is successful for short to medium (i.e. approximately less than 2 hours) horizon lengths, but performance is dependent on the amplitude of the signal;
- The DVR provides a robust and accurate screening and correction layer for further use of sensor data in the monitoring and control applications – especially for anomaly events with a short duration;
- The DVR procedure can be used for other sensor signals using standardised data models.

In order to improve the performance and usability of the DVR, it is proposed to: (i) improve the accuracy of predictions, thereby extending the duration of reconciliation possible, by extending the DVR with another autoencoder specifically trained on slow dynamics, and/or by inclusion of other sensor data signals, (ii) extend the DVR with a technique for classification of anomalies

Acknowledgments

The presented research and development is part of the Fiware4Water project that has received funding from the European Union's Horizon 2020 research and innovation program under grant agreement No. 821036.

References

[1] Ahlgren B, Hidell M and Ngai ECH 2016 Internet of Things for Smart Cities: Interoperability and open data *IEEE Internet Computing* **20**(6) 52–56 doi: 10.1109/MIC.2016.124

[2] Poch M., J. Comas, J. Porro, M. Garrido-Baserba, L. Corominas and M. Pijuan 2014 Where are we in wastewater treatment plants data management? A review and a proposal *Int. Congress on Environmental Modelling and Software* pp 1450–1455 Accessed: May 01, 2022 [Online] Available: https://scholarsarchive.byu.edu/iemssconference

[3] Branisavljević N, D. Prodanović and D. Pavlović 2010 Automatic, semi-automatic and manual validation of urban drainage data *Water Science and Technology* **62**(5) 1013–1021 doi: 10.2166/WST.2010.350

[4] Cirillo F, G. Solmaz, E.L. Berz, M. Bauer, B. Cheng and E. Kovacs 2020 A standard-based open source IoT platform: FIWARE, *IEEE Internet of Things Magazine*, **2**(3) 12–18, doi: 10.1109/IOTM.0001.1800022

[5] van der Helm AWC, M.R. Haas, D. Amoiridis, M.H. Zandvoort, D. Vries and L. Echeverria 2021 Artificial intelligence for wastewater treatment – The development of a digital twin used to train a control optimization agent. Accessed: Feb. 02, 2022. [Online]. Available: https://aiww2021.com/event/smart-water-tools-in-waste-water-treatment/

[6] Echeverria L, N. Nievas, A.W.C. van der Helm, M.H. Zandvoort, M. Ribalta, X. Domingo and C. Fernandez 2021 Artificial intelligence for wastewater treatment – AI-agent-based service for optimal control *Proc. Amsterdam International Water Week*

[7] Privat G, A. Abid, F. le Gall, G. Tropea and J. A. Martinez 2021 Guidelines for modelling with NGSI-LD LD - Guidelines for modelling with NGSI 2 Contributors LD - Guidelines for modelling with NGSI. *Sophia Antipolis, France: ETSI.* Accessed: May 01, 2022. [Online]. Available: www.etsi.org

[8] Abella A and S. Seshan 2022 Data models for wastewater treatment. https://github.com/smart-data-models/dataModel.WasteWater

[9] Bengio Y, Ducharme R and Vincent P 2000 A neural probabilistic language model https://proceedings.neurips.cc/paper/2000/hash/728f206c2a01bf572b5940d7d9a8fa4c-Abstract.html (accessed May 06, 2022)

[10] Provotar OI, Y. M. Linder and M.M. Veres 2019 Unsupervised anomaly detection in time series using LSTM-based autoencoders *IEEE Int. Conf. Advanced Trends in Information Theory, ATIT 2019 - Proceedings*, pp. 513–517. doi: 10.1109/ATIT49449.2019.9030505

[11] Safder U, J. Loy-Benitez, H.T. Nguyen and C.K. Yoo 2022 A hybrid extreme learning machine and deep belief network framework for sludge bulking monitoring in a dynamic wastewater

treatment process *Journal of Water Process Engineering* **46** 102580 doi: 10.1016/J.JWPE.2022.102580

[12] Babiker Y, M. Osman and W. Li 2020 Soft sensor modeling of key effluent parameters in wastewater treatment process based on SAE-NN doi: 10.1155/2020/6347625

[13] Pisa I, A. Morell, J.L. Vicario and R. Vilanova 2020 Denoising autoencoders and LSTM-based artificial neural networks data processing for its application to internal model control in industrial environments – The wastewater treatment plant control case *Sensors* **20**(13) 3743 doi: 10.3390/S20133743

[14] Ba-Alawi AH, P. Vilela, J. Loy-Benitez, S.K. Heo and C.K. Yoo 2021 Intelligent sensor validation for sustainable influent quality monitoring in wastewater treatment plants using stacked denoising autoencoders *Journal of Water Process Engineering* **43** 102206 doi: 10.1016/J.JWPE.2021.102206

[15] Ba-Alawi AH, J. Loy-Benitez, S. Kim and C. Yoo 2021 Missing data imputation and sensor self-validation towards a sustainable operation of wastewater treatment plants via deep variational residual autoencoders *Chemosphere* **288** 132647 1–15 doi: 10.1016/j.chemosphere.2021.132647

[16] Seshan S, D. Vries, A. van der Helm, D. Amoidiris, N. Nievas and L. Echeverria 2022 *D3.3 FIWARE-Enabled Applications for Wastewater Treatment* (Fiware4Water consortium) https://www.fiware4water.eu/sites/default/files/delivrables/F4W-D3.3-WEBSITE-FIWARE-AppWastewaterTreatment_final.pdf

[17] Hinton GE, N. Srivastava, A. Krizhevsky, I. Sutskever and R.R. Salakhutdinov 2012 *Improving Neural Networks by Preventing Co-Adaptation of Feature Detectors* doi: 10.48550/arxiv.1207.0580 Available: https://arxiv.org/pdf/1207.0580.pdf

Enhanced affinity law for centrifugal pumps at variable speed

Romeo Florin Susan-Resiga[1,2], Alin Ilie Bosioc[1] and Adrian Ciprian Stuparu[1]

[1] Research Center for Complex Fluid Systems Engineering; Politehnica University Timişoara, 1 Mihai Viteazu, Timişoara, RO-300222, Romania
[2] Center for Fundamental and Advanced Technical Research, Hydrodynamics and Cavitation Laboratory; Romanian Academy – Timişoara Branch, 22 Mihai Viteazu, Timişoara, RO-300223, Romania

romeo.resiga@upt.ro

Abstract. The paper presents a novel approach for analytical representation of specific energy *E* and torque *T* curves versus discharge *Q* and speed *N*, for centrifugal pumps operated at variable speed both in pump and turbine mode. An enhanced polynomial expression valid for both *E*(*Q*,*N*) and *T*(*Q*,*N*) is introduced and validated against experimental data.

1. Introduction

According to Gibbins [1], "dimensional analysis combines great utility with a demanding intellectual rigour", with applications practically in any field of science and engineering. Essentially, an insightful analysis of the physical phenomenon leads to the dimensionless quantities (numbers) that are not only independent of the particular system of units employed, but also bring the idea of similitude: two phenomena are similar if the relevant dimensionless quantities have the same numerical values. In practice, not all dimensionless quantities can retain the same value because of experimental limitations, thereby usually having an incomplete similarity. The dimensional analysis and similarity have a solid mathematical foundation [2] and are now part of the basic engineering curricula. In particular, for hydraulic turbomachines (pumps, turbines, pump-turbines) the dimensionless groups are now standardized [3, p.477]. For centrifugal pumps, the basic dimensionless quantities are derived and detailed in practically every monograph [4].

In a nutshell, the operation of a centrifugal pump (either in pump or turbine mode) can be described by four quantities: the speed *N*, the volumetric discharge *Q*, the specific hydraulic energy per unit mass *E*, and the shaft torque *T*. The shaft mechanical power is computed as the product of the shaft torque and angular speed, while the hydraulic power is the product of mass flow rate ρQ with the specific energy *E*.

When operating a turbomachine at variable speed, for similar operating points one has the well known (and widely used) affinity laws:

$$\frac{Q_x}{Q_0} = \frac{N_x}{N_0},\ \frac{E_x}{E_0} = \left(\frac{N_x}{N_0}\right)^2,\ \frac{T_x}{T_0} = \left(\frac{N_x}{N_0}\right)^2, \tag{1}$$

where $(\,)_0$ denotes a reference operating speed and $(\,)_x$ an arbitrary speed. The affinity law for the shaft torque follows from the corresponding power law $P_x/P_0 = (N_x/N_0)^3$, and it is employed in this paper because the experimental data examined, available in [5], include the torque measurements. Note from (1) that both specific hydraulic energy and torque share the same affinity law with respect to the variable speed.

The main limitation of the simple affinity laws (1) is that the machine efficiency is not influenced by the operating speed. This is immediately seen from the efficiency definition (e.g. for a pump) $\eta \equiv (\rho Q E)/(\omega T)$, where $\omega = \pi N/30$ is the angular speed with runner speed N in rotations per minute, since the efficiencies ratio is

$$\frac{\eta_x}{\eta_0} = \frac{Q_x}{Q_0}\frac{E_x}{E_0}\frac{N_0}{N_x}\frac{T_0}{T_x} = \frac{N_x}{N_0}\left(\frac{N_x}{N_0}\right)^2 \frac{N_0}{N_x}\left(\frac{N_0}{N_x}\right)^2 = 1 . \tag{2}$$

Experimental data [5], however, show a strong influence of the shaft speed on the efficiency, for centrifugal pumps running either in pump mode or in turbine mode. Some indications of the departure from the above affinity laws are given in [4, §3.10.2], where it is stated for example that the mechanical losses vary with speed, depending on the type of bearings and shaft seals, as $P_m \sim N^{x}$, $x = 1.3 \cdots 1.8$ and not $P_m \sim N^3$ as for scaling the internal power of the pump. Various approaches for improving the pump efficiency estimate at variable speed are presented in [6], [7] and [8], by an ad-hoc inserting the speed ratio or the Reynolds number ratio into a corrected efficiency formula. Similar modifications of the affinity laws are proposed for Pump-as-Turbines in [9], [10] and [11], again by correcting the efficiency ratio (2) with the speed ratio. From our point of view this is not a robust approach, since the efficiency is not a directly measured quantity. As a result, our approach is to improve the estimation of both specific energy $E(Q,N)$ and shaft torque $T(Q,N)$, and the accurate efficiency approximation will follow. Moreover, our approach is equally valid for centrifugal pumps running in pump mode, but also in turbine mode. In fact, we show that the same computer code for processing the experimental data works equally well in both pump and turbine modes.

2. Basic approximation for energy and torque functions of discharge and speed

Thanks to the affinity laws (1), the mathematical expressions for both specific energy $E(Q,N)$ and shaft torque $T(Q,N)$ functions, are identical. As a result, for the remaining of this section we consider only the function $E(Q,N)$ in our development.

Assume that at the pump speed N_0 one measures the specific energy E_0 and torque T_0 values within a range of discharge Q_0 values. Usually, the experimental data are approximated with polynomials $E_0(Q_0)$ or $T_0(Q_0)$ of degree 3 or higher [12]. We will further consider 4-degree polynomials, since higher degree does not improve the data approximation and it may introduce spurious oscillations in the characteristic curves,

$$E_0 = a_1 Q_0^4 + a_2 Q_0^3 + a_3 Q_0^2 + a_4 Q_0 + a_5 . \tag{3}$$

When operating at a different speed N_x, using the affinity laws (1) we can re-write (3) as

$$E_x\left(\frac{N_x}{N_0}\right)^{-2} = a_1\left(\frac{N_x}{N_0}\right)^{-4} Q_x^4 + a_2\left(\frac{N_x}{N_0}\right)^{-3} Q_x^3 + a_3\left(\frac{N_x}{N_0}\right)^{-2} Q_x^2 + a_4\left(\frac{N_x}{N_0}\right)^{-1} Q_x + a_5 , \tag{4}$$

or by explicitly writing the function $E_x(Q_x)$,

$$E_x(Q_x, N_x) = a_1\left(\frac{N_x}{N_0}\right)^{-2} Q_x^4 + a_2\left(\frac{N_x}{N_0}\right)^{-1} Q_x^3 + a_3 Q_x^2 + a_4\left(\frac{N_x}{N_0}\right) Q_x + a_5\left(\frac{N_x}{N_0}\right)^2 . \tag{5}$$

One can see in figure 1 the least squares approximation using (5) of experimental data from [5]. Although the specific energy versus discharge curves seem to be reasonably close to the experimental data, the torque curves significantly depart from measured data at lower speeds. The efficiency curves obviously disagree with experimental data at low speed, where the efficiency is largely overestimated.

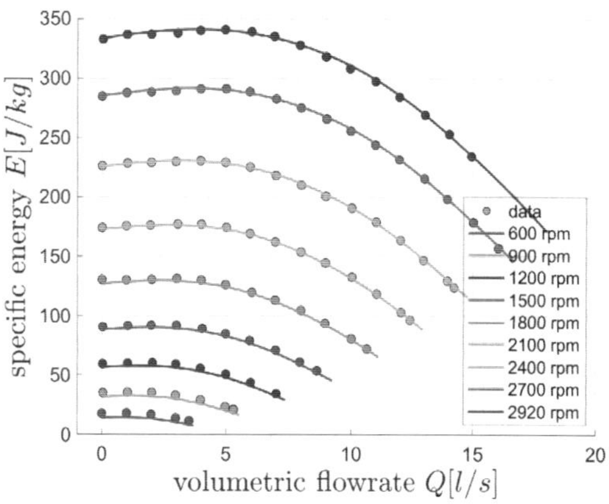

a) Specific energy versus discharge for variable speed, centrifugal pump in pump mode.

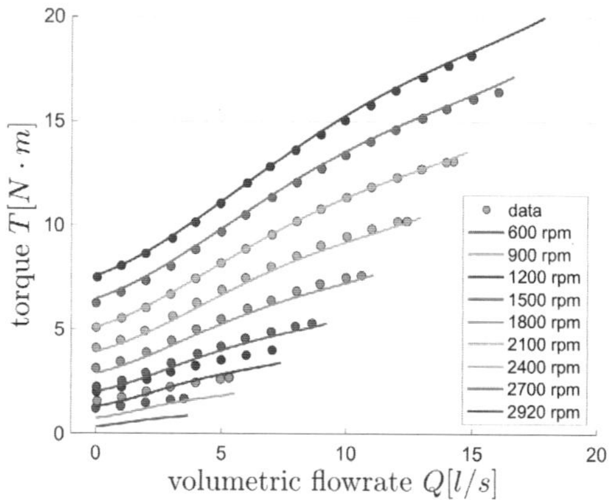

b) Shaft torque versus discharge for variable speed, centrifugal pump in pump mode

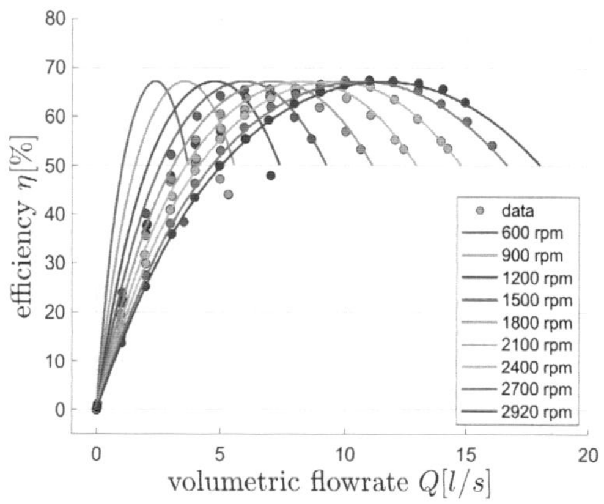

c) Efficiency versus discharge for variable speed, centrifugal pump in pump mode.

Figure 1. Least squares fit of experimental data from [5] for the centrifugal pump NK 40-160/158 in pump mode, using the polynomial (5) for both $E(Q,N)$ and $T(Q,N)$. The maximum efficiency remains the same for all speed values, according to (2). Note the errors at lower speeds, particularly in torque and efficiency curves.

We conclude that (5) is not accurate over a wide range of shaft speeds, and it needs to be enhanced for a better agreement with experiments. In doing so, we use a completely different approach from the various methods in the cited literature by improving the analytical representation of the curves $E(Q,N)$ and $T(Q,N)$, and the efficiency curves $\eta(Q,N)$ will consequently be altered.

The main observation is that for low speeds the friction torque in bearings and shaft seal become large with respect to the hydraulic torque associated with the impeller, and as a result that (5) should be modified accordingly.

3. Enhanced approximation for energy and torque functions of discharge and speed

The main idea put forward in this paper is to augment the polynomial function (5) with extra terms that account in the shaft torque $T(Q,N)$ function for the friction losses in bearings and shaft seal.

$$E_x(Q_x, N_x) = a_1\left(\frac{N_x}{N_0}\right)^{-2} Q_x^4 + a_2\left(\frac{N_x}{N_0}\right)^{-1} Q_x^3 + a_3 Q_x^2 + a_4\left(\frac{N_x}{N_0}\right) Q_x + a_5\left(\frac{N_x}{N_0}\right)^2 + a_6\left(\frac{N_x}{N_0}\right) + a_7 + a_8\left(\frac{N_x}{N_0}\right)^{-1} . \tag{6}$$

The last four terms in (6) do not depend on discharge. In fact, we continue the series in (N_x/N_0), which now includes terms of order 2, 1, 0 and -1 with coefficients $a_5 \ldots a_8$. We prefer this approach since the resulting expression is linear in the unknown coefficients, and as a result the system of equations for the fit coefficients a_i, $i = 1 \ldots 8$ is linear.

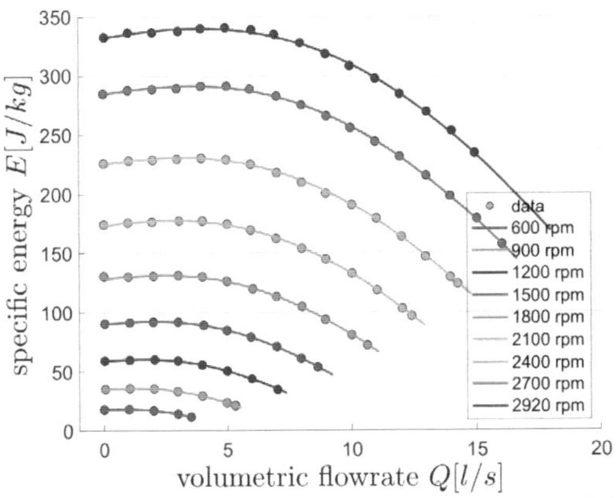

a) Specific energy versus discharge for variable speed, centrifugal pump in pump mode.

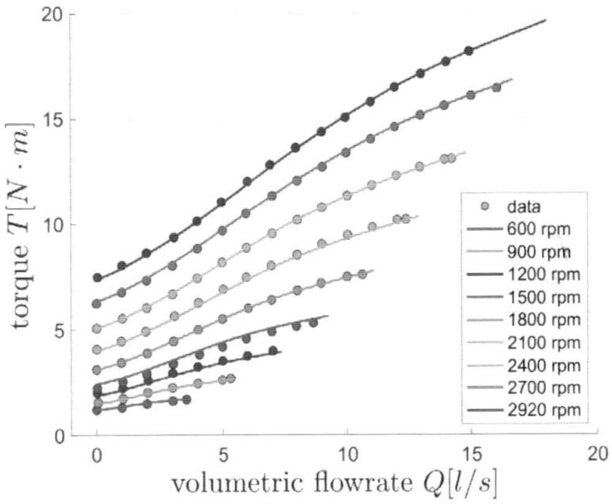

b) Shaft torque versus discharge for variable speed, centrifugal pump in pump mode

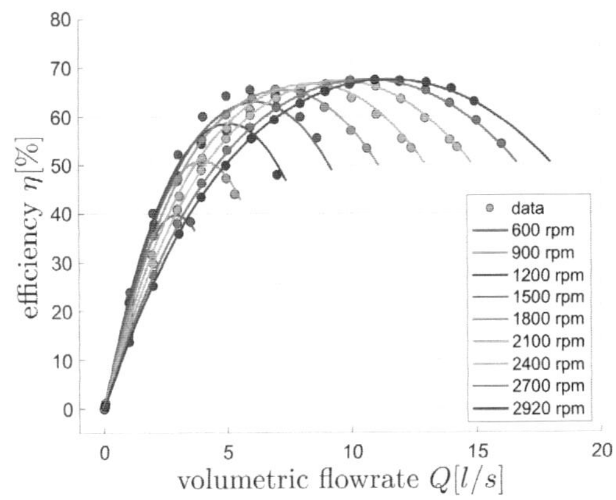

c) Efficiency versus discharge for variable speed, centrifugal pump in pump mode.

Figure 2. Least squares fit of experimental data from [5] for the centrifugal pump NK 40-160/158 in pump mode, using the enhanced polynomial (6) for both $E(Q,N)$ and $T(Q,N)$.

One can see in figure 2 that the approximation of the experimental data at low speeds using the enhanced polynomial (6) is significantly improved, particularly for the torque, figure 2b. As a result, the computed efficiency curves from figure 2c correctly capture the efficiency decrease at low speeds.

The same enhanced polynomial (6) can be used for specific energy and torque curves when the centrifugal pump runs in turbine mode.

Figure 3 shows the data and least squares approximation with (6) for the specific energy and torque when the centrifugal pump is operated in turbine mode. One can notice the excellent approximation of the data, and in particular the very good agreement of the efficiency curves within all range of speeds.

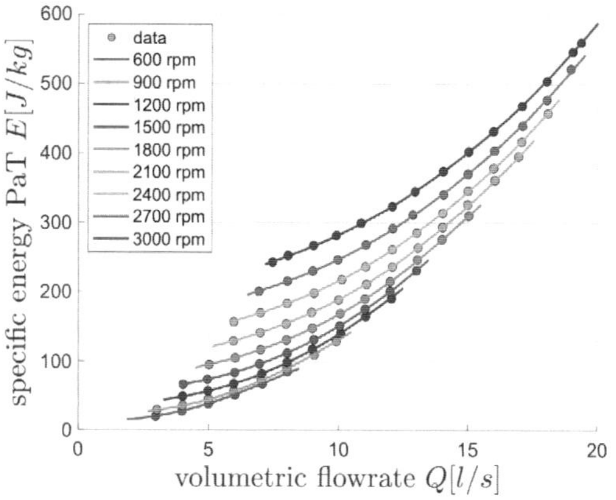

a) Specific energy versus discharge for variable speed, centrifugal pump in turbine mode.

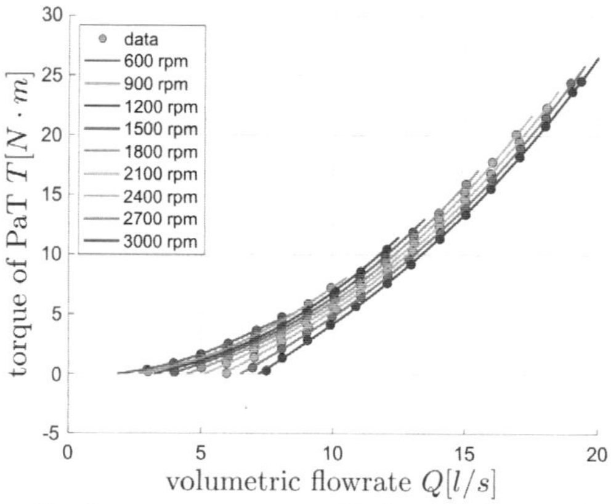

b) Shaft torque versus discharge for variable speed, centrifugal pump in turbine mode

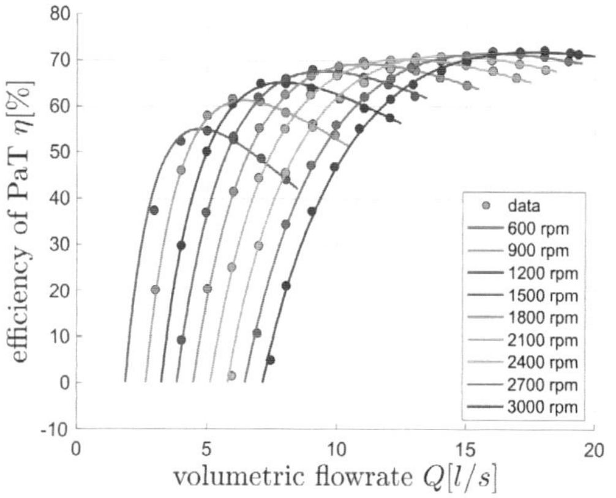

c) Efficiency versus discharge for variable speed, centrifugal pump in turbine mode.

Figure 3. Least squares fit of experimental data from [5] for the centrifugal pump NK 40-160/158 in turbine mode, using the enhanced polynomial (6) for both $E(Q,N)$ and $T(Q,N)$.

4. MATLAB implementation of the least-squares approximation

The fit coefficients in (6) are found by solving an over-determined system of linear equations in the least-squares sense. In doing so, we employ the MATLAB backslash operator, as shown in the MATLAB program excerpt below.

```
%%%% FIT DIMENSIONAL DATA WITH ENHANCED AFFINITY LAW %%%%%%%
% build system matrix (Ndata rows and first 5 columns)
A = zeros(Ndata,8);
for j=1:5
   A(:,j) = ((SPEED/SPEED0).^(j-3)).*(DISCH.^(5-j));
end
% additional 3 columns depending only on speed
A(:,6) = (SPEED/SPEED0);
A(:,7) = ones(Ndata,1);
A(:,8) = (SPEED0./SPEED);
% use backslash for least-squares solving of overdetermined system
pE = A\ENERG; % coefficients a1…a8 for E(Q,N) least-squares fit
pT = A\TRQUE; % coefficients a1…a8 for T(Q,N) least-squares fit
```

The **Ndata** measured points read in the MATLAB program are stored in column vectors for N (variable **SPEED**), Q (variable **DISCH**), E (variable **ENERG**) and T (variable **TRQUE**). The coefficients for the specific energy polynomial are stored in the column vector **pE**, and for the shaft torque in **pT**. Note that the "tall" matrix **A** has 8 columns, but the number of rows, **Ndata**, is much larger and corresponds to the number of operating points measured at variable discharge and speed (106 points in pump mode and 96 points in turbine mode). Details on solving overdetermined linear systems in the least-squares sense are given in [13, §5.7], and what is hidden behind the MATLAB command \ can be found in [13, §5.8]. The reference speed N_0 (variable **SPEED0**) is arbitrary, and it has been set to the value of the maximum speed in the experiments.

5. Conclusions

The paper introduces a novel polynomial expression for specific energy and torque as functions of discharge and speed, to represent experimental data for centrifugal pumps operated at variable speed in both pump and turbine modes. Validation against experimental data shows that this approach correctly captures the efficiency decrease at low speeds, a significant improvement compared with the basic affinity laws commonly used in engineering practice.

Acknowledgments
The first author acknowledges the support from the research program "Hydrodynamics of single- and two-phase flows and applications" within the Romanian Academy – Timişoara Branch.

References

[1] Gibbins J C 2011 *Dimensional Analysis* (London Dordrecht Heidelberg New York: Springer)
[2] Longo SG 2021 *Principles and Applications of Dimensional Analysis and Similarity* (Springer)
[3] International Electrotechnical Commission 1999 *International Standard IEC60193: Hydraulic turbines, storage pumps and pump-turbines – Model acceptance tests* (www.iec.ch)
[4] Gülich JF 2008 *Centrifugal Pumps* (Berlin Heidelberg New York: Springer)
[5] Delgado J, Ferreira JP, Covas DIC and Avellan F 2019 Variable speed operation of centrifugal pumps running as turbines. Experimental investigation *Renewable Energy* **142** pp 437-450
[6] Simpson AR and Marchi A 2013 Evaluating the Approximation of the Affinity Laws and Improving the Efficiency Estimate for Variable Speed Pumps *Journal of Hydraulic Engineering* **139**(12) pp 1314-131 http://doi.org/10.1061/(ASCE)HY.1943-7900.0000776
[7] Coehlo B and Andrade-Campos AG 2016 A new approach for the prediction of speed-adjusted pump efficiency curves *Journal of Hydraulic Research* **54**(5) pp 586-593 http://doi.org/10.1080/00221686.2016.1175521
[8] Agarwal R, Patil A and Morrison G 2020 Efficiency prediction of centrifugal pumps using the modified affinity laws *Journal of Energy Resources Technology* **142**(3) p 032102 http://doi.org/10.1115/1.4044940
[9] Carravetta A, Conte MC, Fecarotta O and Ramos HM 2014 Evaluation of PAT performances by modified affinity law *Procedia Engineering* **89** pp 581-587
[10] Fecarotta O, Carravetta A, Ramos HM and Martino R 2016 An improved model to enhance variable operating strategy for pumps used as turbines Journal of Hydraulic Research **54**(3) pp 332-341 http://doi.org/10.1080/00221686.2016.1141804
[11] Carravetta A, Houreh SD and Ramos HM 2018 *Pumps as Turbines. Fundamentals and Applications* (Springer International Publishing) http://doi.org/10.1007/978-3-319-67507-7
[12] Pöyhönen S, Ahonen T, Ahola J, Punnonen P and Hammo S 2019 Specific speed – based pump flow rate estimator for large scale and long-term energy efficiency auditing *Energy Efficiency* **12** pp 1279-1291 http://doi.org/10.1007/s12053-018-9751-4
[13] Quarteroni A, Saleri F and Gervasio P 2014 *Scientific Computing with MATLAB and Octave* Fourth Edition (Berlin Heidelberg: Springer) https://doi.org/10.1007/978-3-642-45367-0

IOP Conf. Series: Earth and Environmental Science 1136 (2023) 012057 doi:10.1088/1755-1315/1136/1/012057

Effect of start and stop cycles on hydropower plants: modelling the deterioration of the equipment to evaluate the cycling cost

Olivian Savin[1,4,5], Carmen Badina[1], Renaud Pollier[2], Jean-Louis Drommi[3], Julien Baroth[4], Sylvie Charbonnier[5] and Christophe Berenguer[5]

[1] Electricité de France (EDF), General Technical Divison, Saint-Martin Le Vinoux, F-38950, France
[2] EDF, Hydro Industry Division, Saint-Denis, F-93282, France
[3] EDF, Hydro Engineering Center, Le Bourget du Lac, F-73373, France
[4] Université Grenoble Alpes (UGA), CNRS, Grenoble INP, 3SR Laboratory, Saint Martin d'Hères, F-38400, France
[5] UGA, CNRS, Grenoble INP, GIPSA-Lab, Saint Martin d'Hères, F-38400, France

oliviansavin@gmail.com

Abstract. With the high energy demand, renewable energies emerged and the deregulation of the energy market appeared in the late '90. These types of energies are intermittent and can produce high fluctuations in the power grid because they are connected only in certain weather conditions (for example: solar energy can be produced only by day and wind energy only when it is windy). Hydropower energy is recognized for its flexibility in responding to the demands of the network in order to maintain a constant level of energy and to guarantee the security of the electrical network. Nowadays, the hydropower plants are the main provider of energy adjustments thanks to a high number of start and stop cycles (which can sometimes be around 6 per day), the fast variations in load and the use of PSH (pumped-storage hydropower). But these rapid changes in the production conditions of hydroelectric power plants lead to high stresses on the components and consequently to faster degradation (mechanical, electrical, thermal, hydraulic, etc.). This work focuses on the design of a methodology that relies on auditable methods to estimate the cost of start and stop cycles and presents three complementary, representative (by equipment), reproductible (can be applied by different operators) methods based on explicit assumptions and validated by experts and in function of the available data in order to estimate the degradation of the main components of the hydropower plant (shut-off valve, runner, generator, transformer and circuit-breaker). These methods are based on the main degradation phenomena. The first method uses on-site measurements to study the fatigue on runners, the second method uses a reliability law to estimate the lifetime of generators and a predictive model using online temperature measurements and the last method uses expert judgement due to the lack of measurements on the equipment.

1. Introduction

The shift towards green and sustainable energy systems, which has accelerated over the past 30 years, is reflected in the policy decisions of governments worldwide [1]. Nowadays, with the emergence of new energies such as solar or wind, which can produce energy according to climatic conditions (wind and sun), energy fluctuations in the electrical network are fast and high. Solutions were implemented to respond to the intermittent nature of these renewable energies in order to guarantee the security of the electrical network: the energy gaps are filled by hydropower energy. An important role is played by the storage of energy in water dams and the use of PSH (pumped-storage hydropower) to generate electricity when necessary. We can observe the production of energy on the load curve that is the graphical representation of the load (in MW) in chronological order (in hours) by type of energy. The curve is useful for choosing the economic sizes of different generation units and allows for estimating the cost of production [2]. It determines the operating modes of the power plants (system services) and the priority or order in which the different generation units should operate [3]. In figure 1. is represented the load curve for a day in France. At the top of the figure, there is the amount in MW that has been produced, on the right of the image, the percentage of production by type of energy and in the graph, we can see at what time of the day a certain type of energy is produced. It can be seen that in case of need (morning or evening), hydroelectric energy is used to balance the network.

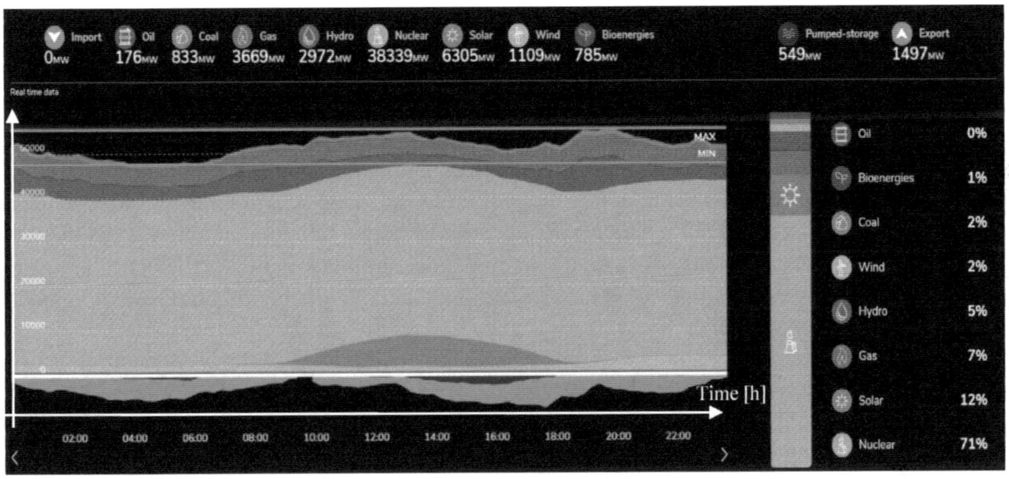

Figure 1. Electricity production by type of energy over one day (23/10/2021) – the load curve (Source: RTE France website).

The high demand of energy and these new modes of functioning (high and partial load operations, rough zone operation, spinning reserve, synchronous condenser) comes with a cost. This increase in the number of cycles has proven effects in terms of degradation and failures of the materials involved, on operational practices and maintenance, and consequently on the overall costs of the installation [4]. Therefore, a more detailed knowledge needs to be acquired for the complete costs generated by these new operating modes with more and more start and stop cycles in order to be able to offer its services at the right cost and to support the economic decisions of equipment operation and maintenance.

This work aims at developing an original approach to evaluate the cost associated with hydroelectric start and stop cycles focusing on the material cost. The details of this work are presented in the articles [5,6,7] and in the thesis written in French [8] and in this paper we summarize and discuss the methodology used. Three methods are proposed to estimate the deterioration of the hydro power equipment taking into consideration the main phenomena of degradation and the available data.

2. The methodologies used in the literature to estimate the start and stop costs

The issue of increasing costs related to start and stop cycles appeared in Europe after the deregulation of the energy market in the late '90 [9]. The need for a short/long term operating strategy for hydropower plants (in terms of operation but especially in terms of cost), has led to the emergence of *top-down* and *bottom-up* models in the hydropower industry to estimate the cost of start and stop cycles [10]. A review of existing work on assessing and modelling costs are described in [5] and shows two types of approaches that have been considered: *economic* and *technical-economic*. In these two categories, six methods are identified (figure 2).

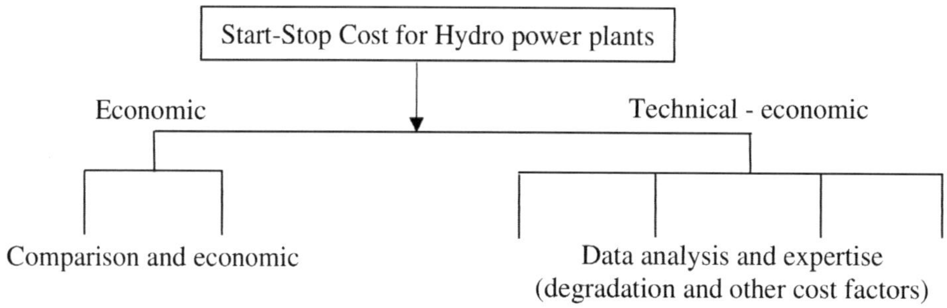

Figure 2. Start and stop cost methodologies [5].

The price range goes from 200$ up to 900$ depending on the equipment, type of turbine and year of estimation. On the one hand, the top-down approach is usually an economic approach, where the overall cost of operating and maintaining of two units are compared: one with high and one with low number of start and stop cycles and assumes that the difference in cost is due to increased start and stop cycles. On the other hand, the bottom-up approach is usually a technical-economic approach and attempts to identify the potential cost factors associated with an individual production unit and to model their relationship with respect to changes in operating conditions. The reliability of hydroelectric plants and the effects of degradation have been discussed and assumptions vary considerably in these models. The cost can be *equipment related* (increasing cost as more frequent and costly maintenance) or *non-equipment related* (reducing revenues as loss opportunity or reduce availability) As observed in [5], these methods do not rely on explicit modelling of deterioration and lifetime consumption due to start and stop cycles. Furthermore, there is no consensus on the influencing factors to be considered or on the methodology for calculating costs.

3. The strategy proposed for estimating the cost of a start and stop cycle

The main research objectives are to develop a methodology that relies on auditable methods for estimating the degradation of hydro power plant equipment and to propose a strategy for estimating the start and stop costs [5,8]. Thus, the methodology is focusing on equipment related costs. These methods are specific, representative (by material) and reproducible (can be applied by different operators) and are based on explicit assumptions validated by EDF (Electricité de France) experts. Priority has been given to the technical-economic (bottom-up) strategy where the main affected equipment and degradation phenomena have been identified. Then, physical and reliability models were proposed with the data available at EDF in order to quantify the impact of transient phases (starts and stops) and to estimate the associated costs. This approach mobilizes skills of analysis, processing and exploitation of data and skills on the engineering of the considered equipment:

- it requires access to a large amount of data on defaults, equipment condition, operating and maintenance costs where the quality of the quantitative results depends on the availability and quality of the existing data;
- this method also requires technical expertise on the materials and on the degradation phenomena involved.

3.1. Data required and used for the study

We systematically search for the impacted equipment and the degradation phenomena involved, as well as their consequences, in terms of malfunction, failures and loss of efficiency. Thus, in-depth research on two axes was carried out: the first one concerns the knowledge of the degradation and the associated phenomena of the equipment of the hydropower plants and the second one synthesizes the information obtained by the interview of the experts, the return of experience and the analysis of the EDF data bases [8]. The interest is to link the approaches based on physics, data and feedback for a cross analysis of the available information. The process is summarized in figure 3.

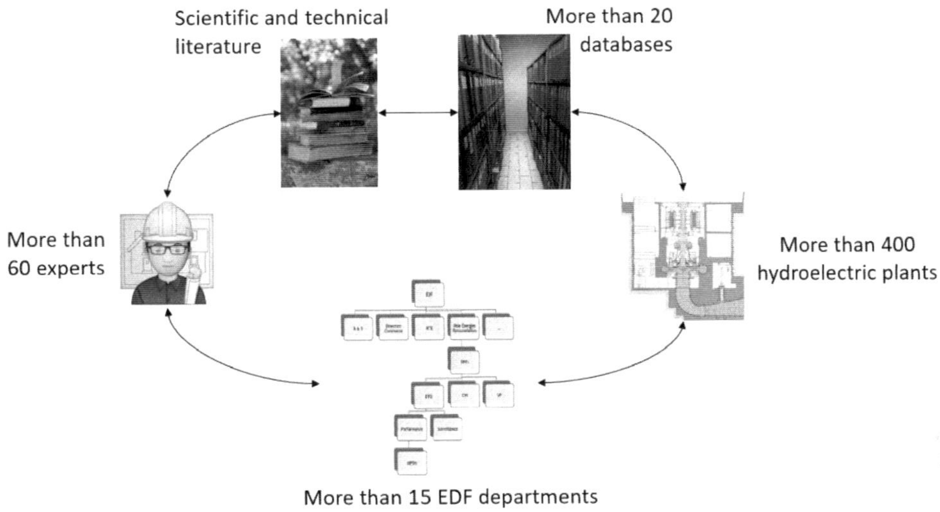

Figure 3. Diagram of the data sources used [8].

3.2. Identification of the impacted equipment and the main phenomena of degradation

In the literature, the generator is one of the most studied equipment. Diagnostic measurements for the remaining lifetime or aging models based on available data associated with the production units on the stator showed that the effect of thermal cycling and mechanical stress (centrifugal force) are the main causes of degradation [11,12]. The turbine was also studied and measurements in the technical literature showed the impacted of fatigue on Pelton and Kaplan turbines [13], [14]. For the circuit breaker, the main causes of failure [15] and methods for estimating remaining life are reviewed [16], but the effect of degradation due to start and stop cycling is not listed. In a methodology where expert judgement is used to validate technical [10], the cost attributable to start and stop cycles is 50% of total replacement cost. In an internal EDF report [17], a correlation between the number of cycles, defaults, and the size of the crack that appeared on the trunnions reveals the impact of start and stop cycles on spherical valves. In the methodologies used to estimate the start and stop costs [5], the authors from the papers reviewed considerer that transformers are not significantly affected by cycling.

Furthermore, the specialized EDF employees (experts, equipment referents, engineers, technicians, maintenance managers and operators) in the field of engineering and reliability were interviewed in several areas (mechanical - turbine, guard valve, electrical - generator, transformer, circuit breaker, etc.) in order to identify the degradation mechanisms of the equipment of a hydroelectric plant. An internal EDF report [18] presents the feedback and a list, as exhaustively as possible, the disorders and pathologies affecting the equipment, the damages and defaults encountered that are possibly related to start and stop cycles. Following the research in the literature and these interviews, the degradation phenomena of materials that are impacted by cycling have been identified.

The system, which in our case is the hydroelectric power plant, was decomposed into sub-components (equipment) and the sequences of start and stop were analyzed in details to identify the main degradation phenomena. In figure 4 is represented a start-up phase in respect with time and the equipment that is impacted [8]. On the vertical axis, one can observe the power in [MW], the pressure of the penstock in [bar] and on the horizontal the speed (which is the rotational speed of frequency) in [Hz] and the opening of the shut-off valve in [mm]. The degradation phenomena are briefly described according to their operating mechanisms. Repeated opening of the shut-off valve to let water through the turbine causes fatigue and subsequently accelerated degradation. The runner is affected by fatigue caused by the mechanical forces of water impact. The torque is transmitted to the generator were start and stop cycles produces heating and cooling due to the electrical energy flowing through the stator and rotor, which then produces thermal cycling that can be damaging. The electrical energy is transmitted to the transformer which is affected by the same thermal cycling degradation forces as the generator, that heats up and cools down during start and stop cycles. Finally, in order to transmit electricity to the grid, the circuit breaker is stressed by mechanical operations which cause accelerated wear to coupling and decoupling the unit from the grid.

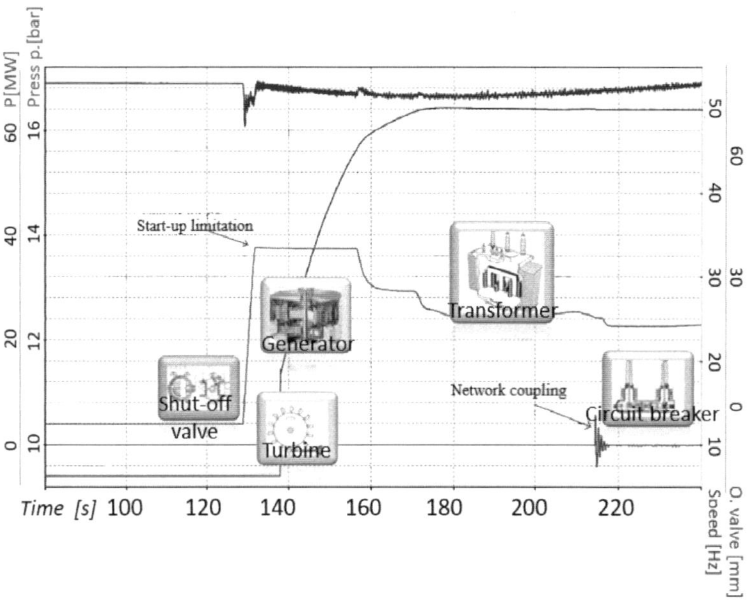

Figure 4. Recordings of a start-up phase of a Francis turbine [8].

The state-of-the-art shows that there are five main materials impacted by start and stop cycles. In table 1 we summarize the main materials and their degradation mechanisms:

Table 1. Equipment impacted by start and stop cycles [8].

Equipment	Sub-component	Mechanism (phenomena) of degradation
Shut-off valve	Spherical and butterfly valves	Mechanical fatigue, wear of sealing parts
Turbine	Runner	Mechanical fatigue
Generator	Stator insulation	Thermal cycling
Transformer	Insulation	Thermal cycling for the transformers with fast power-up
Circuit breaker	Mechanical system	Mechanical fatigue and electrical arc

Once the most impacted equipment has been identified (shut-off valve, runner, generator, transformer and circuit breaker) and their predominant degradation or aging phenomena have been analyzed, physical or stochastic models have been created from the available data and feedbacks from EDF. Thus, the degradation or the decrease in lifetime is estimated from the main degradation phenomenon or the primary end-of-life failure mode. Finally, the lifetime operating cost can be estimated from the purchase and maintenance prices and then divided by the number of cycles.

4. The methods used to estimate the degradation of the equipment

The methods used to estimate the degradation of the equipment are detailed in the articles [6,7] and in the thesis written in French [8]. Three complementary methods are used in function of the available data: comparative damage on runners for fatigue study described in [6], empirical study and lifetime model for thermal cycling on generators described in [7] and estimation of degradation using expert feedback. In the following sub-sections, we describe how the cost was estimated from the estimation of the degradation.

4.1. Comparative damage for runners under high cycle fatigue

The method is presented in details in [6] and uses strain measurements on the runner as a part of a fatigue study. It estimates the degradation of the transient phases (start and stop) and compare them with the normal operation phase to determine a number of normal operating hours for these transient phases. Rain flow analysis [19] is used to extract the numbers and magnitudes of stress cycles from these three sequences.

The study was conducted on regulatory and probabilistic Wöhler curves or S-N curves (cyclic stress "S" against the cycles to failure "N"). In the end, the Palmgren-Miner rule [20] allowed the estimation of the damage rates for each sequence. This method is used when fatigue is the main degradation phenomena, in that case, for Francis and Pelton runners.

Authors obtained results from 6 hours of normal operation (*HNO)* for a start and stop cycle up to several hundred hours depending on the S-N curve used. It should be taken into account that for different materials, the S-N curves can also be different. As an application, an advantage of this method is that it can be used on other materials for which fatigue is the main degradation phenomenon, as for example in the case of a hydroelectric power plant, for guide vanes, servomotors, penstocks, etc.

Once the number of normal operating hours for the start and stop phases (*HNO*) was estimated, it is then multiplied with the number of start and stop cycles of the runner done since commissioning and the results is added to the normal of operating hours (H_t) that the turbine done it in his lifetime. The sum represents the total equivalent hours (*HEOt)* over the lifetime. We can estimate the cost of an operating hour (C_h) by dividing the purchase price and maintenance by the total equivalent hours estimated. Finally, the cost of an operating hour (C_h) is multiplied by the number of normal operating hours estimated and the cost of one runner start and stop (C_R) is obtained.

4.2. Empirical study and lifetime model for thermal cycling on generators

Authors in [7] proposed two methods that can be used together or separately. The main degradation phenomenon of generators was considered the thermal cycling of the stator. In the first method, it was highlighted the effect of cycling on the lifetime of generators by estimating the lifetime for two categories of generators that have experienced different cycling during their lifetime (cycled, with 2 or more cycles per day and non-cycled, with less than 1 cycle per day) using a Weibull reliability law [21] on historical rewind year data. The estimated lifetime for cycled plants (46 years) is lower than the estimated lifetime for less cycled plants (55 years).

The second method consists in estimating an acceleration factor using Coffin-Manson law [22] by integrating the temperature profile characteristics due to cycling into the lifetime law. The advantage of this method is that it allows the use of monitoring measurements over a period of days to years. Another advantage of this method is that it can be used on other materials where the thermal cycling participates in the degradation, as for example in the case of a hydroelectric plant, for *transformers*.

Two scenarios have been tested from the simulated cycles in pump mode for PSH. In both cases, the increase in the number of start and stop cycles leads to a reduction of the lifetime. The method can be tested on different generators and the turbine mode should be added. Finally, the cost of a generator start and stop cycle can be approximated from the difference between cycled and non-cycled plants. The purchase and maintenance price are divided by the total number of cycles performed by a generator over the estimated lifetimes (cycled and non-cycled).

4.3. Expert feedback on the degradation of shut-off valves and circuit breakers

The last method consists in estimating the degradation of shut-off valves and circuit breakers by feedback and expert judgement. For some equipment, such as circuit breakers, experts say that the lifetime and maintenance depend on the number of mechanical operations, which coincide with the number of start and stop cycles. In this case, the number of start and stop cycles can be directly considered as an indicator of degradation and the cost of a circuit breaker start and stop cycle can be estimated based on the number of mechanical operations. For the shut-off valve, the link between the number of start and stop cycles and degradation was not as obvious. For this reason, the authors investigated if there was a correlation between a significant default, the trunnion cracking, and the number of cycles. Indeed, a correlation was found in an internal report [17]. The fact that the valves are mainly used during their opening and closing and that most of the defaults are related to these operations, allowed the experts to attribute 75% of the degradation to the start and stop cycles and the rest to the degradation due to ageing. The start and stop cost for shut-off valves can be approximated by dividing the purchase price and maintenance by the estimated number of cycles over the lifetime and taking into account the degradation assigned.

5. Conclusions

Due to the diversity of the equipment present in a hydroelectric power plant (mechanical and electrical equipment), the realization of the thesis called upon a multidisciplinary approach that combines several fields, each one working in an autonomous frame of reference with specific documentary needs and methods (mechanical and electrical engineering, hydraulics, data processing, reliability, safety and maintenance, etc.) to study the effects of degradation of various natures (hydraulic, mechanical, electrical and thermal). The particularity is represented by the diversity of the methods (probabilistic models, ageing models, degradation models, expert feedback) which are adapted according to each material and which are applied in the field of hydropower (section 4). The multiplicity of degradation phenomena, causes of failure and mobilized competences makes complex the realization of a unique model for all the components of a hydraulic production group.

The start and stop cost (C_{SS}) is the sum of all the costs of the materials studied (equation 1). The details of this study are presented in the confidential EDF report [23].

$$C_{SS} = C_R + C_G + C_T + C_{CB} + C_{SV} \tag{1}$$

where C_R is the cost for the runner, C_G is the cost for the generator, $C_T$is the cost for the transformer, C_{CB} is the cost for the circuit breaker and $C_{SV}$is the cost for the shut-off valve.

It is important to mention that the cost is estimated only from the degradation of the equipment point of view and does not include other costs like unavailability costs or commodity costs.

Finally, the role of hydropower electricity in the network is very important as presented in introduction. While the run-of-rivers plants assures the production of a constant level of energy, more and more hydropower plants, such as PSH and impoundment power plants are solicited by an increased number of start and stop. This paper also emphasizes the role of constant monitoring and informatics in hydropower due to the usage of sensors where the output is used to estimate the degradation of the equipment.

6. Results and discussion

This paper resumes three methods based on real and available data (fatigue tests for runners, on-line monitoring for generators and transformers) in order to estimate the degradation of the hydropower equipment to estimate the start and stop cost. It can also represent a *decision-making tool* to operate the power plants and optimize the operation of the machines according to different scenarios (e.g. optimized start-ups for runners or adjustment of the generators' operating time). It needs to be mentioned that due to the fact that every hydropower plant is unique, the methodology needs to be tested on every plant in function of the site configuration for relevant results. For example, if the methodology is carried out on a 50 MW Francis hydropower plant, making the hypothesis that all five equipment are impacted, the start and stop cost range between 200\$-250\$. This represents only the wear and tear cost, while the other costs are not included (e.g., lost opportunity cost, efficiency loses, etc – costs that depends on the geographical region, grid and on the electrical network in general). For this study, more than 400 equipment were studied for all types of hydropower plants (PSH, impoundment and run-of-rivers) and the average uncertainty for the total cost is ±33% [8].

Acknowledgments
The author wishing to acknowledge the support of the colleagues and experts from EDF and from the two laboratories involved in the project, GIPSA-Lab and 3SR Laboratory, that helped to accomplish this study.

References
[1] Chuang AS and Schwaegerl C 2009 Ancillary services for renewable integration *2009 CIGRE/IEEE PES Joint Symposium Integration of Wide-Scale Renewable Resources Into the Power Delivery System* Calgary, Canada https://ieeexplore.ieee.org/document/5211165
[2] Yumak K, Tosun G, Varlik B and Bağriyanık M 2016 Load curve classification for the evaluation of demand side management programs *IOP Conf. Ser.: Mater. Sci. Eng.* **161** 012111 https://doi.org/10.1088/1757-899X/161/1/012111
[3] Cebulla F, Naegler T and Pohl M 2017 Electrical energy storage in highly renewable European energy systems: Capacity requirements, spatial distribution, and storage dispatch *Journal of Energy Storage* **14** 211–223 https://doi.org/10.1016/j.est.2017.10.004
[4] Nesbitt M 2017 *Flexible Operation of Hydropower Plants* (No. 3002011185)
[5] Savin O, Badina C, Baroth J, Charbonnier S and Bérenguer C 2020 Start and stop costs for hydro power plants: A critical literature review *Proc. 30th European Safety and Reliability Conf. and 15th Probabilistic Safety Assessment and Management Conf.* Presented at *29th European Safety and Reliability Conf. (ESREL 2020)* (Research Publishing Services) Venice, Italy pp 342–349 https://doi.org/10.3850/978-981-14-8593-0_4102-cd
[6] Savin O, Baroth J, Badina C, Charbonnier S and Bérenguer C 2021 Damage due to start-stop cycles of turbine runners under high-cycle fatigue *International Journal of Fatigue* **153** 106458 https://doi.org/10.1016/j.ijfatigue.2021.106458
[7] Savin O, Badina C, Drommi J-L, Baroth J, Charbonnier S and Berenguer C 2021 Influence of starts and stops on the aging of hydroelectric generator stators by thermal cycling: Empirical study and accelerated lifetime model *Proc. 31th European Safety and Reliability Conf.* Presented at *31th European Safety and Reliability Conf. (ESREL 2021)* (Research Publishing) Angers, France p. 8 https://doi.org/10.3850/978-981-18-2016-8_315-cd
[8] Savin O 2021 Effet des cycles de démarrages et d'arrêts sur les centrales hydrauliques: modélisation de la détérioration des matériels pour l'évaluation des coûts de cyclage (in French) PhD Thesis, Université Grenoble Alpes https://theses.hal.science/tel-03606368
[9] Nilsson O and Sjelvgren D 1997 Hydro unit start-up costs and their impact on the short term scheduling strategies of Swedish Power Producers *IEEE Transactions on Power Systems* **12**(1) 38–44 doi: 10.1109/59.574921
[10] Bureau of Reclamation 2014 *Hydrogenerator Start/Stop Costs* Technical Report (U.S. Department of the Interior)
[11] Stone GC, Lyles JF, Braun JM and Kaul CL 1991 A thermal cycling type test for generator stator winding insulation *IEEE Transactions on Energy Conversion* **6**(4) 707–713 https://doi.org/10.1109/60.103645
[12] Kokko V, Ahtiainen J 2014 Estimation de la durée de vie des turbines et générateurs: Vieillissement du aux cycles arrêt-démarrage et à la régulation *Enhancing Hydropower Plants* Presented at the Congrès SHF, Grenoble, p. 16
[13] Ballester JL 1992 *Contribution à l'étude du vieillissement des roues Pelton: L'endommagement comparé* (Technical Report No. H-44201036-2017-001230) EDF Hydro - DTG Service

"Essais hydrauliques", Grenoble, France
[14] Jacquemoud R 1994 Recherche, pour une turbine Kaplan, d'un domaine de fonctionnement fiable, situé au-delà du domaine garanti, à partir d'une approche prenant en compte les processus d'endommagement par fatigue des matériaux *XVII IAHR Symposium* Beijing, China 13
[15] Zhang Z, Zhang J, Gockenbach E and Borsi H 2009 Life management of SF_6 circuit breakers based on monitoring and diagnosis *IEEE Electr. Insul. Mag.* **25** 21–29 https://doi.org/10.1109/MEI.2009.4976899
[16] Huang K and Zhao S 2012 Evaluation of residual service life of high voltage circuit breaker *2012 China Int. Conf. Electricity Distribution (CICED)* Shanghai, China 1–4 https://doi.org/10.1109/CICED.2012.6508698
[17] Savin O 2018 *L'impact des arrêts-démarrages des turbines hydrauliques sur l'état des matériels d'un aménagement hydraulique* Rapport projet de fin d'étude Grenoble INP Ense3 - EDF DTG
[18] Mercier-Allart E 2017 *CR de la réunion de retour d'expérience des experts du DP de l'impact des arrêts-démarrages des aménagements hydrauliques sur les avaries-défauts rencontrés* (Technical Report No. H-44201036-2017-000260-A) EDF
[19] Rychlik I 1987 A new definition of the rainflow cycle counting method *International Journal of Fatigue* **9** 119–121 https://doi.org/10.1016/0142-1123(87)90054-5
[20] Kauzlarich JJ 1989 The Palmgren-Miner rule derived *Tribology Series* Ed Dowson D et al (Elsevier) **14** pp 175–179 https://doi.org/10.1016/S0167-8922(08)70192-5.
[21] Cygan P and Laghari JR 1990 Models for insulation aging under electrical and thermal multistress *IEEE Trans. Elect. Insul.* **25** 923–934 https://doi.org/10.1109/14.59867
[22] Coffin Jr. LF 1953 *A study of the effects of cyclic thermal stresses on a ductile metal* (Technical Report No. KAPL-853) Knolls Atomic Power Lab., United States
[23] Savin O 2021 *Dossier Technique Secret – Thèse "Effet des cycles de démarrages et d'arrêts sur les centrales hydrauliques: modélisation de la détérioration des matériels pour l'évaluation des coûts de cyclage"* (No. H-44201036-2021-000126) EDF Hydro - DTG.

IOP Conf. Series: Earth and Environmental Science 1136 (2023) 012058 doi:10.1088/1755-1315/1136/1/012058

Experimental study of the flow induced by an axial hydrokinetic turbine

Paul Danca[1], Florentina Bunea[1,3], Corina Babutanu[1], Adrian Nedelcu[1] and Ionut Stelian Grecu[2]

[1] Department of Renewable Energy Sources and Energy Efficiency, National Institute for R&D in Electric Engineering, Bucharest, 030138, Romania
[2] Department of Hydraulics, Hydraulic Machinery and Environmental Engineering, University POLITEHNICA of Bucharest, 060042, Romania
[3] Author to whom any correspondence should be addressed

florentina.bunea@icpe-ca.ro

Abstract. The renewable energy policy strongly encourages the use of all renewable sources of energy. The authors started to develop a new hydrokinetic turbine to be implemented in small rivers or water channels with a low investment cost. This paper presents velocity map distribution around a hydrokinetic turbine prototype. Experimental measurements were performed using a Particle Image Velocimetry system in a water channel, while the water velocity and turbine rotation speed were variated. Images are presented in the same rotor position and the rotational flow is captured downstream of the blades, at different turbine operating modes. This paper is part of a larger study on developing an innovative axial hydrokinetic turbine. The resulting velocity distribution will be used to validate the numerical model.

1. Introduction

The energy harvested by hydrokinetic turbines is the kinetic energy of water in rivers, streams, canals, tides and ocean currents. Hydropower plants (HPP) can produce electricity at a low cost without polluting the air, water, or soil. Hydropower plants can be classified according to the power produced as follows: Large (> 10 MW), Small (<10 MW), Mini (<1 MW), Micro (<100 kW), Pico (<5 kW).

Depending on its size, a pico-hydropower plant can provide to a small community, usually isolated, the electricity needed to power appliances such as radios and televisions, light bulbs, etc. The technology used by kinetic hydropower plants differs from traditional hydropower plants that need a water level difference between upstream and downstream of the machine. Hydrokinetic devices, which are placed directly in the flow of running water, take the kinetic energy and convert it into mechanical energy, which drives an electric generator. The major difference from those placed in ocean currents is the one-way direction of the river.

Because they are powered only by kinetic energy, they do not need the head provided by dams to operate, thus the course of a river remains in its natural state. Given many rivers and canals in which traditional hydropower plants cannot be used, few of these HPPs have been put into operation along the rivers [1]. However, they are an interesting alternative for energy harvesting, especially for rural areas [2]. Such systems could be implemented along the rivers in small constructions or without them, having a low impact on the environment [3].

The classic hydrokinetic turbines are constantly improved, and some other new solutions are currently developing to achieve superior efficiency [4,5]. The power coefficient cannot exceed the maximum theoretical value of 0.59 provided by the Betz limit for a certain water velocity. The main ways to increase the power extracted at the same flow rate are to use intubations with different shapes and angles of flare, use rotor blades adapted to a particular flow, and increase the rotor's diameter as much as possible. For the estimation of the power of a real-scale hydrokinetic turbine, experimental models can be used, on a progressive scale, for which the curves of the power coefficient [6,7] are drawn. An example of a hydrokinetic turbine is studied in this paper, developed in an experimental research project (PED478 / 2020 - Hyper).

This paper presents results obtained for the flow induced by an experimental model of a hydrokinetic turbine, at different rotations speeds, and at 1.1 m/s water velocity. A Particle Image Velocimetry (PIV) [8] technique was used to capture the special distribution of the velocity in different planes, at the same position of the studied turbine.

In the first part of the paper, it is presented a short introduction of the actual study. In the second chapter is presented the experimental set-up composed of three subchapters consisting of: impeller details, PIV system presentation and details regarding post-processing. The third chapter is presented and discussed experimental results of the flow field to quantify the effect of the rotor over the water flow. Chapter 4 draws conclusions. This study presented is part of a research project (PED478 / 2020 - Hyper) with the aim to develop an innovative (new) axial hydrokinetic turbine. The design of the new hydrokinetic turbine will be optimized by numerical calculation to obtain the maximum output for an extended operating range. A first prototype was designed and tested in order to validate the numerical flow field. The flow fields obtained in this study will be compared with the flow fields from a numerical study in order to validate it.

2. Experimental setup

2.1. The experimental stand

In this subchapter are presented details regarding the experimental hydrokinetic turbine and the experimental stand. The hydrokinetic turbine has three blades with profiles chosen from NACA 4400 series, using different relative thickness values, from 4% near the hub to 11% at the tip section, disposed at different angles [9]. The design operating point of the turbine runner was set for 1m/s upstream velocity, 165 rpm runner speed and 7W shaft power.

The 3D geometry of the turbine was created in SolidWorks software, then manufactured by acrylonitrile butadiene styrene material with a 3D printer and painted in a matte black color to avoid light reflections. The studied hydrokinetic turbine has a diameter of the rotor $D = 0.2$ m (Figure 1).

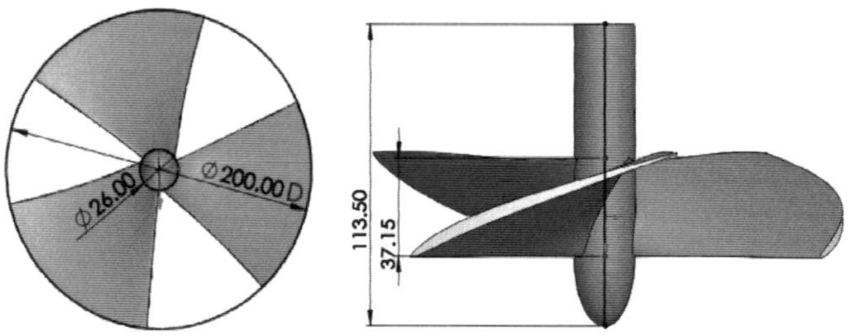

Figure 1. Hydrokinetic turbine model dimensions.

The experimental stand is a closed loop with a pump system that can provide a variety of water velocities from 0.2m/s to 1.1 m/s. This stand has a visualization zone of 375 × 300 × 1015, made from a special Plexiglas in order to allow PIV measurements. Before the measuring area, at a distance of 1 m is a honeycomb structure in order to stabilize the flow. The studied hydrokinetic turbine was placed in the middle of the transparent area at a height of 160 mm from the bottom channel (Figure 2). The use of a limited water channel, with the channel walls in the vicinity of the runner tip, especially the bottom wall, forced the water to enter the rotor, and the extracted power was then higher compared to the free conditions. This effect gives power coefficient values that are higher than in an infinite domain [10].

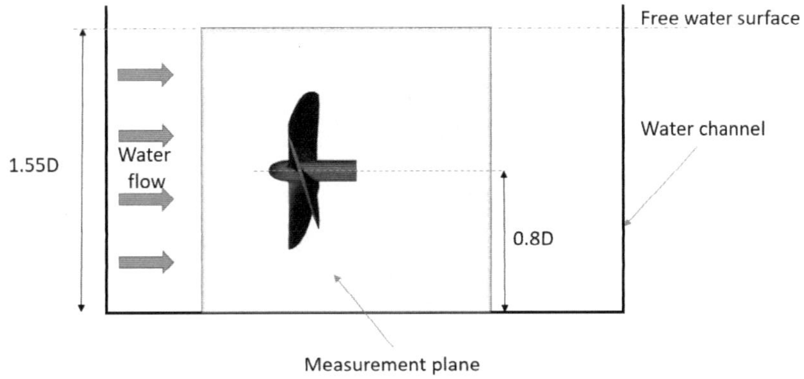

Figure 2. The turbine position in the water channel.

As previously specified, the experimental stand allows varying the water velocity, while a braking system can be controlled to maintain the desired rotation speed of the hydrokinetic turbine. Given these, experimental measurements were performed at 1.1 m/s water mean velocity for 5 different rotation speeds. The experimental settings can be seen in Table 1, resulting in 15 measurement cases. A schematic representation of the three measurement planes can be seen in Figure 3, respectively: in the rotor axis (corresponding to the axis of symmetry of the study area), in the immediate vicinity of the hub and at the end of the rotor blades (the area where detached vortices occur).

Table 1. The analyzed operating regimes.

Studied cases		Case 1	Case 2	Case 3	Case 4	Case 5
Rotation speed	[rpm]	70	100	120	160	210
Water velocity	[m/s]			1.1		
Measurement planes		In the turbine axis (P1)				
		At 14 mm to the turbine axis (P2)				
		At 100 mm from the turbine axis (P3)				

Before the PIV measurements, the velocity field was investigated upstream of the rotor with a Pitot-Prandtl tube. The tube is connected to a differential pressure sensor, which transmits the data to a central unit for data acquisition and storage. The measured differential pressure, Δp, represents the dynamic pressure of the flow and is used to compute the water velocity as $V = \sqrt{2\Delta p/\rho}$, where ρ is the water density. For the studied flow regime, several measurements were conducted in the middle of the channel, moving the Pitot-Prandtl tube on vertical direction, from the free surface to the bottom of the channel. The velocity profile obtained is a typical one for a free surface channel flow, with a uniform spatial steady repartition of the flow field.

The Froude number for each flow velocity is computed in [11], table 2. Starting with the water velocity of 1 m/s, the flow became supercritical. For the velocity of 1.1 m/s, Reynolds number is $Re = 3\cdot 10^5$ and Froude number is $Fr = 1.059$. Due to the wave caused by the turbine in this configuration, the power coefficient values increased compared to the results obtained for lower flow velocities. These aspects are analyzed in [11].

2.2. *Particle Image Velocimetry measurements*

2D flow speed measurements were performed with a Dantec Dynamics PIV system which consists in a Dual Power green (532 nm) Litron laser with an energy of 200 mJ, a synchronizer BNC 575 and a 4M FlowSense EO 4m CCD camera of 2072x2072 pixels resolution combined with a 50 mm lens. The maximum trigger rate of the PIV system between two pairs of images was 5 Hz. The time between pulses was set at 200 μs and 900 μs. We choose these values due to the complex flow from the interest region, as in the area behind the turbine water velocity was significantly decreased, while in the spaces between the wall and the turbine the speeds were much higher. For each experimental set-up were recorded 250 pairs of images.

The flow was seeded with hollow glass spheres with a diameter of 10μm produced by Dantec Dynamics.

Since the flow is with a free surface, waves may appear at certain operating modes, which may reflect and refract the laser beam if the light will come from the top part of the viewing zone. Therefore, the light sheet generator was placed under the viewing zone at 1000mm. At this distance, the entire interest region was lighted.

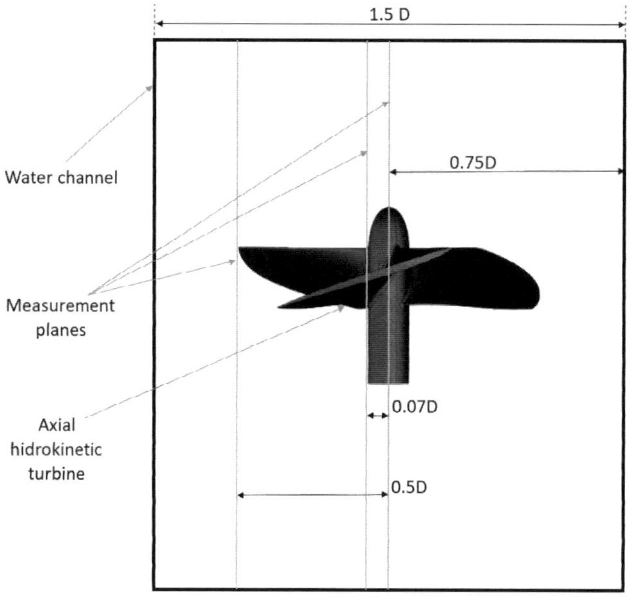

Figure 3. Schematic representation of the measurement planes (upper view).

The CCD camera was placed at a distance of 1500 mm from de measurement plane. We choose this distance in order to capture the flow field from the area of the rotor blades but also upstream and downstream of it. The size of each captured image was 345×345 mm^2 on the axis, 375×375 mm^2 at 14mm and 325×325mm^2 at the distance of 100 mm (see Figure 3).

An optical diffuse-mode laser sensor BANNER QS30LDQ was placed on the other side of the viewing area. This sensor was connected to the BNC 575 synchronizer and its role was to dictate the timing of the trigger for capturing imagers. This sensor has a light emitter and light receptor. When the beam laser emitted meets the shining marker placed on the hub, a part of the light is reflected to the light receptor, these being the moment when the frames are captured. In this way, all the measurement was performed at the same position of the blade.

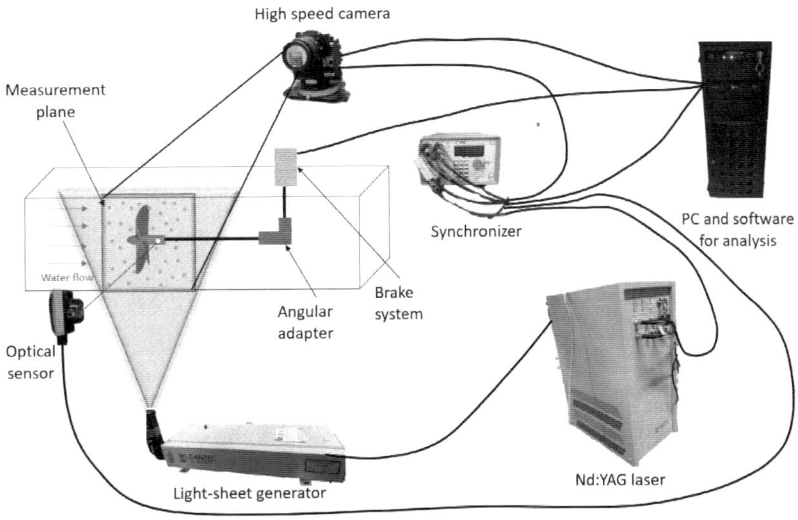

Figure 4. Schematic disposal of the PIV system.

The studied turbine was connected to the rotation speed control system by an angular adapter, through which the rotational movement is translated from the horizontal plane to the vertical plane. The turbine speed control system consists of a torque transducer and a brake, the latter being based on swirling. The type of the brake is FAT 20, produced by Mobac Gmbh and the torque transducer, model DR 2335, is manufactured by Lorenz Messtechnik Gmbh. Both devices are dimensioned for turbine testing with a maximum torque of 2 Nm.

All data taken from the transducers are transmitted to a programmable controller with the role of controlling the converters and acquiring the data. A dedicated software application is installed on a PC connected to the programmable controller.

In accordance with the sketch in figure 4, in figure 5 it can see an image during the experimental measurements and the placement of the equipment.

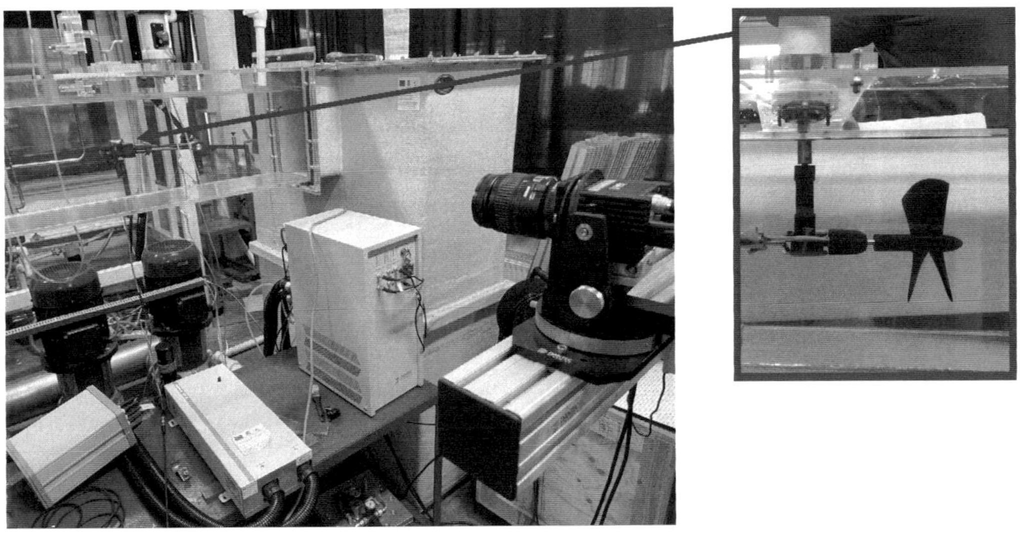

Figure 5. Images of the experimental setup.

A power diagram was drawn to show (black dots) the points at which the PIV measurements were performed (figure 6). The studied turbine has maximum power at 120 rpm, but four other operating points were analyzed.

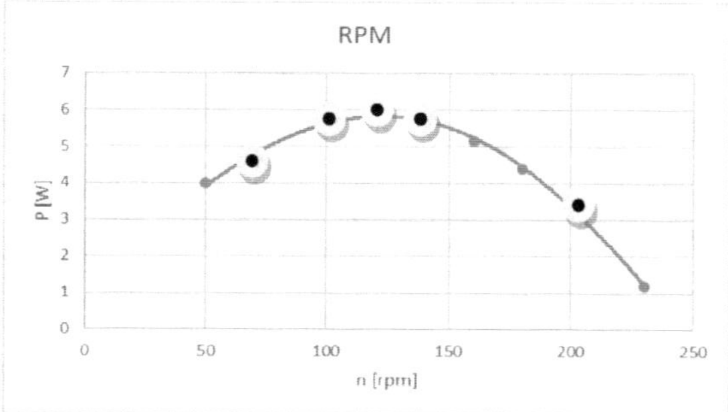

Figure 6. Power diagram at a water velocity of 1.1 m/s.

Uncertainty estimation for PIV velocity vectors were performed using "Peak Height Ratio" [12]. The results show that from 16384, only 1540 were rejected, which is less than 10%.

2.3. PIV image processing

By correlating the 250 image pairs in the Dynamic Studio software, the mean velocity fields for each case studied result. The images correlations were performed through two methods: "Adaptive PIV" and "Average correlation", the results obtained with the last method are presented in this paper. Before the correlation, a mask was applied to the frames to cover the areas with reflections, and other areas that were shaded by the hub or by the blade. Forwards, all the data from the average velocity fields were exported in .dat files. These files were imported to TecPlot software and represented as in the results and discussions section.

3. Results and discussions

The flow induced by a new model of the hydrokinetic turbine is investigated. To characterize the operating range of the turbine, experimental research was performed at five different rotational speeds at an average water velocity of 1.1 m / s (Table 1). The measurements were repeated in three parallel planes: the first plane was placed in the center of the channel, respectively in the turbine axis, the second plane in the immediate vicinity of the hub, and a third plane was chosen so as to cut the tip of the rotor blades, at a distance of 100 mm from the plane 1 (figure 3).

The objective was to measure the main velocity field and the evolution of the velocity distribution at the level of the rotor blades, but also upstream and downstream of them. The resulting velocity vectors maps are presented in the longitudinal plane, on the axis of symmetry (figure 7), at 14 mm of the axis (figure 8), respectively at 100 mm of the axis (figure 9). The blue color regions represent the mask applied to the captured images. These regions are shaded or with high reflections. In Figure 9e the shading effect is diminished and the mask was no longer applied.

The most uniform velocities field were acquired at 210 RPM (figures 7e, 8e, 9e) corresponding with no load cases (the maximum rotation speed of the turbine). As the revolution speed decreases (figure 10), the flow becomes more turbulent behind the turbine area. The upstream flow is uniform and ensures an ideal operating condition for the turbine inlet. Only at the ogive area and in front of the blades these velocities slightly decrease as the RPM number decrease. Downstream, water velocity is highly influenced by the rotation speed. As decreasing the rotation speed, the velocity decrease in the rotor area to 0.3m/s while in other areas increases to 1.5m/s at the top of rotor blades.

The relatively high velocities at the top of the rotor blades can favor the formation of cavitation bubbles, which are difficult to avoid at all points along the blade. The design of the studied experimental model does not favor the formation of cavitation.

However, in the case of hydrokinetic turbines with larger rotor surfaces, attention must be paid to the design of the blades, to avoid the formation of cavitation.

As the measurements were taken for a fixed blade position, by azimuthal phase average in the mean velocity fields, the blade wake and the wingtip vortex are identified (**Error! Reference source not found.** b, c, d) where the value of velocity magnitude is 1.4 m/s. A blade wake can be observed as the farther we go where the velocity values are between 0.5 and 0.9 m/s. The third component of velocity would be useful for wingtip vortex contouring.

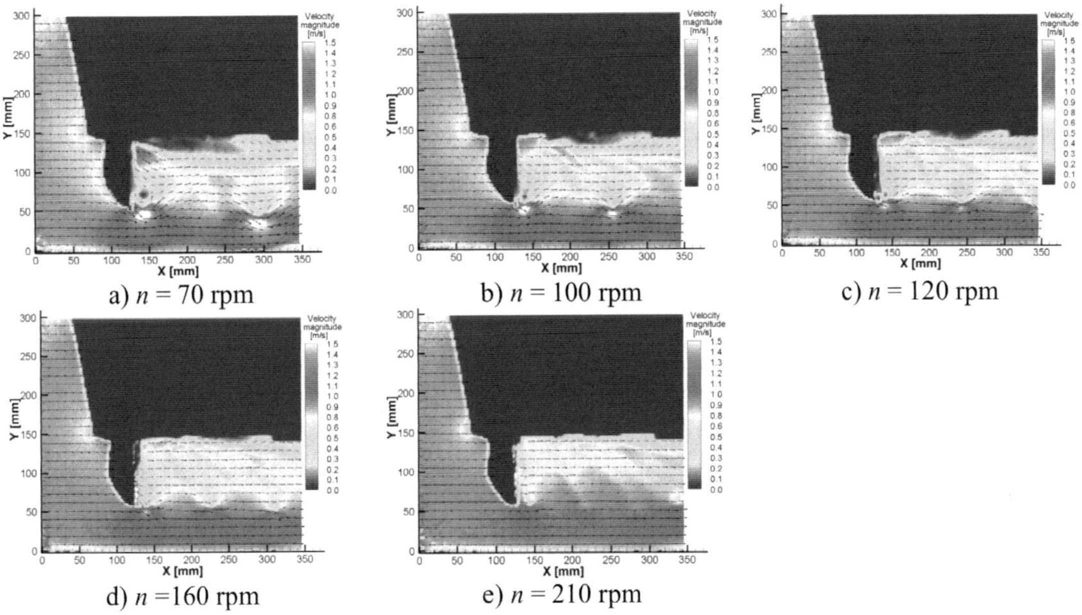

Figure 7. The mean velocity field at five different rotation speed in turbine axis (P1).

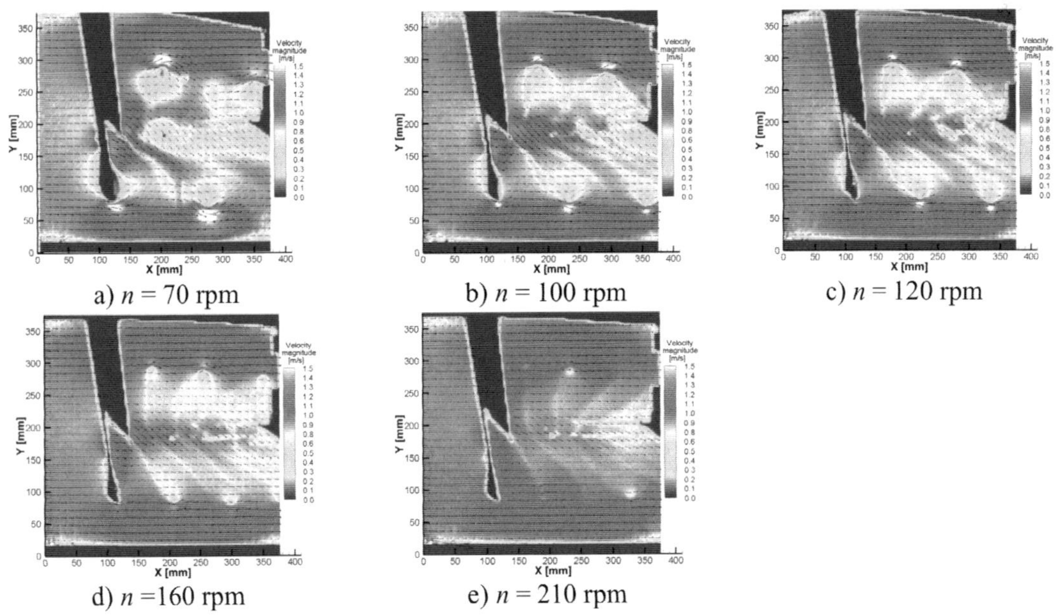

Figure 8. The mean velocity field at five different rotation speed, at 14 mm distance from the turbine axis (in the vicinity of the hub – P2).

IOP Conf. Series: Earth and Environmental Science 1136 (2023) 012058 doi:10.1088/1755-1315/1136/1/012058

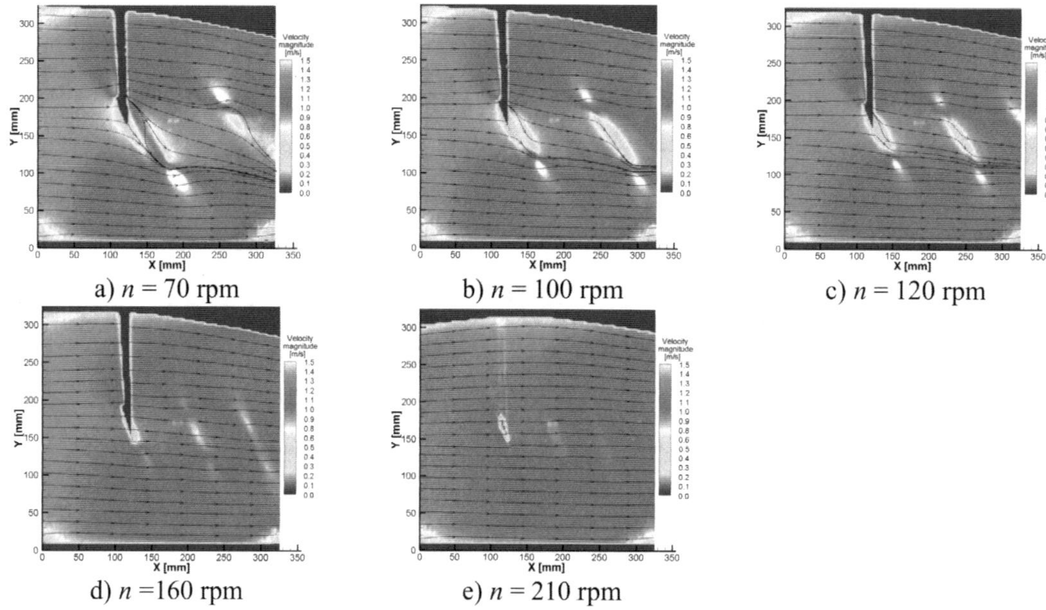

a) $n = 70$ rpm b) $n = 100$ rpm c) $n = 120$ rpm
d) $n = 160$ rpm e) $n = 210$ rpm

Figure 9. The mean velocity field at five different rotation speed, at 100 mm distance from the turbine axis (at the tip of the rotor blades – P3).

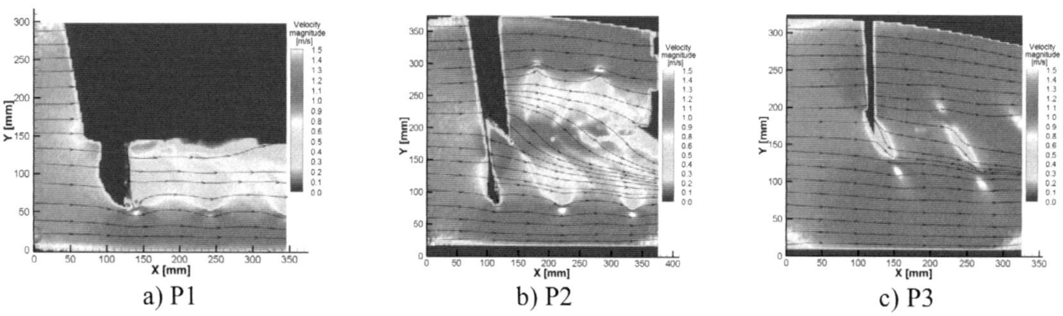

a) P1 b) P2 c) P3

Figure 10. The velocity field at the three measuring planes (P1, P2 and P3), for the rotational speed of 120 rpm.

It is also observed that the level of the free surface decreases due to the increase of the global velocity in the section. This information is important not only for flow analysis but also for implementation purposes due to the additional in situ erosion effects.

4. Conclusions

The results presented in this paper are part of a larger study with the purpose to create a hydrokinetic turbine. The design of the new hydrokinetic turbine will be optimized by numerical calculation to obtain the maximum output for an extended operating range. A first prototype was designed and tested in order to validate the numerical flow field. The flow fields obtained in this study will be compared with the flow fields from the numerical study in order to validate it. Quantitative data – vorticity and turbulence – will be presented in an extended paper. The correlation between the runner speed and the measured velocities, as well as the runner blades wake propagation/dissipation, will be also presented.

In this paper are presented velocity distribution maps around the runner blades of a hydrokinetic turbine model. These measurements were performed using PIV technique in a free surface channel. Using the equipment described in chapter 2.2, the rotor blades were captured in the same position, thus offering the possibility to follow the velocity field inside the rotor. Thus, the velocity field and the evolution of the vortices detached along with the flow were obtained. To have an extended operating range of the turbine model several rotational speeds were investigated. For a better understanding of the phenomenon, the power curve of the analyzed regimes was also presented

The most uniform velocities field was acquired, when the breaking value was zero, this being the maximum rotation speed of the turbine. As the speed rotation decreases the flow becomes more turbulent behind the turbine area. As decreasing the rotation speed, the velocity decrease in the rotor area to 0.3m/s while in other areas the velocity increase to 1.5m/s at the top of rotor blades.

Even though these measurements are 2D can be observed a small vortex in the proximity of the hub where the value of velocity magnitude is 1.4 m/s. A larger vortex can be observed as the farther we go where the velocity values are between 0.5 and 0.9 m/s. The third component of velocity would be useful for vortex contouring. However, the extension of PIV measurements to 3D is being considered.

The level of the free surface decreases due to the increase of the global velocity in the section. This information is important not only for flow analysis but also for implementation purposes due to the additional in situ erosion effects.

Acknowledgments
The work has been funded by the Executive Unit for Financing Higher Education, Research, Development and Innovation (UEFISCDI), PN-III-P2-2.1-PED-2019-3247 ctr. No. 478 PED/2020, HyPER project and by the Romanian Ministry of Education, Research and Digitalization, project number 25PFE/30.12.2021 within PNCDI III, Programme 1.

References

[1] Bhalshankar SS and Thorat CS 2018 Home water-tank pico-hydro power generator modeling and analysis for smart home distributed generation system *International Journal of Electrical Engineering & Technology* **9**(4)

[2] Khomsah A, Sudjito, Wijon and Laksono AS 2019 Pico-hydro as a renewable energy: Local natural resources and equipment availability in efforts to generate electricity *IOP Conf. Ser.: Mater. Sci. Eng.* **462** 012047

[3] Elghali SB, Benbouzid M and Charpentier J 2007 Marine tidal current electric power generation technology: State of the art and current status *IEEE Int. Electric Machines & Drives Conf.* pp 1407–12 doi: 10.1109/IEMDC.2007.383635

[4] Khan MJ, Iqbal MT and Quaicoe JE 2008 River current energy conversion systems: Progress, prospects and challenges *Renewable and sustainable energy reviews* **2**(8) 2177–93, DOI: 10.1016/j.rser.2007.04.016

[5] Miller VB, Ramde EW, Gradoville RT and Schaefer LA 2011 Hydrokinetic power for energy access in rural Ghana *Renewable Energy* **36**(2) 671–675 https://doi.org/10.1016/j.renene.2010.08.014

[6] Chihaia RA, Bunea F, Oprina G and El-Leathey LA 2017 Power prediction method applicable to horizontal axis hydrokinetic turbines *IEEE 2017 Int. Conf. Energy and Environment*, October, Bucharest pp 221–5 doi: 10.1109/CIEM.2017.8120825

[7] Nedelcu A, Bunea F, Danca P, Chihaia RA, Babutanu CA, Marin D and Ciocan GD 2012 Experimental research on a hydrokinetic turbine model *IOP Conf. Ser.: Earth Environ. Sci.* **664** 012061 doi: 10.1088/1755-1315/664/1/01206

[8] Bunea F, Dancă PA and Năstase I 2021 *Determinarea Vitezelor în Curgeri cu Ajutorul Imaginilor de PArticule – PIV. Noțiuni Generale și Aplicații* (in Romanian) (Biblioteca Nationala a Romaniei)

[9] Bucur DM, Grecu IS, Mitruţ R, Drăgoi C, Stroilescu AI and Dunca G 2021 Design and numerical investigation of a small axial hydrokinetic turbine *10th Int. Conf. on Energy and Environment (CIEM)* 1-5 doi: 10.1109/CIEM52821.2021.9614895

[10] Eltayesh A, Hanna MB, Castellani F, Huzayyin AS, El-Batsh HM, Burlando M and Becchetti M 2019 Effect of wind tunnel blockage on the performance of a horizontal axis wind turbine with different blade number *Energies* **12**(10) 1988 https://doi.org/10.3390/en12101988

[11] Bucur DM, Dunca G, Bunea F, Chihaia RA, Grecu IS and Mitruț R 2022 Experimental investigation of small axial hydro-kinetic turbines *IOP Conf. Ser.: Earth Environ. Sci.* **1070** 012020 doi: 10.1088/1755-1315/1079/1/012020

[12] Charonko JJ and Vlachos PP 2013 Estimation of uncertainty bounds for individual particle image velocimetry measurements from cross-correlation peak ratio *Measurement Science and Technology* **24**(6) 065301.

Numerical modelling of the cooling water system of a 106MW Hydro-Power Plant using EPANET

Alexandru-Spiridon Milcă[1,2], Georgiana Dunca[1] and Livia Pitorac[3,4]

[1] Faculty of Energy Engineering, University POLITEHNICA of Bucharest, Bucharest, 060042, Romania
[2] Engineering Department, SC Mecanica-Ind 2004 SRL, Bucharest, 061126, Romania
[3] Civil and Environmental Engineering Department, Norwegian University of Science and Technology, Trondheim, 7031, Norway
[4] Statkraft Energi, Oslo, 0283, Norway

alexandru.milca@mecanicaind.ro

Abstract. The present paper studies the behavior of the new proposed cooling water pumps for a 106 MW Hydro-Power Plant using EPANET. This case-study Hydropower Plant, located in the south of Romania and equipped with two Francis units 53 MW each, processing a 269m gross head, presently goes through a rehabilitation process. The cooling water system of this Hydro-Power Plant consists in four centrifugal pumps, two for each unit, and two additional pumps for the 130 MVA step-up transformer. The rehabilitation process involves the replacement of these pumps. A numerical model following the exact configuration of the cooling water system in what regards pipes diameters, lengths and roughness was designed in EPANET. The focus is to verify if the required pressure is ensured at the cooling water consumers of the plant considering their specific flow demand and if the velocity in the system pipes is within the acceptable limits during the power production process.

1. Introduction

Hydropower Plants (HPP) represent a major energy source in Romania with over 6000 MW installed. Over 10% of these power plants are more than 50 years old [1], therefore, their rehabilitation becomes a necessity. In a rehabilitation process, besides the renewal of the equipment and systems, attention must be paid to the efficiency improvement whenever it is possible, as a rehabilitation process gives this opportunity.

Besides the primary equipment of a HPP, several auxiliary systems ensure the proper conditions for the power producing process. One of the most important of these auxiliary systems is the cooling water system. This system provides water to a series of water/oil or water/air heat exchangers [2] in order to maintain the temperature in the bearings, generator coils, and transformers in the required limits. Also, the cooling water system provides water to the firefighting system of the unit in case of an unwanted event.

The case-study HPP uses two Francis units, 53MW each, at a 269 m gross head and 56 m^3/s nominal flow. Its annually production reaches 262 GWh being a vital power plant in its area. The cooling water system of this plant is an open circuit [3], configured to supply water from each units draft tube using four 55 kW centrifugal pumps (P1 – P4). Each pump must satisfy the cooling water demand for one unit, thus, at nominal, two pumps are running while two are in reserve. The quality of

IOP Conf. Series: Earth and Environmental Science 1136 (2023) 012059 doi:10.1088/1755-1315/1136/1/012059

the cooling water provided to the consumers is ensured by a self-cleaning filter installed on each pump's inlet. At the end of, the cooling water is evacuated to the tailrace.

The power produced is delivered in the system through a step-up 130 MVA transformer. The cooling of this equipment is ensured by two centrifugal pumps (PTr1, PTr2) 5.5 kW each, connected to the filter's outlet. This configuration gives the possibility to select the intake filter for these pumps.

This paper analyses the behaviour of the new proposed cooling water pumps for the described Hydro-Power Plant. The purpose is to verify if these pumps can successfully fulfil the cooling water necessities of the plant for the possible occurring scenarios. Figure 1 shows the configuration of the analysed cooling water system.

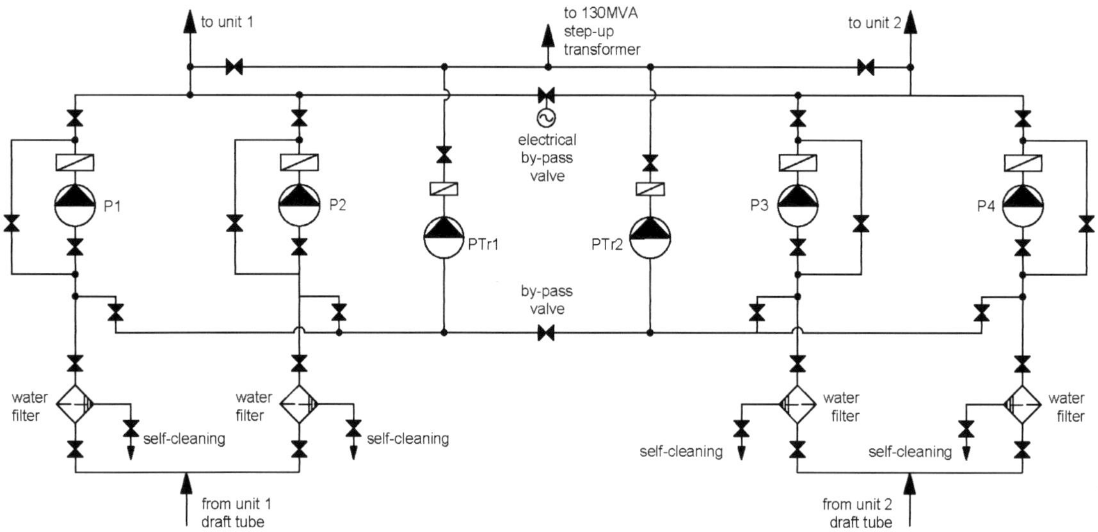

Figure 1. Cooling water system diagram. Legend: P1-P2 – unit 1 cooling water pumps, P3-P4 – unit 2 cooling water pumps, PTr1-PTr2 – step up transformer cooling water pumps.

2. Numerical model

In order to simulate the hydraulic conditions for the case study cooling water system, a numerical model has been built using EPANET [4]. This numerical model follows the exact conditions of the studied system in terms of pipe lengths, diameters and roughness. The valves in the system were also introduced into the model with respect to their type and position. EPANET does not provide filters as predefined components, however, it is possible to substitute a filter with a throttle control valve (TCV) [5]. This is a partially closed valve which introduces a head loss equivalent to the head loss related to the prototype filter.

The head loss introduced by a filter normally depends on the discharge, however, in our case, the discharge is constant as each pump has its own filter. Therefore, knowing that the cooling water demand for one unit of the plant is approximately 320 m^3/h, the head loss value in each filter is around 1.5 mWC or 0.15 bar.

The consumers of the cooling water system are the turbine seals, turbine guide bearing, generator guide bearing, generator air coolers and the heavy duty thrust bearing. The flow demand values for these consumers are presented in Table 1. As observed, the main consumers in terms of the flow required are the generator air coolers, while the lowest cooling water consumers are the turbine seals.

The requirement for our cooling water system is to maintain a minimum pressure of 2.5 bar at the cooling water consumers inlet when delivering their specific flow demand. [3]. The hydraulic network built in EPANET is shown in Figure 2 and it consists of a series of reservoirs, nodes, links, valves, and pumps. Each reservoir simulates the turbines draft tube, namely the supply point for the cooling water.

Table 1. Cooling water consumers flow demand.

Consumer description	No. of heat exchangers	Unitary demand [m^3/h]	Total demand [m^3/h]
Turbine seals	N/A	0.5	0.5
Turbine guide bearing	2	18	36
Generator guide bearing	1	7	7
Generator air coolers	6	40	240
Heavy duty thrust bearing	1	33	33

Each water intake is protected against floating objects intrusion by a metallic grid. The valve inserted downstream the reservoirs simulates this grid which introduces a significant head loss. Upwards the system there are the filters with their inlet and outlet valves, located at the drainage level of the Hydro-Power Plant. The unit cooling water pumps (P1-P4) are located one level higher, at the turbines level, and are also provided with inlet and outlet valves. The transformer cooling water pumps (PTr1, PTr2), located at the same level, only have outlet valves. The unit cooling water pumps are connected to a cooling water bus, separated with a bypass valve between the units. The transformer cooling water pumps are also connected to a common pipe. In case of both PTr1 and PTr2 failure, the transformer can be supplied from the unit pumps.

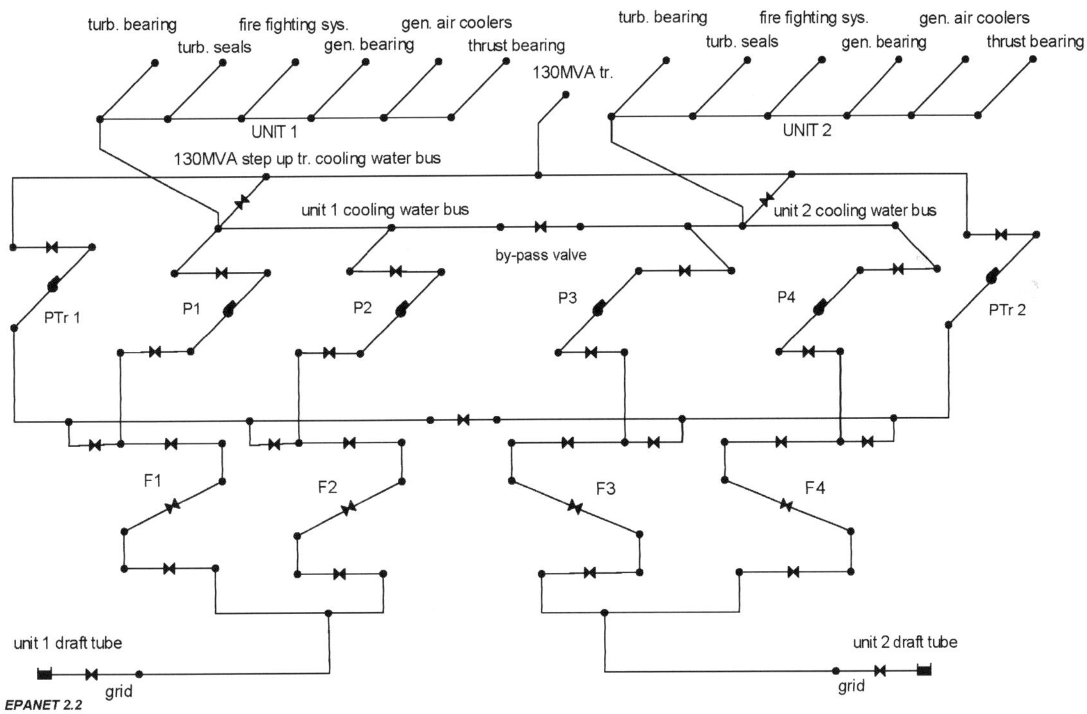

Figure 2. EPANET numerical model for the cooling water system.

Each of the unit pumps must ensure the cooling water flow demand for one unit. Therefore, there are two possible scenarios to be analysed:

- scenario 1 – each unit to be supplied by one of its own pumps.
- scenario 2 – one unit to be supplied by one of the other units pumps.

The second scenario is possible by opening the bypass valve in the middle of the cooling water bus and it is not a normal operation scenario. It occurs when both pumps of a unit are down.

The step-up transformer cooling water flow demand must be ensured by any of its two pumps. Thus, only one of its two pumps is normally running while the other is a reserve in case of failure.

3. Results

Using the numerical model presented and assuming both units of the Hydro-Power Plant running at their nominal power, a series of simulations were performed. In this case, the two possible scenarios for the cooling water system configuration were analysed.

In the first analysed scenario each unit was supplied by one of its own pumps. Thus, P1 supplied unit 1, P3 supplied unit 2 and PTr1 supplied the transformer.

In the second scenario, both units were supplied by one units pumps. Thus, P3 and P4 supplied both units and PTr2 supplied the transformer.

The purpose was to check if the required pressure at the consumers was ensured and if the velocity in the pipes was within the acceptable limits for the analysed scenarios.

3.1. No. 1 scenario – P1, P3 and PTr1 running

In this first scenario, each unit is supplied by one of its own pumps, while the bypass valve is closed, thus, the numerical simulation has been performed for P1, P3 and PTr1 running. The results are presented in Figures 3, 4 and 5.

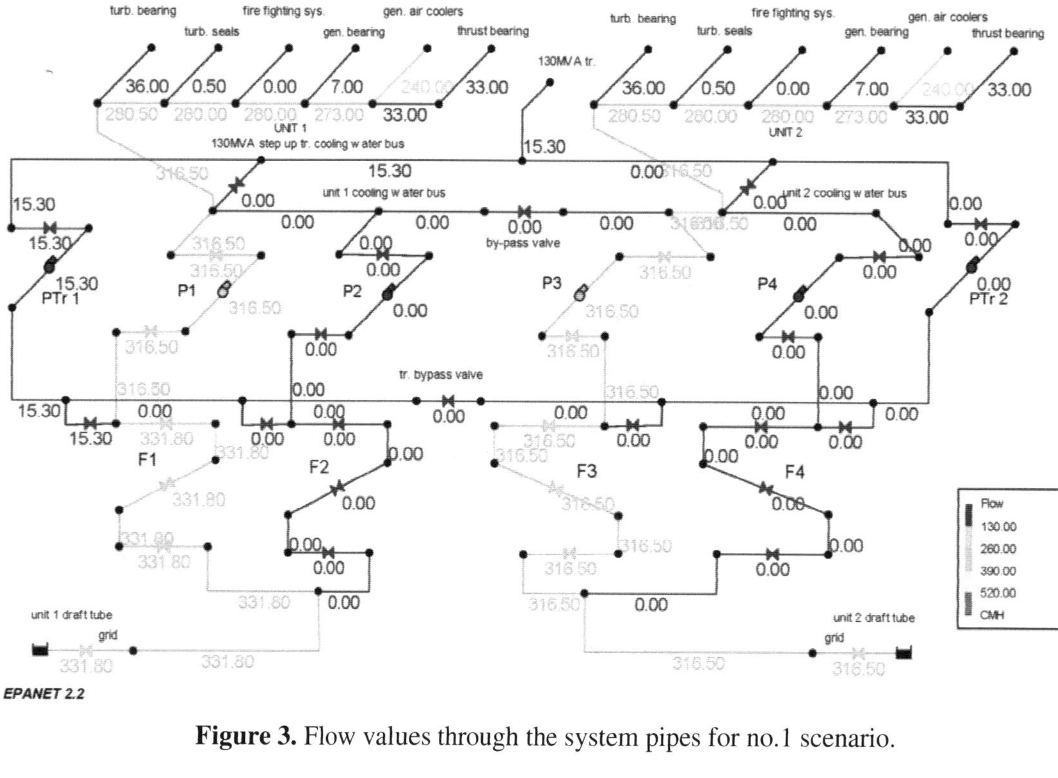

Figure 3. Flow values through the system pipes for no.1 scenario.

3.2. No. 2 scenario – P3, P4 and PTr2 running

In the second scenario, both units are supplied by one units pumps, while the bypass valve is open, thus, the numerical simulation has been performed for P3, P4 and PTr2 running. The results of the are shown in Figures 6, 7 and 8.

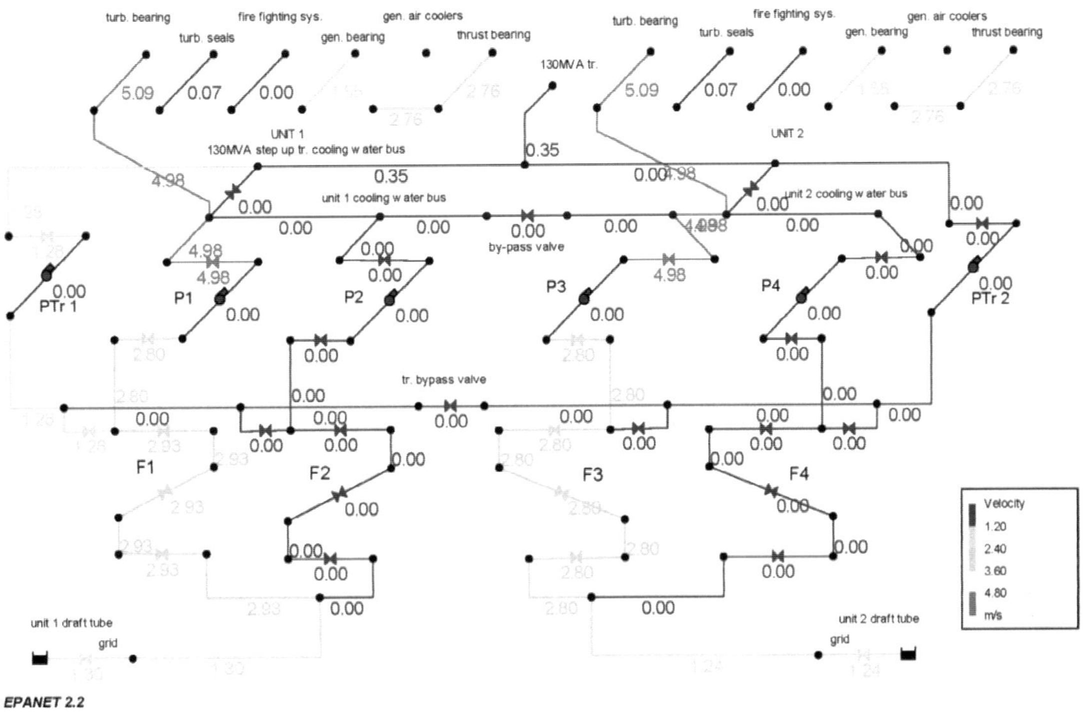

Figure 4. Velocity values through the system pipes for no.1 scenario.

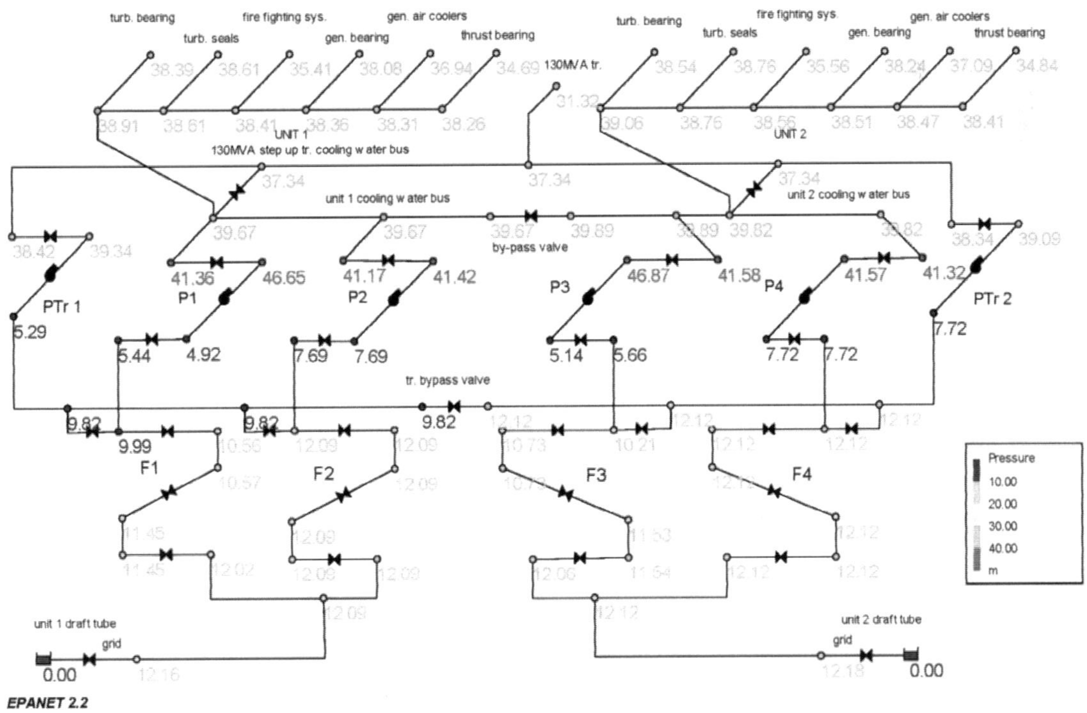

Figure 5. Pressure values in the system for no.1 scenario.

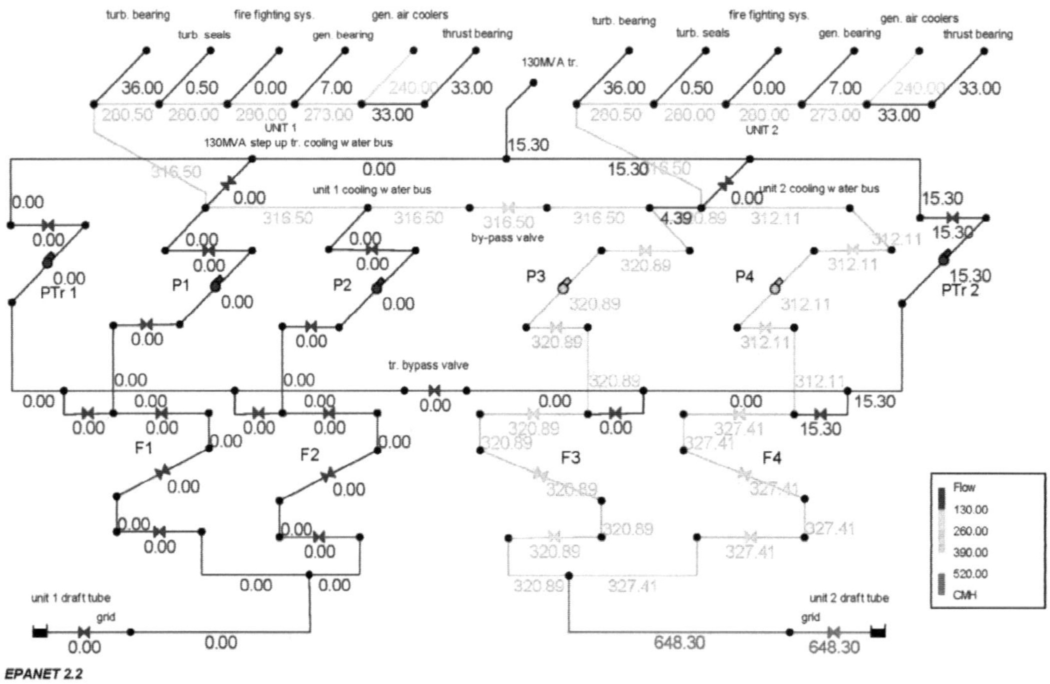

Figure 6. Flow values through the system pipes for no.2 scenario.

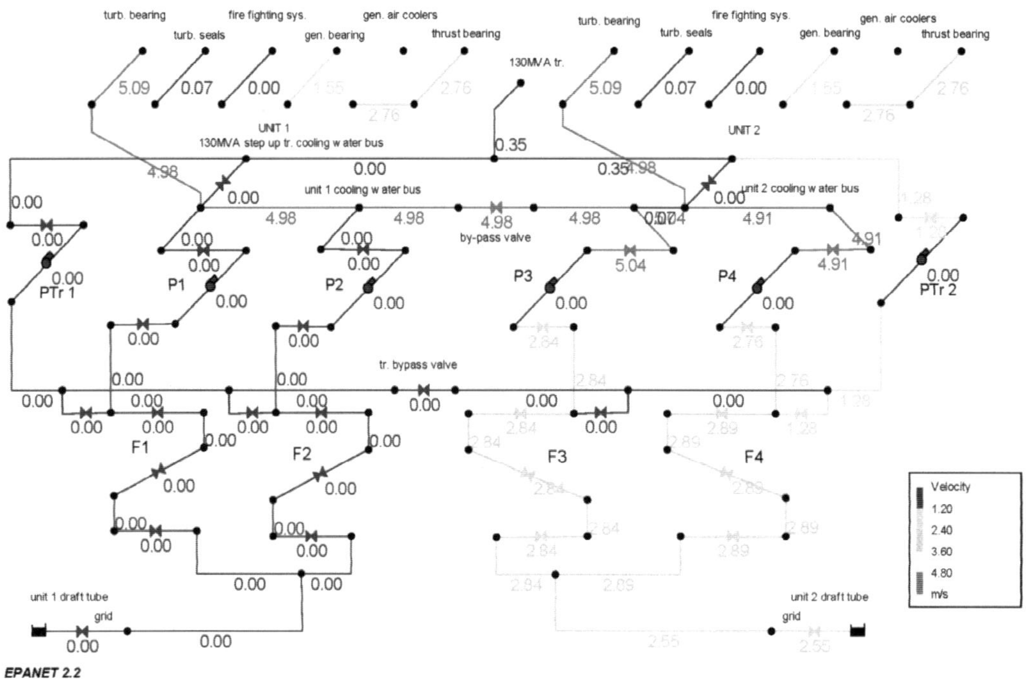

Figure 7. Velocity values through the system pipes for no.2 scenario.

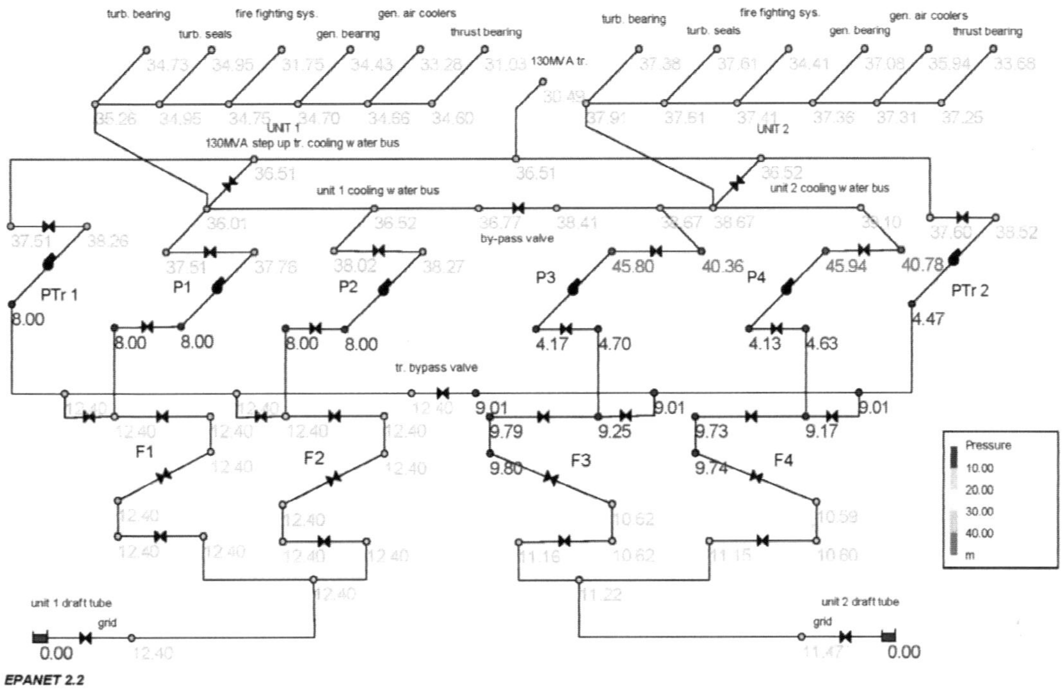

Figure 8. Pressure values in the system for no.2 scenario.

4. Discussion

Similar to the available references [2] and [4], EPANET numerical model has been used to simulate the real conditions in the prototype Hydro-Power Plant cooling water system in order to have an image about the hydraulic parameters in the process.

In the first analysed scenario, for the given flow distribution in the system (see Figure 3), the maximum velocity through the pipes was 5.09 m/s (see Figure 4) and the lowest value for the pressure was 34.69 mWC, encountered at units 1 heavy duty thrust bearing (see Figure 5). The pressure at the transformer cooling unit was 31.32 mWC.

In the second scenario, for the given flow distribution in the system shown in Figure 6, the maximum velocity through the pipes was the same as in the first scenario, 5.09 m/s respectively as can be seen in Figure 7. The lowest pressure value was 31.03 mWC, encountered at unit 1 heavy duty thrust bearing (see Figure 8). The pressure at the transformer cooling unit was 30.49 mWC.

The results retrieved by the numerical model validated the pumps proposed and offered an image of these new cooling water pumps behaviour. The results of this study have been used in the rehabilitation process of the prototype Hydro-Power Plant.

5. Conclusions

The study presented a numerical model built in EPANET for the simulation of a prototype Hydro-Power Plant cooling water system. This study was mandatory as the owner of the plant required a validation of the new pumps. Knowing the results, the following conclusions can be stated:

- the new cooling water pumps proposed for the rehabilitation of the case-study Hydro-Power Plant successfully ensure the cooling water demands;
- the requirement regarding minimum pressure at the cooling water consumers inlet above 2.5 bar is ensured in both analyzed scenarios;
- the velocity in the system pipes for the analyzed scenarios is within the acceptable limits reaching a maximum value of 5.09 m/s;

- the new cooling water pumps proposed for the step-up transformer successfully ensure its cooling water demands.

The results of the current study were as expected, meaning that the new cooling water pumps for the case-study Hydro-Power Plant can be installed. These results were integrated in the rehabilitation documentation and the next step is their validation in the field after the commissioning of the retrofitted cooling water system.

Acknowledgments
Thanks are due to Mecanica-Ind 2004 SRL, which presently performs the rehabilitation process of the case-study Hydropower Plant, for the availability of the data. The results of the present study have been used for the validation of the new pumps provided to the cooling water system of the plant.

References

[1] Hidroelectrica 2019 *Electricity Production* https://www.hidroelectrica.ro/article/de56b7ec-8b2c-5e76-03b8-48e36ae8a533 (accessed: 20 December 2022)
[2] Georgescu S-C, Georgescu A-M, Jumara A, Piraianu V-F and Dunca G 2016 Numerical simulation of the cooling water system of a 115 MW Hydro-Power Plant *Energy Procedia* **85** 228–234 https://doi.org/10.1016/j.egypro.2015.12.230
[3] Mecanica-Ind 2004 SRL 2022 *Case-study Hydro-Power Plant Documentation* (Bucharest: Engineering Department, Mecanica-Ind 2004 SRL)
[4] Bucur DM, Cosoiu CI, Iovanel RG, Nicolae AA and Georgescu S-C 2017 Assessing the operation of the cooling water system of a Hydro-Power Plant using EPANET *Energy Procedia* **112** 51–57 https://doi.org/10.1016/j.egypro.2017.03.1058
[5] Georgescu S-C and Georgescu A-M 2014 *Manual de EPANET* (Bucharest: Editura Printech) https://www.researchgate.net/publication/283268321_Manual_de_EPANET_in_Romanian

IOP Conf. Series: Earth and Environmental Science 1136 (2023) 012060 doi:10.1088/1755-1315/1136/1/012060

Reducing operating costs in district heating systems by optimizing functional parameters. Comparative studies

Daniel Muntean[1] **and Adriana Tokar**[1]

[1] Department of Civil Engineering and Building Services Engineering; Faculty of Civil Engineering; Politehnica University Timişoara, 300223 Timişoara, Romania

adriana.tokar@upt.ro

Abstract. In district heating networks, the influence of conditions: thermohydraulic, reliability, availability and cost of materials, heat demand and energy consumption for pumping, on the operation of the system must be considered in terms of optimizing the design so that the costs of production, transport and maintenance to be reduced as much as possible. These system problems are unavoidable during operation and therefore the paper proposes an analysis based on simulating the operation of the system in two possible situations: reducing the diameters of the heat transmission pipes with the compromise of increasing the energy consumption required for pumping and respectively reducing the maximum pressure losses with the advantage of obtaining an energy saving at pumping. The comparison of the results obtained by the simulation allowed the choice of the optimal solution established on the basis of the minimum total costs (capital and operational) and the minimum heat losses on the pipeline network, taking into account the variation of investment costs of the pipeline and pumping energy depending on the diameter of the pipe.

1. Introduction

The effects of global warming on humanity and ecosystems are already alarming and mitigation of greenhouse gases (GHGs) is essential in limiting their consequences [1-3]. In this regard, special attention is currently being paid to the transition: international, national, regional and local, in the energy sector (electricity and heat) being one of the main sources of GHG emissions worldwide. [4-6].

The key elements of the local energy transition are the energy communities that generate and distribute energy locally to meet the needs of consumers. Given that in the European Union, heating and cooling account for about 50% from primary energy consumption [7], the article focuses on district heating systems that have the potential to help reduce CO2 emissions by addressing solutions that lead to zero carbon emissions by 2050.

Heating systems provide from a central energy center heat and hot water consumption (ACC) to all consumers using the primary thermal network to the thermal point and the secondary thermal network between the thermal point and the final consumers. The study carried out in this article addressed solutions that make the design method more efficient from an ecological and economic point of view.

The study was carried out for a new section designed to supply a newly built neighborhood in Timisoara. The studied section is fed from a thermal point adjacent to the built-up area integrated in the district heating system of Timisoara. The design of a district heating system is a vital step in its construction process so that it can be developed and operated at maximum levels of efficiency. On the other hand, there is the issue of design and construction of new sections or extensions of sections to

IOP Conf. Series: Earth and Environmental Science 1136 (2023) 012060 doi:10.1088/1755-1315/1136/1/012060

supply the new consumer, sections that are powered by the existing network. Concerns about the development of operating techniques in the field of district heating began in the 1980s, focusing on reaching the minimum acceptable temperature, based on heat demand, as a limit condition for optimization.

Over time, various challenges associated with the operational optimization of district heating networks have been addressed in various aspects [7-13]. Even if in recent years, considerable investments have been made to replace the existing pipes with pre-insulated pipes, the problem of reducing the costs of transporting the heating agent in the case of district heating systems in Romania and especially in Timisoara is not completely solved [11,14].

For these reasons, the article conducts a comparative study that aims to determine the water supply temperature and the required flow in the district heating network in relation to the diameter of the pipes, the necessary pumping energy and the thermal energy lost so that consumers are supplied with the required parameters and the cost between the investment value, that of operation and maintenance, to be as low as possible [14].

Considering the fact that some costs such as those represented by the loss of thermal energy on the transmission pipes and the consumption of electricity for pumping are inevitable, the efficiency of the system should be achieved by reducing them to a minimum value, through a design efficiency [14].

In most cases, choosing the right diameter for the path of district heating networks involves a compromise. By choosing pipes with larger diameters, the thermal agent is circulated at a lower speed, which leads to lower pressure losses and therefore to a lower pumping energy. At the same time, the increase in the diameter of the pipe leads to a loss of energy for the transport of the larger thermal agent. In conclusion, the problem of designing district heating networks involves taking into account several aspects, such as reliability, thermohydraulic conditions, availability and cost of the material, energy consumption for pumping and the need for heat [15].

2. Heat supply solution for a new consumer. Case Study

2.1. Presentation of the district heating system in Timisoara

In the third-generation heating systems, such as the one in Timisoara, the primary thermal network has the role of transporting the high temperature thermal agent (90-160°C on the flow pipe and 60-110°C on the return pipe), from the power plants in which it is produced, to the zonal thermal points, in which the thermal energy is transferred to the secondary thermal network of lower temperature, using plate heat exchangers, which supply the final consumers of the locality. The district heating system in Timisoara also works on the same principle, which uses as solid energy source solid fuel with natural gas supply [11]. The functional diagram of the system is presented in figure 1.

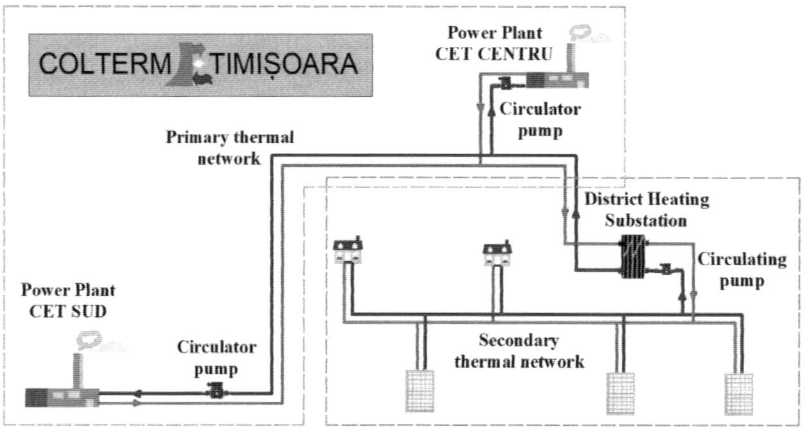

Figure 1. Functional diagram of the district heating network in Timisoara [16].

Considering the continuous development of the locality and the creation of new residential neighborhoods, the problem of finding efficient solutions both from a technical and economic point of view adapted to the current needs of the final consumers appears. The design phase is an essential one in which all the technical and investment aspects must be taken into account so that the extension solution has maximum financial feasibility [11].

2.2. *The proposed power supply solution for a new consumer*

The technical solution for supplying the newly built residential neighborhood, "City of Mara", located in Timisoara, involves supplying the existing thermal point PT 34, by creating a dedicated route of pre-insulated pipes, with a length of 300 meters, through which it is transported thermal agent having a flow temperature of 70°C and a return temperature of 40°C, as can be seen in figure 2 [16].

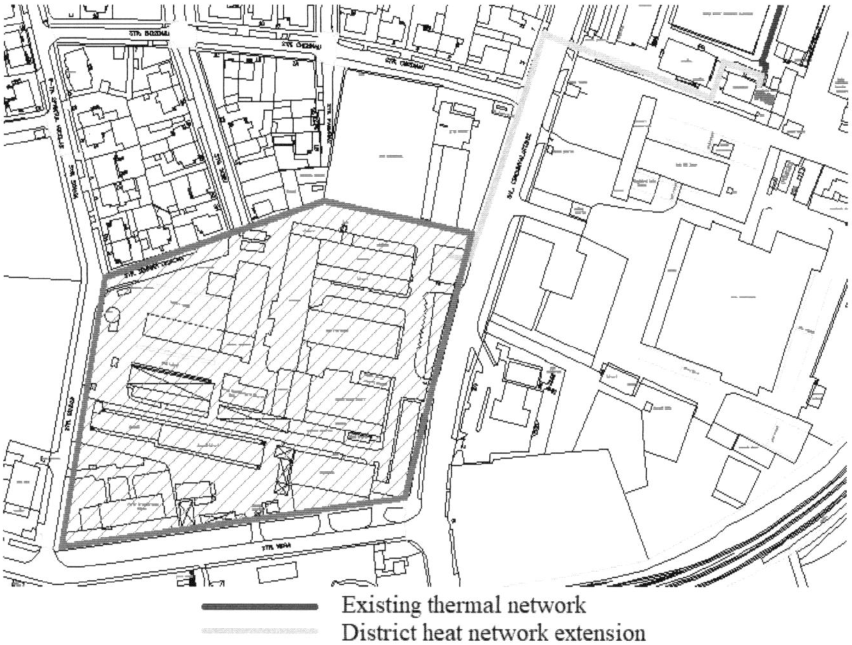

Figure 2. New consumer proposed for connection [16].

The heat exchange between the thermal agent delivered from the thermal point and the thermal agent circulated through the internal installation of the consumer is made with the help of a thermal module.

The simulation for power supply of the newly built residential neighborhood, "City of Mara", was performed using the Polysun calculation program [17] which is based on the analytical relation (1), according to the design regulations for heating installations [18,19].

$$Q_{dot} = m_{dot} \cdot c_p \cdot (t_1 - t_2) \tag{1}$$

where:

Q_{dot} – the amount of heat lost, in J;
m_{dot} – mass flow, in kg/s;
c_p – specific heat, in kJ/(kg·K);
t_1 – internal calculation temperature, in K;
t_2 – external ambient temperature, in K.

The schematic diagram of the consumer power supply installation in the proposed situation is presented in figure 3.

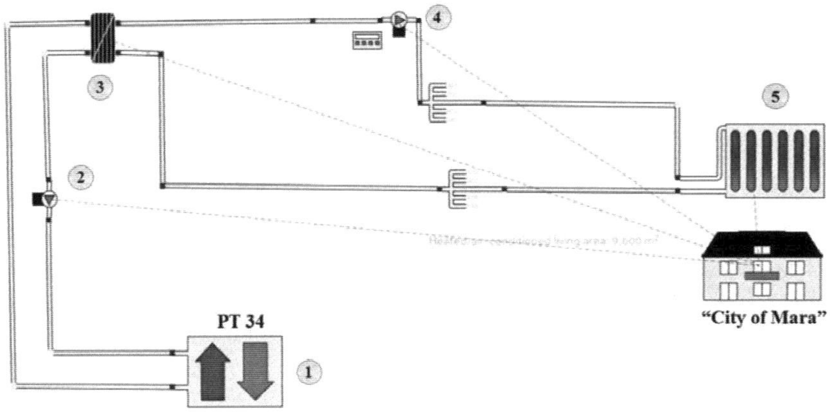

Figure 3. Schematic diagram for connecting the new consumer (1 – District Heating Substation, 2 – Circulating pump DH, 3 – Heat exchanger; 4 – Circulating pump end-user, 5 – End-user).

2.3. *Design optimization by pump height adjustment*

A first optimization possibility studied at the design level takes into account the reduction of the diameters of the heat transmission pipes, with the compromise of increasing the energy consumption required for pumping, or decreasing the pumping energy, increasing the diameter of the pipes, with the compromise of increasing the investment cost and consumption of thermal energy lost through transport.

The study addresses the design of the thermal network of thermal agent transport for the new consumer who wants to be supplied using in the first case a pressure loss considered economical (133 Pa/m), then case 2 was repeated for much higher-pressure losses (397 Pa/m) and in the last case the thermal network was dimensioned for low pressure losses (50 Pa/m).

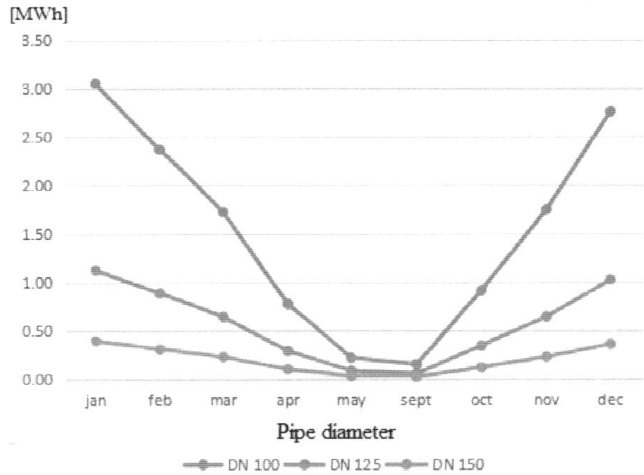

Figure 4. Variation of electricity consumption depending on pipe diameter.

The main simulation parameters for the first optimization variant studied are the flow temperature 70°C, the return temperature 40°C, the heat agent flow rate 62 m^3/h with the maximum consumption of the circulation pump of 7.5 kWh, at the diameter of the pipe DN 100 and the minimum consumption of 1 kWh, at the diameter of the pipe DN 150. Pumping energy was calculated based on the input data regarding the power consumption of the pump [20], and its operating time [17]. The variation of the electric consumption depending on the diameter chosen for the realization of the transport route of the thermal agent to the final consumer is presented in figure 4.

The reduction of the electric energy for pumping by increasing the diameter of the pipes is achieved with the compromise of the increase of the thermal energy loss on the route, calculated using Polysun software, according to EU regulation [17], a loss that increases with the increase of the diameter. Figure 5 shows the variation of heat loss for the 3 diameters of pipes studied.

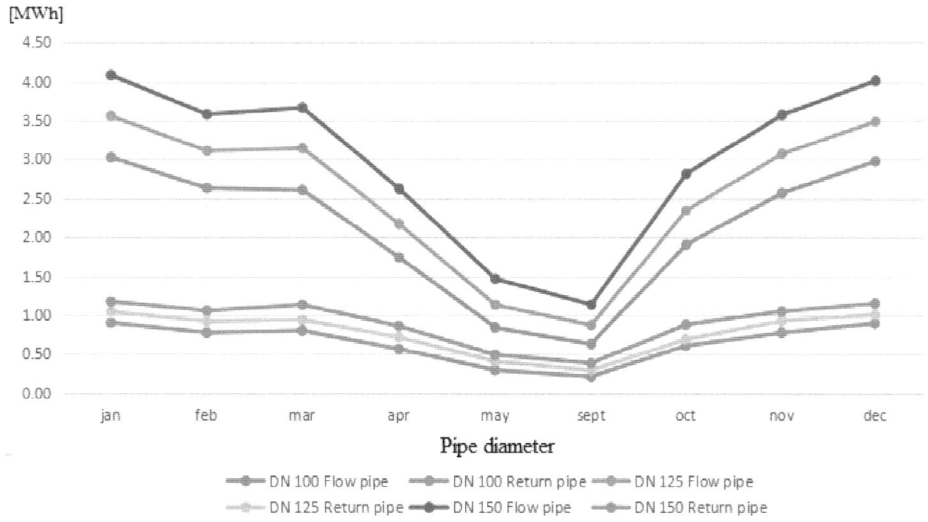

Figure 5. Heat loss variation depending on pipe diameter.

2.4. Design optimization by thermal agent flow temperature adjustment

In the context of the requirements on quality in construction, regarding the sustainable use of natural resources, energy saving and thermal insulation, it is necessary that the temperatures of the transport thermal agent have a decreasing trend so that the district heating system allows the integration of renewable energy sources [11].

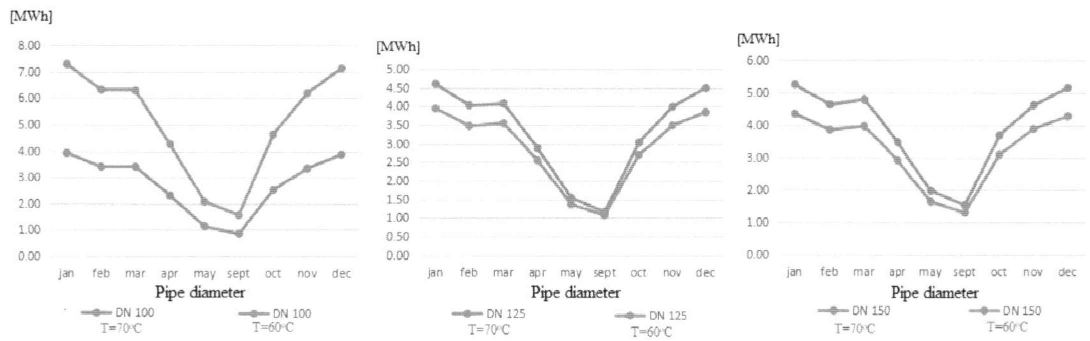

Figure 6. Variation of heat loss depending on pipe diameter and supply temperature.

Considering this aspect, a second possibility of optimizing the studied design implies the decrease of the supply temperature of the thermal agent at the same time with the increase of the flow necessary for the consumption points so that their energy demand is not affected. For this simulation, the supply temperature was reduced from 70°C to 60°C, a change that led to an increase in the required supply flow from 62 m^3/h to 75 m^3/h and to a decrease in the temperature difference between the thermal agent on the pipe flow and the one on the return pipe from 30°C to 20°C. The reduction of the energy loss on the transport pipes achieved by reducing the supply temperature of the thermal agent, for the three studied diameters are presented in figure 6.

The variation of the electric consumption according to the diameter chosen for the realization of the transport route of the thermal agent to the final consumer is presented in figure 7.

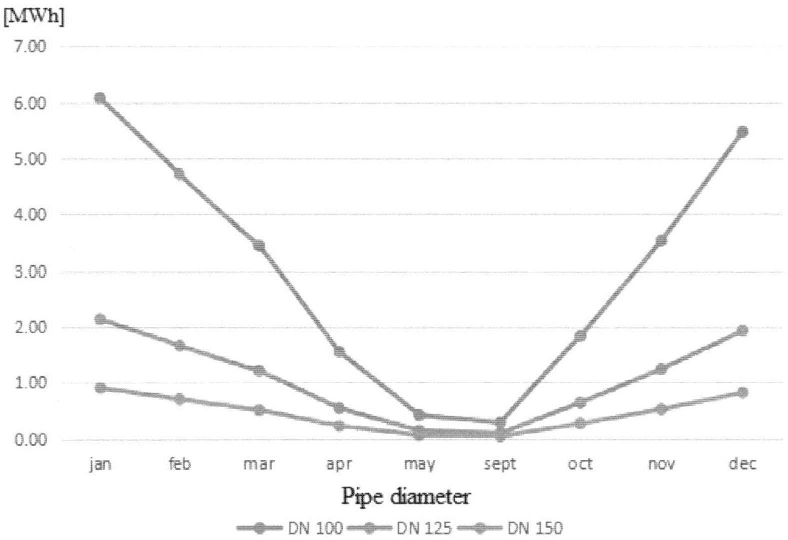

Figure 7. Variation of electricity consumption depending on pipe diameter.

2.5. Thermal energy production through high efficiency cogeneration (CHP)

The use of thermal and electrical energy production through high-efficiency cogeneration has both legislative advantages that encourage the implementation of this solution, as well as technical, economic and pollution reduction advantages for companies that produce these types of energy [19].

The considerable economic advantages of CHP are represented first of all by the reference price of electricity established in the Official Monitor of Romania, Part I, No. 983/14.10.2021, of 308.82 lei/MWh, value to which is added a bonus for a period of 10 years in the amount of 401.79 lei/MWh, excluding VAT, established by the same document. Thus, in the end, the average selling price of electricity produced by CHP is 710.61 lei/MWh, excluding VAT.

Given the current context of climate change that requires reduced CO_2 emissions, the high-efficiency cogeneration system that uses natural gas as fuel is much more environmentally friendly compared to the production of electricity using other conventional fuels, as can be seen in table 1.

Table 1. Specific CO_2 emission values for the primary energy sources used [21].

Primary source of energy	Specific emissions of CO_2 [g/kWh]
Coal	853.76
Natural gas	388.78
Fuel oil	728.84
Other conventional sources	498.07
Renewable sources	0
Sector average	213.37

Another consumption that cannot be neglected in the case of the district heating system is the electricity consumed for the transport of the thermal agent from the power plants where it is produced to the final consumers. The great advantage of cogeneration is that the electricity produced in parallel with the thermal energy covers much of the energy needed for pumping and the surplus is delivered to the national electricity system. The schematic diagram for supplying the consumer with thermal energy produced by high efficiency cogeneration is presented in figure 8.

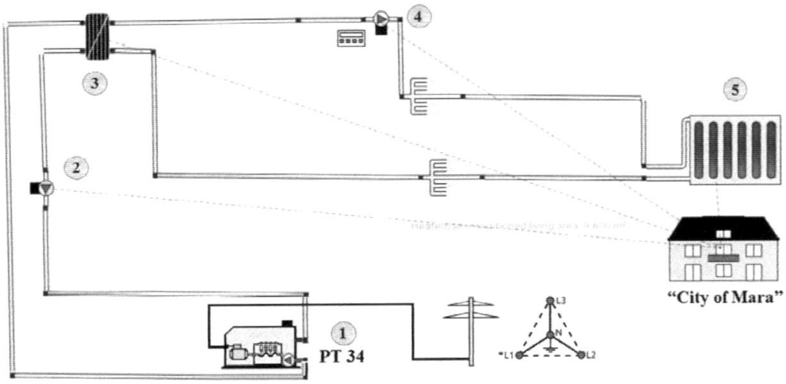

Figure 8. Schematic diagram for supplying the consumer with thermal energy produced by high efficiency cogeneration (1 – District Heating Substation, 2 – Circulating pump DH, 3 – Heat exchanger 4– Circulating pump end-user, 5 – End-user).

3. Results and discussions

For the analysis of the optimization of the design of the district heating system in Timisoara, 3 variants were simulated. The economic calculations were made for the context of the situation in Romania, Timisoara locality, according to the current prices for thermal energy, electricity and the necessary materials, presented in table 2. For the first studied variant of optimizing the design of the district heating network, which proposes reducing the diameters of heat transfer pipes with the compromise of increasing the energy consumption required for pumping or decreasing the pumping energy by increasing the diameter of the pipes with the compromise of increasing investment cost and thermal energy consumption lost during transport, the results highlight the importance of taking into account not only the initial investment cost but also the operating costs for the following years. As can be seen in table 3, if we take into account the cost of investment to supply the new consumer, the costs of electricity for pumping and heat loss along the route, it is found that the best design option involves choosing a smaller diameter (DN 100) of the pipe that reduces the value of thermal energy losses but increases the value of energy required for pumping.

Table 2. Energy and investment costs.

Type of cost	Unit	Price EUR
Electricity price	[MWh]	232.81
Thermal energy price	[MWh]	200
CHP thermal energy price	[MWh]	127.05
Investment cost DN 100 pipe	[m]	146.2
Investment cost DN 125 pipe	[m]	176.8
Investment cost DN 150 pipe	[m]	205.36

Table 3. Economic comparison of costs at flow temperature of 70°C for the first year of operation.

Type of cost/ Pipe diameter	Currency	DN 100	DN 125	DN 150
Investment cost	EUR	87,720.00	106,080.00	123,216.00
Pumping electricity	EUR	3,186.64	1,176.29	407.67
Heat loss	EUR	4,323.62	5,197.59	6,096.90
Total costs year 1	EUR	95,230.26	112,453.87	129,720.56

Carrying out these calculations for a period of 20 years, at the same parameters of the thermal installation, the results change, so that the most economically efficient variant becomes the one

IOP Conf. Series: Earth and Environmental Science 1136 (2023) 012060 doi:10.1088/1755-1315/1136/1/012060

through which a higher diameter is chosen (DN 125). Thus, it can be stated that in the short term the smaller diameter (DN 100) can be considered suitable from a technical-economic point of view, due to lower investment costs and heat loss, but in the long term the most economical design option is using the superior diameter of the pipe (DN 125), which can be seen in table 4.

In the next optimization variant, simulations were performed, taking into account, in addition to the variation of pressure losses depending on the diameter of the pipe, a thermal adjustment, as an influence on heat losses on the pipeline route, reducing temperature supply from 70°C to 60°C. In table 5 it can be seen that for this temperature, out of the 3 proposed diameters of the pipes, the most economical option is the one for which the cost of electricity is reduced, despite the increase of the investment cost and the loss of thermal energy on the route.

Table 4. Economic comparison of costs at a flow temperature of 70°C for the 20th year of operation.

Type of cost/ Pipe diameter	Currency	DN 100	DN 125	DN 150
Investment cost	EUR	87,720.00	106,080.00	123,216.00
Pumping electricity	EUR	3,186.64	1,176.29	407.67
Heat loss	EUR	4,323.62	5,197.59	6,096.90
Total costs year 1	EUR	95,230.26	112,453.87	129,720.56
Total costs year 20	EUR	268,548.04	252,132.96	268,417.74

Table 5. Economic comparison of costs at a flow temperature of 60°C for the 20th year of operation.

Type of cost/ Pipe diameter	Currency	DN 100	DN 125	DN 150
Investment cost	EUR	87,720.00	106,080.00	123,216.00
Pumping electricity	EUR	6,366.98	2,239.88	953.14
Heat loss	EUR	3,632.93	4,369.48	5,078.62
Total costs year 1	EUR	97,719.91	112,689.36	129,247.76
Total costs year 20	EUR	338,796.55	262,478.05	260,653.40

In the last simulated variant, the efficiency is proposed both in terms of reducing the costs of electricity and heat on the supply route with thermal agent and in terms of energy production. The introduction of high-efficiency cogeneration has the great advantage of producing both the thermal energy delivered to the final consumer and the electricity needed to pump the thermal agent. This eliminates from the operating costs, the cost for the consumed electricity, this being realized from own production. In table 6 and table 7 it can be seen that for both heat supply temperatures studied, the most economical solution is the one in which a small diameter is used for the pipes, the costs being influenced by the initial investment and the losses of thermal energy through the transport pipes.

Table 6. Economic comparison of costs at a flow temperature of 70°C.

Type of cost/ Pipe diameter	Currency	DN 100	DN 125	DN 150
Investment cost	EUR	87,720.00	106,080.00	123,216.00
Heat loss	EUR	2,570.54	3,182.34	3,715.07
Total costs year 1	EUR	90,290.54	109,262.34	126,931.07
Total costs year 20	EUR	144,320.74	180,047.15	209,135.00

Table 7. Economic comparison of costs at a flow temperature of 60°C.

Type of cost/ Pipe diameter	Currency	DN 100	DN 125	DN 150
Investment cost	EUR	87,720.00	106,080.00	123,216.00
Heat loss	EUR	2,239.78	2,696.38	3,175.66
Total costs year 1	EUR	89,959.78	108,776.38	126,391.66
Total costs year 20	EUR	137,037.65	168,813.75	196,702.49

4. Conclusions

After simulating the 3 variants of optimizing the design of district heating networks and analyzing the results obtained, it was concluded that in the context of the current situation and prices in Romania for the supply temperature of 70°C, making a cost calculation for a period of 20 years, the most economically efficient diameter is the one that introduces an average pressure drop of 133 Pa/m, namely DN 125. Even if in the first year the lower costs are related to the smaller diameter, which has lower losses of heat, in the long run, the costs necessary to pump the thermal agent, prove to be too high for the DN 100 pipe, this diameter not being the best choice.

Considering the need to reduce the supply temperature of consumers, in order to make it possible to introduce renewable energy sources, simulating the temperature drop on the flow pipe from 70°C to 60°C, for the 3 diameters studied, it turns out that although in the first year DN 100 is more efficient, due to the high costs required for pumping energy, after 20 years it has been proven that, from an economic point of view, DN 150 is more efficient.

The introduction of high-efficiency cogeneration ultimately proves to be the most cost-effective solution, due both to the fact that the pumping energy is provided by its own production and to the fact that the surplus of electricity brings a considerable profit. Of the 3 diameters studied (DN 100, DN 125 and DN 150), for the 2 supply temperatures (70°C and 60°C), it was proved according to the results, that the optimal design is the one by which the diameter of the DN 100 pipe is added, at the temperature of supply to the final consumer of 60°C.

References

[1] Fouladvand J, Ghorbani A, Mouter N and Herber P 2022 Analysing community-based initiatives for heating and cooling: A systematic and critical review *Energy Research & Social Science* **88** https://doi.org/10.1016/j.erss.2022.102507

[2] Lise W and van der Laan J 2015 Investment needs for climate change adaptation measures of electricity power plants in the EU *Energy Sustain. Dev.* **28** 10–20 https://doi.org/10.1016/j.esd.2015.06.003

[3] Van den Broek M, Veenendaal P, Koutstaal P, Turkenburg W and Faaij A 2011 Impact of international climate policies on CO_2 capture and storage deployment. Illustrated in the Dutch energy system *Energy Policy* **39** 2000–19 https://doi.org/10.1016/j.enpol.2011.01.036

[4] Van Der Schoor T and Scholtens B 2015 Power to the people: local community initiatives and the transition to sustainable energy *Renew. Sust. Energ. Rev.* **43** 666–675 https://doi.org/10.1016/j.rser.2014.10.089

[5] Koirala BP, Koliou E, Friege J, Hakvoort RA and Herder PM 2016 Energetic communities for community energy: a review of key issues and trends shaping integrated community energy systems *Renewable and Sustainable Energy Reviews* **56** 722-744 https://doi.org/10.1016/j.rser.2015.11.080

[6] Fouladvand J, Ghorbani A, Mouter N and Herber P 2022 Analysing community-based initiatives for heating and cooling: A systematic and critical review *Energy Research & Social Science* **88** https://doi.org/10.1016/j.erss.2022.102507

[7] van der Zwana S and Pothofab I 2020 Operational optimization of district heating systems with temperature limited sources *Energy and Buildings* **226**(1) 1–21 https://doi.org/10.1016/j.enbuild.2020.110347

[8] Benonysson A, Bøhm B and Ravn H.F 1995 Operational optimization in a district heating system *Energy Conversion and Management* **36**(5) 297–314 https://doi.org/10.1016/0196-8904(95)98895-T

[9] Madsen H, Sejling K, Søgaard HT and Palsson OP 1994 On flow and supply temperature control in district heating systems *Heat Recovery Systems and CHP* **14**(6) 613–620 https://doi.org/10.1016/0890-4332(94)90031-0

[10] Dahl M, Brun A and Andresen GB 2017 Using ensemble weather predictions in district heating operation and load forecasting *Applied Energy* **193** 455–465

https://doi.org/10.1016/j.apenergy.2017.02.066
[11] Muntean D, Mîrza M and Tokar A 2021 Soluții de reducere a pierderilor de căldură în sistemele de termoficare prin zonarea temperaturii de livrare *Ingineria Instalatiilor* **1** 27–31
[12] Muntean D, Mîrza M, Tokar A and Tokar D 2022 Simulations regarding the modernization of the district heating system in Timisoara by controlling the supply and regulating the temperature of the heating agent *Magazine of Hydraulics, Pneumatics, Tribology, Ecology, Sensorics, Mechatronics* **1** 76–84
[13] Muntean D, Mîrza M, Tokar A and Tokar D 2022 Modernization of the heat supply system of the University of Agricultural Sciences and Veterinary Medicine of Banat, from Timisoara *7th World Multidisciplinary Civil Engineering-Architecture-Urban Planning Symposium WMCAUS 2022*, Prague, Czech Republic, 5-9 September
[14] Piroutia M, Bagdanaviciusb A, Wuc J and Ekanayaked J 2012 Optimisation of supply temperature and mass flow rate for a district heating network *Proc. ECOS 2021* pp 26–29
[15] Vesterlund M and Toffolo A 2017 Design optimization of a district heating network expansion, a case study for the town of Kiruna *Appl. Sci.* **7**(5) 488 https://doi.org/10.3390/app7050488
[16] COLTERM S.A. District Heating Company of Timișoara (private communication)
[17] Vela Solaris AG *POLYSUN Software* https://www.velasolaris.com/software/?lang=en
[18] SR 1907-1:2014 *Heating Plants. Design Heat Requirements Computation for Buildings. Computation Method* (Romanian standard)
[19] SR 1907-2:2014 *Heating Plants. Design Heat Requirements Computation for Buildings. Design Conventional Indoor Temperatures* (Romanian standard)
[20] Wilo-Select 4 Online (accessed in 06.04.2022) https://www.wilo-select.com/Region.aspx
[21] Romanian Energy Regulatory Authority *Annual Report ANRE 2020* (accessed in 06.04.2022) https://www.anre.ro/ro/despre-anre/rapoarte-anuale

Carbon footprint of Vidraru hydropower development

Liana-Ioana Vuta[1], Gabriela-Elena Dumitran[1], Eliza-Isabela Tica[1] and Bogdan Popa[1]

[1] Department of Hydraulics, Hydraulic Machinery and Environmental Engineering, University Politehnica of Bucharest, Bucharest, 060041, Romania

eliza.tica@upb.ro

Abstract. Large hydropower developments are not considered to be environmentally friendly due to large landscape flooded by their reservoirs. There is a certain carbon footprint related to this, in the phase of construction and of the operation of the reservoirs. This paper presents GHG emissions related to an iconic reservoir from Romania, Vidraru, and demonstrate that values are close to those determined around the world and that comparing with other conventional power plants producing electricity is the best environmental option. Furthermore, must be considered the other water uses related to reservoirs as flood mitigation, water supply for population, irrigation, and industry, where reservoirs cannot be replaced by anything else.

1. Introduction

Waterpower is the most important and the oldest renewable source of energy. Compared with other renewable energy sources (RES) such as wind and solar, large hydropower developments (HPD) impact the environment during the construction and operational phases, mainly due to their most important component, the reservoir.

A lifecycle assessment (LCA) related to HPDs must consider three main sources of greenhouse gas (GHG) emissions: construction of the dam and the other components of the development, decaying biomass from the land covered by the water, and the thermal backup power. The last one is part of the assessment for run-of-river (ROR) hydropower plants (HPPs) as this type of scheme has no possibilities for flow regulation and produces electricity only using the available flows. Main GHG considered for research were generally: carbon dioxide (CO_2), methane (CH_4), nitrous oxide (N_2O), and carbon tetrafluoride (CF_4) [1]. The main results related to the state of research in 1996 were that for hydropower can be considered a mean value for the emission factors of 20 gCO_2e/kWh, for fossil fuel generation is 720 gCO_2e/kWh and that hydropower generation in that year (1996) decreased annual worldwide emissions of GHG from energy generation by about 8% [1].

Related to small hydropower plants (SHPPs), in 2012 was published a paper presenting correlations for GHG emissions for three different types of SHPPs schemes: ROR, canal-based and dam-toe, in India [2]. The authors analyzed 145 SHPPs, and they concluded that GHG emissions depend on the type of scheme, head, and installed capacity, determining correlations among these characteristics. Their findings show GHG emissions ranging from 10 to 70 gCO_2e/kWh. Including large hydropower, the authors indicate for GHG emissions 3.7 to 237 gCO_2e/kWh, comparing with other RES: wind 9.7–123.7 gCO_2e/kWh, Solar PV 9.4–300 gCO_2e/kWh, Biomass 35–178 gCO_2e/kWh, and with conventional power plants: coal fired 870–975.3 gCO_2e/kWh, oil fired 742.1 gCO_2e/kWh, gas fired 607.6 gCO_2e/kWh, nuclear 24.2 gCO_2e/kWh.

In 2014, Li and Zhang [3] reviewed all the findings related to GHG emissions from hydropower reservoirs at the global level by considering reservoir surface area, drawdown zone, and reservoir downstream. They estimated around 301.3 $MtCO_2$/year and 18.7 $MtCH_4$/year from global hydropower reservoirs, drawing attention that drawdown and downstream emissions counted together represent 42% CO_2 and 67% CH_4 of the total emissions. Related to energy generation from HPPs, the global average emissions were estimated at 92 gCO_2e/kWh and 5.7 gCH_4/kWh, representing a global reduction concerning fossil fuel power plants of about 2,351 $MtCO_2eq$/year.

From the global to regional level, some studies are devoted to the presentation of GHG emissions at river basin or country level. A study related to Ecuador [4] presents the analysis of only two hydropower schemes, one ROR and one reservoir type. For the HPD with reservoir GHG emissions were estimated at 547 gCO_2e/kWh, of which the largest part is due to the reservoir, and for ROR only 2.6 gCO_2e/kWh. The authors mentioned that social aspects were not considered and that this is a very important aspect of every analysis.

In Brazil, 8 representative HPDs having distinct technical features were the subject of a two-year field campaign to estimate GHG emissions and to compare them with those from thermal power plants producing electricity [5]. The authors consider their results as an argument that emissions of GHG from hydropower reservoirs are very variable and are site-dependent; results obtained for one reservoir cannot be extrapolated to another one, and all net emissions values per unit of electricity generation are lower than those from an equivalent thermal power using gas or coal, ranging from 2 to 60 gCO_2e/kWh, only one HPD being far from this range with 690 gCO_2e/kWh.

Kumar started in 2015 to study GHG emissions of hydropower reservoirs with the case of a large reservoir from India, 4 bm^3 and 52 km^2 surface area concluding that it can be considered down to medium risks associated with HPDs [6]. Then, for another large reservoir from India, 0.88 bm^3 and 29 km^2 surface area they estimated GHG emissions at 13.87 gCO_2e/kWh [7] and continued with a state-of-the art for all India [8]. The review reveals that most Indian hydropower reservoirs are coming under the low to medium GHG risk of reservoirs, that SHPPs are good alternative for the sustainable management of hydropower projects even if there are still many issues associated with them which cannot be neglected.

There are studies related to huge river basins which are under extensive hydropower development, as that of Mekong River [9], for which GHG emissions estimates were done for the 141 reservoirs in operation or in project phase. Findings show that the emission range of hydropower is 0.2–1994 gCO_2e/kWh over a 100-year lifetime, with a median of 26 (1–114) gCO_2e/kWh; 82% of hydropower reservoirs (119) and 45% of reservoirs (22) also having irrigation as water use have emissions comparable to other RES, meaning lower than 190 gCO_2e/kWh. The other reservoirs have emissions at the scale of those from fossil fuel power plants, meaning greater than 380 gCO_2e/kWh. Among their conclusions most important ones are that each hydropower project must be analyzed related to GHG emissions, that a minimization of these emissions can be done by removing vegetation and other easily degradable organic matter from the landscape flooded by the reservoir and that related to huge rivers hydropower cannot be considered as low GHG emissions technology.

Some authors consider that because the water-carbon nexus of a HPP and vegetation are similar; it is interesting to compare the water-carbon nexus of hydropower with that of biomass power [10]. Thus, they developed a new method to solve the problem of water allocation at the scale of a river basin for maximizing the GHG emissions reduction. Their results show that, for the first three years of utilization, the GHG reduction efficiency is bigger for the biomass power plant then becoming lower than that of the HPP.

Related to the place that GHG emissions must have when choosing the type of the hydropower development, ROR or reservoir type, some authors concluded that this criterion is not strong enough to discourage the construction of new reservoir type HPDs [11]. They recommend being developed as many large-scale hydropower projects as possible to capture all possible hydropower potential in the world and, for all new HPDs, detailed studies to determine the need of a reservoir, case by case, as the impact of flooding, melting snow and ice vary dramatically for each location.

An excellent paper considered when conducting this research related to 220 MW installed capacity Vidraru HPD, presented results related to GHG emissions of four HPDs from China (95, 160, 201 and 500 MW installed capacities), assessed using GHG Reservoir Tool (G-res) [12]. For the first 10 years GHG intensity varied from 45 to 75 gCO_2e/kWh, after 20 years from 45 to 75 gCO_2e/kWh, halving after 20 years, and becoming 7 to 12 gCO_2e/kWh, larger values corresponding to larger installed capacities. The authors indicate as average for GHG emissions for the four analyzed HPDs 13.6 gCO_2e/kWh for lifetime of 50 years and 8.13 gCO_2e/kWh for 100 years. They mentioned that previous studies for other five large HPDs from China show 11.11 gCO_2e/kWh for lifetime of 44 years and an average of 7.25 gCO_2e/kWh for 100 years, suggesting that the increase in installed capacity from 50 MW to 16,000 MW does not lead to major increase of efficiency related to GHG emissions. The authors mentioned for comparison other studies as for 24 reservoir type SHPPs in India having the average GHG emissions of 21.05 gCO_2e/kWh; or the case of Itaipu HPD in Brazil, with an installed capacity of 14,000 MW and carbon intensity of 4.33 gCO_2e/kWh.

The second largest Romanian reservoir, Stanca-Costesti, was subject of analysis related to hydrological variability impact on eutrophication [13] and even of the evaluation of GHG emissions using 7 models from literature, among which G-res [14]. The results show GHG emissions varying from 567.34 gCO_2e/kWh to 19.06 gCO_2e/kWh using G-res.

In this paper, we used G-res tool to evaluate GHG emissions related to Vidraru reservoir, and we compare our results with those from other studies.

2. Materials and Methods

2.1. The GHG Reservoir Tool (G-res)

Usually, data from direct measurements for GHG emissions for reservoirs are not available and estimation are done comparing with similar systems where results from measurements are available [12]. This approach seems not to be appropriate as GHG emissions measurements are done where it is supposed that there are potential high emissions; because it does not consider specific environmental conditions of the target reservoir; because it ignore the drop in GHG emissions after impoundment [15]; because it ignores the GHG emissions from aquatic systems and from the covered landscape and attributes all to the reservoir. These are some of the reasons which make the assessment of GHC footprint of reservoirs a very complex task.

GHG emissions (footprint) of a reservoir are the difference between the landscape and the atmosphere before and after the realization of the reservoir. To estimate GHG emissions associated with the construction phase, the consumption of materials, plants, and transport associated with the construction of the infrastructure must be known [16]. In some cases, this information is known in detail; however, in other cases, such as early design, this information is not known [17]. The calculation of GHG emissions is based on: GHG Emissions = amount of consumption or activity × emission rate per unit of activity [12].

Consumption is specified by the user as volume or mass depending on the material. Transport of materials during construction is specified in terms of t km (tons-kilometers), combining the amount of material to be transported over a given distance. The user can use activities defined in G-res and add additional activities using the 'Own assessment' option [18,19,20].

Emissions related to transport of materials and construction of hydrotechnical works are considered based on input values, for example [20]:

$$\begin{aligned} GHG = {} & concrete\ (\mathrm{m^3}) \times concrete\ material\ EF\ (\mathrm{kgCO_2/m^3}) \\ & + mass\ (\mathrm{t}) \times distance\ (\mathrm{km}) \times transport\ EF\ (\mathrm{kgCO_2/t\ km}) \\ & + concrete\ (\mathrm{m^3}) \times plant\ EF\ (\mathrm{kgCO_2/m^3}). \end{aligned} \tag{1}$$

Equation (1) is applied for all the components of a hydrotechnical development, and the total footprint is the sum of the results.

When applying, for example, G-res tool related to the realization of a reservoir, the components of the carbon footprint depending on different moments in time are [20]:

Net carbon footprint =
– [Pre-impoundment carbon balance of the reservoir area before reservoir creation]
+ [Post-impoundment carbon balance of the reservoir] (2)
– [Emissions from the reservoir due to unrelated anthropogenic sources (UAS)]
+ [GHG due to construction]

Where the signification of terms is:

- pre-impoundment considers emissions of CH_4 and CO_2 on the land that will be covered with water by the reservoir;
- post-impoundment refers to GHG flux. it comprises CO_2 and CH_4 diffusive fluxes as well as bubbling and degassing for CH_4. The semi-empirical model of G-res use values established by observations on GHG flux at 223 reservoirs around the world [14,20]. Emissions are given as areal emissions ($gCO_2e/m^2/yr$) and as reservoir wide emissions (tCO_2e/yr) and can also be expressed just as CO_2 or CH_4. A global warming potential of 34, for 100 years (GWP100), was used to obtain CH_4 emissions and CO_2e [21,22];
- unrelated anthropogenic sources (UAS) represent the carbon emissions related to other anthropogenic sources which lead to sewage and influence the carbon cycle in the area. Those emission must be removed [23];
- the construction term covers all the activities and materials needed to complete the dam and associated structures. Each activity and type of material used have a corresponding emission factor [24].

2.2. *Vidraru hydropower development*

Arges River, long of 340 km, has its hydrographic basin situated in the central-southern part of the country and covers an area of 12,550 km^2. The spring area of the Arges river is represented by the Fagaras Mountains where the density of the hydrographic network is high, often exceeding 1.4 km/km^2. The average altitude in this mountain area is 1000-1200 MASL, so the average slope has high values (150–80$^0/_{00}$). The upper hydrographic basin of the Arges river stretches in the north-south direction from the strongly crenelated ridge of the Fagaras Mountains to the south of the Pitesti municipality and is composed of 62 cadastral hydrographic subbasins in which are present 25 natural lakes and 15 storage lakes, with a total volume at the normal retention level (NRL) of 781 Mm^3 and an area of 4095 ha, occupying 25.2% of the hydrographic basin of the Arges river [25].

The mountain hydrographic basins tributary to the Arges river have important water resources, due to their positioning at high altitudes, but also to the cantonment at their level of an important layer of snow for a longer period, of about 6 months. Due to the uneven leakage and distribution of water resources during a year and for their efficient management, numerous accumulations have been built in the mountain area that store a large volume of water redistributed during the dry periods of a year or used in the production of electricity. These reservoirs in the Arges river basin are on Arges, Cumpana, Valsan, Doamnei, and Târgului rivers.

The Vidraru storage is the largest in the Arges River catchment and has as main functions the production of electricity (installed capacity of 220 MW and energy generation in the hydrological average year of 400 GWh/year) and the flood mitigation (attenuation coefficient 39% / attenuation degree 61%, corresponding to a volume of 24.56 Mm^3). Also, regularizing the flows and compensating with the rest of the basin downstream of the Vidraru reservoir allows the realization of an additional flow of about 8 m^3/s for downstream water uses (localities, industries, irrigation) [26].

The multiannual average water resources in Vidraru reservoir hydrographical basin are evaluated at 614 Mm^3, and the theoretical resources are evaluated at 921 Mm^3. The hydrographical basin of

Vidraru reservoir has an area of 745 km^2 and the related mean flow is 19.7 m^3/s (figure 1). The location of the accumulation in a more open area behind a tight gorge determines the retention of a very large volume of water, which could not be ensured only by the hydrographical basin of the Arges River, with an area of 282 km^2 in the dam section; this is why numerous secondary intakes with a total length of 29 km were built on the rivers Valsan, Dobroneagu, Cernat, Izvorul Bradului, Draghina, Doamnei, Baciu, Cumpana, Topolog, Valea lui Stan and Limpedea; thus, the tributary flow of Vidraru reservoir increases by 34% [27].

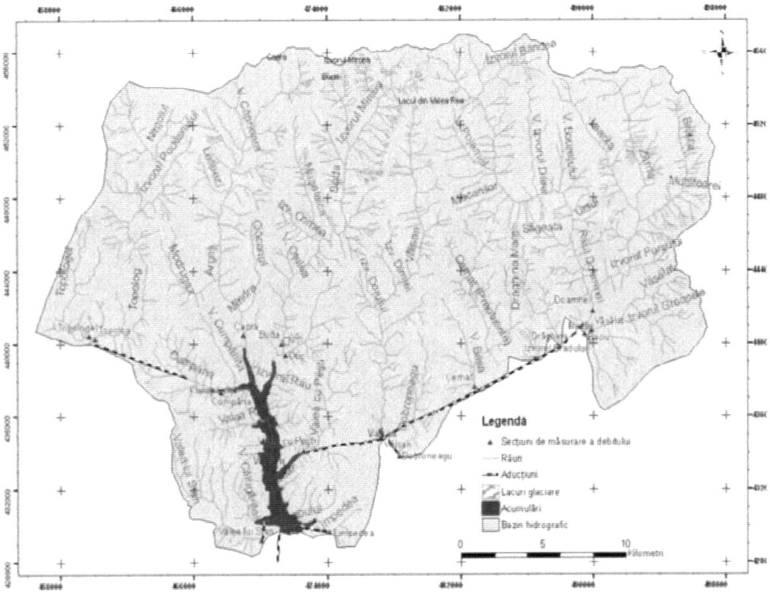

Figure 1. Catchment area related to Vidraru reservoir [3].

Vidraru development falls into category "A" of exceptional importance and consists of a large reservoir, a concrete dam in the double arch, a derivation, an underground power plant, and a tailrace gallery. The construction of the arrangement began in 1959, and at the date of entry into operation in 1965, the executed works totalized: drilling of underground galleries on 42 km, rock excavations of 1.768 Mm^3, of which about 1 Mm^3 underground, pouring of 0.93 Mm^3 of concrete, of which 0.4 Mm^3 underground and installation of 6,300 tons of electromechanical equipment [28]. The work cost 1.47 billion lei at that time, which is equivalent to about 964 MRON (~193 MEUR) and the investment was amortized in 28 years.

The dam is a double curvature concrete type, made of 22 vertical plots, with a concrete volume of 480,000 m^3, a height of 166.6 m, a length at the crown of 307 m and contains nine horizontal interior galleries. During the construction of the dam, the river was diverted through a 5.2 m diameter gallery, used later as bottom outlet. The reservoir created by damming Arges river upstream the gorge sector in the Ghitu Mountains, has a length of 14 km, a maximum water depth of 155 m, an area of 870 ha, and a volume of 465 Mm^3 at the normal retention level of 830 MASL (table 1). The diversion gallery, with a length of 2130 m and a diameter of 5.15 m, ensures the 90 m^3/s of the water for the operation of the hydropower plant and can withstand a pressure of 12-14 atm. The underground power plant is located 103 m below the level of the Arges riverbed.

Because in the initial project there was not foreseen the minimum flow required in the downstream riverbed, at present on the Vidraru – Oiesti storage sector, the easement flow is variable with a minimum value of about 100 l/s immediately downstream of the dam (due to the infiltrations from the dam and the losses related to the bottom overtures – 150 l/s at the NRL quota). This value is increased by the difference in the basin up to the Oiesti section.

Table 1. Characteristics of Vidraru reservoir.

Parameter	Value	Unit
Impoundment year	1965	-
Normal operating level	830	MASL
Maximum dam height	167	m
Top length of the dam	307	m
Catchment area	745	km^2
Total reservoir volume	496.93	Mm^3
Regulated storage capacity	462.20	Mm^3
Net head	280	m
Flow rate	90	m^3/s
Installed capacity	220	MW
Total annual generation	400	GWh/year
Investments	193	MEUR
Longitude	45.22	DD
Latitude	24.37	DD
Catchment area	745	km^2

Nowadays, exist several studies investigating the carbon footprint of some hydropower stations around the world but very few for Romania. Thus, this study aims to calculate the carbon footprint of the Vidraru arrangement using as a working tool the G-res program.

2.3. *Input data for G-res*

Considering this purpose, as input data necessary for the calculation of the carbon footprint of the Vidraru development were the Regulations for the exploitation of the arrangement (Hidroelectrica), the water resources management reports – NARW, but also the meteorological data from the HPD area.

For the annual mean meteorological data of the Vidraru site, needed as inputs for the G-res tool, were used the public data available at www.meteoblue.com: data related to average monthly temperatures, average wind speed, and mean global horizontal radiance (figure 2).

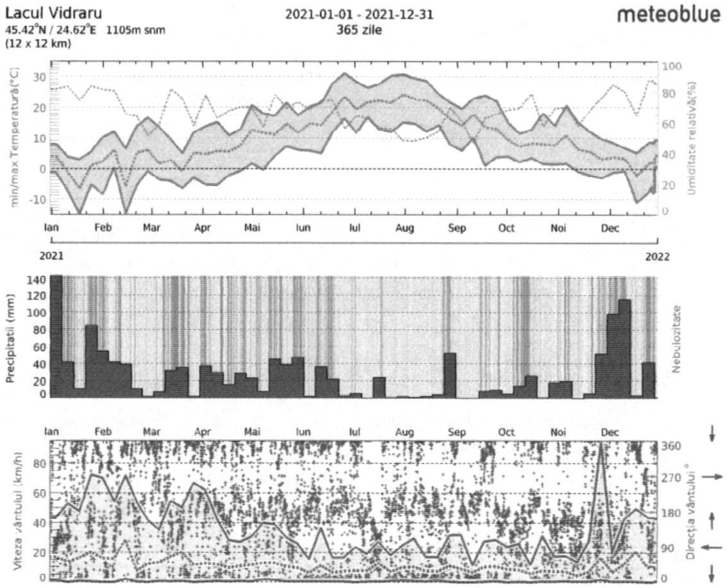

Figure 2. Annual mean meteorological data of the Vidraru reservoir site (www.meteoblue.com).

The main information about Vidraru reservoir used in this paper comes from the construction planning documents, which contain dam's characteristics, details regarding the materials and the quantities used, aspects related to different construction stages, as well as information regarding the transportation, all being necessary as input data for G-res tool. Figure 3 presents a data input window of G-res.

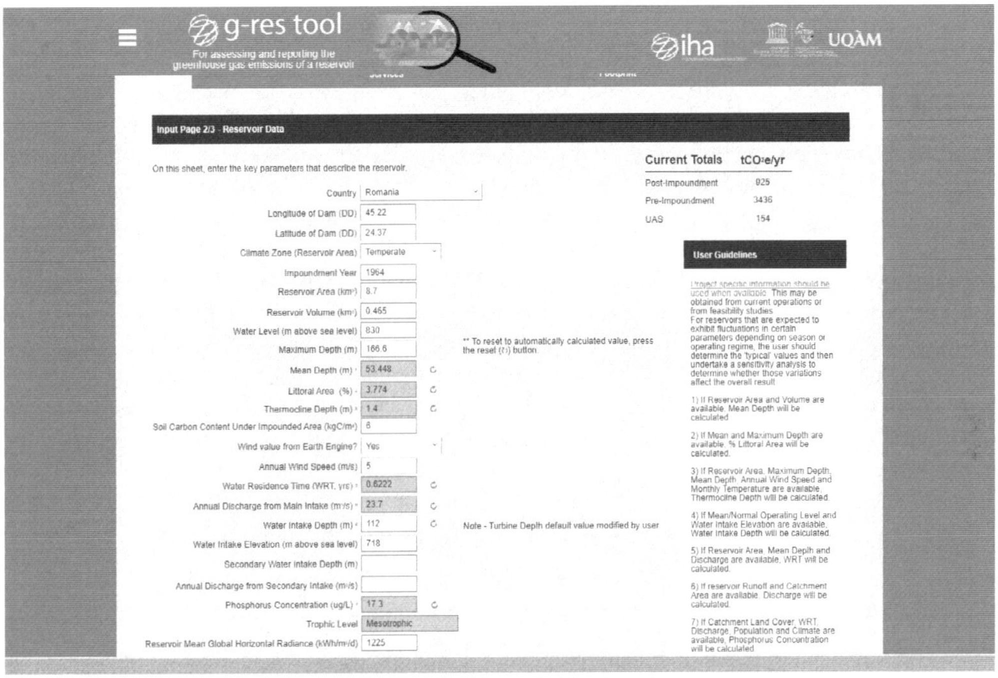

Figure 3. G-res: data input window.

3. Results and Discussion

3.1. Pre-Impoundment GHG emissions

Because the land cover or land use types have an impact on the emission factors, based on the available area catchment data, it was estimated the land cover types for the flooded area by the reservoir. Thus, in Vidraru catchment exist mainly non-waterlogged, mineral soils with few organic soils. Forest represents an important part in the Vidraru catchment (33%), grassland represents 30%, both playing an important role as carbon source. There are also important rocky bare areas (30%).

The GHG balance for the area flooded by Vidraru reservoir is:

- landscape CO_2 emission rate = 377 $gCO_2e/m^2/year$;
- landscape CH_4 emission rate = 10 $gCO_2e/m^2/year$.

3.2. Post-Impoundment GHG emissions

The impoundment of a reservoir changes the environment in terms of physical and chemical characteristics and favours the growth of biomass, leading to variation in carbon transformation processes. The balance between the carbon from the water and the atmosphere is assured by:

1. Diffusive fluxes at the water-air interface, as main emission source; in case of CH_4 is influenced by temperature and spatial parameters of the reservoir; the CO_2 diffusive flux is also controlled by the content in carbon of the surface soil;
2. CH_4 bubbling is a influenced by the surface horizontal radiance and spatial parameters of the reservoir;
3. Release of GHG downstream as a result of power plant operation (e.g. degassing).

In case of Vidraru reservoir, the values of CO_2 and CH_4 in the post-impoundment stage, are:

- CO_2 release rate = 188 CO_2 gCO2e/m^2/year
- CH_4 release rate = 36 gCH_4/m^2/year in which diffusive flux 82% and degassing 18%.

3.3. *Emissions associated with other anthropogenic activities*

All anthropogenic activities lead to some sewage which eventually reach the reservoir and perturb the chemical characteristics of water (e.g., higher values of phosphorus which favour the algal bloom), thus being necessary to consider it as part of the total emission of post-impoundment. Even if in the Arges-Vedea hydrographic basin the total population in 2018 was 3,984,686 inhabitants, with a density of 185 inhabitants/km^2 for the hydrographic basin related to Vidraru reservoir, the population density is much lower, given that most of it are unpopulated due to hard-to-reach mountainous areas.

At the same time, the urban agglomerations of Bucharest and Pitesti are in the Arges-Vedea catchment areas, which substantially change the values of the population density. Therefore, for the reception basin related to Vidraru reservoir, it was considered a population of 10,000 inhabitants with a density of about 13.4 inhabitants/km^2, with only 0.4% settlements. Besides, because the Vidraru hydrographic basin is placed in the mountain region, the croplands in all catchments are rainfed. Thus, the managed levels of all land type inputs for the G-res tool are set to low.

Table 2 presents the CH_4 emissions associated with other anthropogenic sources. Given the fact that the water residence times for Vidraru reservoir is less than one year, the CH_4 emissions associated with other anthropogenic sources is high and has an important share from the total CH_4 emission of post-impoundment.

Table 2. CH_4 values associated to other anthropogenic sources.

Water residence time (year)	0.8
CH_4 emissions (gCO_2e/m^2/year)	30
- due to land use (%)	2
- due to sewage (%)	98
UAS/Post-impoundment CH_4 emissions (%)	83

3.4. *Emissions from construction*

The G-res tools can estimate the GHG emissions due to construction at different stages. It considers the full construction cycle: from the production of the materials used until the finalization of the entire infrastructure. For each activity and material used, a specific GHG emission factor is assigned and by using equation (1) the GHG emissions are computed. This implies the necessity of knowing in detail all aspects related to the construction, including the quantity of materials and their transportation. Unfortunately, for Vidraru reservoir, this information is not fully known.

3.5. *Net GHG footprint*

Net GHG footprint of Vidraru reservoir is obtained by applying equation (2), the results being presented in table 3.

Table 3. Estimate emissions for Vidraru reservoir with G-res tool.

Pre-impoundment (tCO_2e/year)	–3362
Post-impoundment (tCO_2e/year)	1333
Unrelated Anthropogenic Source (tCO_2e/year)	–264
Net GHG footprint (tCO_2e/year)	–2293

3.6. *Discussion*

Based on the designed annual generation, the software shows that the Vidraru hydropower development has a total carbon footprint of 120,884 tCO_2e, with 336,249 tCO_2e before impoundment,

133,222 tCO_2e after impoundment, 26,348 tCO_2e Unrelated Anthropogenic Source, and 350,294 tCO_2e construction. The GHG intensity of a hydropower station relates the GHG emission per unit of electricity generated to the lifetime of a hydropower station (table 4).

Table 4. Parameters and carbon footprint for Vidraru HPD.

Installed capacity (MW)	220
Total carbon footprint (tCO_2e)	120,884
Total annual generation (GWh/year)	400
Designed lifetime (year)	100
Carbon intensity (tCO_2e/GWh)	3.02

GHG intensity over the lifespan for Vidraru HPD (from 10th to 100th year) reveals a fast decrease, from 30 to 7.5 gCO_2e/kWh, between 10th and 40th year, followed by a slow decrease, approaching 3.02 gCO_2e/kWh, when the lifespan is around 100 years (figure 4).

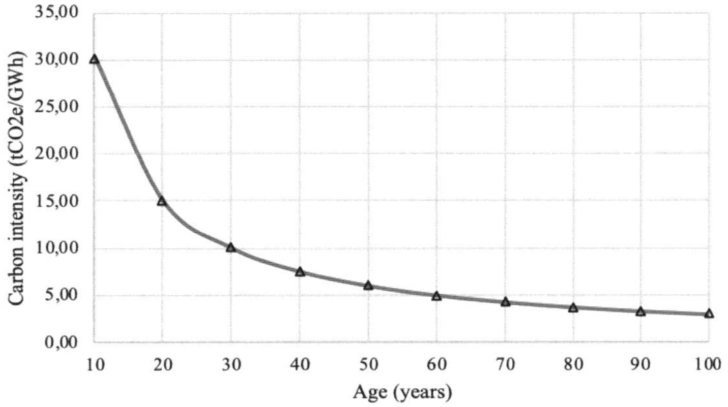

Figure 4. Variation of GHG intensity over the lifespan for Vidraru HPD.

4. Conclusions

This paper determined carbon footprint related to an iconic reservoir from Romania, Vidraru, using GHG Reservoir Tool (G-res). Obtained results, which show a GHG intensity exponentially decreasing from 30 gCO_2e/kWh for 10 years lifespan to 3.02 gCO_2e/kWh for 100 years lifespan, are in good agreement with those obtained for the second largest Romanian border reservoir, Stanca-Costesti, namely 19.06 gCO_2e/kWh for 100 years lifespan, also using G-res. Results are also in good agreement compared with the literature presenting results related to large reservoirs from abroad, where values of 26 gCO_2e/kWh (for HPDs of Mekong River), 8.13 gCO_2e/kWh (for four HPDs in China), or 7.25 gCO_2e/kWh (for other five large HPDs from China), or 4.33 gCO_2e/kWh (for Itaipu HPD), related to 100 years lifespan.

Comparing with other conventional power plants producing electricity hydropower is certainly the best environmental option. Furthermore, must be considered the other water uses related to reservoirs as flood mitigation, water supply for population, irrigation, and industry, where reservoirs cannot be replaced by anything else.

References

[1] Gagnon L and van de Vate JF 1997 Greenhouse gas emissions from hydropower *Energy Policy* **25**(1) pp 7–13 https://doi.org/10.1016/S0301-4215(96)00125-5
[2] Varun G, Prakash R and Bhat IK 2012 Life cycle greenhouse gas emissions estimation for small hydropower schemes in India *Energy* **44** pp 498–508 http://dx.doi.org/10.1016/j.energy.2012.05.052

[3] Li S and Zhang Q 2014 Carbon emission from global hydroelectric reservoirs revisited *Environ Sci Pollut Res* **21** pp 13636–13641 http://dx.doi.org/10.1007/s11356-014-3165-4
[4] Hidrovo AB, Uche J and Martinez-Garcia A 2017 Accounting for GHG net reservoir emissions of hydropower in Ecuador *Renewable Energy* **112** pp 209–221 http://dx.doi.org/10.1016/j.renene.2017.05.047
[5] Dos Santos MA et al 2017 Estimates of GHG emissions by hydroelectric reservoirs: The Brazilian case *Energy* **133** pp 99–107 http://dx.doi.org/10.1016/j.energy.2017.05.082
[6] Kumar A and Sharma MP 2016 Assessment of risk of GHG emissions from Tehri hydropower reservoir, India *Human and Ecological Risk Assessment: An International Journal* **22**(1) http://dx.doi.org/10.1080/10807039.2015.1055708
[7] Kumar A, Sharma MP and Yang T 2018 Estimation of carbon stock for greenhouse gas emissions from hydropower reservoirs *Stochastic Environmental Research and Risk Assessment* **32** pp 3183–3193 https://doi.org/10.1007/s00477-018-1608-z
[8] Kumar A, Yu ZG, Klemes JJ and Bokhari A 2021 A state-of-the-art review of greenhouse gas emissions from Indian hydropower reservoirs *Journal of Cleaner Production* **320** 128806 https://doi.org/10.1016/j.jclepro.2021.128806
[9] Rasanen TA, Varis O, Scherer L and Kummu M 2018 Greenhouse gas emissions of hydropower in the Mekong River Basin *Environ. Res. Lett.* **13** 034030 https://doi.org/10.1088/1748-9326/aaa817
[10] Zhang J, Xu L and Cai Y 2018 Water-carbon nexus of hydropower: The case of a large hydropower plant in Tibet, China *Ecological Indicators* **92** pp 107–112 http://dx.doi.org/10.1016/j.ecolind.2017.06.019
[11] Amanda HAH, Wijayapala WADS, Lucas JR and Fernando WJLS 2018 The economic comparison of reservoir type and run of the river type hydropower plants: A case study for Upper Kotmale Hydropower Plant *Engineer: Journal of the Institution of Engineers, Sri Lanka* **51**(4) pp 75–84 http://doi.org/10.4038/engineer.v51i4.7315
[12] Jiang T, Shen Z, Liu Y and Hou Y 2018 Carbon footprint assessment of four normal size hydropower stations in China *Sustainability* **10**(6) 1–14 http://doi.org/10.3390/su10062018
[13] Dumitran GE, Vuta LI, Popa B and Popa F 2020 Hydrological variability impact on eutrophication in a large Romanian border reservoir, Stanca–Costesti *Water* **12** 3065 http://doi.org/10.3390/w12113065
[14] Ion VI and Ene A 2021 Evaluation of greenhouse gas emissions from reservoirs: A review *Sustainability* **13** 11621 https://doi.org/10.3390/su132111621
[15] Teodoru CR et al 2012 The net carbon footprint of a newly created boreal hydroelectric reservoir *Global Biogeochem. Cycles* **26** GB2016 doi:10.1029/2011GB004187
[16] Pandey D, Agrawal M and Pandey JS 2011 Carbon footprint: Current methods of estimation *Environ. Monit. Assess.* **178** pp 135–160 https://doi.org/10.1007/s10661-010-1678-y
[17] Barros N, Cole JJ, Tranvik LJ, Prairie YT, Bastviken D, Huszar VLM, del Giorgio P and Roland F 2011 Carbon emission from hydroelectric reservoirs linked to reservoir age and latitude *Nat. Geosci.* **4**(9) pp 593–596 https://doi.org/10.1038/ngeo1211
[18] Fearnside PM 2004 Greenhouse gas emissions from hydroelectric dams: Controversies provide a springboard for rethinking a supposedly 'clean' energy source. An editorial comment. *Climatic Change* **66** pp 1–8 https://doi.org/10.1023/B:CLIM.0000043174.02841.23
[19] Prairie YT et al 2018 Greenhouse gas emissions from freshwater reservoirs: What does the atmosphere see? *Ecosystems* **21** pp 1058–1071 https://doi.org/10.1007/s10021-017-0198-9
[20] Prairie YT, Alm J, Harby A, Mercier-Blais S and Nahas R 2017 The GHG Reservoir Tool (G-res) Technical documentation. Updated version 3.0 (2021-10-27), UNESCO/IHA Research Project on the GHG status of freshwater reservoirs, Joint publication of the UNESCO Chair in Global Environmental Change and the International Hydropower Association
[21] Zhao Y, Wu BF and Zeng Y 2013 Spatial and temporal patterns of greenhouse gas emissions from Three Gorges Reservoir of China *Biogeosciences* **10**(2) pp 1219–1230

https://doi.org/10.5194/bg-10-1219-2013

[22] St. Louis VL, Kelly CA, Duchemin É, Rudd JWM and Rosenberg DM 2000 Reservoir surfaces as sources of greenhouse gases to the atmosphere: A global estimate *BioScience* **50**(9) pp 766-775 https://doi.org/10.1641/0006-3568(2000)050[0766:RSASOG]2.0.CO;2

[23] Kadiyala A, Kommalapati R and Huque Z 2016 Evaluation of the life cycle greenhouse gas emissions from hydroelectricity generation systems *Sustainability* **8**(6) 539 https://doi.org/10.3390/su8060539

[24] Gallagher J, Styles D, McNabola A and Williams AP 2015 Life cycle environmental balance and greenhouse gas mitigation potential of micro-hydropower energy recovery in the water industry *J. Clean Prod.* **99** pp 152–159 https://doi.org/10.1016/j.jclepro.2015.03.011

[25] Mailat E 2010 Water resources in the superior hydrographic basin of Argeş River *Proc. Conf. Water resources from Romania. Vulnerability to the pressure of man's activities* ed P Gâştescu and P Breţcan (11-13 June 2010, Targoviste, Romania) (in Romanian)

[26] Mititelu LA 2011 Water quality in the upper and middle hydrographic basin of the Arges River, *Lakes, reservoirs and ponds* **5**(1) pp 27–38 https://www.limnology.ro/Lakes/2011/201105102.pdf

[27] *** 2017 *The Plan of Flood Risk Management Argeş-Vedea Water Basin Administration* (Bucharest, Romania: National Administration "Apele Romane") (in Romanian)

[28] Marinescu AM, Vasilescu M, Ştefănescu M, Ion A and Dumitru M 2021 *The Project "2021 Updated Management Plan of the Hydrographic Space Argeş-Vedea"* (Bucharest, Romania: National Administration "Apele Romane") (in Romanian).

IOP Conf. Series: Earth and Environmental Science 1136 (2023) 012062 doi:10.1088/1755-1315/1136/1/012062

A way to develop a micro-grid using a Siemens S7-1500 PLC

Oana-Carmen Niculescu-Faida[1], Adrian Ion Minea[2] and Adrian Niculescu-Faida[3]

[1] Faculty of Hydrotechnics, Technical University of Civil Engineering Bucharest, 020396, Romania
[2] Department of Automation and Industrial Informatics, Faculty of Automatic Control and Computer Science, University POLITEHNICA of Bucharest, 060042, Romania
[3] Engine Department, Renault Technologie Roumanie, Bucharest North Gate, Voluntari, 077190, Romania

oanacarmen2000@yahoo.com

Abstract. The main objective of the paper is to present a way to develop a micro grid that will help minimize electricity costs. A controller can monitor all subsystems of the micro grid, thus helping the operator to manage the key performance indicators for generation, storage and consumption in an efficient way. The micro grid generators are: a micro hydropower plant, photovoltaic panels, wind turbines. This proposal comes in support of energy developers and producers to use their investments more efficiently.

1. Introduction

Microgrids are local electrical networks that produce, consume and distribute electrical energy. They usually integrate several local renewable energy sources which serve a small group of consumers. The energy of a microgrid is cheaper, clean and the supply is not affected if the main grid fails. On the other hand the reliability of the energy sources may be an issue but with smart management and batteries used for energy storage the problem can be overcome.

Microgrids are recognized as the basic solution for supplying electricity to various locations where renewable energy sources are available [1]÷[12]. The possible locations for these micro grids are the industrial parks around the cities, hospitals, universities that all have in common large available spaces to install PV panels.

The use of renewable energy resources is a solution to the problem of electricity shortage. Renewable energy sources will become the main pillar of development of the Romanian energy system, as it is in Europe and as it happens everywhere in the world.

The microgrid pilot presented in the paper is part of the Romanian energy industry, with the ambition to actively participate in the transformation of the Romanian energy sector by developing renewable energy projects like hydropower, wind power and solar power. This management is no longer based solely on transport and distribution infrastructure but solves local problems if a wind or photovoltaic park and a battery are installed next to a village, it is no longer needed to inject energy into the system, there is no need for line upgrades, but the energy can be used at the locality level.

The novelty of the paper is the development of an efficient control solution for a microgrid that will help minimize electricity costs. The aim is to increase the number of such micro-hydro networks to be implemented at national level as this reduces the load on transmission lines, increasing operational safety and feeding the area inside the micro-hydro network.

According to IRENA (International Renewable Energy Agency), the renewable energy, especially wind energy, today offers not only the cleanest and most sustainable way to produce energy, but also the cheapest [13] (figure 1).

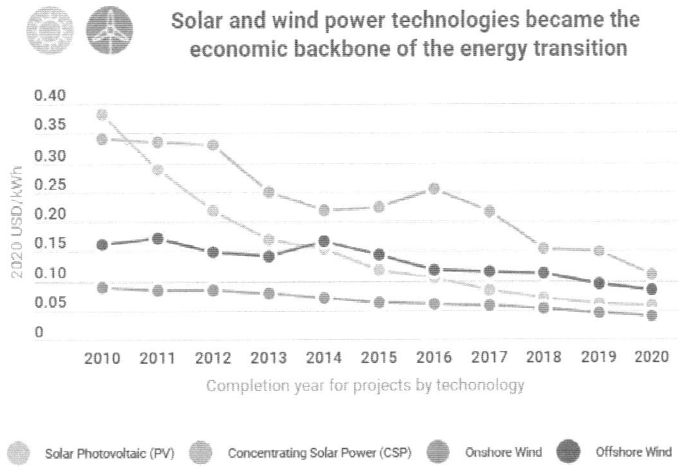

Figure 1. The production cost of renewable energy systems [13].

The use of microgrid systems offers many benefits to both the consumer and the utility provider (figure 2).

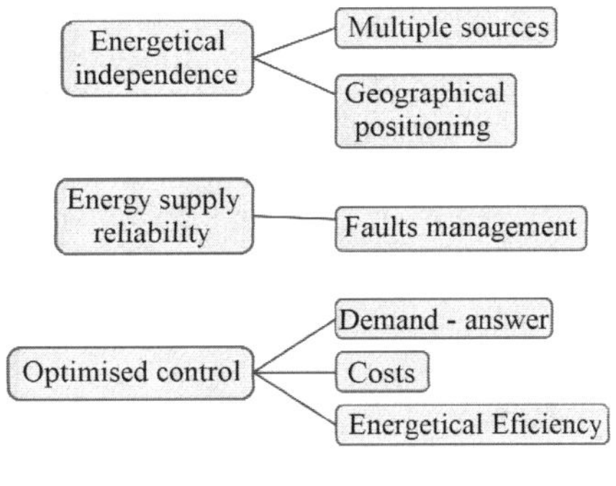

Figure 2. Advantages of using the microgrid.

The controller simplifies the way power sources integrate into the consumer network. It distributes and controls the flow of energy between the public electricity grid, energy sources, loads and energy storage elements.

Depending on the user's consumption curve, the importance of consumers, the price of electricity, the price of gas and the weather forecast, the use of energy sources can be optimized to know when to store, when to sell or when to consume available energy.

The principles an intelligent micro-hydro-grid:

- It must be reliable under normal conditions and durable under unusual conditions;
- extensible in terms of functionality;
- adaptable to future circumstances.

Although the adoption of microgrids is constantly growing, there are still some obstacles to competently designing, regulating, and operating microgrids when connected to the network (figure 3). From the analyzed data, a series of barriers were observed in the development of policies for the creation of micro-networks that appeared due to the lack of education, the persistence of personal beliefs, the lack of awareness of the need to develop micro-networks.

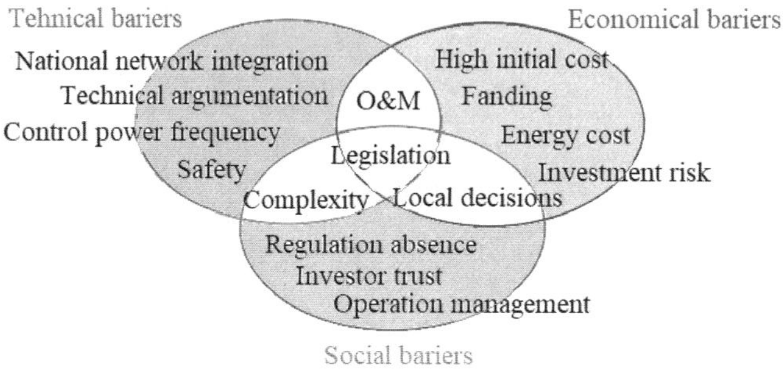

Figure 3. Barriers in the wide scale adoption of the microgrid.

The aim of the paper is to present an efficient way to develop a microgrid using the Siemens S7-1500 PLC.

2. Microgrid components

The microgrid includes photovoltaic solar panels, wind turbines, hydro generation (figure 4).

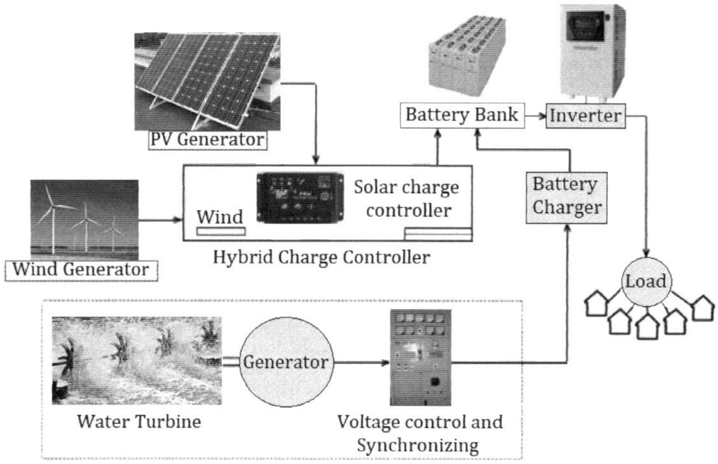

Figure 4. Microgrid components.

Each of this renewable energy production devices like solar panels, wind turbines produce energy that is stored into the battery bank. From the battery bank the current must pass through an inverter in order to be usable by the consumers.

3. The analyzed microgrid

The active reference power will be established by the controller based on the operation strategy of the microgrid (figure 5).

Battery operating modes are set by the controller (figure 6). It sends reference values to the two DC-DC converters. The aim is to maintain the same state of charge, which means that the reference value of P sent by the controller is set in proportion to the capacity of the battery. SBB aims to maximize the power produced (through MPPT).

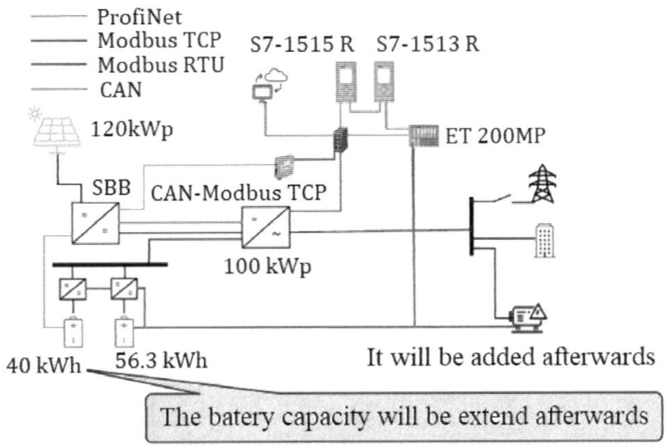

Figure 5. Microgrid pilot – phase 1.

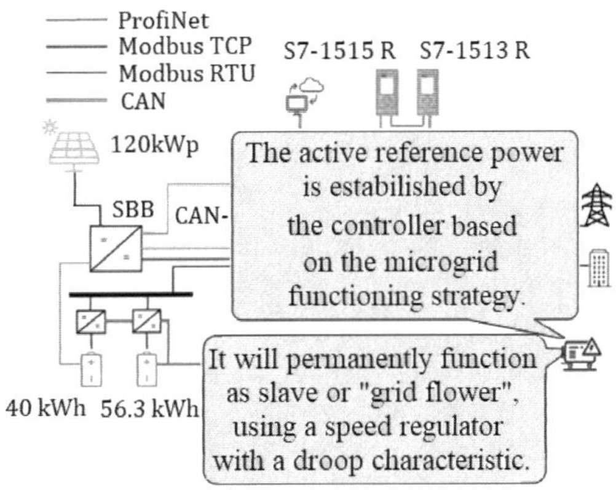

Figure 6. Microgrid pilot – phase 2.

In normal mode (synchronized), the converter is in slave mode (grid follower), and in insular mode it becomes master (grid former), it becomes the main element that regulates the frequency and forms the voltage wave (figure7). The converter provides the function of synchronizing the microgrid with the public network and the activation or deactivation of some functions is done communicating with the controller. It also identifies the conditions of islanding and commands the disconnection of the circuit breaker. Technical data of the pilot are listed below:
Photovoltaic panels:

- PV Generator Output: 127.26 kW;
- surface: 607.3 sqm;
- number of PV modules: 252;
- SBB: 1 (WSTECH);
- inverter: 1 (WSTECH);

Batteries:

- Li-Ion;
- Model 1: capacity 42 kWh@500V, Pmax = 120 kW;
- Model 2: capacity 56.3 kWh @ 563V, 4000 cycles, Pmax = 30 kW;
- connected via DC-DC converter (Siemens SINAMICS DCP).

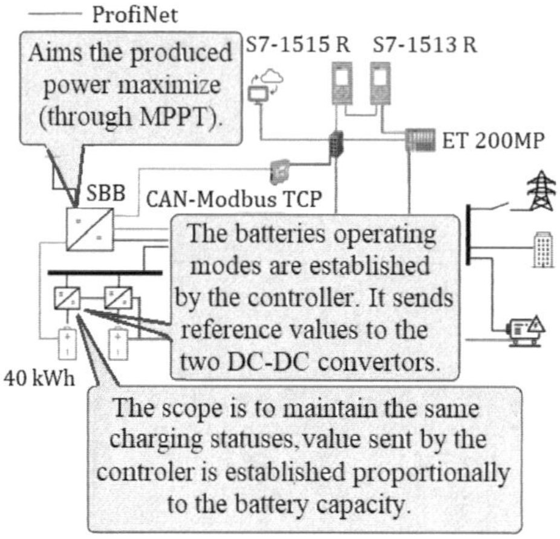

Figure 7. Microgrid pilot – phase 3.

Distributed energy storage plays an important role in the operation and control of the micro-grid, as it can improve local reliability and resilience, reduce operating costs and mitigate the challenges posed by renewable energy generation. The control solution includes:

PLC Siemens S7-1500

- State-of-the-art redundant CPU;
- flexibility in implementing control logic;
- high flexibility in the integration and control of various equipment;
- remote monitoring and operation;

PcVue monitoring application (figure 8)

- client-server configuration;
- the server is the only connection point to the process;
- the client station retrieves data from the server based on the IP address or computer name;
- there may be multiple client stations.

Optimizing the control strategy means obtaining the best results, considering the specific limitations of each application:

- local control - quick reaction, reactive to a concrete situation;
- control based on predefined rules;
- model-based control.

Depending on the availability of existing energy sources, the energy stored in the batteries and the consumer demand, the microgrid controller distributes and controls the energy between all the components of the microgrid. The controller will monitor the production and power consumption inside the microgrid to decide the optimal operating strategy based on the defined indicators and the user's market strategy.

For the control strategy, a Siemens S7-1500 controller in redundant configuration will be used, which will communicate with the SCADA application, PcVue through the ISO-S7 protocol. The SCADA application will include the control screens of the installation with the display of the graphs. The data will be archived and analyzed; thus, the system performance can be evaluated in different operating modes. It is desired to achieve a 61% reduction in emissions through a micro-hydro-grid with intelligent controller.

Intelligent grid control helps minimize electricity costs. The controller monitors all subsystems of the micro-network, thus helping the operator to manage the key performance indicators for generation, storage and consumption in an efficient way.

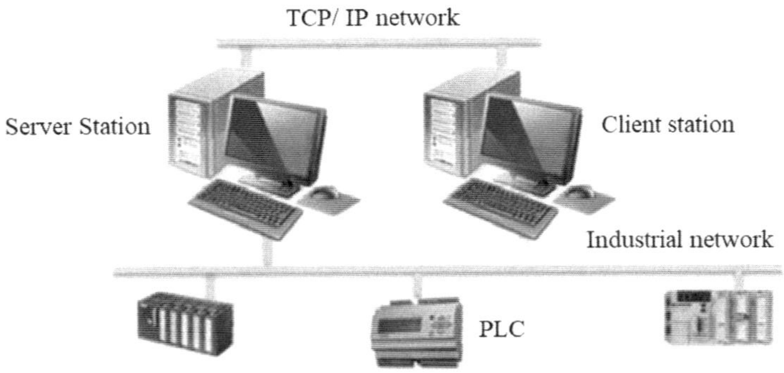

Figure 8. PcVue monitoring application.

The top line of the diagram in figure 9 illustrates the generators, storage and consumers of a micro-hydro-grid.

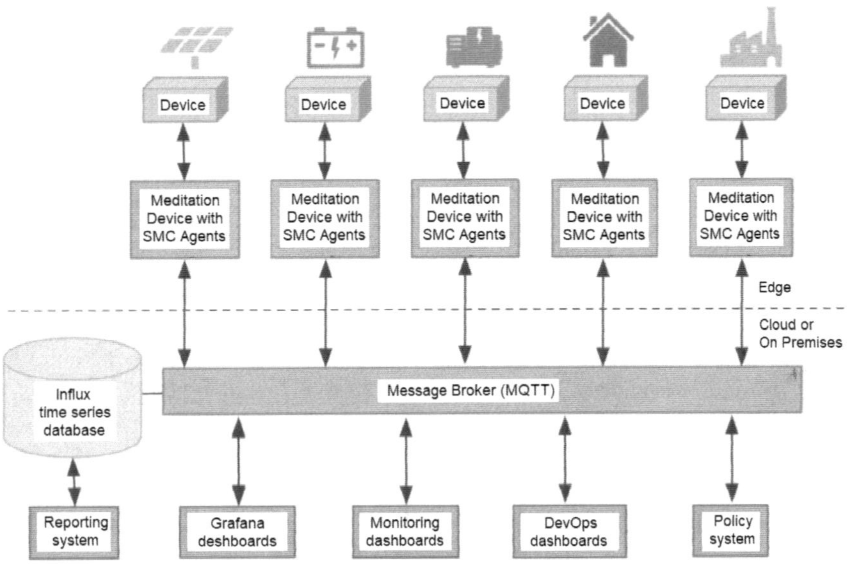

Figure 9. Simplified controller diagram.

For each mediation device, services are offered that allow efficient functionality for micro-networks, namely:

- automation service to announce, identify, register and configure all new devices that are added to the system;
- monitoring and control services that are easy to install and change;
- the messaging system exploits the features of the MQTT industry standard, which is a modern network protocol designed specifically for network connections to remote locations where network bandwidth is limited and connections are unsafe. MQTT is the ideal choice to support micro-hydro-grids where the WAN connection could become temporarily unavailable due to extreme local weather or geological events.

The charge state of the batteries is determined by the formula:

$$SOC_i = \frac{E^{(i)}_{BESS,av}(t)}{E^{(i)}_{BESS,max}} 100 \quad (1)$$

where $E_{BESS,av}$ represents the energy stored inside the battery at the give time t, and $E_{BESS,max}$ represents the maximum energy stored inside the battery.

Battery charge status at time t depending on the previously considered moment in time $(t-1)$:

$$SOC_i(t) = SOC_i(t-1) + \frac{P^{(i)}_{BESS}(t)\cdot \Delta t}{E^{(i)}_{BESS,max}} \left(\eta_i \, S_{ch,i} + \frac{1}{\eta_i} S_{disch,i} \right) \quad (2)$$

where P_{BESS} represents the absorbed power at the time t; η_i represents the charging efficiency; S_{ch} represents the charging current, and S_{disch} represents the discharging current.

Limits imposed on charge status are defined as:

$$SOC^{(i)}_{BESS,min} \leq SOC^{(i)}_{BESS}(t) \leq SOC^{(i)}_{BESS,max} \quad (3)$$

Limits imposed on power produced / absorbed are defined as:

$$-P^{(i)}_{BESS,max} \leq P^{(i)}_{BESS}(t) \leq P^{(i)}_{BESS,max} \quad (4)$$

where $P_{BESS,max}$ represents the maximum absorbed power from the battery specifications.

One hour sampling simulations considers that:

- the production of electricity from the photovoltaic plant comes first;
- battery systems are mainly used to take variations from a reference profile; batteries may be required to have a certain amount of energy available in case of an emergency (islanding);
- the synchronous generator (on natural gas) covers consumption when there is no availability from the photovoltaic plant, thus minimizing the cost of fuel and the amount of CO_2;
- at the micro-grid level, the control strategy correlates with the strategies related to the electricity market.

In addition to the devices for generating, storing and consuming, inside the micro-hydro-grid other components are integrated through the messaging system:

- database from generators, storage and consumers
- reports that provide history of generation, storage and consumption, used to create a more efficient smart hydro-grid.

4. Barriers in the wide scale adoption of the microgrid

In figure 10 is the outline of the daily load curve. In figure 11 is the distribution of energy value. Figure 12 illustrates that the maximum number of sampling is obtained for the 50-60 kW range.

Figure 13 shows the distribution of energy values – it shows that most consumption values are recorded at $P_m = 58$kW.

One minute sampling simulations gives the following results: the single-wire diagram of the microgrid is outlined in figure 14. The graphs show that maintaining the power exchange with the public network at a predefined value (the balance within the microgrid) is influenced by the quality of the forecast and market strategy.

The sampling interval becomes important depending on market strategy. Islanding operation requires 1 second interval sampling. The control strategy is customized for the beneficiary's market strategy.

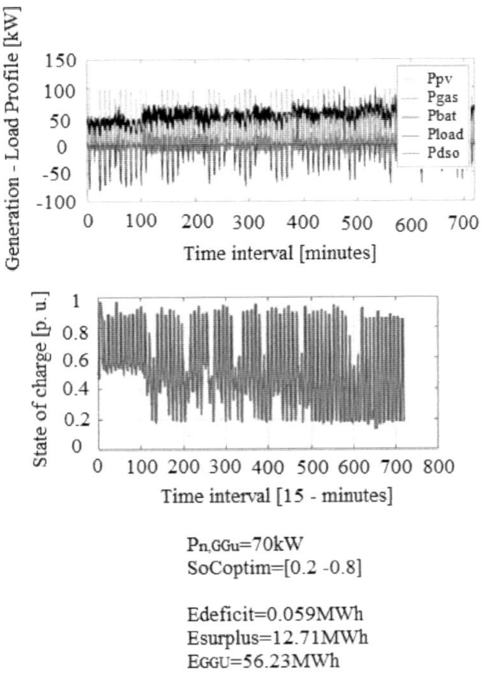

Figure 10. Applying an optimal operating range for batteries, respectively [50% - 60%].

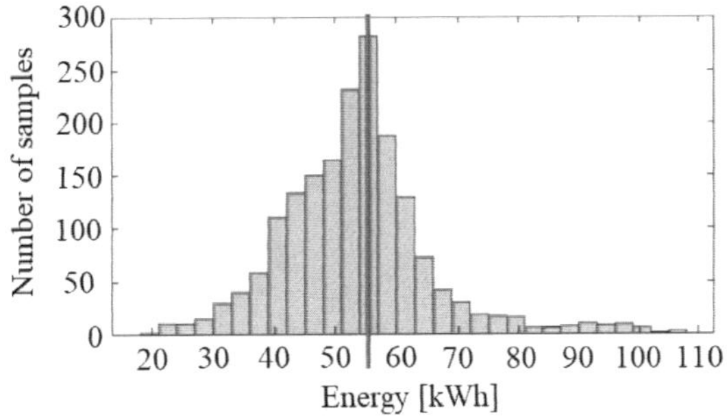

Figure 11. Distribution of energy values.

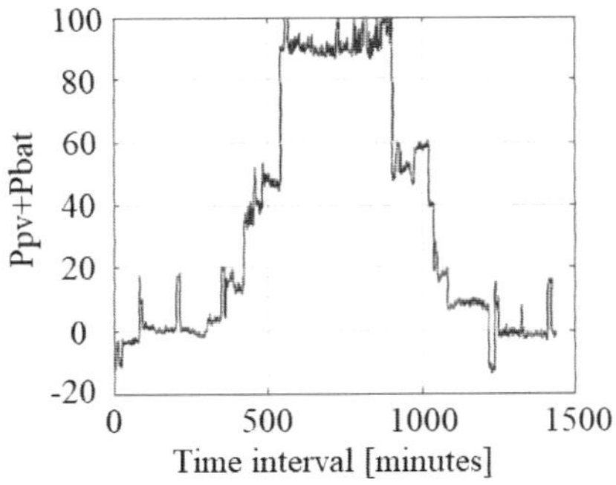

Figure 12. Powers transferred through the DC/AC converter.

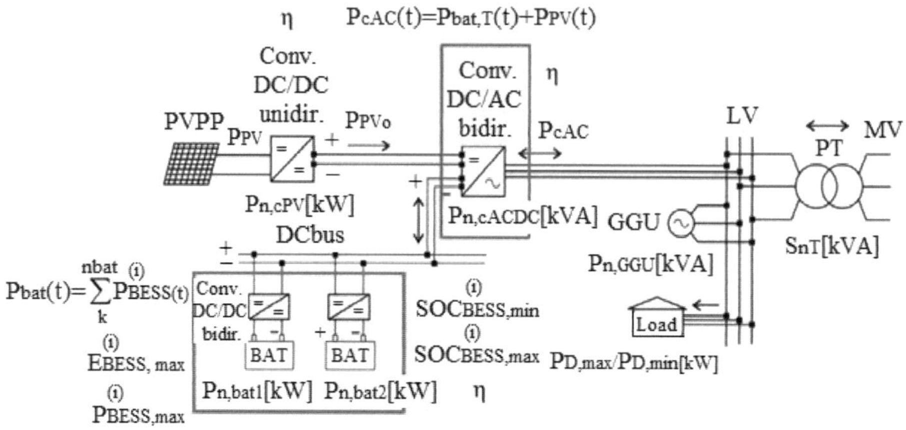

Figure 13. Monofilament diagram of the microgrid.

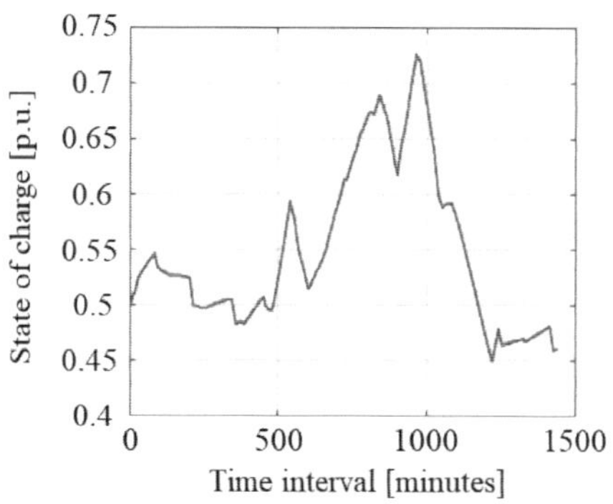

Figure 14. Battery charge status.

5. Conclusions and futures perspectives

In conclusion, the set in place of a microgrid is perfectly achievable and all the necessary devices are already available.

For optimal results, the microgrid developer must find the best control logic for each facility depending on its particularities and configuration (type of renewable source available, energetic needs, day-night energy consumption graphic etc).

This proposal aims to increase the number of micro-hydro-grid systems to be implemented at national level, thus supporting sustainability in the use of environmentally friendly technologies by increasing efficiency in energy production and use locally, facilitating transactions between production and consumer. This should reduce the load on the transmission lines, increasing the operational safety and power supply of the area inside the micro-hydro-grid.

The presented solution is dedicated to the developers of renewable energy parks, municipalities that want to implement micro-hydro-grid islands in order to reduce CO_2 emissions.

References

[1] Rehman AU, Zeb S, Khan HU, Shah SSU and Ullah A 2017 Design and operation of microgrid with renewable energy sources and energy storage system: A case study *Proc. 2017 IEEE 3rd Int. Conf. Engineering Technologies and Social Sciences (ICETSS)* 1-6 doi: 10.1109/ICETSS.2017.8324151

[2] Solanki A 2012 Generation capacity design for a microgrid for measurable power quality indexes *Proc. 2012 IEEE PES Innovative Smart Grid Technologies*, doi: 10.1109/ISGT.2012.6175742
[3] Kroposki B et al 2008 Microgrid standards and technologies *Proc. 2017 IEEE Power and Energy Society General Meeting - Conversion and Delivery of Electrical Energy in the 30st Century* doi: 10.1109/PES.2008.4596703
[4] Zhao B 2010 Design and implementation of an integrated micro-grid system *Proc. CICED 2010* ISBN 978-1-4577-0066-8
[5] Arjun K et al 2013 An intelligent microgrid with distributed generation *Proc. 2013 IEEE Innovative Smart Grid Technologies-Asia* doi: 10.1109/ISGT-Asia.2013.6698755
[6] Borani P et al 2020 A review on microgrid control techniques *Proc. 2020 IEEE Innovative Smart Grid Technologies*, Bulgaria
[7] Shrivastwa R et al 2021 Understanding microgrids and their future trends *Proc. 2021 IEEE Int. Conf. Industrial Technology* doi: 10.1109/ICIT.2019.8754952; doi: 10.1109/ISGT-Germany.2021.6873886
[8] Kumar V et al 2018 Overview on microgrid system *Proc. 2018 Fifth Int. Conf. Parallel, Distributed and Grid Computing* doi: 10.1109/PDGC.2018.8745849
[9] Zhou Y et al 2019 A review on microgrid architectures and control methods *Proc. 2019 IEEE 8th Int. Power Electronics and Motion Control Conf.* doi: 10.1109/IPEMC.2016.7512799
[10] Chen Z et al 2017 A review on control strategies of AC/DC micro grid *Proc. 2017 IEEE Int. Conf. Environment and Electrical Engineering and 2017 IEEE Industrial and Commercial Power Systems Europe (EEEIC / I&CPS Europe)* 1-6 doi: 10.1109/EEEIC.2017.7977807
[11] Minshew Z et al 2017 DC micro-grid pricing and market model *Proc. 2017 IEEE Global Humanitarian Technology Conf.* doi: 10.1109/GHTC.2017.8239282
[12] Rajab Z et al 2020 Optimum microgrid planning and operation for improving reliability and power quality: Review *Proc. 2020 11th Int. Renewable Energy Congress* doi: 10.1109/IREC48820.2020.9310384, 2020.
[13] IRENA – Int. Renewable Energy Agency https://www.irena.org/news/articles/2021/Jun/Low-Renewable-Costs-Allow-To-Power-Past-Coal

AUTHOR INDEX